Workbook to Accompany *Anatomy & Physiology | Revealed*® Version 3.2

Robert B. Broyles, Jr.
Butler County Community College

WORKBOOK TO ACCOMPANY ANATOMY & PHYSIOLOGY REVEALED VERSION 3.2

Published by McGraw-Hill Education, 2 Penn Plaza, New York, NY 10121. Copyright © 2019 by McGraw-Hill Education. All rights reserved. Printed in the United States of America. Previous editions © 2012 and 2009. No part of this publication may be reproduced or distributed in any form or by any means, or stored in a database or retrieval system, without the prior written consent of McGraw-Hill Education, including, but not limited to, in any network or other electronic storage or transmission, or broadcast for distance learning.

Some ancillaries, including electronic and print components, may not be available to customers outside the United States.

This book is printed on acid-free paper.

1 2 3 4 5 6 7 8 9 LMN 21 20 19 18

ISBN 978-1-260-17014-6
MHID 1-260-17014-4

Portfolio Manager: *Matt Garcia*
Product Developers: *Michelle Gaseor/Lora Neyens*
Content Project Managers: *Mary Jane Lampe/Sandy Wille*
Buyer: *Sandy Ludovissy*
Cover Design: *SPi Global*
Content Licensing Specialist: *Lori Hancock*
Cover Image: *©McGraw-Hill Education*
Compositor: *SPi Global*

All credits appearing on page or at the end of the book are considered to be an extension of the copyright page.

The Internet addresses listed in the text were accurate at the time of publication. The inclusion of a website does not indicate an endorsement by the authors or McGraw-Hill Education, and McGraw-Hill Education does not guarantee the accuracy of the information presented at these sites.

mheducation.com/highered

PREFACE

Anatomy & Physiology | Revealed® has proven to be an invaluable tool for the student of the human body. Whether you are taking your first anatomy and physiology course, or you are a med-school student, or you are somewhere in between, this is the ultimate hands-on learning tool for you. Through the use of layer-by-layer dissection photographs of actual cadavers, students will gain invaluable insight into the intricate design of the human body. In addition, histological and radiological images and animations round out the students' learning experience with their unique perspectives. Not every student has access to a cadaver, and even those who do, cannot take it home for review. Now, however, students *can* take a cadaver home with them, through this accurate and detailed study of the human body.

The updated version of *Anatomy & Physiology | Revealed®* now includes new interactive animations and revised histology images as well as the 11 Body Systems with a powerful search engine that quickly locates information across all modules.

Audience

This workbook is designed to complement *Anatomy & Physiology | Revealed®* 3.2, a powerful learning tool. Regardless of their computer skills, students will be up and running in a short time as they learn how to master the concepts within the user-friendly interface. Developed to support one-semester or two-semester anatomy and physiology courses, human anatomy, human physiology courses, and a variety of health related career programs, this workbook, combined with *Anatomy & Physiology | Revealed®* 3.2, will ensure students the very best opportunity for getting the most out of their classes. For many schools, the combination of the two will take the place of their laboratory manual.

Key Features

- *NEW* Enhancements for Accessibility. New updates meet a number of WCAG 2.0 AA guidelines for keyboard navigation and low vision/colorblindness. This workbook reflects these changes.
- *NEW* Over 50 Clinical 3D Animations were added to APR 3.2, spread across all of APRs modules. This workbook incorporates those new animations.
- Customize your online version of *Anatomy & Physiology | Revealed®* 3.2 for your class.
- *NEW* 7 Concept Overview Interactives cover key concepts like Glomerular filtration and regulation and Neuron physiology enhance the APR experience and are now covered in this workbook.

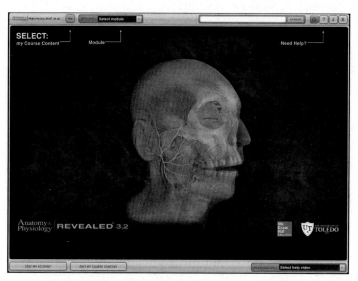

From Anatomy and Physiology | Revealed

- The Introduction walks students through the format and important sections of *Anatomy & Physiology | Revealed®* 3.2.
- Exercises follow the arrangement of *Anatomy & Physiology | Revealed®* 3.2 requiring students to complete steps and answer follow-up questions.
- Perforated pages allow exercises to be easily removed and handed in to instructors.
- "Heads-Up!" sections offer tips and reminders about the program.
- "Check-Point" questions serve as brief self-tests to check comprehension.
- Useful tables include bonus questions on related topics.

Acknowledgments

I offer my heartfelt thanks to colleagues who've offered suggestions and/or reviewed this and previous versions of *Workbook to Accompany Anatomy & Physiology | Revealed®*. Their contributions are much appreciated.

I would like to give a special thank-you to my developmental editor Amy Oline. This revision would not have been possible without her guidance and counsel.

A special thank-you to my colleagues at Butler Community College, especially Michael Heffron, who has been my sounding board throughout all editions of this workbook. Thank you Miguel!

I would be remiss if I did not thank Dr. Ann Stalheim Smith, Professor Emeritus of Human Body at Kansas State University. Ann not only taught me anatomy and physiology while an undergrad those many years ago, but she also taught me, by her example, how to teach. Thank you, Ann, for being such a remarkable role model.

The third edition of the workbook was dedicated to the loving memory of my dear friend, Dr. Norm Dillman. Even as a retired professor from Kansas State University, Norm modeled the concept of a lifelong learner. His wide-eyed, childlike curiosity about the natural world was contagious—he was never embarrassed to ask questions and seek out answers. His example is exemplary!

This edition of the workbook is dedicated to Martin "Skip" Pickens, USMC Retired, father figure and mentor, who saw my potential when I didn't, and urged me on to excel in my endeavors. Thank you, Skipper!

Finally, a special thank-you to my beloved bride, Patricia for her loving support and her many sacrifices that were made as this edition came into being. This project could never have even begun, much less come to fruition, without her continued support and encouragement. You are indeed my Bride and Joy!

Reviewers

Shaheem Abrahams
Thomas Nelson Community College

Isaac Barjis
New York City College of Technology

Jerry Barton
Tarrant County College—South Campus

David Bastedo
San Bernardino Valley College

Valerie A. Bennett
Clarion University of Pennsylvania

J. Gordon Betts
Tyler Junior College

Lois Brewer Borek
Georgia State University

Scott Dunham
Illinois Central College

Kathryn Durham
Lorain County Community College

Adam Eiler
San Jacinto College South

Deanna Ferguson
Gloucester County College

Cynthia R. Hartzog
Catawba Valley Community College

Amy Harwell
Oregon State University

Chris Herman
Eastern Michigan University

Mark F. Hoover
Penn State Altoona

Julie Huggins
Arkansas State University

Susanne Kalup
Westmoreland County Community College

Michael Kopenits
Amarillo College

Ellen Lathrop-Davis
Community College of Baltimore County Essex

Elisabeth C. Martin
College of Lake County

Alice McAfee
University of Toledo

Patrick McArthur
Okaloosa Walton College

Elizabeth M. Meyer
College of Lake County

Alfredo Munoz
University of Texas at Brownsville

Margaret (Betsy) Ott
Tyler Junior College

Mark Paternostro
Pennsylvania College of Technology

Robert L. Pope
Miami Dade College

Gregory K. Reeder
Broward Community College

Dawn E. Roberts
University of Massachusetts Amherst

Walied Samarrai
New York City College of Technology

A Special Note to Students

Anatomy & Physiology | Revealed® 3.2 is not a learning tool that is exhausted with a single use, nor is it any less valuable after repeated use. The high-quality cadaver photos in *Anatomy & Physiology | Revealed*® 3.2 are designed to provide you a unique opportunity to match visuals of the human body with what you are learning from your instructor. With each use, you will glean more information to add to your knowledge base, and as you do, you will also gain the confidence necessary at exam time. With this in mind, use *Anatomy & Physiology | Revealed*® 3.2 on a daily basis to hone your skills, refresh your memory, and stay sharp.

Be sure to take advantage of the self-tests in *Anatomy & Physiology | Revealed*® 3.2. These allow you to determine how well you understand the information presented and will reveal areas that you may need to revisit.

This workbook is written for YOU, the student. If you have any comments on how it may be improved, please let me know—I am always open to suggestions! My desire is that this workbook, in combination with *Anatomy & Physiology | Revealed*®, will help you to excel in your study of the human body.

Wishing you the best,

Bob Broyles
bbroyles@butlercc.edu

ABOUT THE AUTHOR

©Bob Broyles

Bob embarked on his teaching career while in the third grade, bringing critters from his rural home to share with the kindergarten classroom. While in graduate school, he began teaching undergraduate anatomy and physiology labs. This set the stage for his current position at Butler County Community College, where he teaches or has taught anatomy and physiology lectures and labs, cadaver dissection, and general biology and chemistry review for biology students.

Bob has a passion for teaching; desiring to guide others in understanding the intricacies of the natural world as experienced in their daily lives. His other passions include organic gardening (he is a Master Gardener), landscaping for wildlife (primarily birds), prairie restoration, photography, and bird studies.

Bob's previous projects with McGraw-Hill Publishing include preparing the instructor's PowerPoint presentations for Kenneth S. Saladin's *Human Anatomy* text, writing multiple textbook reviews, and offering online tutoring for Saladin's *Anatomy and Physiology* textbook.

Bob resides with his wife Patricia on a small farm, where they raise heritage poultry and grow prairie wildflowers, as well as trees and shrubs that benefit birds, bees, and butterflies.

CONTENTS

Introduction: Becoming Familiar with *Anatomy and Physiology | Revealed*® 1

Overview: Introduction: 1

The DISSECTION Study Area 5
    The DISSECTION Study Area: *my* COURSE CONTENT 13

The ANIMATIONS Study Area 15
    The ANIMATIONS Study Area: *my* COURSE CONTENT 15

The HISTOLOGY Study Area 18
    The HISTOLOGY Study Area: *my* COURSE CONTENT 19

The IMAGING Study Area 21
    The IMAGING Study Area: *my* COURSE CONTENT 25
    The IMAGING Study Area: A Side-Trip to the Skeletal System 25

The QUIZ Study Area 26
    The QUIZ Study Area: *my* COURSE CONTENT 30

IN REVIEW 32
    What Have I Learned? 32

Module 1: Body Orientation 33

Overview: Body Orientation 33

    Exercise 1.1: Body Position 34
    Exercise 1.2: Planes of Section 35

In Review
    What Have I Learned? 36
    Exercise 1.3: Directional Terms 36

In Review
    What Have I Learned? 37
    Exercise 1.4: Body Regions 38

In Review
    What Have I Learned? 45
    Exercise 1.5: Body Cavities 45

In Review
    What Have I Learned? 46
    Exercise 1.6: Abdominal Quadrants and Regions 46

In Review
    What Have I Learned? 48

In Review
    What Have I Learned? 50
    Exercise 1.7: Peritoneum 50

In Review
    What Have I Learned? 51
    Exercise 1.8: Organ Systems 51

In Review
    What Have I Learned? 52

In Review
    What Have I Learned? 55

In Review
    What Have I Learned? 57

In Review
    What Have I Learned? 58

In Review
    What Have I Learned? 60

In Review
    What Have I Learned? 62

In Review
    What Have I Learned? 64

In Review
    What Have I Learned? 67

In Review
    What Have I Learned? 69

In Review
    What Have I Learned? 72

In Review
    What Have I Learned? 73

In Review
    What Have I Learned? 75
    Self-Quiz 75

Module 2: Cells and Chemistry 77

Overview: Cells and Chemistry 77

    Self-Quiz 79
    Self-Quiz 79
    Self-Quiz 79
    Self-Quiz 79
    Self-Quiz 80
    Self-Quiz 81
    Self-Quiz 81
    Self-Quiz 82
    Self-Quiz 82
    Exercise 2.1: Generalized Cell 82

Histology: Cilium 83

Histology: Flagellum of Sperm 84

Histology: Microvillus (longitudinal section) 86

Histology: Microvillus (cross-section) 86

Histology: Cilium and Microvillus 87

Histology: Plasma Membrane 88

Histology: Cytoplasm 88

Histology: Cytoskeleton (microfilament) 89

Histology: Cytoskeleton (microfilament and microtubule) 89
Histology: Cytoskeleton (neurofilament and microtubule) 90
   Self-Quiz 92
Histology: Lysosome 92
Histology: Mitochondrion 93
Histology: Peroxisome 94
Histology: Rough Endoplasmic Reticulum 95
Histology: Ribosome 96
Histology: Smooth Endoplasmic Reticulum 97
Histology: Golgi Apparatus 98
Histology: Nucleus 99
Histology: Chromosome 100
Histology: Centrosome 101
Histology: Centriole 102
In Review
   What Have I Learned? 104
   Exercise 2.2: Plasma Membrane 105
   Structure Identification: Plasma Membrane 108
In Review
   What Have I Learned? 109
Membrane Transport 109
   Self-Quiz 110
   Self-Quiz 110
   Self-Quiz 110
   Self-Quiz 111
   Self-Quiz 111
   Self-Quiz 111
   Self-Quiz 111
   Exercise 2.3: Chemistry 112
   Self-Quiz 112
   Self-Quiz 112
   Self-Quiz 114
   Self-Quiz 114

Module 3: Tissues 115

Overview: Tissues 115

Epithelial Tissues 115
   Self-Quiz 117
   Exercise 3.1: Epithelial Tissue—Simple 117
Histology: Epithelial Tissue—Simple Squamous Epithelium 117
Histology: Epithelial tissue—Simple Cuboidal Epithelium 118
Histology: Epithelial Tissue—Simple Columnar Epithelium (ciliated) 119
Histology: Epithelial Tissue—Simple Columnar Epithelium (nonciliated) 120
Histology: Epithelial Tissue—Pseudostratified Columnar Epithelium 121
In Review
   What Have I Learned? 122
   Exercise 3.2: Epithelial Tissue—Stratified 123
Histology: Epithelial Tissue—Stratified Squamous Epithelium (keratinized) 123
Histology: Epithelial Tissue—Stratified Squamous Epithelium (nonkeratinized) 124
Histology: Epithelial Tissue—Stratified Cuboidal Epithelium 124
Histology: Epithelial Tissue—Stratified Columnar Epithelium 125
Histology: Epithelial Tissue—Transitional Epithelium 126
In Review
   What Have I Learned? 127
Connective Tissues 127
   Self-Quiz 128
   Exercise 3.3: Connective Tissue Proper—Loose 128
Histology: Connective Tissue—Areolar Connective Tissue 128
Histology: Connective Tissue—Adipose Connective Tissue 129
Histology: Connective Tissue—Reticular Connective Tissue 130
In Review
   What Have I Learned? 131
   Exercise 3.4: Connective Tissue Proper—Dense 131
Histology: Connective Tissue—Dense Regular Connective Tissue 131
Histology: Connective Tissue—Dense Irregular Connective Tissue 132
Histology: Connective Tissue—Elastic Connective Tissue 133
In Review
   What Have I Learned? 134
   Exercise 3.5: Supporting Connective Tissue—Cartilage 134
Histology: Connective Tissue—Hyaline Cartilage 134
Histology: Connective Tissue—Fibrocartilage 135
Histology: Connective Tissue—Elastic Cartilage 136
In Review
   What Have I Learned? 137
   Exercise 3.6: Supporting Connective Tissue—Bone 137
Histology: Connective Tissue—Compact Bone 137
Histology: Connective Tissue—Spongy Bone 138
In Review
   What Have I Learned? 139
   Exercise 3.7: Fluid Connective Tissue—Blood 139
Histology: Connective Tissue—Blood 139
In Review
   What Have I Learned? 140
Muscle Tissue 140
   Self-Quiz 141
   Exercise 3.8: Muscle Tissue 141
Histology: Muscle Tissue—Skeletal Muscle 141
Histology: Muscle Tissue—Cardiac Muscle 142

Histology: Muscle Tissue—Smooth Muscle 142

In Review
What Have I Learned? 144

Nervous Tissue 144
Self-Quiz 145
Exercise 3.9: Nervous Tissue 145

Histology: Nervous Tissue 145

In Review
What Have I Learned? 146
Self-Quiz 146

Module 4: The Integumentary System 147

Overview: The Integumentary System 147

Exercise 4.1: Integumentary System—Thin Skin and Subcutaneous Tissues 148

In Review
What Have I Learned? 151
Exercise 4.2: Fingernail—Sagittal View 152

Fingernail: Sagittal View 152

In Review
What Have I Learned? 153
Exercise 4.3a: Thick Skin—Histology 154
Exercise 4.3b: Thick Skin—Histology 154
Exercise 4.4a: Thin Skin—Histology 155
Exercise 4.4b: Thin Skin—Histology 155
Exercise 4.5: Hair—Histology 156
Exercise 4.6a: Hair Follicle—Histology 156
Exercise 4.6b: Hair Follicle—Histology 157
Exercise 4.7a: Sebaceous Gland—Histology 157
Exercise 4.7b: Sebaceous Gland—Histology 158
Exercise 4.8a: Merocrine Sweat Gland—Histology 158
Exercise 4.8b: Merocrine Sweat Gland—Histology 159
Exercise 4.9: Duct of Sweat Gland—Histology 159
Exercise 4.10a: Apocrine Sweat Gland—Histology 160
Exercise 4.10b: Apocrine Sweat Gland—Histology 160

In Review
What Have I Learned? 163
Self-Quiz 163

Module 5: The Skeletal System 165

Overview: Skeletal System 165

Naming Bony Processes and Other Landmarks 165
Exercise 5.1: Coloring Exercise 166
Exercise 5.2: Appositional Bone Growth 168

Skeleton (Regions) 168
Exercise 5.3: Skeleton—Anterior View 168
Exercise 5.4: Skeleton—Posterior View 169
Exercise 5.5: Locating Structures of the Head and Neck 170
Exercise 5.6: Head and Neck—Anterior View 173
Exercise 5.7: Head and Neck—Lateral View 174
Exercise 5.8: Thorax—Anterior View 175
Exercise 5.9: Abdomen—Anterior View 175
Exercise 5.10: Back—Posterior View 176

Skull and Associated Bones 177
Exercise 5.11: Skull—Anterior View 178
Exercise 5.12: Skull—Superior View 179
Exercise 5.13: Skull—Lateral View 180
Exercise 5.14: Skull—Posterior View 181
Exercise 5.15: Skull—Midsagittal View 182
Exercise 5.16: Skull—Inferior View 184
Exercise 5.17: Skull—Cranial Cavity—Superior View 185
Exercise 5.18: Orbit 186
Exercise 5.19: Ethmoid Bone 187
Exercise 5.20: Frontal Bone 187
Exercise 5.21: Mandible 188
Exercise 5.22: Maxilla 188
Exercise 5.23: Occipital Bone 189
Exercise 5.24: Parietal Bone 189
Exercise 5.25: Sphenoid Bone 190
Exercise 5.26: Temporal Bone 190
Exercise 5.27: Teeth 191
Exercise 5.28: Hyoid Bone 192
Exercise 5.29a: Imaging—Skull 193
Exercise 5.29b: Imaging—Skull 193
Self-Quiz 193

In Review
What Have I Learned? 194

Vertebral Column 194
Exercise 5.30: Vertebral Column 194
Exercise 5.31: Imaging—Vertebral Column (with Scoliosis) 195
Exercise 5.32: Cervical Vertebra 196
Exercise 5.33a: Imaging—Cervical Region 197
Exercise 5.33b: Imaging—Cervical Region 197
Exercise 5.33c: Imaging—Cervical Region 198
Exercise 5.34: Atlas (C1 Vertebra) 198
Exercise 5.35: Imaging—Atlas (C1 Vertebra) 199
Exercise 5.36: Axis (C2 Vertebra) 200
Exercise 5.37: Thoracic Vertebra 200
Exercise 5.38: Lumbar Vertebra 201
Exercise 5.39: Imaging—Lumbar Region 201
Exercise 5.40: Sacrum and Coccyx 202
Self-Quiz 203

In Review
What Have I Learned? 204

Thoracic Cage 205
Exercise 5.41: Thoracic Cage 205
Exercise 5.42a: Imaging—Thorax 206
Exercise 5.42b: Imaging—Thorax 206
Exercise 5.42c: Imaging—Thorax 207

In Review
What Have I Learned? 207

Pectoral Girdle and Upper Limb 208
Exercise 5.43: Shoulder and Arm—Anterior View 208
Exercise 5.44: Shoulder and Arm—Posterior View 208
Exercise 5.45: Imaging—Shoulder and Arm 210
Exercise 5.46: Clavicle 210
Exercise 5.47: Scapula 211
Exercise 5.48: Humerus 212
Exercise 5.49: Forearm and Hand—Anterior View 213
Exercise 5.50: Forearm and Hand—Posterior View 214
Exercise 5.51a: Imaging—Elbow 215
Exercise 5.51b: Imaging—Elbow 215
Exercise 5.52: Radius and Ulna 216
Exercise 5.53: Wrist and Hand—Anterior View 217
Exercise 5.54: Wrist and Hand—Posterior View 218
Exercise 5.55: Wrist and Hand—Anterior and Posterior View 218
Exercise 5.56a: Imaging—Hand (adult) 221
Exercise 5.56b: Imaging—Hand (teenaged) 222
Self-Quiz 222

In Review
   What Have I Learned? 222
Pelvic Girdle and Lower Limb 223
   Exercise 5.57: Hip and Thigh—Anterior View 223
   Exercise 5.58: Hip and Thigh—Posterior View 223
   Exercise 5.59: Imaging—Hip and Thigh 224
   Exercise 5.60: Pelvis 225
   Exercise 5.61: Pelvic Girdle—Female 225
   Exercise 5.62: Pelvic Girdle—Male 227
   Exercise 5.63: Hip Bone 228
   Exercise 5.64: Femur 229
   Exercise 5.65: Femur—Anterior and Coronal 230
   Exercise 5.66: Patella 231
   Exercise 5.67a: Imaging—Knee 231
   Exercise 5.67b: Imaging—Knee 232
   Exercise 5.67c: Imaging—Knee 232
   Exercise 5.68: Leg and Foot—Anterior View 233
   Exercise 5.69: Leg and Foot—Posterior View 234
   Exercise 5.70: Tibia and Fibula 234
   Exercise 5.71: Ankle and Foot 236
   Exercise 5.72a: Imaging—Foot 237
   Exercise 5.72b: Imaging—Foot 238
   Exercise 5.73: Imaging—Bone Scan 238
   Self-Quiz 239
In Review
   What Have I Learned? 239
   Exercise 5.74: Temporomandibular Joint 242
   Exercise 5.75: Glenohumeral (Shoulder) Joint 245
   Exercise 5.76: Elbow Joint 248
   Exercise 5.77: Knee Joint—Anterior View 251
   Exercise 5.78: Knee Joint—Posterior View 253
   Exercise 5.79: Ankle Joint—Lateral View 256
   Exercise 5.80: Ankle Joint—Medial View 258
   Self-Quiz 259
In Review
   What Have I Learned? 260
   Exercise 5.81a: Elastic Cartilage—Histology 260
   Exercise 5.81b: Elastic Cartilage—Histology 261
   Exercise 5.82a: Fibrocartilage—Histology 261
   Exercise 5.82b: Fibrocartilage—Histology 261
   Exercise 5.83a: Hyaline Cartilage—Histology 262
   Exercise 5.83b: Hyaline Cartilage—Histology 262
   Exercise 5.84a: Compact Bone—Histology 264
   Exercise 5.84b: Compact Bone—Histology 264
   Exercise 5.85a: Spongy Bone—Histology 264
   Exercise 5.85b: Spongy Bone—Histology 265
   Exercise 5.85c: Spongy Bone—Histology 265
   Self-Quiz 265
In Review
   What Have I Learned? 265
   Exercise 5.86: Coloring Exercise 266

Module 6: The Muscular System 269

Overview: Muscular System 269

Animations: Anatomy and Physiology 269

Muscular System: Head and Neck 277
   Exercise 6.1: Skeletal Muscle—Head and Neck—Anterior View 277

Animations: Muscle Actions 279
   Exercise 6.2: Skeletal Muscle—Head and Neck—Lateral View 280

   Exercise 6.3: Skeletal Muscles—Head and Neck—Midsagittal View 282
   Exercise 6.4: Skeletal Muscle—Head and Neck—Posterior View 284
   Self-Quiz 285
In Review
   What Have I Learned? 286
Muscular System: Trunk, Shoulder Girdle, and Upper Limb 287
   Exercise 6.5: Skeletal Muscle—Thorax—Anterior View 287
   Self-Quiz 289
In Review
   What Have I Learned? 289
   Exercise 6.6: Skeletal Muscle—Abdomen—Anterior View 290
   Self-Quiz 292
In Review
   What Have I Learned? 292
   Exercise 6.7: Skeletal Muscle—Pelvis—Superior View 293
   Self-Quiz 293
In Review
   What Have I Learned? 293
   Exercise 6.8: Skeletal Muscle—Back—Posterior View 294
   Exercise 6.9: Skeletal Muscle—Shoulder and Arm—Anterior View 296
   Self-Quiz 296
   Exercise 6.10: Skeletal Muscle—Shoulder and Arm—Posterior View 299
   Self-Quiz 300
In Review
   What Have I Learned? 301
   Exercise 6.11: Skeletal Muscles—Forearm and Hand—Anterior View 301
   Exercise 6.12: Skeletal Muscles—Forearm and Hand—Posterior View 304
   Self-Quiz 305
In Review
   What Have I Learned? 306
   Exercise 6.13: Skeletal Muscles—Wrist and Hand—Anterior View 306
   Exercise 6.14: Skeletal Muscles—Wrist and Hand—Posterior View 308
   Self-Quiz 310
In Review
   What Have I Learned? 310
   Exercise 6.15: Skeletal Muscles—Hip and Thigh—Anterior View 310
   Exercise 6.16: Skeletal Muscles—Hip and Thigh—Posterior View 312
   Self-Quiz 315
In Review
   What Have I Learned? 315
   Exercise 6.17: Skeletal Muscles—Leg and Foot—Anterior View 315
   Exercise 6.18: Skeletal Muscles—Leg and Foot—Posterior View 317
   Self-Quiz 319
In Review
   What Have I Learned? 319
   Exercise 6.19: Skeletal Muscles—Foot—Plantar View 320

In Review
What Have I Learned? 322
Self-Quiz 322

Muscle Attachments (on Skeleton) 322
Exercise 6.20: Skull—Anterior View 323
Exercise 6.21: Skull—Lateral View 323
Exercise 6.22: Skull—Midsagittal View 324
Exercise 6.23: Skull—Posterior View 324
Exercise 6.24: Skull—Inferior View 325
Exercise 6.25: Orbit—Anterolateral View 325
Exercise 6.26: Hyoid Bone—Anterior and Lateral View 326
Exercise 6.27: Clavicle 326
Self-Quiz 326
Exercise 6.28: Scapula 327
Exercise 6.29: Humerus 327
Exercise 6.30: Radius and Ulna 328
Exercise 6.31: Wrist and Hand 328
Self-Quiz 328
Self-Quiz 329
Self-Quiz 329
Exercise 6.32: Thoracic Cage 329
Exercise 6.33: Vertebral Column 330
Exercise 6.34: Pelvic Girdle—Female 330
Self-Quiz 330
Self-Quiz 331
Exercise 6.35: Pelvic Girdle—Male 331
Exercise 6.36: Sacrum and Coccyx 331
Exercise 6.37: Hip Bone 332
Exercise 6.38: Femur 332
Self-Quiz 333
Exercise 6.39: Tibia and Fibula 333
Exercise 6.40: Ankle and Foot 333
Exercise 6.41a: Skeletal Muscle—Histology 334
Exercise 6.41b: Skeletal Muscle—Histology 334
Self-Quiz 334
Exercise 6.42: Skeletal Muscle (Striations)—Histology 335
Exercise 6.43: Sarcomere—Histology 335
Exercise 6.44a: Neuromuscular Junction—Histology 335
Exercise 6.44b: Neuromuscular Junction—Histology 336
Exercise 6.45a: Smooth Muscle—Histology 336
Exercise 6.45b: Smooth Muscle—Histology 336
Self-Quiz 337
Exercise 6.46: Cardiac Muscle—Histology 337
Exercise 6.47: Cardiac Muscle (Intercalated Disc)—Histology 337

In Review
What Have I Learned? 338
Exercise 6.48: Coloring Exercise 339

Module 7: The Nervous System 341

Overview: Nervous System 341
Exercise 7.1: Multipolar Neuron—Golgi Stain—Histology 349
Exercise 7.2: Axon Hillock—Histology 349
Exercise 7.3: Unmyelinated Axon—Histology 350
Exercise 7.4: Synapse—Histology 350
Exercise 7.5a: Neuromuscular Junction—Histology 351
Exercise 7.5b: Nervous System—Neuromuscular Junction—Histology 351
Exercise 7.6a: Schwann Cell—Histology 352
Exercise 7.6b: Nervous System—Schwann Cell—Histology 352
Self-Quiz 353

In Review
What Have I Learned? 353

The Brain 355
Exercise 7.7: Nervous System—Brain—Coronal View 356
Self-Quiz 359
Exercise 7.8: Nervous System—Brain—Lateral View 359
Self-Quiz 362
Exercise 7.9: Nervous System—Brain—Superior View 362
Self-Quiz 365
Exercise 7.10: Nervous System—Brain—Inferior View 366
Self-Quiz 367
Exercise 7.11: Nervous System—Brain—Inferior View (close-up) 368
Self-Quiz 369
Exercise 7.12a: Imaging—Brain 369
Exercise 7.12b: Imaging—Brain 370
Exercise 7.12c: Imaging—Brain 370
Exercise 7.12d: Imaging—Brain 371

In Review
What Have I Learned? 372

Cranial Nerves 372
Exercise 7.13: Nervous System—Cranial Nerves—Inferior Brain (CN I-XII) 372
Exercise 7.14: Nervous System—Cranial Nerves—CN I Olfactory 373
Exercise 7.15: Nervous System—Cranial Nerves—CN II Optic 374
Exercise 7.16: Nervous System—Cranial Nerves—CN III Oculomotor 375
Exercise 7.17: Nervous System—Cranial Nerves—CN IV Trochlear 376
Exercise 7.18: Nervous System—Cranial Nerves—CN V Trigeminal 376
Exercise 7.19: Nervous System—Cranial Nerves—CN VI Abducens 377
Exercise 7.20: Nervous System—Cranial Nerves—CN VII Facial 378
Exercise 7.21: Nervous System—Cranial Nerves—CN VIII Vestibulocochlear 379
Exercise 7.22: Nervous System—Cranial Nerves—CN IX Glossopharyngeal 380
Exercise 7.23: Nervous System—Cranial Nerves—CN X Vagus 381
Exercise 7.24: Cranial Nerves—CN XI Accessory 382
Exercise 7.25: Nervous System—Cranial Nerves—CN XII Hypoglossal 383

In Review
What Have I Learned? 384

Spinal Cord 385
Exercise 7.26: Nervous System—Spinal Cord—Overview 385
Exercise 7.27: Nervous System—Spinal Cord—Typical Spinal Nerve 388
Exercise 7.28: Nervous System—Spinal Cord—Cervical Region 389
Exercise 7.29: Nervous System—Spinal Cord—Thoracic Region 391
Exercise 7.30: Nervous System—Spinal Cord—Lumbar Region 392
Self-Quiz 393

In Review
    What Have I Learned? 393

Peripheral Nerves 394
    Exercise 7.31: Nervous System—Peripheral Nerves—Cervical Plexus 394
    Self-Quiz 396
    Exercise 7.32: Nervous System—Peripheral Nerves—Brachial Plexus 396
    Self-Quiz 399
    Exercise 7.33: Nervous System—Peripheral Nerves—Upper Limb—Anterior View 399
    Self-Quiz 400
    Exercise 7.34: Nervous System—Peripheral Nerves—Upper Limb—Posterior View 400
    Self-Quiz 401
    Exercise 7.35: Nervous System—Peripheral Nerves—Shoulder and Arm—Anterior View 401
    Exercise 7.36: Nervous System—Peripheral Nerves—Shoulder and Arm—Posterior View 404
    Exercise 7.37: Nervous System—Peripheral Nerves—Arm, Forearm, and Hand—Anterior View 407
    Exercise 7.38: Nervous System—Peripheral Nerves—Arm, Forearm, and Hand—Posterior View 411
    Exercise 7.39: Nervous System—Peripheral Nerves—Hand—Palmar View 414
    Exercise 7.40: Nervous System—Peripheral Nerves—Hand—Dorsum View 417
    Exercise 7.41: Nervous System—Peripheral Nerves—Trunk 419
    Self-Quiz 420
    Exercise 7.42: Nervous System—Peripheral Nerves—Lumbosacral Plexus 421
    Self-Quiz 422
    Exercise 7.43: Nervous System—Peripheral Nerves—Lower Limb—Anterior View 422
    Self-Quiz 424
    Exercise 7.44: Nervous System—Peripheral Nerves—Lower Limb—Posterior View 424
    Exercise 7.45: Nervous System—Peripheral Nerves—Hip and Thigh—Anterior View 425
    Exercise 7.46: Nervous System—Peripheral Nerves—Hip and Thigh—Posterior View 429
    Exercise 7.47: Nervous System—Peripheral Nerves—Knee—Anterior View 433
    Exercise 7.48: Nervous System—Peripheral Nerves—Knee—Posterior View 434
    Exercise 7.49: Nervous System—Peripheral Nerves—Leg and Foot—Anterior View 438
    Exercise 7.50: Nervous System—Peripheral Nerves—Leg and Foot—Posterior View 440
    Exercise 7.51: Nervous System—Peripheral Nerves—Foot—Dorsum View 444
    Exercise 7.52: Nervous System—Peripheral Nerves—Foot—Plantar View 446
    Self-Quiz 449

Autonomic Nervous System 449
    Exercise 7.53: Nervous System—Sympathetic (ANS)—Overview 449
    Exercise 7.54: Nervous System—Sympathetic (ANS)—Thoracic Region 450
    Self-Quiz 451
    Exercise 7.55: Nervous System—Parasympathetic (ANS)—Overview 451
    Exercise 7.56: Nervous System—Parasympathetic (ANS)—Inferior Brain 452

In Review
    What Have I Learned? 452

The Senses 453
    Exercise 7.57: Nervous System—Taste—Inferior Brain 453
    Exercise 7.58: Nervous System—Taste—Tongue—Superior View 453
    Exercise 7.59: Nervous System—Taste—Vallate Papilla—Histology 454
    Exercise 7.60: Nervous System—Taste—Taste Bud—Histology 454
    Exercise 7.61: Nervous System—Smell—Inferior Brain 455
    Exercise 7.62: Nervous System—Smell—Nasal Cavity—Lateral View 455
    Exercise 7.63: Nervous System—Smell—Olfactory Mucosa—Histology 456
    Exercise 7.64: Nervous System—Hearing/Balance—Inferior Brain 457
    Exercise 7.65: Nervous System—Hearing/Balance—Ear—Anterior View 457
    Exercise 7.66: Imaging—Tympanic Membrane 458
    Exercise 7.67: Imaging—Tympanic Membrane with Otitis Media 458
    Exercise 7.68a: Nervous System—Hearing/Balance—Cochlea—Histology 459
    Exercise 7.68b: Nervous System—Hearing/Balance—Cochlea—Histology 459
    Exercise 7.69: Nervous System—Hearing/Balance—Spiral Organ—Histology 460
    Exercise 7.70: Nervous System—Vision—Inferior Brain 461
    Exercise 7.71: Nervous System—Vision—Orbit—Lateral View 461
    Exercise 7.72: Nervous System—Vision—Eye—Lateral View 463

In Review
    What Have I Learned? 464
    Exercise 7.73: Nervous System—Vision—Retina—Histology 464
    Exercise 7.74: Nervous System—Vision—Retinal Rods and Cones—Histology 465
    Exercise 7.75: Imaging—Retina 465

In Review
    What Have I Learned? 466
    Exercise 7.76: Coloring Exercise 471
    Exercise 7.77: Coloring Exercise 472

Module 8: The Endocrine System 473

Overview: The Endocrine System 473
    Self-Quiz 474
    Self-Quiz 474
    Self-Quiz 474
    Self-Quiz 475

The Hypothalamus, Pituitary, and Pineal Glands 475
    Exercise 8.1: Endocrine System—Hypothalamus/Pituitary/Pineal—Lateral View 475
    Exercise 8.2: Imaging—Hypothalamus and Pituitary Gland 476

The Pituitary Gland 477
    Exercise 8.3: Endocrine System—Pituitary—Histology 477

Exercise 8.4: Endocrine System—Anterior Pituitary—Histology 477
Exercise 8.5: Endocrine System—Posterior Pituitary—Histology 478
Exercise 8.6: Endocrine System—Posterior Pituitary—Histology 478

The Thyroid Gland 479
Self-Quiz 480
Exercise 8.7: Endocrine System—Thyroid Gland—Anterior View 480
Exercise 8.8a: Endocrine System—Thyroid Gland—Histology 482
Exercise 8.8b: Endocrine System—Thyroid Gland—Histology 482

The Parathyroid Glands 483
Self-Quiz 483
Exercise 8.9a: Endocrine System—Parathyroid Gland—Histology 483
Exercise 8.9b: Endocrine System—Parathyroid Gland—Histology 484
Exercise 8.9c: Endocrine System—Parathyroid Gland—Histology 484

The Pancreas 485
Self-Quiz 485
Exercise 8.10: Endocrine System—Pancreas—Anterior View 488
Exercise 8.11a: Endocrine System—Pancreas—Histology 489
Exercise 8.11b: Endocrine System—Pancreas—Histology 490
Exercise 8.12: Endocrine System—Pancreas Alpha Cell—Histology 490
Exercise 8.13: Endocrine System—Pancreas Beta Cell—Histology 491
Exercise 8.14: Endocrine System—Pancreas (Endocrine)—Histology 491
Exercise 8.15: Imaging—Pancreas 492

The Suprarenal (Adrenal) Gland 492
Self-Quiz 493
Exercise 8.16: Endocrine System—Suprarenal (Adrenal) Gland—Anterior View 493
Exercise 8.17a: Endocrine System—Suprarenal Gland—Histology 494
Exercise 8.17b: Endocrine System—Suprarenal Gland—Histology 494
Exercise 8.18: Endocrine System—Suprarenal Gland, Zona Glomerulosa, and Fasciculata—Histology 495
Exercise 8.19: Endocrine System—Suprarenal Gland, Zona Reticularis—Histology 495

The Ovary 496
Exercise 8.20: Endocrine System—Ovary—Superior View 496

The Testis 496
Exercise 8.21: Endocrine System—Testis and Spermatic Cord—Anterior View 496
Exercise 8.22: Endocrine System—Testis and Spermatic Cord (Isolated)—Lateral View 497

The Seminiferous Tubule 499
Exercise 8.23: Endocrine System—Interstitial (Leydig) Cell of Seminiferous Tubule—Histology 499

In Review
What Have I Learned? 500
Self-Quiz 500

Module 9: The Cardiovascular System 501

Overview: Cardiovascular System 501
Self-Quiz 501

Blood 501
Self-Quiz 502
Self-Quiz 502

In Review
What Have I Learned? 503
Exercise 9.1: Cardiovascular System—Blood (Peripheral Smear)—Histology 503
Exercise 9.2a: Cardiovascular System—Erythrocyte—Histology 504
Exercise 9.2b: Cardiovascular System—Erythrocyte—Histology 504
Exercise 9.2c: Cardiovascular System—Erythrocyte—Histology 505
Exercise 9.3: Cardiovascular System—Neutrophil—Histology 505
Exercise 9.4: Cardiovascular System—Eosinophil—Histology 506
Exercise 9.5: Cardiovascular System—Basophil—Histology 506
Exercise 9.6: Cardiovascular System—Lymphocyte—Histology 507
Exercise 9.7: Cardiovascular System—Monocyte—Histology 507
Exercise 9.8: Cardiovascular System—Megakaryocyte—Histology 508
Self-Quiz 511

Elastic Artery—The Aorta 512
Exercise 9.9a: Cardiovascular System—Elastic Artery (Aorta)—Histology 512
Exercise 9.9b: Cardiovascular System—Elastic Artery (Aorta)—Histology 513

In Review
What Have I Learned? 513

Large Vein—The Inferior Vena Cava 514
Exercise 9.10: Cardiovascular System—Large Vein (Inferior Vena Cava)—Histology 514

In Review
What Have I Learned? 514

Muscular Artery and Medium-sized Vein 515
Exercise 9.11: Cardiovascular System—Muscular Artery and Medium-sized Vein—Histology 515

In Review
What Have I Learned? 515

Arteriole and Venule 516
Exercise 9.12: Cardiovascular System—Arteriole and Venule—Histology 516

In Review
What Have I Learned? 516

Neurovascular Bundle 517
Exercise 9.13: Cardiovascular System—Neurovascular Bundle—Histology 517

In Review
What Have I Learned? 517
Self-Quiz 518

The Heart 518
    Exercise 9.14: Cardiovascular System—
        Heart—Internal Features—
        Anterior View 518
    Exercise 9.15: Cardiovascular System—Heart—
        Vasculature—Anterior View 522
    Exercise 9.16: Cardiovascular System—Heart—
        Vasculature—Posterior View 523
    Exercise 9.17: Imaging—Heart—Left Ventricle 525
    Exercise 9.18: Imaging—Heart—Right Ventricle 525
    Self-Quiz 526

Cardiac Muscle 526
    Exercise 9.19: Cardiovascular System—Cardiac
        Muscle—Histology 526

In Review
    What Have I Learned? 527
    Self-Quiz 527

In Review
    What Have I Learned? 537

The Thorax 537
    Exercise 9.20: Cardiovascular System—Thorax—
        Arteries—Anterior View 537
    Exercise 9.21a: Imaging—Aortic Arch 541
    Exercise 9.21b: Imaging—Aortic Arch 541
    Exercise 9.22: Cardiovascular System—Thorax—
        Veins—Anterior View 542
    Self-Quiz 545
    Exercise 9.23: Imaging—Thorax 546
    Self-Quiz 546

In Review
    What Have I Learned? 546

The Head and Neck 547
    Exercise 9.24: Cardiovascular System—
        Head and Neck—Vasculature—
        Lateral View 547
    Exercise 9.25: Imaging—Carotid Artery 550
    Self-Quiz 550
    Self-Quiz 551
    Exercise 9.26a: Imaging—Head and Neck 551
    Exercise 9.26b: Imaging—Head and Neck 552
    Exercise 9.26c: Imaging—Head and Neck 552
    Exercise 9.26d: Imaging—Head and Neck 553
    Self-Quiz 553

In Review
    What Have I Learned? 553

The Brain 554
    Exercise 9.27: Cardiovascular System—Brain—
        Arteries—Inferior View 554
    Exercise 9.28: Cardiovascular System—Brain—Veins—
        Lateral View 555
    Self-Quiz 557

In Review
    What Have I Learned? 558

The Shoulder 559
    Exercise 9.29: Cardiovascular System—Shoulder—
        Arteries—Anterior View 559
    Exercise 9.30: Cardiovascular System—Shoulder—
        Veins—Anterior View 561
    Self-Quiz 562

In Review
    What Have I Learned? 562

The Shoulder and Arm 563
    Exercise 9.31: Cardiovascular System—Shoulder and
        Arm—Vasculature—Anterior View 563
    Exercise 9.32: Cardiovascular System—Shoulder and
        Arm—Vasculature—Posterior View 566
    Exercise 9.33: Cardiovascular System—Shoulder and
        Arm—Arteries—Anterior View 567
    Exercise 9.34: Cardiovascular System—Shoulder and
        Arm—Veins—Anterior View 568
    Self-Quiz 569

In Review
    What Have I Learned? 569
    Exercise 9.35: Cardiovascular System—Arm, Forearm,
        and Hand Vasculature—Anterior View 570
    Exercise 9.36: Cardiovascular System—Arm, Forearm,
        and Hand Vasculature—Posterior View 573

The Forearm and Hand 576
    Exercise 9.37: Cardiovascular System—Forearm and
        Hand—Arteries—Anterior View 576
    Exercise 9.38: Cardiovascular System—Forearm and
        Hand—Veins—Anterior View 577
    Exercise 9.39: Imaging—Hand 578
    Exercise 9.40: Cardiovascular System—Hand
        Vasculature—Dorsum 578
    Exercise 9.41: Cardiovascular System—Hand
        Vasculature—Palmar 580
    Self-Quiz 583

In Review
    What Have I Learned? 583

The Abdomen 584
    Exercise 9.42: Cardiovascular System—Abdomen—
        Celiac Trunk—Anterior View 584
    Exercise 9.43: Imaging—Abdominal Aorta and
        Branches 587
    Exercise 9.44: Imaging—Abdominal Aorta and Iliac
        Arteries 588
    Exercise 9.45a: Imaging—Aortic Aneurysm 589
    Exercise 9.45b: Imaging—Aortic Aneurysm 589
    Exercise 9.46: Cardiovascular System—Abdomen—
        Mesenteric Arteries—Anterior View 590
    Self-Quiz 594
    Exercise 9.47: Cardiovascular System—Abdomen—
        Veins—Anterior View 594
    Self-Quiz 597

In Review
    What Have I Learned? 597

The Pelvis 598
    Exercise 9.48: Cardiovascular System—Pelvis—
        Female—Vasculature—Anterior View 598
    Self-Quiz 600
    Exercise 9.49: Cardiovascular System—Pelvis—Male—
        Vasculature—Anterior View 600
    Self-Quiz 602

In Review
    What Have I Learned? 603
    Exercise 9.50: Cardiovascular System—Hip and Thigh
        Vasculature—Anterior View 604
    Exercise 9.51: Cardiovascular System—Hip and Thigh
        Vasculature—Posterior View 606
    Exercise 9.52: Cardiovascular System—Hip and Thigh—
        Arteries—Anterior View 610
    Exercise 9.53: Cardiovascular System—Hip and Thigh—
        Arteries—Posterior View 611

Exercise 9.54: Cardiovascular System—Hip and Thigh—Veins—Anterior View   612
Exercise 9.55: Cardiovascular System—Hip and Thigh—Veins—Posterior View   613
Self-Quiz   614

In Review
What Have I Learned?   615

The Knee   615
Exercise 9.56: Cardiovascular System—Knee Vasculature—Anterior View   615
Exercise 9.57: Cardiovascular System—Knee Vasculature—Posterior View   617
Exercise 9.58: Cardiovascular System—Knee—Arteries—Anterior View   620
Exercise 9.59: Cardiovascular System—Knee—Arteries—Posterior View   621
Exercise 9.60: Cardiovascular System—Knee—Veins—Anterior View   623
Exercise 9.61: Cardiovascular System—Knee—Veins—Posterior View   624
Exercise 9.62: Imaging—Knee   625
Self-Quiz   625

In Review
What Have I Learned?   626

The Leg and Foot   626
Exercise 9.63: Cardiovascular System—Leg and Foot Vasculature—Anterior View   626
Exercise 9.64: Cardiovascular System—Leg and Foot Vasculature—Posterior View   628
Exercise 9.65: Cardiovascular System—Leg and Foot—Arteries—Anterior View   631
Exercise 9.66: Cardiovascular System—Leg and Foot—Arteries—Posterior View   633
Exercise 9.67: Cardiovascular System—Leg and Foot—Veins—Anterior View   634
Exercise 9.68: Cardiovascular System—Leg and Foot—Veins—Posterior View   635
Exercise 9.69: Cardiovascular System—Foot Vasculature—Dorsum   636
Exercise 9.70: Cardiovascular System—Foot Vasculature—Plantar   638
Self-Quiz   640

In Review
What Have I Learned?   640
Exercise 9.71: Coloring Exercise   641
Exercise 9.72: Labeling Exercise   642
Exercise 9.73: Labeling Exercise   643

Module 10: The Lymphatic System   645

Overview: Lymphatic System   645
Self-Quiz   646
Exercise 10.1: Lymphatic System—B Lymphocyte—Histology   652
Exercise 10.2: Lymphatic System—Plasma Cell—Histology   652
Exercise 10.3: Lymphatic System—Antigen-presenting Cells—Histology   653
Exercise 10.4: Lymphatic System—Macrophage—Histology   653
Self-Quiz   653
Exercise 10.5: Lymphatic System—Lymph Node—Histology   654
Exercise 10.6: Lymphatic System—Lymph Node—Histology   654
Exercise 10.7a: Lymphatic System—Lymph Node (Lymphoid Nodule)—Histology   655
Exercise 10.7b: Lymphatic System—Lymph Node (Lymphoid Nodule)—Histology   655
Exercise 10.8: Lymphatic System—Lymph Node (Medullary Sinus)—Histology   656
Exercise 10.9: Imaging—Pelvis—Lymphangiogram—Anterior–Posterior   656

In Review
What Have I Learned?   657
Exercise 10.10: Lymphatic System—Tonsils   657

In Review
What Have I Learned?   659

Palatine Tonsil   660
Exercise 10.11a: Lymphatic System—Palatine Tonsil—Histology   660
Exercise 10.11b: Lymphatic System—Palatine Tonsil—Histology   660

In Review
What Have I Learned?   661

Peyer's Patch   661
Exercise 10.12: Lymphatic System—Peyer's Patch—Histology   661

In Review
What Have I Learned?   662

The Thorax   662
Exercise 10.13: Lymphatic System—Thorax—Anterior View   662

In Review
What Have I Learned?   663
Exercise 10.14: Lymphatic System—Breast and Axillary Nodes—Female   664

The Spleen   664
Exercise 10.15: Lymphatic System—Spleen—Anterior View   664
Exercise 10.16: Lymphatic System—Spleen—Histology   667
Exercise 10.17: Lymphatic System—Spleen—Lymphoid Nodule—Histology   668
Exercise 10.18: Lymphatic System—Spleen (Unstimulated White Pulp)   668

In Review
What Have I Learned?   669

The Thymus   669
Exercise 10.19: Lymphatic System—Thymus—Adult—Anterior View   669
Exercise 10.20: Lymphatic System—Thymus—Fetus—Anterior View   671
Exercise 10.21a: Lymphatic System—Fetal Thymus—Histology   671
Exercise 10.21b: Lymphatic System—Fetal Thymus—Histology   672
Self-Quiz   672
Exercise 10.22: Lymphatic System—Thymus (Hassall's Corpuscle)—Histology   672

In Review
What Have I Learned?   674

Module 11: The Respiratory System 675

Overview: The Respiratory System 675
Self-Quiz 676

The Upper Respiratory System 677
Exercise 11.1: Respiratory System—Upper Respiratory—Lateral View 677
Exercise 11.2: Respiratory System—Nasal Cavity—Coronal View 679
Exercise 11.3a: Imaging—Upper Respiratory System 682
Exercise 11.3b: Imaging—Upper Respiratory System 682
Self-Quiz 683

In Review
What Have I Learned? 683

The Lower Respiratory System 684
Exercise 11.4: Respiratory System—Lower Respiratory—Anterior View 684
Self-Quiz 686
Self-Quiz 687
Self-Quiz 687
Self-Quiz 688
Exercise 11.5a: Imaging—Lower Respiratory System 688
Exercise 11.5b: Imaging—Lower Respiratory—Pneumonia 689
Exercise 11.5c: Imaging—Lower Respiratory—Pneumonia 689

In Review
What Have I Learned? 690

The Trachea 690
Exercise 11.6a: Respiratory System—Trachea—Histology 690
Exercise 11.6b: Respiratory System—Trachea—Histology 691
Exercise 11.7: Respiratory System—Respiratory Epithelium 691

In Review
What Have I Learned? 692

The Alveolus and Alveolar Duct 692
Exercise 11.8: Respiratory System—Alveolar Duct—Histology 692
Exercise 11.9a: Respiratory System—Alveolus—Histology 693
Exercise 11.9b: Respiratory System—Alveolus—Histology 693
Self-Quiz 693

In Review
What Have I Learned? 693
Exercise 11.10: Respiratory System—Larynx—Anterior View 694
Exercise 11.11: Respiratory System—Larynx—Lateral View 695
Exercise 11.12: Respiratory System—Larynx—Posterior View 697

In Review
What Have I Learned? 699

Module 12: The Digestive System 713

Overview: The Digestive System 713
Self-Quiz 714

The Oral Cavity 715
Exercise 12.1: Digestive System—Oral Cavity and Pharynx—Lateral View 715

In Review
What Have I Learned? 716

The Salivary Glands and Teeth 717
Exercise 12.2: Digestive System—Salivary Glands—Lateral View 717
Exercise 12.3: Digestive System—Teeth—Superior and Inferior Views 719

In Review
What Have I Learned? 720

The Esophagus 721
Exercise 12.4: Digestive System—Esophagus—Anterior View 721
Exercise 12.5a: Digestive System—Esophagus—Histology 723
Exercise 12.5b: Digestive System—Esophagus 724
Exercise 12.5c: Digestive System—Gastro-esophageal Junction 724

The Abdominal Cavity 725
Exercise 12.6: Digestive System—Abdominal Cavity—Anterior View 725

In Review
What Have I Learned? 728

The Stomach 729
Self-Quiz 729
Exercise 12.7a: Digestive System—Stomach—Histology 729
Exercise 12.7b: Digestive System—Stomach—Histology 730
Exercise 12.7c: Digestive System—Stomach—Histology 730
Exercise 12.7d: Digestive System—Stomach—Histology 731
Exercise 12.8: Digestive System—Stomach and Duodenum—Anterior View 731
Self-Quiz 733
Self-Quiz 733

In Review
What Have I Learned? 734

The Small Intestine 734
Exercise 12.9: Digestive System—Duodenum—Histology 734
Exercise 12.10a: Digestive System—Jejunum—Histology 735
Exercise 12.10b: Digestive System—Jejunum—Histology 735
Exercise 12.11a: Digestive System—Ileum—Histology 736
Exercise 12.11b: Digestive System—Ileum—Histology 736
Exercise 12.11c: Digestive System—Ileum—Histology 737
Exercise 12.12: Digestive System—Small Intestine—Histology 737
Exercise 12.13: Digestive System—Peyer's Patch—Histology 738
Exercise 12.14: Digestive System—Intestinal Microvilli—Histology 738

Exercise 12.15: Imaging—Stomach and Small Intestine 739
Exercise 12.16: Imaging—Digestive—Small Intestine 739

In Review
What Have I Learned? 740

The Colon 741
Exercise 12.17a: Digestive System—Colon—Histology 741
Exercise 12.17b: Digestive System—Colon—Histology 741
Exercise 12.18a: Imaging—Digestive—Colon 742
Exercise 12.28b: Imaging—Colon 742

In Review
What Have I Learned? 743

The Peritoneum 743
Exercise 12.19: Digestive System—Peritoneum—Midsagittal View 743
Self-Quiz 744

In Review
What Have I Learned? 744

The Liver 744
Self-Quiz 744
Exercise 12.20: Digestive System—Liver—Anterior and Postero-inferior Views 745
Exercise 12.21a: Digestive System—Liver—Histology 745
Exercise 12.21b: Digestive System—Liver—Histology 746
Exercise 12.21c: Digestive System—Liver—Histology 746
Exercise 12.21d: Digestive System—Liver—Histology 747
Exercise 12.22: Digestive System—Hepatocyte—Histology 747
Exercise 12.23: Digestive System—Biliary Ducts—Anterior View 748

In Review
What Have I Learned? 752

The Pancreas 752
Exercise 12.24: Digestive System—Pancreas (Exocrine)—Histology 752
Exercise 12.25a: Digestive System—Pancreas—Histology 753
Exercise 12.25b: Digestive System—Pancreas—Histology 753

In Review
What Have I Learned? 754

ATP Synthesis 754
Self-Quiz 755
Exercise 12.26: Coloring Exercise 757

Module 13: The Urinary System 759

Overview: The Urinary System 759
Self-Quiz 760

The Upper Urinary System 761
Exercise 13.1: Urinary System—Upper Urinary—Anterior View 761
Self-Quiz 763

In Review
What Have I Learned? 763

The Kidney 764
Self-Quiz 764
Exercise 13.2: Urinary System—Kidney—Anterior View 765
Exercise 13.3a: Imaging—Kidney 767
Exercise 13.3b: Imaging—Kidney 767
Self-Quiz 768

In Review
What Have I Learned? 768

The Renal Corpuscle 768
Exercise 13.4: Urinary System—Renal Corpuscle—Histology 768
Self-Quiz 769
Self-Quiz 770
Exercise 13.5: Urinary System—Renal Cortex—Histology 770
Exercise 13.6: Urinary System—Renal Medulla—Histology 771
Exercise 13.7: Urinary System—Renal Medulla (Longitudinal Section) 771
Exercise 13.8: Urinary System—Podocyte—Histology 772
Exercise 13.9: Urinary System—Distal Convoluted Tubule—Histology 772
Exercise 13.10: Urinary System—Proximal Convoluted Tubule—Histology 773

In Review
What Have I Learned? 773

The Ureter 783
Exercise 13.11a: Urinary System—Ureter—Histology 783
Exercise 13.11b: Urinary System—Ureter—Histology 783
Exercise 13.12a: Imaging—Kidney, Ureter, and Urinary Bladder 784
Exercise 13.12b: Imaging—Kidney, Ureter, and Urinary Bladder 784

The Female Lower Urinary System 785
Exercise 13.13: Urinary System—Lower Urinary—Female—Sagittal View 785
Self-Quiz 785

The Male Lower Urinary System 786
Exercise 13.14: Urinary System—Lower Urinary—Male—Sagittal View 786
Self-Quiz 786

In Review
What Have I Learned? 787

The Urinary Bladder 787
Exercise 13.15: Urinary System—Urinary Bladder—Histology 787
Exercise 13.16: Urinary System—Urinary Bladder—Histology 788
Self-Quiz 788

In Review
What Have I Learned? 789
Self-Quiz 789
Exercise 13.17: Coloring Exercise 790

Module 14: The Reproductive System   791

Overview: The Reproductive System   791

The Female Breast   792
  Exercise 14.1: Reproductive System—Breast—Female—Anterior View   792
  Exercise 14.2: Imaging—Reproductive System—Mammogram   795

In Review
  What Have I Learned?   797

The Female Pelvis   797
  Exercise 14.3: Reproductive System—Pelvis—Female—Sagittal View   797
  Exercise 14.4: Reproductive System—Pelvis—Female—Superior View   798
  Exercise 14.5: Imaging—Uterus and Vagina   799

In Review
  What Have I Learned?   800

The Uterus   801
  Exercise 14.6a: Reproductive System—Uterus—Histology   801
  Exercise 14.6b: Reproductive System—Uterus—Histology   801
  Exercise 14.7: Reproductive System—Ovary—Histology   802

The Ovarian Follicle   802
  Exercise 14.8: Reproductive System—Primordial Follicle—Histology   804
  Exercise 14.9: Reproductive System—Primary Follicle—Histology   805
  Exercise 14.10: Reproductive System—Secondary Follicle—Histology   806
  Exercise 14.11: Reproductive System—Corpus Luteum—Histology   807
  Exercise 14.12: Reproductive System—Corpus Albicans—Histology   808

The Uterine Tube   808
  Exercise 14.13a: Reproductive System—Uterine Tube—Histology   808
  Exercise 14.13b: Reproductive System—Uterine Tube—Histology   809
  Exercise 14.13c: Reproductive System—Uterine Tube—Histology   809

The Female Perineum   810
  Exercise 14.14: Reproductive System—Perineum—Female—Inferior View   810
  Exercise 14.15: Reproductive System—Vagina—Histology   811

In Review
  What Have I Learned?   812

The Male Pelvis   812
  Exercise 14.16: Reproductive System—Pelvis—Male—Sagittal View   813

In Review
  What Have I Learned?   815

The Prostate   815
  Exercise 14.17: Reproductive System—Prostate—Histology   815

The Male Perineum   817
  Exercise 14.18: Reproductive System—Perineum—Male—Inferior View   817

In Review
  What Have I Learned?   818

The Penis and Scrotum   819
  Exercise 14.19: Reproductive System—Penis and Scrotum—Anterior View   819

In Review
  What Have I Learned?   822

The Seminiferous Tubule   822
  Exercise 14.20a: Reproductive System—Seminiferous Tubule—Histology   822
  Exercise 14.20b: Reproductive System—Seminiferous Tubule—Histology   823

In Review
  What Have I Learned?   823
  Self-Quiz   823
  Exercise 14.21: Coloring Exercise   824
  Exercise 14.22: Coloring Exercise   825

INTRODUCTION

Becoming Familiar with *Anatomy and Physiology | Revealed*®

Overview: Introduction

Welcome to *Anatomy and Physiology | Revealed*®, the premier learning tool for the Anatomy and Physiology student. *Anatomy and Physiology | Revealed*® is a powerful software tool to guide you through 11 body systems, plus Body Orientation, Chemistry, Cells, and Tissues. As I previewed this newly revised version 3.2 of *Anatomy and Physiology | Revealed*®, one thought kept coming to my mind: "Wow, is there anything that could be added or improved on beyond what you have available in this edition?" My answer, a resounding "not that I'm aware of!" I must tell you how excited I am to be a part of this project. My students have used the first three editions of *Anatomy and Physiology | Revealed*® with great success over the years, and I can't wait to engage them with this newest edition. I trust you will experience similar results!

So, let's dive right into this tutorial module so that you can become familiar with the ins and outs of this robust tool.

First we must note that there are two different methods by which you can experience *Anatomy and Physiology | Revealed*®, and this Workbook is designed to complement each one.

The first option is the standard online version of *Anatomy and Physiology | Revealed*®, where you subscribe for a period of time and have access to the complete software 24/7 during that time. All you need is an Internet connection and an access code and you are up and running.

The second option is the online version of *Anatomy and Physiology | Revealed*® with specific content selected for you by your instructor. This content is referred to as *my* **Course Content**. The **Dissection Study Area** alone contains more than 11,000 structures in 260 views. The *my* **Course Content** option allows your instructor to pare this number down to just the specific structures you are required to know.

This workbook has been written with a focus on the standard online access option. This approach is not intended to neglect those of you using the *my* **Course Content** option. Rather, as the structures your instructor has selected are included in the overall list of structures incorporated into *Anatomy and Physiology | Revealed*®, you will simply skip those exercises that your instructor does not require you to complete.

So, let's get started, by logging on to the *Anatomy and Physiology | Revealed*® website with the information supplied by your instructor.

Upon opening *Anatomy and Physiology | Revealed*®, you will see the **HOME** screen, shown here.

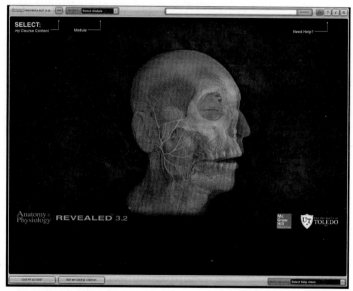

©McGraw-Hill Education

Navigate around the **HOME** screen, starting at the top left. The software version of *Anatomy and Physiology | Revealed*® is listed in the top left corner (as well as just below and to the left of the central image). To the right of this is the *my* button. Your instructor may have selected specific structures and other information for you to learn, and the *my* button allows you access to this material. We discuss this in more detail later.

To the right of the *my* button is the **MODULE** menu. **Select module** is visible in the drop-down menu. Click the downward arrow and you will see the 14 modules covered in *Anatomy and Physiology | Revealed*®. Click the downward arrow again to close this menu.

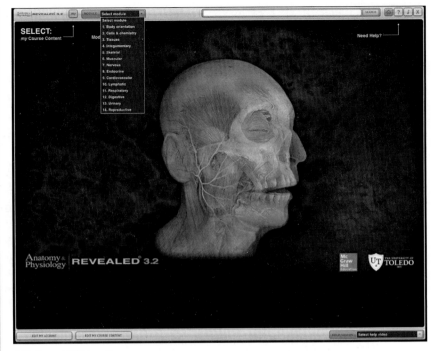
©McGraw-Hill Education

> **HEADS UP!**
>
> *You will find, as you go through this workbook, that all terms referring to items on the screen of* Anatomy and Physiology | Revealed® *are in bold type. For example, the opening screen is referred to as the* **HOME** *screen. Later, you will see the names of the different buttons in* Anatomy and Physiology | Revealed® *listed in bold in the same case in this workbook as they appear on your screen. An example of this is the* **DISSECTION** *button. This is intended to make it easier for you to refer to these terms in this workbook while viewing* Anatomy and Physiology | Revealed® *on your computer screen.*

To the right of the **MODULE** menu is the **SEARCH** window. Type in a term or structure name such as "blood" in the blank space on the left and click **SEARCH.**

In response, you will see three tabs entitled **Structures, Animations,** and **General Topics/Views** from within *Anatomy and Physiology | Revealed*® pertaining to blood. Each tab has a list of links that contain information concerning blood. Click through the three tabs to view the links that they contain.

Mouse-over the text and it will become highlighted. Click the text and you will be linked to that location in *Anatomy and Physiology | Revealed*®.

Return to the **HOME** screen by clicking the **X** button to the right of the **SEARCH** window.

Speaking of the **HOME** screen, we will now look at the four **SUB-NAVIGATION** buttons at the top right of the **HOME** screen. If you mouse-over them, their identity is revealed. They are, in sequence from left to right, the **HOME** button, the **HELP** button, the **INFO** button, and the **EXIT** button. The **HOME** button will return you to the **HOME** screen from any location within *Anatomy and Physiology | Revealed*® where it is active.

The **HELP** button takes you to a list of help videos.

Click this button and notice that each help video has the specific icon for that topic when applicable.

Click the **X** in the top right corner to close the **HELP** screen and click the **INFO** button.

This screen displays information about the publisher, McGraw-Hill, including a tech-support email address, the authors and creators of *Anatomy and Physiology | Revealed*®, and acknowledgments of professionals who aided in the development of *Anatomy and Physiology | Revealed*®. Click the **X** in the top right corner to close the information screen.

The **EXIT** button allows you to end your *Anatomy and Physiology | Revealed*® session.

In the lower left corner of the **HOME** screen are the **SET UP** menus. The left button, labeled **EDIT MY ACCOUNT**, allows you to set up access to *Anatomy and Physiology | Revealed*®. The **EDIT MY COURSE CONTENT** button allows your instructor to select the structures and other materials you will be required to know.

In the lower right corner is the **HELP VIDEOS** menu. Click on the down arrow and you will see a list of videos designed to get you off to a quick start with *Anatomy and Physiology | Revealed*®. As you can see, these videos will walk you through the basic steps of navigating the software and show you how the information is presented.

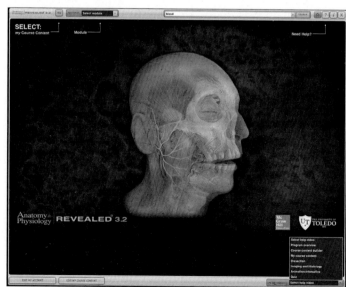

©McGraw-Hill Education

Before continuing on, click on and view each of these videos to get an overview of *Anatomy and Physiology | Revealed*®. As you do, note the **PAUSE** and **CLOSE** buttons near the top of the screen.

> **HEADS UP!**
> Terms used in this Workbook:
> *Click*—Click once with the left mouse button.
> *Mouse-over*—Place the mouse cursor over an object.
> *Select*—Click on a particular item in a menu.

We will use the **DIGESTIVE SYSTEM** to walk through the many facets of this powerful learning tool called *Anatomy and Physiology | Revealed®*. After you have mastered the techniques described here, you will be able to apply them to each of the modules.

Return to the **HOME** screen, if you are not there already. In the **Module** menu, click the **Select module** drop-down box.

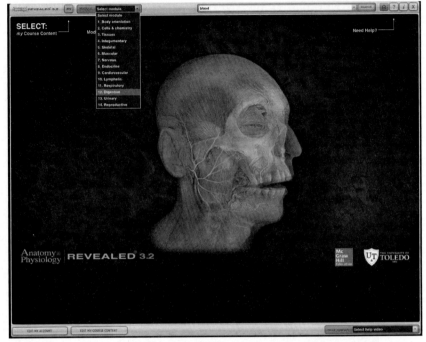
©McGraw-Hill Education

Click **Digestive,** and you will see the **DIGESTIVE SYSTEM** opening screen. We will explore the five buttons at the top center of the screen.

These five buttons, the **STUDY AREAS,** are the heart and soul of *Anatomy and Physiology | Revealed®*. It is through the use of these buttons that we can access the five areas of focus for *Anatomy and Physiology | Revealed®*. It should be pointed out that depending on the structures contained within each module, not all modules will have all five of these **STUDY AREAS** available to you.

The five **STUDY AREAS** and their respective buttons are:

DISSECTION   **ANIMATIONS**   **HISTOLOGY**

IMAGING   **QUIZ**

We will look at these one at a time, in the next five sections of this module.

CHECK POINT

Becoming Familiar with *Anatomy and Physiology | Revealed*®

1. What is the name for the opening screen of *Anatomy and Physiology | Revealed®*?
2. Why is the text of this workbook often in bold letters?
3. What are the functions of each of the four **SUB-NAVIGATION** buttons?
4. What button would you click for a helpful guided tour of *Anatomy and Physiology | Revealed®*?
5. What are the five **STUDY AREAS** in *Anatomy and Physiology | Revealed®*?

The DISSECTION Study Area

The first **STUDY AREA** button or icon is shaped like a scalpel. Mouse-over this button and it will identify itself as **DISSECTION,** as seen in the image shown here.

Click the **DISSECTION** button, and you will see the image shown here.

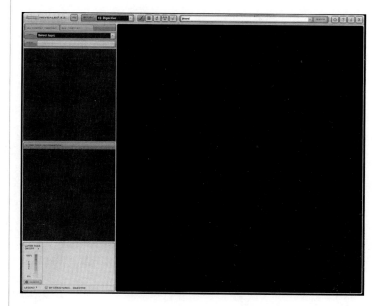

As you can see, there is not a lot of information available in this screen yet. Directly beneath the *Anatomy and Physiology | Revealed*® 3.2 at the top left are two tabs. If your instructor has selected specific structures for you to learn, they will be listed under the **my COURSE CONTENT** tab. Otherwise, you will learn from the list of all available structures under the **ALL CONTENT** tab. We will discuss the **my COURSE CONTENT** tab later in this module, so for now select the **ALL CONTENT** tab. Directly under this tab is the **Select topic** menu. Click here to see a list of the topics available for the digestive system.

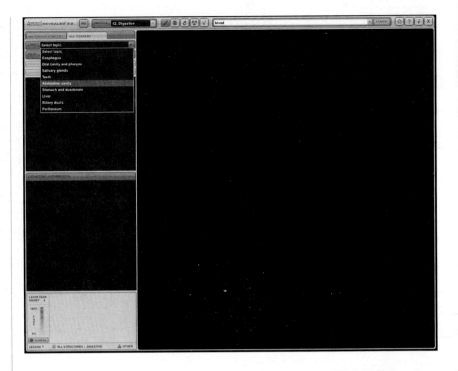

Select **Abdominal cavity** from the list. The **VIEW** menu defaults to **Anterior.** If we had chosen a topic that had multiple views, you would have the option of selecting a view from the **Select view** menu. An image of the anterior view of the abdomen is now visible in the large window to the right: the **IMAGE AREA.** The **Select structure type** menu is now available. Click the down arrow and select **Gastrointestinal tract** from the list.

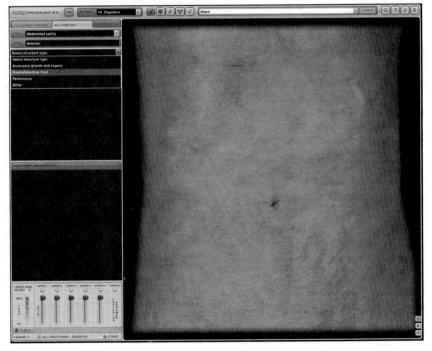

©McGraw-Hill Education

The **Select group** menu is now available. Click this menu and select **Small and large intestines** from the list.

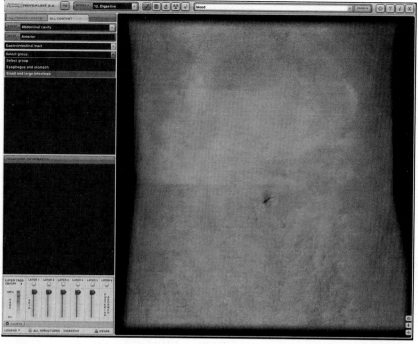
©McGraw-Hill Education

A list of the structures visible in the anterior view of the abdomen will appear. These are often accompanied by a number in brackets. This number refers to the layer of dissection where the image of that structure is located. Select **Ascending colon [4]** from the list. The ascending colon is now highlighted blue in the **IMAGE AREA**.

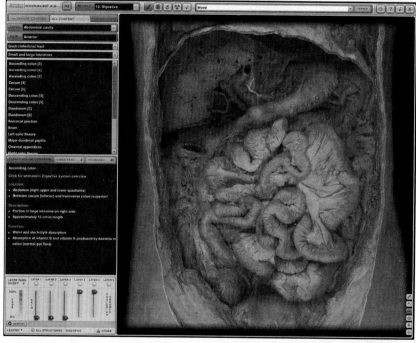
©McGraw-Hill Education

Notice that information pertaining to the structures highlighted in the **IMAGE AREA** is displayed in the **STRUCTURE INFORMATION** window. This window displays information pertaining to the highlighted structure, including location, description, and function. Other pertinent information is often included for various structures.

Notice also that *Anatomy and Physiology | Revealed*® also contains an animation that includes the ascending colon, as indicated in green type in the **INFORMATION** window. By clicking on this link, you will be taken to that animation.

Return to the list of structures and click through them to view them highlighted in the **IMAGE AREA** and to see the information for them displayed in the **STRUCTURE INFORMATION** window.

When finished, select **Transverse colon [3]** from the list before continuing.

There are two buttons located next to the title of the **STRUCTURE INFORMATION** window: **OTHER VIEWS** and **PRONOUNCE**.

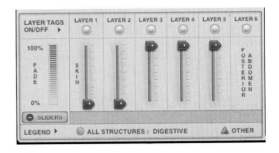

When the **OTHER VIEWS** button is available, it indicates that there are alternate views of the highlighted structure. To preview these images, click this button. A preview with a different view of the transverse colon appears in the **IMAGE AREA**, and the **STRUCTURE INFORMATION** window remains visible. At the top right of the new image is a number indicating which number this image is of the number of previews available. When the number of previews is 2 or more, you can use the **NEXT** and **BACK** buttons to move through the different images. If you want to fully interact with that image, click the **GO TO VIEW** button at the bottom right corner of the **PREVIEW IMAGE**. To return to the original image, close the **PREVIEW IMAGE** by clicking on the **X** in the top right corner.

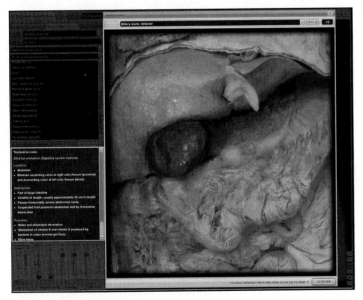

©McGraw-Hill Education

The **PRONOUNCE** button, when clicked, will give the correct pronunciation of the highlighted structure. Select several structures from the list and listen to the pronunciation of those terms.

Located at the bottom left of the screen, the **LAYER CONTROLS** window consists of **LAYER TAGS ON/OFF** buttons and several sliders.

The **LAYER TAGS ON/OFF** buttons allow you to quickly reveal specific layers of the selected image, and place pins (referred to as **TAGS**) on structures in the list above. Click the **LAYER 4** radio button and you will see the following image:

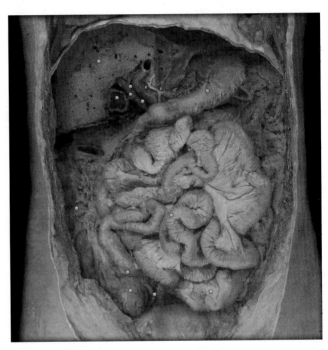

©McGraw-Hill Education

The pins come in two colors: green and blue. Green pins indicate structures that are components of the current system or module—in this case, those that *are* part of the digestive system. The blue pins indicate structures that are *not* part of the current system. In case you don't remember which pins are which, the **LEGEND** in the **LAYER CONTROLS** window will clarify this for you. The blue pins in *Anatomy and Physiology | Revealed*® version 3.2 are triangle shaped, as well as blue in color to make distinguishing the nonsystem structures even more obvious. This workbook contains images from both version 3.0, with round blue pins and version 3.2, with triangle-shaped blue pins.

Mouse-over the pins in the image to identify the structures.

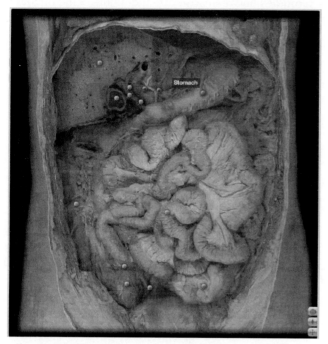
©McGraw-Hill Education

Clicking on a pin will highlight that structure, and the information for that structure appears in the **STRUCTURE INFORMATION** window.

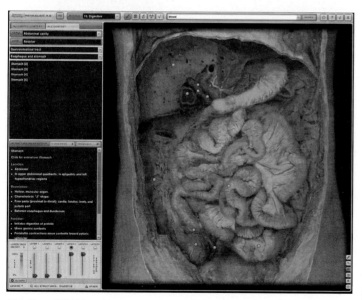
©McGraw-Hill Education

Notice also that the sliders have moved downward for the first two layers, indicating that this view is from the third layer. The structure name is also highlighted in the list above the **STRUCTURE INFORMATION** window. Click the layer buttons to reveal the various views provided for the small and large intestine.

The sliders are arranged from left to right to allow views from superficial to deep—in this example, from the **SKIN** to the **POSTERIOR ABDOMEN.** Move the sliders up and down to see layers of anatomy melt away. Note also that when a slider is in the downward position, it can be partially raised to reveal a "ghost" image of the structures in that layer. This technique will allow you to see the relationships between structures of different layers.

Six buttons are located at the bottom right of the image when a structure is highlighted, while three are visible otherwise.

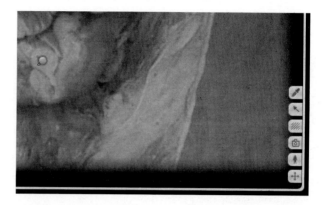

The top three buttons are only available when a structure is highlighted, which occurs when you click on a **TAG** or pin. We will look at those three first.

The top button is the **HIGHLIGHT COLOR** button:

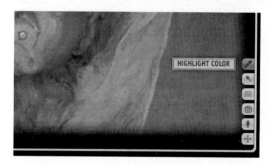

Click this button to view a selection of colors to choose from for the highlighted structure:

In this case, if you choose the default radio button, the highlight color will be turquoise. Click through the choices to see the different colors available. Note that only the default color allows you to see the texture and other details of the highlighted structure.

The next button is the **ARROW ON/OFF** button:

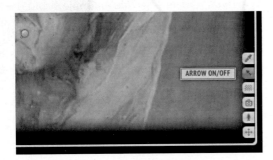

By clicking this button, an arrow appears pointing to the highlighted structure.

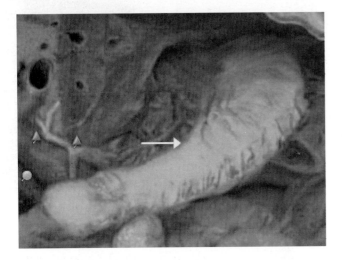

This is a helpful addition for those who are color blind or visually impaired.

The next button is the **HIGHLIGHT ON/OFF** button which toggles between structure highlights being on and off. Again, these first three buttons are only available when a structure has been selected by clicking an image pin.

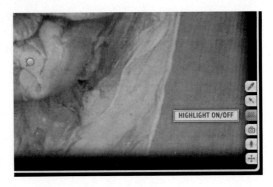

Three buttons are available on **DISSECTION** and **IMAGING** views. The **SAVE IMAGE** button . . .

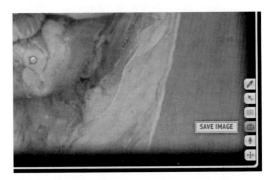

. . . allows you to save the image in the **IMAGE AREA** to your hard drive or other memory device. This is a handy option for studying the images on flash cards or making your own quizzes. Simply follow the directions on the pop-up window to save the image.

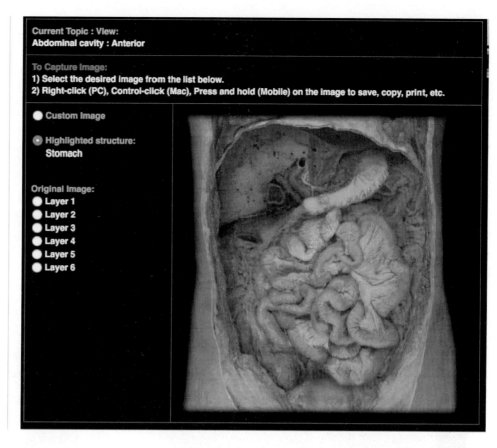

The **CURRENT VIEW** button . . .

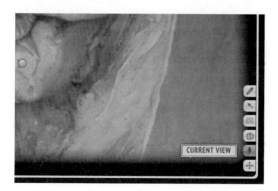

. . . illustrates the **TOPIC** and **VIEW** for that image. This image can be closed by clicking the **X** located in the top right corner.

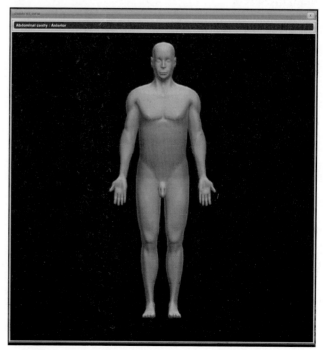

©McGraw-Hill Education

The **DIRECTIONAL LABELS** button . . .

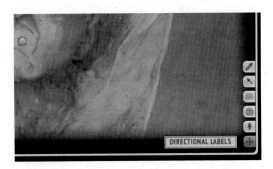

. . . toggles on and off helpful labels indicating the directional references for that image (see arrows).

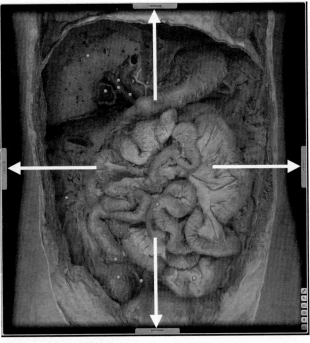

©McGraw-Hill Education

The **IMAGE INFORMATION** button is unique to **HISTOLOGY** images when viewed in the **VIEW AREA**:

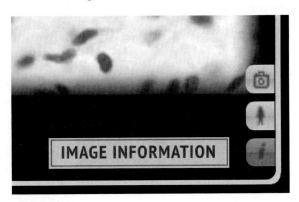

By clicking this button, a screen appears with the technical information on how that image was produced.

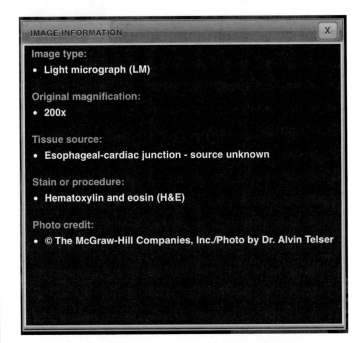

CHECK POINT

The DISSECTION Study Area: ALL CONTENT

1. What icon or button do you click on to enter the **DISSECTION Study Area** of *Anatomy and Physiology | Revealed*®?
2. What is the name for the area of the screen where the images are located?
3. When clicking the **LAYER** buttons, pins of different colors are visible on the dissected cadaver image. What are those colors, and what do they signify?
4. What happens when you mouse-over one of the pins? What happens when you left-click them?
5. What button do you select to change the topic you are viewing in *Anatomy and Physiology | Revealed*®?

The DISSECTION Study Area: *my* COURSE CONTENT

If your instructor has selected specific structures for you to learn, then you will have the ***my* COURSE CONTENT** option available to you. Return to the **MODULE** menu and select **Digestive**. Click the ***my*** icon next to the **MODULE** menu. You will see the image shown here:

This is the ***my* COURSE CONTENT** window. A list of structures of the digestive system selected from the Saladin's *Anatomy and Physiology* textbook occupies the large window. Above this window, from left to right, are the **Study Area** buttons for ***my* COURSE CONTENT**. The ***my* DISSECTION** button is activated by default and highlighted in green. We will discuss the other **Study Area** buttons later in this module. Also selected by default is the **ALPHABETICAL** radio button in the **SORT BY:** menu. Scroll down the list of structures to get a feel for how they are listed. Select **Transverse colon** from the list, and you will see the identical screen we viewed earlier in this module.

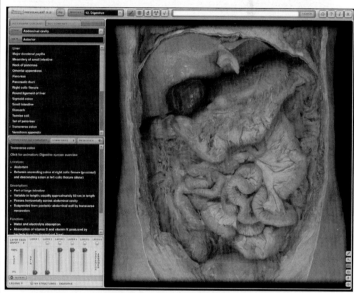

©McGraw-Hill Education

Notice the yellow check mark beside the **Transverse colon** in the list. Every time you select a structure from the ***my* COURSE CONTENT** list, a check mark will appear next to that term so you can keep track of which structures you have already viewed. If you are not confident that you have mastered that structure, you can click the check mark to remove it.

Click the green ***my*** button at the top of the screen to return to the ***my* COURSE CONTENT** window.

Select the **TOPIC: VIEW** button from the **SORT BY:** menu and you will see the image shown here:

Click **Abdominal cavity: Anterior** and you will see the image shown here:

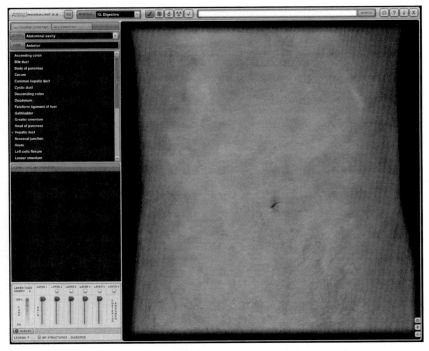

©McGraw-Hill Education

Does this screen look familiar? This is the same screen we saw earlier. So, as you can see, using the *my* **COURSE CONTENT** window, you have a variety of methods available to access the information in *Anatomy and Physiology | Revealed*®!

Click the green *my* button at the top of the screen to return to the *my* **COURSE CONTENT** window. The last button on the **SORT BY:** menu is the **REMAINING** button. The number in parentheses is the number of structures that you have yet to view. By clicking on this button, you will see a list of those structures.

The top row of the *my* **COURSE CONTENT** window consists of the **LIST** window, and three large yellow buttons to the right of this window. The **LIST** window shows the name of the list available in the *my* **COURSE CONTENT** view. If more than one list is available, you can click here and select from those lists. The **ADD/DELETE** button allows you to enter *my* **COURSE CONTENT** codes for other lists or to remove lists from the menu. The **PRINT LIST** button allows you to print out the *my* **COURSE CONTENT** lists by module. The **CLEAR ✓** button removes the check marks associated with the listed structures you have already viewed.

Finally, to the right of the *my* **Study Area** tabs is the **TAKE *my* QUIZ** button. Click this button and you will see a screen similar to this one:

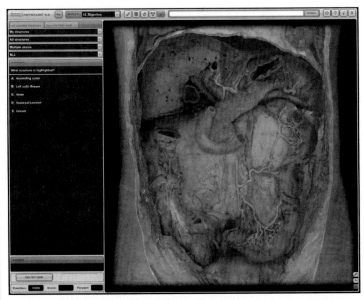

©McGraw-Hill Education

We will be discussing the **QUIZ Study Area** later in this module but it is important for you to see that there are quizzes in *Anatomy and Physiology | Revealed®* designed to help you review the specific structures that your instructor has selected for you. Click through the various options listed here to become familiar with them.

Now would be a good time for you to review the video over the **DISSECTION Study Area.** Click the **HOME** button, and from the **HELP VIDEOS** menu at the bottom right of the **HOME** page, select **Dissection.** You may want to review this video several times to help you become proficient in this study area.

CHECK POINT

The DISSECTION Study Area: *my* COURSE CONTENT

1. What is the function of the *my* button?
2. How are the **STUDY AREAS** listed in the *my* **COURSE CONTENT** window?
3. How are you able to keep track of the structures you have viewed from the *my* **COURSE CONTENT** window?
4. How are you able to keep track of the structures you have yet to view from the *my* **COURSE CONTENT** window? How do you know how many remain?
5. What are the options available with the **SORT BY:** button?
6. How would you print a list of structures from the *my* **COURSE CONTENT** window?
7. What information is found in the **LIST** menu in the *my* **COURSE CONTENT** window?

The ANIMATIONS Study Area

From the **HOME** screen, select **Digestive** from the **MODULE** menu. Click the **ANIMATIONS** icon, and the **ALL CONTENT** tab, and you will see the following screen:

Select **Digestive system anatomy and physiology (3D)** from the list. Click the **PLAY** button and the following image will be seen in the **IMAGE AREA.**

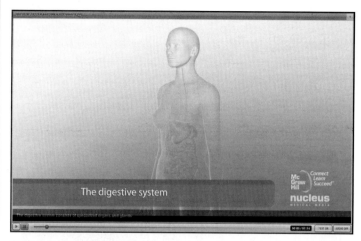

©Nucleus Medical Media, 2019

The control buttons at the bottom of the screen include **PLAY** (▶), **PAUSE** (❙❙), and the red **SLIDER.** The red **SLIDER** moves from left to right as the animation plays and can be grasped with the mouse and moved right and left to advance the animation or to replay a part of it. The **AUDIO ON** and **TEXT ON** buttons toggle these features on and off. The default is on for both. Click the **PLAY** button, and view the animation. After viewing this animation, familiarize yourself with the **ANIMATIONS** Study Area of *Anatomy and Physiology | Revealed®* by viewing several other animations.

The ANIMATIONS Study Area: *my* COURSE CONTENT

From the **HOME** screen, click the green *my* icon. From the new screen select **Digestive** from the **MODULE** menu. Click the *my* **ANIMATIONS** tab, and you will see the following screen:

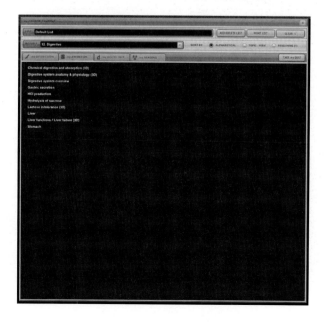

From here on, the sequence is the same for viewing the animations as previously discussed for the **ALL CONTENT** tab.

A new feature added to *Anatomy and Physiology | Revealed®* is a series of interactive videos for many of the modules. As there are no interactive animations for the digestive system, we will go to the first system where they are available—the muscular system. In the drop-down **MODULE** menu, select **Muscular** to access the muscular system. Click on the **ANIMATION** button and you will see the following screen:

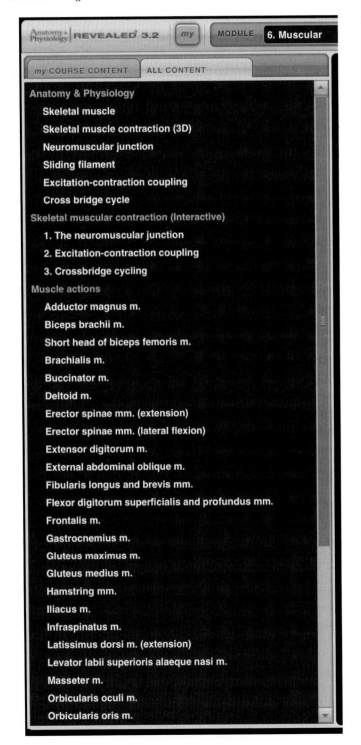

Note that the list of animations is divided under three headings in blue text. These are: **Anatomy & Physiology,** **Skeletal muscle contraction (Interactive),** and **Muscle actions.** Many modules will also include **Clinical application** animations as well, where commonly encountered physical conditions are explained in reference to normally functioning systems.

Under the heading **Skeletal muscle contraction (Interactive)**, select **1. The neuromuscular junction.** You will see the following image:

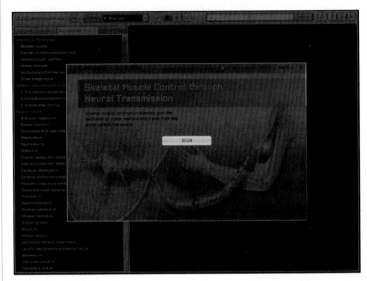

©McGraw-Hill Education

Note that the screen is grayed-out for the most part, with a white **BEGIN** box in the center, and nerve impulses flowing down the nerve cell. Click the **BEGIN** button, and the narration will begin:

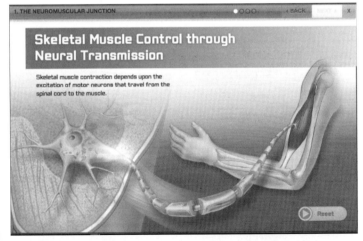

©McGraw-Hill Education

There are no interactions with this slide, but notice some components of this first slide that will be common with all of the interactive animations. Across the top of the slide, from left to right, are the animation title; followed by four circles with the first one filled with white, indicating that this is the first of four slides in this animation; a **BACK** button, where you can return to a previous slide; a **NEXT** button, that flashes when the narration is finished; and an **X,** where you can click to close the animation. Also, notice the blue **Reset** button in the bottom right corner. By clicking this button, the narration and any animations on this slide will repeat.

Click the **NEXT** button to view the next slide:

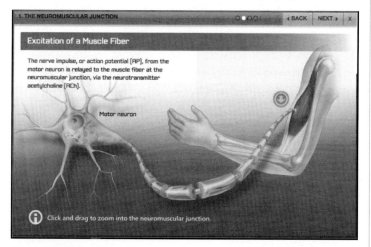
©McGraw-Hill Education

The details are the same for this slide as they were for the first one, except now the white-filled circle is the second one, and there is a blue arrow at the neuromuscular junction at the arm. At the bottom of the slide there is an "i" with a circle around it followed by instructions. Follow the instructions and click on the blue arrow and drag it downward. As you drag the arrow, the image will zoom in on the neuromuscular junction so that you can see the details:

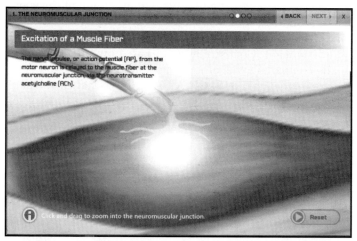
©McGraw-Hill Education

After interacting with this slide, either click the **Reset** button to view the animation again, or click **NEXT** to see the following slide:

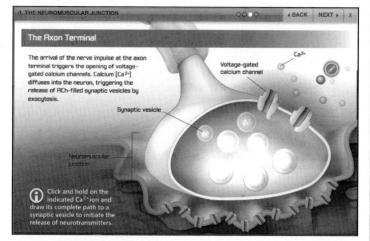
©McGraw-Hill Education

On this slide, the instructions indicate that you are to move a calcium ion to a synaptic vesicle. When you do:

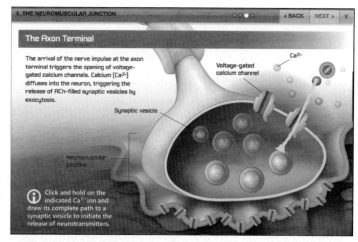
©McGraw-Hill Education

You will see this confirmation of your correct response:

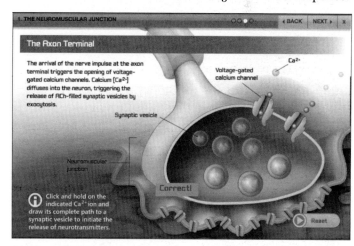
©McGraw-Hill Education

Click the **NEXT** button to view the final slide in this interactive animation:

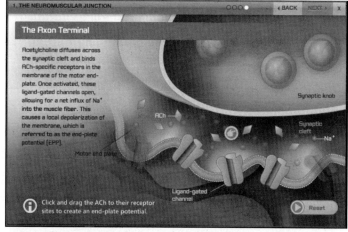
©McGraw-Hill Education

Once again, the instructions indicate that you are to interact with the slide, this time by dragging the ACh to their receptor sites where your correct actions will be affirmed:

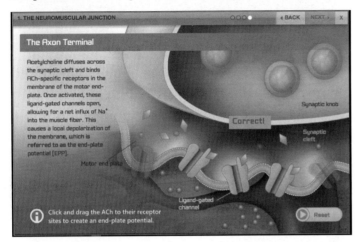
©McGraw-Hill Education

When you are finished interacting with the animation, click the **X** in the top right corner to close the animation.

CHECK POINT
The ANIMATIONS Study Area

1. What steps must be completed to access the **ALL CONTENT** list of animations from the **HOME** page?
2. What steps must be completed to access *my* **ANIMATIONS** from the **HOME** page?
3. Describe the function of the buttons and slider on the **ANIMATION** window.

The HISTOLOGY Study Area

Just like the previous study areas of *Anatomy and Physiology | Revealed®*, the **HISTOLOGY Study Area** can be accessed from any screen within a specific module. From the **HOME** screen, select **Digestive** from the **MODULE** menu. Select the **HISTOLOGY** icon. Click the **ALL CONTENT** tab. From the **TOPIC** menu, click **Select topic**. You will see the following screen:

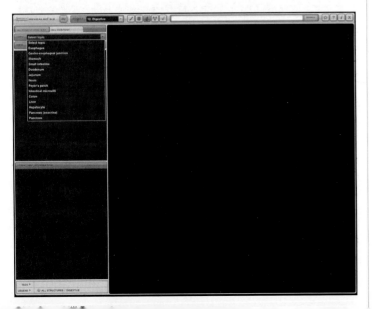

Select **Stomach** from the list. Click **Select view,** and you will see the following list:

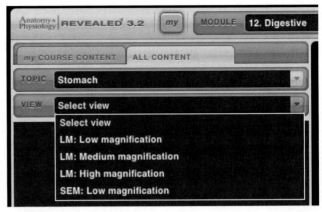
©McGraw-Hill Education

Anatomy and Physiology | Revealed® provides numerous histological views from which to learn. The **VIEW** of an image is listed by its source and relative magnification. Views with the abbreviation **LM** are produced with a **L**ight **M**icroscope, much like those that you may have access to in your lab. The abbreviation **SEM** refers to an image produced with a **S**canning **E**lectron **M**icroscope. You will also see the abbreviation **TEM** when a **T**ransmission **E**lectron **M**icroscope produces the image. The relative magnification of the specimen in the image follows the source, and ranges from **Low magnification** through **Medium magnification** and **High magnification**. Compare how an **LM: Low magnification** image compares to the others by clicking each one and viewing the image in the **IMAGE AREA**. Notice that as you click each **VIEW**, a list of structures appears underneath, as seen here for **LM: Low magnification.**

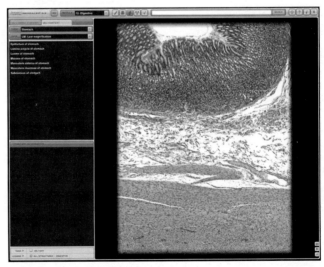
©McGraw-Hill Education/Al Telser

Click on each structure in the list, and just as with the **DISSECTION Study Area,** each structure is highlighted in the image and information concerning that structure appears in the **STRUCTURE INFORMATION** window, as seen here for **Epithelium of stomach.**

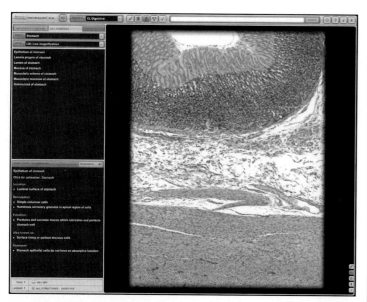
©McGraw-Hill Education/Al Telser

Click each of the structures listed and observe the highlighted image and the information provided for each.

For each view, the **TAGS ON/OFF** button is provided at the bottom left of the screen. Return to **Epithelium of stomach.** Click the **TAGS ON/OFF** button, and you will see the following image:

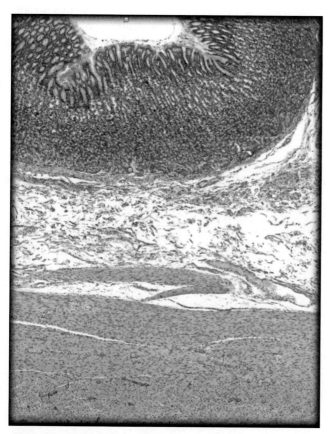
©McGraw-Hill Education/Al Telser

Just as you did in the **DISSECTION Study Area,** you can mouse-over the pins to activate the identification labels.

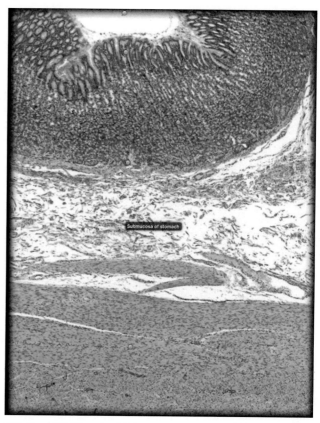
©McGraw-Hill Education/Al Telser

You can also left-click the pins to highlight that structure and activate the **STRUCTURE IDENTIFICATION** window.

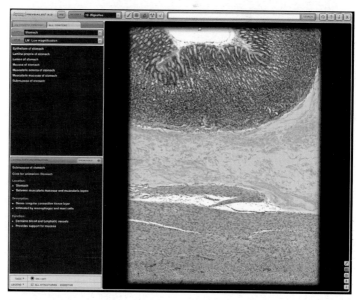
©McGraw-Hill Education/Al Telser

The HISTOLOGY Study Area: *my* COURSE CONTENT

If your instructor has chosen specific histology images for you to learn, then they will be available under the ***my* COURSE CONTENT** tab. To access these images, either click the green ***my*** button at the top of the screen, or when previously instructed to select the **ALL CONTENT** tab, select the ***my* COURSE CONTENT** tab instead.

Click the *my* button and you will see the following screen:

Select the *my* **HISTOLOGY** tab and you will see the following window:

Click on **Brush border of ileum** and you will see the following screen:

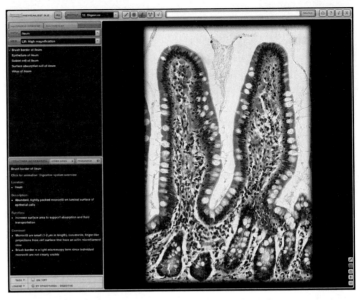

©Image Source/Getty Images

As you can see, this view is located under the **TOPIC: Ileum** and the **LM: High magnification** view. The brush border is highlighted, and the **STRUCTURE INFORMATION** window contains pertinent information about this structure. A check mark indicates that this structure is now listed as a previously viewed structure. Click each of the other structures listed. For each one, that structure is highlighted, with information available to you in the **STRUCTURE INFORMATION** window. As you go through the list, a check mark appears beside each term.

Click the **TAGS ON/OFF** button, and you will see the following image:

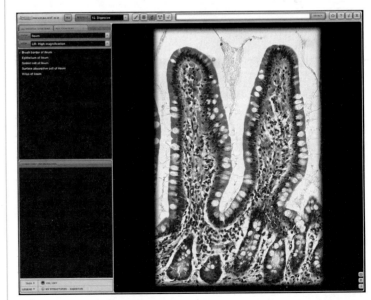

©Image Source/Getty Images

As discussed earlier, if you mouse-over the pins, the structure identity will become visible, and if you click a pin, the structure becomes highlighted with information available in the **STRUCTURE INFORMATION** window. Click the *my* button to return to the *my* **HISTOLOGY** tab. Notice that the structures you viewed have a check mark next to them.

CHECK POINT

The HISTOLOGY Section

1. Describe the sequence of steps you would take to access the **HISTOLOGY Study Area** of the **Digestive System**. Assume you are starting from the **HOME** screen.
2. What steps would you then take to view the **Epithelium of stomach** tissue?
3. What two different methods could you use to access the **STRUCTURE IDENTIFICATION** window for the **Epithelium of stomach** tissue?
4. What steps would you take to save the image of the **Epithelium of stomach** tissue?
5. What steps would you take to view the **Villus of duodenum** tissue? Include steps to highlight structures and activate the **STRUCTURE IDENTIFICATION** window.

HEADS UP!

The imaging techniques used in *Anatomy and Physiology | Revealed®* include the following:

CT (**C**omputer **T**omography) scan is a medical procedure where a machine called a scanner takes a series of X ray images in cross-section around the body's circumference—much like taking sliced images—as it travels down the body's length. A computer is then used to convert the scans into a picture.

X ray is the radiograph you are familiar with which is commonly used to take images of broken bones.

Nuclear scan uses a special camera to detect energy emitted by radioactive tracers.

MRI (**M**agnetic **R**esonance **I**maging) is a procedure used to provide high-contrast images of body soft tissues without the use of radiation, such as used in CT scans and X rays.

Angiogram is a type of X ray used with a special dye. It is commonly employed to take images of the blood flow in a blood vessel.

CTA is an angiogram done using a **CT** scanner.

Lymphangiogram is an angiogram conducted on a lymphatic vessel instead of a blood vessel.

Bronchogram like an angiogram, is an X ray used with a special dye, which is employed to take images of the bronchial tree.

Pyelogram or intravenous pyelogram (**IVP**) is an X ray of the urinary tract. The image is taken after the injection of a radiopaque contrast material, which is filtered out by the kidneys.

The IMAGING Study Area

The **IMAGING Study Area,** when available, can be accessed from any screen of a module. We have been using the digestive system to demonstrate the many facets of *Anatomy and Physiology | Revealed®*, so we will continue to do so with a side trip to the skeletal system to explore the images available there.

From the **HOME** screen, select **Digestive** from the **MODULE** menu. Select the **IMAGING** icon. Click the **ALL CONTENT** tab. From the **TOPIC** menu, click **Select topic.** You will see this screen:

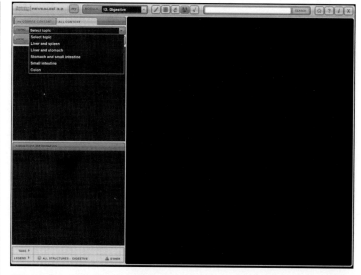
©Image Source/Getty Images

From the list, select **Colon.** Click **Select view** in the **VIEW** menu. Notice that there are two views, a **CT** scan and an **X ray.** Select **CT: Axial.** You will see this image:

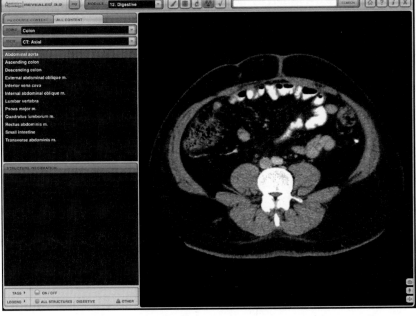
©McGraw-Hill Education

As we have seen with the previous **STUDY AREAS,** by clicking the structures listed, the structure will become highlighted in the **IMAGE AREA,** and the **STRUCTURE INFORMATION window** will contain information concerning that structure. Click **Ascending colon** and you will see this screen:

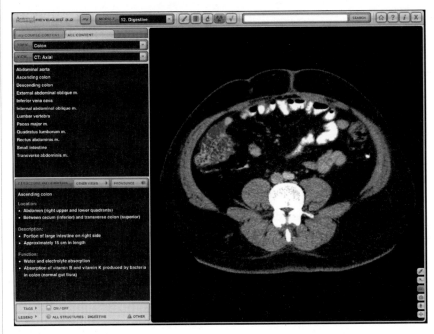
©McGraw-Hill Education

Click **OTHER VIEWS** and you will see this image:

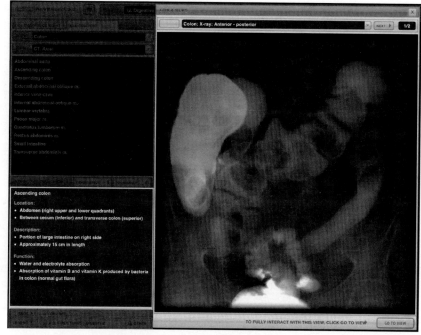
©McGraw-Hill Education

Click the **NEXT** button, and you will see the second view:

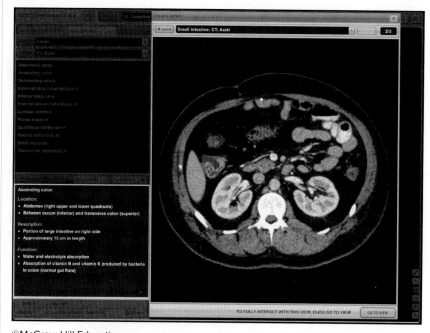
©McGraw-Hill Education

If you want to go to either of the **OTHER VIEWS,** you can click the **GO TO VIEW** button in the bottom right corner. Click on the **X** in the top right corner to return to the original screen.

Click the **TAGS ON/OFF** button to see the image with location pins.

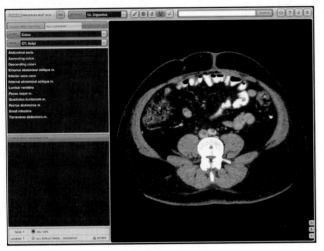
©McGraw-Hill Education

Notice the green pins indicating structures included in the digestive system, and blue pins indicating structures not included in the digestive system. As we have seen previously, you can mouse-over the pins to reveal the structure identity, or you can click them to highlight the structure and open the **STRUCTURE INFORMATION** window.

Return to the **VIEW** menu, and select **X ray Anterior—posterior.** You will see this screen:

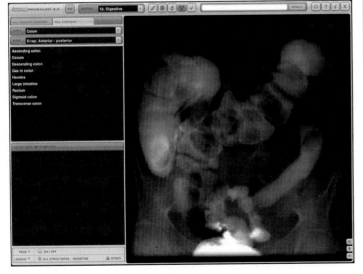
©McGraw-Hill Education

From the list, select **Large intestine.** You will see the following screen:

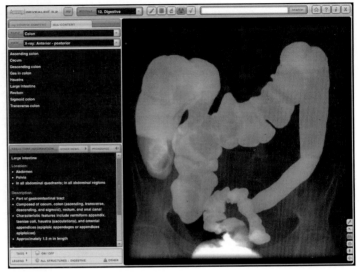
©McGraw-Hill Education

Click the **TAGS ON/OFF** button to see the image with location pins.

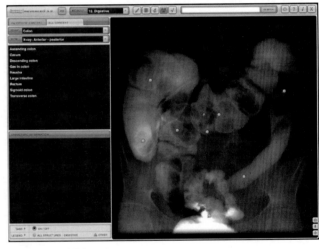
©McGraw-Hill Education

Again, you can mouse-over the pins to reveal the structure identity . . .

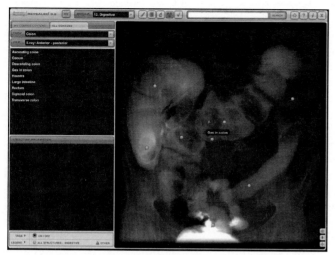
©McGraw-Hill Education

. . . or you can click them to highlight the structure and open the **STRUCTURE INFORMATION** window.

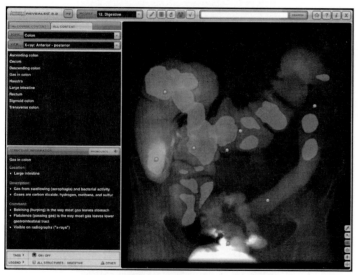
©McGraw-Hill Education

The IMAGING Study Area: *my* COURSE CONTENT

If your instructor has chosen specific **IMAGING** images for you to learn, then they will be available under the *my* **COURSE CONTENT** tab. To access these images, either click the green *my* button at the top of the screen, or when previously instructed to select the **ALL CONTENT** tab, select the *my* **COURSE CONTENT** tab instead. Click the green *my* button at the top left, and you will see the following screen:

Click the *my* **IMAGING** tab. The list will be populated with the structures your instructor has selected for you to learn. By clicking any of the listed structures, you will see a screen similar to the following one:

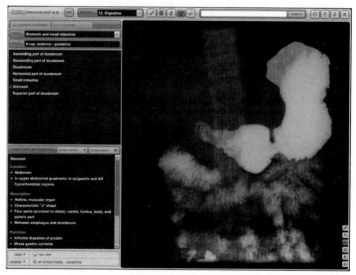
©McGraw-Hill Education

The results are similar to what we viewed previously, with the exception of the check mark next to the structure selected. Continue through the different aspects of the digestive system as you did with the **ALL CONTENT** tab. You will see that the *my* **IMAGING Study Area** functions very much like we just discussed.

The IMAGING Study Area: A Side-Trip to the Skeletal System

Click the **HOME** button to return to the **HOME** page. From the **MODULE** menu, select **Skeletal**. Click the **IMAGING** button. Click the **ALL CONTENT** tab. From the **TOPIC** menu, select **Thorax**. From the **VIEW** menu, select **CT: Sagittal**. You will see the following screen:

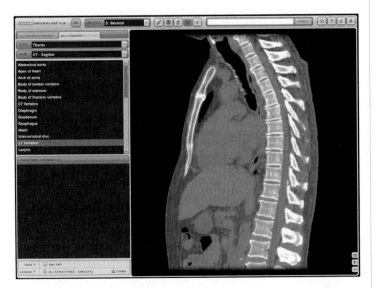

Select **Body of thoracic vertebra**. The structures are now highlighted, and the **STRUCTURE INFORMATION** window is now available. Spend some time here, selecting the various structures from the list, and toggling the **TAGS ON/OFF** button to activate the structure identification pins. Be sure to mouse-over the pins to identify the structures and click on them to highlight the structures and to activate the **STRUCTURE IDENTIFICATION** windows.

Return to the **VIEW** menu and select **X ray: Posterior—anterior**. Select **Ribs**. The ribs are now highlighted in the **IMAGE** area, and the **STRUCTURE IDENTIFICATION** window is activated. Spend some time selecting the various structures and reading the information presented for each one. Be sure to click the **TAGS ON/OFF** button as well, and mouse-over and click the identification pins.

CHECK POINT

The IMAGING Section

1. List the images available in the **Skeletal System** for the **Skull**.
2. List the images available in the **Skeletal System** for the **Hip and thigh**.
3. List the images available in the **Skeletal System** for the **Knee**.

Take this opportunity to view the **Imaging and Histology** video in the **HELP VIDEOS** menu on the **HOME** page.

The QUIZ Study Area

The **QUIZ Study Area** of *Anatomy and Physiology | Revealed*® provides an opportunity for you to assess your understanding of the structures and concepts that have been presented. These quizzes are valuable for you to use in preparation for upcoming classroom examinations or quizzes. They allow you to find your strengths *and* your weaknesses, and thus you will be able to fine-tune your study time as needed. Be sure to use this study area regularly before you move on to new material, so that you can increase your proficiency of the subject information.

Just like the previous study areas of *Anatomy and Physiology | Revealed*®, the **QUIZ Study Area** can be accessed from any screen once you have entered a module.

Begin from the **HOME** screen. From the **MODULE** menu, select **Digestive**. Select the **QUIZ** icon. Click the **ALL CONTENT** tab. From the **TOPIC** menu, click **Select topic**. You will see the following screen:

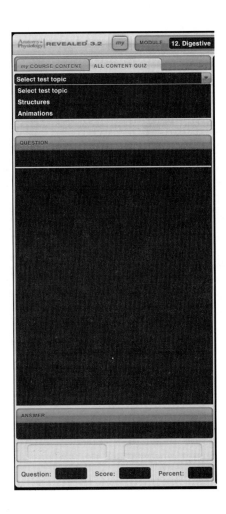

You have the option to select a quiz from the **Structures** you have learned, or from the **Animations** you have viewed. For now, select **Structures.** Click the **Select region** menu, and you will see the following list:

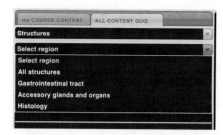

Your options now include quizzes covering **All structures** included in the current module, or groups of structures from the current module. For this module, these include the **Gastrointestinal tract,** its associated **Accessory glands and organs,** or **Histology.** Select **Gastrointestinal tract.** Click the **Select test type** menu, and you will see the following list:

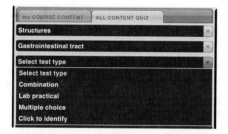

The list includes **Combination,** which is a sampling of all available test types, **Lab Practical,** which consists of labeled structures for you to identify, **Multiple choice,** which lists a series of options for you to choose from for the correct answer, and **Click to identify,** which gives you a structure name, and you are to click on the image to identify that structure. We will explore each of these test types, but for now select **Lab practical.**

Next select the number of questions you would like on your quiz. Click the **Select number of questions** menu. If you have a limited amount of time, choosing **10** or **25** questions will give you a brief quiz. But, for the most benefit from this study area, choosing the **ALL** option will ask you questions from every image available. Either way, the self-quizzes will appear identical.

Select **All** from the list. Your quiz, with its first question, will appear:

Note: Due to the random order of the questions, your results may not be identical to these examples.

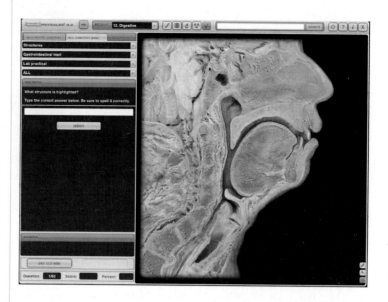

The screen now contains the following components:

- The **QUESTION** window, with the current question in colored text.

- The answer space, with a **SUBMIT** button. After typing your answer, click the **SUBMIT** button. Note that your answer must be spelled correctly.

- The **ANSWER** window, where the correct answer is displayed after you submit your answer. If your submission is correct, then the word **CORRECT** will appear in this window. If your submission is incorrect, then **INCORRECT will appear in red, along with the correct answer.**

- The **END TEST NOW** button, which will terminate your quiz.

- The **QUESTION** window, where the current question number is listed as a fraction of the total number of questions.

- Your **Score,** with the number of correct submissions as a fraction of your total submissions.

- Your **Percent,** which is the percent of correct submissions for this quiz.

Answer the question, and click the **SUBMIT** button. Your results will be listed in this screen:

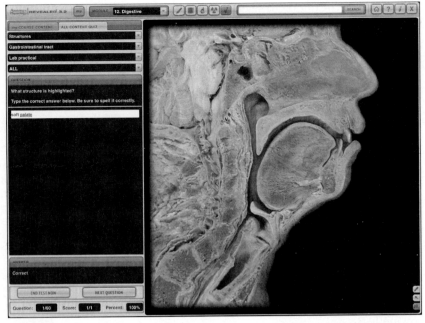

©McGraw-Hill Education

You will now be given the option of answering the **NEXT QUESTION.** Click this button and the next question will appear:

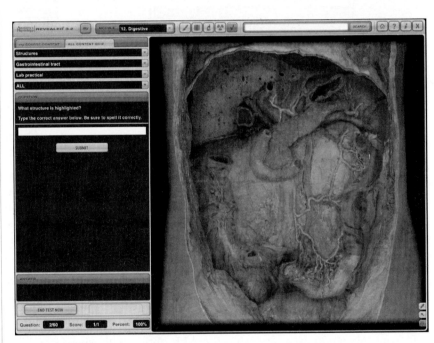

©McGraw-Hill Education

You can continue answering questions until you complete the quiz, or you can end the quiz at this point. Click **END TEST NOW** and you will see the following screen:

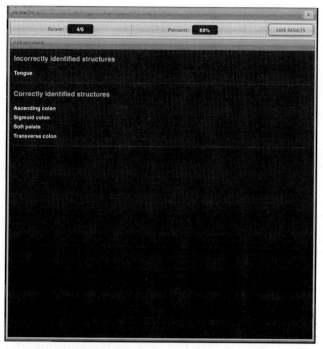

©McGraw-Hill Education

Across the top of this **RESULTS** window are your **Score**, your **Percent**, and a **SAVE RESULTS** button.

In the **STRUCTURES** window, you will see a breakdown of all **Incorrectly identified structures** as well as **Correctly identified structures**. You can save your results from as many quizzes as you desire, and this allows you to chart your progress as you learn the material. You will be given a pdf file of your results, which you can save on your computer.

Close the results pdf file and you will return to your **RESULTS**. Note that the structures listed as correct and incorrect are hyperlinked to the appropriate study area of *Anatomy and Physiology | Revealed*®. Click on the listed terms, and you will be taken to the area where you can review those structures.

Here I clicked on the structure that was misidentified—tongue—so that I can review it again before retaking the quiz:

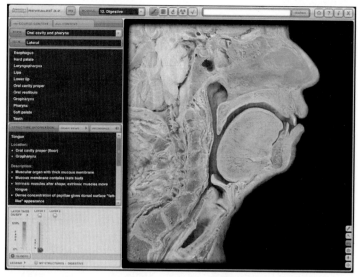

©McGraw-Hill Education

Note that a check mark appears beside the structures that were misidentified. This makes learning from your mistakes much easier.

Close your **RESULTS** window by clicking the **X** in the top right corner. This returns you to the Quiz menus. From the **Select test topic** menu, select **Structures**. From the **Select region** menu, select **Gastrointestinal tract**. From the **Select test type** menu, select **Multiple choice**. From the **Select number of questions** menu, select **ALL**. The first question then appears on the following screen:

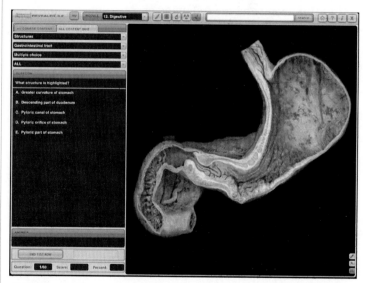

©McGraw-Hill Education

This quiz is similar to the previous example with the exception of the **QUESTION** window. Here the question is listed in colored text, with the possible answers listed below it. Click the answer you think is correct, and you will receive the same feedback as you did with the **Lab practical** example. You can continue on with the quiz or you can terminate it by clicking the **END TEST NOW** button.

Return to the **Select test type** menu, and select **Click to identify.** Select **ALL** from the **Number of questions** menu. You will see a screen similar to the following one:

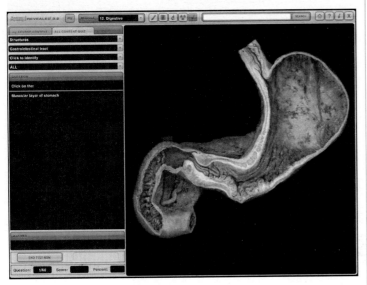
©McGraw-Hill Education

Click the location of the structure on the image. If correct, the structure will highlight and **Correct** appears in the **ANSWER** window. If you incorrectly answered the question, the structure will be highlighted on the image, and **Incorrect** will appear in red letters, followed by **correct structure shown** to indicate the highlighted structure. You now have the option of continuing this quiz, or terminating it as before.

Return to the quiz menus above and select **Animations** from the **Select test topic** menu. Click the **Select animation** menu and the following choices appear:

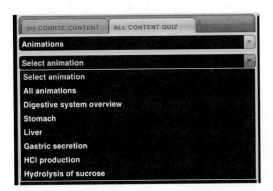

You can select from any one of the animations pertaining to the digestive system or all of them. Select **All animations.** You will see a screen similar to the following one:

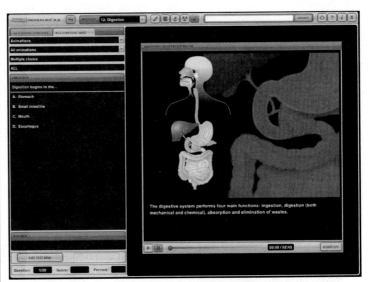
©McGraw-Hill Education

By default, the **Test type** is **Multiple choice** and the **Number of questions** is **ALL**. The **QUESTION** and **ANSWER** windows are the same as the **Multiple choice** quiz discussed above, but notice that the corresponding animation appears in the **IMAGE AREA.** You can click the **Play** button and view the animation to find the answer to the question. When you correctly answer the question, **Correct** appears in the **ANSWER** window. And when you answer incorrectly, **Incorrect** appears in red text in the **ANSWER** window, along with the correct answer.

Continue to explore the quiz options available to you. These assessments will prove invaluable to you as you gain proficiency with *Anatomy and Physiology | Revealed*®.

Take this opportunity to view the **Quiz** video in the **HELP VIDEOS** menu on the **HOME** page.

The QUIZ Study Area: *my* COURSE CONTENT

If your instructor has chosen specific **Study Area** information for you to learn, then quizzes over this information will be available under the ***my* COURSE CONTENT** tab. To access these quizzes, either click the green ***my*** button at the top of the screen, or when previously instructed to select the **ALL CONTENT** tab, select the ***my* COURSE CONTENT** tab instead.

Click the green, square *my* button at the top of the screen. To the right of the *my* tabs is the **TAKE *my* QUIZ** button. Click this button, and a screen similar to the following one will appear:

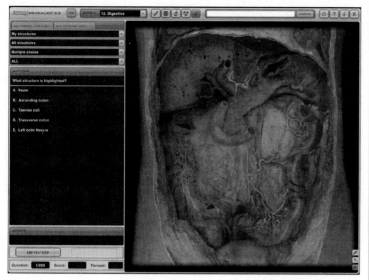

©McGraw-Hill Education

Click the **Select test topic** menu, and notice that the options are now *my* **structures** and *my* **animations**. Click the **Select region** menu and the options are identical to those discussed previously with the **ALL CONTENT** tab. Click the **Select test type** menu, and again the choices are the same as we discussed previously. And when you click the **Select number of questions** menu, the list is the same as before. The only difference now is that the quizzes will only cover the information that your instructor has selected for you to learn. Therefore, if this is the case for your class, you can explore the quiz examples in the *my* **COURSE CONTENT** using the same directions and examples as previously discussed for the **ALL CONTENT** tab.

> ## CHECK POINT
>
> **The QUIZ Section**
>
> 1. Describe the sequence of steps that you would take to access the **QUIZ Learning Area** for the **Nervous System.** Assume that you are starting from the **HOME** screen.
> 2. List the **Test topics** available for the **Muscular System.**
> 3. What are the advantages of selecting **ALL** from the **Select number of questions** menu?
> 4. Describe the process of saving your results from a **QUIZ.**

IN REVIEW

What Have I Learned?

1. Name the five **study areas** in *Anatomy and Physiology | Revealed®*.

2. List the steps you would go through from the *Anatomy and Physiology | Revealed®* **HOME** screen to view a deep dissection of the posterior abdomen.

3. List the steps you would go through from the *Anatomy and Physiology | Revealed®* **HOME** screen to view the animation **Gastric secretion**.

4. Name the function of each of the following buttons:

   (a)  (b)   (c)

   (d)  (e)

5. Name the function of each of these five buttons:

   (a) (b) (c) (d) (e)

   (a)
   (b)
   (c)
   (d)
   (e)

6. Using *Anatomy and Physiology | Revealed®*, define **Umbilical region**.

7. Using *Anatomy and Physiology | Revealed®*, define **Oblique plane**.

8. Using *Anatomy and Physiology | Revealed®*, define **Anatomical position**.

9. Using *Anatomy and Physiology | Revealed®*, define **Proximal** and **Distal**.

10. Using *Anatomy and Physiology | Revealed®*, define **Right upper quadrant**.

MODULE 1

Body Orientation

Overview: Body Orientation

Imagine, if you will, what the world would be like if we did not all agree on some basic tenets of everyday life, such as which direction is at the top of a map, or which side of the road to drive on, or just how much gasoline is in a gallon or a liter. Fortunately, we have standards or conventions for these and so many more everyday experiences. We know that north is always at the top of a map that when in a particular country we drive on a predetermined side of the road, and that when you buy a gallon or liter of gasoline in one location it will contain the same volume as it would if purchased in any other. These standards prevent confusion, frustration, and even ill will. Now consider the medical field. Is it not just as important for all surgeons, first responders, nurses, and scientists to agree on basic directions and locations in and on the human body? We have all heard the horror stories of a patient having the wrong limb amputated or surgery conducted in the wrong location. This is why we must have a basic understanding of **body orientation,** and the terms associated with it. In this manner, a paramedic responding in the field can communicate with an emergency room team awaiting their arrival, which can communicate with a team of specialists on call for their specialized procedures. Communication is the key! And this communication goes beyond international borders, allowing medical personnel from various locations worldwide to communicate with each other.

So, it cannot be overemphasized just how important this terminology is to the bigger picture. But it also has relevance to you, the student, as you learn anatomy and physiology. Once you learn that brachial refers to the arm, and the arm, contrary to popular opinion, is only the portion of your upper limb from your shoulder to your elbow, then when you are looking for a brachial artery, brachial vein, or biceps brachii muscle, you will know immediately where to look. Once you master these terms, they will aid you for the remainder of your career. If you do not learn them now, they will haunt you until you do.

The first step to understanding this terminology is to remember that these terms are always used in reference to a person standing in what is termed **anatomical position.** As shown in the figure below anatomical position consists of standing erect with the arms at the sides, palms facing forward with fingers pointing downward, feet parallel to each other and flat on the floor, with eyes directed forward. All references to left or right, for example, are those of the subject, and not your own.

To assist your learning, the body orientation terms have been divided into **Planes of section, Directional terms, Body regions, Pleura and pericardium,** and **Organ systems.** Let's begin at the top, with **Planes of section,** and discover the wealth of information contained in the terminology of **Body orientation.**

> ### HEADS UP!
> *As we discussed in the introductory module, these exercises will cover all of the options available with Anatomy and Physiology | Revealed®. Your instructor may have selected particular exercises for you to complete, while omitting others.*

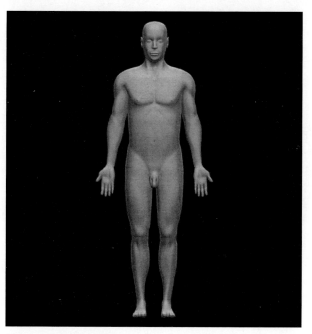

©McGraw-Hill Education

MODULE 1 Body Orientation

EXERCISE 1.1:
Body Position

Select the **MODULE** *drop-down menu in the* **HOME** *screen of* Anatomy and Physiology | Revealed®. *Choose* **1. Body orientation.** *Click the* **DISSECTION** *icon. From the* **Select topic** *menu, choose* **Body position.** *Select the* **my COURSE CONTENT** *tab if your instructor has selected specific exercises for you to complete. Otherwise select the* **ALL CONTENT** *tab. You will see the following image:*

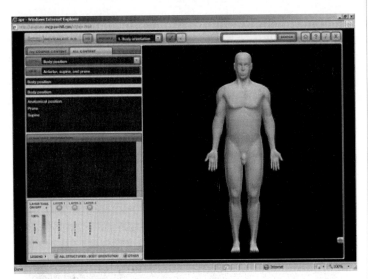

©McGraw-Hill Education

Click on each body position—**Anatomical position, Prone,** *and* **Supine**—*and view the image in the* **IMAGE** *screen as well as the* **STRUCTURE INFORMATION.**

1. Which of these is the position of the body when lying face up?

2. Which of these is the reference position for anatomical description?

3. Which of these is the position of the body when lying face down?

CHECK POINT
Body Position
1. Describe anatomical position.
2. Describe the prone position.
3. Describe the supine position.

Click **LAYER 1** *in the* **LAYER CONTROLS** *window, and you will find the following image:*

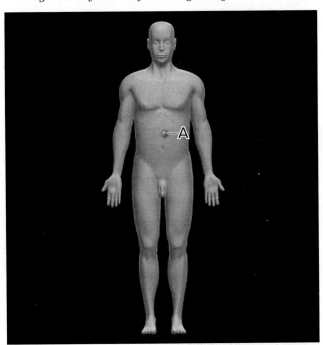

©McGraw-Hill Education

Mouse-over the pin on the screen to find the information to fill in the following blank.

A. _____

Click **LAYER 2** *in the* **LAYER CONTROLS** *window, and you will find the following image:*

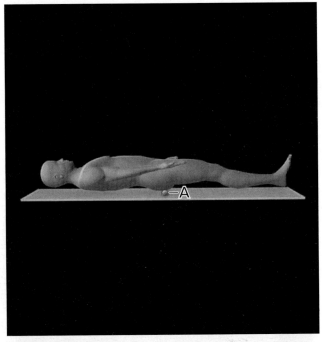

©McGraw-Hill Education

Mouse-over the pin on the screen to find the information to fill in the following blank.

A. _____

Click **LAYER 3** *in the* **LAYER CONTROLS** *window, and you will find the following image:*

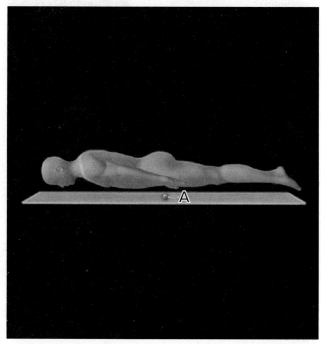

©McGraw-Hill Education

Mouse-over the pin on the screen to find the information to fill in the following blank.

A. _____

EXERCISE 1.2:
Planes of Section

Return to the **Topic** *drop-down menu, and select* **Planes of section.** *You will see the following image:*

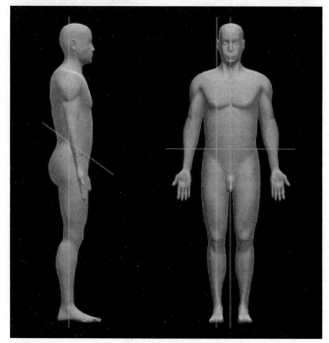

©McGraw-Hill Education

Select the **my COURSE CONTENT** *tab if your instructor has selected specific exercises for you to complete. Otherwise select the* **ALL CONTENT** *tab.*

Click through the five **Planes of section** *to answer the following questions.*

1. Which of the planes passes through the body at a slant or an angle?

2. Which of the planes is also called a cross-section?

3. Which of the planes passes side-to-side through the body, dividing it into anterior and posterior portions?

4. Which of the planes passes from front to back through the body, dividing it into equal or unequal right and left portions?

5. Which of the planes passes from front to back through the body, dividing it into equal right and left halves?

Click **LAYER 1** *in the* **LAYER CONTROLS** *window, and you will find the following image:*

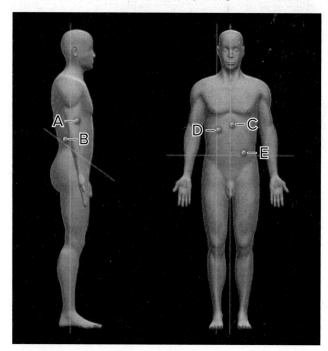

©McGraw-Hill Education

Mouse-over the pins on the screen to find the information necessary to identify the following structures.

A. _____
B. _____
C. _____
D. _____
E. _____

CHECK POINT

Planes of Section

1. Which plane is also known as a frontal plane?
2. Which plane is also known as a cross-section?
3. Which plane is also known as a parasagittal plane?
4. Which plane is also known as a median plane?

IN REVIEW

What Have I Learned?

The following questions cover the material that you have just learned, planes of section. Apply what you have learned to answer these questions on a separate piece of paper.

1. Define coronal plane.
2. Define midsagittal plane.
3. Define oblique plane.
4. Define sagittal plane.
5. Define transverse plane.

EXERCISE 1.3:
Directional Terms

Return to the **TOPIC** *drop-down menu and select* **Directional terms.** *Notice* **Anterior, Lateral,** *and* **Midsagittal** *appear in the* **VIEW** *window.*

Select the **my COURSE CONTENT** *tab if your instructor has selected specific exercises for you to complete. Otherwise select the* **ALL CONTENT** *tab.*

Click through the **Directional terms** *to find the term that matches each of the following descriptions.*

1. Closer to trunk or origin of a structure.
2. Downward or below.
3. Toward the front of the body.
4. Farther from the trunk or origin of a structure.
5. Away from the surface of the body.
6. Upward or above.
7. Toward the back of the body, or relating to the back.
8. Toward the midline.
9. Toward the surface of the body.
10. Away from the midline of the body.

Click **LAYER 1** *in the* **LAYER CONTROLS** *window, and you will find the following image:*

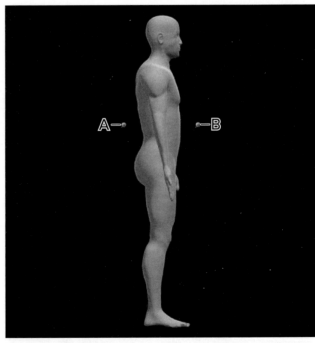

©McGraw-Hill Education

Mouse-over the pins on the screen to find the information necessary to identify the following structures.

A. _____

B. _____

Click **LAYER 2** *in the* **LAYER CONTROLS** *window, and you will find the following image:*

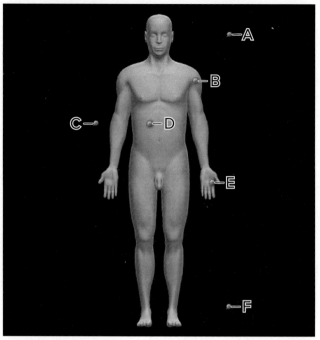

©McGraw-Hill Education

Mouse-over the pins on the screen to find the information necessary to identify the following structures.

A. _____

B. _____

C. _____

D. _____

E. _____

F. _____

Click **LAYER 3** *in the* **LAYER CONTROLS** *window, and you will find the following image:*

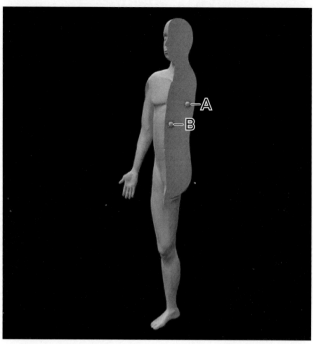

©McGraw-Hill Education

Mouse-over the pins on the screen to find the information necessary to identify the following structures.

A. _____

B. _____

CHECK POINT

Directional Terms

1. Which term is synonymous with dorsal?
2. Which term is synonymous with caudal?
3. Which term is synonymous with ventral?
4. Which term is synonymous with cranial?

IN REVIEW

What Have I Learned?

The following questions cover the material that you have just learned, directional terms. Apply what you have learned to answer these questions on a separate piece of paper.

1. Define anterior.
2. Define deep.
3. Define distal.
4. Define inferior.
5. Define lateral.
6. Define medial.
7. Define posterior.
8. Define proximal.
9. Define superficial.
10. Define superior.

EXERCISE 1.4:
Body Regions

Return to the **TOPIC** *drop-down menu and select* **Body Regions.** *Notice* **Anterior and posterior** *appear in the* **VIEW** *window.*

Select the **my COURSE CONTENT** *tab if your instructor has selected specific exercises for you to complete. Otherwise select the* **ALL CONTENT** *tab.*

From the **Select group** *menu, select* **Abdomen and pelvis.**

Click through the terms to find the one that matches each of the following descriptions.

1. Which region consists of the trunk, inferior to the thoracic region?

2. Which cavity is bounded by the diaphragm and the pelvic brim?

3. Which cavity is bounded by the pelvic inlet and the pelvic outlet?

4. Which of the nine abdominal regions is the median region?

5. Which of the nine abdominal regions is the left upper lateral region?

6. Which of the nine abdominal regions is the right lateral region?

7. Which of the nine abdominal regions is the right lower lateral region?

8. Which of the nine abdominal regions is the right upper lateral region?

9. Which of the nine abdominal regions is the left lateral region?

10. Which of the nine abdominal regions is the lower median region?

11. Which of the nine abdominal regions is the left lower lateral region?

12. Which of the nine abdominal regions is the upper median region?

13. Which of the four abdominal quadrants is left of the midline and superior to the transverse plane through the umbilicus?

14. Which of the four abdominal quadrants is right of the midline and inferior to the transverse plane through the umbilicus?

15. Which of the four abdominal quadrants is right of the midline and superior to the transverse plane through the umbilicus?

16. Which of the four abdominal quadrants is left of the midline and inferior to the transverse plane through the umbilicus?

CHECK POINT

Body Regions: Abdomen and Pelvis

1. Which cavities combine to form a continuous abdominopelvic cavity?
2. List the nine regions used in clinical practice to describe the location of abdominal organs.
3. List the four quadrants used in clinical practice to describe the location of abdominal organs and abdominal pain.
4. List the major organs of the abdominal cavity.
5. List the major organs of the pelvic cavity.

Return to the **Select group** *menu and select* **Back.**

1. Identify the location where the absence of extrinsic back musculature allows respiratory sounds to be heard more clearly with a stethoscope.

2. Identify the term for the posterior trunk.

3. Identify the region that includes the sacrum and the attached muscles.

4. Identify the region of the posterior midline.

5. Identify the region also known as the shoulder blade.

6. Identify the region also known as the lower back.

CHECK POINT

Body Regions: Back

1. Identify the function of the intrinsic and extrinsic muscle groups of the back.
2. Identify the common site of back pain and intervertebral disc herniation.
3. Describe the location of the triangle of auscultation.
4. List the vertebrae by region, from superior to inferior.

Return to the **Select group** *menu and select* **Head.**

1. Identify the region surrounding the brain.

2. Identify the region of the forehead.

3. Identify the part of the facial region that includes the nose.

4. Identify the region of the eye.

5. Identify the region of the external ear.

6. Identify the region of the posterior head.

7. Identify the region of the head inferior to the auricular region.

8. Identify the region of the lateral superior head.

9. Identify the region of the lateral head inferior to the orbital region.

10. Identify the region of the lateral head superior to the zygomatic and auricular regions.

11. Identify the region of the cheek.

12. Identify the region of the anterior inferior head.

13. Identify the region of the mouth.

14. Identify the structure superior to the neck region.

15. Identify the region of the chin.

CHECK POINT

Body Regions: Head

1. Identify the part of the skull containing the brain.
2. List the facial regions.
3. The facial region is related to the _____ bone.
4. The mental region is related to the _____.
5. Identify the region related to the parotid salivary gland and ramus of mandible.

Return to the **Select group** *menu and select* **Lower limb.**

1. Identify the region of the anterior leg.

2. Identify the dorsal aspect of the foot.

3. Identify the region also known as the hip and buttocks.

4. Identify the middle region of the lower limb.

5. Identify the region of the thigh.

6. Identify the lateral proximal region of the lower limb.

7. Identify the structure with the function of locomotion, balance, and support of the body.

8. Identify the region known as the talocrural region.

9. Identify the posterior knee region.

10. Identify the calcaneal region.

11. Identify the plantar region.

12. Identify the distal end of the lower limb.

13. Identify the sural region.

14. Identify the distal subdivision of the foot.

CHECK POINT
Body Regions: Lower Limb

1. Identify the region also known as the talocrural region.
2. Identify the region known as hip and buttocks.
3. The calcaneal region is also known as the _____.
4. Identify the region with the function of locomotion, balance, and body support.
5. Identify the region of the posterior knee.
6. Identify the region known as the sural region.

Return to the **Select group** *menu and select* **Neck.**

1. Identify the structure that connects the head with the upper limb and trunk.

2. Identify the subdivision of the neck related to the sternocleidomastoid muscle.

3. Identify the anterior subdivision of the neck.

4. Identify the posterior aspect of the neck.

5. Identify the lateral subdivision of the neck.

CHECK POINT
Body Regions: Neck

1. Identify the region also known as the anterior cervical triangle.
2. Identify the region also known as the posterior cervical triangle.
3. Identify the subdivision of the neck deep to the trapezius.
4. Identify the subdivision of the neck related to the sternocleidomastoid muscle.

Return to the **Select group** *menu and select* **Perineum.**

Describe the location of the perineum.

Return to the **Select group** *menu and select* **Thoracic.**

1. Identify the region known as the clavipectoral triangle.

2. Identify the subdivision of the thoracic region over the sternum.

3. Identify the region known as the armpit.

4. Identify the anterior thoracic region.

5. Identify the region of the superior part of the trunk.

CHECK POINT
Body Regions: Thoracic

1. Identify the region between upper arm and lateral thoracic wall.
2. Identify the region containing the cephalic vein.
3. Identify the organs of the thoracic cavity.

Return to the **Select group** *menu and select* **Upper limb.**

1. Identify the region between the glenohumeral and elbow joints.

2. Identify the region known as the shoulder.

3. Identify the region known as the wrist.

4. Identify the distal subdivision of the hand.

5. Identify the anterior aspect of the carpal and metacarpal bones.

6. Identify the posterior portion of the hand.

7. Identify the distal subdivision of the upper limb.

8. Identify the region between the elbow and wrist joints.

9. Identify the limb that is suspended from the shoulder.

10. Identify the region known as the elbow.

CHECK POINT

Body Regions: Upper Limb

1. Identify the region known as the forearm.
2. Identify the region known as the arm.
3. Identify the region known as the wrist.
4. Identify the region sometimes referred to as the shoulder.
5. Identify the distal subdivision of the hand.
6. Identify the non-weight-bearing limb used primarily to position the hand.

Click **LAYER 1** *in the* **LAYER CONTROLS** *window, and you will find the following image:*

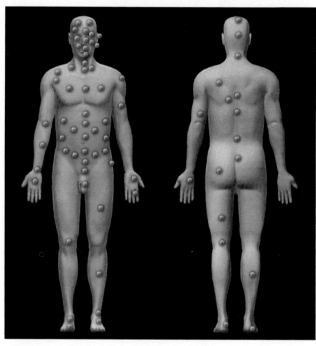

©McGraw-Hill Education

Due to the multitude of pins to identify, we will divide this exercise into separate regions.

Mouse-over the pins on the screen to find the information necessary to identify the following structures.

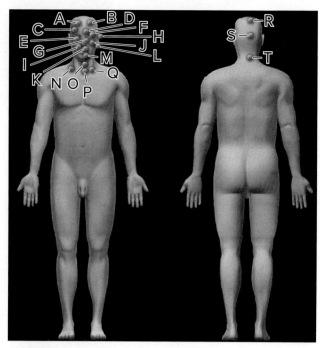

©McGraw-Hill Education

A. _____
B. _____
C. _____
D. _____
E. _____
F. _____
G. _____
H. _____
I. _____
J. _____
K. _____
L. _____
M. _____
N. _____
O. _____
P. _____
Q. _____
R. _____
S. _____
T. _____

42 MODULE 1 Body Orientation

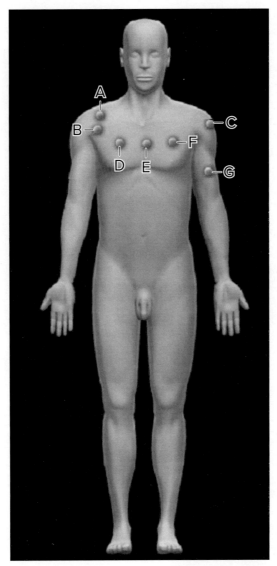
©McGraw-Hill Education

A. _____   E. _____
B. _____   F. _____
C. _____   G. _____
D. _____

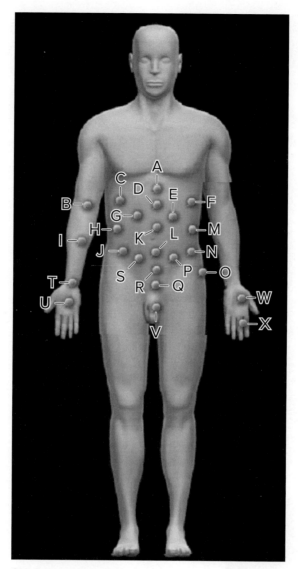

©McGraw-Hill Education

A. _____ M. _____

B. _____ N. _____

C. _____ O. _____

D. _____ P. _____

E. _____ Q. _____

F. _____ R. _____

G. _____ S. _____

H. _____ T. _____

I. _____ U. _____

J. _____ V. _____

K. _____ W. _____

L. _____ X. _____

44 MODULE 1 Body Orientation

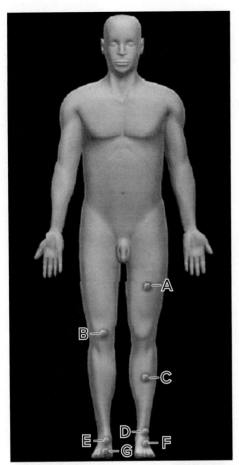

©McGraw-Hill Education

A. _____
B. _____
C. _____
D. _____
E. _____
F. _____
G. _____

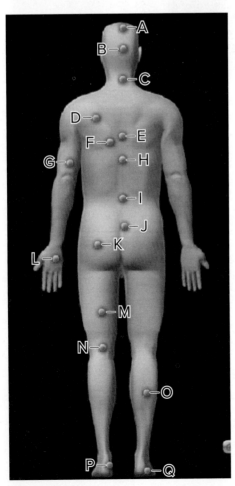

©McGraw-Hill Education

A. _____
B. _____
C. _____
D. _____
E. _____
F. _____
G. _____
H. _____
I. _____
J. _____
K. _____
L. _____
M. _____
N. _____
O. _____
P. _____
Q. _____

IN REVIEW

What Have I Learned?

The following questions cover the material that you have just learned, body regions. Apply what you have learned to answer these questions on a separate piece of paper.

1. Describe the abdominal cavity.
2. Describe the abdominopelvic region.
3. Describe the pelvic cavity.
4. Describe the auricular region.
5. Describe the buccal region.
6. Describe the facial region.
7. Describe the mental region.
8. Describe the orbital region.
9. Describe the temporal region.
10. Describe the zygomatic region.
11. Describe the perineum region.
12. Describe the axillary region.
13. Describe the thoracic cavity.
14. Describe the antebrachial region.
15. Describe the brachial region.
16. Describe the cubital region.

EXERCISE 1.5:

Body Cavities

Return to the **TOPIC** *drop-down menu and select* **Body cavities.** *Notice* **Anterior and lateral** *appear in the* **VIEW** *window.*

Select the **my COURSE CONTENT** *tab if your instructor has selected specific exercises for you to complete. Otherwise select the* **ALL CONTENT** *tab.*

From the **Select group** *menu, select* **Dorsal cavity (posterior aspect).**

Select **Cranial cavity.** *Answer the following questions.*

1. In which bony structure is this cavity located?

2. What structures are located within this cavity?

3. Which bones form this cavity?

Select **Vertebral canal.** *Answer the following questions.*

1. In which bony structure is this cavity located?

2. What structures combine to form the vertebral canal?

3. What structures are located within this cavity?

Return to the **Select group** *menu and select* **Ventral cavity.** *Click through the terms to find the one that matches each of the following descriptions.*

1. The primary muscle of respiration.

2. Latin: middle septum.

3. Major organs include: stomach, intestines, liver, gallbladder, spleen, pancreas, kidneys, ureters, suprarenal glands, aorta, inferior vena cava, lumbar nerve plexus, urinary bladder, rectum, and reproductive organs.

4. Bounded by abdominal walls, thoracic diaphragm, and pelvic brim.

5. Major organs include: urinary bladder, loops of small intestine, inferior part of sigmoid colon, rectum, and reproductive organs.

6. Bounded by the sternum, ribs, costal cartilages, intercostal muscles, thoracic vertebrae, and diaphragm.

7. Located in the trunk, between the thoracic and pelvic diaphragms.

46 MODULE 1 Body Orientation

8. Lined by parietal plurae.

9. Three divisions include central mediastinum and bilateral pulmonary cavities.

10. Major organs include: stomach, intestines, liver, gallbladder, spleen, pancreas, kidneys, ureters, suprarenal glands, aorta, inferior vena cava, and lumbar nerve plexus.

11. Bilateral subdivision of thoracic cavity.

12. Bounded by pelvic inlet and pelvic outlet.

13. Contraction of this muscle increases vertical dimension of thoracic cavity.

14. One of the three parts of the thoracic cavity.

CHECK POINT

Body Regions: Ventral Cavity

1. What pressure changes occur upon contraction of the diaphragm?
2. Identify the region that separates right and left pulmonary cavities?
3. Identify the cavity located in the pelvic region.
4. Identify the structures located in the pulmonary cavity.
5. Identify the cavity of the chest.
6. Identify the boundaries of the thoracic cavity.

Click **LAYER 1** *in the* **LAYER CONTROLS** *window, and you will find the following image:*

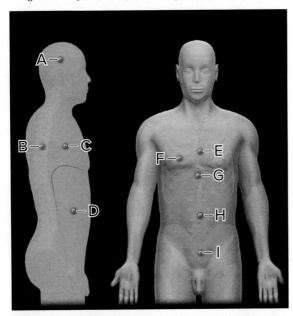

©McGraw-Hill Education

Mouse-over the pins on the screen to find the information necessary to identify the following structures.

A. _____
B. _____
C. _____
D. _____
E. _____
F. _____
G. _____
H. _____
I. _____

IN REVIEW

What Have I Learned?

The following questions cover the material that you have just learned, body cavities. Apply what you have learned to answer these questions on a separate piece of paper.

1. Describe the cranial cavity.
2. Describe the vertebral canal.
3. Describe the diaphragm.
4. Describe the zygomatic mediastinum.
5. Describe the pulmonary cavity.

EXERCISE 1.6:

Abdominal Quadrants and Regions

Return to the **TOPIC** *drop-down menu and select* **Abdominal quadrants and regions.** *Notice* **Anterior** *appears in the* **VIEW** *window.*

Select the **my COURSE CONTENT** *tab if your instructor has selected specific exercises for you to complete. Otherwise select the* **ALL CONTENT** *tab.*

From the **Select group** *menu, select* **Abdominal surface quadrants.**

Click through the terms to find the one that matches each of the following descriptions.

1. Left of midline, inferior to transverse plane through umbilicus.

2. Right of midline, inferior to transverse plane through umbilicus.

3. Left of midline, superior to transverse plane through umbilicus.

4. Right of midline, superior to transverse plane through umbilicus.

CHECK POINT

Abdominal Surface Quadrants

1. Which quadrant is abbreviated RLQ?
2. Which quadrant is abbreviated LUQ?
3. Which quadrant is abbreviated RUQ?
4. Which quadrant is abbreviated LLQ?

CHECK POINT

Abdominal Surface Regions

1. What structures delineate the abdominal surface regions?
2. What is the function of the abdominal surface regions?
3. Which abdominal surface region is sometimes referred to as left lumbar region?
4. Which two abdominal surface regions have the Greek meaning "under cartilage"?
5. Which abdominal surface region is also known as the hypogastric region?

Return to the **Select group** *menu and select* **Abdominal surface regions.**

Click through the terms to find the one that matches each of the following descriptions.

1. Contains cecum and vermiform appendix.

2. Contents include suprarenal glands, and parts of stomach, large intestine, liver, gallbladder, and pancreas.

3. Contents include the spleen.

4. Contents include parts of small and large intestines and right kidney.

5. Median region containing parts of small and large intestine.

6. Contents include parts of large intestine, liver, gallbladder, and right kidney.

7. Contents include parts of small and large intestines and left kidney.

8. Contents include urinary bladder when distended.

9. Left lower lateral region containing parts of small and large intestines.

Click **LAYER 1** *in the* **LAYER CONTROLS** *window, and you will find the following image:*

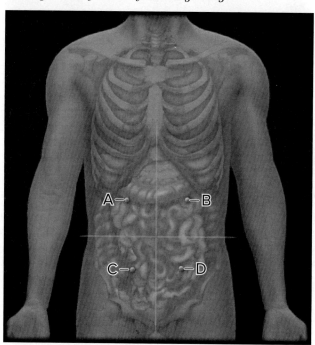

©McGraw-Hill Education

Mouse-over the pins on the screen to find the information necessary to identify the following structures.

A. _____
B. _____
C. _____
D. _____

48 MODULE 1 Body Orientation

Click **LAYER 2** *in the* **LAYER CONTROLS** *window, and you will find the following image:*

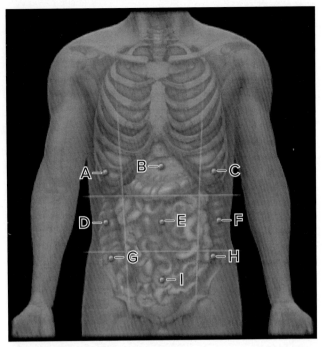

©McGraw-Hill Education

Mouse-over the pins on the screen to find the information necessary to identify the following structures.

A. _____
B. _____
C. _____
D. _____
E. _____
F. _____
G. _____
H. _____
I. _____

IN REVIEW

What Have I Learned?

The following questions cover the material that you have just learned, abdominal quadrants and regions. Apply what you have learned to answer these questions on a separate piece of paper.

1. List the contents of the left lower quadrant.
2. List the contents of the left upper quadrant.
3. List the contents of the right lower quadrant.
4. List the contents of the right upper quadrant.
5. List the contents of the epigastric region.
6. List the contents of the left flank region.
7. List the contents of the left hypochondriac region.
8. List the contents of the left inguinal region.
9. List the contents of the pubic region.
10. List the contents of the right flank region.
11. List the contents of the right hypochondriac region.
12. List the contents of the right inguinal region.
13. List the contents of the umbilical region.

Return to the **TOPIC** *drop-down menu and select* **Pleura and pericardium.** *Notice* **Anterior** *appears in the* **VIEW** *window.*

Select the **my COURSE CONTENT** *tab if your instructor has selected specific exercises for you to complete. Otherwise select the* **ALL CONTENT** *tab.*

From the **Select structure type** *menu, select* **Pleura and pericardium.**

From the **Select group** *menu, select* **Pericardium.**

Click through the terms to find the one that matches each of the following descriptions.

1. Inner limit of pericardial cavity.

2. Outer limit of pericardial cavity.

3. Potential space between layers of serous pericardium.

Return to the **Select group** *menu and select* **Pleura.**

Click through the terms to find the one that matches each of the following descriptions.

1. Thin, serous membrane fused to internal walls of thoracic cavity and lateral surface of mediastinum.

2. Thin, serous membrane fused to the surface of the lung.

3. Bilateral potential space between layers of pleura.

CHECK POINT
Pleura and Pericardium

1. What structures form the pericardial cavity?
2. Where is the visceral layer of the serous pericardium located?
3. Where is the parietal layer of the serous pericardium located?
4. Identify the term for the muscular surface of the heart.
5. The epicardium is also known as _____.
6. What structures line the pleural cavity?
7. Identify the three divisions of the thoracic cavity.

Return to the **Select structure type** *menu and select* **Other.** *From the* **Select group** *menu select* **Cardiovascular.** *Select* **Heart.** *Review the information presented in the* **STRUCTURE INFORMATION** *window and answer the following questions.*

1. Where is the heart located?

2. Describe the heart.

3. What is the heart's function?

Return to the **Select group** *menu and select* **Respiratory.** *Select* **Lungs.** *Review the information presented in the* **STRUCTURE INFORMATION** *window and answer the following questions.*

1. Describe the location of the lungs.

2. On what structure do the lungs rest?

3. How many lobes does each lung have?

4. What is the function of the lungs?

Select **Trachea and main bronchi** *from the list of Respiratory structures. Review the information presented in the* **STRUCTURE INFORMATION** *window and answer the following questions.*

1. What structures hold the trachea open?

2. Identify the structure of the posterior aspect of the cartilaginous rings.

3. What structures are the paired branches of the trachea?

4. What is another term for the structures in question 3?

Click **LAYER 1** *in the* **LAYER CONTROLS** *window, and you will find the following image:*

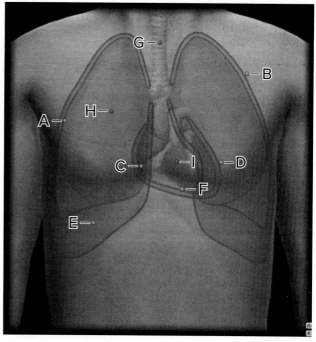

©McGraw-Hill Education

Mouse-over the pins on the screen to find the information necessary to identify the following structures.

A. _____
B. _____
C. _____
D. _____
E. _____
F. _____

Nonbody Cavity Structures (blue pins)

G. _____
H. _____
I. _____

IN REVIEW

What Have I Learned?

The following questions cover the material that you have just learned, pleura and pericardium. Apply what you have learned to answer these questions on a separate piece of paper.

1. Describe the parietal layer of the serous pericardium.
2. Describe the pericardial cavity.
3. Describe the visceral layer of the serous pericardium.
4. Describe the parietal pleura.
5. Describe the pleural cavity.
6. Describe the visceral pleura.

EXERCISE 1.7:
Peritoneum

Return to the **TOPIC** *drop-down menu and select* **Peritoneum.** *Notice* **Midsagittal** *appears in the* **VIEW** *window.*

Select the **my COURSE CONTENT** *tab if your instructor has selected specific exercises for you to complete. Otherwise select the* **ALL CONTENT** *tab.*

From the **Select structure type** *menu, select* **Peritoneum.**

Click through the terms to find the one that matches each of the following descriptions.

1. Located between the posterior abdominal wall and small intestine.
2. Consists of a structure containing the hepatic artery proper, bile duct, and hepatic portal vein.
3. Single layer of serous membrane lining wall of abdomen.
4. Single layer of serous membrane coating outer surface of many abdominal organs.
5. Potential space between parietal and visceral layers of peritoneum.
6. Double-layered fold of peritoneum suspended from greater curvature of stomach.
7. Located between the lesser curvature of the stomach, the duodenum, and the liver.
8. Capable of storing large amounts of fat and limits the spread of peritoneal infection.
9. Two structures that secrete and absorb serous fluid within peritoneal cavity.
10. Provides support for, and contains blood, nerve, and lymphatic supply for small intestine.

CHECK POINT
Peritoneum

1. Identify the structure that limits the spread of peritoneal infection.
2. Identify the structure contained in the hepatoduodenal ligament.
3. What are the functions of the mesentery of the small intestine?
4. What is the function of the parietal peritoneum?
5. From the comments section, identify the functions of the visceral and parietal layers of peritoneum.
6. What is the function of the visceral peritoneum?

Return to the **Select structure type** *menu and select* **Other.** *From the* **Select group** *menu select* **Digestive.** *Click through the terms to find the one that matches each of the following descriptions.*

1. Two organs that store feces.
2. Modifies hormones and phagocytizes bacteria.
3. Most mobile part of large intestine.
4. Primary site for digestion.
5. Peristaltic contractions move contents toward pyloric sphincter.
6. Abdominal organs that lie along posterior abdominal wall, posterior to parietal peritoneum.
7. Releases hormones (insulin, glucagon, and somatostatin) into blood.
8. Secretes and absorbs serous fluid within peritoneal cavity.

9. Mixes gastric contents.

10. Stores glycogen, minerals, and vitamins.

11. Blood, lymphatic, and nerve supply to transverse colon lie between double layer of peritoneum.

12. Receives bile and pancreatic secretions.

13. Located between sigmoid colon and anal canal.

14. Detoxifies drugs and alcohol.

15. Has both an endocrine and exocrine function.

16. Primary site for absorption of nutrients.

17. Include esophagus, parts of duodenum, ascending and descending colon, pancreas, kidneys, suprarenal glands, ureters, abdominal aorta, and inferior vena cava.

18. Produces and secretes bile and plasma proteins.

19. Two organs that absorb water and electrolytes.

20. Initiates digestion of proteins.

21. Produces digestive enzymes and bicarbonate ions.

22. Largest visceral organ.

Return to the **Select group** *menu and select* **Muscular.** *Click on* **Diaphragm** *and read the information in the* **STRUCTURE INFORMATION** *window.*

Return to the **Select group** *menu and select* **Skeletal.** *Click through the terms to find the one that matches each of the following descriptions.*

1. Composed of 33 vertebrae and intervertebral disks.

2. Joint formed by two pubic bones and intervening fibrocartilage disc.

3. Distributed in five regions.

4. In female, fibrocartilage softens in late pregnancy to allow slight separation of bones.

Return to the **Select group** *menu and select* **Urinary.** *Click on* **Urinary bladder** *and read the information in the* **STRUCTURE INFORMATION** *window.*

IN REVIEW

What Have I Learned?

The following questions cover the material that you have just learned, peritoneum. Apply what you have learned to answer these questions on a separate piece of paper.

1. Describe the greater omentum.

2. Describe the lesser omentum.

3. Describe the mesentery of the small intestine.

4. Describe the parietal peritoneum.

5. Describe the peritoneal cavity.

6. Describe the visceral peritoneum.

EXERCISE 1.8:

Organ Systems

Return to the **TOPIC** *drop-down menu and select* **Organ systems.** *Notice* **Anterior** *appears in the* **VIEW** *window.*

Select the **my COURSE CONTENT** *tab if your instructor has selected specific exercises for you to complete. Otherwise select the* **ALL CONTENT** *tab.*

From the **Select group** *menu, select* **Cardiovascular.** *Click through the terms to find the one that matches each of the following descriptions.*

1. Consists of two divisions: pulmonary and systemic circulations.

2. Drain capillaries.

3. Blood vessels that carry blood away from the heart.

4. Blood vessels that carry blood toward the heart.

5. Consists of two atria and two ventricles.

52 MODULE 1 Body Orientation

6. Three basic types: large, medium-small, and venules.

7. Pulmonary circulation carries deoxygenated blood.

8. Conical muscular organ.

9. Consists of blood, blood vessels, and heart.

10. Pulmonary circulation carries oxygenated blood.

11. Three basic types: elastic, muscular, and arterioles.

12. Pumps blood to the body.

13. Systemic circulation carries oxygenated blood.

14. Systemic circulation carries deoxygenated blood.

Click **LAYER 6—CARDIOVASCULAR** in the **LAYER CONTROLS** window, and you will find the following image:

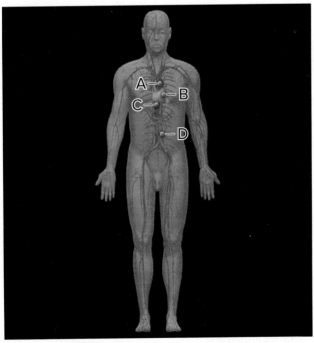

©McGraw-Hill Education

Mouse-over the pins to find the information necessary to fill in the following blanks:

A. _____

B. _____

C. _____

D.

CHECK POINT

Organ Systems: Cardiovascular

1. Identify the three basic types of arteries.
2. Identify the structure that connects arterioles to venules.
3. Contrast arteries of pulmonary circulation to those of systemic circulation.
4. Identify the two circulatory divisions of the cardiovascular system.
5. Compare the functions of the two circulatory divisions of the cardiovascular system.
6. What causes the release of atriopeptin?
7. What is the function of atriopeptin?
8. Identify the three basic types of veins.
9. Contrast veins of pulmonary circulation to those of systemic circulation.

IN REVIEW

What Have I Learned?

The following questions cover the material that you have just learned, cardiovascular system. Apply what you have learned to answer these questions on a separate piece of paper.

1. Blood vessels that carry blood away from the heart.
2. Blood vessels that carry blood toward the heart.
3. System consisting of blood, blood vessels, and heart.
4. Identify the structure that connects arterioles to venules.
5. Identify the two circulatory divisions of the cardiovascular system.

*Return to the **Select group** menu and select **Digestive**. Click through the terms to find the one that matches each of the following descriptions.*

1. Initiates digestion of proteins.

2. Modifies hormones and phagocytizes bacteria.

3. Mastication only.

4. Muscular tube continuous with nasal cavity, oral cavity, and larynx.

5. Composed of cecum, colon, rectum, and anal canal.

6. Consists of teeth, tongue, salivary glands, liver, gallbladder, and pancreas.

7. Mastication and chemical digestion.

8. Largest visceral organ.

9. Two parts: oral vestibule and oral cavity proper.

10. Consists of cervical part, thoracic part, and abdominal part.

11. Has both an endocrine and exocrine function.

12. A muscular tube lined with epithelium.

13. Stores, concentrates, and releases bile.

14. Consists of the gastrointestinal tract and accessory digestive organs.

15. Stores glycogen, minerals, and vitamins.

16. Produces digestive enzymes and bicarbonate ions.

17. Also known as alimentary tract or canal.

18. Produce and secrete components of saliva.

19. Continuous with pylorus of stomach and cecum of large intestine.

20. Absorption of vitamin B and vitamin K produced by normal gut flora.

21. Functions include mastication, taste, and phonation.

22. Four lobes: right, left, quadrate, and caudate.

23. Receives bile and pancreatic secretions.

24. Functions include ingestion, digestion, propulsion, secretion, and absorption, and elimination of wastes.

25. Characteristic features include vermiform appendix, taeniae coli, haustra, and omental appendices.

26. Functions include secretion of lysozyme.

27. Also known as buccal cavity.

28. Pear-shaped, hollow muscular organ connected to cystic duct.

29. Detoxifies drugs and alcohol.

30. Primary site for digestion.

31. Water and electrolyte absorption.

32. Located in alveolar processes of mandible and maxilla.

33. Mixes gastric contents.

34. Passes through diaphragm at esophageal hiatus.

35. Muscular organ with thick mucous membrane containing taste buds.

36. Primary site for absorption of nutrients.

37. Ventilation and phonation.

38. Functions include moistening and lubricating food before swallowing.

39. Elongated modular gland divided into head, neck, body, and tail.

40. Consists of oral cavity, pharynx, esophagus, stomach, small intestine, and large intestine.

41. Produces and secretes bile and plasma proteins.

42. Inferior to liver in shallow fossa on quadrate lobe.

43. Conveys food from pharynx to stomach.

44. Storage and defecation of feces.

45. Floor is tongue, roof is hard palate.

46. Peristaltic contractions move contents toward pyloric sphincter.

47. Two of these whose functions include ingestion, digestion, propulsion, secretion, absorption, and elimination of wastes.

48. Releases hormones (insulin, glucagon, and somatostatin) into blood.

Click **LAYER 9—DIGESTIVE** *in the* **LAYER CONTROLS** *window, and you will find the following image:*

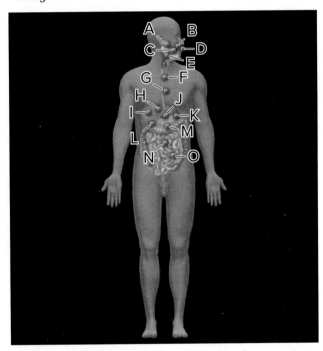

©McGraw-Hill Education

Mouse-over the pins to find the information necessary to fill in the following blanks:

A. _____
B. _____
C. _____
D. _____
E. _____
F. _____
G. _____
H. _____
I. _____
J. _____
K. _____
L. _____
M. _____
N. _____
O. _____

CHECK POINT

Organ Systems: Digestive

1. List the accessory digestive organs.
2. The GI tract is also known as _____.
3. List the organs of the digestive system from proximal to distal.
4. What is "heartburn"?
5. List the structures of the large intestine.
6. What are the causes of cirrhosis?
7. List the three subdivisions of the pharynx.
8. List the three salivary glands.
9. List the three parts of the small intestine from proximal to distal.
10. List the four parts of the stomach from proximal to distal.
11. The first part of mechanical digestion is _____.
12. List the teeth of the mandible and the maxilla.

IN REVIEW

What Have I Learned?

The following questions cover the material that you have just learned, digestive system. Apply what you have learned to answer these questions on a separate piece of paper.

1. Identify the muscular tube continuous with nasal cavity, oral cavity, and laryngopharynx.

2. Identify the structure composed of cecum, colon, rectum, and anal canal.

3. Identify the organ group consisting of teeth, tongue, salivary glands, liver, gallbladder, and pancreas.

4. Identify the organ that stores, concentrates, and releases bile.

5. Identify the structure with functions including mastication, deglutition, and phonation.

6. Identify the organ that detoxifies drugs and alcohol.

7. Identify the primary site for digestion.

8. Identify the location for water and electrolyte absorption.

9. Identify the primary site for absorption of nutrients.

10. Identify the organ that produces and secretes bile and plasma proteins.

11. Identify the largest visceral organ.

12. Define mastication.

13. Define deglutition.

14. Define phonation.

Return to the Select group menu and select Endocrine. Click through the terms to find the one that matches each of the following descriptions.

1. Regulates body temperature, food and water intake, and emotional behavior.

2. Paired female gonad.

3. In children, a large bilobed gland in the anterior and superior mediastinum.

4. Two functional lobes: anterior and posterior.

5. Urine formation.

6. Collection of organs and cells that secrete hormones.

7. Has both an endocrine and exocrine function.

8. Small nodules, partially embedded in posterior surface of thyroid gland.

9. Produces progesterone, estrogen, and inhibin.

10. Controls anterior pituitary gland with releasing and inhibiting hormones.

11. Paired, oval male gonad.

12. Secretes melatonin.

13. Removes excess water, electrolytes, and wastes of protein metabolism from blood.

14. Medulla secretes epinephrine and norepinephrine.

15. Releases hormones (insulin, glucagon, and somatostatin) into blood.

16. Site for maturation and differentiation of T-lymphocytes.

17. Functions: regulate body functions by secreting hormones into bloodstream.

18. Anterior lobe produces TSH, PRL, ACTH, GH, LH, MSH, and FSH.

19. Synthesize and secrete triiodothyronine and thyroxin.

20. Produces hormones that are transported to and stored in posterior pituitary.

21. Secretions stimulate and promote differentiation, growth, and maturation of T-lymphocytes.

22. Produces sperm and androgen.

23. Produces the enzyme rennin.

24. Secrete PTH.

25. Produces digestive enzymes and bicarbonate ions.

26. Synthesizes and secretes calcitonin.

27. Controls autonomic nervous system.

28. Oogenesis and ovulation.

29. Cortex secretes corticosteroids and androgen.

30. When enlarged, known as goiter.

31. Glands are ductless.

32. Elongated modular gland divided into head, neck, body, and tail.

33. Regulates body functions.

34. Considered master control center for endocrine system.

35. Bilobed endocrine gland of the neck, lobes lie lateral to trachea and larynx.

36. Posterior lobe stores and releases ADH and OT.

37. Synthesizes calcitrol, a form of vitamin D.

38. Begin development near kidneys, and later descend to scrotum.

39. Atrophies during adolescence, remnant in adults consists primarily of fibrous and adipose tissue.

40. Maintains sleep/wake cycle.

41. Size decreases after menopause.

Click **LAYER 5—ENDOCRINE** *in the* **LAYER CONTROLS** *window, and you will find the following image:*

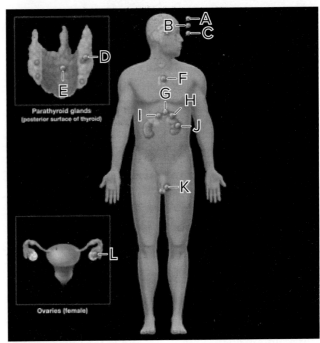

©McGraw-Hill Education

Mouse-over the pins to find the information necessary to fill in the following blanks:

A. _____

B. _____

C. _____

D. _____

E. _____

F. _____

G. _____

H. _____

I. _____

J. _____

K. _____

L. _____

CHECK POINT

Organ Systems: Endocrine

1. List the organs and tissues of the endocrine system.
2. Where is calcitrol synthesized? What is its function?
3. Where is erythropoietin released? What is its function?
4. Where is renin produced? What is its function?
5. What hormone is synthesized by the chief cells of the parathyroid gland? What is the function of this hormone?
6. What is the adenohypophysis? What is the neurohypophysis?
7. List the functions of the thymus.
8. How does the thymus compare from childhood through adolescence and adulthood?

IN REVIEW

What Have I Learned?

The following questions cover the material that you have just learned, endocrine system. Apply what you have learned to answer these questions on a separate piece of paper.

1. Identify the organ system that regulates body functions by secreting hormones into the bloodstream.

2. Identify the collection of organs and cells that secrete hormones.

3. Identify the organ that regulates body temperature, food and water intake, and emotional behavior.

4. In children, a large bilobed endocrine gland in the anterior and superior mediastinum.

5. Identify the visceral organ that has both an endocrine and exocrine function.

6. Identify the endocrine gland that controls anterior pituitary gland with releasing and inhibiting hormones.

7. Identify the endocrine gland that secretes melatonin.

8. Identify the site for maturation and differentiation of T-lymphocytes.

9. Identify the endocrine gland that produces hormones that are transported to and stored in posterior pituitary gland.

10. Identify the endocrine gland that controls autonomic nervous system.

11. Identify the endocrine gland considered the master control center for endocrine system.

12. Identify the endocrine gland that maintains sleep/wake cycle.

13. What is the adenohypophysis? What is the neurohypophysis?

14. List the hormones produced by the suprarenal cortex.

15. List the hormones produced by the suprarenal medulla.

Return to the **Select group** *menu and select* **Female reproductive.** *Click through the terms to find the one that matches each of the following descriptions.*

1. The paired female gonad.

2. Function: milk production and fat storage.

3. Female organ of copulation.

4. The site of fertilization.

5. Consists of female gonads, accessory glands, uterus, external genitalia, and mammary glands.

6. Also known as Fallopian tube or oviduct.

7. Implantation site for blastocyst.

8. Contains oocytes.

9. Duct for menstrual fluid.

10. Bisected during tubal ligation.

11. Contracts to expel fetus and placenta during parturition.

12. Pyramidal structure composed of skin with areola and nipple; fat and connective tissue; mammary glands and associated ducts.

13. Receives cervix of uterus and opens into vestibule.

14. Common site for ectopic implantation of fertilized egg.

15. Supplies nutrients and oxygen, via placenta, to embryo and fetus.

16. Conduit for movement of germ cells.

17. Produces progesterone, estrogen, and inhibin.

18. Base located over pectoralis major muscle, extending into axilla as "tail."

Click **LAYER 12—FEMALE REPRODUCTIVE** *in the* **LAYER CONTROLS** *window, and you will find the following image:*

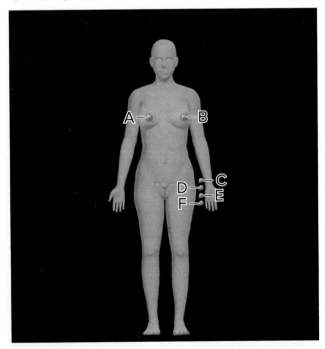

©McGraw-Hill Education

Mouse-over the pins to find the information necessary to fill in the following blanks:

A. _____
B. _____
C. _____
D. _____
E. _____
F. _____

CHECK POINT

Organ Systems: Female Reproductive

1. List the organs and tissues of the female reproductive system.
2. Describe the perineum in the female.
3. How does the uterine position vary in reference to nearby organs or pregnancy?
4. What is another name for the perineum?

IN REVIEW

What Have I Learned?

The following questions cover the material that you have just learned, female reproductive system. Apply what you have learned to answer these questions on a separate piece of paper.

1. Identify the organ system consisting of female gonads, accessory organs, uterus, external genitalia, and mammary glands.

2. Identify the female organ of copulation.

3. Identify the site of fertilization.

4. Identify the implantation site for the blastocyst.

5. Identify the organ that contracts to expel fetus and placenta during parturition.

6. Identify the organ that receives the cervix of uterus and opens into vestibule.

7. Identify the common site for ectopic implantation of fertilized egg.

8. Identify the conduit for movement of female and male germ cells.

9. Identify the organ that produces progesterone, estrogen, and inhibin.

10. Identify the structure whose base is located over the pectoralis major muscle, extending into the axilla as a "tail."

11. What are fimbriae? Where are they located?

12. What is another name for the perineum?

Return to the **Select group** *menu and select* **Integumentary.** *Click through the terms to find the one that matches each of the following descriptions.*

1. Merocrine sweat glands included in these.

2. Not found in thick skin of palms or soles.

3. Projects from epidermal surface.

4. Major cell type is keratinocyte.

5. Forms as invagination from epidermis.

6. Accessory organ of skin consisting of fine, keratinized filament.

7. Between hair bulb and mature hair shaft.

8. Apocrine sweat glands included in these.

9. Contains appendages of skin.

10. Functions in hair formation and growth.

11. The external surface of the body.

12. Grows from oblique tube in skin.

13. Also called hypodermis.

14. Contains sensory nerve endings and dense network of blood and lymphatic vessels.

15. Portion of hair within follicle.

16. Functions include protection, immune defense, secretion, thermoregulation.

17. Keratinized stratified squamous epithelium of variable thickness.

18. Site of hair elongation.

19. Functions include fat storage, thermoregulation, and padding.

20. Skin (between epidermis and hypodermis)

21. Location of vitamin D synthesis.

22. Two structures that function in heat retention, cutaneous sensation, and protection.

23. Also called superficial fascia.

24. Thick skin has five layers while thin skin has only four.

25. Sebaceous glands included in these.

26. Contains skin, hair, nails, and exocrine glands.

27. Apocrine glands release here.

Click **LAYER 1—INTEGUMENTARY** *in the* **LAYER CONTROLS** *window, and you will find the following image:*

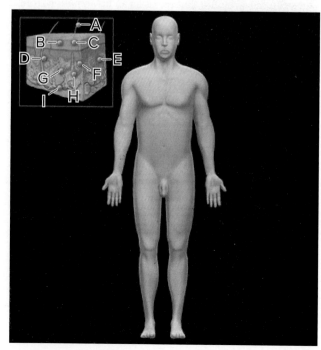

©McGraw-Hill Education

Mouse-over the pins to find the information necessary to fill in the following blanks:

A. _____
B. _____
C. _____
D. _____
E. _____
F. _____
G. _____
H. _____
I. _____

CHECK POINT

Organ Systems: Integumentary

1. Where are the different exocrine glands of the skin located?
2. Describe the various exocrine glands of the skin. What are their functions?
3. List the other integumentary glands. What are their functions?
4. Where are hair follicles located?
5. Describe the hair follicle. What are its functions?
6. How do hair follicles develop?
7. What structures are associated with the hair follicle?
8. Where are hair roots located?
9. Describe the hair root. What is its function?

IN REVIEW

What Have I Learned?

The following questions cover the material that you have just learned, integumentary system. Apply what you have learned to answer these questions on a separate piece of paper.

1. Merocrine sweat glands are included in these.
2. Identify a structure not found in thick skin of palms or soles.
3. Identify a structure that projects from epidermal surface.
4. Identify the integumentary layer whose major cell type is the keratinocyte.
5. Identify the structure that forms as an invagination from the epidermis.
6. Identify the integumentary layer containing appendages of the skin.
7. Identify the structure that functions in hair formation and growth.
8. The external surface of the body.
9. Also called hypodermis.
10. Contains sensory nerve endings and dense network of blood and lymphatic vessels.
11. The portion of a hair within the follicle.
12. The location of vitamin D synthesis.
13. Thick skin has five layers here while thin skin has only four.
14. Identify the organ system containing skin, hair, nails, and exocrine glands.
15. Apocrine glands release their products here.
16. What is the arrector pili muscle?
17. Where is hair located?
18. Where is hair *not* located?
19. What determines body hair distribution?
20. Where is the integumentary system located?
21. Describe the integumentary system. What are its functions?
22. Where is the hair shaft located?
23. Describe the hair shaft. What are its functions?

Return to the **Select group** menu and select **Lymphatic.** Click through the terms to find the one that matches each of the following descriptions.

1. Contains MALT.

2. Small, irregular-shaped lymph sac, the origin of thoracic duct.

3. Located in the left hypochondriac region.

4. Located along pathways of lymphatic vessels throughout the body except the brain and spinal cord.

5. Located in medullary cavity of all bones.

6. Consists of capillaries, vessels, trunks, and ducts.

7. Detects antigens and initiates immune response.

8. System that produces lymphocytes.

9. Receives lymph from body inferior to diaphragm.

10. Largest lymphatic organ.

11. Large bilobed gland in children.

12. Receives lymph from abdomen, pelvis, and lower limbs.

13. Pharyngeal also known as "adenoids" when infected or inflamed.

14. A large collection of lymphatic nodules.

15. Phagocytizes old and damaged blood cells, bacteria, and foreign material.

16. Small, usually bean-shaped lymphatic organs.

17. Atrophies during adolescence.

18. Function to transport lymph.

19. Receives lymph from right upper limb.

20. Soft pulpy tissue in two forms: red and yellow.

21. Trap foreign material and facilitate identification by lymphocytes.

22. Initiates immune response to blood antigens.

23. Site for maturation and differentiation of T-lymphocytes.

24. Functions include filtering blood.

25. Receives lymph from right side of head and neck.

26. Mucosa-associated lymphatic tissue.

27. Lingual, palatine, and pharyngeal.

28. Reservoir for red and white blood cells.

29. Found in prominent clusters.

30. System that initiates immune response.

Click **LAYER 7—LYMPHATIC** *in the* **LAYER CONTROLS** *window, and you will find the following image:*

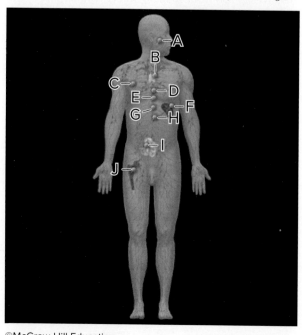

©McGraw-Hill Education

Mouse-over the pins to find the information necessary to fill in the following blanks:

A. _____
B. _____
C. _____
D. _____
E. _____
F. _____
G. _____
H. _____
I. _____
J. _____

CHECK POINT

Organ Systems: Lymphatic

1. Where is the bone marrow located?
2. Where are lymph nodes located?
3. What situations can cause lymph nodes to become enlarged?
4. Where is the lymphatic system located?
5. Where does lymph return to the venous system?
6. Lymphatic vessels typically follow _____.
7. Where is the spleen located?
8. Where is the thoracic duct located?
9. Where is the thymus located?

IN REVIEW

What Have I Learned?

The following questions cover the material that you have just learned, lymphatic system. Apply what you have learned to answer these questions on a separate piece of paper.

1. The organ system that initiates immune response.
2. Identify the structures located along pathways of lymphatic vessels throughout the body except the brain and spinal cord.
3. Located in medullary cavity of all bones.
4. Consists of capillaries, vessels, trunks, and ducts.
5. System that produces lymphocytes.
6. Receives lymph from body inferior to diaphragm.
7. Largest lymphatic organ.
8. Large bilobed gland in children.
9. Atrophies during adolescence.
10. Functions to transport lymph.
11. Receives lymph from right upper limb.
12. Site for maturation and differentiation of T-lymphocytes.
13. Reservoir for red and white blood cells.
14. Describe bone marrow. What are its functions?
15. Over time, how does the type and location of bone marrow change?
16. What is lymph? How does it originate?
17. Where are lymph nodes located?
18. How do lymph nodes compare to the spleen?
19. Describe lymphatic system. What are its functions?
20. Describe lymph nodes. What are their functions?
21. What are adenoids?
22. Describe the thymus. What are its functions?
23. Describe the tonsils. What are their functions?
24. Where is mucosa-associated lymphatic tissue located?
25. Describe mucosa-associated lymphatic tissue. What are its functions?
26. List the organs and tissues of the lymphatic system.
27. How does the spleen compare to lymph nodes?
28. Describe lymphatic vessels. What is their function?

*Return to the **Select group** menu and select **Male reproductive**. Click through the terms to find the one that matches each of the following descriptions.*

1. Contributes 60 percent of semen volume.

2. Male organ of copulation.

3. Convoluted muscular tube along posterolateral surface of testis.

4. Transmits urine and semen.

5. Produces androgens.

6. Secretes thick, alkaline fluid containing fructose and prostaglandins.

7. Contains prostatic urethra, ejaculatory ducts, and prostatic ducts.

8. Longer in males than in females.

9. Unites with vas deferens to form ejaculatory duct.

10. Divided into head, body, and tail.

11. Also known as ductus deferens.

12. Termed flaccid when nonerect.

13. Preprostatic, prostatic, membranous, and spongy.

14. Structures include gonads, accessory glands, and external genitals.

15. Secretes thin, milky, slightly acid fluid.

16. Produces sperm cells.

17. Site of maturation and storage of sperm.

18. Scrotal part dissected during vasectomy.

19. Contributes 30 percent of semen volume.

20. Transport sperm and fluid from epididymis during emission.

21. Produces androgens, gametes, and semen.

22. Transports semen and urine.

23. Body contains dense network of sensory nerve endings.

24. Paired, oval male gonad.

*Click **LAYER 11—MALE REPRODUCTIVE** in the **LAYER CONTROLS** window, and you will find the following image:*

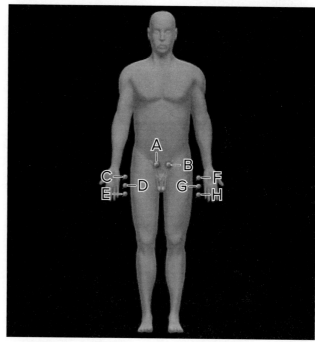

©McGraw-Hill Education

Mouse-over the pins to find the information necessary to fill in the following blanks:

A. _____

B. _____

C. _____

D. _____

E. _____

F. _____

G. _____

H. _____

64 MODULE 1 Body Orientation

CHECK POINT

Organ Systems: Male Reproductive

1. Where is the epididymis located?
2. Where is the male reproductive system located?
3. Describe the perineum in the male.
4. Define flaccid.
5. Where are the seminal vesicles located?
6. The vas deferens is also known as the _____.

IN REVIEW

What Have I Learned?

The following questions cover the material that you have just learned, male reproductive system. Apply what you have learned to answer these questions on a separate piece of paper.

1. The organ system that produces androgens, gametes, and semen.
2. The convoluted muscular tube along posterolateral surface of testis.
3. Secretes thick, alkaline fluid containing fructose and prostaglandins.
4. Longer in males than females.
5. Unites with vas deferens to form ejaculatory duct.
6. Termed flaccid when nonerect.
7. Preprostatic, prostatic, membranous, and spongy.
8. Structures include gonads, accessory glands, and external genitals.
9. Secretes thin, milky, slightly acid fluid.
10. Produces sperm cells.
11. Site of maturation and storage of sperm.
12. Contributes 30 percent of semen volume.
13. Transports sperm and fluid from epididymis during emission.
14. Paired, oval male gonad.
15. How long can sperm be stored in the epididymis?
16. List the organs and tissues of the male reproductive system.
17. What two structures unite to form the ejaculatory duct?
18. Where do the testes begin development? How do they arrive at their destination?
19. How does scrotal temperature compare to body temperature?
20. How is the temperature of the testes regulated?
21. Describe the testis. What are its functions?
22. Describe the male reproductive system. What are its functions?
23. Describe the penis. What are its functions?
24. What is BPH? How common is it?
25. Describe the urethra. What are its functions?
26. What is emission?
27. What is a vasectomy?

Return to the **Select group** *menu and select* **Muscular.** *Click through the terms to find the one that matches each of the following descriptions.*

1. Unilateral: rotation of head.
2. Longest muscle in body.
3. Movement of pectoral girdle and upper limb.
4. Around muscles and some organs.
5. Function in nonverbal communication.

6. Muscles of respiration.

7. Origin: tibia and interosseous membrane.

8. Lateral part: abduction of arm.

9. Encapsulates and defines shape of some organs.

10. At right angle to internal abdominal oblique.

11. Site for intramuscular injections.

12. Functions include maintenance of posture.

13. Flexion of forearm.

14. Active during forced inspiration.

15. Function in verbal communication.

16. Layer of dense irregular connective tissue devoid of fat.

17. Produces up to 85 percent of body heat.

18. Movement of pelvic girdle and lower limb.

19. Functions include temperature regulation.

20. One of the four muscles of this group flex thigh.

21. Anterior part: flexion and medial rotation of arm.

22. Actions include flexion of trunk during sit-ups.

23. Retinacula hold tendons close to joints.

24. Inversion of foot.

25. Two muscles important in abdominal breathing.

26. Underlies hypodermis.

27. Twenty-three muscles innervated by facial nerve.

28. Important in assuming "cross-legged" position.

29. Bilateral: flexion of head.

30. Muscles of facial expression.

31. System with the function of body movement.

32. Tissue with the function of body movement.

33. Origin: clavicle and manubrium of sternum.

34. Actions include compression of anterior abdominal wall.

35. Muscles of mastication.

36. Separates and supports individual muscles and neurovascular bundles.

37. Weak flexion of arm.

38. Group of four muscles extending the leg.

39. Extrinsic eye muscles.

40. Intrinsic back muscles.

41. Dorsiflexion of foot.

42. Two muscles important in straining.

43. Also known as striated or voluntary muscle.

44. Supination of forearm.

45. Posterior part: extension and lateral rotation of arm.

Click **LAYER 3—MUSCULAR** *in the* **LAYER CONTROLS** *window, and you will find the following image:*

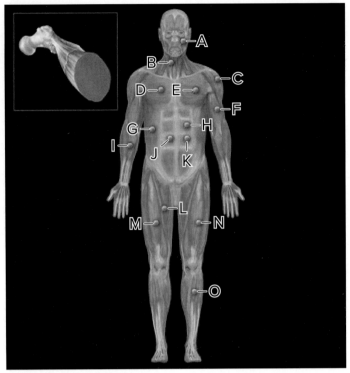

©McGraw-Hill Education

Mouse-over the pins to find the information necessary to fill in the blanks:

A. _____
B. _____
C. _____
D. _____
E. _____
F. _____
G. _____
H. _____
I. _____
J. _____
K. _____
L. _____
M. _____
N. _____
O. _____

CHECK POINT

Organ Systems: Muscular

1. Where are the appendicular muscles located?
2. Where are the axial muscles located?
3. Where is the deep fascia located?
4. Fascia refers to _____.
5. How many muscles are included in the muscles of facial expression? What is their innervation?
6. Which muscles are under voluntary control? Which muscles are under involuntary control?
7. List the four muscles that comprise the quadriceps femoris muscle. Which is the only one to act on the thigh?
8. Which is the longest muscle in the body?
9. Where are the skeletal muscles located?
10. Skeletal muscle is also known as _____ or _____.
11. Approximately _____ muscles are formed by skeletal muscle in the body.

IN REVIEW

What Have I Learned?

The following questions cover the material that you have just learned, muscular system. Apply what you have learned to answer these questions on a separate piece of paper.

1. Identify the muscle with unilateral contraction yields rotation of head.
2. The longest muscle in the body.
3. Involved with movement of pectoral girdle and upper limb.
4. Located around muscles and some organs.
5. The site for intramuscular injections.
6. Produces up to 85 percent of body heat.
7. Actions include flexion of trunk during sit-ups.
8. Two muscles important in abdominal breathing.
9. Underlies hypodermis.
10. Separates and supports individual muscles and neurovascular bundles.
11. Group of four muscles extending the leg.
12. Also known as striated or voluntary muscle.
13. How many muscles are included in the muscles of facial expression?
14. Describe the appendicular muscles. What are their functions?
15. Describe the axial muscles. What are their functions?
16. Describe deep fascia. What are its functions?
17. Describe the muscles of facial expression. What are their functions?
18. Describe the muscular system. What are its functions?
19. Describe skeletal muscle. What are its functions?

*Return to the **Select group** menu and select **Nervous**. Click through the terms to find the one that matches each of the following descriptions.*

1. Comprised of axons connecting sympathetic ganglia.
2. Includes the brain and spinal cord.
3. Mediates some reflexes.
4. Hemispheres connected by corpus callosum.
5. Major organ of CNS.
6. Consists of CNS and PNS.
7. Receives extensive sensory input from body and CNS.
8. Contributions from rami of C5-T1 spinal nerves.
9. Ventral rami of C1-4 spinal nerves.
10. Located in vertebral canal.
11. Vertical stalklike portion of the brain.
12. Also known as sympathetic chain ganglia.
13. Includes two large hemispheres separated by longitudinal fissure.
14. Seat of consciousness, intelligence, learning, emotion, and memory.
15. Consists of the brain, spinal cord, nerves, and ganglia.

16. Vascular tunic composed of choroid, ciliary body, and iris.

17. Ends at level of L2 vertebra in adults.

18. Monitors muscles to ensure fluid movements.

19. Rostral portion of the brain.

20. Includes midbrain, pons, and medulla oblongata.

21. Innervates all muscles of pelvic floor, perineum, and lower limb.

22. Composed of inner core of gray matter and outer coat of white matter.

23. Located in cranial cavity and vertebral canal.

24. Neural tunic composed of retina and pigment epithelium.

25. Transmits information to and receives information from CNS.

26. Caudal portion of the brain.

27. Ventral rami of L1-S4 spinal nerves.

28. Coordinates complex movements.

29. Gives rise to 31 pairs of spinal nerves.

30. Motor nerves distribute to muscles of upper limb.

31. General sensation nerves distribute to skin over shoulder, anterior and lateral neck, and parts of head.

32. Innervates skin on lower anterior abdominal wall, lower back, perineum, and lower limb.

33. Functional divisions include afferent and efferent.

34. Located in cranial cavity.

35. Fibrous tunic composed of cornea and sclera.

36. Twelve pair of nerves; most connected to brainstem.

37. Integrates and processes nervous information.

38. Located adjacent to vertebral bodies from base of skull to coccyx.

39. Motor nerves include phrenic nerve.

40. General sensation nerves distribute to skin of upper limb.

41. Located in orbit of skull.

42. Located outside cranial cavity and vertebral canal.

Click **LAYER 4—NERVOUS** *in the* **LAYER CONTROLS** *window, and you will find the following image:*

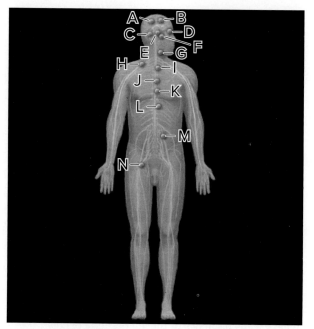

©McGraw-Hill Education

Mouse-over the pins to find the information necessary to fill in the following blanks:

A. _____
B. _____
C. _____
D. _____
E. _____
F. _____
G. _____
H. _____
I. _____
J. _____
K. _____
L. _____
M. _____
N. _____

CHECK POINT

Organ Systems: Nervous

1. Where is the brachial plexus located?
2. Which nerves contribute to the brachial plexus?
3. Where is the brain located?
4. Where is the brainstem located?
5. Where is the central nervous system located?
6. Where is the cerebellum located?
7. Where is the cerebrum located?
8. Where is the cervical plexus located?
9. Where are the cranial nerves located?
10. Where is the lumbosacral plexus located?
11. Where is the nervous system located?
12. Where is the peripheral nervous system located?
13. Where is the spinal cord located?
14. Where are the sympathetic trunks located?

IN REVIEW

What Have I Learned?

The following questions cover the material that you have just learned, nervous system. Apply what you have learned to answer these questions on a separate piece of paper.

1. Comprised of axons connecting sympathetic ganglia.
2. Includes the brain and spinal cord.
3. Receives extensive sensory input from the body and the CNS.
4. Contributions from rami of C5-T1 spinal nerves.
5. Ventral rami of C1-4 spinal nerves.
6. Located in vertebral canal.
7. Vertical stalklike portion of brain.
8. Includes two large hemispheres separated by longitudinal fissure.
9. Seat of consciousness, intelligence, learning, emotion, and memory.
10. Consists of the brain, spinal cord, nerves, and ganglia.
11. Vascular tunic composed of choroid, ciliary body, and iris.
12. Ends at level of L2 vertebra in adults.
13. Monitors muscles to ensure fluid movements.
14. Rostral portion of the brain.

15. Includes midbrain, pons, and medulla oblongata.
16. Composed of inner core of gray matter and outer coat of white matter.
17. Located in cranial cavity and vertebral canal.
18. Ventral rami of L1-S4 spinal nerves.
19. Coordinates complex movements.
20. Gives rise to 31 pairs of spinal nerves.
21. Twelve pair of nerves; most connected to brainstem.
22. Integrates and processes nervous information.
23. Located adjacent to vertebral bodies from base of skull to coccyx.
24. The plexus with motor nerves including the phrenic nerve.
25. Located outside cranial cavity and vertebral canal.
26. What is a plexus?
27. Define somatic motor.
28. Define visceral motor.
29. List the organs and tissues of the nervous system.
30. What is a spinal nerve?
31. What is the motor composition of the brachial plexus?
32. What is the general sensation composition of the brachial plexus?
33. What are the divisions of the brachial plexus?
34. How many total branches comprise the brachial plexus? What are the five terminal branches?
35. Describe the brain. What traits are attributed to the brain?
36. Describe the brainstem.
37. Describe the central nervous system. What is its function?
38. Define rostral.
39. Which nerves contribute to the cervical plexus?
40. What is the motor composition of the cervical plexus?
41. What is the general sensation composition of the cervical plexus?
42. Describe the cranial nerves. What are their functions?
43. Which general sensations are included by the cranial nerves?
44. Which special sensations are included by the cranial nerves?
45. Describe the cerebrum.
46. Which nerves contribute to the lumbosacral plexus?
47. What are the branches of the lumbosacral plexus?
48. What are the innervations of the lumbosacral plexus?
49. Describe the nervous system. What are its functions?
50. What is the Latin for "to bring to"? What is the Latin for "to bring out"?
51. Describe the peripheral nervous system. What are its functions?
52. Where is the spinal cord continuous with the brain?
53. Describe the spinal cord. What are its functions?
54. How many spinal nerves arise from the spinal cord?
55. Describe the sympathetic trunk.
56. What other names refer to the sympathetic trunk?
57. What structures give the sympathetic trunk a "beaded" appearance?

Return to the **Select group** *menu and select* **Respiratory.** *Click through the terms to find the one that matches each of the following descriptions.*

1. Latin = throat.
2. Primary organ of respiration.
3. The respiratory tract superior to the larynx.
4. The respiratory tract from the larynx inferiorly.
5. Involved with air passage, air cleaning, phonation, gas exchange, and acid-base balance.

6. Sound production.

7. Hollow cavities within some skull bones.

8. Right has three lobes, while left has two.

9. Rigid tube held open by series of "C-shaped" rings.

10. Continuation of trachea into right and left lungs.

11. Communicates with paranasal sinuses and nasolacrimal duct.

12. Decrease weight of anterior skull.

13. Located in pulmonary cavity.

14. Muscular tube continuous with nasal cavity, oral cavity, and larynx.

15. Contribute to voice resonance.

16. Located between pharynx and trachea.

17. Cavity with functions including olfaction.

18. Main branch into lobar, which branches into segmental.

19. Medial surface has hilum.

20. Posterior aspect closed by trachealis muscle.

21. Located between hard palate and base of skull.

22. Surface covered with visceral pleura.

23. Conducts air between trachea and lungs.

24. Three subdivisions include naso-, oro-, and laryngo.

25. Prevents swallowed food from entering lower respiratory tract.

Click **LAYER 8—RESPIRATORY** *in the* **LAYER CONTROLS** *window, and you will find the following image:*

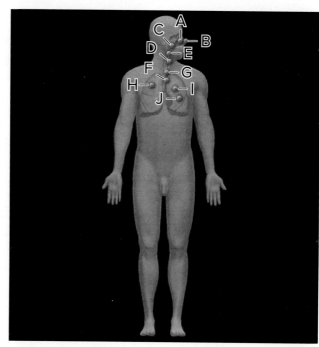

©McGraw-Hill Education

Mouse-over the pins to find the information necessary to fill in the following blanks:

A. _____
B. _____
C. _____
D. _____
E. _____
F. _____
G. _____
H. _____
I. _____
J. _____

CHECK POINT

Organ Systems: Respiratory

1. Where are the bronchi located?
2. Where is the larynx located?
3. Where are the lungs located?
4. Where is the nasal cavity located?
5. Where are the nasal sinuses located?
6. Where is the pharynx located?
7. Where is the respiratory system located?
8. Where is the trachea located?
9. Where is the upper respiratory tract located?

IN REVIEW

What Have I Learned?

The following questions cover the material that you have just learned, respiratory system. Apply what you have learned to answer these questions on a separate piece of paper.

1. Involved with air passage, air cleaning, phonation, gas exchange, and acid-base balance.
2. Primary organ of respiration.
3. The respiratory tract superior to the larynx.
4. The respiratory tract from the larynx inferiorly.
5. Communicates with paranasal sinuses and nasolacrimal duct.
6. Responsible for sound production.
7. Hollow cavities within some skull bones.
8. Decreases weight of the anterior skull.
9. Located in the pulmonary cavity.
10. Muscular tube continuous with nasal cavity, oral cavity, and larynx.
11. Contribute to voice resonance.
12. Surface covered with visceral pleura.
13. Conducts air between trachea and lungs.
14. Where does gas exchange occur?
15. List the laryngeal cartilages.
16. List the three divisions of the thoracic cavity.
17. Describe the bronchi. What is their function?
18. Describe the larynx. What is its function?
19. Describe the lower respiratory tract. What is its function?
20. Describe the lungs. What is their function?
21. Describe the nasal cavity. What is its function?
22. Describe the nasal sinuses. What are their functions?
23. How are the paranasal sinuses named?
24. Describe the pharynx.
25. List the three subdivisions of the pharynx.
26. List the organs and tissues of the respiratory system.
27. Describe the respiratory system. What is its function?
28. Describe the trachea.
29. What is a tracheotomy?
30. Describe the upper respiratory tract. What is its function?

*Return to the **Select group** menu and select **Skeletal**. Click through the terms to find the one that matches each of the following descriptions.*

1. Provides support for the body.
2. Bones of hip, thigh, leg, and foot.
3. Provides storage of calcium and phosphorus.
4. Contains cranial and facial bones.
5. The bones of the limbs, shoulder, and pelvis.
6. Provides protection for the body.
7. Located at the posterior midline of axial skeleton.
8. Provides site of muscle attachment.
9. Has series of curvatures along its length.
10. Include humerus, scapula, femur, and hip bones.
11. The bones of the head, neck, and trunk.
12. Consists of cervical, thoracic, lumbar, sacral, and coccygeal vertebrae.
13. Comprised of axial and appendicular skeleton.

14. Consists of 12 ribs and sternum.

15. Bones of the head.

16. Includes bones of skull, vertebral column, and thoracic cage.

17. Site of hemopoiesis.

18. Skeletal framework of chest.

19. Bones of shoulder, arm, forearm, wrist, and hand.

20. Consists of bone, cartilage, ligament, and tendon.

21. Provides movement of the body via joints.

Click **LAYER 2—SKELETAL** in the **LAYER CONTROLS** window, and you will find the following image:

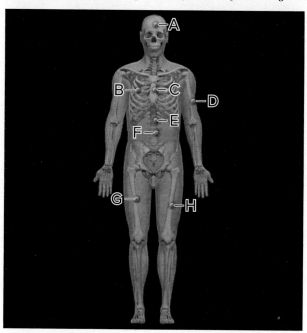

©McGraw-Hill Education

Mouse-over the pins to find the information necessary to fill in the following blanks:

A. _____
B. _____
C. _____
D. _____
E. _____
F. _____
G. _____
H.

CHECK POINT

Organ Systems: Skeletal

1. Where is the appendicular skeleton located?
2. Where is the axial skeleton located?
3. Where is the skeleton of the lower limb located?
4. Where is the skeleton of the upper limb located?
5. Where is the thoracic cage located?
6. Where do all ribs articulate?
7. Where is the vertebral column located?

IN REVIEW

What Have I Learned?

The following questions cover the material that you have just learned, skeletal system. Apply what you have learned to answer these questions on a separate piece of paper.

1. Provides support for the body.
2. Provides storage of calcium and phosphorus.
3. Located at the posterior midline of axial skeleton.
4. Includes bones of skull, vertebral column, and thoracic cage.
5. The bones of the limbs, shoulder, and pelvis.
6. Has a series of curvatures along its length.
7. Which ribs are considered true ribs?
8. Which ribs are considered false ribs?
9. Which ribs are considered floating ribs?
10. Describe the appendicular skeleton.
11. Describe the axial skeleton.
12. Describe the skeletal system. What is its function?
13. Describe the skeleton of the upper limb.
14. Describe the skeleton of the lower limb.
15. Describe the skull.
16. Describe the thoracic cage.
17. Describe the vertebral column.
18. The vertebral column is also known as _____.

Return to the **Select group** *menu and select* **Urinary**. *Click through the terms to find the one that matches each of the following descriptions.*

1. Hollow organ with smooth muscle wall.

2. Muscular tube runs from hilum of kidney to urinary bladder.

3. Consists of kidneys, ureters, urinary bladder, and urethra.

4. Transports urine in female.

5. Retroperitoneal paired organ.

6. Stores urine.

7. Fibromuscular tube running from urinary bladder to urethral orifice.

8. Transport urine from kidney to urinary bladder.

9. Filters waste products from blood and eliminates them in urine.

10. Contains cortex, medulla, and renal sinus.

11. Ureters transport urine from kidneys to urinary bladder, urinary bladder stores and, with urethra, expels urine from the body.

12. Transports semen and urine in male.

13. Removes excess water, electrolytes, and wastes from blood.

14. Paired, bean-shaped organ.

15. Detrusor muscle makes up smooth muscle wall.

Click **LAYER 10—URINARY** *in the* **LAYER CONTROLS** *window, and you will find the following image:*

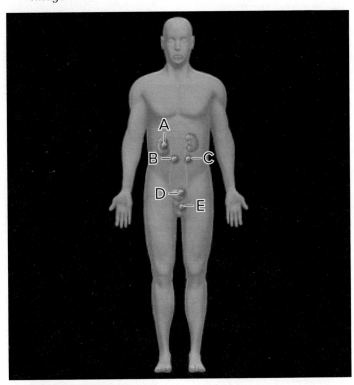

©McGraw-Hill Education

Mouse-over the pins to find the information necessary to fill in the following blanks:

A. _____

B. _____

C. _____

D. _____

E. _____

CHECK POINT

Organ Systems: Urinary

1. Where is the kidney located?
2. Where is the ureter located?
3. Where is the urethra located?
4. Where is the urinary bladder located?
5. Where is the urinary system located?

IN REVIEW

What Have I Learned?

The following questions cover the material that you have just learned, urinary system. Apply what you have learned to answer these questions on a separate piece of paper.

1. Hollow organ with smooth muscle wall.
2. Muscular tube running from the hilum of the kidney to the urinary bladder.
3. Consists of kidneys, ureters, urinary bladder, and urethra.
4. Paired, bean-shaped organ of the urinary system.
5. Detrusor muscle makes up the smooth muscle wall.
6. Contains cortex, medulla, and renal sinus.
7. List the flow of urine from the kidney distally.
8. List the organs and tissues of the urinary system.
9. Why are urinary tract infections more common in females?
10. Describe the kidney. What is its function?
11. What structures surround and support the kidney?
12. Describe the ureter. What is its function?
13. Describe the urethra. What is its function?
14. Describe the urinary bladder. What is its function?
15. What aspect of the urinary bladder affects its size and position?
16. What aspect of the urinary bladder affects the position of surrounding organs?
17. Describe the urinary system. What is its function?

Self-Quiz

Select the **QUIZ** icon from the main selections of the Menu Bar. From the **Select region** menu, select the following self-quizzes and complete them to assess your understanding of the material presented. Be sure to select each of the different test types from the **Select test type** menu. From the **Select number of questions** menu, select the appropriate number of questions for the time that you have available. This allows you to return often and challenge yourself to answer more questions as time allows.

- *All structures*
- *General terms and regions*
- *Cavities and membranes*
- *Organ systems*

MODULE 2

Cells and Chemistry

Overview: Cells and Chemistry

An understanding of cells and chemistry is foundational to your study of anatomy and physiology. From the cell theory, we know that "an organism's structure and all of its functions are ultimately due to the activities of its cells."[1] Moreover these activities are biochemical reactions that produce, for example, ATP (adenosine triphosphate) as the currency of energy exchange, or the synthesis of proteins for tissue growth and repair. Therefore, in order to gain a more complete understanding of the structure and function of the various systems of the body, one must first understand these basic concepts. We will begin by approaching these concepts through a variety of perspectives, including animations, histology, and virtual dissections of a generalized cell. We will conclude with a discussion of chemistry through animations, explaining in simple step-by-step illustrations, the basic chemical processes of the cell.

[1] Saladin, Kenneth S. (2009). *Anatomy and Physiology: The Unity of Form and Function*, 5th ed. New York: McGraw-Hill.

We will begin the animations with a series of foundational concepts that the more specific animations to come will build upon. It is imperative that you understand these key concepts before you move on.

SELECT ANIMATION
Structural Hierarchy of the Human Body (3D) — PLAY

After viewing the animation, answer the following questions.

1. The _____ is the structural and functional unit of the body.

2. Together, cells construct _____, and _____ combine to form _____.

3. Together, various _____ make up _____ _____ which work collectively to create an _____.

SELECT ANIMATION
Cells and Organelles (3D) — PLAY

After viewing the animation, answer the following questions.

1. A human cell contains a _____ and specialized structures called _____.

2. Inside the nucleus, the cell's _____ orchestrates activities integral to the cells _____ and _____ in the body.

3. Organelles, suspended in a gel-like solution known as _____ perform specific functions to assist with the cell's _____ and _____.

4. For example, _____ synthesize _____.

5. And the _____ _____ assembles and transports _____ for use by the cell.

6. _____ produce energy for the cell in the form of _____ powering activities such as _____.

7. The _____ _____ encloses the cell and serves a site of exchange for _____, a point of attachment for _____ and other _____, and a conduit for _____ between cells and their surroundings.

SELECT ANIMATION
Mitosis (3D) — PLAY

After viewing the animation, answer the following questions.

1. Mitosis is a type of _____ with many vital functions.

2. These functions include _____ _____, promoting _____ _____ after birth, and replacing _____ or _____ cells in the body.

3. How many divisions occur in mitosis?

77

78   MODULE 2   Cells and Chemistry

4. What are the results?

5. How are the two daughter cells identical to the parental cell?

6. These cells are called _____ cells.

7. How many pairs of chromosomes do these daughter cells contain?

8. Each pair of chromosomes contains one _____ and one _____ chromosome.

9. What process precedes mitosis?

10. What is the first phase of mitosis?

11. What occurs in the cell during this phase?

12. During prophase _____ condenses into _____, each consisting of two identical _____ _____.

13. During prophase, the _____ _____ dissolves and _____ _____ begin to grow from the cell's _____.

14. During metaphase, the _____ _____ pull the _____ into alignment in the center of the cell.

15. In anaphase, each chromosome consisting of two genetically identical _____ splits in two.

16. Each _____ now considered a single stranded _____ _____ migrates to the opposite end of the cell from its twin.

17. During telophase _____ _____ reform around the chromosomes as the cell finishes dividing.

Now that you have this foundation, we will build on it with these animations, as they reveal the more intricate anatomy and physiology of the human cell.

SELECT ANIMATION
Cell Cycle and Mitosis   PLAY

After viewing the animation, answer the following questions.

1. Define the cell cycle.

2. These events include a _____ stage, _____ or _____ _____ and _____ or _____ of the _____.

3. What events occur during interphase?

4. List the three discrete phases of interphase. What events occur in each?

5. What events occur during prophase?

6. What are sister chromatids?

7. What changes occur to the nuclear envelope during prophase?

8. What changes occur with the chromosomes during metaphase?

9. What interaction occurs between the chromosomes and the microtubules of the spindle?

10. During anaphase, what changes occur in the centromeres and the chromosomes?

11. Where do the two groups of chromosomes arrive during telophase?

12. What structure forms around each group of chromosomes?

13. What changes then occur in the chromosomes and the spindle?

14. Define cytokinesis. What other term refers to cytokinesis?

15. What is the result of mitosis and cytokinesis?

16. In summary, which phases occur during interphase?

17. What is occurring in the cell during interphase?

18. Mitosis, or _____, consists of four phases.

19. List sequentially the four phases of mitosis.

20. What is the result of the four phases of mitosis?

21. Finally, the cell's cytoplasm divides during _____, resulting in the formation of _____.

Self-Quiz

Take this opportunity to quiz yourself by taking the **QUIZ** over this animation.

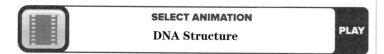

After viewing the animation, answer the following questions.

1. Describe the structure of DNA.

2. Describe the nucleotides.

3. If the DNA molecule unwinds, its structure resembles a _____.

4. Which molecules form the sides of this structure and which molecules form the rungs?

5. Which structures of the nucleotides run in opposite directions?

6. How are the ends of each strand labeled? What are these labels?

7. How do the ends of the complementary strands connect?

8. What bonds are facilitated by this antiparallel orientation of the two nucleotide strands?

9. What are the specific combinations for the pairing of the four different nitrogenous bases?

10. How do the number of hydrogen bonds correspond with each nitrogenous base pairing?

11. What is the function of the hydrogen bonds?

Self-Quiz

Take this opportunity to quiz yourself by taking the **QUIZ** over this animation.

After viewing the animation, answer the following questions.

1. Define DNA replication.

2. What sequence must occur for DNA replication to commence?

3. What function do the two original DNA strands serve during DNA replication?

4. What is the key feature that allows a DNA molecule to produce two DNA molecules with the same DNA sequence?

5. With complementary base pairing, which bases bind with which other bases?

6. What type of bonds bind the base pairs?

7. As a result of DNA replication, each copy contains one _____ _____ _____ and one _____ _____.

8. What is this mechanism called? Why?

Self-Quiz

Take this opportunity to quiz yourself by taking the **QUIZ** over this animation.

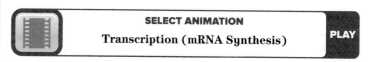

After viewing the animation, answer the following questions.

1. Define transcription.

2. What steps begin transcription?

3. What is the function of RNA polymerase?

4. Which direction does this synthesis occur?

5. Which end of the growing mRNA molecule receives the added nucleotides?

6. What is the function of the transcription terminator?

7. What does the hairpin loop cause to occur?

Self-Quiz

Take this opportunity to quiz yourself by taking the **QUIZ** over this animation.

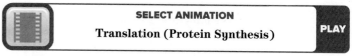

After viewing the animation, answer the following questions.

1. What structures constitute an initiation complex?

2. What other proteins, among those not shown in the animation, are involved?

3. How many binding sites for transfer RNA are located on the 70s ribosome?

4. Identify those sites, using both names for each.

5. Where does the initiating transfer RNA bind to the 70s ribosome?

6. Where does the next transfer RNA bind?

7. What reaction occurs between the two amino acids transported by the two transfer RNAs?

8. What occurs next for the ribosome and the first transfer RNA?

9. Where does the next transfer RNA bind? What then occurs for the ribosome?

10. What occurs next for the now three amino acids?

11. What occurs next for the ribosome and the growing polypeptide chain?

12. What is a stop codon? What occurs to the elongating polypeptide when the ribosome reaches the stop codon?

13. What occurs next for the ribosome, the messenger RNA and the newly formed protein?

Self-Quiz
Take this opportunity to quiz yourself by taking the **QUIZ** over this animation.

SELECT ANIMATION
Protein Synthesis (3D)
PLAY

After viewing the animation, answer the following questions.

1. Protein synthesis is the process by which the body creates _____.

2. Proteins consist of chains of _____ _____.

3. What determines each particular protein?

4. Where does the assembly of amino acids into proteins take place?

5. What is the first stage?

6. Where does the first stage occur?

7. What is the second stage?

8. Where does the second stage occur?

9. _____ is the process of converting instructions for assembling a protein located in the cell's _____ into _____ _____.

10. The template for building messenger RNA is a _____ along a section of the _____ strand.

11. Each strand of DNA contains _____ with complimentary _____.

12. _____ pairs with _____, and _____ pairs with _____.

13. To start transcription, an _____ called _____ _____ attaches to the beginning of the DNA template.

14. A sequence of three DNA bases called a _____ _____ contains information for assembling each _____ _____ of a _____.

15. _____ _____ reads the base triplets to build _____ using free _____.

16. Corresponding _____ _____ _____ are called _____.

17. In mRNA codons, _____ replaces _____.

18. Once the mRNA is built, certain _____ remove _____, or sections that will not be used to build the protein.

19. _____ splice the remaining ends or _____ together.

20. What does the functional mRNA do next?

21. _____ is the process of using _____ _____ to assemble _____ _____ into a _____.

22. What is the structure called that will read the mRNA?

23. How is it able to read the mRNA strand?

24. Initiated by a _____ _____, the ribosome reads each subsequent codon, which signals a _____ _____ molecule that has the matching _____ sequence and specific _____ _____.

25. What occurs as this process continues?

26. What signals that the protein is completely assembled?

27. What does the assembled protein do next?

28. What happens to the ribosome after this occurs?

SELECT ANIMATION — NADH/Oxidation–reduction Reactions — PLAY

After viewing the animation, answer the following questions.

1. Cells obtain energy through the process of _____.

2. How is this energy obtained? What is the energy molecule that results from this process?

3. What is oxidation?

4. A hydrogen atom consists of _____.

5. Whenever a molecule is oxidized, another molecule must be _____, which means _____.

6. What is the enzyme's role in the oxidation–reduction reaction in the cell?

7. What structures does the enzyme have that facilitate this reaction?

8. In this reaction, the substrate is _____ (loses a hydrogen) and the _____ is reduced to _____.

9. What two things occur after the reaction is complete?

10. What is NADH? What is this molecule available to do now?

Self-Quiz
Take this opportunity to quiz yourself by taking the **QUIZ** over this animation.

SELECT ANIMATION — Glycolysis — PLAY

After viewing the animation, answer the following questions.

1. Cells derive energy from the _____ of nutrients, such as _____.

2. The oxidation of _____ to _____ occurs through a series of steps called _____.

3. How many carbons are in a molecule of glucose?

4. The energy released during these _____ reactions is used to form _____ (_____), the _____ of the cell.

5. Name the two initial steps in glycolysis.

6. What are the three molecules that result?

7. What then occurs to the 6-carbon molecule?

8. These 3-carbon molecules are converted to _____.

9. What happens to the electrons in this reaction? What two molecules are formed?

10. What happens to the pyruvate under aerobic conditions?

11. What happens to the pyruvate under anaerobic conditions?

Self-Quiz
Take this opportunity to quiz yourself by taking the **QUIZ** over this animation.

SELECT ANIMATION — Kreb's Cycle — PLAY

After viewing the animation, answer the following questions.

1. Name the product of glycolysis that enters the Krebs cycle. How many of these molecules are produced per glucose molecule?

2. Name the 2-carbon fragment that enters the Krebs cycle.

3. What is another name for the Krebs cycle? Where in the cell does it occur? Where in the cell did glycolysis occur?

4. What two products are formed during the conversion of pyruvate to acetyl-CoA?

5. What reaction occurs to release the CoA carrier molecule?

6. What two products are formed as the 6-carbon molecule is converted to a 5-carbon molecule?

7. What three products are produced during the second oxidation and decarboxylation? This process forms a _____-carbon molecule.

8. Finally, the _____-carbon molecule is further oxidized, and the _____ removed are used to form _____ and _____. What is regenerated by these reactions?

9. Each glucose molecule is broken down into _____ _____ molecules during glycolysis. Then each _____ molecule is converted to _____ and enters the Krebs cycle.

10. For each glucose molecule, how many circuits of the Krebs cycle are completed to completely break down the two pyruvate molecules?

11. For each circuit of the Krebs cycle, how many ATP are produced? Therefore, how many ATP are produced in the Krebs cycle for every glucose molecule? (Remember, the products from each glucose molecule require two circuits of the Krebs cycle.)

Self-Quiz
Take this opportunity to quiz yourself by taking the **QUIZ** over this animation.

SELECT ANIMATION
Electron Transport and ATP Synthesis
PLAY

After viewing the animation, answer the following questions.

1. In the mitochondrion, the energy stored in NADH is _____.

2. This energy of the _____ is used to form _____.

3. What changes occur to NAD^+ and FAD during glycolysis and the Krebs cycle?

4. Where are the electrons from NADH transferred? How? What happens to the protons?

5. Where are the electrons carried? By what molecule? What else occurs as this is happening?

6. Where are the electrons from $FADH_2$ transferred? And the protons?

7. Where do the electrons travel next? And the protons?

8. What is the terminal electron acceptor? What molecule forms when the electrons are transferred (accepted)?

9. What force is generated by the transfer of protons to the intermembrane space? Where is this force generated?

10. Name the special proton channel proteins that allow the protons to re-enter the matrix of the mitochondrion.

11. What energy is used to synthesize the ATP molecules? What products are used to make the ATP?

12. What name is given to this method of ATP production?

Self-Quiz
Take this opportunity to quiz yourself by taking the **QUIZ** over this animation.

EXERCISE 2.1:
Generalized Cell

Select the **MODULE** *drop-down menu in the* **HOME** *screen of APR. Choose* **2. Cells and chemistry.** *Click the* **DISSECTION** *icon. From the* **Select topic** *menu, choose* **Generalized cell.** *Select the* **my COURSE CONTENT** *tab if your instructor has selected specific exercises for you to complete. Otherwise select the* **ALL CONTENT** *tab. You will see the following image:*

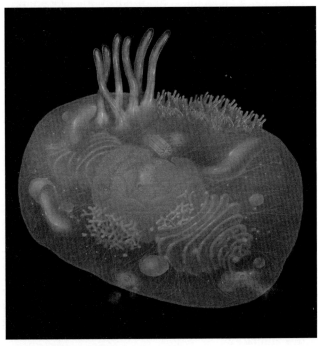
©McGraw-Hill Education

Click the **Select structure type** *menu and choose* **Cell surface.** *Click the* **Select group** *menu and choose* **Cell surface extensions.** *Select* **Cilium [1]** *from the list of structures. This image reveals the superficial view of a generalized cell and a highlighted group of cilia (plural for cilium). The number* **[1]** *correlates with the* **LAYER 1 (SURFACE)** *radio button in the* **LAYER CONTROLS** *window. Use the information in the* **STRUCTURE INFORMATION** *window to answer the following questions.*

1. Where on the cell are the cilia located?

2. Which group of tissue has cilia?

3. Where are the cilia evident?

4. Describe a cilium.

5. What is the axoneme?

6. Describe the function of cilia.

7. How do cilia compare to microvilli?

8. Define micrometer. What is a micrometer also known as?

Return to the list under **Cell surface extensions** *and select* **Cilium [2]**. *This image reveals the deep view of generalized cell and a highlighted group of cilia with other cell organelles. The number* **[2]** *correlates with the* **LAYER 2 (deep)** *radio button in the* **LAYER CONTROLS** *window. Note also that the* **LAYER 1: SURFACE** *slider in the* **LAYER CONTROLS** *window has moved downward. This indicates that* **LAYER 2** *is the active view.*

Histology: Cilium

Select the **MODULE** *drop-down menu in the* **HOME** *screen of APR. Choose* **2. Cells and chemistry.** *Click the* **HISTOLOGY** *icon. Select the* **my COURSE CONTENT** *tab if your instructor has selected specific exercises for you to complete. Otherwise select the* **ALL CONTENT** *tab. From the* **Select topic** *menu, choose* **Cilium. TEM: High magnification** *appears in the* **VIEW** *window. To the right in the* **IMAGE AREA** *you will see a high-magnification Transmission Electron Micrograph (TEM) of the cilium. Click on* **Axoneme** *in the* **STRUCTURE LIST** *and use the information in the* **STRUCTURE INFORMATION** *window to answer the following questions.*

1. Where is the axoneme located?

2. Describe the axoneme.

3. List the functions of the axoneme.

Return to the **STRUCTURE LIST** *and click on* **Cilium.** *Review the information presented in the* **STRUCTURE INFORMATION** *window.*

Return to the **STRUCTURE LIST** *and click on* **Dynein arm of axoneme.** *Use the information in the* **STRUCTURE INFORMATION** *window to answer the following questions.*

1. Where are the dynein arms of axoneme located?

2. Describe the dynein arms of axoneme.

3. What is the function of the dynein arms of axoneme?

Return to the **STRUCTURE LIST** *and click on* **Microvillus.** *Use the information in the* **STRUCTURE INFORMATION** *window to answer the following questions.*

1. Where are the microvilli located?

2. Describe the microvilli.

3. List the functions of the microvilli.

4. Compare microvilli to cilia and flagella.

Click the **TAGS ON/OFF** *radio button located below the* **STRUCTURE INFORMATION** *window to activate the green pins on the image. You will see the following image.*

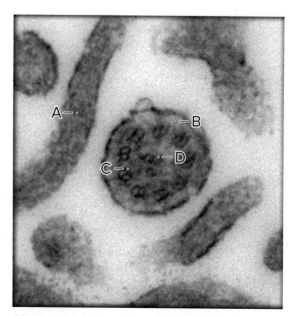

©EM Research Services, Newcastle University

84 MODULE 2 Cells and Chemistry

Mouse-over the pins on the screen to find the information to identify the following structures.

A. _____

B. _____

C. _____

D. _____

Histology: Flagellum of Sperm

Return to the Select topic menu, choose Flagellum of sperm. TEM: Medium magnification appears in the VIEW window. To the right in the IMAGE AREA you will see a Medium magnification Transmission Electron Micrograph (TEM) of the sperm. Click on Acrosome in the STRUCTURE LIST and use the information in the STRUCTURE INFORMATION window to answer the following questions.

1. Where is the acrosome located?

2. Describe the acrosome.

3. List the functions of the acrosome.

4. The acrosome is considered a special type of _____.

Return to the STRUCTURE LIST and click on Axoneme. Review the information presented in the STRUCTURE INFORMATION window.

Return to the STRUCTURE LIST and click on Centriole of sperm. Use the information in the STRUCTURE INFORMATION window to answer the following questions.

1. Where is the centriole of the sperm located?

2. Describe the centriole of the sperm.

3. What is the function of the centriole of the sperm?

Return to the STRUCTURE LIST and click on Condensed nucleus of sperm. Use the information in the STRUCTURE INFORMATION window to answer the following questions.

1. Where is the condensed nucleus of sperm located?

2. Describe the condensed nucleus of sperm.

3. What is the function of the condensed nucleus of sperm?

Return to the STRUCTURE LIST and click on Flagellum of sperm. Use the information in the STRUCTURE INFORMATION window to answer the following questions.

1. Where is the flagellum of sperm located?

2. Describe the flagellum of sperm.

3. What are the four regions of the tail of sperm?

4. What structural arrangement constitutes the core of the flagellum?

5. What is the function of the flagellum of sperm?

Return to the STRUCTURE LIST and click on Head of sperm. Use the information in the STRUCTURE INFORMATION window to answer the following questions.

1. Where is the head of sperm located?

2. Describe the head of sperm.

3. List the functions of the head of sperm.

Return to the STRUCTURE LIST and click on Middle piece of sperm. Use the information in the STRUCTURE INFORMATION window to answer the following questions.

1. Where is the middle piece of sperm located?

2. Describe the middle piece of sperm.

3. List the functions of the middle piece of sperm.

Return to the STRUCTURE LIST and click on Neck of sperm. Use the information in the STRUCTURE INFORMATION window to answer the following questions.

1. Where is the neck of sperm located?

2. Describe the neck of sperm.

3. List the functions of the neck of sperm.

Return to the STRUCTURE LIST and click on Principal piece of sperm. Use the information in the

STRUCTURE INFORMATION *window to answer the following questions.*

1. Where is the principal piece of sperm located?

2. Describe the principal piece of sperm.

3. What is the function of the principal piece of sperm?

Return to the **STRUCTURE LIST** *and click on* **Sperm cell.** *Use the information in the* **STRUCTURE INFORMATION** *window to answer the following questions.*

1. Where is the sperm cell located?

2. Describe the sperm cell.

3. What is the function of the sperm cell?

Return to the **STRUCTURE LIST** *and click on* **Spiral mitochondrion in sperm.** *Use the information in the* **STRUCTURE INFORMATION** *window to answer the following questions.*

1. Where is the spiral mitochondrion in sperm located?

2. Describe the spiral mitochondrion in sperm.

3. What is the function of the spiral mitochondrion in sperm?

Click the **TAGS ON/OFF** *radio button located below the* **STRUCTURE INFORMATION** *window to activate the green pins on the image. You will see the following image:*

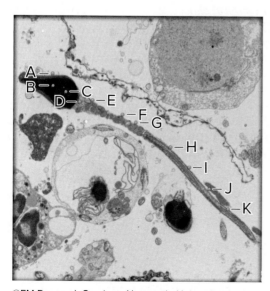

©EM Research Services, Newcastle University

Mouse-over the pins on the screen to find the information to identify the following structures:

A. _____

B. _____

C. _____

D. _____

E. _____

F. _____

G. _____

H. _____

I. _____

J. _____

K. _____

Return to the **DISSECTION** *menu. Select* **ALL CONTENT.** *From the* **Select topic** *menu, choose* **Generalized cell.** *Click the* **Select structure type** *menu and choose* **Cell surface.** *Click the* **Select group** *menu and choose* **Cell surface extensions.** *From the list under* **Cell surface extensions** *select* **Microvillus [1].** *This image reveals the superficial view of a generalized cell and a highlighted group of microvilli. The number* **[1]** *correlates with the* **LAYER 1 (SURFACE)** *radio button in the* **LAYER CONTROLS** *window. Use the information in the* **STRUCTURE INFORMATION** *window to answer the following questions.*

1. Where on the cell are the microvilli located?

2. Where are microvilli most developed?

3. Describe microvilli.

4. What is the function of microvilli?

5. How do microvilli differ from cilia and flagella?

6. What are stereocilia?

7. What are villi?

Return to the list under **Cell surface extensions** *and select* **Microvillus [2].** *This image reveals the deep view of generalized cell and a highlighted group of microvilli with other cell organelles. The number* **[2]** *correlates with the* **LAYER 2 (deep)** *radio*

86 MODULE 2 Cells and Chemistry

button in the **LAYER CONTROLS** *window. Note also that the* **LAYER 1: SURFACE** *slider in the* **LAYER CONTROLS** *window has moved downward. This indicates that* **LAYER 2** *is the active view.*

Histology: Microvillus (longitudinal section)

Return to the **HISTOLOGY** *menu. Select the* **my COURSE CONTENT** *tab if your instructor has selected specific exercises for you to complete. Otherwise select the* **ALL CONTENT** *tab. From the* **Select topic** *menu, choose* **Microvillus (longitudinal section)**. **TEM: Medium magnification** *appears in the* **VIEW** *window. To the right in the* **IMAGE AREA** *you will see a Medium magnification Transmission Electron Micrograph (TEM) of the microvillus in longitudinal section. Click on* **Glycocalyx** *in the* **STRUCTURE LIST** *and use the information in the* **STRUCTURE INFORMATION** *window to answer the following questions.*

1. Where is the glycocalyx located?

2. Describe the glycocalyx.

3. List the functions of the glycocalyx.

4. What other term refers to the glycocalyx?

Return to the **STRUCTURE LIST** *and click* **Microfilament.** *Use the information in the* **STRUCTURE INFORMATION** *window to answer the following questions.*

1. Where are the microfilaments located?

2. Describe the microfilaments.

3. List the functions of the microfilaments.

Return to the **STRUCTURE LIST** *and click* **Microvillus.** *Use the information in the* **STRUCTURE INFORMATION** *window to answer the following questions.*

1. Where are the microvilli located?

2. Describe the microvilli.

3. List the functions of the microvilli.

Click the **TAGS ON/OFF** *radio button located below the* **STRUCTURE INFORMATION** *window to activate the green pins on the image. You will see the following image:*

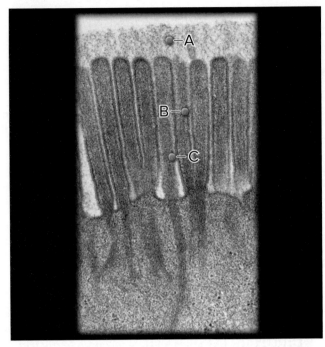

©Dr. S. Ito, Harvard Medical School

Mouse-over the pins on the screen to find the information to identify the following structures:

A. _____

B. _____

C. _____

Histology: Microvillus (cross-section)

Return to the **HISTOLOGY** *menu. From the* **Select topic** *menu, choose* **Microvillus (cross-section)**. **TEM: Medium magnification** *appears in the* **VIEW** *window. To the right in the* **IMAGE AREA** *you will see a medium magnification Transmission Electron Micrograph (TEM) of the microvillus in cross-section. We have already covered the structures on the* **STRUCTURE LIST,** *so click each one to review the information presented in the* **STRUCTURE INFORMATION** *window and to view them in the* **IMAGE AREA.**

Click the **TAGS ON/OFF** *radio button located below the* **STRUCTURE INFORMATION** *window to activate the green pins on the image. You will see the following image:*

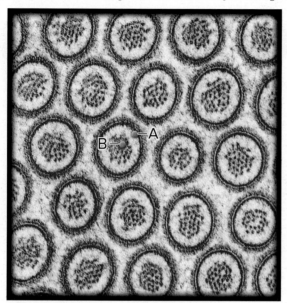

©Biophoto Associates/Photo Researchers, Inc.

Mouse-over the pins on the screen to find the information to identify the following structures:

A. _____

B. _____

Histology: Cilium and Microvillus

Return to the **HISTOLOGY** *menu. From the* **Select topic** *menu, choose* **Cilium and microvillus. SEM: Low magnification** *appears in the* **VIEW** *window. To the right in the* **IMAGE AREA** *you will see a medium magnification Transmission Electron Micrograph (TEM) of the microvillus in cross-section. We have already covered the structures on the* **STRUCTURE LIST,** *so click each one to review the information presented in the* **STRUCTURE INFORMATION** *window and to view them in the* **IMAGE AREA.**

Click the **TAGS ON/OFF** *radio button located below the* **STRUCTURE INFORMATION** *window to activate the green pins on the image. You will see the following image:*

©Science Photo Library/Alamy Stock Photo

Mouse-over the pins on the screen to find the information to identify the following structures:

A. _____

B. _____

CHECK POINT

Cell Surface Extensions

1. What are the dimensions of a cilium?
2. What are the dimensions of a microvillus?
3. What are oscillations?
4. How are the microvilli visible in a light microscope? How can individual microvilli be distinguished?

Histology: Plasma Membrane

Return to the HISTOLOGY menu. From the Select topic menu, choose Plasma membrane. To the right in the IMAGE AREA you will see a high-magnification Transmission Electron Micrograph (TEM) of the plasma membrane, as indicated in the IMAGE VIEW. We have already covered most of the structures in the STRUCTURE LIST, so click through each one to see it highlighted in the IMAGE AREA. When you click on Extracellular matrix, use the information in the STRUCTURE INFORMATION window to answer the following questions.

1. Where is the extracellular matrix located?

2. Describe the extracellular matrix.

3. What other terms refer to the extracellular matrix?

Click the TAGS ON/OFF radio button located below the STRUCTURE INFORMATION window to activate the green pins on the image. You will see the following image:

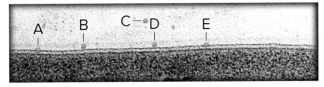

©Biophoto/Science Source

Mouse-over the pins on the screen to find the information to identify the following structures:

A. _____
B. _____
C. _____
D. _____
E. _____

Histology: Cytoplasm

Return to the HISTOLOGY menu. From the Select topic menu, choose Cytoplasm. To the right in the IMAGE AREA you will see a medium magnification Transmission Electron Micrograph (TEM) of the cytoplasm, as indicated in the IMAGE VIEW. We have already covered the structures in the STRUCTURE LIST, so click through each one to see it highlighted in the IMAGE AREA.

Click the TAGS ON/OFF radio button located below the STRUCTURE INFORMATION window to activate the green pins on the image. You will see the following image:

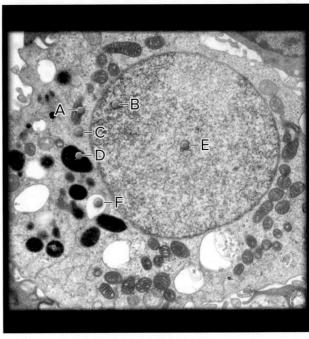

©EM Research Services, Newcastle University

Mouse-over the pins on the screen to find the information to identify the following structures:

A. _____
B. _____
C. _____
D. _____
E. _____
F. _____

Return to the DISSECTION menu. Click the Select structure type menu and choose Cytoplasm. Click the Select group menu and choose Cytoskeleton. Select Intermediate filament from the list of structures. Use the information in the STRUCTURE INFORMATION window to answer the following questions.

1. Where are the intermediate filaments located?

2. Describe the intermediate filaments.

3. List the functions of the intermediate filaments.

4. List the three different filaments that compose the cytoskeleton.

5. Define nanometer.

Return to the list under **Cytoskeleton** *and select* **Microfilament** *from the list of structures. Use the information in the* **STRUCTURE INFORMATION** *window to answer the following questions.*

1. Where are the microfilaments located?

2. Describe the microfilaments.

3. List the functions of the microfilaments.

4. What other terms refer to the microfilaments?

Return to the list under **Cytoskeleton** *and select* **Microtubule** *from the list of structures. Use the information in the* **STRUCTURE INFORMATION** *window to answer the following questions.*

1. Where are the microtubules located?

2. Describe the microtubules.

3. List the functions of the microtubules.

Histology: Cytoskeleton (microfilament)

Return to the **Select topic** *menu and choose* **Cytoskeleton (microfilament). SEM: High magnification** *appears in the* **VIEW** *window. To the right in the* **IMAGE AREA** *you will see a high-magnification Scanning Electron Micrograph (SEM) of the microfilament.*

Click the **TAGS ON/OFF** *radio button located below the* **STRUCTURE INFORMATION** *window to activate the green pin on the image. You will see the following image:*

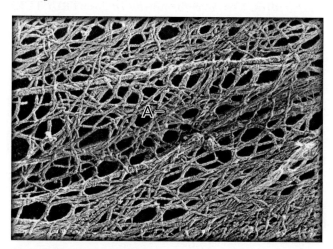

©Don W. Fawcett/Science Source

Mouse-over the pin on the screen to find the information to identify the following structure:

A. _____

Click on **Intermediate filament** *in the* **STRUCTURE LIST** *and use the information in the* **STRUCTURE INFORMATION** *window to answer the following questions.*

1. Where is the nuclear Intermediate filament?

2. Describe the Intermediate filament.

3. List the functions of the Intermediate filament.

Histology: Cytoskeleton (microfilament and microtubule)

Return to the **Select topic** *menu and choose* **Cytoskeleton (microfilament and microtubule). FM: Low magnification** *appears in the* **VIEW** *window. To the right in the* **IMAGE AREA** *you will see a low-magnification Fluorescence Micrograph (FM) of the microfilament and microtubule.*

We have already covered the structures on the **STRUCTURE LIST**, *so click each one to review the information presented in the* **STRUCTURE INFORMATION** *window and to view them in the* **IMAGE AREA**.

90 MODULE 2 Cells and Chemistry

Click the **TAGS ON/OFF** radio button located below the **STRUCTURE INFORMATION** window to activate the green pins on the image. You will see the following image:

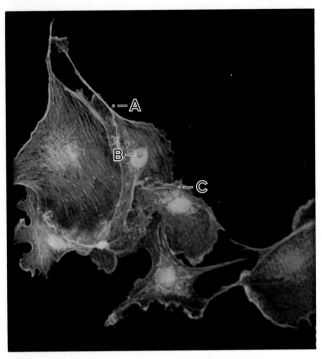
©defun/Getty Images

Mouse-over the pins on the screen to find the information to identify the following structures:

A. _____
B. _____
C. _____

Histology: Cytoskeleton (neurofilament and microtubule)

Return to the **Select topic** menu and choose **Cytoskeleton (neurofilament and microtubule)**. **TEM: High magnification** appears in the **VIEW** window. To the right in the **IMAGE AREA** you will see a high-magnification Transmission Electron Micrograph (TEM) of the neurofilament and microtubule.

Click the **TAGS ON/OFF** radio button located below the **STRUCTURE INFORMATION** window to activate the green pins on the image. You will see the following image:

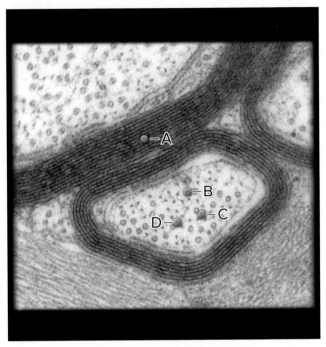
©EM Research Services, Newcastle University

Mouse-over the pins on the screen to find the information to identify the following structures:

A. _____
B. _____
C. _____
D. _____

Return to the **DISSECTION** menu. Select **ALL CONTENT**. From the **Select topic** menu, choose **Generalized cell**. Click the **Select structure type** menu and choose **Cytoplasm**. Click the **Select group** menu and choose **Membrane-bound organelles**. Select **cis-face of Golgi apparatus** from the list of structures. Use the information in the **STRUCTURE INFORMATION** window to answer the following questions.

1. Where is the cis-face of the Golgi apparatus located?

2. Describe the cis-face of the Golgi apparatus.

3. What is the function of the cis-face of the Golgi apparatus?

4. What other term refers to the cis-face of the Golgi apparatus?

Return to the list of **Membrane-bound organelles** *and select* **Cristae of mitochondrion** *from the list of structures. Use the information in the* **STRUCTURE INFORMATION** *window to answer the following questions.*

1. Where are the cristae of mitochondrion located?

2. Describe the cristae of mitochondrion.

3. What is the function of the cristae of mitochondrion?

Return to the list of **Membrane-bound organelles** *and select* **Endocytic vesicle [1]** *from the list of structures. The number* **[1]** *correlates with the* **LAYER 1 (SURFACE)** *radio button in the* **LAYER CONTROLS** *window. Use the information in the* **STRUCTURE INFORMATION** *window to answer the following questions.*

1. Where are the endocytic vesicles located?

2. Describe the endocytic vesicles.

3. What are the two different endocytic vesicles? How do they differ?

4. What are receptor-mediated endocytic vesicles?

5. What is the function of the endocytic vesicles?

Return to the list of **Membrane-bound organelles** *and select* **Golgi apparatus [2]** *from the list of structures. The number* **[2]** *correlates with the* **LAYER 2 (deep)** *radio button in the* **LAYER CONTROLS** *window. Use the information in the* **STRUCTURE INFORMATION** *window to answer the following questions.*

1. Where is the Golgi apparatus located?

2. Describe the Golgi apparatus.

3. Describe the three subdivisions of the Golgi apparatus.

4. List the functions of the Golgi apparatus.

5. What other term refers to the Golgi apparatus?

Return to the list of **Membrane-bound organelles** *and select* **Golgi apparatus [3]** *from the list of structures. The number* **[3]** *correlates with the* **LAYER 3 (CROSS SECTION)** *radio button in the* **LAYER CONTROLS** *window. Compare the structures visible in* **LAYER 3** *with those visible in* **LAYER 2.**

Return to the list of **Membrane-bound organelles** *and select* **Lysosome [2]** *from the list of structures. The number* **[2]** *correlates with the* **LAYER 2 (deep)** *radio button in the* **LAYER CONTROLS** *window. Use the information in the* **STRUCTURE INFORMATION** *window to answer the following questions.*

1. Where are lysosomes located?

2. Describe lysosomes.

3. Describe the two types of lysosomes.

4. What is the function of lysosomes?

5. How do lysosomes compare with peroxisomes?

SELECT ANIMATION: Lysosomes — PLAY

After viewing the animation, answer the following questions.

1. Lysosomes are _____ _____ vesicles that contain _____ enzymes.

2. What is the function of these enzymes?

3. What do these enzymes degrade?

4. Where are these enzymes formed?

5. Where are they transported after formation? How are they transported?

6. From where do the lysosomes arise?

7. What is a phagosome? How does one form?

8. What occurs after the lysosome fuses with the phagosome?

9. What other structure do lysosomes fuse with? What are the results?

92 MODULE 2 Cells and Chemistry

Self-Quiz
Take this opportunity to quiz yourself by taking the **QUIZ** over this animation.

Histology: Lysosome

Return to the **HISTOLOGY** *menu. From the* **Select topic** *menu, choose* **Lysosome. TEM: High magnification** *appears in the* **VIEW** *window. To the right in the* **IMAGE AREA** *you will see a high-magnification Transmission Electron Micrograph (TEM) of a lysosome. Some of these structures we have already covered, so click through all of the structures to view them on the image. Then click on* **Primary lysosome** *in the* **STRUCTURE LIST** *and use the information in the* **STRUCTURE INFORMATION** *window to answer the following questions.*

1. Where are primary lysosomes located?

2. Describe primary lysosomes.

3. Describe how primary lysosomes form.

4. What is the function of primary lysosomes?

Return to the **STRUCTURE LIST** *and click on* **Secondary lysosome.** *Use the information in the* **STRUCTURE INFORMATION** *window to answer the following questions.*

1. Where are secondary lysosomes located?

2. Describe secondary lysosomes.

3. Describe how secondary lysosomes form.

4. What is the function of secondary lysosomes?

5. What other terms refer to secondary lysosomes?

Click the **TAGS ON/OFF** *radio button located below the* **STRUCTURE INFORMATION** *window to activate the green pins on the image. You will see the following image:*

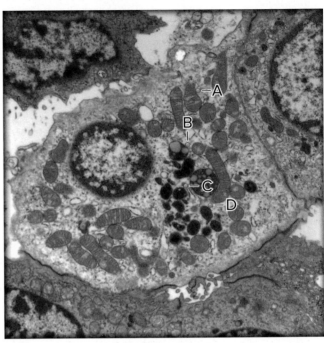

©EM Research Services, Newcastle University

Mouse-over the pins on the screen to find the information to identify the following structures:

A. _____

B. _____

C. _____

D. _____

Return to the **DISSECTION** *menu. Select* **Cytoplasm** *from the* **Select structure type** *menu. Click the* **Select group** *menu and from the list of* **Membrane-bound organelles** *select* **Mitochondrion [2].** *The number* **[2]** *correlates with the* **LAYER 2 (deep)** *radio button in the* **LAYER CONTROLS** *window. Use the information in the* **STRUCTURE INFORMATION** *window to answer the following questions.*

1. Where are mitochondria located?

2. Describe mitochondria.

3. What is the function of mitochondria?

Histology: Mitochondrion

Return to the HISTOLOGY menu. From the Select topic menu, choose Mitochondrion. TEM: High magnification appears in the VIEW window. To the right in the IMAGE AREA you will see a high-magnification Transmission Electron Micrograph (TEM) of a mitochondrion. Click on Cristae of mitochondrion in the STRUCTURE LIST and use the information in the STRUCTURE INFORMATION window to answer the following questions.

1. Where are the cristae of mitochondrion located?

2. Describe the cristae of mitochondrion.

3. What is the function of the cristae of mitochondrion?

Return to the STRUCTURE LIST and click on Inner mitochondrial membrane. Use the information in the STRUCTURE INFORMATION window to answer the following questions.

1. Where is the inner mitochondrial membrane located?

2. Describe the inner mitochondrial membrane.

3. List the functions of the inner mitochondrial membrane.

4. How does the mitochondrial membrane compare to the plasma membrane?

Return to the STRUCTURE LIST and click on Mitochondrial matrix. Use the information in the STRUCTURE INFORMATION window to answer the following questions.

1. Where is the mitochondrial matrix located?

2. Describe the mitochondrial matrix.

3. List the functions of the mitochondrial matrix.

4. What other terms refer to the tricarboxylic acid cycle?

Return to the STRUCTURE LIST and click on Mitochondrial matrix granule. Use the information in the STRUCTURE INFORMATION window to answer the following questions.

1. Where are the mitochondrial matrix granules located?

2. Describe the mitochondrial matrix granules.

3. List the functions of the mitochondrial matrix granules.

Return to the STRUCTURE LIST and click on Mitochondrion. Review the information presented in the STRUCTURE INFORMATION window.

Return to the STRUCTURE LIST and click on Outer mitochondrial membrane. Use the information in the STRUCTURE INFORMATION window to answer the following questions.

1. Where is the outer mitochondrial membrane located?

2. Describe the outer mitochondrial membrane.

3. List the functions of the outer mitochondrial membrane.

Click the TAGS ON/OFF radio button located below the STRUCTURE INFORMATION window to activate the green pins on the image. You will see the following image:

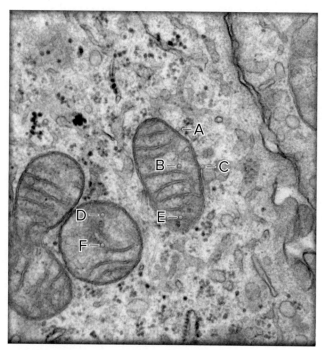

©EM Research Services, Newcastle University

94 MODULE 2 Cells and Chemistry

Mouse-over the pins on the screen to find the information to identify the following structures:

A. _____
B. _____
C. _____
D. _____
E. _____
F. _____

Return to the DISSECTION menu. From the list of Membrane-bound organelles select Peroxisome. Use the information in the STRUCTURE INFORMATION window to answer the following questions.

1. Where are peroxisomes located?

2. Describe peroxisomes.

3. What is a nucleoid? What substance composes a nucleoid?

4. What is the function of peroxisomes?

5. How do peroxisomes and lysosomes compare?

Histology: Peroxisome

Return to the HISTOLOGY menu. From the Select topic menu, choose Peroxisome. TEM: High magnification appears in the VIEW window. To the right in the IMAGE AREA you will see a high-magnification Transmission Electron Micrograph (TEM) of a peroxisome. We have already covered the structures in the STRUCTURE LIST, so click through the list to view them on the image.

Click the TAGS ON/OFF radio button located below the STRUCTURE INFORMATION window to activate the green pins on the image. You will see the following image:

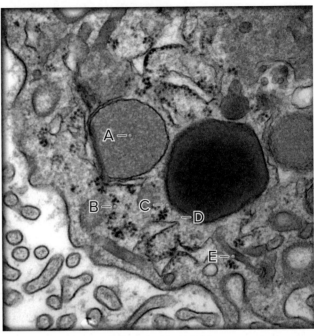

©EM Research Services, Newcastle University

Mouse-over the pins on the screen to find the information to identify the following structures:

A. _____
B. _____
C. _____
D. _____
E. _____

Return to the DISSECTION menu. From the list of Membrane-bound organelles select Rough endoplasmic reticulum (RER) [2]. The number [2] correlates with the LAYER 2 (deep) radio button in the LAYER CONTROLS window. Use the information in the STRUCTURE INFORMATION window to answer the following questions.

1. Where is the rough endoplasmic reticulum located?

2. Describe the rough endoplasmic reticulum.

3. What structures are attached to the rough endoplasmic reticulum?

4. What structure is continuous with the rough endoplasmic reticulum?

5. What is the function of the rough endoplasmic reticulum?

6. How does the rough endoplasmic reticulum compare to the smooth endoplasmic reticulum?

Return to the list of **Membrane-bound organelles** *and select* **Rough endoplasmic reticulum (RER) [3]** *from the list of structures. The number* **[3]** *correlates with the* **LAYER 2 (CROSS SECTION)** *radio button in the* **LAYER CONTROLS** *window. Compare the structures visible in* **LAYER 3** *with those visible in* **LAYER 2**.

Histology: Rough Endoplasmic Reticulum

Return to the **HISTOLOGY** *menu. From the* **Select topic** *menu, choose* **Rough endoplasmic reticulum (RER)**. *From the* **Select view** *menu, choose* **TEM: Low magnification**. *To the right in the* **IMAGE AREA** *you will see a low-magnification Transmission Electron Micrograph (TEM) of the rough endoplasmic reticulum. We have already covered most of the structures in the* **STRUCTURE LIST**, *so click through each one to see it highlighted in the* **IMAGE AREA**. *When you click on* **Plasma cell**, *use the information in the* **STRUCTURE INFORMATION** *window to answer the following questions.*

1. Where are plasma cells located?

2. Describe plasma cells.

3. List the functions of plasma cells.

4. What other terms refer to plasma cells?

5. Where are antibodies secreted? Where do they travel from there?

Click the **TAGS ON/OFF** *radio button located below the* **STRUCTURE INFORMATION** *window to activate the green pins on the image. You will see the following image:*

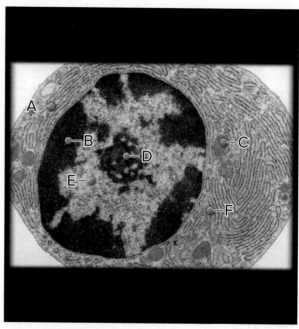

©Don W. Fawcett/Science Source

Mouse-over the pins on the screen to find the information to identify the following structures:

A. _____

B. _____

C. _____

D. _____

E. _____

F. _____

Return to the **Select view** *menu and choose* **TEM: High magnification**. *To the right in the* **IMAGE AREA** *you will see a high-magnification Transmission Electron Micrograph (TEM) of the rough endoplasmic reticulum. Click on* **Free ribosome** *in the* **STRUCTURE LIST** *and use the information in the* **STRUCTURE INFORMATION** *window to answer the following questions.*

1. Where are free ribosomes located?

2. Describe free ribosomes.

3. What is the function of free ribosomes?

4. What are polyribosomes?

96 MODULE 2 Cells and Chemistry

Return to the **STRUCTURE LIST** *and click on* **Membrane-bound ribosome.** *Use the information in the* **STRUCTURE INFORMATION** *window to answer the following questions.*

1. Where are membrane-bound ribosomes located?

2. Describe membrane-bound ribosomes.

3. What is the function of membrane-bound ribosomes?

Return to the **STRUCTURE LIST** *and click on* **Ribosome.** *Use the information in the* **STRUCTURE INFORMATION** *window to answer the following questions.*

1. Where are ribosomes located?

2. Describe ribosomes.

3. Distinguish between the two categories of ribosomes.

4. List the functions of ribosomes.

Click the **TAGS ON/OFF** *radio button located below the* **STRUCTURE INFORMATION** *window to activate the green pins on the image. You will see the following image:*

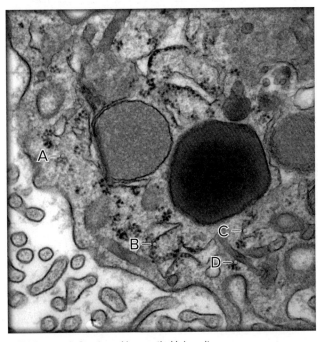

©EM Research Services, Newcastle University

Mouse-over the pins on the screen to find the information to identify the following structures:

A. _____

B. _____

C. _____

D. _____

Histology: Ribosome

Return to the **Select topic** *menu and choose* **Ribosome. TEM: High magnification** *appears in the* **VIEW** *window. To the right in the* **IMAGE AREA** *you will see a high-magnification Transmission Electron Micrograph (TEM) of a ribosome.*

We have already covered the structures on the **STRUCTURE LIST**, *so click each one to review the information presented in the* **STRUCTURE INFORMATION** *window and to view them in the* **IMAGE AREA.**

Click the **TAGS ON/OFF** *radio button located below the* **STRUCTURE INFORMATION** *window to activate the green pins on the image. You will see the following image:*

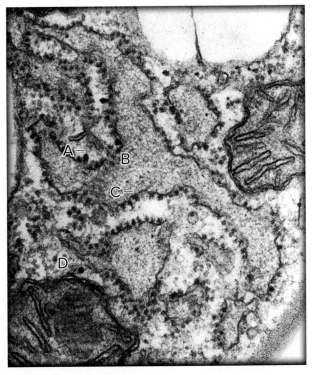

©Biophoto/Science Source

Mouse-over the pins on the screen to find the information to identify the following structures:

A. _____

B. _____

C. _____

D. _____

Return to the DISSECTION menu. From the list of Membrane-bound organelles select Secretory vesicle of Golgi apparatus [1] from the list of structures. The number [1] correlates with the LAYER 1 (SURFACE) radio button in the LAYER CONTROLS window. Use the information in the STRUCTURE INFORMATION window to answer the following questions.

1. Where are secretory vesicles of Golgi apparatus located?

2. Describe secretory vesicles of Golgi apparatus.

3. What is the function of secretory vesicles of Golgi apparatus?

Return to the list of Membrane-bound organelles and select Smooth endoplasmic reticulum (SER) [2] from the list of structures. The number [2] correlates with the LAYER 2 (deep) radio button in the LAYER CONTROLS window. Use the information in the STRUCTURE INFORMATION window to answer the following questions.

1. Where is the smooth endoplasmic reticulum located?

2. Describe the smooth endoplasmic reticulum.

3. What applies to the smooth endoplasmic reticulum in skeletal muscle tissue?

4. What structure is continuous with the smooth endoplasmic reticulum?

5. What is the function of the smooth endoplasmic reticulum?

6. How does the smooth endoplasmic reticulum compare to the rough endoplasmic reticulum?

Return to the list of Membrane-bound organelles and select Smooth endoplasmic reticulum (SER) [3] from the list of structures. The number [3] correlates with the LAYER 2 (CROSS SECTION) radio button in the LAYER CONTROLS window. Compare the structures visible in LAYER 3 with those visible in LAYER 2.

Histology: Smooth Endoplasmic Reticulum

Return to the HISTOLOGY menu. From the Select topic menu, choose Smooth endoplasmic reticulum (SER). TEM: High magnification appears in the VIEW window. To the right in the IMAGE AREA you will see a high-magnification Transmission Electron Micrograph (TEM) of the smooth endoplasmic reticulum. Click on Inclusion in the STRUCTURE LIST and use the information in the STRUCTURE INFORMATION window to answer the following questions.

1. Where are inclusions located?

2. Describe inclusions.

3. List the functions of inclusions.

4. Describe glycogen storage disease.

Click the TAGS ON/OFF radio button located below the STRUCTURE INFORMATION window to activate the green pins on the image. You will see the following image:

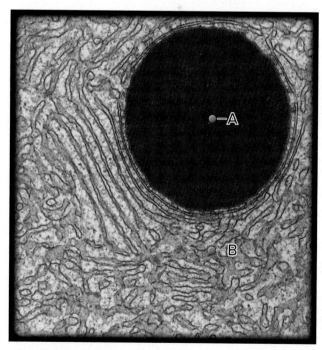

©Don W. Fawcett/Photo Researchers, Inc.

Mouse-over the pins on the screen to find the information to identify the following structures:

A. _____

B. _____

98 MODULE 2 Cells and Chemistry

Return to the **DISSECTION** *menu. From the list of* **Membrane-bound organelles** *select* **trans-face of Golgi apparatus.** *Use the information in the* **STRUCTURE INFORMATION** *window to answer the following questions.*

1. Where is the trans-face of the Golgi apparatus located?

2. Describe the trans-face of the Golgi apparatus.

3. What is the function of the trans-face of the Golgi apparatus?

4. What other terms refer to the trans-face of the Golgi apparatus?

CHECK POINT

Cytoplasm

1. What is endocytosis?
2. What are lysosomes?
3. What are secretory vesicles?
4. What does mitochondrial DNA code for?
5. Define perinuclear.
6. Where are membrane phospholipids synthesized?

Histology: Golgi Apparatus

Return to the **HISTOLOGY** *menu. From the* **Select topic** *menu, choose* **Golgi apparatus. TEM: High magnification** *appears in the* **VIEW** *window. To the right in the* **IMAGE AREA** *you will see a high-magnification Transmission Electron Micrograph (TEM) of the Golgi apparatus. Some of these structures we have already covered, so click through all of the structures to view them on the image. Then click on* **Transport vesicle of Golgi apparatus** *in the* **STRUCTURE LIST** *and use the information in the* **STRUCTURE INFORMATION** *window to answer the following questions.*

1. Where are transport vesicles of Golgi apparatus located?

2. Describe transport vesicles of Golgi apparatus.

3. What is the function of transport vesicles of Golgi apparatus?

Click the **TAGS ON/OFF** *radio button located below the* **STRUCTURE INFORMATION** *window to activate the green pins on the image. You will see the following image:*

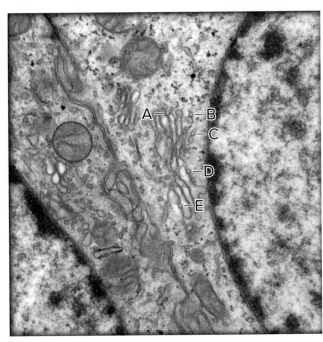

©EM Research Services, Newcastle University

Mouse-over the pins on the screen to find the information to identify the following structures:

A. _____
B. _____
C. _____
D. _____
E. _____

Return to the **DISSECTION** *menu. Click the* **Select structure type** *menu and choose* **Nucleus. Nucleus** *appears in the* **Select group** *menu. Select* **Nuclear pore [2]** *from the list of structures. The number [2] correlates with the* **LAYER 2 (deep)** *radio button in the* **LAYER CONTROLS** *window. Use the information in the* **STRUCTURE INFORMATION** *window to answer the following questions.*

1. Where are the nuclear pores located?

2. Describe the nuclear pores.

3. What is the function of the nuclear pores?

Return to the list of **Nucleus** *structures and select* **Nucleolus.** *Use the information in the* **STRUCTURE INFORMATION** *window to answer the following questions.*

1. Where is the nucleolus located?

2. Describe the nucleolus.

3. What determines the number of nucleoli?

4. What is the function of the nucleolus?

Return to the list of **Nucleus** *structures and select* **Nucleus [2].** *The number* **[2]** *correlates with the* **LAYER 2 (deep)** *radio button in the* **LAYER CONTROLS** *window. Use the information in the* **STRUCTURE INFORMATION** *window to answer the following questions.*

1. Where is the nucleus located?

2. Describe the nucleus.

3. What is the function of the nucleus?

Return to the list of **Nucleus** *structures and select* **Nucleus [3].** *The number* **[3]** *correlates with the* **LAYER 3 (CROSS SECTION)** *radio button in the* **LAYER CONTROLS** *window. Compare the structures visible in* **LAYER 3** *with those visible in* **LAYER 2.**

CHECK POINT

Cytoplasm

1. Define eccentric.
2. What is the largest organelle?
3. Define anucleate. Give an example of an anucleate cell.
4. Define multinucleate. Give an example of a multinucleate cell.
5. What is euchromatin?
6. What is heterochromatin?

Histology: Nucleus

Return to the **HISTOLOGY** *menu. From the* **Select topic** *menu, choose* **Nucleus. TEM: High magnification** *appears in the* **VIEW** *window. To the right in the* **IMAGE AREA** *you will see a high-magnification Transmission Electron Micrograph (TEM) of the microvillus in longitudinal section. Click* **Heterochromatin** *in the* **STRUCTURE LIST** *and use the information in the* **STRUCTURE INFORMATION** *window to answer the following questions.*

1. Where is heterochromatin located?

2. Describe heterochromatin.

3. What is the function of heterochromatin?

4. What other term refers to the heterochromatin?

5. Compare chromatin, euchromatin, and heterochromatin.

Return to the **STRUCTURE LIST** *and click* **Nuclear envelope.** *Use the information in the* **STRUCTURE INFORMATION** *window to answer the following questions.*

1. Where is the nuclear envelope located?

2. Describe the nuclear envelope.

3. List the functions of the nuclear envelope.

4. Each nuclear envelope is a _____.

We have already covered the remaining structures on the **STRUCTURE LIST,** *so click each one to review the information presented in the* **STRUCTURE INFORMATION** *window and to view them in the* **IMAGE AREA.**

100 MODULE 2 Cells and Chemistry

Click the **TAGS ON/OFF** *radio button located below the* **STRUCTURE INFORMATION** *window to activate the green pins on the image. You will see the following image:*

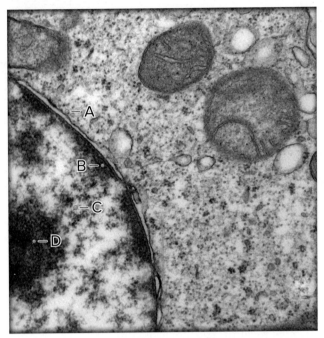

©EM Research Services, Newcastle University

Mouse-over the pins on the screen to find the information to identify the following structures:

A. _____

B. _____

C. _____

D. _____

Histology: Chromosome

Return to the **Select topic** *menu and choose* **Chromosome. SEM: Medium magnification** *appears in the* **VIEW** *window. To the right in the* **IMAGE AREA** *you will see a medium-magnification Transmission Electron Micrograph (TEM) of the microvillus in longitudinal section. Click on* **Centromere** *in the* **STRUCTURE LIST** *and use the information in the* **STRUCTURE INFORMATION** *window to answer the following questions.*

1. Where is the centromere located?

2. Describe the centromere.

3. List the functions of the centromere.

Return to the **STRUCTURE LIST** *and click on* **Chromatid of metaphase chromosome.** *Use the information in the* **STRUCTURE INFORMATION** *window to answer the following questions.*

1. Where are the chromatid of metaphase chromosome located?

2. Describe the chromatid of metaphase chromosome.

3. What is the function of the chromatid of metaphase chromosome?

4. What are sister chromatids?

Return to the **STRUCTURE LIST** *and click on* **Long arm of metaphase chromosome.** *Use the information in the* **STRUCTURE INFORMATION** *window to answer the following questions.*

1. Where are the long arms of metaphase chromosomes located?

2. Describe the long arms of metaphase chromosomes.

3. What other term refers to the long arms of metaphase chromosomes?

Return to the **STRUCTURE LIST** *and click on* **Metaphase chromosome.** *Use the information in the* **STRUCTURE INFORMATION** *window to answer the following questions.*

1. Where are the metaphase chromosomes located?

2. Describe metaphase chromosomes.

3. What is the function of the metaphase chromosome?

4. How many chromosomes are found in human somatic cells?

5. When are chromosomes visible in light microscopes?

Return to the **STRUCTURE LIST** *and click on* **Short arm of metaphase chromosome.** *Use the information in the* **STRUCTURE INFORMATION** *window to answer the following questions.*

1. Where are the short arms of metaphase chromosomes located?

2. Describe the short arms of metaphase chromosomes.

3. What other term refers to the short arms of metaphase chromosomes?

Click the **TAGS ON/OFF** *radio button located below the* **STRUCTURE INFORMATION** *window to activate the green pins on the image. You will see the following image:*

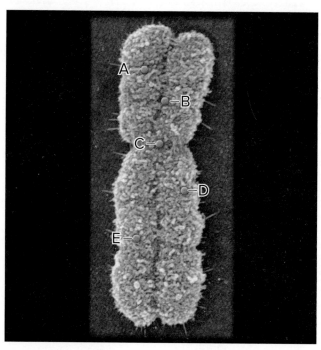

©Biophoto/Science Source

Mouse-over the pins on the screen to find the information to identify the following structures:

A. _____

B. _____

C. _____

D. _____

E. _____

Histology: Centrosome

Return to the **Select topic** *menu and choose* **Centrosome. TEM: High magnification** *appears in the* **VIEW** *window. To the right in the* **IMAGE AREA** *you will see a high-magnification Transmission Electron Micrograph (TEM) of the centrosome. Click on* **Centrosome** *in the* **STRUCTURE LIST** *and use the information in the* **STRUCTURE INFORMATION** *window to answer the following questions.*

1. Where is the centrosome located?

2. Describe the centrosome.

3. List the functions of the centrosome.

4. What other term refers to the centrosome?

Return to the **STRUCTURE LIST** *and click on* **Microtubule.** *Review the information presented in the* **STRUCTURE INFORMATION** *window.*

Click the **TAGS ON/OFF** *radio button located below the* **STRUCTURE INFORMATION** *window to activate the green pins on the image. You will see the following image:*

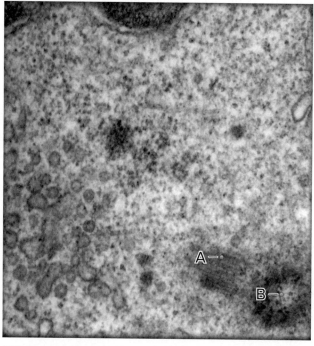

©EM Research Services, Newcastle University

Mouse-over the pins on the screen to find the information to identify the following structures:

A. _____

B. _____

Histology: Centriole

Return to the Select topic menu and choose Centriole. TEM: High magnification appears in the VIEW window. To the right in the IMAGE AREA you will see a high-magnification Transmission Electron Micrograph (TEM) of the centriole. Click on Centriole in the STRUCTURE LIST and use the information in the STRUCTURE INFORMATION window to answer the following questions.

1. Where is the centriole located?

2. Describe the centriole.

3. List the functions of the centriole.

4. What structure is formed by a pair of centrioles?

5. What other term refers to the centrioles?

Return to the STRUCTURE LIST and click on Microtubule triplet. Use the information in the STRUCTURE INFORMATION window to answer the following questions.

1. Where are the microtubule triplets located?

2. Describe the microtubule triplets.

3. What is the function of the microtubule triplets?

4. What structure is formed by microtubule doublets?

Click the TAGS ON/OFF radio button located below the STRUCTURE INFORMATION window to activate the green pins on the image. You will see the following image:

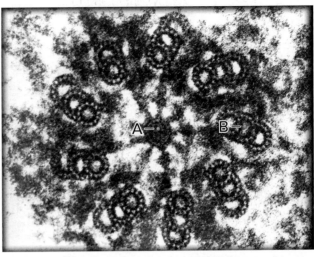

©Don W. Fawcett/Science Source

Mouse-over the pins on the screen to find the information to identify the following structures:

A. _____

B. _____

Structure Identification: Generalized Cell

Return to the **DISSECTION** *menu. From the* **Select topic** *menu, choose* **Generalized cell.**

Click **LAYER 1** *in the* **LAYER CONTROLS** *window and you will find the following image:*

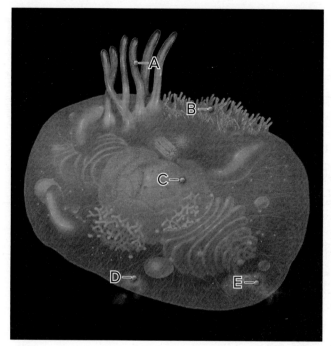

©McGraw-Hill Education

Mouse-over the pins on the screen to find the information to identify the following structures:

A. _____

B. _____

C. _____

D. _____

E. _____

Click **LAYER 2** *in the* **LAYER CONTROLS** *window and you will find the following image:*

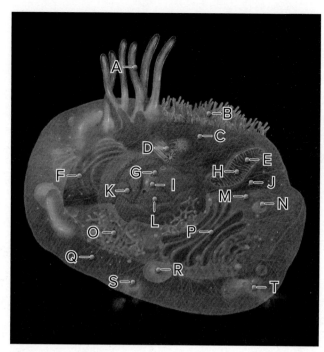

©McGraw-Hill Education

Mouse-over the pins on the screen to find the information to identify the following structures:

A. _____

B. _____

C. _____

D. _____

E. _____

F. _____

G. _____

H. _____

I. _____

J. _____

K. _____

L. _____

M. _____

N. _____

O. _____

P. _____

Q. _____

R. _____

S. _____

T. _____

Click **LAYER 3** *in the* **LAYER CONTROLS** *window and you will find the following image:*

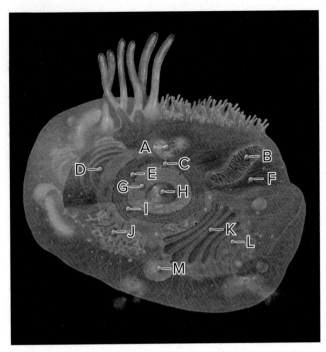

©McGraw-Hill Education

Mouse-over the pins on the screen to find the information to identify the following structures:

A. _____
B. _____
C. _____
D. _____
E. _____
F. _____
G. _____
H. _____
I. _____
J. _____
K. _____
L. _____
M. _____

IN REVIEW

What Have I Learned?

The following questions cover the material that you have just learned, Generalized Cell. Apply what you have learned to answer these questions on a separate piece of paper.

1. Where on the cell are the cilia located?
2. Which group of tissue has cilia?
3. Where are the cilia evident?
4. Describe the function of cilia.
5. Define micrometer. What is a micrometer also known as?
6. Where on the cell are the microvilli located?
7. Where are microvilli most developed?
8. Compare microvilli, cilia, and flagella.
9. What are stereocilia?
10. What are oscillations?
11. How are the microvilli visible in a light microscope? How can individual microvilli be distinguished?
12. List the functions of the plasma membrane.
13. Describe the phospholipid bilayer.
14. List the three different filaments that compose the cytoskeleton.
15. List the functions of the intermediate filaments.
16. Define nanometer.
17. List the functions of the microfilaments.
18. List the functions of the microtubules.
19. What is the function of the cis-face of the Golgi apparatus?
20. What is the function of the cristae of mitochondrion?
21. What are the two different endocytic vesicles? How do they differ?
22. What is the function of the endocytic vesicles?
23. Describe the three subdivisions of the Golgi apparatus.
24. List the functions of the Golgi apparatus.

Continued

25. Describe the two types of lysosomes.
26. What is the function of lysosomes?
27. What is the function of peroxisomes?
28. How do peroxisomes and lysosomes compare?
29. Where are mitochondria located?
30. Describe mitochondria.
31. What is the function of mitochondria?
32. What structures are attached to the rough endoplasmic reticulum?
33. What structure is continuous with the rough endoplasmic reticulum?
34. What is the function of the rough endoplasmic reticulum?
35. How does the rough endoplasmic reticulum compare to the smooth endoplasmic reticulum?
36. What is the function of secretory vesicles of the Golgi apparatus?
37. What applies to the smooth endoplasmic reticulum in skeletal muscle tissue?
38. What is the function of the trans-face of the Golgi apparatus?
39. What is endocytosis?
40. What is the function of the nuclear pores?
41. What determines the number of nucleoli?
42. What is the function of the nucleus?
43. What is the largest organelle?
44. Define anucleate. Give an example of an anucleate cell.
45. Define multinucleate. Give an example of a multinucleate cell.

EXERCISE 2.2:
Plasma Membrane

Select the **Module** drop-down menu in the **Home Screen** of APR. Choose **2. Cells and chemistry**. Click the **DISSECTION** icon. From the **Select topic** menu, choose **Plasma membrane**. Select the **my COURSE CONTENT** tab if your instructor has selected specific exercises for you to complete. Otherwise select the **ALL CONTENT** tab. Notice that **Cross-section** is displayed in the **VIEW** menu. You will see the following image:

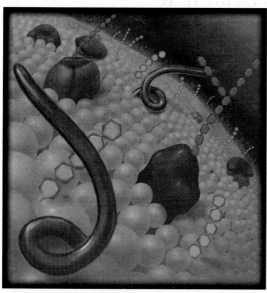

©McGraw-Hill Education

Click the **Select group** menu and choose **Membrane carbohydrates**. Select **Glycocalyx [1]** from the list of structures. This image reveals the superficial view of a generalized plasma membrane and a highlighted group of glycocalyces (plural for glycocalyx). Use the information in the **STRUCTURE INFORMATION** window to answer the following questions.

1. Where on the cell are the glycocalyces located?

   outer surface of plasma membrane

2. Describe the glycocalyces.

   fuzzy coat on extracellular surface of plasma membrane, comprised of carbohydrates attached to membrane glycolipids and glycoproteins

3. List the functions of the glycocalyces.

4. What other term refers to the glycocalyces?

   cell coat

Return to the list of **Membrane carbohydrates** and select **Glycocalyx [2]**. Compare the structures visible in **LAYER 2** with those visible in **LAYER 1**.

CHECK POINT
Plasma Membrane: Membrane Carbohydrates

1. How are the glycocalyces made visible under the light microscope?

MODULE 2 Cells and Chemistry

Return to the **Select group** *menu and select* **Membrane lipids.**

Select **Cholesterol molecules in lipid bilayer.** *Use the information in the* **STRUCTURE INFORMATION** *window to answer the following questions.*

1. Where are the cholesterol molecules in the lipid bilayer located?

   hydrophobic region of lipid bilayer

2. Describe the cholesterol molecules in the lipid bilayer.
   - steroid molecules
   - accounts for approx 20% of plasma memb lipids

3. List the functions of the cholesterol molecules in the lipid bilayer. strengthens
   prevent other denye

Return to the list of **Membrane lipids** *and select* **Fatty acid tails of lipid bilayer.** *Use the information in the* **STRUCTURE INFORMATION** *window to answer the following questions.*

1. Where are the fatty acid tails of the lipid bilayer located?

   hydrophobic region of lipid bilayer of plasma membrane

2. Describe the fatty acid tails of the lipid bilayer.

   3 pic

3. What is the function of the fatty acid tails of the lipid bilayer?

   length and saturation

Return to the list of **Membrane lipids** *and select* **Glycolipid [1].** *Use the information in the* **STRUCTURE INFORMATION** *window to answer the following questions.*

1. Where are the glycolipids located?

   outer surface of plasma mem

2. Describe the fatty acid tails of the lipid bilayer.

   sugar containing
   may account for ~ 5%

3. What is the function of the fatty acid tails of the lipid bilayer?

Return to the list of **Membrane lipids** *and select* **Glycolipid [2].** *Compare the structures visible in* **LAYER 2** *with those visible in* **LAYER 1.**

Return to the list of **Membrane lipids** *and select* **Inner leaflet of lipid bilayer.** *Use the information in the* **STRUCTURE INFORMATION** *window to answer the following questions.*

1. Where is the inner leaflet of lipid bilayer located?

   cytoplasmic (inner) side of lipid bilayer

2. Describe the inner leaflet of lipid bilayer.
   - inner part of lipid bilayer
   - comprised of polar heads and

3. What is the function of the inner leaflet of the lipid bilayer?

   interacts with cytoplasmic compartment of cells

4. What other term refers to the "plasma" membrane? What other membrane has similar structure?

   pic

Return to the list of **Membrane lipids** *and select* **Lipid bilayer.** *Use the information in the* **STRUCTURE INFORMATION** *window to answer the following questions.*

1. Where is the lipid bilayer located?

   pic

2. Describe the lipid bilayer.   ↓

3. List the functions of the lipid bilayer.   ↓

4. What other term refers to the lipid bilayer?   ↓

Return to the list of **Membrane lipids** *and select* **Outer leaflet of lipid bilayer.** *Use the information in the* **STRUCTURE INFORMATION** *window to answer the following questions.*

1. Where is the outer leaflet of the lipid bilayer located?

   extracellular side of lipid bila

2. Describe the outer leaflet of the lipid bilayer.
   outer part of lipid bilayer
   - comprised of polar heads & non polar fatty acid tail

3. What is the function of the outer leaflet of the lipid bilayer?

   interacts with extracellular environment

Return to the list of **Membrane lipids** *and select* **Polar heads of lipid bilayer [1].** *Use the information in the* **STRUCTURE INFORMATION** *window to answer the following questions.*

1. Where are the polar heads of the lipid bilayer located?

   hydrophillic region of lipid bilayer of plasma membrane

2. Describe the polar heads of the lipid bilayer.

3. List the functions of the polar heads of the lipid bilayer.

Return to the list of **Membrane lipids** *and select* **Polar heads of lipid bilayer [2].** *Compare the structures visible in* **LAYER 2** *with those visible in* **LAYER 1.**

CHECK POINT

Plasma Membrane: Membrane Lipids
1. Define hydrophilic.
2. Define hydrophobic.
3. Define unsaturated as it refers to fatty acids.
4. Define amphipathic.

Return to the **Select group** *menu and select* **Membrane proteins**.

Select **Glycoprotein [1]**. *Use the information in the* **STRUCTURE INFORMATION** *window to answer the following questions.*

1. Where are glycoproteins located?

    plasma membrane

2. Describe glycoproteins.

3. What is the function of glycoproteins?

Return to the list of **Membrane proteins** *and select* **Glycoprotein [2]**. *Compare the structures visible in* **LAYER 2** *with those visible in* **LAYER 1**.

Return to the list of **Membrane proteins** *and select* **Integral membrane protein [1]**. *Use the information in the* **STRUCTURE INFORMATION** *window to answer the following questions.*

1. Where are integral membrane proteins located?

    plasma membrane

2. Describe integral membrane proteins. *plasma membrane*

3. List the functions of integral membrane proteins.

    receptors, enzymes etc

4. What other term refers to integral membrane proteins?

    intrinsic membrane protein

5. How do integral membrane proteins compare to peripheral membrane proteins?

Return to the list of **Membrane proteins** *and select* **Integral membrane protein [2]**. *Compare the structures visible in* **LAYER 2** *with those visible in* **LAYER 1**.

Return to the list of **Membrane proteins** *and select* **Membrane channel pore [1]**. *Use the information in the* **STRUCTURE INFORMATION** *window to answer the following questions.*

1. Where are membrane channel pores located?

    membrane channel protein

2. Describe membrane channel pores.

    narrow pore

3. What is the function of membrane channel pores?

    permits selective passage of it

Return to the list of **Membrane proteins** *and select* **Membrane channel pore [2]**. *Compare the structures visible in* **LAYER 2** *with those visible in* **LAYER 1**. *hydrophilic solu*

Return to the list of **Membrane proteins** *and select* **Membrane channel protein [1]**. *Use the information in the* **STRUCTURE INFORMATION** *window to answer the following questions.*

1. Where are membrane channel proteins located?

    plasma mem

2. Describe membrane channel proteins.

    transmembr forms narrow

3. List the functions of membrane channel proteins.

Return to the list of **Membrane proteins** *and select* **Membrane channel protein [2]**. *Compare the structures visible in* **LAYER 2** *with those visible in* **LAYER 1**.

Return to the list of **Membrane proteins** *and select* **Membrane channel [1]**. *Use the information in the* **STRUCTURE INFORMATION** *window to answer the following questions.*

1. Where are membrane channel proteins located?

    plasma membrane

2. Describe membrane channel proteins.

    transmembrane protein that forms narrow pore

3. List the functions of membrane channel proteins.

Return to the list of **Membrane proteins** *and select* **Membrane channel [2]**. *Compare the structures visible in* **LAYER 2** *with those visible in* **LAYER 1**.

Return to the list of **Membrane proteins** *and select* **Peripheral membrane protein [1]**. *Use the information in the* **STRUCTURE INFORMATION** *window to answer the following questions.*

1. Where are peripheral membrane proteins located?

    plasma mem

2. Describe peripheral membrane proteins.

3. List the functions of peripheral membrane proteins.

Return to the list of **Membrane proteins** and select **Peripheral membrane protein [2]**. Compare the structures visible in **LAYER 2** with those visible in **LAYER 1**.

> ### CHECK POINT
> **Plasma Membrane: Membrane Proteins**
> 1. List the structures that contribute to the glycocalyx.
> 2. What are transmembrane proteins?
> 3. Compare the hydrophobic regions of integral membrane proteins with the hydrophilic regions.

Structure Identification: Plasma Membrane

Click **LAYER 1** in the **LAYER CONTROLS** window and you will find the following image:

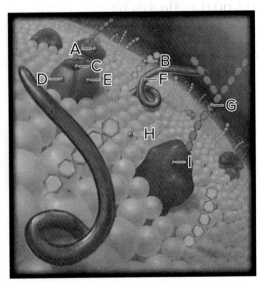

©McGraw-Hill Education

Mouse-over the pins on the screen to find the information to identify the following structures:

A. peripheral membrane protein
B. glycolipid
C. membrane channel pore
D. membrane channel
E. membrane channel protein
F. integral membrane protein
G. glycocalyx
H. polar heads of lipid bilayer
I. glycoprotein

Click **LAYER 2** in the **LAYER CONTROLS** window and you will find the following image:

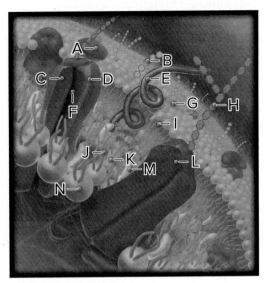

©McGraw-Hill Education

Mouse-over the pins on the screen to find the information to identify the following structures:

A. peripheral membrane protein
B. glycolipid
C. membrane channel
D. membrane channel protein
E. integral membrane protein
F. membrane channel pore
G. outer leaflet of lipid bilayer
H. glycocalyx
I. lipid bilayer
J. fatty acid tails of lipid bilayer
K. inner leaflet of lipid bilayer
L. glycoprotein
M. cholesterol molecules in lipid bilayer
N. polar heads of lipid bilayer

IN REVIEW

What Have I Learned?

The following questions cover the material that you have just learned, Plasma Membrane. Apply what you have learned to answer these questions on a separate piece of paper.

1. Where on the cell are the glycocalyces located?
2. Describe the glycocalyces.
3. List the functions of the glycocalyces?
4. Where are the cholesterol molecules in the lipid bilayer located?
5. List the functions of the cholesterol molecules in the lipid bilayer.
6. Where are the fatty acid tails of the lipid bilayer located?
7. What is the function of the fatty acid tails of the lipid bilayer?
8. Where are the glycolipids located?
9. What is the function of the fatty acid tails of the lipid bilayer?
10. What is the function of the inner leaflet of the lipid bilayer?
11. Where is the lipid bilayer located?
12. Describe the lipid bilayer.
13. List the functions of the lipid bilayer.
14. What is the function of the outer leaflet of the lipid bilayer?
15. Where are the polar heads of the lipid bilayer located?
16. List the functions of the polar heads of the lipid bilayer.
17. Define hydrophilic.
18. Define hydrophobic.
19. Define unsaturated as it refers to fatty acids.
20. Define amphipathic.
21. Where are glycoproteins located?
22. Describe glycoproteins.
23. What is the function of glycoproteins?
24. List the functions of integral membrane proteins.
25. How do integral membrane proteins compare to peripheral membrane proteins?
26. Where are membrane channel pores located?
27. Describe membrane channel pores.
28. What is the function of membrane channel pores?
29. List the functions of membrane channel proteins.
30. List the functions of peripheral membrane proteins.
31. List the structures that contribute to the glycocalyx.
32. What are transmembrane proteins?
33. Compare the hydrophobic regions of integral membrane proteins with the hydrophilic regions.

Membrane Transport

In keeping with the theme of the plasma membrane, we will now discuss methods cells employ to transport substances into and out of this selectively permeable membrane. These methods both prevent unwanted materials from entering the cell and allow desired materials to pass through unhindered.

The following animations explain these phenomena. View each animation as often as necessary until you are comfortable with the concepts covered, and then test your understanding by taking the **QUIZ** for each specific animation.

SELECT ANIMATION: Diffusion — PLAY

After viewing the animation, answer the following questions.

1. Molecules dissolved in a solution are in constant _____ _____ _____ due to their _____.

2. One result of this motion is _____.

3. This tendency of molecules to spread out is an example of _____.

4. Even as a solid lump, the individual sugar molecules are _____ _____.

5. What happens to the lump of sugar when it is dropped into the water?

6. How do the individual sugar molecules move?

7. How does this movement define diffusion?

8. How long does diffusion continue?

9. What factors affect the rate of diffusion?

Self-Quiz
Take this opportunity to quiz yourself by taking the **QUIZ** over this animation.

SELECT ANIMATION
Osmosis
PLAY

After viewing the animation, answer the following questions.

1. What is diffusion?

2. What does this process allow?

3. Do most polar molecules freely cross the lipid cell membrane? Name two groups of polar molecules.

4. What is the name for the special case of diffusion that involves the movement of water molecules across a membrane?

5. Why is a molecule of urea unable to diffuse across the membrane?

6. How does a urea molecule interact with water molecules? Why?

7. Why is there now a net movement of water molecules? Which direction do they move?

8. What happens to the water level on the side of the beaker where the water molecules are moving into?

9. Define isotonic, hypertonic, and hypotonic.

Self-Quiz
Take this opportunity to quiz yourself by taking the **QUIZ** over this animation.

SELECT ANIMATION
Osmosis and Tonicity
PLAY

After viewing the animation, answer the following questions.

1. Define osmosis.

2. Define tonicity. What determines tonicity in cell biology?

3. List the three ranges of tonicity with reference to cells.

4. What changes can occur in a cell's volume with changes in tonicity?

5. How does an isotonic solution compare to the tonicity of the intracellular fluid?

6. Do erythrocytes placed in an isotonic solution show any effects?

7. What effects are demonstrated by erythrocytes placed in a hypertonic solution?

8. What is the process of crenation?

9. What effects are demonstrated by erythrocytes placed in a hypotonic solution?

10. What is the process of hemolysis?

Self-Quiz
Take this opportunity to quiz yourself by taking the **QUIZ** over this animation.

SELECT ANIMATION
Facilitated Diffusion
PLAY

After viewing the animation, answer the following questions.

1. What occurs in the process of facilitated diffusion?

2. What is unique about the carrier molecules and the molecules to which they bind?

3. Once the molecule binds to the carrier protein, the protein will facilitate the diffusion process by _____.

4. Facilitated diffusion and simple diffusion are similar in that both _____.

5. How is facilitated diffusion different from simple diffusion?

6. What determines the direction in which facilitated diffusion occurs?

Self-Quiz
Take this opportunity to quiz yourself by taking the **QUIZ** over this animation.

SELECT ANIMATION
Sodium-potassium Pump
PLAY

After viewing the animation, answer the following questions.

1. The sodium-potassium exchange pump is an example of an _____ process.

2. Active transport processes can move substances _____ a concentration gradient.

3. What do active transport processes require?

4. What ions are moved out of the cell by the sodium-potassium exchange pump?

5. What ions are moved into the cell by the sodium-potassium exchange pump?

6. Where is the carrier protein for the sodium-potassium exchange pump located?

7. What items can bind to the carrier protein on the inside of the cell membrane?

8. What changes occur to the ATP molecule? What effect does this have on the carrier protein?

9. What occurs to the carrier protein in this new configuration?

10. What events occur when the carrier protein returns to its original conformation?

Self-Quiz
Take this opportunity to quiz yourself by taking the **QUIZ** over this animation.

SELECT ANIMATION
Cotransport
PLAY

After viewing the animation, answer the following questions.

1. Which direction can small molecules, such as sugars and amino acids, be transported?

2. How does the sugar move? How does the concentration of sugar compare inside and outside the cell?

3. How is this transport of sugar driven through a coupled transport protein? Are these counterions moving from a higher to a lower concentration or from a lower to a higher concentration?

4. What is a symport? What occurs there?

5. How is a low concentration of sodium maintained inside the cell? How is it powered?

6. What is counter-transport?

7. What is an antiport? What occurs there? How is this different than what occurs at a symport?

8. How does the sodium-potassium pump come into play in this process?

Self-Quiz
Take this opportunity to quiz yourself by taking the **QUIZ** over this animation.

SELECT ANIMATION
Endocytosis and Exocytosis

PLAY

After viewing the animation, answer the following questions.

1. Define endocytosis.

2. How do these large particles enter the cell?

3. List the three types of endocytosis.

4. When is phagocytosis employed?

5. When is pinocytosis employed?

6. When is receptor-mediated endocytosis employed?

7. Describe the process of receptor-mediated endocytosis.

8. Define exocytosis.

9. Describe the process of exocytosis.

Self-Quiz
Take this opportunity to quiz yourself by taking the **QUIZ** over this animation.

EXERCISE 2.3:
Chemistry

As promised, we will conclude this module with an overview of concepts of chemistry that are vital to an understanding of physiology. If one were to summarize physiology, it would include all of the cell processes that maintain homeostasis—keeping the body environment on an even keel. Moreover, as stated before, these processes are chemical, specifically biochemical processes.

So, without any further ado, let's delve into this fascinating world of biochemistry!

SELECT ANIMATION — Atomic Structure — **PLAY**

After viewing the animation, answer the following questions.

1. Define an atom.

2. List the smaller particles that compose an atom.

3. Where are these smaller particles located?

4. Define an electron shell or energy level.

5. Compare the charges present on protons, electrons, and neutrons.

6. How are each element uniquely defined?

7. Define atomic number.

8. The number of _____ equals the number of _____, making atoms _____ _____.

9. How massive are electrons?

10. Define atomic mass.

11. The number of _____ in a particular atom, and thus _____ _____ can vary.

12. Define isotopes.

13. The first shell holds a maximum of _____ electrons.

14. The second shell holds a maximum of _____ electrons.

15. The third shell can hold up to _____ electrons.

16. Define the valence shell. The valence shell holds a maximum of _____ electrons.

17. What specifically determines an atom's chemical bonding properties?

18. What tendency do atoms with partially filled valence shells demonstrate? Why? What does this produce?

19. What determines the number and kind of bonds an atom will form?

Self-Quiz
Take this opportunity to quiz yourself by taking the **QUIZ** over this animation.

SELECT ANIMATION — Bonds — **PLAY**

After viewing the animation, answer the following questions.

1. How do atoms tend to achieve a stable number of valence electrons?

2. How are ionic bonds formed?

3. How are covalent bonds formed?

4. Define ionic bond.

5. How is a nonpolar covalent bond formed between two chlorine atoms?

6. How are electrons shared in nonpolar covalent bonds?

7. Define polar covalent bond.

8. How is a polar covalent bond formed between a Hydrogen atom and a Chlorine atom?

9. What charges are associated with a polar covalent bond? How do they compare to an ionic bond and a nonpolar covalent bond?

Self-Quiz
Take this opportunity to quiz yourself by taking the **QUIZ** over this animation.

MODULE 2 Cells and Chemistry

SELECT ANIMATION
Organic Molecules (3D) PLAY

1. What are organic molecules?

2. Each organic molecule contains _____ _____ _____ _____, including _____.

3. Carbon has _____ _____ _____ and is able to bond with other elements that can contribute another _____ _____ to complete its outer shell.

4. Carbon can form long chains or _____ _____, as a base for a variety of _____ _____.

5. Carbon's ability to bond with _____ and _____ _____ allow it to form complex molecules necessary for _____.

6. These molecules include _____, _____, and _____ _____.

7. _____ are organic molecules made up of _____, _____, and _____.

8. Carbohydrates have a two-to-one ratio of _____ to _____ and _____.

9. For example, glucose has _____ _____ atoms, _____ _____, and _____ _____ atoms.

10. Why are carbohydrates important?

11. _____ are chains of _____ _____.

12. What is the structure of all amino acids?

13. What differentiates each amino acid?

14. Amino acids join together to form _____.

15. Longer chains of amino acids are called _____.

16. Groups of _____ join to form _____.

17. Describe the structure of proteins.

18. What determines the function of a protein?

19. List the functions of proteins.

20. Lipids are organic molecules composed mainly of _____, _____, and _____.

21. Fatty acids are lipids consisting of a _____ _____, a chain of _____, and a _____ _____.

22. What are triglycerides?

23. What are the functions of lipids?

24. Nucleic acids are composed of repeating units called _____.

25. What are the three segments of a nucleotide?

26. What is the function of DNA?

SELECT ANIMATION
Electrolytes PLAY

After viewing the animation, answer the following questions.

1. Define an electrolyte.

2. Sodium chloride is a _____ _____ when dissolved in water.

3. Explain how the polar water molecules interact with ions of the salt.

4. How is carbonic acid formed?

5. Does carbonic acid ionize in water? How strongly? Why?

6. "That makes this acid a _____ _____."

7. How does the number of ions in carbonic acid affect the conduction of electricity?

8. How does glucose react with water?

9. Glucose is a _____. A _____ solution does not contain _____. Therefore, the solution _____ _____ conduct electricity.

10. What is a dipole? Give an example from the animation.

11. How does water interact with glucose molecules? What are these particular reactions called?

12. How does glucose dissolve in water?

13. How does dissolved glucose compare with dissolved sodium chloride?

Self-Quiz
Take this opportunity to quiz yourself by taking the **QUIZ** over this animation.

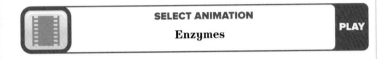

After viewing the animation, answer the following questions.

1. Define enzymes.

2. Describe the active site of an enzyme.

3. How does an enzyme work?

4. What are reactants or substrates?

5. Where does binding between reactants and enzymes occur?

6. The enzyme and substrates form an _____.

7. What does this interaction between the substrate and the enzyme cause?

8. What is the end result of this interaction?

9. What occurs to the enzyme after the product is released?

Self-Quiz
Take this opportunity to quiz yourself by taking the **QUIZ** over this animation.

MODULE 3

Tissues

Overview: Tissues

As you may recall, one of the major themes in *Anatomy and Physiology | Revealed* is that structure determines function. The realm of tissues is no exception! Here we will explore the various tissues that build the structures of your body. As you examine these tissues through animations and histology, watch for the key structural components that give each tissue its unique function. And as you practice and review these images, you will begin to see the patterns that form to give you clues to their identification and the role they play in the body.

Epithelial Tissues

Select the **MODULE** drop-down menu in the **HOME** screen of *Anatomy and Physiology | Revealed*®. Choose **3. Tissues.** Click the **ANIMATIONS** icon. Select the *my* **COURSE CONTENT** tab if your instructor has selected specific exercises for you to complete. Otherwise select the **ALL CONTENT** tab.

We will begin the animations with a series of foundational concepts that the more specific animations to come will build upon. It is imperative that you understand these key concepts before you move on.

SELECT ANIMATION
Structural Hierarchy of the Human Body (3D) PLAY

After viewing the animation, answer the following questions:

1. The _____ is the structural and functional unit of the body.

2. Together, cells construct _____, and _____ combine to form _____.

3. Together, various _____ make up _____ _____ which work collectively to create an _____.

SELECT ANIMATION
Four primary tissues (3D) PLAY

1. What is a tissue?

2. List the four basic types of tissues in the body.

3. Which tissue type covers and lines body structures and exhibits specialized shapes and functions?

4. In the lining of the small intestine, or _____, a single layer of cells called _____ absorbs nutrients through surface extensions called _____.

5. Where is nervous tissue located?

6. What are the two specialized cells in nervous tissue?

7. Which of these two are the neuron supporting cells?

8. Inside the brain, _____ _____ contains _____ _____ _____ and their extensions known as _____ and various _____ such as _____ and _____.

9. _____ _____ contains _____ which are long fibers extending away from _____ _____ _____.

10. Axons carry _____ and _____ impulses to body tissues such as _____.

11. List the classifications of muscle tissue.

12. _____ _____ _____ form whole _____ that attach to _____ and move the body.

115

13. Cells called _____ containing contractile _____ fibers comprise _____.

14. List the connective tissues.

15. _____ is comprised of _____ molecules in a _____ matrix surrounding a mesh work of _____ fibrils that impart tensile strength and _____ that help maintain the tissue's _____ and _____.

16. What is the definition of an organ?

17. In the heart, _____ _____ tissue in the _____ contracts and pushes blood through the _____ tissue valves into vessels lined by vascular epithelial tissue called _____.

SELECT ANIMATION
Epithelial Tissue Overview
PLAY

After viewing the animation, answer the following questions:

1. What body structures are formed of epithelia?

2. Any substance that enters or leaves the body must cross an _____.

3. What features characterize all epithelia?

4. List the two distinguishing features of epithelia.

5. What structures join adjacent cells of epithelia?

6. What structure separates the basal surface of epithelia from underlying the connective tissue layers?

7. Although epithelia lack _____ _____, many _____ _____ _____ are present.

8. What characteristic of epithelia allows them to maintain their function?

9. List the primary functions of epithelia.

10. What are the two basic criteria for classifying epithelia?

11. Describe a simple epithelium.

12. Describe a stratified epithelium.

13. Describe a squamous epithelium.

14. Describe a cuboidal epithelium.

15. Describe a columnar epithelium.

16. Describe simple squamous epithelium.

17. Describe stratified squamous epithelium.

18. Describe simple cuboidal epithelium.

19. Describe stratified cuboidal epithelium.

20. Describe simple columnar epithelium.

21. Describe stratified columnar epithelium.

22. Describe transitional epithelium.

23. Describe a pseudostratified epithelium.

24. Give one location for a pseudostratified epithelium.

25. List the subclassifications of simple columnar epithelium.

26. What structures are located on the free surface of epithelial cells involved in absorption?

27. Are microvilli motile or nonmotile?

28. What is the function of microvilli?

29. How do microvilli appear in the light microscope? Therefore, cells with numerous microvilli are said to have a _____ _____.

30. Describe the two cell types of the two different stratified squamous epithelia.

31. What protein is contained in dead stratified squamous epithelium cells?

32. Describe the two classifications of stratified squamous epithelium.

33. List the four types of intercellular junctions.

34. Describe tight junctions. What are their functions?

35. How are tight junctions involved in the primary function of epithelium?

36. Describe adhering junctions.

37. What is located between the cell membranes of adjacent cells at adhering junctions?

38. Describe desmosomes. Where are they located? How do they differ from tight and adhering junctions?

39. Describe gap junctions. What substances pass through them?

Self-Quiz
Take this opportunity to quiz yourself by taking the **QUIZ** over this animation.

EXERCISE 3.1:
Epithelial Tissue—Simple

Click the **HISTOLOGY** icon. Select the **my COURSE CONTENT** tab if your instructor has selected specific exercises for you to complete. Otherwise select the **ALL CONTENT** tab.

Histology: Epithelial Tissue—Simple Squamous Epithelium

From the **Select topic** menu, choose **Simple squamous epithelium. LM: Medium magnification** appears in the **VIEW** window. To the right in the

IMAGE AREA *you will see a medium magnification Light Microscope image of this tissue. Use the information in the* **STRUCTURE INFORMATION** *window to answer the following questions.*

1. What structures may be found on the apical surface of cuboidal epithelial cells?

2. List the functions of cuboidal epithelial cells.

3. Where is embryonic loose connective tissue located?

4. List and describe the tissues included in embryonic loose connective tissue.

5. List the functions of embryonic loose connective tissue.

6. Describe the nucleus of the cuboidal epithelial cell. Where is it located?

7. List the functions of nucleus of the cuboidal epithelial cell.

8. What is euchromatin?

9. What is heterochromatin?

10. Describe the nucleus of the squamous cell. Where is it located?

11. List the functions of the nucleus of the squamous cell.

12. Describe simple cuboidal epithelium. Where is it located?

13. What is the function of simple cuboidal epithelium?

14. Describe simple squamous epithelium. Where is it located?

15. List the functions of simple squamous epithelium.

16. Define mesothelium.

17. Describe a squamous cell. Where are squamous cells located?

18. What is the function of squamous cells?

118 MODULE 3 Tissues

Click the **TAGS ON/OFF** *radio button located below the* **STRUCTURE INFORMATION** *window to activate the green pins on the image. You will see the following image:*

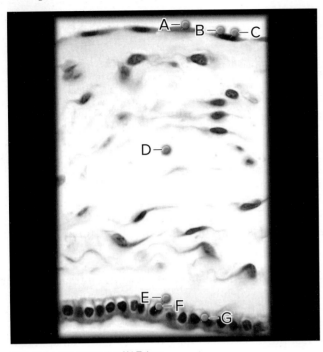

©McGraw-Hill Education/Al Telser

Mouse-over the pins on the screen to find the information to identify the following structures:

A. _____

B. _____

C. _____

D. _____

E. _____

F. _____

G. _____

Histology: Epithelial tissue—Simple Cuboidal Epithelium

From the **Select topic** *menu, choose* **Simple cuboidal epithelium. LM: Medium magnification** *appears in the* **VIEW** *window. To the right in the* **IMAGE AREA** *you will see a medium magnification Light Microscope image of this tissue. Click through the listed structures and use the information in the* **STRUCTURE INFORMATION** *window to answer the following questions.*

1. What structures may be found on the apical surface of cuboidal epithelial cells?

2. List the functions of cuboidal epithelial cells.

3. Where are cuboidal epithelial cells located?

4. Describe the lumen of the collecting duct. Where is it located?

5. What is the function of the lumen of the collecting duct?

6. Describe the nucleus of the cuboidal epithelial cell. Where is it located?

7. List the functions of the nucleus of the cuboidal epithelial cell.

8. Describe a simple cuboidal epithelium. Where is it located?

9. What is the function of a simple cuboidal epithelium?

10. What is a tangential plane of section?

Click the **TAGS ON/OFF** *radio button located below the* **STRUCTURE INFORMATION** *window to activate the green pins on the image. You will see the following image:*

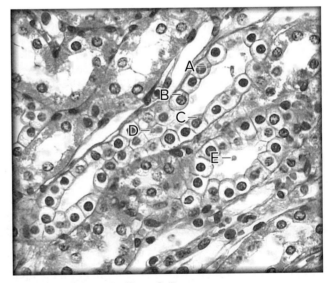

©McGraw-Hill Education/Steve Sullivan

Mouse-over the pins on the screen to find the information to identify the following structures:

A. _____

B. _____

C. _____

D. _____

E. _____

Histology: Epithelial Tissue—Simple Columnar Epithelium (ciliated)

From the **Select topic** *menu, choose* **Simple columnar epithelium (ciliated). LM: Medium magnification** *appears in the* **VIEW** *window. To the right in the* **IMAGE AREA** *you will see a medium magnification Light Microscope image of this tissue. Click through the listed structures and use the information in the* **STRUCTURE INFORMATION** *window to answer the following questions.*

1. Describe the basement membrane of simple columnar epithelium. Where is it located?

2. List the functions of the basement membrane of simple columnar epithelium.

3. Describe the cilia on simple columnar epithelium. Where are they located?

4. List the functions of the cilia on simple columnar epithelium.

5. What are oscillations of cilia? What substances are moved by these oscillations?

6. What relationship do the cilia have with the ovum and with the sperm?

7. Describe the ciliated columnar epithelial cells. Where are they located?

8. What is the function of the ciliated columnar epithelial cells?

9. Describe the lamina propria of ciliated simple columnar epithelium. Where is it located?

10. List the functions of the lamina propria of ciliated simple columnar epithelium.

11. Describe the lumen of the uterine tube. Where is it located?

12. List the functions of the lumen of the uterine tube.

13. Describe the nucleus of the ciliated columnar epithelial cell. Where is it located?

14. List the functions of the nucleus of the ciliated columnar epithelial cell.

15. Describe the nonciliated peg cell. Where is it located?

16. List the functions of the nonciliated peg cell.

17. Why does the apical surface of the uterine tube epithelium typically appear uneven?

18. Describe ciliated simple columnar epithelium. Where is it located?

19. List the functions of ciliated simple columnar epithelium.

20. What effect does estrogen have on the cilia of the uterine tube?

21. Cilia are not present on which group of epithelia?

Click the **TAGS ON/OFF** *radio button located below the* **STRUCTURE INFORMATION** *window to activate the green pins on the image. You will see the following image:*

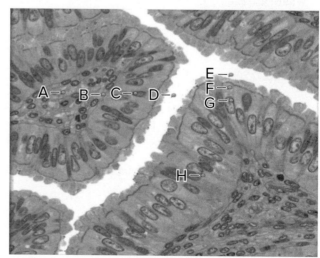

©McGraw-Hill Education/Al Telser

Mouse-over the pins on the screen to find the information to identify the following structures:

A. _____
B. _____
C. _____
D. _____
E. _____
F. _____
G. _____
H. _____

Histology: Epithelial Tissue—Simple Columnar Epithelium (nonciliated)

From the **Select topic** *menu, choose* **Simple columnar epithelium (nonciliated). LM: Medium magnification** *appears in the* **VIEW** *window. To the right in the* **IMAGE AREA** *you will see a medium magnification Light Microscope image of this tissue. Click through the listed structures and use the information in the* **STRUCTURE INFORMATION** *window to answer the following questions.*

1. Describe the basement membrane of simple columnar epithelium. Where is it located?

2. List the functions of the basement membrane of simple columnar epithelium.

3. Describe the nonciliated columnar epithelial cells. Where are they located?

4. What is the function of the nonciliated columnar epithelial cells?

5. Describe the goblet cells in nonciliated simple columnar epithelium. Where are they located?

6. What is the function of the goblet cells in nonciliated simple columnar epithelium?

7. What other tissue contains goblet cells? Where is it located? What is the function of these goblet cells?

8. Describe the lamina propria of nonciliated simple columnar epithelium. Where is it located?

9. List the functions of the lamina propria of nonciliated simple columnar epithelium.

10. Describe the lumen of the jejunum. Where is it located?

11. List the functions of the lumen of the jejunum.

12. Describe the microvilli (brush border). Where are they located?

13. List the functions of the microvilli (brush border).

14. Describe the nucleus of the nonciliated columnar epithelial cell. Where is it located?

15. List the functions of the nucleus of the nonciliated columnar epithelial cell.

16. Describe nonciliated simple columnar epithelium. Where is it located?

17. List the functions of nonciliated simple columnar epithelium.

Click the **TAGS ON/OFF** *radio button located below the* **STRUCTURE INFORMATION** *window to activate the green pins on the image. You will see the following image:*

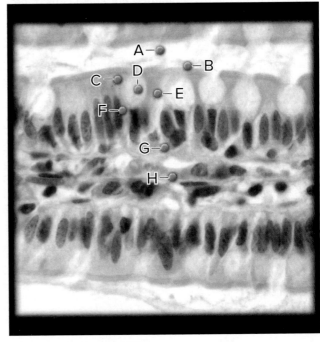

©McGraw-Hill Education/Al Telser

Mouse-over the pins on the screen to find the information to identify the following structures:

A. _____
B. _____
C. _____
D. _____
E. _____
F. _____
G. _____
H. _____

Histology: Epithelial Tissue— Pseudostratified Columnar Epithelium

*From the **Select topic** menu, choose **Pseudostratified columnar epithelium. LM: Medium magnification** appears in the **VIEW** window. To the right in the **IMAGE AREA** you will see a medium magnification Light Microscope image of this tissue. Click through the listed structures and use the information in the **STRUCTURE INFORMATION** window to answer the following questions.*

1. Describe the cilia on pseudostratified columnar epithelium. Where are they located?

2. List the functions of the cilia on pseudostratified columnar epithelium.

3. Describe the ciliated columnar epithelial cells. Where are they located?

4. What is the function of the ciliated columnar epithelial cells?

5. Describe the goblet cells in pseudostratified columnar epithelium. Where are they located?

6. What is the function of the goblet cells in pseudostratified columnar epithelium?

7. Describe the lamina propria of pseudostratified columnar epithelium. Where is it located?

8. List the functions of the lamina propria of pseudostratified columnar epithelium.

9. Describe the lumen of the trachea. Where is it located?

10. List the functions of the lumen of the trachea.

11. Describe the nucleus of the basal cell. Where is it located?

12. List the functions of the nucleus of the basal cell.

13. Describe pseudostratified columnar epithelium. Where is it located?

14. List the functions of pseudostratified columnar epithelium.

*Click the **TAGS ON/OFF** radio button located below the **STRUCTURE INFORMATION** window to activate the green pins on the image. You will see the following image:*

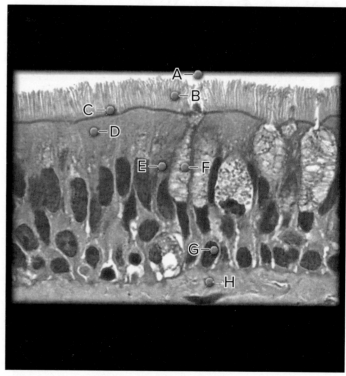

©McGraw-Hill Education/Dennis Strete

Mouse-over the pins on the screen to find the information to identify the following structures:

A. _____
B. _____
C. _____
D. _____
E. _____
F. _____
G. _____
H. _____

CHECK POINT

Epithelial Tissue—Simple

1. List the functions of simple cuboidal epithelium.
2. Embryonic loose connective tissue provides an environment for the diffusion of what materials?
3. What is mesothelium?
4. Identify the site of ribosomal RNA synthesis and ribosome subunit assembly.
5. Define sperm capacitation.
6. Identify the secretions that inhibit microorganisms in the uterine tubes.
7. What are the two parts of the basement membrane of simple columnar epithelium?
8. What is the composition of each of the two parts of the basement membrane of simple columnar epithelium?
9. Identify the cells that produce each of the two parts of the basement membrane of simple columnar epithelium.
10. Identify the structures responsible for increasing cell surface area in simple columnar epithelium.
11. Individual cells may have up to _____ cilia.

IN REVIEW

What Have I Learned?

The following questions cover the material that you have just learned, simple epithelial tissues. Apply what you have learned to answer these questions on a separate piece of paper.

1. What body structures are formed of epithelia?
2. List the primary functions of epithelia.
3. Describe a simple epithelium.
4. Describe a stratified epithelium.
5. Describe simple cuboidal epithelium. Where is it located?
6. List the functions of cuboidal epithelial cells.
7. What is euchromatin?
8. What is heterochromatin?
9. List the functions of the nucleus of the squamous cell.
10. List the functions of simple squamous epithelium.
11. Describe a squamous cell. Where are squamous cells located?
12. What is the function of squamous cells?
13. What is a tangential plane of section?
14. Describe the basement membrane of simple columnar epithelium. Where is it located?
15. List the functions of the basement membrane of simple columnar epithelium.
16. Describe the cilia on simple columnar epithelium. Where are they located?
17. List the functions of the cilia on simple columnar epithelium.
18. What are oscillations of cilia? What substances are moved by these oscillations?
19. What is the function of the goblet cells?
20. What tissues contain goblet cells? Where are they located?
21. Describe the microvilli (brush border). Where are they located?
22. List the functions of the microvilli (brush border).
23. Describe nonciliated simple columnar epithelium. Where is it located?
24. List the functions of nonciliated simple columnar epithelium.
25. Describe pseudostratified columnar epithelium. Where is it located?
26. List the functions of pseudostratified columnar epithelium.

EXERCISE 3.2:
Epithelial Tissue—Stratified

Click the **HISTOLOGY** *icon. Select the* **my COURSE CONTENT** *tab if your instructor has selected specific exercises for you to complete. Otherwise select the* **ALL CONTENT** *tab.*

Histology: Epithelial Tissue—Stratified Squamous Epithelium (keratinized)

From the **Select topic** *menu, choose* **Stratified squamous epithelium (keratinized). LM: Medium magnification** *appears in the* **VIEW** *window. To the right in the* **IMAGE AREA** *you will see a medium magnification Light Microscope image of this tissue. Use the information in the* **STRUCTURE INFORMATION** *window to answer the following questions.*

1. Describe the dermis. Where is the dermis located?

2. List the functions of the dermis.

3. Describe the epidermis. Where is the epidermis located?

4. List the functions of the epidermis.

5. Compare the epidermal layers of thick skin and thin skin.

6. Describe the squamous cell (without nucleus). Where is this cell located?

7. What is the function of the squamous cell (without nucleus)?

8. Describe keratinized stratified squamous epithelium. Where is this tissue located?

9. What is the function of keratinized stratified squamous epithelium?

10. Describe the stratum basale. Where is this layer located?

11. What is the function of the stratum basale?

12. What other terms refer to the stratum basale?

13. Describe the stratum corneum. Where is this layer located?

14. What is the function of the stratum corneum?

15. What other term refers to the stratum corneum?

16. How are cells sloughed from the stratum corneum replaced?

17. Describe the stratum granulosum. Where is this layer located?

18. What is the function of the stratum granulosum?

19. What other term refers to the stratum granulosum?

20. Describe the stratum spinosum. Where is this layer located?

21. List the functions of the stratum spinosum.

22. What other term refers to the stratum spinosum?

Click the **TAGS ON/OFF** *radio button located below the* **STRUCTURE INFORMATION** *window to activate the green pins on the image. You will see the following image:*

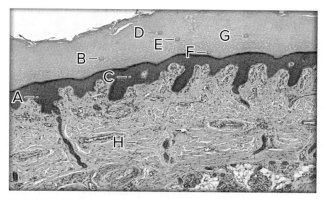

©McGraw-Hill Education/Al Telser

Mouse-over the pins on the screen to find the information to identify the following structures:

A. _____
B. _____
C. _____
D. _____
E. _____
F. _____
G. _____
H. _____

Histology: Epithelial Tissue— Stratified Squamous Epithelium (nonkeratinized)

*From the **Select topic** menu, choose **Stratified squamous epithelium (nonkeratinized)**. LM: Medium magnification appears in the **VIEW** window. To the right in the **IMAGE AREA** you will see a medium magnification Light Microscope image of this tissue. Use the information in the **STRUCTURE INFORMATION** window to answer the following questions.*

1. Describe the lamina propria of stratified squamous epithelium. Where is it located?

2. List the functions of the lamina propria of stratified squamous epithelium.

3. Describe the nucleus of the squamous cell. Where is it located?

4. List the functions of the nucleus of the squamous cell.

5. Describe the squamous cell (with nucleus). Where is it located?

6. What is the function of the squamous cell (with nucleus)?

7. Describe nonkeratinized stratified squamous epithelium. Where is this tissue located?

8. What is the function of nonkeratinized stratified squamous epithelium?

*Click the **TAGS ON/OFF** radio button located below the **STRUCTURE INFORMATION** window to activate the green pins on the image. You will see the following image:*

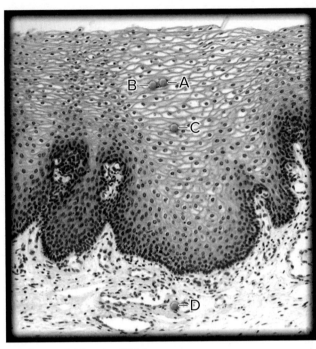

©McGraw-Hill Education/Al Telser

Mouse-over the pins on the screen to find the information to identify the following structures:

A. _____
B. _____
C. _____
D. _____

Histology: Epithelial Tissue— Stratified Cuboidal Epithelium

*From the **Select topic** menu, choose **Stratified cuboidal epithelium**. LM: Medium magnification appears in the **VIEW** window. To the right in the **IMAGE AREA** you will see a medium magnification Light Microscope image of this tissue. Use the information in the **STRUCTURE INFORMATION** window to answer the following questions.*

1. Describe a cuboidal epithelial cell. Where is it located?

2. What is the function of a cuboidal epithelial cell?

3. Describe the interlobular duct of parotid salivary gland. Where is it located?

4. What is the function of the interlobular duct of parotid salivary gland.

5. Describe the nucleus of a cuboidal epithelial cell. Where is it located?

6. List the functions of the nucleus of a cuboidal epithelial cell.

7. Describe stratified cuboidal epithelium. Where is it located?

8. What is the function of stratified cuboidal epithelium?

9. Describe a venule. Where is it located?

10. What is the function of a venule?

Click the **TAGS ON/OFF** *radio button located below the* **STRUCTURE INFORMATION** *window to activate the green pins on the image. You will see the following image:*

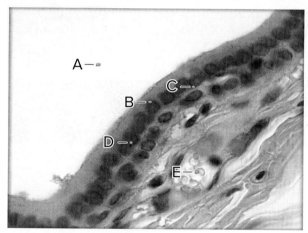

©McGraw-Hill Education/Steve Sullivan

Mouse-over the pins on the screen to find the information to identify the following structures:

A. _____
B. _____
C. _____
D. _____
E. _____

Histology: Epithelial Tissue—Stratified Columnar Epithelium

From the **Select topic** *menu, choose* **Stratified columnar epithelium. LM: Medium magnification** *appears in the* **VIEW** *window. To the right in the* **IMAGE AREA** *you will see a medium magnification Light Microscope image of this tissue. Use the information in the* **STRUCTURE INFORMATION** *window to answer the following questions.*

1. Describe the basement membrane of stratified columnar epithelium. Where is it located?

2. List the functions of the basement membrane of stratified columnar epithelium.

3. Describe a columnar epithelial cell. Where is it located?

4. What is the function of a columnar epithelial cell?

5. Describe the lamina propria of stratified columnar epithelium. Where is it located?

6. List the functions of the lamina propria of stratified columnar epithelium.

7. Describe the lumen of the spongy urethra. Where is it located?

8. What is the function of the lumen of the spongy urethra.

9. Describe stratified columnar epithelium. Where is it located?

10. What is the function of stratified columnar epithelium?

Click the **TAGS ON/OFF** *radio button located below the* **STRUCTURE INFORMATION** *window to activate the green pins on the image. You will see the following image:*

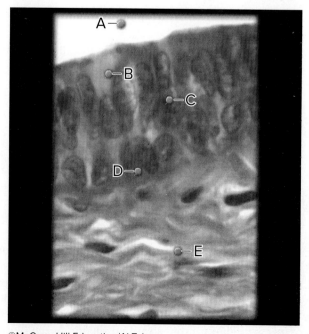

©McGraw-Hill Education/Al Telser

Mouse-over the pins on the screen to find the information to identify the following structures:

A. _____

B. _____

C. _____

D. _____

E. _____

Histology: Epithelial Tissue— Transitional Epithelium

From the **Select topic** *menu, choose* **Transitional epithelium. LM: Medium magnification** *appears in the* **VIEW** *window. To the right in the* **IMAGE AREA** *you will see a medium magnification Light Microscope image of this tissue. Use the information in the* **STRUCTURE INFORMATION** *window to answer the following questions.*

1. Describe a capillary. Where is it located?

2. What is the function of a capillary?

3. Describe dome-shaped luminal cells of transitional epithelium. Where is it located?

4. What is the function of dome-shaped luminal cells of transitional epithelium?

5. Describe loose connective tissue. Where is it located?

6. List the functions of loose connective tissue.

7. Describe the lumen of the allantoic duct. Where is it located?

8. What is the function of the lumen of the allantoic duct?

9. Describe transitional epithelium. Where is it located?

10. What is the function of transitional epithelium?

11. What other term refers to transitional epithelium?

Click the **TAGS ON/OFF** *radio button located below the* **STRUCTURE INFORMATION** *window to activate the green pins on the image. You will see the following image:*

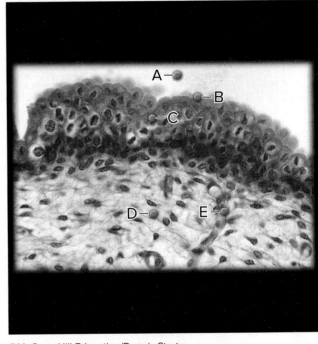

©McGraw-Hill Education/Dennis Strete

Mouse-over the pins on the screen to find the information to identify the following structures:

A. _____

B. _____

C. _____

D. _____

E. _____

CHECK POINT

Epithelial Tissue—Stratified

1. Identify the location of insertion for the muscles of facial expression.
2. Where are stem cells, melanocytes, Merkel cells, and dendritic cells located?
3. Where would one find keratinocytes as the major cell type?
4. Discuss the changes that occur as epidermal cells reach the apical surface in keratinized stratified squamous epithelium.
5. Cells in which layers are responsible for turnover of epidermal keratinocytes?
6. What is the function of melanocytes? Where are they located?
7. What is the major component of household dust?
8. Which epidermal layer is not a distinct layer in thin skin?
9. List the epidermal layers found in thick skin from deep to superficial.
10. List the epidermal layers found in thin skin from deep to superficial.
11. How is nonkeratinized stratified squamous epithelium kept moist?

IN REVIEW

What Have I Learned?

The following questions cover the material that you have just learned, stratified epithelial tissues. Apply what you have learned to answer these questions on a separate piece of paper.

1. Describe the dermis. Where is the dermis located?
2. List the functions of the dermis.
3. Describe the epidermis. Where is the epidermis located?
4. List the functions of the epidermis.
5. Compare the epidermal layers of thick skin and thin skin.
6. Describe the squamous cell (without nucleus). Where is this cell located?
7. What is the function of the squamous cell (without nucleus)?
8. Describe keratinized stratified squamous epithelium. Where is this tissue located?
9. What is the function of keratinized stratified squamous epithelium?
10. Describe nonkeratinized stratified squamous epithelium. Where is this tissue located?
11. What is the function of nonkeratinized stratified squamous epithelium?
12. Describe stratified cuboidal epithelium. Where is it located?
13. What is the function of stratified cuboidal epithelium?
14. Describe stratified columnar epithelium. Where is it located?
15. What is the function of stratified columnar epithelium?
16. Describe transitional epithelium. Where is it located?
17. What is the function of transitional epithelium?
18. List the epidermal layers found in thick skin from deep to superficial.
19. List the epidermal layers found in thin skin from deep to superficial.
20. How is nonkeratinized stratified squamous epithelium kept moist?

Connective Tissues

SELECT ANIMATION
Connective Tissue Overview
PLAY

After viewing the animation, answer the following questions:

1. List the primary functions of connective tissue.
2. List the three components shared by all connective tissues.
3. Which components form the extracellular matrix?
4. How are the various categories of connective tissue defined?
5. List the three broad categories of connective tissue in adults.
6. List the two basic subcategories of connective tissue proper.
7. List the characteristics of loose connective tissue.
8. List the characteristics of dense connective tissue.
9. List the subdivisions of loose connective tissue.
10. Where is areolar connective tissue found?
11. Describe areolar connective tissue.
12. Where is adipose tissue found?
13. Describe adipose tissue.
14. Where is reticular connective tissue found?
15. Define stroma.
16. Describe reticular connective tissue.
17. List the subdivisions of dense connective tissue.
18. Which structures are formed of dense regular connective tissue?

19. Describe dense regular connective tissue.

20. Where is dense irregular connective tissue found?

21. Describe dense irregular connective tissue.

22. Where is elastic tissue found?

23. Describe elastic tissue.

24. What protein forms the elastic fibers?

25. How do elastic fibers function?

26. List the two subdivisions of supporting connective tissue.

27. Compare the structure of cartilage to that of bone.

28. List the three types of cartilage.

29. Which is the most common type of cartilage? Where is it found?

30. What are chondrocytes? Where are they found?

31. What are lacunae?

32. What fibrils are present in the matrix of hyaline cartilage?

33. What is a perichondrium? Where is it located?

34. Where is elastic cartilage found?

35. Describe elastic cartilage.

36. How is elastic cartilage similar to hyaline cartilage?

37. Where is fibrocartilage found?

38. Describe fibrocartilage.

39. Does fibrocartilage have a perichondrium?

40. List the two forms of bone connective tissue.

41. What is another term for bone tissue?

42. Where is compact bone located?

43. Describe compact bone.

44. What are osteocytes?

45. Where does spongy bone form?

46. What is another term for spongy bone tissue?

47. What structures form a bony network in spongy bone?

48. Describe spongy bone.

49. Fluid connective tissue is represented by _____.

50. Describe the formed elements of blood.

51. Describe the ground substance of blood.

Self-Quiz
Take this opportunity to quiz yourself by taking the **QUIZ** over this animation.

EXERCISE 3.3:
Connective Tissue Proper—Loose

Click the **HISTOLOGY** *icon. Select the* **my COURSE CONTENT** *tab if your instructor has selected specific exercises for you to complete. Otherwise select the* **ALL CONTENT** *tab.*

Histology: Connective Tissue—Areolar Connective Tissue

From the **Select topic** *menu, choose* **Areolar connective tissue. LM: Medium magnification** *appears in the* **VIEW** *window. To the right in the* **IMAGE AREA** *you will see a medium magnification Light Microscope image of this tissue. Use the information in the* **STRUCTURE INFORMATION** *window to answer the following questions.*

1. Describe a capillary. Where is it located?

2. What is the function of a capillary?

3. Describe a collagen fiber in areolar connective tissue. Where is it located?

4. What is the function of a collagen fiber in areolar connective tissue?

5. Describe an elastic fiber in elastic cartilage. Where is it located?

6. What is the function of an elastic fiber in elastic cartilage?

7. Describe the ground substance of areolar connective tissue? Where is it located?

8. List the functions of the ground substance of areolar connective tissue.

9. Describe a mast cell and granules. Where are they located?

10. List the functions of a mast cell and granules.

11. Describe the nucleus of an endothelial cell. Where is it located?

12. List the functions of the nucleus of an endothelial cell.

13. Describe the nucleus of a fibroblast. Where is it located?

14. List the functions of a fibroblast.

Click the **TAGS ON/OFF** *radio button located below the* **STRUCTURE INFORMATION** *window to activate the green pins on the image. You will see the following image:*

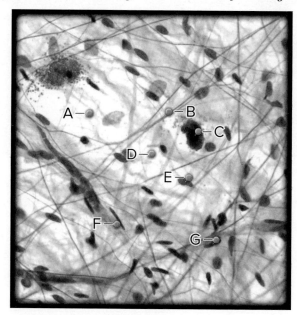

©McGraw-Hill Education/Al Telser

Mouse-over the pins on the screen to find the information to identify the following structures:

A. _____
B. _____
C. _____
D. _____
E. _____
F. _____
G. _____

Histology: Connective Tissue— Adipose Connective Tissue

From the **Select topic** *menu, choose* **Adipose connective tissue. LM: Medium magnification** *appears in the* **VIEW** *window. To the right in the* **IMAGE AREA** *you will see a medium magnification Light Microscope image of this tissue. Use the information in the* **STRUCTURE INFORMATION** *window to answer the following questions.*

1. Describe an adipocyte. Where is it located?

2. What is the function of an adipocyte?

3. What is another term for an adipocyte?

4. Describe a lipid inclusion in an adipocyte. Where is it located?

5. What is the function of a lipid inclusion in an adipocyte?

6. Describe the nucleus of an adipocyte. Where is it located?

7. What is the function of the nucleus of an adipocyte?

130 MODULE 3 Tissues

Click the **TAGS ON/OFF** *radio button located below the* **STRUCTURE INFORMATION** *window to activate the green pins on the image. You will see the following image:*

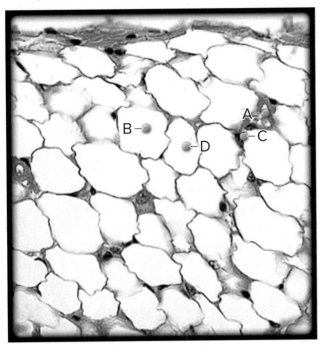

©McGraw-Hill Education/Dennis Strete

Mouse-over the pins on the screen to find the information to identify the following structures:

A. _____
B. _____
C. _____
D. _____

Histology: Connective Tissue—Reticular Connective Tissue

From the **Select topic** *menu, choose* **Reticular connective tissue. LM: Medium magnification** *appears in the* **VIEW** *window. To the right in the* **IMAGE AREA** *you will see a medium magnification Light Microscope image of this tissue. Use the information in the* **STRUCTURE INFORMATION** *window to answer the following questions.*

1. Describe the ground substance of reticular tissue. Where is it located?

2. List the functions of the ground substance of reticular tissue?

3. Describe a leukocyte. Where is it located?

4. What is the function of a leukocyte?

5. Describe a macrophage. Where is it located?

6. List the functions of a macrophage.

7. What is another term for a macrophage?

8. Describe a reticular fiber. Where is it located?

9. What is the function of a reticular fiber?

10. Where are reticular fibers produced?

Click the **TAGS ON/OFF** *radio button located below the* **STRUCTURE INFORMATION** *window to activate the green pins on the image. You will see the following image:*

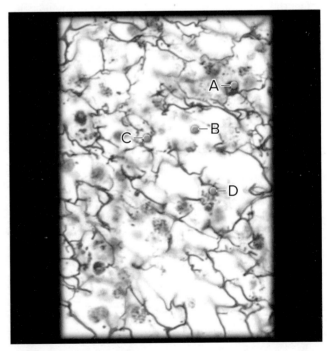

©McGraw-Hill Education/Al Telser

Mouse-over the pins on the screen to find the information to identify the following structures:

A. _____
B. _____
C. _____
D. _____

CHECK POINT

Connective Tissue—Loose

1. What is the diameter of a capillary? How does it compare to the diameter of a red blood cell?
2. In connective tissue, what structures form the extracellular matrix?
3. List the specific structures where elastic cartilage is present.
4. What is the function of heparin?
5. What is the function of histamine?
6. What is anaphylaxis?
7. How does the structure of white adipose tissue compare with the structure of brown adipose tissue?
8. What causes the classic "signet ring" appearance of adipocytes in histological preparations?
9. Describe the process of phagocytosis.
10. List the antigen presenting cells.

IN REVIEW

What Have I Learned?

The following questions cover the material that you have just learned, loose connective tissues. Apply what you have learned to answer these questions on a separate piece of paper.

1. List the primary functions of connective tissue.
2. List the three components shared by all connective tissues.
3. Which components form the extracellular matrix?
4. How are the various categories of connective tissue defined?
5. List the three broad categories of connective tissue in adults.
6. Describe a collagen fiber in areolar connective tissue. Where is it located?
7. What is the function of a collagen fiber in areolar connective tissue?
8. Describe an elastic fiber in elastic cartilage. Where is it located?
9. What is the function of an elastic fiber in elastic cartilage?
10. Describe the ground substance of areolar connective tissue? Where is it located?
11. List the functions of the ground substance of areolar connective tissue.
12. Describe an adipocyte. Where is it located?
13. What is the function of an adipocyte?
14. Describe a macrophage. Where is it located?
15. List the functions of a macrophage.

EXERCISE 3.4:

Connective Tissue Proper—Dense

Click the **HISTOLOGY** icon. Select the **my COURSE CONTENT** tab if your instructor has selected specific exercises for you to complete. Otherwise select the **ALL CONTENT** tab.

Histology: Connective Tissue—Dense Regular Connective Tissue

From the **Select topic** menu, choose **Dense regular connective tissue. LM: Medium magnification** appears in the VIEW window. To the right in the IMAGE AREA you will see a medium magnification Light Microscope image of this tissue. Use the information in the **STRUCTURE INFORMATION** window to answer the following questions.

1. Describe the collagen fibers in dense regular connective tissue. Where are they located?

2. What is the function of the collagen fibers in dense regular connective tissue?

3. Describe the ground substance of dense regular connective tissue. Where is it located?

4. List the functions of the ground substance of dense regular connective tissue.

5. Describe the nucleus of the fibroblast. Where is it located?

6. List the functions of the nucleus of the fibroblast.

Click the **TAGS ON/OFF** *radio button located below the* **STRUCTURE INFORMATION** *window to activate the green pins on the image. You will see the following image:*

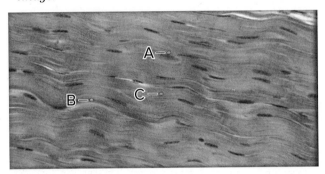

©Deagostini/Science Source

Mouse-over the pins on the screen to find the information to identify the following structures:

A. _____

B. _____

C. _____

Histology: Connective Tissue—Dense Irregular Connective Tissue

From the **Select topic** *menu, choose* **Dense irregular connective tissue. LM: Medium magnification** *appears in the* **VIEW** *window. To the right in the* **IMAGE AREA** *you will see a medium magnification Light Microscope image of this tissue. Use the information in the* **STRUCTURE INFORMATION** *window to answer the following questions.*

1. Describe the collagen fibers in dense irregular connective tissue. Where are they located?

2. What is the function of the collagen fibers in dense irregular connective tissue?

3. Describe the ground substance of dense irregular connective tissue. Where is it located?

4. List the functions of the ground substance of dense irregular connective tissue.

5. Describe the nucleus of the fibroblast. Where is it located?

6. List the functions of the nucleus of the fibroblast.

Click the **TAGS ON/OFF** *radio button located below the* **STRUCTURE INFORMATION** *window to activate the green pins on the image. You will see the following image:*

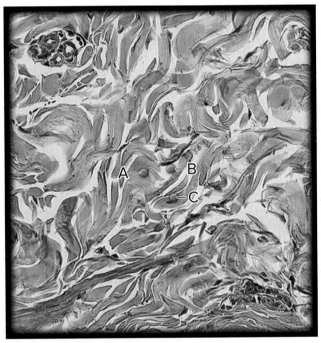

©McGraw-Hill Education/Dennis Strete

Mouse-over the pins on the screen to find the information to identify the following structures:

A. _____

B. _____

C. _____

Histology: Connective Tissue—Elastic Connective Tissue

From the **Select topic** *menu, choose* **Elastic connective tissue. LM: Medium magnification** *appears in the* **VIEW** *window. To the right in the* **IMAGE AREA** *you will see a medium magnification Light Microscope image of this tissue. Use the information in the* **STRUCTURE INFORMATION** *window to answer the following questions.*

1. Describe the elastic lamellae. Where are they located?

2. What is the function of the elastic lamellae?

3. Describe the endothelium. Where is it located?

4. List the functions of the endothelium.

5. Describe the internal elastic lamellae. Where is it located?

6. Describe the lumen of elastic artery. Where is it located?

7. What is the function of the lumen of elastic artery?

8. Describe the nucleus of the fibroblast. Where is it located?

9. List the functions of the nucleus of the fibroblast.

10. Describe the nucleus of the smooth muscle cell. Where is it located?

11. List the functions of the nucleus of the smooth muscle cell.

12. Describe the tunica intima of the elastic artery. Where is it located?

13. List the functions of the tunica intima of the elastic artery.

14. What is another term for the tunica intima of the elastic artery?

15. Describe the tunica media of the elastic artery. Where is it located?

16. What is the function of the tunica media of the elastic artery?

Click the **TAGS ON/OFF** *radio button located below the* **STRUCTURE INFORMATION** *window to activate the green pins on the image. You will see the following image:*

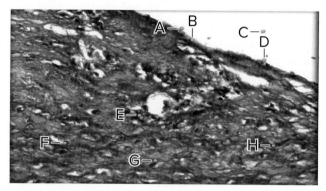

©McGraw-Hill Education/Steve Sullivan

Mouse-over the pins on the screen to find the information to identify the following structures:

A. _____

B. _____

C. _____

D. _____

E. _____

F. _____

G. _____

H. _____

CHECK POINT

Connective Tissue—Dense

1. How many elastic lamellae are contained in the human aorta?
2. Describe the thickness of the tunica intima of elastic arteries.
3. Describe the thickness of the tunica intima in the aorta.
4. What is the vasa vasorum? What is its function?
5. Which is the thickest layer of arterial walls? Which is the thickest layer of venous walls?

IN REVIEW

What Have I Learned?

The following questions cover the material that you have just learned, dense connective tissues. Apply what you have learned to answer these questions on a separate piece of paper.

1. Describe the collagen fibers in dense regular connective tissue. Where are they located?
2. What is the function of the collagen fibers in dense regular connective tissue?
3. Describe the ground substance of dense regular connective tissue. Where is it located?
4. List the functions of the ground substance of dense regular connective tissue.
5. Describe the collagen fibers in dense irregular connective tissue. Where are they located?
6. What is the function of the collagen fibers in dense irregular connective tissue?
7. Describe the ground substance of dense irregular connective tissue. Where is it located?
8. List the functions of the ground substance of dense irregular connective tissue.
9. Describe the elastic lamellae. Where are they located?
10. What is the function of the elastic lamellae?
11. Describe the endothelium. Where is it located?
12. List the functions of the endothelium.

EXERCISE 3.5:
Supporting Connective Tissue—Cartilage

Click the **HISTOLOGY** *icon. Select the my* **COURSE CONTENT** *tab if your instructor has selected specific exercises for you to complete. Otherwise select the* **ALL CONTENT** *tab.*

Histology: Connective Tissue—Hyaline Cartilage

From the **Select topic** *menu, choose* **Hyaline cartilage. LM: Medium magnification** *appears in the* **VIEW** *window. To the right in the* **IMAGE AREA** *you will see a medium magnification Light Microscope image of this tissue. Use the information in the* **STRUCTURE INFORMATION** *window to answer the following questions.*

1. Describe the chondrocytes (in lacunae) in hyaline cartilage. Where are they located?
2. What is the function of chondrocytes (in lacunae) in hyaline cartilage?
3. Describe the extracellular matrix of hyaline cartilage. Where are they located?
4. What is the function of the extracellular matrix of hyaline cartilage?
5. What are other terms for the extracellular matrix?

6. Describe the lacunae (with chondrocytes) in hyaline cartilage. Where are they located?

7. Describe the nucleus of the chondrocyte. Where is it located?

8. List the functions of the nucleus of the chondrocyte.

9. Describe the perichondrium of hyaline cartilage. Where is it located?

10. List the functions of the perichondrium of hyaline cartilage.

Click the **TAGS ON/OFF** *radio button located below the* **STRUCTURE INFORMATION** *window to activate the green pins on the image. You will see the following image:*

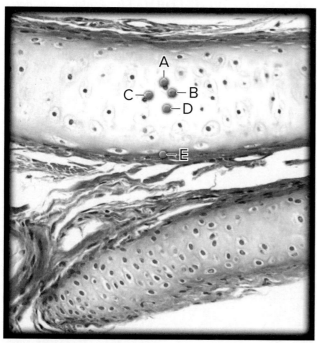

©McGraw-Hill Education/Al Telser

Mouse-over the pins on the screen to find the information to identify the following structures:

A. _____

B. _____

C. _____

D. _____

E. _____

Histology: Connective Tissue—Fibrocartilage

From the **Select topic** *menu, choose* **Fibrocartilage**. **LM: Medium magnification** *appears in the* **VIEW** *window. To the right in the* **IMAGE AREA** *you will see a medium magnification Light Microscope image of this tissue. Use the information in the* **STRUCTURE INFORMATION** *window to answer the following questions.*

1. Describe the chondrocytes (in lacunae) in fibrocartilage. Where are they located?

2. What is the function of chondrocytes (in lacunae) in fibrocartilage?

3. Describe the collagen fibers in fibrocartilage. Where are they located?

4. What is the function of the collagen fibers in fibrocartilage?

5. Describe the ground substance in fibrocartilage. Where is it located?

6. List the functions of the ground substance in fibrocartilage.

7. Describe the lacunae (with chondrocytes) in fibrocartilage. Where are they located?

8. Describe the nucleus of the chondrocyte. Where is it located?

9. List the functions of the nucleus of the chondrocyte.

136 MODULE 3   Tissues

Click the **TAGS ON/OFF** *radio button located below the* **STRUCTURE INFORMATION** *window to activate the green pins on the image. You will see the following image:*

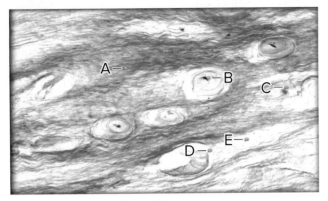

©McGraw-Hill Education/Steve Sullivan

Mouse-over the pins on the screen to find the information to identify the following structures:

A. _____

B. _____

C. _____

D. _____

E. _____

Histology: Connective Tissue—Elastic Cartilage

From the **Select topic** *menu, choose* **Elastic cartilage. LM: Medium magnification** *appears in the* **VIEW** *window. To the right in the* **IMAGE AREA** *you will see a medium magnification Light Microscope image of this tissue. Use the information in the* **STRUCTURE INFORMATION** *window to answer the following questions.*

1. Describe the chondrocytes (in lacunae) in elastic cartilage. Where are they located?

2. What is the function of chondrocytes (in lacunae) in elastic cartilage?

3. Describe the elastic fibers in elastic cartilage. Where are they located?

4. What is the function of the elastic fibers in elastic cartilage?

5. Describe the ground substance of elastic cartilage. Where is it located?

6. List the functions of the ground substance of elastic cartilage.

7. Describe the lacunae (with chondrocytes) in elastic cartilage. Where are they located?

8. Describe the nucleus of the chondrocyte. Where is it located?

9. List the functions of the nucleus of the chondrocyte.

10. Describe the perichondrium of elastic cartilage. Where is it located?

11. List the functions of the perichondrium of elastic cartilage.

Click the **TAGS ON/OFF** *radio button located below the* **STRUCTURE INFORMATION** *window to activate the green pins on the image. You will see the following image:*

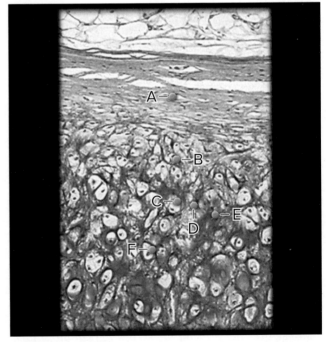

©McGraw-Hill Education/Dennis Strete

Mouse-over the pins on the screen to find the information to identify the following structures:

A. _____

B. _____

C. _____

D. _____

E. _____

F. _____

CHECK POINT

Connective Tissue—Cartilage

1. What is an isogenous group of chondrocytes?
2. What is a lacuna? What is the plural for lacuna?
3. Which structures or organs contain hyaline cartilage?
4. How does the extracellular matrix of bone compare to that of cartilage?
5. What is the function of fibroblasts?
6. Which cartilage types have a perichondrium and which do not?
7. When does synovial fluid nourish hyaline cartilage?
8. Which structures contain fibrocartilage?
9. Which structures or organs contain elastic cartilage?

IN REVIEW

What Have I Learned?

The following questions cover the material that you have just learned, cartilage tissues. Apply what you have learned to answer these questions on a separate piece of paper.

1. Describe the chondrocytes (in lacunae) in hyaline cartilage. Where are they located?
2. What is the function of chondrocytes (in lacunae) in hyaline cartilage?
3. Describe the extracellular matrix of hyaline cartilage. Where are they located?
4. What is the function of the extracellular matrix of hyaline cartilage?
5. Describe the lacunae (with chondrocytes) in hyaline cartilage. Where are they located?
6. Describe the perichondrium of hyaline cartilage. Where is it located?
7. List the functions of the perichondrium of hyaline cartilage.
8. Describe the elastic fibers in elastic cartilage. Where are they located?
9. What is the function of the elastic fibers in elastic cartilage?

EXERCISE 3.6:

Supporting Connective Tissue—Bone

Click the **HISTOLOGY** *icon. Select the* **my COURSE CONTENT** *tab if your instructor has selected specific exercises for you to complete. Otherwise select the* **ALL CONTENT** *tab.*

Histology: Connective Tissue—Compact Bone

From the **Select topic** *menu, choose* **Compact bone**. **LM: Medium magnification** *appears in the* **VIEW** *window. To the right in the* **IMAGE AREA** *you will see a medium magnification Light Microscope image of this tissue. Use the information in the* **STRUCTURE INFORMATION** *window to answer the following questions.*

1. Describe the cement line of the osteon. Where is it located?
2. What is another term for the cement line of the osteon?
3. Describe the central canal of the osteon. Where is it located?
4. What is another term for the central canal of the osteon?
5. Describe the interstitial lamella. Where is it located?
6. Describe the osteocytes in lacunae. Where are they located?

7. List the functions of the osteocytes in lacunae.

8. Describe the osteon. Where is it located?

9. List the functions of the osteon.

Click the **TAGS ON/OFF** *radio button located below the* **STRUCTURE INFORMATION** *window to activate the green pins on the image. You will see the following image:*

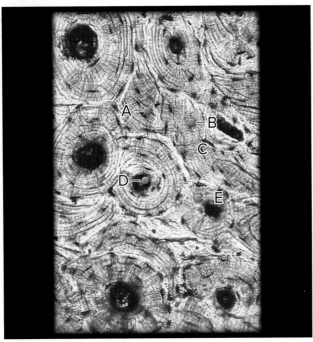

©McGraw-Hill Education/Dennis Strete

Mouse-over the pins on the screen to find the information to identify the following structures:

A. _____

B. _____

C. _____

D. _____

E. _____

Histology: Connective Tissue—Spongy Bone

From the **Select topic** *menu, choose* **Spongy bone.** **LM: Medium magnification** *appears in the* **VIEW** *window. To the right in the* **IMAGE AREA** *you will see a medium magnification Light Microscope image of this tissue. Use the information in the* **STRUCTURE INFORMATION** *window to answer the following questions.*

1. Describe the marrow cavity in spongy bone. Where is it located?

2. List the functions of the marrow cavity in spongy bone.

3. Describe the trabeculae of spongy bone. Where are they located?

4. List the functions of the trabeculae of spongy bone.

Click the **TAGS ON/OFF** *radio button located below the* **STRUCTURE INFORMATION** *window to activate the green pins on the image. You will see the following image:*

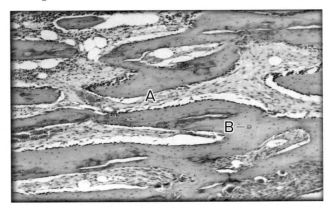

©McGraw-Hill Education/Steve Sullivan

Mouse-over the pins on the screen to find the information to identify the following structures:

A. _____

B. _____

CHECK POINT

Connective Tissue Bone

1. What structure marks the limit of bone erosion?
2. What is another term for the osteon?
3. What are other terms for compact bone?
4. What is another term for red marrow?
5. What is another term for yellow marrow?
6. What are other terms for spongy bone?
7. At birth, all marrow is _____. With age, this marrow is converted to _____ _____.
8. Define hematopoiesis.
9. What cells dominate red marrow?
10. What cells dominate yellow marrow?
11. What is another term for the shaft of long bones?
12. What is another term for the ends of long bones?

IN REVIEW

What Have I Learned?

The following questions cover the material that you have just learned, bone tissues. Apply what you have learned to answer these questions on a separate piece of paper.

1. Describe the cement line of the osteon. Where is it located?

2. Describe the central canal of the osteon. Where is it located?

3. Describe the interstitial lamella. Where is it located?

4. Describe the osteocytes in lacunae. Where are they located?

5. List the functions of the osteocytes in lacunae.

6. Describe the osteon. Where is it located?

7. List the functions of the osteon.

8. Describe the marrow cavity in spongy bone. Where is it located?

9. List the functions of the marrow cavity in spongy bone.

10. At birth, all marrow is _____. With age, this marrow is converted to _____ _____.

11. Define hematopoiesis.

12. What cells dominate red marrow?

13. What cells dominate yellow marrow?

14. What is another term for the shaft of long bones?

15. What is another term for the ends of long bones?

EXERCISE 3.7:

Fluid Connective Tissue—Blood

Click the **HISTOLOGY** *icon. Select the* **my COURSE CONTENT** *tab if your instructor has selected specific exercises for you to complete. Otherwise select the* **ALL CONTENT** *tab.*

Histology: Connective Tissue—Blood

From the **Select topic** *menu, choose* **Blood. LM: Medium magnification** *appears in the* **VIEW** *window. To the right in the* **IMAGE AREA** *you will see a medium magnification Light Microscope image of this tissue. Use the information in the* **STRUCTURE INFORMATION** *window to answer the following questions.*

1. Describe erythrocytes. Where are they located?

2. What is the function of erythrocytes?

3. What is another term for erythrocytes?

4. Describe the ground substance of fluid connective tissue.

5. What is the function of the ground substance of fluid connective tissue?

6. Describe leukocytes. Where are they located?

7. What is the function of leukocytes?

8. What is another term for leukocytes?

9. Describe platelets. Where are they located?

10. What is the function of platelets?

11. What is another term for platelets?

Click the **TAGS ON/OFF** *radio button located below the* **STRUCTURE INFORMATION** *window to activate the green pins on the image. You will see the following image:*

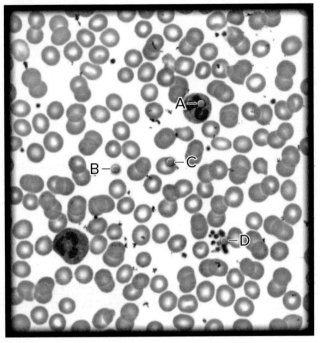

©McGraw-Hill Education/Al Telser

Mouse-over the pins on the screen to find the information to identify the following structures:

A. _____

B. _____

C. _____

D. _____

CHECK POINT
Connective Tissue—Blood

1. What is another term for the extracellular matrix of blood?
2. What percentage of blood consists of the extracellular matrix?
3. What percentage of blood consists of the formed elements?
4. What are the three most abundant plasma proteins? What are their percentages?
5. List the formed elements and their numbers per cubic millimeter.
6. List the various leukocytes and their relative abundance in blood.
7. What is thrombocytopenia?

IN REVIEW
What Have I Learned?

The following questions cover the material that you have just learned, blood tissue. Apply what you have learned to answer these questions on a separate piece of paper.

1. Describe erythrocytes. Where are they located?
2. What is the function of erythrocytes?
3. Describe leukocytes. Where are they located?
4. What is the function of leukocytes?
5. Describe platelets. Where are they located?
6. What is the function of platelets?

Muscle Tissue

Select the **MODULE** *drop-down menu in the* **Home Screen** *of APR. Choose* **3. Tissues.** *Click the* **ANIMATIONS** *icon. Select the* **my COURSE CONTENT** *tab if your instructor has selected specific exercises for you to complete. Otherwise select the* **ALL CONTENT** *tab.*

 SELECT ANIMATION Muscle Tissue Overview **PLAY**

After viewing the animation, answer the following questions:

1. What is the primary function of muscle tissue?

2. What are the results of this function?

3. What are muscle fibers?

4. List the five properties of muscle tissue.

5. List the three types of muscle tissue.

6. Describe skeletal muscle tissue.

7. Describe the skeletal muscle fibers.

8. Identify the connective tissue that surrounds the muscle fibers.

9. What are fascicles? Identify the connective tissue surrounding the fascicles.

10. What structures are contained in this connective tissue?

11. Numerous fascicles are collected together and form a _____, which is covered by an external sheath known as the _____.

12. What is the name for the cytoplasm of a skeletal muscle fiber?

13. Describe the myofibrils and myofilaments located in the cytoplasm.

14. Describe the two types of myofilaments, including their composition.

15. How are the myofilaments arranged? How are they responsible for striations?

16. Describe the A bands.

17. Describe the I bands.

18. Describe the Z discs.

19. What is a sarcomere? How is it defined?

20. How many sarcomeres are located in each myofibril?

21. Where is cardiac muscle found?

22. Is cardiac muscle under voluntary or involuntary control?

23. Describe cardiac muscle. How is it similar to skeletal muscle?

24. What are intercalated discs? How do they function?

25. How are cardiac muscle cells able to contract as a unit?

26. How are cardiac muscle cells prevented from being pulled apart during contraction?

27. Where is smooth muscle located?

28. Describe smooth muscle cells.

29. Describe dense bodies. What is their function?

30. How are smooth muscle cells typically arranged?

31. How are smooth muscle cells similar to cardiac muscle cells?

Self-Quiz

Take this opportunity to quiz yourself by taking the **QUIZ** over this animation.

EXERCISE 3.8:
Muscle Tissue

Click the **HISTOLOGY** *icon. Select the* **my COURSE CONTENT** *tab if your instructor has selected specific exercises for you to complete. Otherwise select the* **ALL CONTENT** *tab.*

Histology: Muscle Tissue—Skeletal Muscle

From the **Select topic** *menu, choose* **Skeletal muscle. LM: Medium magnification** *appears in the* **VIEW** *window. To the right in the* **IMAGE AREA** *you will see a medium magnification Light Microscope image of this tissue. Use the information in the* **STRUCTURE INFORMATION** *window to answer the following questions.*

1. Describe the A band. Where is it located?

2. What is another term for the A band?

3. Describe the I band. Where is it located?

4. What is another term for the I band?

5. Describe the nucleus of skeletal muscle fiber. Where is it located?

6. What is the function of the nucleus of skeletal muscle fiber?

Click the **TAGS ON/OFF** *radio button located below the* **STRUCTURE INFORMATION** *window to activate the green pins on the image. You will see the following image:*

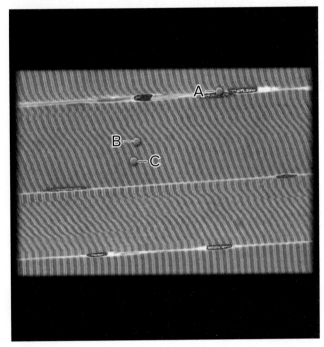

©Victor Eroschenko

Mouse-over the pins on the screen to find the information to identify the following structures:

A. _____
B. _____
C. _____

Histology: Muscle Tissue—Cardiac Muscle

From the **Select topic** *menu, choose* **Cardiac muscle. LM: Medium magnification** *appears in the* **VIEW** *window. To the right in the* **IMAGE AREA** *you will see a medium magnification Light Microscope image of this tissue. Use the information in the* **STRUCTURE INFORMATION** *window to answer the following questions.*

1. Describe the branched cardiac muscle fiber. Where is it located?

2. List the functions of the branched cardiac muscle fiber.

3. Describe the intercalated discs. Where are they located?

4. List the functions of the intercalated discs.

5. Describe the nucleus of cardiac muscle fiber. Where is it located?

6. What is the function of the nucleus of cardiac muscle fiber?

Click the **TAGS ON/OFF** *radio button located below the* **STRUCTURE INFORMATION** *window to activate the green pins on the image. You will see the following image:*

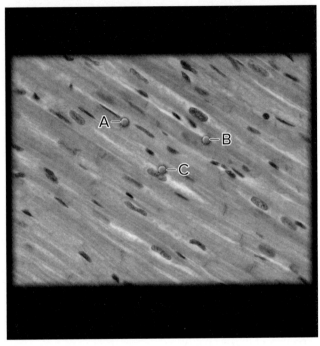

©McGraw-Hill Education/Al Telser

Mouse-over the pins on the screen to find the information to identify the following structures:

A. _____
B. _____
C. _____

Histology: Muscle Tissue—Smooth Muscle

From the **Select topic** *menu, choose* **Smooth muscle.** *Select* **LM: Low magnification** *from the* **VIEW** *menu. To the right in the* **IMAGE AREA** *you will see a low-magnification Light Microscope image of this tissue. Use the information in the* **STRUCTURE INFORMATION** *window to answer the following questions.*

1. Describe the longitudinal smooth muscle fiber. Where is it located?

2. What is another term for smooth muscle?

3. Describe smooth muscle fiber. Where is it located?

Click the **TAGS ON/OFF** *radio button located below the* **STRUCTURE INFORMATION** *window to activate the green pins on the image. You will see the following image:*

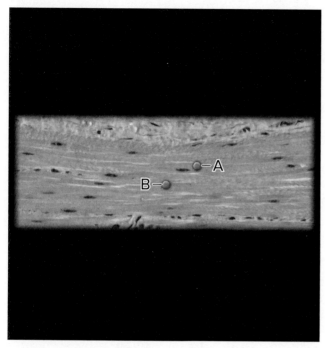

©McGraw-Hill Education/Al Telser

Mouse-over the pins on the screen to find the information to identify the following structures:

A. _____

B. _____

Return to the **VIEW** *menu and select* **LM: High magnification.** *To the right in the* **IMAGE AREA** *you will see a high-magnification Light Microscope image of this tissue. Use the information in the* **STRUCTURE INFORMATION** *window to answer the following questions.*

1. Describe the nucleus of smooth muscle fiber. Where is it located?

2. What is the function of the nucleus of smooth muscle fiber?

3. Describe the smooth muscle fiber. Where is it located?

Click the **TAGS ON/OFF** *radio button located below the* **STRUCTURE INFORMATION** *window to activate the green pins on the image. You will see the following image:*

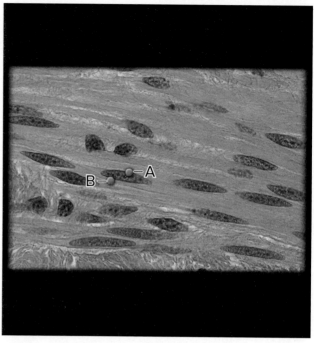

©McGraw-Hill Education/Dennis Strete

Mouse-over the pins on the screen to find the information to identify the following structures:

A. _____

B. _____

CHECK POINT

Muscle Tissue

1. How is the "A" band designation derived?
2. How is the "I" band designation derived?
3. What are other terms for skeletal muscle fiber?
4. What are other terms for cardiac muscle fiber?
5. What is another term for adhering junctions?
6. How does the location of the nucleus in cardiac muscle tissue compare to the location in skeletal muscle tissue?

IN REVIEW

What Have I Learned?

The following questions cover the material that you have just learned, muscle tissues. Apply what you have learned to answer these questions on a separate piece of paper.

1. What is the primary function of muscle tissue?
2. What are the results of this function?
3. List the five properties of muscle tissue.
4. List the three types of muscle tissue.
5. Describe the A bands.
6. Describe the I bands.
7. Describe the Z discs.
8. What is a sarcomere? How is it defined?
9. Describe the branched cardiac muscle fiber. Where is it located?
10. List the functions of the branched cardiac muscle fiber.
11. Describe the intercalated discs. Where are they located?
12. List the functions of the intercalated discs.
13. Describe the smooth muscle fiber. Where is it located?

Nervous Tissue

Select the **Module** *drop-down menu in the* **Home Screen** *of APR. Choose* **3. Tissues.** *Click the* **ANIMATIONS** *icon. Select the* **my COURSE CONTENT** *tab if your instructor has selected specific exercises for you to complete. Otherwise select the* **ALL CONTENT** *tab.*

SELECT ANIMATION
Nervous Tissue Overview   **PLAY**

After viewing the animation, answer the following questions:

1. What is the function of the nervous system?
2. What are the two main types of cells found in nervous tissue?
3. Describe a neuron.
4. What are the functional properties of neurons?
5. Describe the basic features of neurons.
6. Describe the two types of processes that project from the cell body. What are their functions?
7. How do most axons terminate? Describe the structures at their terminus.
8. Describe dendrites.
9. What is a synapse?
10. What are glial cells? What percentage of the nervous system is made up of glial cells?
11. List the four types of glial cells found in the central nervous system.
12. List the two additional types of glial cells found in the peripheral nervous system.
13. Which of the glial cells is most abundant? Describe these glial cells. What is their function?
14. How do glial cells form the blood-brain barrier? What is its function?
15. What other important functions do astrocytes have?
16. What property of astrocytes is unlike neurons, and allows them to form a glial scar?
17. Describe ependymal cells. What is their function?
18. Describe microglia. What is their function?

19. What response do microglia demonstrate when activated?

20. Describe oligodendrocytes. What is their function?

21. What is the myelin sheath? How many axons can an oligodendrocyte myelinate?

22. What is the function of Schwann cells? How do they differ from oligodendrocytes?

23. Describe satellite cells. What is their function?

Self-Quiz
Take this opportunity to quiz yourself by taking the **QUIZ** over this animation.

EXERCISE 3.9:
Nervous Tissue

Click the **HISTOLOGY** icon. Select the **my COURSE CONTENT** tab if your instructor has selected specific exercises for you to complete. Otherwise select the **ALL CONTENT** tab.

Histology: Nervous Tissue

From the **Select topic** menu, choose **Nervous tissue. LM: High magnification** appears in the **VIEW** window. To the right in the **IMAGE AREA** you will see a high-magnification Light Microscope image of this tissue. Use the information in the **STRUCTURE INFORMATION** window to answer the following questions.

1. Describe an axon. Where is it located?

2. What is the function of the axon?

3. What is another term for the axon?

4. Describe a dendrite. Where is it located?

5. What is the function of the dendrite?

6. Describe a neuron. Where are they located?

7. What is the function of a neuron?

8. Describe Nissl substance. Where is it located?

9. What is the function of Nissl substance?

10. Describe nucleolus in the nucleus of the neuron. Where is it located?

11. What is the function of the nucleus of the neuron?

12. Describe the nucleus of the glial cell. Where is it located?

13. List the functions of the nucleus of the glial cell.

14. Describe the nucleus of the neuron. Where is it located?

15. List the functions of the nucleus of the neuron.

Click the **TAGS ON/OFF** radio button located below the **STRUCTURE INFORMATION** window to activate the green pins on the image. You will see the following image:

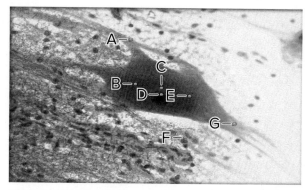

©McGraw-Hill Education/Steve Sullivan

Mouse-over the pins on the screen to find the information to identify the following structures:

A. _____

B. _____

C. _____

D. _____

E. _____

F. _____

G. _____

CHECK POINT

Nervous Tissue

1. Where does the axon arise on a neuron?
2. What is another term for the soma?
3. How does an axon terminate?
4. Which neurons are usually myelinated?
5. What is the neuron's response to peripheral axon damage?
6. What is another term for Nissl substance?
7. What is another term for glial cells?
8. List the various glial cells.

IN REVIEW

What Have I Learned?

The following questions cover the material that you have just learned, nervous tissue. Apply what you have learned to answer these questions on a separate piece of paper.

1. What is the function of the nervous system?
2. What are the two main types of cells found in the nervous system?
3. Describe a neuron.
4. What are the functional properties of neurons?
5. Describe the basic features of neurons.
6. Describe an axon. Where is it located?
7. What is the function of the axon?
8. What is another term for the axon?
9. Describe a dendrite. Where is it located?
10. What is the function of the dendrite?

Self-Quiz
Take this opportunity to quiz yourself by taking the **QUIZ** over this module.

MODULE 4
The Integumentary System

Overview: The Integumentary System

Your skin (or integument) is your body's largest organ, weighing in at around 15 percent of your body weight. The **Integumentary System** consists not only of your skin, but also your hair, nails, sweat, and sebaceous glands—the so-called accessory organs. In our exploration of the **Integumentary System**, we will focus our attention on the nails and the layers of the skin in the dissection views and the detailed structure of the skin and its accessory organs in the histology views.

From the **HOME** screen, click the drop-down box on the **MODULE** menu.

From the list, click **Integumentary**.

Select the **ANIMATIONS** icon. Select the **my COURSE CONTENT** tab if your instructor has selected specific exercises for you to complete. Otherwise select the **ALL CONTENT** tab.

SELECT ANIMATION
Homeostasis: Thermoregulation (3D) — PLAY

After viewing the animation, answer the following questions.

1. All structures in the body function together to maintain _____. **homeostasis**

2. What is homeostasis? **process by which the body maintains to its internal environ in response to extern env**

3. How does the liver function to maintain homeostasis? **metabolism of drugs & toxins in liver**

4. How do the kidneys function to maintain homeostasis? **regulation of water and solutes in blood**

5. How does the pancreas function to maintain homeostasis? **reg of blood glucose**

6. What is thermoregulation? **maintenance of normal body temp**

7. If the body's skin or core temperature drops, **thermal receptors** in the skin or internal environment send impulses to the _____, which acts as the body's thermostat. **hypothalamus**

8. The hypothalamus responds through the **SNS** _____ _____ by **constricting** blood vessels in the skin.

9. **vasoconstriction** diverts blood away from the skin and extremities to the warmer interior of the body to prevent further **loss** of **heat** to the surroundings and prevent the body's core temperature from dropping further.

10. **arrector pili** muscles contract, causing **piloerections** in which **hair follicles** stand up in an attempt to trap warm air next to the skin.

11. What causes shivering? **continued drop in temp prompts hypothalmus to send impulses or that elicit a shivering reflex**

12. What is the function of shivering? **generating additional heat to increase body temp**

13. If the body's skin or core temperature increases, **thermoreceptors** in the skin or internal organs prompt the **hypothalmus** to halt **sympathetic stimulation** of **blood vessels** in the skin.

14. The vessels **dilate** and allow **warm blood** to distribute **heat** through the skin.

147

15. What affect does this have on the arrector pili muscles and hair follicles? erector pili muscles relax and hair follicles lie flat against skin

16. _Sweat_ glands produce _sweat_, allowing _heat_ loss through _evaporation_

17. The _stimulus_ or _decrease_ in body temperature causes the _brain_ to act as a _thermostat_ and _dissipate_ heat throughout the body.

18. What happens to the thermostat once normal temperature is reached? thermostat shuts off

19. What are negative feedback loops? allows body to maintain homeostasis

20. How do negative feedback loops allow the body to maintain homeostasis? examples of body's response to contract stimuli

This Instructional Text Box is directing you to:

Click the **DISSECTION Study Area** *icon.*

Click the **ALL CONTENT** *tab.*

From the **Select Topic** *menu, select* **Thin skin and subcutaneous tissues.**

Layers *will appear in the* **Select View** *menu.*

In the **Select structure type** *menu, select* **Skin and subcutaneous tissue.**

Click the **LAYER** *radio buttons as directed.*

EXERCISE 4.1

Integumentary System—Thin Skin and Subcutaneous Tissues

SELECT TOPIC	SELECT VIEW
Thin Skin and Subcutaneous Tissue	Layers

Click **LAYER 1** *in the* **LAYER CONTROLS** *window, and you will see the following image:*

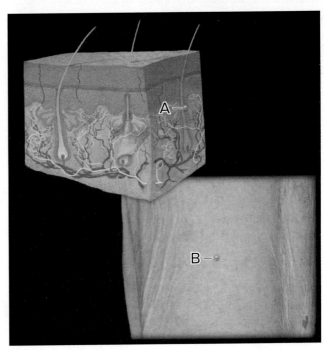

©McGraw-Hill Education

Mouse-over the pins on the screen to find the information necessary to fill in the following blanks:

A. _hair_
B. _epidermis_

CHECK POINT

Thin Skin and Subcutaneous Tissues

1. Name the two layers of the epidermis consisting of cells without nuclei.
2. Name the major cell type of the epidermis.
3. Name the accessory organ of the skin that consists of a filament of keratinized cells.

Click **LAYER 2** *in the* **LAYER CONTROLS** *window, and you will see the following image:*

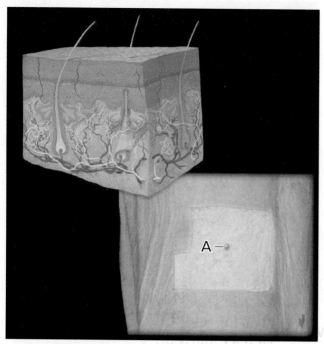

©McGraw-Hill Education

Mouse-over the pin on the screen to find the information necessary to fill in the following blank:

A. _____ dermis _____

CHECK POINT

Thin Skin and Subcutaneous Tissues, *continued*

4. Name the skin layer that lies deep to the epidermis.
5. What tissue types constitute this layer?
6. This layer consists of two structural layers. What are they?

Click **LAYER 3** *in the* **LAYER CONTROLS** *window, and you will see the following image:*

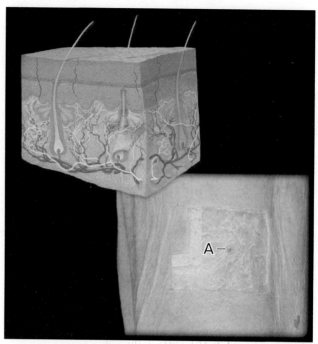

©McGraw-Hill Education

Mouse-over the pin on the screen to find the information necessary to fill in the following blank:

A. _____ hypodermis _____

CHECK POINT

Thin Skin and Subcutaneous Tissues, *continued*

7. Name the integumentary layer located deep to the skin.
8. What tissue types are found in this layer?

Click **LAYER 4** *in the* **LAYER CONTROLS** *window, and you will see the following image:*

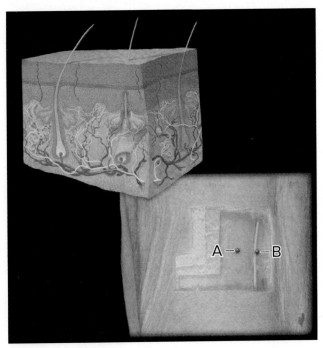

©McGraw-Hill Education

Click **LAYER 5** *in the* **LAYER CONTROLS** *window, and you will see the following image:*

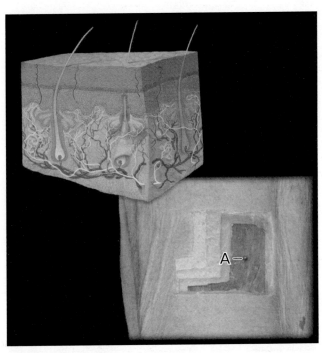

©McGraw-Hill Education

Mouse-over the pins on the screen to find the information necessary to fill in the following blanks:

Nonintegumentary System Structures (blue pins)

A. _fascia lata_
B. _cutaneous nerve in thigh_

Mouse-over the pin on the screen to find the information necessary to fill in the following blank:

Nonintegumentary System Structure (blue pin)

A. _skeletal muscle_

CHECK POINT

Thin Skin and Subcutaneous Tissues, *continued*

9. What structures are located in the integumentary layer located deep to the skin?
10. Name the deep fascia of the thigh.

IN REVIEW

What Have I Learned?

The following questions cover the material that you just learned—the **Thin Skin and Subcutaneous Tissues**. Use the information in the **STRUCTURE INFORMATION** window for these structures to answer the following questions on a separate piece of paper.

1. Name the two layers of the epidermis consisting of cells without nuclei.

2. Name the major cell type of the epidermis.

3. Other than the cells listed in question 2, what other cells are found in the epidermis?

4. Name the accessory organ of the skin that consists of a filament of keratinized cells.

5. What skin type contains these structures?

6. What is the name of the oblique tube in the skin where these structures are located?

7. Where on the body surface are these structures not found?

8. Name the skin layer that lies deep to the epidermis.

9. What tissue type does this layer consist of?

10. This layer consists of two structural layers. What are they?

11. What fibers located in the dermis give strength to the skin?

12. What general sensory reception occurs through the dermis?

13. How are these stimuli received?

14. Name two other functions of the dermis not mentioned in these questions.

15. Name the integumentary layer located deep to the skin.

16. What tissue types are found in this layer?

17. What structures are located in this layer?

18. Name three functions of this integumentary layer.

19. What are two other names for this layer?

20. Name the deep fascia of the thigh.

This Instructional Text Box is directing you to:

Click the **DISSECTION Study Area** icon.

Click the **ALL CONTENT** tab.

From the **Select Topic** menu, select **Fingernail**.

The default view for the **Fingernail** is **Sagittal**. If there is more than one option, you will be directed in this text box which one to choose.

MODULE 4 The Integumentary System

EXERCISE 4.2:
Fingernail—Sagittal View

SELECT TOPIC: Fingernail → SELECT VIEW: Sagittal

Fingernail: Sagittal View

The **View** menu displays **Sagittal**. This indicates that this is the only view available for the fingernail. Otherwise, as you will recall from the introduction, the **Select view** menu would have a drop-down box with different views from which to choose.

Click **LAYER 1** in the **LAYER CONTROLS** window, and you will see the following image:

Click on **LAYER 2** in the **LAYER CONTROLS** window, and you will see the following image:

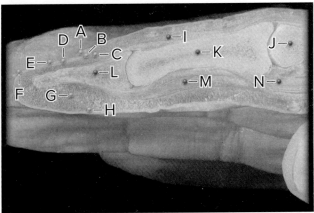

©McGraw-Hill Education

Mouse-over the pins on the screen to find the information necessary to fill in the following blanks:

A. eponychium
B. nail root
C. nail matrix
D. nail bed
E. nail body
F. free edge of nail
G. dermis
H. epidermis

HEADSUP!
Some of the structures in the dissections are not part of the integumentary system, but are important structures in the vicinity. These nonintegumentary structures are tagged with blue pins to distinguish them from the green pins of the integumentary system.

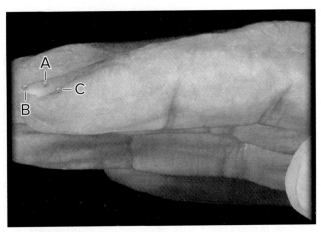

©McGraw-Hill Education

Mouse-over the pins on the screen to find the information necessary to fill in the following blanks:

A. nail body
B. free edge of nail
C. lateral nail fold

CHECK POINT

Fingernail
- Left-click the pins on the screen to find the information necessary to answer the following questions:
1. Name the structures that are scalelike modifications of epidermis.
2. What are the two functions of these structures?
3. What is the name for the distal edge of these structures?

nail plate

free edge of nail nail body
nail bed / nail matrix
↓   ↓
            • protection for tip of finger and toe
            • aids in grasping function of finger

MODULE 4   The Integumentary System   153

Nonintegumentary System Structures (blue pins)

I. extensor expansion
J. proximal phalanx of finger
K. middle phalanx of finger
L. distal phalanx of finger
M. " " " " profundus
N. tendon of flexor digitorum superficialis m.

CHECK POINT

Fingernail, *continued*

4. Name the structure that consists of the stratum corneum of the proximal nail fold. *[handwritten: eponychium / nail body? cuticle]*
5. What is another name for this structure? *[handwritten: lateral nail fol]*
6. What structure does this structure overlie?

IN REVIEW

What Have I Learned?

The following questions cover the material that you just learned—the **Integumentary System.** Use the information in the **STRUCTURE INFORMATION** window for these structures to answer the following questions on a separate piece of paper.

1. Name the structures that are scalelike modifications of the epidermis.

2. What are the two functions of these structures?

3. What is the name for the distal edge of these structures?

4. Name the part of the nail plate that consists of layers of compacted, highly keratinized epithelial cells.

5. What layer of the epidermis does this structure correspond to?

6. What is the name for the epidermal fold along the lateral edge of the nail plate?

7. Name the structure that consists of the stratum corneum of the proximal nail fold.

8. What is another name for this structure?

9. What structure does this structure overlie?

10. What is the name for the proximal part of the nail plate?

11. What is its function?

12. Name the growth zone of the nail that contains mitotic cells.

13. Normal nail growth is _____ mm/day.

14. Which grow faster, fingernails or toenails?

15. Where do the muscles of facial expression insert?

16. Name the four functions of this structure.

17. Name the skin layer deep to the epidermis.

18. Name the two functions of the epidermis.

19. Name the five epidermal layers of thick skin, from deep to superficial.

20. Which of these five layers is absent in thin skin?

21. Name the structure that provides nutrients for the epidermis.

154 MODULE 4 The Integumentary System

EXERCISE 4.3a:
Thick Skin—Histology

SELECT TOPIC	SELECT VIEW
Thick Skin	LM: Low Magnification

Click the **TAGS ON/OFF** button in the lower left panel, and you will see the following image:

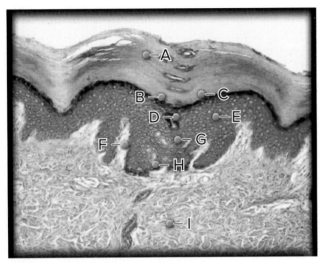

©Victor Eroschenko

Mouse-over the pins on the screen to find the information necessary to fill in the following blanks:

A. Stratum corneum
B. Stratum granulosum
C. Stratum lucidum
D. duct of merocrine sweat gland
E. epidermis
F. dermal papilla
G. Stratum spinosum
H. Stratum basale
I. dermis

CHECK POINT
Thick Skin, Histology (Low Magnification) Stratum lucidum
1. Which epidermal layer is present only in the skin of the palm of the hand and the sole of the foot?
2. What is the major cell type of the epidermis? Keratinocyte
3. Which epidermal layers contain cells responsible for the turnover of epidermal keratinocytes?

stratum basale
stratum spinosum

EXERCISE 4.3b:
Thick Skin—Histology

SELECT TOPIC	SELECT VIEW
Thick Skin	LM: High Magnification

Click the **TAGS ON/OFF** button in the lower left panel, and you will see the following image:

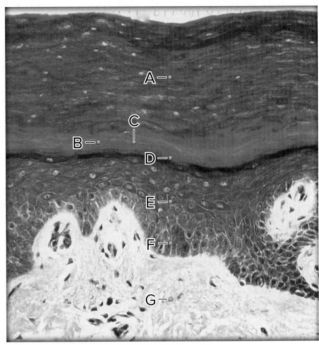

©Lutz Slomianka

Mouse-over the pins on the screen to find the information necessary to fill in the following blanks:

A. Stratum corneum
B. epidermis
C. Stratum lucidum
D. Stratum granulosum
E. Stratum spinosum
F. Stratum basale
G. dermis

CHECK POINT
Thick Skin, Histology (High Magnification)
1. Name the layer of the epidermis that creates a barrier to fluids. Stratum corneum
2. What relationship does your skin have with household dust? sloughed cells form household dust
3. Where do the replacement cells for the sloughed stratum corneum cells originate?

MODULE 4   The Integumentary System   155

EXERCISE 4.4a:
Thin Skin—Histology

SELECT TOPIC: Thin Skin ▸ SELECT VIEW: LM: Low Magnification

Click the **TAGS ON/OFF** button in the lower left panel, and you will see the following image:

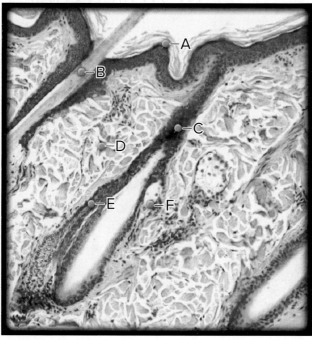

©Victor Eroschenko

Mouse-over the pins on the screen to find the information necessary to fill in the following blanks:

A. epidermis
B. hair
C. hair follicle
D. dermis
E. duct of sebaceus gland
F. sebaceous gland

CHECK POINT
Thin Skin, Histology (Low Magnification)

1. Name the layer of the skin consisting of stratified squamous epithelium. epidermis
2. The cells from which epidermal layers lack nuclei?
3. What is sebum? Where is it produced?

stratum lucidum & stratum corneum

oily substance — sebaceous gland

EXERCISE 4.4b:
Thin Skin—Histology

SELECT TOPIC: Thin Skin ▸ SELECT VIEW: LM: High Magnification

Click the **TAGS ON/OFF** button in the lower left panel, and you will see the following image:

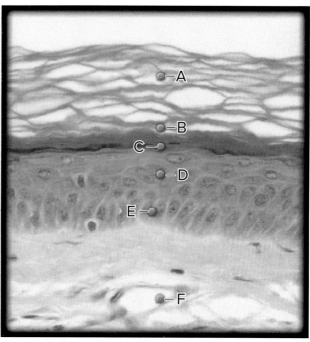

©Lutz Slomianka

Mouse-over the pins on the screen to find the information necessary to fill in the following blanks:

A. stratum corneum
B. epidermis
C. stratum granulosum
D. stratum spinosum
E. stratum basale
F. dermis

CHECK POINT
Thin Skin, Histology (High Magnification)

1. Name the layer of the epidermis that is not a distinct layer in thin skin. stratum granulosum
2. List the appendages of the skin. Where are they located? hair follicle, glands → dermis
3. Where is the hair follicle located? How does the hair follicle develop? dermis

develope as invagination from epidermis

156 MODULE 4   The Integumentary System

EXERCISE 4.5:
Hair—Histology

SELECT TOPIC	SELECT VIEW
Hair	SEM: Low Magnification

*Click the **TAGS ON/OFF** button in the lower left panel, and you will see the following image:*

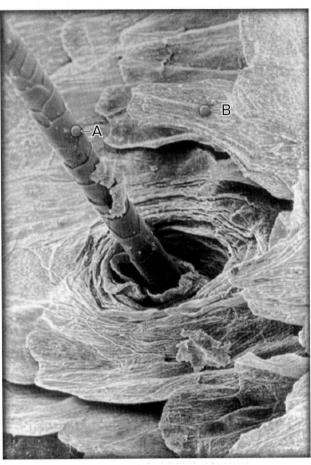

©SPL/Photo Researchers, Inc.

Mouse-over the pins on the screen to find the information necessary to fill in the following blanks:

A. _____

B. _____

CHECK POINT

Hair, Histology (SEM: Low Magnification)

1. Describe the hair shaft. Where is it located?
2. What is the function of the hair shaft?

EXERCISE 4.6a:
Hair Follicle—Histology

SELECT TOPIC	SELECT VIEW
Hair Follicle	LM: Medium Magnification

*Click the **TAGS ON/OFF** button in the lower left panel, and you will see the following image:*

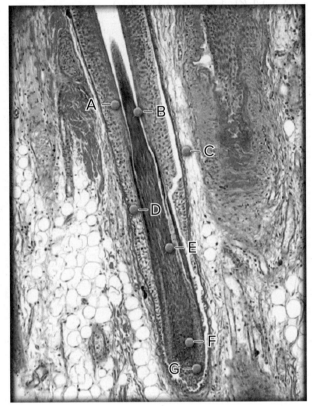

©McGraw-Hill Education/Al Telser

Mouse-over the pins on the screen to find the information necessary to fill in the following blanks:

A. _____

B. _____

C. _____

D. _____

E. _____

F. _____

G. _____

CHECK POINT

Hair Follicle, Histology (LM: Low Magnification)

1. List the layers of hair from deep to superficial.
2. Where is the site for differentiation of cells in the hair root?
3. Describe the external root sheath. What is its function?

EXERCISE 4.6b:
Hair Follicle—Histology

SELECT TOPIC	SELECT VIEW
Hair Follicle	LM: High Magnification

Click the **TAGS ON/OFF** *button in the lower left panel, and you will see the following image:*

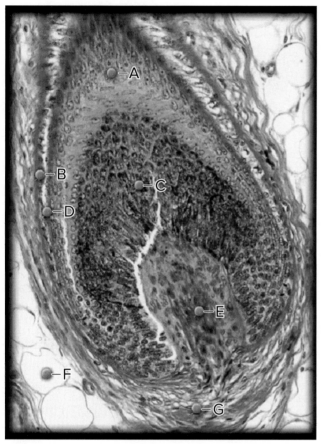

©McGraw-Hill Education/Al Telser

Mouse-over the pins on the screen to find the information necessary to fill in the following blanks:

A. _____
B. _____
C. _____
D. _____
E. _____
F. _____
G. _____

CHECK POINT

Hair Follicle, Histology (LM: Low Magnification)

1. Identify the blood supply for the hair follicle.
2. Describe how hair color is determined.
3. What activity accounts for hair growth?

EXERCISE 4.7a:
Sebaceous Gland—Histology

SELECT TOPIC	SELECT VIEW
Sebaceous Gland	LM: Low Magnification

Click the **TAGS ON/OFF** *button in the lower left panel, and you will see the following image:*

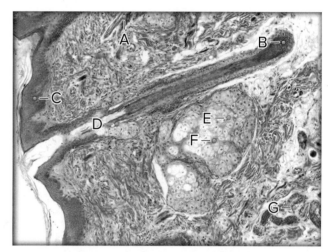

©McGraw-Hill Education/Al Telser

Mouse-over the pins on the screen to find the information necessary to fill in the following blanks:

A. _____
B. _____
C. _____
D. _____
E. _____
F. _____
G. _____

CHECK POINT

Hair Follicle, Histology (LM: Low Magnification)

1. Describe the sebaceous gland. What is its function?
2. What skin does not contain sebaceous glands?
3. What is the function of sebum?
4. How do holocrine glands function?
5. Where does the duct of the sebaceous gland most frequently open?
6. The duct of the sebaceous gland opens directly on the skin surface in which locations?
7. Describe the arrector pili muscle. What is its function?

158 MODULE 4 The Integumentary System

EXERCISE 4.7b:
Sebaceous Gland—Histology

SELECT TOPIC	SELECT VIEW
Sebaceous Gland	LM: High Magnification

Click the **TAGS ON/OFF** *button in the lower left panel, and you will see the following image:*

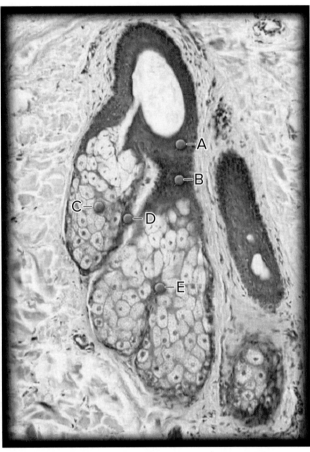

©Lutz Slomianka

Mouse-over the pins on the screen to find the information necessary to fill in the following blanks:

A. _____
B. _____
C. _____
D. _____
E. _____

CHECK POINT

Hair Follicle, Histology (LM: High Magnification)

1. Describe the secretory cell sebaceous gland. What is its function?
2. How long is the complete process for sebum production?
3. Describe the basal cell sebaceous gland. What is its function?

EXERCISE 4.8a:
Merocrine Sweat Gland—Histology

SELECT TOPIC	SELECT VIEW
Merocrine Sweat Gland	LM: Low Magnification

Click the **TAGS ON/OFF** *button in the lower left panel, and you will see the following image:*

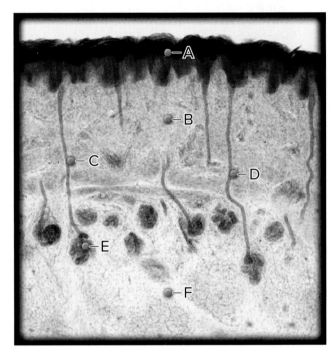

©Lutz Slomianka

Mouse-over the pins on the screen to find the information necessary to fill in the following blanks:

A. _____
B. _____
C. _____
D. _____
E. _____
F. _____

CHECK POINT

Merocrine Sweat Gland, Histology (LM: Low Magnification)

1. Describe the merocrine sweat gland. What is its function?
2. Describe the sweat from the merocrine sweat gland.
3. What is the most numerous and widely distributed sweat gland?
4. Where are the highest concentrations of merocrine sweat glands located?

MODULE 4 The Integumentary System 159

EXERCISE 4.8b:
Merocrine Sweat Gland—Histology

SELECT TOPIC	SELECT VIEW
Merocrine Sweat Gland	LM: High Magnification

Click the **TAGS ON/OFF** *button in the lower left panel, and you will see the following image:*

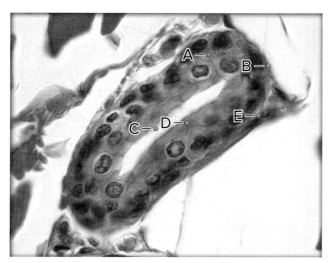

©McGraw-Hill Education/Steve Sullivan

Mouse-over the pins on the screen to find the information necessary to fill in the following blanks:

A. _____
B. _____
C. _____
D. _____
E. _____

CHECK POINT

Merocrine Sweat Gland, Histology (LM: High Magnification)

1. Describe the epithelial cell of the merocrine sweat gland. What is its function?
2. Describe the myoepithelial cell of the merocrine sweat gland. What is its function?
3. Describe the lumen of the merocrine sweat gland. What is its function?

EXERCISE 4.9:
Duct of Sweat Gland—Histology

SELECT TOPIC	SELECT VIEW
Duct of Sweat Gland	LM: High Magnification

Click the **TAGS ON/OFF** *button in the lower left panel, and you will see the following image:*

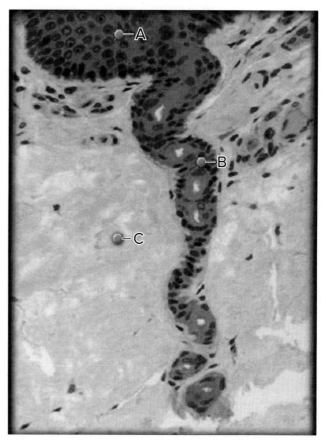

©Dr. Thomas Caceci, Virginia-Maryland Regional College of Verinary Medicine

Mouse-over the pins on the screen to find the information necessary to fill in the following blanks:

A. _____
B. _____
C. _____

CHECK POINT

Merocrine Sweat Gland, Histology (LM: High Magnification)

1. Describe the duct of the sweat gland. What is its function?

EXERCISE 4.10a:
Apocrine Sweat Gland—Histology

SELECT TOPIC	SELECT VIEW
Apocrine Sweat Gland	LM: Low Magnification

Click the **TAGS ON/OFF** *button in the lower left panel, and you will see the following image:*

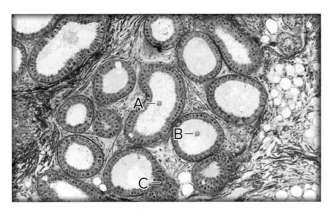

©McGraw-Hill Education/Dennis Strete

Mouse-over the pins on the screen to find the information necessary to fill in the following blanks:

A. _____
B. _____
C. _____

CHECK POINT

Merocrine Sweat Gland, Histology (LM: High Magnification)

1. Where are the secretory products of the apocrine sweat gland secreted?
2. Compare the secretory products of the apocrine sweat gland when released and after being metabolized by bacteria.
3. Describe the composition of the secretory products of the apocrine sweat gland.

EXERCISE 4.10b:
Apocrine Sweat Gland—Histology

SELECT TOPIC	SELECT VIEW
Apocrine Sweat Gland	LM: High Magnification

Click the **TAGS ON/OFF** *button in the lower left panel, and you will see the following image:*

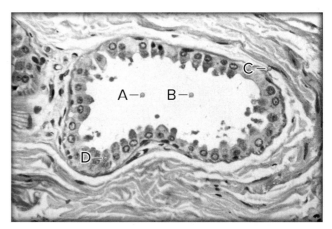

©Biophoto/Science Source

Mouse-over the pins on the screen to find the information necessary to fill in the following blanks:

A. _____
B. _____
C. _____
D. _____

CHECK POINT

Merocrine Sweat Gland, Histology (LM: High Magnification)

1. Describe the epithelial cell of the apocrine sweat gland. What is its function?
2. What is the function of the myoepithelial cell of the merocrine sweat gland?

We follow up the study of the integumentary system with two clinical application animations covering burns and wound healing. As you view these animations, keep in mind the normal structures of the skin and compare how they appear when damaged, and how your body repairs itself in an attempt to return to pre-injury conditions.

SELECT ANIMATION: Burns (3D) — PLAY

After viewing the animation, answer the following questions.

1. What are the three main layers of the skin?

2. What are burns?

3. What affect do burns have on the skin's framework and function?

4. What are superficial burns?

5. What is erythema?

6. What are the two types of partial-thickness burns?

7. Describe the tissues involved and the damage caused by superficial partial-thickness burns.

8. Describe the tissues involved and the damage caused by deep partial-thickness burns.

9. Describe the tissues involved and the damage caused by full-thickness burns.

10. Burn can also be evaluated by determining the _____ of _____ _____ area they cover.

11. How is this percentage quickly estimated?

12. Severe burns involving a large surface area increase _____ _____ and lead to two stages of _____.

13. In _____ shock, _____, _____, and _____ _____ leak from the bloodstream into _____ _____, creating widespread _____.

14. In _____ shock, the lower _____ _____ increases _____ _____ and _____ _____.

15. How does the body compensate for this?

16. What are the results of burn shock?

17. This tissue and organ death results from a lack of _____.

18. What is the aim of burn treatment?

19. How does it accomplish this aim?

20. How does this intervention affect vascular volume?

21. How does wound care reduce complications?

22. _____ control the pain and inflammation of _____ burns.

23. How are superficial partial-thickness burns treated?

24. In _____ _____ _____ burns, dead skin or _____ is routinely _____ or _____ to a healthy level.

25. The _____ _____ is kept clean and moist to allow _____ _____ and to accept _____ _____ called a _____ _____.

26. How are full-thickness burns treated?

SELECT ANIMATION — Wound Healing (3D) — PLAY

1. An injury to the skin, such as a _____, _____, or _____ _____, kills nearby _____ and damages underlying _____ and triggers the complex process of _____ the skin.

2. How many steps are included in the process of wound healing?

3. The _____ phase begins immediately upon injury.

4. _____ _____ constrict to reduce blood loss.

5. Then _____ arrive to plug the leak.

6. The _____ _____ initiates the clotting mechanism by facilitating the reactions of _____ _____ - called _____ _____, which interact to form a _____ _____.

7. After the clot forms, the blood vessels _____ and become more _____ to allow _____ _____ _____ to leave the blood vessel and populate at the site of injury.

8. During this process, called _____ white blood cells eat _____ and kill _____, reducing the risk of _____.

9. The _____ phase begins _____ days to _____ weeks after injury.

10. The first step in the _____ phase is _____.

11. Connective tissue cells called _____ lay a matrix of _____ that reinforces the wound and provides _____ for other cells.

12. _____ then contracts to pull together the _____ of the wound.

13. _____, or the growth of new blood vessels, begins _____ _____ and supplies _____ to the repairing cells.

14. _____ is the restoration of the _____ _____ _____.

15. _____ cells migrate from the margins of the wound, protected by the _____ until they meet.

16. What eventually happens to the scab?

17. The _____ phase begins _____ weeks after the injury and can continue _____ _____.

18. During this phase, a new, more organized _____ _____ forms in the wound bed and _____ disappear, leaving an avascular _____.

19. What is one possible complication of wound healing?

20. How does this structure form?

21. What are keloids composed of?

22. How rapidly do they grow?

23. Do keloids regress spontaneously?

24. How are keloids treated?

25. Is an excision a permanent treatment?

26. What is a common initial treatment for keloids?

27. What do these treatments accomplish?

IN REVIEW

What Have I Learned?

The following questions cover the material that you just learned—**Histology of the Integumentary System.** Use the information in the **STRUCTURE INFORMATION** window for these structures to answer the following questions on a separate piece of paper.

1. Name the structure responsible for hair growth.
2. Where is this structure located?
3. Name the filamentous, pigmented, and keratinized structure that projects from the epidermal surface.
4. What is the function of this structure?
5. Name the angulated tubular invagination of the epidermis containing an inner epidermic and an outer dermic coat.
6. What is its function?
7. List the characteristic parts of this structure.
8. What structures are associated with a hair follicle?
9. What is another name for the arrector muscle of the hair?
10. What is included with the apocrine gland secretions?
11. Name the simple, saccular holocrine gland with ducts opening into the hair follicle or onto the skin.
12. Name two locations where these glands are *not* found.
13. Where on the body surface would you find thick skin? Thin skin?
14. Name the structures found at the interface between the dermis and the epidermis.
15. What is the function of these structures?
16. Describe these structures.
17. Define avascular.
18. Name the layer of the epidermis that creates a barrier to liquids.
19. What relationship does your skin have with household dust?
20. What is the meaning of the Latin word *cornu*?
21. Where do the replacement cells for the sloughed stratum corneum cells originate?
22. The stratum corneum consists of _____ layers of cornified dead cells.
23. Name the layer of the epidermis present only in thick skin.
24. What is keratohyalin? Where is it located?
25. What is the function of the stratum spinosum?
26. The cells from which two layers of the epidermis are responsible for the turnover of epidermal cells?
27. Name the deepest layer of the epidermis.
28. What is another name for this layer?
29. What is the function of this layer?
30. Name the layer of the skin consisting of stratified squamous epithelium.
31. The cells from which epidermal layers lack nuclei?
32. Name the two types of sweat glands. What is the difference between the two?

Self-Quiz

Take this opportunity to quiz yourself by taking the **QUIZ** over the **Integumentary system.** See the introduction for a reminder on how to access the **QUIZ** for this module.

MODULE 5

The Skeletal System

Overview: Skeletal System

We are born with 270 bones in our bodies, and even more bones form during childhood. By the time we reach adulthood though, several separate bones have fused together so that the number of our bones has decreased to around 206,[1] which make up the adult skeletal system. An example of this reduction occurs in each half of our pelvis, where three separate bones—the ilium, the ischium, and the pubis—fuse into one single bone called the *os coxa*.

The skeletal system is further divided into the **axial skeleton**, consisting of the bones of the skull, hyoid bone, inner ear ossicles, vertebral column, and the thoracic cage; and the **appendicular skeleton**, which consists of the bones of the upper and lower extremities along with their associated girdles (Table 5.1).

[1]Around 206—some people develop varying numbers of miscellaneous bones, either **sesamoid bones**, which form within some tendons as a response to stress (such as the patella) or **sutural bones**, which develop within the sutures of the skull.

CHECK POINT

Overview: Skeletal System

1. The average human adult has _____ bones in their body, whereas the average newborn has _____ bones in theirs.
2. Explain the difference.

Naming Bony Processes and Other Landmarks

Bony landmarks are various ridges, spines, depressions, pores, bumps, grooves, and articulating structures on the surface of bones. These structures allow for the passage of blood vessels and nerves; for joints between bones; and for the attachment of ligaments, muscles, and tendons. Therefore, a working knowledge of the names of these structures will be a valuable asset when considering the structure and function of these surface features.

Some of the most commonly encountered bony landmarks are listed in Table 5.2.

Table 5.1 Summary of the Bones of the Adult Skeletal System

Axial Skeleton—80 Bones		Appendicular Skeleton—126 Bones	
Skull and Hyoid	23 bones	Pectoral Girdle	4 bones
Inner Ear Ossicles	6 bones	Upper Extremities	60 bones
Vertebral Column	26 bones	Pelvic Girdle	2 bones
Sternum and Ribs	25 bones	Lower Extremities	60 bones
Total — 206 Bones			

Bonus Question: Using this information as a reference, how would this table appear if it was a list of the bones for a newborn?

Table 5.2 Common Bony Landmarks

FEATURE	DESCRIPTION	EXAMPLE
Landmarks of Articulation:		
Condyle:	Smooth, rounded knob	Occipital condyle of skull
Facet:	Smooth, flat, slightly concave or convex articular surface	Articular facets of vertebrae
Head:	Prominent expanded end of a bone, sometimes rounded	Head of the femur
Elevated Landmarks:		
Process:	Any bony prominence	Mastoid process of the skull
Spine:	Sharp, slender, or narrow process	Spine of the scapula

Continued

MODULE 5 The Skeletal System

Table 5.2 Continued

FEATURE	DESCRIPTION	EXAMPLE
Elevated Landmarks: (continued)		
Crest:	Narrow ridge	Iliac crest of the pelvis
Line:	Slightly raised, elongated ridge	Nuchal lines of the skull
Tuberosity:	Rough elevated surface	Tibial tuberosity
Tubercle:	Small, rounded process	Greater tubercle of the humerus
Trochanter:	Massive processes unique to the femur	Greater trochanter of the femur
Epicondyle:	Projection superior to a condyle	Medial epicondyle of the femur
Depressions or Flat Surfaces:		
Alveolus:	Pit or socket	Tooth socket
Fossa:	Shallow; broad, or elongated basin	Mandibular fossa
Fovea:	Small pit	Fovea capitis of the femur
Sulcus:	Groove for a tendon, nerve, or blood vessel	Intertubercular sulcus of the humerus
Spaces or openings:		
Foramen:	Hole through a bone, usually round	Foramen magnum of the skull
Fissure:	Slit through a bone	Orbital fissure behind the eye
Meatus or canal:	Tubular passage or tunnel through a bone	Auditory meatus of the ear
Sinus:	Space or cavity within a bone	Frontal sinus of the skull

Bonus Question: Alveolus is a common term in human anatomy. How many other examples of alveoli can you find in *Anatomy & Physiology | Revealed®*?

CHECK POINT

Naming Bone Processes and Other Landmarks

1. Explain the difference between a crest and a line.
2. Explain the difference between a condyle and an epicondyle.
3. Explain the difference between a foramen and a fissure.

EXERCISE 5.1:

Coloring Exercise

Color in the structures with colored pens or pencils.

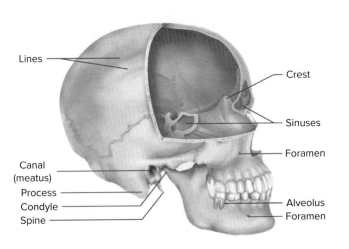

©McGraw-Hill Education

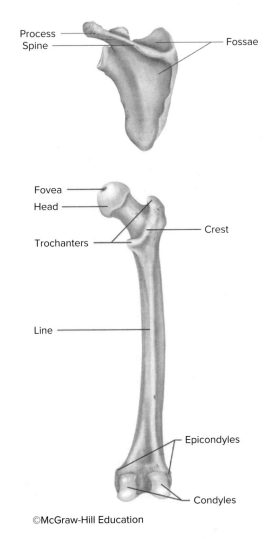

©McGraw-Hill Education

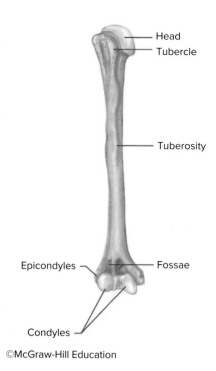

A Few Notes About Naming Processes

A process is any bony prominence—that is, a piece of bone that sticks out from the rest of the bone. When it comes to naming these processes, there are a few rules that need to be followed to minimize confusion. For example, let's consider when a process articulates with another bone, such as the zygomatic process of the temporal bone. This process is a structure on the **temporal bone** that articulates with the **zygomatic bone**. The formula for naming one of these processes is as follows:

The *x* process of the *y* bone

where *x* = the name of bone articulated with
and *y* = the name of the bone it is part of

So, with the ZYGOMATIC process of the TEMPORAL bone, x = ZYGOMATIC (the bone articulated with) and y = TEMPORAL (the bone it is part of). Now, how does the *zygomatic* process of the *temporal* bone compare to the *temporal* process of the *zygomatic* bone? If you are not sure, don't worry, we will cover these structures shortly.

Let's look at the styloid process of the temporal bone. *This* styloid process does *not* articulate with any other bone, but it shares its name with the styloid processes of both the ulna and radius bones of the forearm. Therefore, it must be named in reference to the bone that it is part of to prevent confusion—hence the name the styloid process of the temporal bone. What problems would you predict could occur if this distinction is not made in an emergency room scenario?

Some processes, the mastoid process for example, are unique in name and do not articulate with any other bones. These require no further clarification when naming them.

CHECK POINT

A Few Notes About Naming Processes

1. Consider the *temporal* process of the *zygomatic* bone.
   a) This process articulates with which bone?
   b) This process is part of which bone?
2. What is the correct way to say the two styloid processes of the forearm?
3. Why is it correct to refer to the mastoid process as the mastoid process and not the mastoid process of the temporal bone?

From the **HOME** *screen, click the drop-down box on the* **MODULE** *menu.*

From the systems listed, click on **Skeletal**.

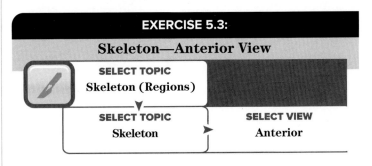

EXERCISE 5.2:
Appositional Bone Growth

SELECT ANIMATION
Appositional Bone Growth — PLAY

After viewing the animation, answer the following questions:

1. Define appositional bone growth.

2. Which cells produce bone material?

3. How is a tunnel formed around a blood vessel?

4. How is the tunnel filled in to produce a new osteon?

5. What is the name for the concentric rings that form the osteon?

*When you are finished with the animation, click on the **DISSECTION** button at the top of the screen to begin the following exercises, or click on the **EXIT** button at the top-right of the screen to exit* Anatomy & Physiology | Revealed®.

HEADS UP!

Anatomy & Physiology | Revealed® *has several animations available to aid your study of different systems. Watch for the green-highlighted **Animation** title in the **INFORMATION** window, which indicates that an animation is available for the specific structure(s) you are viewing.*

Skeleton (Regions)

EXERCISE 5.3:
Skeleton—Anterior View

SELECT TOPIC
Skeleton (Regions)

SELECT TOPIC: Skeleton ▶ SELECT VIEW: Anterior

*Click **LAYER 1** in the **LAYER CONTROLS** window, and you will see the following image:*

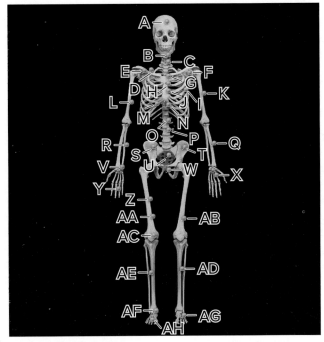

©McGraw-Hill Education

Mouse-over the pins on the screen to find the information necessary to identify the following structures:

A. skull
B. hyoid bone
C. cervical vertebra
D. clavicle
E. pectoral girdle
F. scapula
G. axial skeleton
H. thoracic cage
I. ribs
J. sternum
K. humerus
L. skeleton of upper limb
M. costal cartilages
N. thoracic vertebrae
O. vertebral column
P. lumbar vertebra
Q. radius
R. ulna
S. pelvic girdle
T. hip bone
U. sacrum
V. carpal bones
W. coccyx
X. metacarpals
Y. phalanges of fingers
Z. femur
AA. skeleton of lower limb
AB. appendicular skeleton
AC. patella
AD. tibia
AE. fibula
AF. tarsal bone
AG. metatarsal
AH. phalanges of toes

CHECK POINT

Skeleton, Anterior View

1. Name the bones that make up the cranial group. How many of each bone are in this group?
2. Name the bones that make up the facial group. How many of each bone are in this group?
3. Name the bones that form the pelvic girdle.
4. Name the bones that form the wrist.
5. The vertebral column is composed of _____ vertebrae. How many of each group of vertebrae are there?

EXERCISE 5.4:

Skeleton—Posterior View

SELECT TOPIC: Skeleton ▶ SELECT VIEW: Posterior

Click **LAYER 1** *in the* **LAYER CONTROLS** *window, and you will see the following image:*

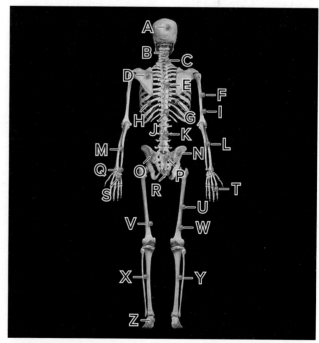

©McGraw-Hill Education

Mouse-over the pins on the screen to find the information necessary to identify the following structures:

A. skull
B. cervical vertebra
C. vertebra prominens
D. scapula
E. thoracic vertebrae
F. humerus
G. costal cartilages
H. ribs
I. skeleton of upper limb
J. vertebral column
K. lumbar vertebra
L. ulna
M. radius
N. pelvic girdle
O. hip bone
P. sacrum
Q. carpal bones
R. coccyx
S. metacarpals
T. phalanges of fingers
U. femur
V. appendicular skeleton
W. skeleton of lower limb
X. fibula
Y. tibia
Z. tarsal bones

CHECK POINT

Skeleton, Posterior View

1. Name the group of bones that form the palm of the hand. What structures form the knuckles?
2. Name the group of bones that form the foot. How many are there?
3. Name the group of bones that form the fingers. How many compose each finger?
4. Name the lateral bone of the forearm. Name the medial bone of the forearm.
5. What is another name for the tailbone?

EXERCISE 5.5:

Locating Structures of the Head and Neck

There are 30 bones in the human adult head and neck, which can present a daunting task when it comes to learning each bone. With this in mind, the following (Tables 5.3–5.6) are designed to allow you to discover which bones are visible in each view of *Anatomy & Physiology | Revealed*®. These tables are not meant to be tedious, but rather to help you become more familiar with each bone.

On the left of the following tables are names of structures found on the head and neck. The names of specific bones are aligned to the left, and the structures found on those bones are listed under them indented to the right. Using *Anatomy & Physiology | Revealed*®, open each dissection view listed across the top-right of the table and put an "X" in the columns under the views where you find the structures listed in the left column. Not all bones or structures will be visible in all views.

Table 5.3 Structures of the Head and Neck—Cranial Bones

Structures	Dissection Views								
	Individual Bone	Anterior	Inferior	Lateral	Mid-sagittal	Posterior	Superior	Skull Cranial Cavity	Orbit
Frontal bone									
Frontal sinus									
Supraorbital notch									
Parietal bone									
Occipital bone									
External occipital protuberance									
Foramen magnum									
Occipital condyle									
Temporal bone									
Carotid canal									
External acoustic meatus									
Internal acoustic meatus									
Mastoid process									
Squamous part of temporal bone									
Styloid process of temporal bone									
Zygomatic process of temporal bone									
Sphenoid bone									
Body									
Foramen ovale									
Foramen spinosum									
Greater wing									
Sella turcica									
Sphenoidal sinus									
Ethmoid bone									
Crista galli									
Ethmoid air cells									
Nasal concha—middle									
Nasal concha—superior									
Perpendicular plate of the ethmoid bone									

Bonus Question: What is the name for the bone shaped like a butterfly?

172 MODULE 5 The Skeletal System

Table 5.4 Structures of the Head and Neck—Facial Bones

Structures	Dissection Views								
	Individual Bone	Anterior	Inferior	Lateral	Mid-sagittal	Posterior	Superior	Skull Cranial Cavity	Orbit
Maxilla									
Alveolar process of maxilla									
Infraorbital foramen									
Palatine bone									
Zygomatic bone									
Temporal process of zygomatic bone									
Nasal bone									
Vomer bone									
Mandible									
Alveolar process of mandible									
Angle of mandible									
Body of mandible									
Condylar process of mandible									
Coronoid process of mandible									
Mandibular foramen									
Mental foramen									
Ramus of mandible									

Bonus Question: What is the anatomical term that refers to the chin area? How do you suppose it received this name?

Table 5.5 Structures of the Head and Neck—Other Skull Structures

Structures	Dissection Views								
	Individual Bone	Anterior	Inferior	Lateral	Mid-sagittal	Posterior	Superior	Skull Cranial Cavity	Orbit
Coronal suture									
Cranial fossa—anterior									
Cranial fossa—middle									
Cranial fossa—posterior									
Foramen lacerum									
Hard palate									
Hyoid bone									
Jugular foramen									
Lambda									
Lambdoid suture									
Nasal concha (inferior)									
Pterion									
Sagittal suture									
Septal cartilage									
Temporomandibular joint (TMJ)									
Articular disk of the TMJ									
Zygomatic arch									

Bonus Question: What is the name for the cheekbone?

MODULE 5 The Skeletal System 173

Table 5.6 Structures of the Head and Neck—Cervical Vertebrae

Structures	Dissection Views								
	Individual Bone	Anterior	Inferior	Lateral	Mid-sagittal	Posterior	Superior	Skull Cranial Cavity	Orbit
Atlas (C1 vertebra)									
Axis (C2 vertebra)									
Cervical vertebrae									
Spinous process (cervical)									
Transverse process (cervical)									
Vertebral body (cervical)									
Intervertebral disk									

Bonus Question: Which bones are characterized by the presence of transverse foramina?

EXERCISE 5.6:
Head and Neck—Anterior View

SELECT TOPIC: Head and Neck ▸ SELECT VIEW: Anterior

Click **LAYER 1** in the **LAYER CONTROLS** window, and you will see the following image:

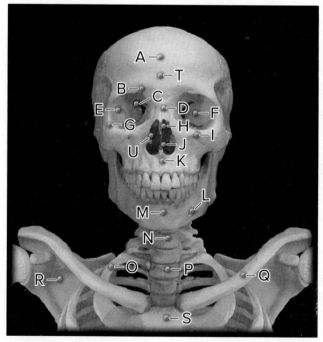

©McGraw-Hill Education

Mouse-over the pins on the screen to find the information necessary to identify the following structures:

A. _____
B. _____
C. _____
D. _____
E. _____
F. _____
G. _____
H. _____
I. _____
J. _____
K. _____
L. _____
M. _____
N. _____
O. _____
P. _____
Q. _____
R. _____
S. _____
T. _____
U. _____

CHECK POINT
Head and Neck, Anterior View

1. Name the bones contributing to the orbit.
2. Name the bone known as the collar bone.
3. What structures are formed by the frontal bone?
4. Which is the shortest rib?
5. Name the superior part of the sternum.

174 MODULE 5   The Skeletal System

EXERCISE 5.7:
Head and Neck—Lateral View

SELECT TOPIC	SELECT VIEW
Head and Neck	Lateral

Click **LAYER 1** in the **LAYER CONTROLS** *window, and you will see the following image:*

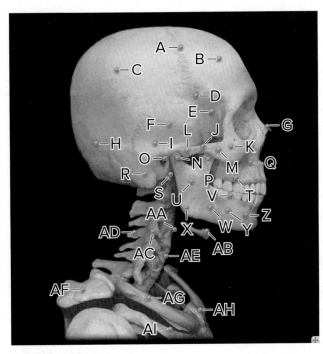

©McGraw-Hill Education

Mouse-over the pins on the screen to find the information necessary to identify the following structures:

A. _____
B. _____
C. _____
D. _____
E. _____
F. _____
G. _____
H. _____
I. _____
J. _____
K. _____
L. _____
M. _____
N. _____
O. _____
P. _____
Q. _____
R. _____
S. _____
T. _____
U. _____
V. _____
W. _____
X. _____
Y. _____
Z. _____
AA. _____
AB. _____
AC. _____
AD. _____
AE. _____
AF. _____
AG. _____
AH. _____
AI. _____

CHECK POINT

Head and Neck, Lateral View

1. Name the only bone in the body that does not articulate with any other bone.
2. Name the two bone processes that make up the zygomatic arch.
3. Name the "sockets" for the teeth.
4. Describe the pterion.
5. What structures of which bones form the temporomandibular joint?

MODULE 5   The Skeletal System   175

EXERCISE 5.8:
Thorax—Anterior View

SELECT TOPIC	SELECT VIEW
Thorax	Anterior

Click **LAYER 1** *in the* **LAYER CONTROLS** *window, and you will see the following image:*

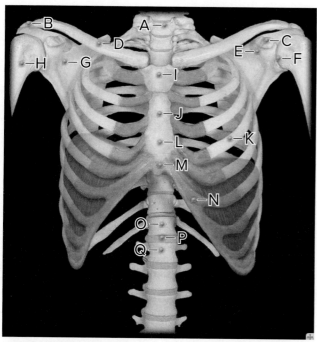

©McGraw-Hill Education

Mouse-over the pins on the screen to find the information necessary to identify the following structures:

A. _____
B. _____
C. _____
D. _____
E. _____
F. _____
G. _____
H. _____
I. _____
J. _____
K. _____
L. _____
M. _____
N. _____
O. _____
P. _____
Q. _____

CHECK POINT
Thorax, Anterior View

1. Name the three parts of the sternum.
2. Name a landmark for intramuscular injections.
3. Name the structures that attach the ribs to the sternum.
4. Name the two bones that form the glenohumeral joint.
5. Name the structure found between the bodies of all but two vertebrae. This structure is lacking between which two vertebrae?

EXERCISE 5.9:
Abdomen—Anterior View

SELECT TOPIC	SELECT VIEW
Abdomen	Anterior

Click **LAYER 1** *in the* **LAYER CONTROLS** *window, and you will see the following image:*

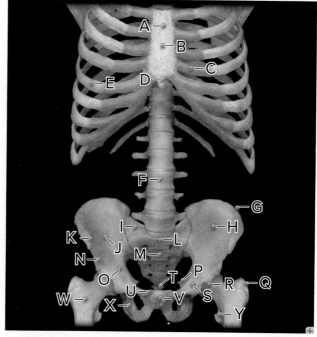

©McGraw-Hill Education

Mouse-over the pins on the screen to find the information necessary to identify the following structures:

A. _____
B. _____
C. _____
D. _____
E. _____
F. _____
G. _____

176 MODULE 5 The Skeletal System

H. _____
I. _____
J. _____
K. _____
L. _____
M. _____
N. _____
O. _____
P. _____
Q. _____
R. _____
S. _____
T. _____
U. _____
V. _____
W. _____
X. _____
Y. _____

CHECK POINT

Abdomen, Anterior View

1. What process occurs to the pubic symphysis during late pregnancy?
2. What Latin term means "wing"? Where is a structure with this name located?
3. Name a landmark for intramuscular injections.
4. What term refers to the hip joint socket? Which bones contribute to this structure?
5. Name the structure formed by five fused vertebrae.

EXERCISE 5.10:

Back—Posterior View

SELECT TOPIC	SELECT VIEW
Back	Posterior

Click **LAYER 1** *in the* **LAYER CONTROLS** *window, and you will see the following image:*

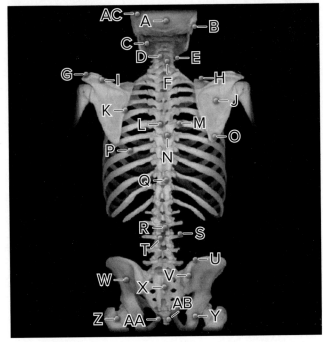

©McGraw-Hill Education

Mouse-over the pins on the screen to find the information necessary to identify the following structures:

A. _____
B. _____
C. _____
D. _____
E. _____
F. _____
G. _____
H. _____
I. _____
J. _____
K. _____
L. _____
M. _____
N. _____
O. _____
P. _____

Q. _____
R. _____
S. _____
T. _____
U. _____
V. _____
W. _____
X. _____
Y. _____
Z. _____
AA. _____
AB. _____
AC. _____

CHECK POINT

Back, Posterior View

1. Name the large, triangular flat bone of the superior back.
2. Name the three coxal bones.
3. Name the vertebrae of the lower back. How many are there?
4. Which intervertebral discs most commonly herniate?
5. Name the structure marked by a "dimple" on the lower back.

Skull and Associated Bones

SELECT ANIMATION
Skull—Exploded (3D)   PLAY

After viewing the animation, answer the following questions:

1. The bones that surround and protect the brain are referred to as the _____ bones.

2. How many of these bones are there?

3. The bones that form the underlying structure of the face are referred to as the _____ bones.

4. How many of these bones are there?

5. With one exception, the bones of the skull articulate with each other through joints known as _____. The exception is the _____.

6. There are numerous holes in the skull known as _____.

7. What are the functions of these holes?

8. What are the functions of the bones on the surface of the skull?

9. List these bones.

10. Identify the butterfly-shaped bone that makes up the anterior base of the cranium.

11. The wings of this bone form portions of what structure?

12. Identify the structure that houses the pituitary gland.

13. What structure of the ethmoid bone forms part of the nasal septum?

14. What structure of the ethmoid bone serves as the attachment point for the falx cerebri?

15. What is the function of the cribriform plate?

16. What features of the skull are formed by the palatine bones?

17. What is the function of the lacrimal groove?

178 MODULE 5 The Skeletal System

18. What bone forms the inferior and posterior part of the nasal septum?

19. How do the skull bones form an intricate structure perfectly suited to its functions?

EXERCISE 5.11:
Skull—Anterior View

SELECT TOPIC
Skull and Associated Bones

SELECT TOPIC
Skull

SELECT VIEW
Anterior

HEADS UP!
Many of the views we will see of the Skeletal System will indicate the locations of muscle attachments. These attachments are of two types—**Origins** and **Insertions**. **Origins**, as indicated by red shading, are the relatively stationary or immobile points of skeletal muscle attachment. The **Insertions**, indicated by blue shading, are the points where a muscle attaches to a bone and produces movement. One way to remember the distinction between the two is to consider your life. Your **Origin**, or location of your birth, never changes or moves as you go about your life. But, your **Insertion**, where you are inserted on the earth at this time may be different than your origin. Thus, your **Origin never moves**, but when you move, your **Insertion moves** to another location.

Click **LAYER 1** in the **LAYER CONTROLS** window, and you will see the following image:

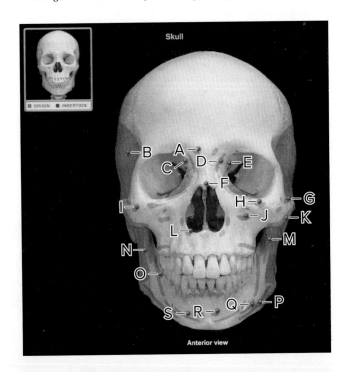

Mouse-over the pins on the screen to find the information necessary to identify the following structures:

A. _____
B. _____
C. _____
D. _____
E. _____
F. _____
G. _____
H. _____
I. _____
J. _____
K. _____
L. _____
M. _____
N. _____
O. _____
P. _____
Q. _____
R. _____
S. _____

Click **LAYER 2** in the **LAYER CONTROLS** window, and you will see the following image:

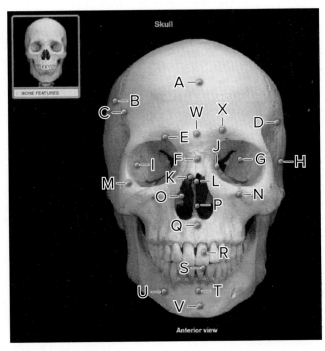

©McGraw-Hill Education

Mouse-over the pins on the screen to find the information necessary to identify the following structures:

A. _____
B. _____
C. _____
D. _____
E. _____
F. _____
G. _____
H. _____
I. _____
J. _____
K. _____
L. _____
M. _____
N. _____
O. _____
P. _____
Q. _____
R. _____
S. _____
T. _____
U. _____
V. _____
W. _____
X. _____

CHECK POINT

Skull, Anterior View

1. Name the structures transmitted through the mental foramen.
2. Name a feature of the skull that can be either a notch or a foramen.
3. What two bones contain teeth?
4. What is the name for the sockets of the teeth? (You may have to revisit earlier parts of this module to answer this one.)
5. The nasal septum consists of what specific bones or structures of bones?

EXERCISE 5.12:
Skull—Superior View

SELECT TOPIC	SELECT VIEW
Skull	Superior

*Click **LAYER 1** in the **LAYER CONTROLS** window, and you will see the following image:*

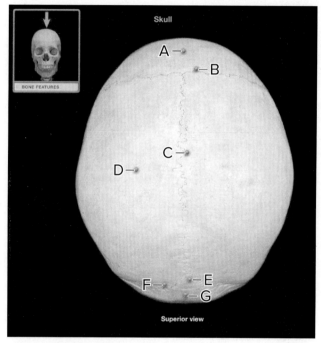

©McGraw-Hill Education

Mouse-over the pins on the screen to find the information necessary to identify the following structures:

A. _____
B. _____
C. _____
D. _____
E. _____
F. _____
G. _____

CHECK POINT

Skull, Superior View

1. Name the joint between the parietal bones.
2. Name the joint between the frontal and the parietal bones.
3. Name the bone type found in or near the sutures of the skull.
4. Where are these bones most often found?
5. Name the skull bone that articulates with the vertebral column.

EXERCISE 5.13:
Skull—Lateral View

SELECT TOPIC: Skull ▸ **SELECT VIEW:** Lateral

Click **LAYER 1** in the **LAYER CONTROLS** window, and you will see the following image:

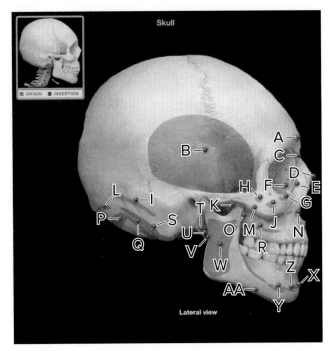

©McGraw-Hill Education

Mouse-over the pins on the screen to find the information necessary to identify the following structures:

A. _____
B. _____
C. _____
D. _____
E. _____
F. _____
G. _____
H. _____
I. _____
J. _____
K. _____
L. _____
M. _____
N. _____
O. _____
P. _____
Q. _____
R. _____
S. _____
T. _____
U. _____
V. _____
W. _____
X. _____
Y. _____
Z. _____
AA. _____

Click **LAYER 2** in the **LAYER CONTROLS** window, and you will see the following image:

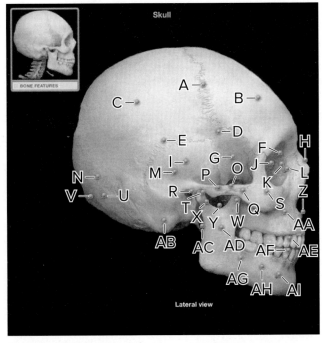

©McGraw-Hill Education

Mouse-over the pins on the screen to find the information necessary to identify the following structures:

A. _____
B. _____
C. _____
D. _____
E. _____
F. _____
G. _____

MODULE 5 The Skeletal System 181

H. _____
I. _____
J. _____
K. _____
L. _____
M. _____
N. _____
O. _____
P. _____
Q. _____
R. _____
S. _____
T. _____
U. _____
V. _____
W. _____
X. _____
Y. _____
Z. _____
AA. _____
AB. _____
AC. _____
AD. _____
AE. _____
AF. _____
AG. _____
AH. _____
AI. _____

CHECK POINT

Skull, Lateral View

1. Name the specific bony structures that form the temporomandibular joint.
2. What two bones make up most of the lateral skull (one on each side)?
3. What suture is their point of articulation?

EXERCISE 5.14:

Skull—Posterior View

SELECT TOPIC	SELECT VIEW
Skull	Posterior

Click **LAYER 1** *in the* **LAYER CONTROLS** *window, and you will see the following image:*

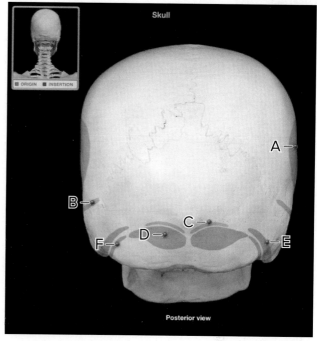

©McGraw-Hill Education

Mouse-over the pins on the screen to find the information necessary to identify the following structures:

A. _____
B. _____
C. _____
D. _____
E. _____
F. _____

CHECK POINT

Skull, Posterior View

1. What suture forms the joint between the parietal and occipital bones?
2. What bone forms most of the posterior skull?
3. What is the attachment site on the skull for the nuchal ligament?

182 MODULE 5 The Skeletal System

Click **LAYER 2** *in the* **LAYER CONTROLS** *window, and you will see the following image:*

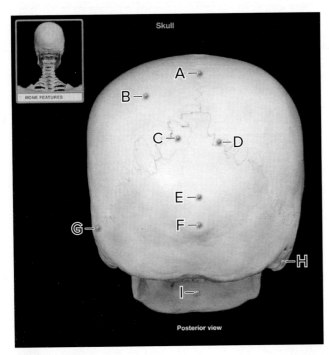

©McGraw-Hill Education

Mouse-over the pins on the screen to find the information necessary to identify the following structures:

A. _____
B. _____
C. _____
D. _____
E. _____
F. _____
G. _____
H. _____
I. _____

EXERCISE 5.15:
Skull—Midsagittal View

SELECT TOPIC	SELECT VIEW
Skull	Midsagittal

Click **LAYER 1** *in the* **LAYER CONTROLS** *window, and you will see the following image:*

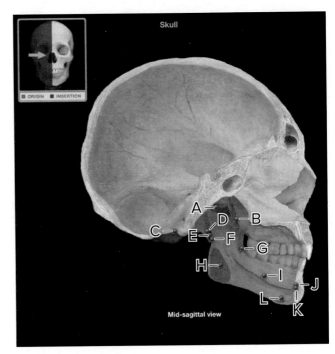

©McGraw-Hill Education

Mouse-over the pins on the screen to find the information necessary to identify the following structures:

A. _____
B. _____
C. _____
D. _____
E. _____
F. _____
G. _____
H. _____
I. _____
J. _____
K. _____
L. _____

Click **LAYER 2** *in the* **LAYER CONTROLS** *window, and you will see the following image:*

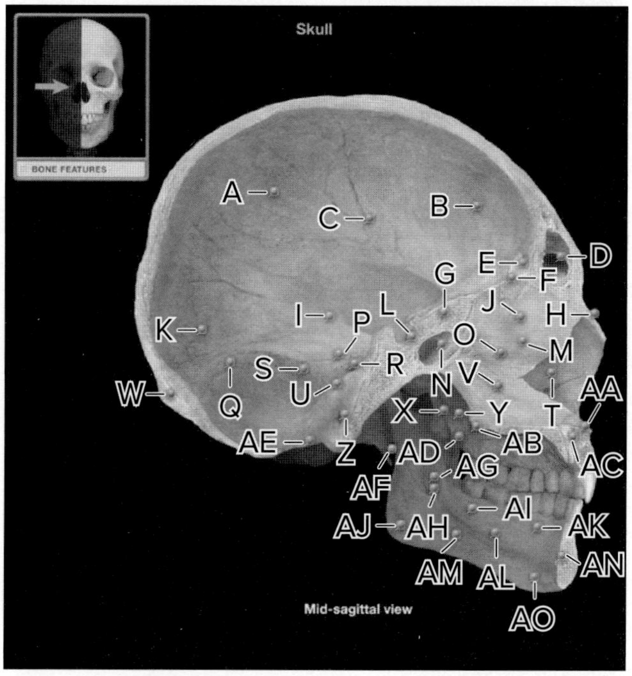

©McGraw-Hill Education

Mouse-over the pins on the screen to find the information necessary to identify the following structures:

A. _____
B. _____
C. _____
D. _____
E. _____
F. _____
G. _____
H. _____
I. _____
J. _____
K. _____
L. _____
M. _____
N. _____
O. _____
P. _____
Q. _____
R. _____
S. _____
T. _____
U. _____
V. _____
W. _____
X. _____
Y. _____
Z. _____
AA. _____
AB. _____
AC. _____
AD. _____
AE. _____
AF. _____
AG. _____
AH. _____
AI. _____
AJ. _____
AK. _____
AL. _____
AM. _____
AN. _____
AO. _____

CHECK POINT

Skull, Mid-sagittal View

1. When your dentist wants to numb your lower jaw by anesthetizing the nerves that serve the teeth and skin, what "hole" in what bone would be used to access those nerves?
2. Name the paranasal sinuses.
3. Name the structure that increases the surface area of the nasal cavity and plays an important role in warming inhaled air.
4. Name the structure that contains the sublingual salivary gland.
5. Name the muscles that attach to the styloid process of the temporal bone.

184 MODULE 5   The Skeletal System

EXERCISE 5.16:
Skull—Inferior View

SELECT TOPIC	SELECT VIEW
Skull	Inferior

Click **LAYER 1** *in the* **LAYER CONTROLS** *window, and you will see the following image:*

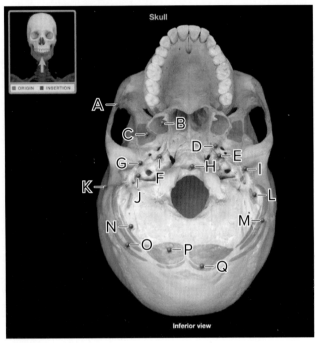

©McGraw-Hill Education

Mouse-over the pins on the screen to find the information necessary to identify the following structures:

A. _____
B. _____
C. _____
D. _____
E. _____
F. _____
G. _____
H. _____
I. _____
J. _____
K. _____
L. _____
M. _____
N. _____
O. _____
P. _____
Q. _____

Click **LAYER 2** *in the* **LAYER CONTROLS** *window, and you will see the following image:*

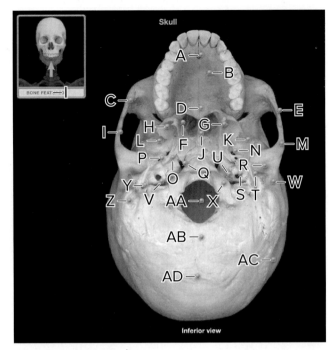

©McGraw-Hill Education

Mouse-over the pins on the screen to find the information necessary to identify the following structures:

A. _____
B. _____
C. _____
D. _____
E. _____
F. _____
G. _____
H. _____
I. _____
J. _____
K. _____
L. _____
M. _____
N. _____
O. _____
P. _____
Q. _____
R. _____
S. _____
T. _____
U. _____

V. _____
W. _____
X. _____
Y. _____
Z. _____
AA. _____
AB. _____
AC. _____
AD. _____

CHECK POINT

Skull, Inferior View

1. The zygomatic arch is made up of what two specific structures of what two bones?
2. What is the name for the large foramen on the inferior side of the occipital bone?
3. Name the irregular-shaped opening formed by the sphenoid, temporal, and occipital bones.

EXERCISE 5.17:

Skull—Cranial Cavity—Superior View

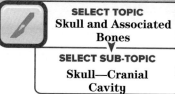

SELECT TOPIC: Skull and Associated Bones
SELECT SUB-TOPIC: Skull—Cranial Cavity
SELECT VIEW: Superior

Click **LAYER 1** in the **LAYER CONTROLS** window, and you will see the following image:

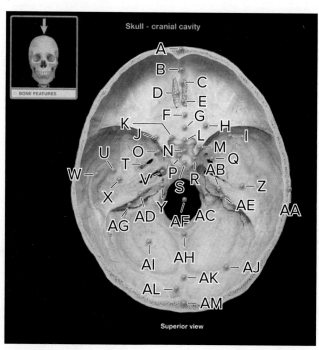

Mouse-over the pins on the screen to find the information necessary to identify the following structures:

A. _____
B. _____
C. _____
D. _____
E. _____
F. _____
G. _____
H. _____
I. _____
J. _____
K. _____
L. _____
M. _____
N. _____
O. _____
P. _____
Q. _____
R. _____
S. _____
T. _____
U. _____
V. _____
W. _____
X. _____
Y. _____
Z. _____
AA. _____
AB. _____
AC. _____
AD. _____
AE. _____
AF. _____
AG. _____
AH. _____
AI. _____
AJ. _____
AK. _____
AL. _____
AM. _____

©McGraw-Hill Education

186 MODULE 5 The Skeletal System

CHECK POINT
Skull—Cranial Cavity, Superior View
1. Describe the cribriform plate.
2. Name the structures that traverse through the jugular foramen.
3. Name the structure that passes through the foramen ovale.
4. Name the structure contained in the hypophyseal fossa.
5. What structure is transmitted by the foramen rotundum?

EXERCISE 5.18:
Orbit

SELECT TOPIC	SELECT VIEW
Orbit	Anterolateral

Click **LAYER 1** *in the* **LAYER CONTROLS** *window, and you will see the following image:*

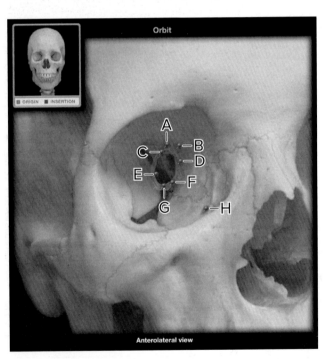

©McGraw-Hill Education

Mouse-over the pins on the screen to find the information necessary to identify the following structures:

A. _____
B. _____
C. _____
D. _____
E. _____
F. _____
G. _____
H. _____

Click **LAYER 2** *in the* **LAYER CONTROLS** *window, and you will see the following image:*

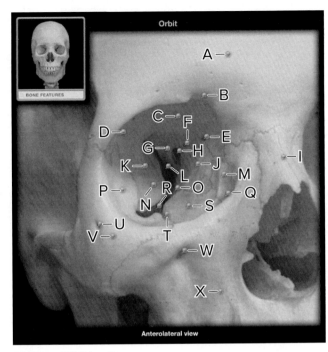

©McGraw-Hill Education

Mouse-over the pins on the screen to find the information necessary to identify the following structures:

A. _____
B. _____
C. _____
D. _____
E. _____
F. _____
G. _____
H. _____
I. _____
J. _____
K. _____
L. _____
M. _____
N. _____
O. _____
P. _____
Q. _____
R. _____
S. _____
T. _____
U. _____
V. _____
W. _____
X. _____

CHECK POINT

Orbit

1. Name the seven bones that constitute the orbit.
2. Name the structure that transmits the optic nerve (CN II) and the ophthalmic artery.
3. Name the nerves and blood vessels transmitted by the superior orbital fissure.
4. Name the nerves and blood vessels transmitted by the inferior orbital fissure.
5. Name the nerves and blood vessels transmitted by the anterior ethmoidal foramen.

EXERCISE 5.19:
Ethmoid Bone

SELECT TOPIC: Ethmoid Bone ▸ SELECT VIEW: Superior–Posterior

Click **LAYER 1** *in the* **LAYER CONTROLS** *window, and you will see the following image:*

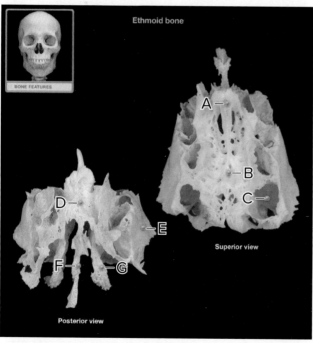

©McGraw-Hill Education

Mouse-over the pins on the screen to find the information necessary to identify the following structures:

A. _____
B. _____
C. _____
D. _____
E. _____
F. _____
G. _____

CHECK POINT

Ethmoid Bone

1. Name the characteristic features of the ethmoid bone.
2. Describe the cribriform plate of the ethmoid bone.
3. Name the structure that forms the superior part of the nasal septum.

EXERCISE 5.20:
Frontal Bone

SELECT TOPIC: Frontal Bone ▸ SELECT VIEW: Anterior

Click **LAYER 1** *in the* **LAYER CONTROLS** *window, and you will see the following image:*

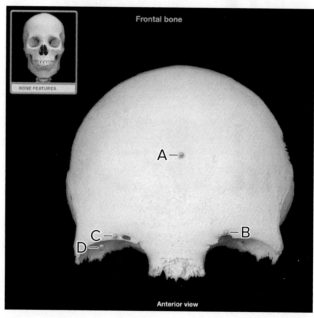

©McGraw-Hill Education

Mouse-over the pins on the screen to find the information necessary to identify the following structures:

A. _____
B. _____
C. _____
D. _____

CHECK POINT

Frontal Bone

1. Name the structures that traverse the supraorbital notch.
2. Describe the supraorbital margin.
3. Name the sutures associated with the frontal bone. Name the bones that articulate the frontal bone at each suture.

188 MODULE 5 The Skeletal System

EXERCISE 5.21:
Mandible

SELECT TOPIC	SELECT VIEW
Mandible	Lateral

Click **LAYER 1** *in the* **LAYER CONTROLS** *window, and you will see the following image:*

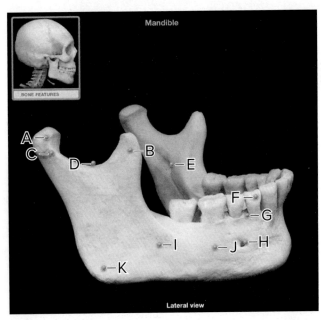

©McGraw-Hill Education

Mouse-over the pins on the screen to find the information necessary to identify the following structures:

A. _____
B. _____
C. _____
D. _____
E. _____
F. _____
G. _____
H. _____
I. _____
J. _____
K. _____

CHECK POINT

Mandible

1. Name the bone structures that constitute the temporomandibular joint.
2. What structure is also known as the mandibular incisure?

EXERCISE 5.22:
Maxilla

SELECT TOPIC	SELECT VIEW
Maxilla	Medial and Lateral

Click **LAYER 1** *in the* **LAYER CONTROLS** *window, and you will see the following image:*

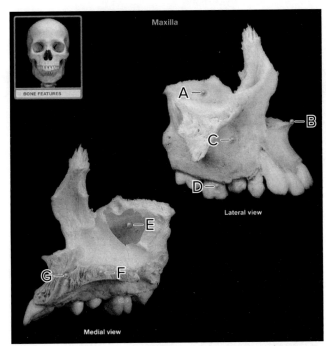

©McGraw-Hill Education

Mouse-over the pins on the screen to find the information necessary to identify the following structures:

A. _____
B. _____
C. _____
D. _____
E. _____
F. _____
G. _____

CHECK POINT

Maxilla

1. Name a structure of the maxilla bone that contributes to voice resonance.
2. Name the function of the incisive canal.
3. Name the structure that forms the anterior three-quarters of the hard palate.

MODULE 5 The Skeletal System 189

EXERCISE 5.23:
Occipital Bone

SELECT TOPIC	SELECT VIEW
Occipital Bone	Superior and Inferior

Click **LAYER 1** *in the* **LAYER CONTROLS** *window, and you will see the following image:*

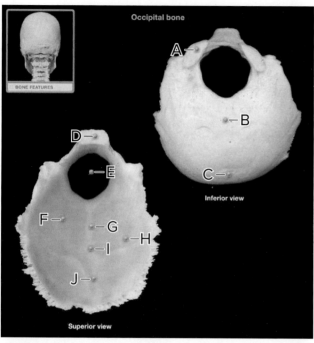

©McGraw-Hill Education

Mouse-over the pins on the screen to find the information necessary to identify the following structures:

A. _____
B. _____
C. _____
D. _____
E. _____
F. _____
G. _____
H. _____
I. _____
J. _____

CHECK POINT
Occipital Bone

1. Name the structures that pass through the foramen magnum.
2. What is the function of the occipital condyles?
3. What structures are contained within the cerebellar fossa?

EXERCISE 5.24:
Parietal Bone

SELECT TOPIC	SELECT VIEW
Parietal Bone	Lateral

Click **LAYER 1** *in the* **LAYER CONTROLS** *window, and you will see the following image:*

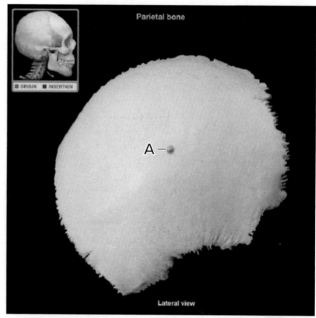

©McGraw-Hill Education

Mouse-over the pin on the screen to find the information necessary to identify the following structure:

A. _____

CHECK POINT
Parietal Bone

1. Name the bones that articulate with the parietal bone.
2. Name the sutures involved with each articulation listed in question 1.

190 MODULE 5   The Skeletal System

EXERCISE 5.25:
Sphenoid Bone

SELECT TOPIC: Sphenoid Bone
SELECT VIEW: Superior and Posterior

Click **LAYER 1** *in the* **LAYER CONTROLS** *window, and you will see the following image:*

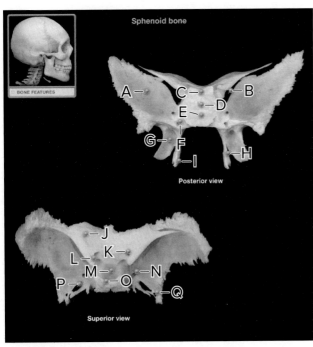

©McGraw-Hill Education

Mouse-over the pins on the screen to find the information necessary to identify the following structures:

A. _____
B. _____
C. _____
D. _____
E. _____
F. _____
G. _____
H. _____
I. _____
J. _____
K. _____
L. _____
M. _____
N. _____
O. _____
P. _____
Q. _____

CHECK POINT
Sphenoid Bone

1. What structure is contained in the sella turcica?
2. Name the two structures transmitted through the optic canal.
3. Name the paired lateral projections of the sphenoid bone.

EXERCISE 5.26:
Temporal Bone

SELECT TOPIC: Temporal Bone
SELECT VIEW: Medial and Lateral

Click **LAYER 1** *in the* **LAYER CONTROLS** *window, and you will see the following image:*

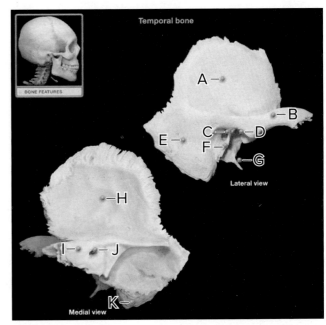

©McGraw-Hill Education

Mouse-over the pins on the screen to find the information necessary to identify the following structures:

A. _____
B. _____
C. _____
D. _____
E. _____
F. _____
G. _____
H. _____
I. _____
J. _____
K. _____

CHECK POINT

Temporal Bone

1. Name the bony canal that ends at the tympanic membrane.
2. Name the bony canal traversed by the facial and vestibulocochlear nerves.
3. Name the bony structure containing the organs of hearing and equilibrium.

EXERCISE 5.27:
Teeth

SELECT TOPIC	SELECT VIEW
Teeth	Superior and Inferior

Click **LAYER 1** *in the* **LAYER CONTROLS** *window, and you will see the following image:*

©McGraw-Hill Education

Mouse-over the pins on the screen to find the information necessary to identify the following structures:

A. _____
B. _____
C. _____
D. _____
E. _____
F. _____
G. _____
H. _____
I. _____
J. _____
K. _____
L. _____
M. _____
N. _____
O. _____
P. _____
Q. _____
R. _____
S. _____
T. _____
U. _____
V. _____
W. _____
X. _____
Y. _____
Z. _____
AA. _____
AB. _____

CHECK POINT

Teeth

1. What four teeth are also known as wisdom teeth?
2. What structures are transmitted through the lesser palatine foramen?
3. What structures are transmitted through the greater palatine foramen?

EXERCISE 5.28:
Hyoid Bone

SELECT TOPIC	SELECT VIEW
Hyoid Bone	Anterior and Lateral

Click **LAYER 1** *in the* **LAYER CONTROLS** *window, and you will see the following image:*

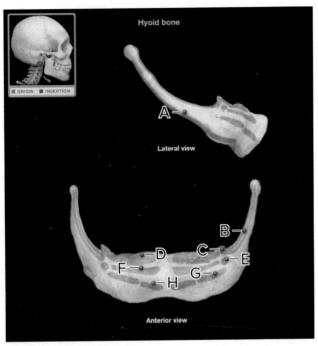

©McGraw-Hill Education

Mouse-over the pins on the screen to find the information necessary to identify the following structures:

A. _____
B. _____
C. _____
D. _____
E. _____
F. _____
G. _____
H. _____

Click **LAYER 2** *in the* **LAYER CONTROLS** *window, and you will see the following image:*

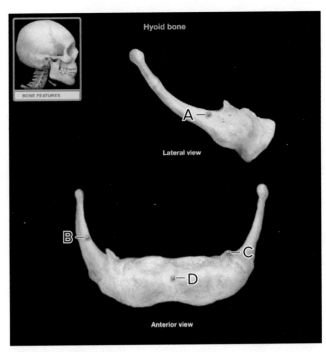

©McGraw-Hill Education

Mouse-over the pins on the screen to find the information necessary to identify the following structures:

A. _____
B. _____
C. _____
D. _____

CHECK POINT

Hyoid Bone

1. Name the function of the greater horn of the hyoid bone.
2. Name the function of the lesser horn of the hyoid bone.
3. What muscles attach to the body of the hyoid bone?

EXERCISE 5.29a:
Imaging—Skull

SELECT TOPIC: Skull ▸ **SELECT VIEW:** X ray: Anterior–Posterior

Click the **TAGS ON/OFF** button, and you will see the following image:

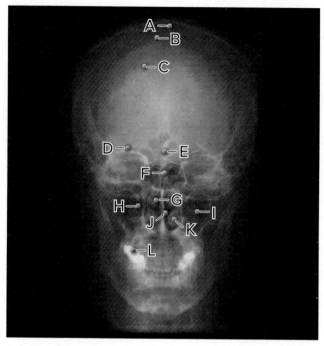

©McGraw-Hill Education

Mouse-over the pins on the screen to find the information necessary to identify the following structures:

A. _____
B. _____
C. _____
D. _____
E. _____
F. _____
G. _____
H. _____
I. _____
J. _____
K. _____

Nonskeletal System Structure (blue pin)

L. _____

EXERCISE 5.29b:
Imaging—Skull

SELECT TOPIC: Skull ▸ **SELECT VIEW:** X ray—Lateral

Click the **TAGS ON/OFF** button and you will see the following image:

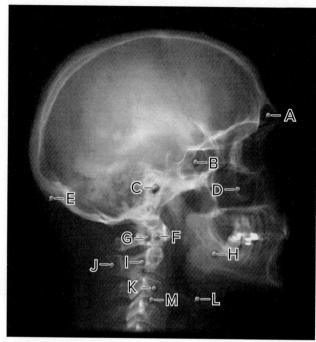

©McGraw-Hill Education

Mouse-over the pins on the screen to find the information necessary to identify the following structures:

A. _____
B. _____
C. _____
D. _____
E. _____
F. _____
G. _____
H. _____
I. _____
J. _____
K. _____
L. _____
M. _____

Self-Quiz
Take this opportunity to check your progress by taking the **QUIZ**. See the **Introduction Module** for a reminder on how to access the **QUIZ** for this Study Area.

194 MODULE 5 The Skeletal System

IN REVIEW

What Have I Learned?

The following questions cover the material that you just read, the introduction to the skeleton and the skeletal structures of the head and neck. Apply what you have learned in answering these questions on a separate piece of paper.

1. The hard palate consists of which bones?

2. What is the name for the large foramen on the inferior side of the occipital bone?

3. What passes through this foramen?

4. Name the two bones that form the bridge of the nose.

5. At the junction of the temporal and occipital bones, the internal jugular vein passes through which foramen?

6. The greater and lesser wings are parts of what bone?

7. Name the bony canal of the external ear. What bone is it a part of?

8. What is the attachment site for the sternocleidomastoid muscle on the skull?

9. What bony landmark is defined as "any bony prominence"?

10. What bony landmark is defined as "a pit or socket"?

11. What bony landmark is defined as "a hole through a bone, usually round"?

12. What bony landmark is defined as "a smooth, rounded knob"?

Vertebral Column

EXERCISE 5.30:
Vertebral Column

SELECT TOPIC
Vertebral Column

SELECT VIEW
Anterior, Posterior, and Lateral

Click **LAYER 1** in the **LAYER CONTROLS** window, and you will see the following image:

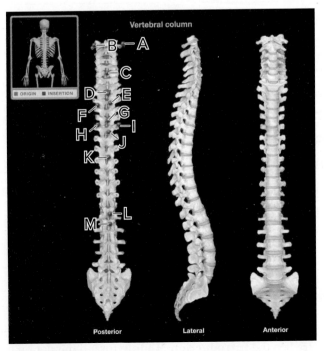

©McGraw-Hill Education

Mouse-over the pins on the screen to find the information necessary to identify the following structures:

A. _____

B. _____

C. _____

D. _____

E. _____

F. _____

G. _____

H. _____

I. _____

J. _____

K. _____

L. _____

M. _____

Click **LAYER 2** *in the* **LAYER CONTROLS** *window, and you will see the following image:*

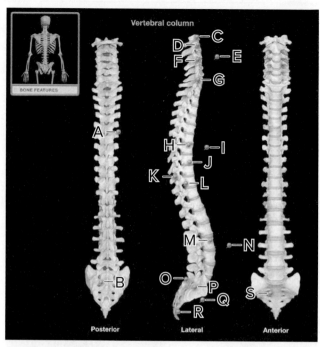

©McGraw-Hill Education

Mouse-over the pins on the screen to find the information necessary to identify the following structures:

A. _____
B. _____
C. _____
D. _____
E. _____
F. _____
G. _____
H. _____
I. _____
J. _____
K. _____
L. _____
M. _____
N. _____
O. _____
P. _____
Q. _____
R. _____
S. _____

EXERCISE 5.31:
Imaging—Vertebral Column (with Scoliosis)

 SELECT TOPIC: Vertebral Column (with scoliosis) ▶ **SELECT VIEW**: X ray—Anterior–Posterior

Click the **TAGS ON/OFF** *button, and you will see the following image:*

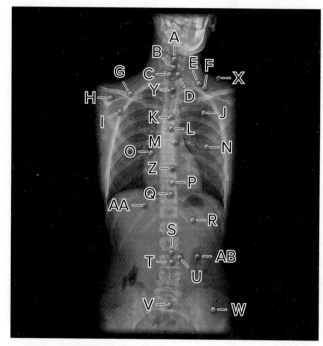

©McGraw-Hill Education

Mouse-over the pins on the screen to find the information necessary to identify the following structures:

A. _____
B. _____
C. _____
D. _____
E. _____
F. _____
G. _____
H. _____
I. _____
J. _____
K. _____
L. _____
M. _____
N. _____

MODULE 5 The Skeletal System

O. _____
P. _____
Q. _____
R. _____
S. _____
T. _____
U. _____
V. _____
W. _____

Nonskeletal System Structures (blue pins)

X. _____
Y. _____
Z. _____
AA. _____
AB. _____

CHECK POINT

Vertebral Column

1. Name the different groups of vertebrae. How many of each in a typical skeleton?
2. What structure passes through each intervertebral foramen?
3. Which vertebra has a dens?
4. Which vertebra has no vertebral body?

EXERCISE 5.32:
Cervical Vertebra

SELECT TOPIC: Cervical Vertebra ▶ SELECT VIEW: Superior and Lateral

*Click **LAYER 1** in the **LAYER CONTROLS** window, and you will see the following image:*

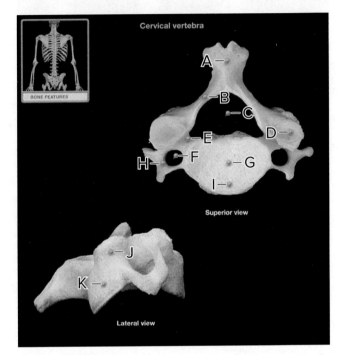

©McGraw-Hill Education

Mouse-over the pins on the screen to find the information necessary to identify the following structures:

A. _____
B. _____
C. _____
D. _____
E. _____
F. _____
G. _____
H. _____
I. _____
J. _____
K. _____

MODULE 5   The Skeletal System   197

EXERCISE 5.33a:
Imaging—Cervical Region

SELECT TOPIC	SELECT VIEW
Cervical Region	X ray—Lateral

Click the **TAGS ON/OFF** *button, and you will see the following image:*

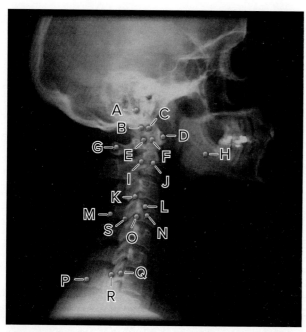

©McGraw-Hill Education

Mouse-over the pins on the screen to find the information necessary to identify the following structures:

A. _____
B. _____
C. _____
D. _____
E. _____
F. _____
G. _____
H. _____
I. _____
J. _____
K. _____
L. _____
M. _____
N. _____
O. _____
P. _____
Q. _____
R. _____

Nonskeletal System Structure (blue pin)

S. _____

EXERCISE 5.33b:
Imaging—Cervical Region

SELECT TOPIC	SELECT VIEW
Cervical Region	CT—Coronal

Click the **TAGS ON/OFF** *button, and you will see the following image:*

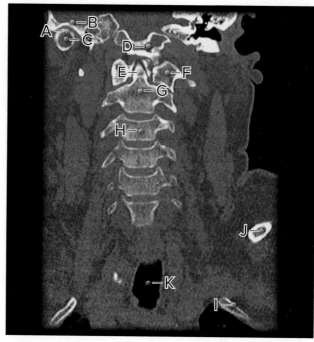

©McGraw-Hill Education

Mouse-over the pins on the screen to find the information necessary to identify the following structures:

A. _____
B. _____
C. _____
D. _____
E. _____
F. _____
G. _____
H. _____
I. _____
J. _____

Nonskeletal System Structure (blue pin)

K. _____

MODULE 5 The Skeletal System

EXERCISE 5.33c:
Imaging—Cervical Region

SELECT TOPIC: Cervical Region
SELECT VIEW: CT—Sagittal

Click the **TAGS ON/OFF** button, and you will see the following image:

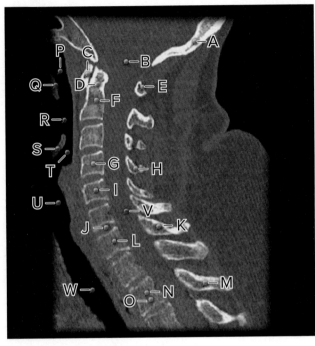

©McGraw-Hill Education

Mouse-over the pins on the screen to find the information necessary to identify the following structures:

A. _____
B. _____
C. _____
D. _____
E. _____
F. _____
G. _____
H. _____
I. _____
J. _____
K. _____
L. _____
M. _____
N. _____
O. _____

Nonskeletal System Structures (blue pins)

P. _____
Q. _____
R. _____
S. _____
T. _____
U. _____
V. _____
W. _____

CHECK POINT

Cervical Vertebra

1. Name a characteristic unique to the transverse processes of the cervical vertebrae.
2. What structure passes through this opening in C1–C6?
3. What is unique about the spinous process of most cervical vertebrae?
4. Which cervical vertebra does not have a spinous process?

EXERCISE 5.34:
Atlas (C1 Vertebra)

SELECT TOPIC: Atlas (C1 Vertebra)
SELECT VIEW: Superior

Click **LAYER 1** in the **LAYER CONTROLS** window, and you will see the following image:

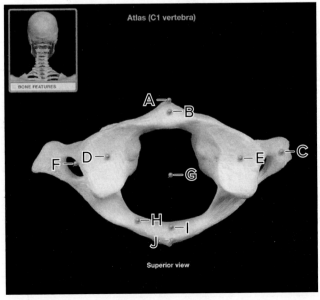

©McGraw-Hill Education

Mouse-over the pins on the screen to find the information necessary to identify the following structures:

A. _____
B. _____
C. _____
D. _____
E. _____
F. _____
G. _____
H. _____
I. _____
J. _____

EXERCISE 5.35:
Imaging—Atlas (C1 Vertebra)

SELECT TOPIC	SELECT VIEW
Atlas (C1 Vertebra)	CT—Axial

Click the **TAGS ON/OFF** *button, and you will see the following image:*

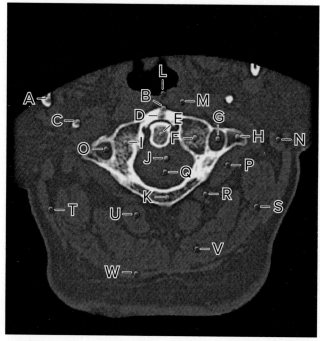

©McGraw-Hill Education

Mouse-over the pins on the screen to find the information necessary to identify the following structures:

A. _____
B. _____
C. _____
D. _____
E. _____
F. _____
G. _____
H. _____
I. _____
J. _____
K. _____

Nonskeletal System Structures (blue pins)

L. _____
M. _____
N. _____
O. _____
P. _____
Q. _____
R. _____
S. _____
T. _____
U. _____
V. _____
W. _____

CHECK POINT

Atlas (C1 Vertebra)

1. What is the most posterior structure of the Atlas (C1 vertebra)?
2. What is its function?
3. What structure is lacking on the atlas and coccygeal vertebrae that is present on all other vertebrae?

EXERCISE 5.36:
Axis (C2 Vertebra)

SELECT TOPIC: Axis (C2 Vertebra) ▶ **SELECT VIEW:** Posteriosuperior

Click **LAYER 1** in the **LAYER CONTROLS** *window, and you will see the following image:*

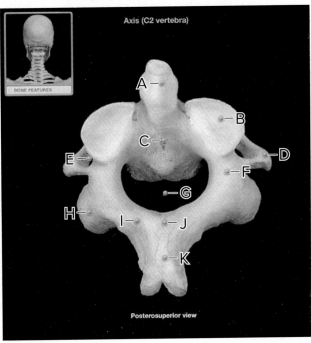

©McGraw-Hill Education

Mouse-over the pins on the screen to find the information necessary to identify the following structures:

A. _____
B. _____
C. _____
D. _____
E. _____
F. _____
G. _____
H. _____
I. _____
J. _____
K. _____

CHECK POINT

Axis (C2 Vertebra)

1. Name a structure unique to the axis.
2. What is another name for this structure?
3. What does this structure represent?

EXERCISE 5.37:
Thoracic Vertebra

SELECT TOPIC: Thoracic Vertebra ▶ **SELECT VIEW:** Superior and Lateral

Click **LAYER 1** in the **LAYER CONTROLS** *window, and you will see the following image:*

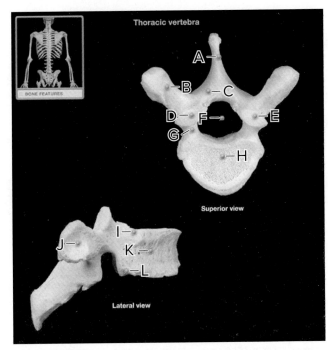

©McGraw-Hill Education

Mouse-over the pins on the screen to find the information necessary to identify the following structures:

A. _____
B. _____
C. _____
D. _____
E. _____
F. _____
G. _____
H. _____
I. _____
J. _____
K. _____
L. _____

CHECK POINT

Thoracic Vertebra

1. List the characteristic features of thoracic vertebrae.
2. How many thoracic vertebrae do you have?
3. Where are they located?

MODULE 5 The Skeletal System 201

EXERCISE 5.38:
Lumbar Vertebra

SELECT TOPIC	SELECT VIEW
Lumbar Vertebra	Superior–Lateral

Click **LAYER 1** in the **LAYER CONTROLS** window, and you will see the following image:

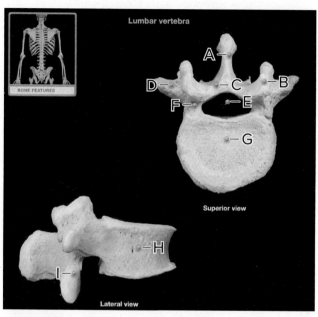

©McGraw-Hill Education

Mouse-over the pins on the screen to find the information necessary to identify the following structures:

A. _____
B. _____
C. _____
D. _____
E. _____
F. _____
G. _____
H. _____
I. _____

EXERCISE 5.39:
Imaging—Lumbar Region

SELECT TOPIC	SELECT VIEW
Lumbar Region	X ray—Lateral

Click the **TAGS ON/OFF** button, and you will see the following image:

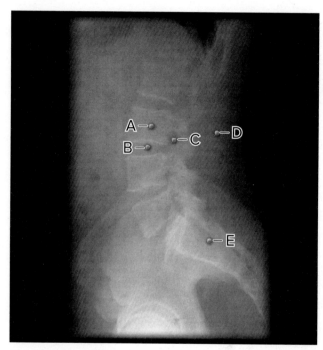

©McGraw-Hill Education

Mouse-over the pins on the screen to find the information necessary to identify the following structures:

A. _____
B. _____
C. _____
D. _____
E. _____

CHECK POINT
Lumbar Vertebra

1. List the characteristic features of lumbar vertebrae.
2. How many lumbar vertebrae do you have?
3. Where are they located?

SELECT ANIMATION
Joint Movements: Intervertebral Joints   PLAY

After viewing the animation, answer the following questions:

1. How are intervertebral joints formed?

2. Where on the vertebral column is there no intervertebral disc?

3. Describe flexion of a joint. In what direction does flexion of a joint usually occur? Is there an exception?

4. Describe flexion of intervertebral joints. Give an example of this flexion.

5. Describe extension of a joint. In what direction does extension of a joint usually occur? Is there an exception?

6. Describe extension of intervertebral joints. Give an example of this extension.

7. Describe lateral flexion of intervertebral joints. Give an example of this lateral flexion.

8. Describe rotation of a joint.

9. Describe rotation of intervertebral joints.

SELECT ANIMATION
Joint Movements: Cervical Intervertebral Joints   PLAY

After viewing the animation, answer the following questions:

1. How are intervertebral joints formed?

2. Where on the vertebral column is there no intervertebral disc?

3. Describe flexion of a joint. In what direction does flexion of a joint usually occur? Is there an exception?

4. Describe flexion of cervical vertebrae.

5. Describe extension of a joint. In what direction does extension of a joint usually occur? Is there an exception?

6. Describe extension of cervical vertebrae.

7. Describe hyperextension of cervical vertebrae.

8. Describe lateral flexion of cervical vertebrae. Give an example of this lateral flexion.

EXERCISE 5.40
Sacrum and Coccyx

SELECT TOPIC	SELECT VIEW
Sacrum and Coccyx	Anterior–Posterior

Click **LAYER 1** *in the* **LAYER CONTROLS** *window, and you will see the following image:*

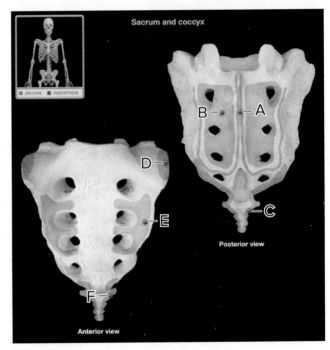

©McGraw-Hill Education

Mouse-over the pins on the screen to find the information necessary to identify the following structures:

A. _____

B. _____

C. _____

D. _____

E. _____

F. _____

Click **LAYER 2** *in the* **LAYER CONTROLS** *window, and you will see the following image:*

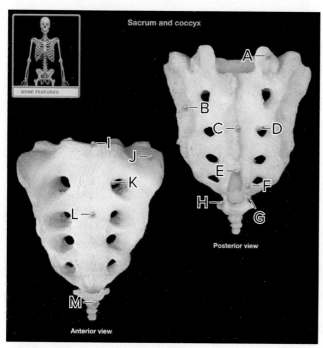

Mouse-over the pins on the screen to find the information necessary to identify the following structures:

A. _____

B. _____

C. _____

D. _____

E. _____

F. _____

G. _____

H. _____

I. _____

J. _____

K. _____

L. _____

M. _____

CHECK POINT

Sacrum and Coccyx

1. Name the landmark for establishing female pelvic dimensions.
2. Name the route for injection in caudal epidural anesthesia.
3. Describe the coccyx.

Self-Quiz

Take this opportunity to check your progress by taking the **QUIZ**. See the **Introduction Module** for a reminder on how to access the **QUIZ** for this Study Area.

IN REVIEW

What Have I Learned?

The following questions cover the material that you just read, the vertebral column. Apply what you have learned in answering these questions on a separate piece of paper.

1. Name the different groups of vertebrae. How many of each in a typical skeleton?

2. What structure passes through each intervertebral foramen?

3. Which vertebra has a dens?

4. Which vertebra has no vertebral body?

5. Name a characteristic unique to the transverse processes of the cervical vertebrae.

6. What structure passes through this opening in C1–C7?

7. What is unique about the spinous process of most cervical vertebrae?

8. Which vertebra does not have a spinous process?

9. What is the most posterior structure of the Atlas (C1 vertebra)?

10. What is its function?

11. What structure is lacking on the atlas and coccygeal vertebrae that is present on all other vertebrae?

12. Name a structure unique to the axis.

13. What is another name for this structure?

14. What does this structure represent?

15. List the characteristic features of thoracic vertebrae.

16. How many thoracic vertebrae do you have?

17. Where are they located?

18. List the characteristic features of lumbar vertebrae.

19. How many lumbar vertebrae do you have?

20. Where are they located?

21. Name the landmark for establishing female pelvic dimensions.

22. Name the route for injection in caudal epidural anesthesia.

23. Describe the coccyx.

Thoracic Cage

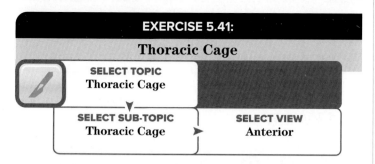

EXERCISE 5.41:
Thoracic Cage

SELECT TOPIC: Thoracic Cage
SELECT SUB-TOPIC: Thoracic Cage
SELECT VIEW: Anterior

Click **LAYER 1** *in the* **LAYER CONTROLS** *window, and you will see the following image:*

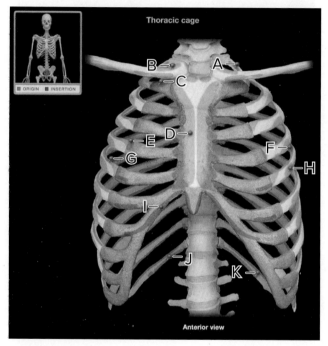

©McGraw-Hill Education

Mouse-over the pins on the screen to find the information necessary to identify the following structures:

A. _____
B. _____
C. _____
D. _____
E. _____
F. _____
G. _____
H. _____
I. _____
J. _____
K. _____

Click **LAYER 2** *in the* **LAYER CONTROLS** *window, and you will see the following image:*

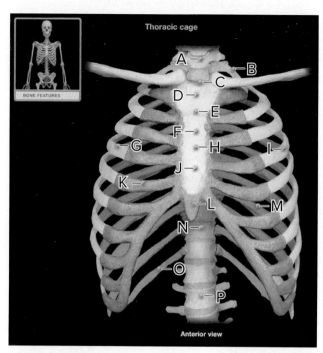

©McGraw-Hill Education

Mouse-over the pins on the screen to find the information necessary to identify the following structures:

A. _____
B. _____
C. _____
D. _____
E. _____
F. _____
G. _____
H. _____
I. _____
J. _____
K. _____
L. _____
M. _____
N. _____
O. _____
P. _____

CHECK POINT
Thoracic Cage

1. Describe the thoracic cage.
2. Define a true rib. How many are there?
3. Define a false rib. How many are there?
4. Define a floating rib. How many are there?

206 MODULE 5 The Skeletal System

EXERCISE 5.42a:
Imaging—Thorax

SELECT TOPIC	SELECT VIEW
Thorax	X ray: Posterior–Anterior

*Click the **TAGS ON/OFF** button, and you will see the following image:*

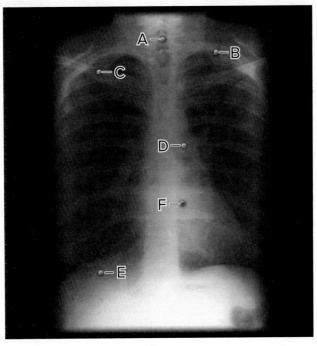

©McGraw-Hill Education

Mouse-over the pins on the screen to find the information necessary to identify the following structures:

A. _____
B. _____
C. _____
D. _____
E. _____

Nonskeletal System Structure (blue pin)

F. _____

EXERCISE 5.42b:
Imaging—Thorax

SELECT TOPIC	SELECT VIEW
Thorax	CT—Axial

*Click the **TAGS ON/OFF** button, and you will see the following image:*

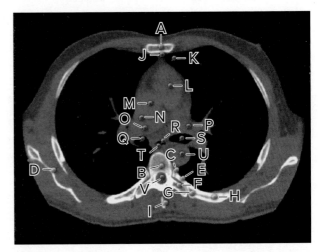

©McGraw-Hill Education

Mouse-over the pins on the screen to find the information necessary to identify the following structures:

A. _____
B. _____
C. _____
D. _____
E. _____
F. _____
G. _____
H. _____
I. _____

Nonskeletal System Structures (blue pins)

J. _____
K. _____
L. _____
M. _____
N. _____
O. _____
P. _____
Q. _____
R. _____
S. _____
T. _____
U. _____
V. _____

MODULE 5   The Skeletal System   207

EXERCISE 5.42c:
Imaging—Thorax

SELECT TOPIC: **Thorax** ▸ SELECT VIEW: **CT—Sagittal**

Click the **TAGS ON/OFF** *button, and you will see the following image:*

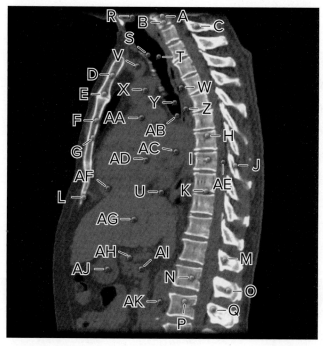

©McGraw-Hill Education

Mouse-over the pins on the screen to find the information necessary to identify the following structures:

A. _____
B. _____
C. _____
D. _____
E. _____
F. _____
G. _____
H. _____
I. _____
J. _____
K. _____
L. _____
M. _____
N. _____
O. _____
P. _____
Q. _____

Nonskeletal System Structures (blue pins)

R. _____
S. _____
T. _____
U. _____
V. _____
W. _____
X. _____
Y. _____
Z. _____
AA. _____
AB. _____
AC. _____
AD. _____
AE. _____
AF. _____
AG. _____
AH. _____
AI. _____
AJ. _____
AK. _____

IN REVIEW

What Have I Learned?

1. Describe the thoracic cage.

2. Define a true rib. How many are there?

3. Define a false rib. How many are there?

4. Define a floating rib. How many are there?

Pectoral Girdle and Upper Limb

EXERCISE 5.43:
Shoulder and Arm—Anterior View

SELECT TOPIC: Pectoral Girdle and Upper Limb
SELECT SUB-TOPIC: Shoulder and Arm
SELECT VIEW: Anterior

Click **LAYER 1** in the **LAYER CONTROLS** window, and you will see the following image:

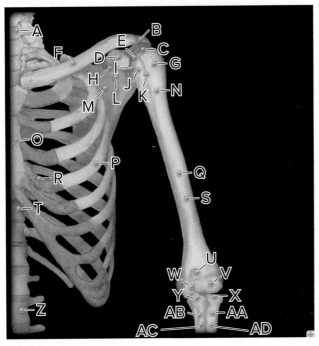

©McGraw-Hill Education

Mouse-over the pins on the screen to find the information necessary to identify the following structures:

A. ___
B. ___
C. ___
D. ___
E. ___
F. ___
G. ___
H. ___
I. ___
J. ___
K. ___
L. ___
M. ___
N. ___
O. ___
P. ___
Q. ___
R. ___
S. ___
T. ___
U. ___
V. ___
W. ___
X. ___
Y. ___
Z. ___
AA. ___
AB. ___
AC. ___
AD. ___

EXERCISE 5.44:
Shoulder and Arm—Posterior View

SELECT TOPIC: Shoulder and Arm
SELECT VIEW: Posterior

Click **LAYER 1** in the **LAYER CONTROLS** window, and you will see the following image:

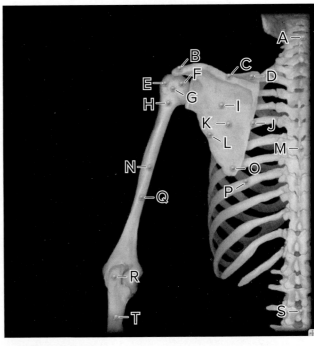

©McGraw-Hill Education

Mouse-over the pins on the screen to find the information necessary to identify the following structures:

A. _____
B. _____
C. _____
D. _____
E. _____
F. _____
G. _____
H. _____
I. _____
J. _____
K. _____
L. _____
M. _____
N. _____
O. _____
P. _____
Q. _____
R. _____
S. _____
T. _____

CHECK POINT

Shoulder and Arm

1. Which bony structures form the glenohumeral joint?
2. How many ribs do you have?
3. Both true and false ribs articulate with _____.
4. Name the characteristic features of the scapula.

SELECT ANIMATION
Joint Movements: Scapula

 PLAY

After viewing the animation, answer the following questions:

1. The scapula does not have _____ with the axial skeleton.

2. The scapula's movement affects _____.

3. The scapula's movement is controlled by _____.

4. Describe elevation of a body part.

5. Describe elevation of the scapula. Give an example of this elevation.

6. Describe depression of a body part.

7. Describe depression of the scapula. Give an example of this depression.

8. Describe protraction of a body part. What is protrusion?

9. Describe protraction of the scapula. Give an example of this protraction.

10. Describe retraction of a body part.

11. Describe retraction of the scapula. Give an example of this retraction.

12. Describe upward rotation of a body part.

13. Describe retraction of the scapula. Give an example of this retraction.

14. Describe upward rotation of the scapula. Give an example of this upward rotation.

15. Upward rotation is also known as _____. Why?

16. Describe downward rotation of the scapula. Give an example of this upward rotation.

17. Downward rotation is also known as _____. Why?

EXERCISE 5.45:
Imaging—Shoulder and Arm

SELECT TOPIC	SELECT VIEW
Shoulder	X ray—Anterior–Posterior

Click the **TAGS ON/OFF** button, and you will see the following image:

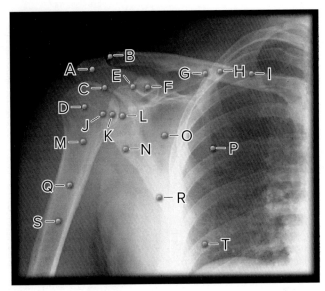

©McGraw-Hill Education

Mouse-over the pins on the screen to find the information necessary to identify the following structures:

A. _____
B. _____
C. _____
D. _____
E. _____
F. _____
G. _____
H. _____
I. _____
J. _____
K. _____
L. _____
M. _____
N. _____
O. _____
P. _____
Q. _____
R. _____
S. _____
T. _____

EXERCISE 5.46:
Clavicle

SELECT TOPIC	SELECT VIEW
Clavicle	Superior and Inferior

Click **LAYER 1** in the **LAYER CONTROLS** window, and you will see the following image:

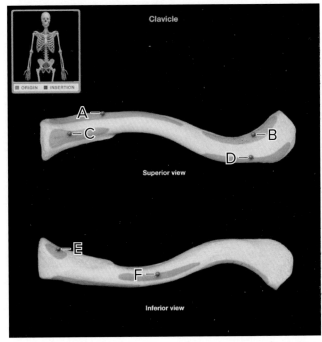

©McGraw-Hill Education

Mouse-over the pins on the screen to find the information necessary to identify the following structures:

A. _____
B. _____
C. _____
D. _____
E. _____
F. _____

Click **LAYER 2** *in the* **LAYER CONTROLS** *window, and you will see the following image:*

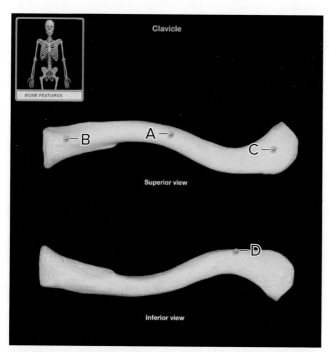

©McGraw-Hill Education

Mouse-over the pins on the screen to find the information necessary to identify the following structures:

A. _____

B. _____

C. _____

D. _____

CHECK POINT

Clavicle

1. The medial end of the clavicle articulates _____.
2. The lateral end of the clavicle articulates _____.
3. Another name for the clavicle is _____.

EXERCISE 5.47:

Scapula

 | SELECT TOPIC: Scapula ▸ SELECT VIEW: Anterior, Posterior and Lateral

Click **LAYER 1** *in the* **LAYER CONTROLS** *window, and you will see the following image:*

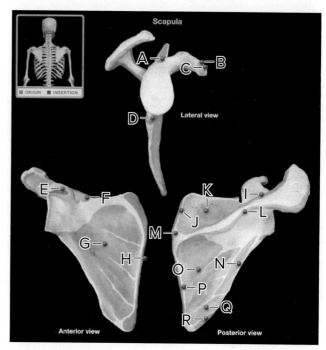

©McGraw-Hill Education

Mouse-over the pins on the screen to find the information necessary to identify the following structures:

A. _____
B. _____
C. _____
D. _____
E. _____
F. _____
G. _____
H. _____
I. _____
J. _____
K. _____
L. _____
M. _____
N. _____
O. _____
P. _____
Q. _____
R. _____

212 MODULE 5 The Skeletal System

Click **LAYER 2** *in the* **LAYER CONTROLS** *window, and you will see the following image:*

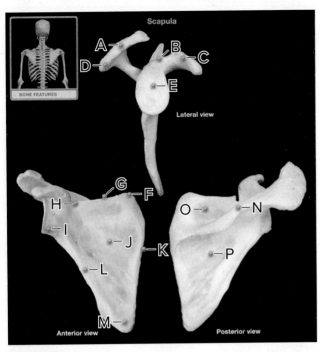

©McGraw-Hill Education

Mouse-over the pins on the screen to find the information necessary to identify the following structures:

A. _____
B. _____
C. _____
D. _____
E. _____
F. _____
G. _____
H. _____
I. _____
J. _____
K. _____
L. _____
M. _____
N. _____
O. _____
P. _____

CHECK POINT

Scapula

1. Name a landmark for intramuscular injections.
2. Name another visible subcutaneous landmark.

EXERCISE 5.48:

Humerus

 SELECT TOPIC: Humerus ▶ SELECT VIEW: Anterior and Posterior

Click **LAYER 1** *in the* **LAYER CONTROLS** *window, and you will see the following image:*

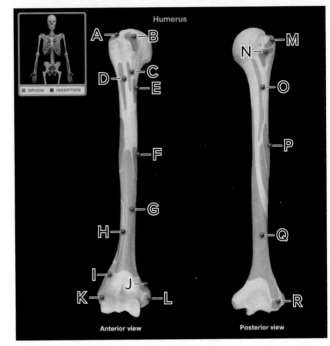

©McGraw-Hill Education

Mouse-over the pins on the screen to find the information necessary to identify the following structures:

A. _____
B. _____
C. _____
D. _____
E. _____
F. _____
G. _____
H. _____
I. _____
J. _____
K. _____
L. _____
M. _____
N. _____
O. _____
P. _____
Q. _____
R. _____

Click **LAYER 2** in the **LAYER CONTROLS** *window, and you will see the following image:*

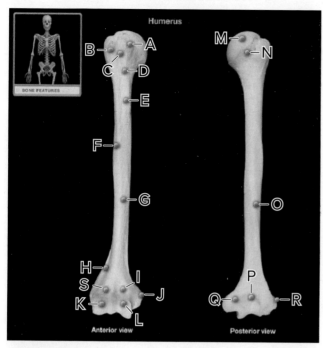

©McGraw-Hill Education

Mouse-over the pins on the screen to find the information necessary to identify the following structures:

A. _____
B. _____
C. _____
D. _____
E. _____
F. _____
G. _____
H. _____
I. _____
J. _____
K. _____
L. _____
M. _____
N. _____
O. _____
P. _____
Q. _____
R. _____
S. _____

CHECK POINT

Humerus

1. How many necks are on the proximal end of the humerus? Where are they located?
2. Which neck is a common site for fractures?

EXERCISE 5.49:
Forearm and Hand—Anterior View

SELECT TOPIC: Forearm and Hand ▸ **SELECT VIEW**: Anterior

Click **LAYER 1** in the **LAYER CONTROLS** *window, and you will see the following image:*

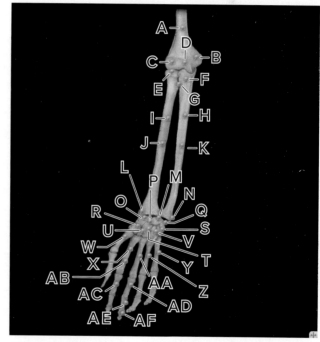

©McGraw-Hill Education

Mouse-over the pins on the screen to find the information necessary to identify the following structures:

A. _____
B. _____
C. _____
D. _____
E. _____
F. _____
G. _____

214 MODULE 5 The Skeletal System

H. _____
I. _____
J. _____
K. _____
L. _____
M. _____
N. _____
O. _____
P. _____
Q. _____
R. _____
S. _____
T. _____
U. _____
V. _____
W. _____
X. _____
Y. _____
Z. _____
AA. _____
AB. _____
AC. _____
AD. _____
AE. _____
AF. _____

EXERCISE 5.50:
Forearm and Hand—Posterior View

SELECT TOPIC	SELECT VIEW
Forearm and Hand	Posterior

Click **LAYER 1** in the **LAYER CONTROLS** window, and you will see the following image:

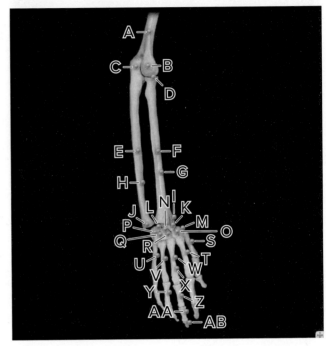

©McGraw-Hill Education

Mouse-over the pins on the screen to find the information necessary to identify the following structures:

A. _____
B. _____
C. _____
D. _____
E. _____
F. _____
G. _____
H. _____
I. _____
J. _____
K. _____
L. _____
M. _____
N. _____
O. _____
P. _____
Q. _____
R. _____

S. _____
T. _____
U. _____
V. _____
W. _____
X. _____
Y. _____
Z. _____
AA. _____
AB. _____

CHECK POINT
Forearm and Hand

1. List the proximal carpal bones from lateral to medial.
2. List the distal carpal bones from lateral to medial.
3. List the phalanges of fingers II through V from proximal to distal.
4. List the phalanges of finger I from proximal to distal.

Mouse-over the pins on the screen to find the information necessary to identify the following structures:

A. _____
B. _____
C. _____
D. _____
E. _____
F. _____
G. _____
H. _____
I. _____
J. _____
K. _____
L. _____
M. _____
N. _____
O. _____
P. _____
Q. _____
R. _____

EXERCISE 5.51a:
Imaging—Elbow

SELECT TOPIC: Elbow → SELECT VIEW: X ray—Anterior–Posterior

Click the **TAGS ON/OFF** *button, and you will see the following image:*

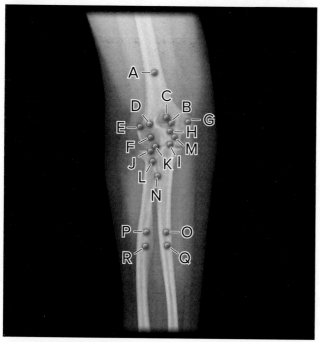

©McGraw-Hill Education

EXERCISE 5.51b:
Imaging—Elbow

SELECT TOPIC: Elbow → SELECT VIEW: X ray—Lateral

Click the **TAGS ON/OFF** *button, and you will see the following image:*

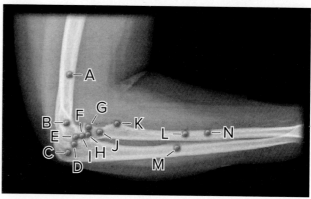

©McGraw-Hill Education

Mouse-over the pins on the screen to find the information necessary to identify the following structures:

A. _____
B. _____
C. _____
D. _____

E. _____
F. _____
G. _____
H. _____
I. _____
J. _____
K. _____
L. _____
M. _____
N. _____

J. _____
K. _____
L. _____
M. _____
N. _____
O. _____
P. _____
Q. _____
R. _____
S. _____
T. _____

EXERCISE 5.52:
Radius and Ulna

SELECT TOPIC: Radius and Ulna ▸ SELECT VIEW: Anterior and Posterior

Click **LAYER 1** *in the* **LAYER CONTROLS** *window, and you will see the following image:*

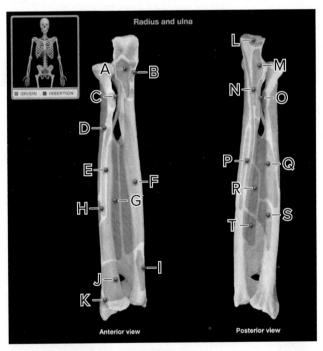

©McGraw-Hill Education

Mouse-over the pins on the screen to find the information necessary to identify the following structures:

A. _____
B. _____
C. _____
D. _____
E. _____
F. _____
G. _____
H. _____
I. _____

Click **LAYER 2** *in the* **LAYER CONTROLS** *window, and you will see the following image:*

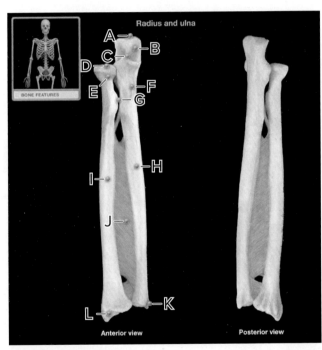

©McGraw-Hill Education

Mouse-over the pins on the screen to find the information necessary to identify the following structures:

A. _____
B. _____
C. _____
D. _____
E. _____
F. _____
G. _____
H. _____
I. _____
J. _____
K. _____
L. _____

CHECK POINT

Radius and Ulna

1. Describe the interosseous membrane of the forearm.
2. What structure holds the radio-ulnar joint in place?
3. Name the distal pointed projection of the radius. Of the ulna?

SELECT ANIMATION
Joint Movements: Radio-ulnar Joint   PLAY

After viewing the animation, answer the following questions:

1. How is the radio-ulnar joint formed?

2. Pronation is a movement unique to _____.

3. Describe pronation of the forearm with the elbow extended.

4. Supination is a movement unique to _____.

5. Describe supination of the forearm with the elbow extended.

6. Describe pronation of the forearm with the elbow flexed.

7. Describe supination of the forearm with the elbow flexed.

EXERCISE 5.53:
Wrist and Hand—Anterior View

SELECT TOPIC	SELECT VIEW
Wrist and Hand ▶	Anterior

*Click **LAYER 1** in the **LAYER CONTROLS** window and you will see the following image:*

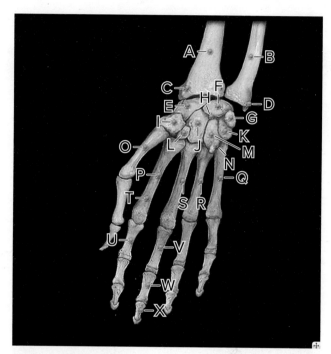

©McGraw-Hill Education

Mouse-over the pins on the screen to find the information necessary to identify the following structures:

A. _____
B. _____
C. _____
D. _____
E. _____
F. _____
G. _____
H. _____
I. _____
J. _____
K. _____
L. _____
M. _____
N. _____
O. _____
P. _____

218 MODULE 5   The Skeletal System

Q. _____
R. _____
S. _____
T. _____
U. _____
V. _____
W. _____
X. _____

F. _____
G. _____
H. _____
I. _____
J. _____
K. _____
L. _____
M. _____
N. _____
O. _____
P. _____
Q. _____
R. _____
S. _____
T. _____
U. _____

EXERCISE 5.54:
Wrist and Hand—Posterior View

SELECT TOPIC: Wrist and Hand ▶ SELECT VIEW: Posterior

Click **LAYER 1** in the **LAYER CONTROLS** window, and you will see the following image:

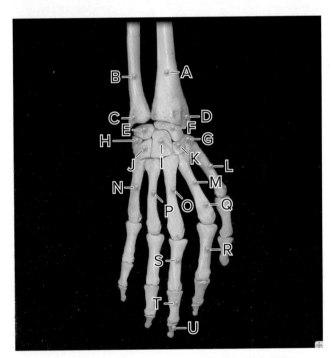

©McGraw-Hill Education

Mouse-over the pins on the screen to find the information necessary to identify the following structures:

A. _____
B. _____
C. _____
D. _____
E. _____

EXERCISE 5.55:
Wrist and Hand—Anterior and Posterior View

SELECT TOPIC: Wrist and Hand ▶ SELECT VIEW: Anterior and Posterior

Click **LAYER 1** in the **LAYER CONTROLS** window, and you will see the following image:

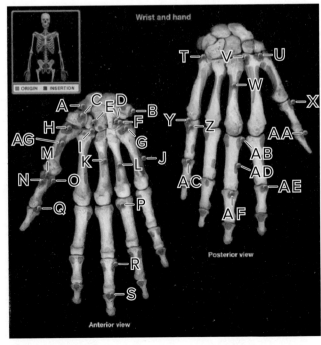

©McGraw-Hill Education

MODULE 5 The Skeletal System

Mouse-over the pins on the screen to find the information necessary to identify the following structures:

A. _____
B. _____
C. _____
D. _____
E. _____
F. _____
G. _____
H. _____
I. _____
J. _____
K. _____
L. _____
M. _____
N. _____
O. _____
P. _____
Q. _____
R. _____
S. _____
T. _____
U. _____
V. _____
W. _____
X. _____
Y. _____
Z. _____
AA. _____
AB. _____
AC. _____
AD. _____
AE. _____
AF. _____
AG. _____

Click **LAYER 2** *in the* **LAYER CONTROLS** *window, and you will see the following image:*

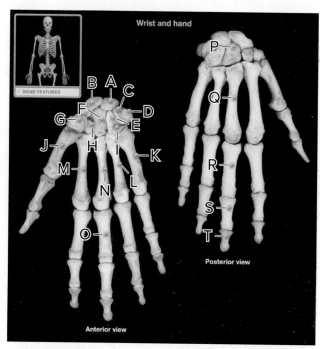

©McGraw-Hill Education

Mouse-over the pins on the screen to find the information necessary to identify the following structures:

A. _____
B. _____
C. _____
D. _____
E. _____
F. _____
G. _____
H. _____
I. _____
J. _____
K. _____
L. _____
M. _____
N. _____
O. _____
P. _____
Q. _____
R. _____
S. _____
T. _____

CHECK POINT

Wrist and Hand

1. Name the lateral bone of the forearm.
2. Name the medial bone of the forearm.
3. How are the fingers numbered?

SELECT ANIMATION
Joint Movements: Radiocarpal Joint (wrist) — PLAY

After viewing the animation, answer the following questions:

1. How is the radiocarpal joint formed?

2. Describe flexion of a joint. In what direction does flexion of a joint usually occur? Is there an exception?

3. Describe flexion of the radiocarpal joint. Give an example of this flexion.

4. Describe extension of a joint. In what direction does extension of a joint usually occur? Is there an exception?

5. Describe extension of the radiocarpal joint. Give an example of this extension.

6. Describe abduction of a joint.

7. Describe abduction of the radiocarpal joint.

8. Describe adduction of a joint.

9. Describe adduction of the radiocarpal joint.

SELECT ANIMATION
Joint Movements: Carpometacarpal (CMC) Joint (Joint of Thumb) — PLAY

After viewing the animation, answer the following questions:

1. How is the carpometacarpal joint of the thumb formed?

2. Describe flexion of a joint.

3. Describe flexion of the carpometacarpal joint. What other joint of the thumb flexes in the same plane?

4. Describe extension of a joint.

5. Describe extension of the carpometacarpal joint. What other joint of the thumb flexes in the same plane?

6. Describe abduction of a joint.

7. Describe abduction of the carpometacarpal joint. Give an example of this abduction.

8. Describe adduction of a joint.

9. Describe adduction of the carpometacarpal joint. Give an example of this adduction.

10. Describe opposition of the carpometacarpal joint.

11. Describe reposition of the carpometacarpal joint.

SELECT ANIMATION
Joint Movements: Metacarpophalangeal (MP) Joints of Fingers — PLAY

After viewing the animation, answer the following questions:

1. How are the metacarpophalangeal joints of the fingers formed?

2. Describe flexion of a joint.

3. Describe flexion of the metacarpophalangeal joints (digits II to V). Give an example of this flexion.

4. Describe extension of a joint.

5. Describe extension of the metacarpophalangeal joints (digits II to V). Give an example of this extension.

6. Describe abduction of a joint.

7. Describe abduction of the metacarpophalangeal joints (digits II, IV, V).

8. Describe adduction of a joint.

9. Describe adduction of the metacarpophalangeal joints (digits II, IV, V).

10. Describe abduction of the metacarpophalangeal joint (digit III).

11. Describe adduction of the metacarpophalangeal joint (digit III).

MODULE 5 The Skeletal System 221

SELECT ANIMATION
Joint Movements: Interphalangeal (IP) Joints of Fingers II to V   **PLAY**

After viewing the animation, answer the following questions:

1. How are the interphalangeal joints of fingers II to V formed?

2. How does the interphalangeal joint of the thumb compare to the interphalangeal joints of fingers II to V?

3. Describe flexion of a joint.

4. Describe flexion of the interphalangeal joints of fingers II to V.

5. Describe extension of a joint.

6. Describe extension of the interphalangeal joints of fingers II to V.

EXERCISE 5.56a:

Imaging—Hand (adult)

SELECT TOPIC	SELECT VIEW
Hand (adult)	X ray— Anterior–Posterior

Click the **TAGS ON/OFF** *button, and you will see the following image:*

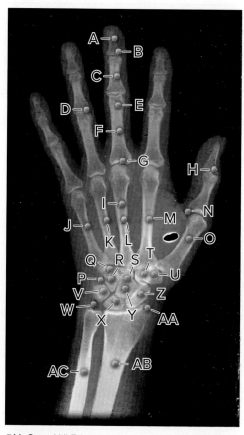

©McGraw-Hill Education

Mouse-over the pins on the screen to find the information necessary to identify the following structures:

A. _____
B. _____
C. _____
D. _____
E. _____
F. _____
G. _____
H. _____
I. _____
J. _____
K. _____
L. _____
M. _____
N. _____
O. _____
P. _____
Q. _____
R. _____
S. _____
T. _____
U. _____
V. _____
W. _____
X. _____
Y. _____
Z. _____
AA. _____
AB. _____
AC. _____

222 MODULE 5 The Skeletal System

EXERCISE 5.56b:
Imaging—Hand (teenaged)

SELECT TOPIC	SELECT VIEW
Hand (teenaged)	X ray—Anterior–Posterior

Click the **TAGS ON/OFF** button, and you will see the following image:

Mouse-over the pins on the screen to find the information necessary to identify the following structures:

A. _____

B. _____

C. _____

D. _____

E. _____

F. _____

G. _____

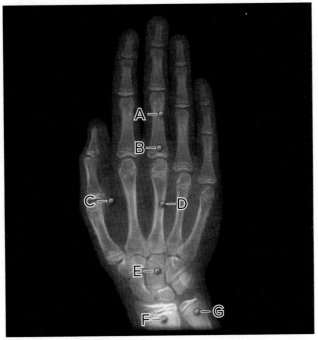

©McGraw-Hill Education

Self-Quiz
Take this opportunity to check your progress by taking the **QUIZ**. See the **Introduction Module** for a reminder on how to access the **QUIZ** for this Study Area.

IN REVIEW

What Have I Learned?

The following questions cover the material that you just learned: the pectoral girdle and upper limb. Apply what you have learned in answering these questions on a separate piece of paper.

1. Which bony structures form the glenohumeral joint?

2. Name the characteristic features of the scapula.

3. The lateral end of the clavicle articulates _____.

4. Name a landmark of the scapula for intramuscular injections.

5. Name another visible subcutaneous landmark of the scapula.

6. How many necks are on the proximal end of the humerus? Where are they located?

7. Which neck is a common site for fractures?

8. List the proximal carpal bones from lateral to medial.

9. List the distal carpal bones from lateral to medial.

10. List the phalanges of fingers II through V from proximal to distal.

11. List the phalanges of finger I from proximal to distal.

12. Describe the interosseous membrane of the forearm.

13. What structure holds the radio-ulnar joint in place?

14. Name the distal pointed projection of the radius. Of the ulna.

15. Name the lateral bone of the forearm.

16. Name the medial bone of the forearm.

17. Name the bones and the bony structures that form the shoulder joint.

18. Name the wrist bone with a prominent hook (hamulus).

19. What term describes a bone embedded in a tendon? Which wrist bone is embedded in a tendon? Which tendon?

20. Name the prominent ridge on the posterior scapula.

21. Name the eight wrist bones.

Pelvic Girdle and Lower Limb

EXERCISE 5.57:
Hip and Thigh—Anterior View

SELECT TOPIC: Pelvic Girdle and Lower Limb
SELECT SUB-TOPIC: Hip and Thigh
SELECT VIEW: Anterior

Click **LAYER 1** in the **LAYER CONTROLS** window, and you will see the following image:

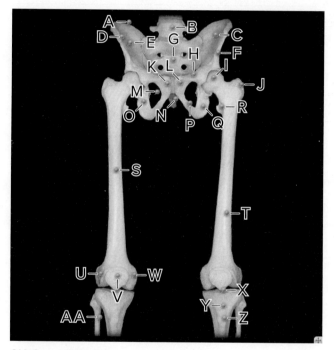

©McGraw-Hill Education

Mouse-over the pins on the screen to find the information necessary to identify the following structures:

A. _____
B. _____
C. _____
D. _____
E. _____
F. _____
G. _____
H. _____
I. _____
J. _____
K. _____
L. _____
M. _____
N. _____
O. _____
P. _____
Q. _____
R. _____
S. _____
T. _____
U. _____
V. _____
W. _____
X. _____
Y. _____
Z. _____
AA. _____

EXERCISE 5.58:
Hip and Thigh—Posterior View

SELECT TOPIC: Hip and Thigh
SELECT VIEW: Posterior

Click **LAYER 1** in the **LAYER CONTROLS** window, and you will see the following image:

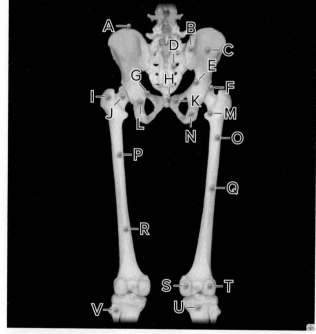

©McGraw-Hill Education

Mouse-over the pins on the screen to find the information necessary to identify the following structures:

A. _____
B. _____
C. _____
D. _____

224 MODULE 5 The Skeletal System

E. _____
F. _____
G. _____
H. _____
I. _____
J. _____
K. _____
L. _____
M. _____
N. _____
O. _____
P. _____
Q. _____
R. _____
S. _____
T. _____
U. _____
V. _____

EXERCISE 5.59:

Imaging—Hip and Thigh

SELECT TOPIC: Hip and Thigh ▸ **SELECT VIEW**: X ray—Anterior–Posterior

Click the **TAGS ON/OFF** button, and you will see the following image:

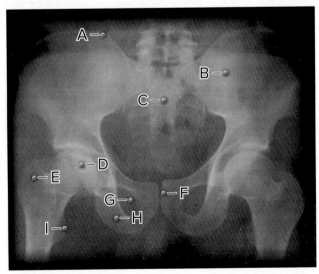

©McGraw-Hill Education

Mouse-over the pins on the screen to find the information necessary to identify the following structures:

A. _____
B. _____
C. _____
D. _____
E. _____
F. _____
G. _____
H. _____
I. _____

CHECK POINT

Hip and Thigh

1. Describe the patella.
2. Describe the sacrum.
3. Identify the enlarged distal ends of the femur.

 SELECT ANIMATION
Joint Movements: Hip Joint **PLAY**

After viewing the animation, answer the following questions:

1. Describe the hip joint.

2. Describe flexion of a joint.

3. Describe flexion of the hip joint. Give an example of this flexion.

4. Describe extension of a joint.

5. Describe extension of the hip joint. Give an example of this extension.

6. Describe abduction of a joint.

7. Describe abduction of the hip joint. Give an example of this abduction.

8. Describe adduction of a joint.

9. Describe adduction of the hip joint. Give an example of this adduction.

10. Describe circumduction of a joint.

11. Describe circumduction of the hip joint.

12. Describe medial rotation of a joint. Which joints are capable of medial rotation?

13. Describe medial rotation of the hip joint.

14. Describe lateral rotation of a joint. Which joints are capable of medial rotation?

15. Describe lateral rotation of the hip joint.

EXERCISE 5.60:
Pelvis

SELECT TOPIC Pelvis ▸ **SELECT VIEW** Superior

Click **LAYER 1** *in the* **LAYER CONTROLS** *window, and you will see the following image:*

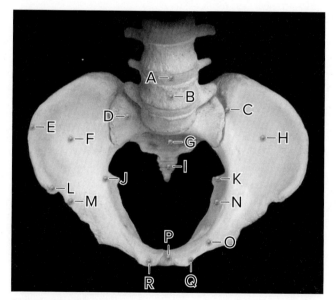

©McGraw-Hill Education

Mouse-over the pins on the screen to find the information necessary to identify the following structures:

A. _____
B. _____
C. _____
D. _____
E. _____
F. _____
G. _____
H. _____
I. _____
J. _____
K. _____
L. _____
M. _____
N. _____
O. _____
P. _____
Q. _____
R. _____

CHECK POINT
Pelvis

1. Name the synovial joint between the sacrum and the ilium.
2. How much movement does this joint allow? Why?
3. Name the landmark for administering anesthetic during childbirth.

EXERCISE 5.61:
Pelvic Girdle—Female

SELECT TOPIC Pelvic Girdle—Female ▸ **SELECT VIEW** Anterior

Click **LAYER 1** *in the* **LAYER CONTROLS** *window, and you will see the following image:*

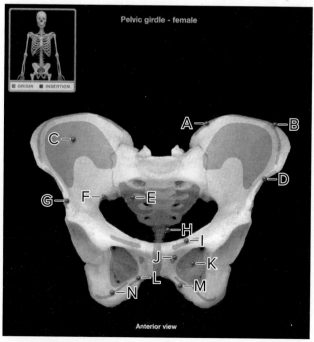

©McGraw-Hill Education

Mouse-over the pins on the screen to find the information necessary to identify the following structures:

A. _____
B. _____
C. _____
D. _____
E. _____
F. _____
G. _____
H. _____
I. _____
J. _____
K. _____
L. _____
M. _____
N. _____

*Click **LAYER 2** in the **LAYER CONTROLS** window, and you will see the following image:*

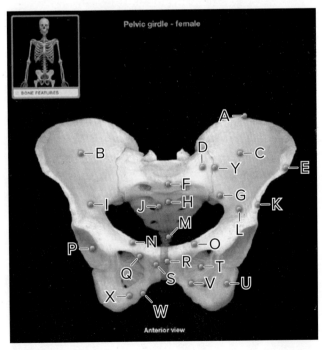

©McGraw-Hill Education

Mouse-over the pins on the screen to find the information necessary to identify the following structures:

A. _____
B. _____
C. _____
D. _____
E. _____
F. _____
G. _____
H. _____
I. _____
J. _____
K. _____
L. _____
M. _____
N. _____
O. _____
P. _____
Q. _____
R. _____
S. _____
T. _____
U. _____
V. _____
W. _____
X. _____
Y. _____

CHECK POINT

Pelvic Girdle—Female

1. In the pregnant female, what events occur concerning the pubic symphysis?
2. Describe the subpubic angle.
3. The subpubic angle in females is usually _____ degrees.

MODULE 5 The Skeletal System 227

EXERCISE 5.62:
Pelvic Girdle—Male

SELECT TOPIC	SELECT VIEW
Pelvic Girdle—Male ▸	Anterior

Click **LAYER 1** *in the* **LAYER CONTROLS** *window, and you will see the following image:*

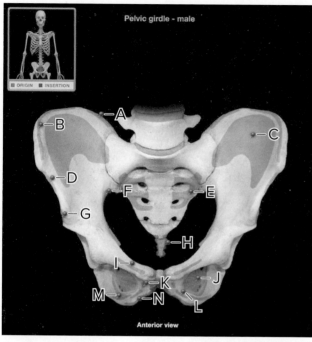

©McGraw-Hill Education

Mouse-over the pins on the screen to find the information necessary to identify the following structures:

A. _____
B. _____
C. _____
D. _____
E. _____
F. _____
G. _____
H. _____
I. _____
J. _____
K. _____
L. _____
M. _____
N. _____

Click **LAYER 2** *in the* **LAYER CONTROLS** *window, and you will see the following image:*

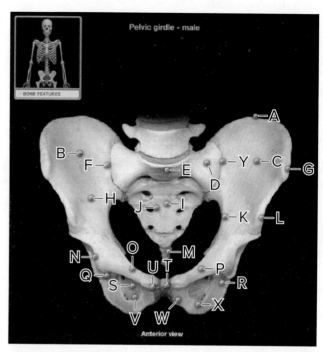

©McGraw-Hill Education

Mouse-over the pins on the screen to find the information necessary to identify the following structures:

A. _____
B. _____
C. _____
D. _____
E. _____
F. _____
G. _____
H. _____
I. _____
J. _____
K. _____
L. _____
M. _____
N. _____
O. _____
P. _____
Q. _____
R. _____
S. _____
T. _____
U. _____
V. _____
W. _____
X. _____
Y. _____

CHECK POINT

Pelvic Girdle—Male

1. Describe the difference between the male and female obturator foramena.
2. What structures are found in the right iliac fossa? The left iliac fossa?
3. The subpubic angle in males is usually _____ degrees.

EXERCISE 5.63:
Hip Bone

SELECT TOPIC: Hip Bone → SELECT VIEW: Medial and Lateral

Click **LAYER 1** in the **LAYER CONTROLS** window, and you will see the following image:

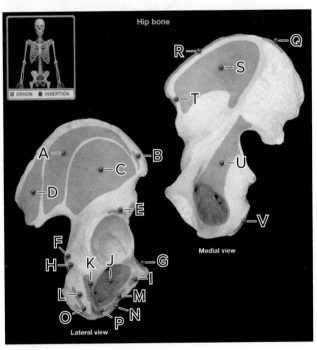

©McGraw-Hill Education

Mouse-over the pins on the screen to find the information necessary to identify the following structures:

A. _____
B. _____
C. _____
D. _____
E. _____
F. _____
G. _____
H. _____
I. _____
J. _____
K. _____
L. _____
M. _____
N. _____
O. _____
P. _____
Q. _____
R. _____
S. _____
T. _____
U. _____
V. _____

Click **LAYER 2** in the **LAYER CONTROLS** window, and you will see the following image:

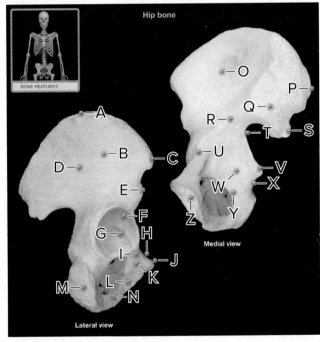

©McGraw-Hill Education

Mouse-over the pins on the screen to find the information necessary to identify the following structures:

A. _____
B. _____
C. _____
D. _____
E. _____
F. _____

G. _____
H. _____
I. _____
J. _____
K. _____
L. _____
M. _____
N. _____
O. _____
P. _____
Q. _____
R. _____
S. _____
T. _____
U. _____
V. _____
W. _____
X. _____
Y. _____
Z. _____

CHECK POINT

Hip Bone

1. Describe the acetabulum.
2. Name a hip landmark for intramuscular injections.
3. Name a hip landmark for administering anesthetic during childbirth.

EXERCISE 5.64:

Femur

SELECT TOPIC	SELECT VIEW
Femur	Anterior and Posterior

Click **LAYER 1** *in the* **LAYER CONTROLS** *window, and you will see the following image:*

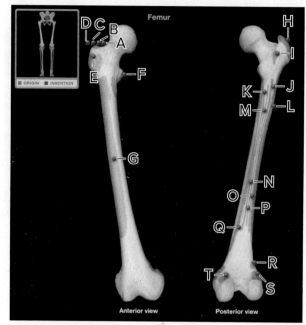

©McGraw-Hill Education

Mouse-over the pins on the screen to find the information necessary to identify the following structures:

A. _____
B. _____
C. _____
D. _____
E. _____
F. _____
G. _____
H. _____
I. _____
J. _____
K. _____
L. _____
M. _____
N. _____
O. _____
P. _____
Q. _____
R. _____
S. _____
T. _____

230 MODULE 5 The Skeletal System

Click **LAYER 2** *in the* **LAYER CONTROLS** *window, and you will see the following image:*

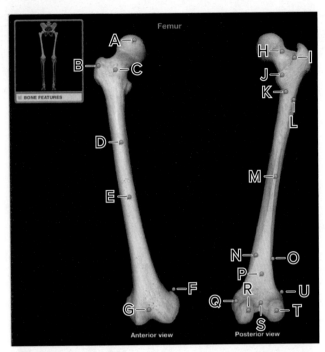

©McGraw-Hill Education

Mouse-over the pins on the screen to find the information necessary to identify the following structures:

A. _____
B. _____
C. _____
D. _____
E. _____
F. _____
G. _____
H. _____
I. _____
J. _____
K. _____
L. _____
M. _____
N. _____
O. _____
P. _____
Q. _____
R. _____
S. _____
T. _____
U. _____

CHECK POINT

Femur

1. Name a common site of femur fractures, especially in the elderly.
2. What correlation exists between the length of the femur and body height?
3. What is the fovea capitis? What is its function?
4. Describe the epiphyseal line. How does it relate to bone growth?
5. Describe the proximal epiphysis. Where is it located?
6. Describe the distal epiphysis. Where is it located?
7. Describe the medullary cavity. Where is it located? What is its function?

EXERCISE 5.65:

Femur—Anterior and Coronal

 SELECT TOPIC: Femur ▸ SELECT VIEW: Anterior and Coronal

Click **LAYER 1** *in the* **LAYER CONTROLS** *window, and you will see the following image:*

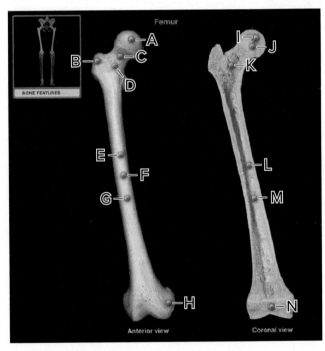

©McGraw-Hill Education

Mouse-over the pins on the screen to find the information necessary to identify the following structures:

A. _____
B. _____
C. _____
D. _____
E. _____
F. _____

G. _____
H. _____
I. _____
J. _____
K. _____
L. _____
M. _____
N. _____

EXERCISE 5.66:
Patella

SELECT TOPIC Patella ▸ **SELECT VIEW** Anterior–Posterior

Click **LAYER 1** *in the* **LAYER CONTROLS** *window, and you will see the following image:*

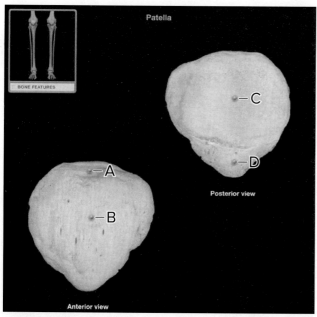

©McGraw-Hill Education

Mouse-over the pins on the screen to find the information necessary to identify the following structures:

A. _____
B. _____
C. _____
D. _____

CHECK POINT
Patella

1. Describe the location of the patella.
2. What structures form the knee joint?
3. What affect does the patella have on the tendon of the quadriceps femoris muscle?

MODULE 5   The Skeletal System   **231**

EXERCISE 5.67a:
Imaging—Knee

SELECT TOPIC Knee ▸ **SELECT VIEW** X ray—Anterior–Posterior

Click the **TAGS ON/OFF** *button, and you will see the following image:*

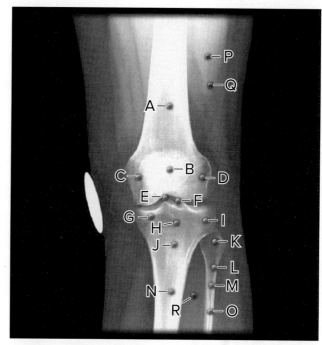

©McGraw-Hill Education

Mouse-over the pins on the screen to find the information necessary to identify the following structures:

A. _____
B. _____
C. _____
D. _____
E. _____
F. _____
G. _____
H. _____
I. _____
J. _____
K. _____
L. _____
M. _____
N. _____
O. _____

232 MODULE 5 The Skeletal System

Nonskeletal System Structures (blue pins)

P. _____
Q. _____
R. _____

EXERCISE 5.67b:
Imaging—Knee

SELECT TOPIC: Knee ▶ SELECT VIEW: X ray—Posterior–Anterior

Click the **TAGS ON/OFF** button, and you will see the following image:

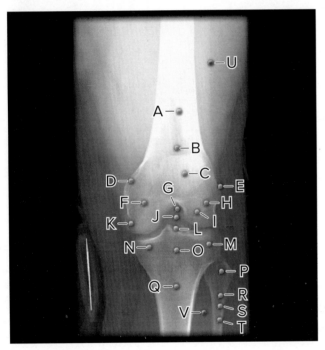

©McGraw-Hill Education

Mouse-over the pins on the screen to find the information necessary to identify the following structures:

A. _____
B. _____
C. _____
D. _____
E. _____
F. _____
G. _____
H. _____
I. _____
J. _____
K. _____
L. _____
M. _____
N. _____
O. _____
P. _____
Q. _____
R. _____
S. _____
T. _____

Nonskeletal System Structures (blue pins)

U. _____
V. _____

EXERCISE 5.67c:
Imaging—Knee

SELECT TOPIC: Knee ▶ SELECT VIEW: X ray—Lateral

Click the **TAGS ON/OFF** button, and you will see the following image:

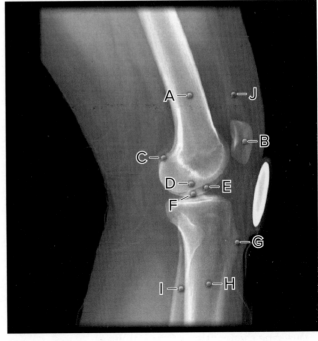

©McGraw-Hill Education

MODULE 5 The Skeletal System 233

Mouse-over the pins on the screen to find the information necessary to identify the following structures:

A. _____
B. _____
C. _____
D. _____
E. _____
F. _____
G. _____
H. _____
I. _____

Nonskeletal System Structure (blue pin)

J. _____

EXERCISE 5.68:
Leg and Foot—Anterior View

SELECT TOPIC	SELECT VIEW
Leg and Foot	Anterior

Click **LAYER 1** *in the* **LAYER CONTROLS** *window, and you will see the following image:*

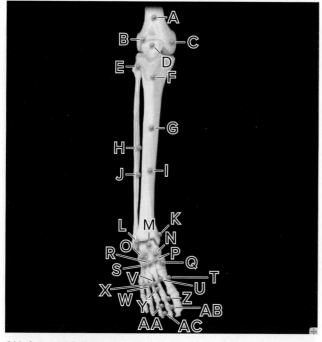

©McGraw-Hill Education

Mouse-over the pins on the screen to find the information necessary to identify the following structures:

A. _____
B. _____
C. _____
D. _____
E. _____
F. _____
G. _____
H. _____
I. _____
J. _____
K. _____
L. _____
M. _____
N. _____
O. _____
P. _____
Q. _____
R. _____
S. _____
T. _____
U. _____
V. _____
W. _____
X. _____
Y. _____
Z. _____
AA. _____
AB. _____
AC. _____

234 MODULE 5 The Skeletal System

EXERCISE 5.69:
Leg and Foot—Posterior View

SELECT TOPIC	SELECT VIEW
Leg and Foot	Posterior

Click **LAYER 1** *in the* **LAYER CONTROLS** *window, and you will see the following image:*

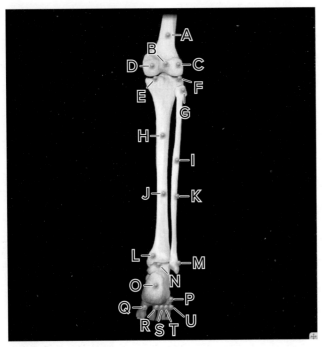

©McGraw-Hill Education

Mouse-over the pins on the screen to find the information necessary to identify the following structures:

A. _____
B. _____
C. _____
D. _____
E. _____
F. _____
G. _____
H. _____
I. _____
J. _____
K. _____
L. _____
M. _____
N. _____
O. _____
P. _____
Q. _____
R. _____
S. _____
T. _____
U. _____

CHECK POINT
Leg and Foot

1. What bone forms the connecting link between the foot and the leg?
2. What bone contributes to the knee and ankle joints?
3. Name the largest tarsal bone. Where is it located?

EXERCISE 5.70:
Tibia and Fibula

SELECT TOPIC	SELECT VIEW
Tibia and Fibula	Anterior and Posterior

Click **LAYER 1** *in the* **LAYER CONTROLS** *window, and you will see the following image:*

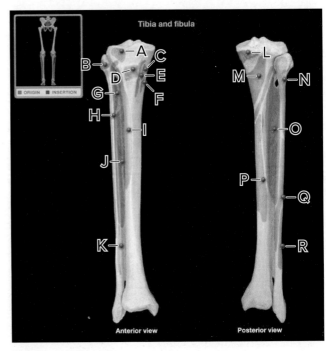

©McGraw-Hill Education

Mouse-over the pins on the screen to find the information necessary to identify the following structures:

A. _____
B. _____
C. _____
D. _____
E. _____
F. _____
G. _____
H. _____
I. _____
J. _____
K. _____
L. _____
M. _____
N. _____
O. _____
P. _____
Q. _____
R. _____

Mouse-over the pins on the screen to find the information necessary to identify the following structures:

A. _____
B. _____
C. _____
D. _____
E. _____
F. _____
G. _____
H. _____
I. _____
J. _____
K. _____
L. _____
M. _____
N. _____
O. _____
P. _____
Q. _____

Click **LAYER 2** *in the* **LAYER CONTROLS** *window, and you will see the following image:*

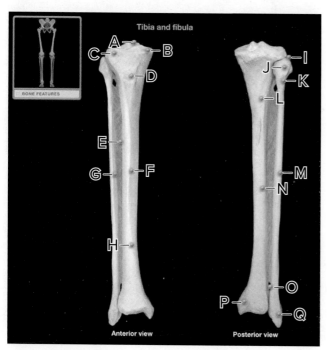

CHECK POINT

Tibia and Fibula

1. What structure on the tibia contributes to the ankle joint?
2. What structure on the fibula contributes to the ankle joint?
3. What is unique about the anterior border of the tibia?

EXERCISE 5.71:
Ankle and Foot

SELECT TOPIC: Ankle and Foot ▶ **SELECT VIEW:** Superior–Inferior

Click **LAYER 1** *in the* **LAYER CONTROLS** *window, and you will see the following image:*

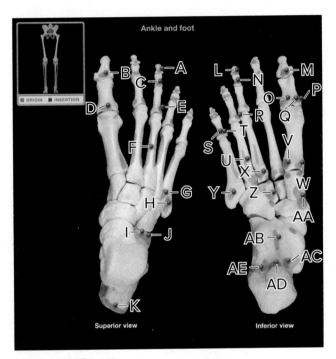

©McGraw-Hill Education

Mouse-over the pins on the screen to find the information necessary to identify the following structures:

A. _____
B. _____
C. _____
D. _____
E. _____
F. _____
G. _____
H. _____
I. _____
J. _____
K. _____
L. _____
M. _____
N. _____
O. _____
P. _____
Q. _____
R. _____
S. _____
T. _____
U. _____
V. _____
W. _____
X. _____
Y. _____
Z. _____
AA. _____
AB. _____
AC. _____
AD. _____
AE. _____

Click **LAYER 2** *in the* **LAYER CONTROLS** *window, and you will see the following image:*

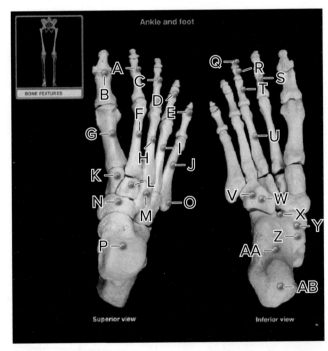

©McGraw-Hill Education

Mouse-over the pins on the screen to find the information necessary to identify the following structures:

A. _____
B. _____
C. _____
D. _____
E. _____
F. _____
G. _____
H. _____
I. _____

J. _____
K. _____
L. _____
M. _____
N. _____
O. _____
P. _____
Q. _____
R. _____
S. _____
T. _____
U. _____
V. _____
W. _____
X. _____
Y. _____
Z. _____
AA. _____
AB. _____

EXERCISE 5.72a:
Imaging—Foot

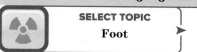

SELECT TOPIC: Foot ▸ SELECT VIEW: X ray—Dorsal

Click the **TAGS ON/OFF** button, and you will see the following image:

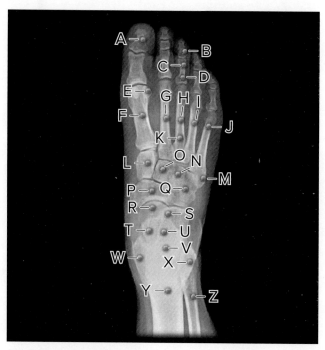

©McGraw-Hill Education

Mouse-over the pins on the screen to find the information necessary to identify the following structures:

A. _____
B. _____
C. _____
D. _____
E. _____
F. _____
G. _____
H. _____
I. _____
J. _____
K. _____
L. _____
M. _____
N. _____
O. _____
P. _____
Q. _____
R. _____
S. _____
T. _____
U. _____
V. _____
W. _____
X. _____
Y. _____
Z. _____

238 MODULE 5 The Skeletal System

EXERCISE 5.72b:
Imaging—Foot

SELECT TOPIC	SELECT VIEW
Foot	X ray—Lateral

Click the **TAGS ON/OFF** *button, and you will see the following image:*

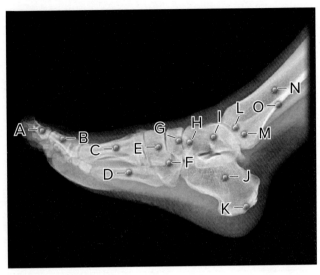

©McGraw-Hill Education

Mouse-over the pins on the screen to find the information necessary to identify the following structures:

A. _____
B. _____
C. _____
D. _____
E. _____
F. _____
G. _____
H. _____
I. _____
J. _____
K. _____
L. _____
M. _____
N. _____
O. _____

CHECK POINT

Ankle and Foot

1. Name the structure commonly called the "ball of the foot."
2. Name the structures that are the surface contact points on the plantar foot.
3. List three common names for digit I of the foot.

EXERCISE 5.73:
Imaging—Bone Scan

SELECT TOPIC	SELECT VIEW
Bone Scan	Nuclear Scan—Anterior–Posterior

Click the **TAGS ON/OFF** *button, and you will see the following image:*

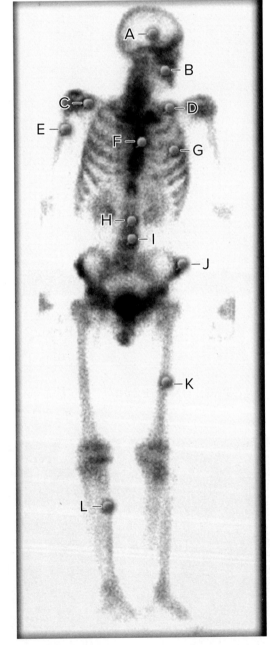

©Scott Camazine/Photo Researchers, Inc.

Mouse-over the pins on the screen to find the information necessary to identify the following structures:

A. _____
B. _____
C. _____
D. _____
E. _____
F. _____
G. _____
H. _____
I. _____
J. _____
K. _____
L. _____

Self-Quiz

Take this opportunity to check your progress by taking the **QUIZ**. See the **Introduction Module** for a reminder on how to access the **QUIZ** for this Study Area.

IN REVIEW

What Have I Learned?

The following questions cover the material that you have just learned: the pelvic girdle and lower limb. Apply what you have learned in answering these questions on a separate piece of paper.

1. Where on the femur does the lateral collateral ligament attach?
2. How many tarsal bones are there on each foot? Name them.
3. The head of which bone is commonly referred to as the "ball of the foot"?
4. Name the small bones of the toes. How many of these bones make up each toe?
5. Name the two enlargements of the distal femur. What structures do they articulate with?
6. Describe the patella.
7. Describe the sacrum.
8. Name the synovial joint between the sacrum and the ilium.
9. How much movement does this joint allow? Why?
10. Name the landmark for administering anesthetic during childbirth.
11. In the pregnant female, what events occur concerning the pubic symphysis?
12. The subpubic angle in females is usually _____ degrees.
13. The subpubic angle in males is usually _____ degrees.
14. Describe the difference between the male and female obturator foramena.
15. Name a hip landmark for intramuscular injections.
16. Name a hip landmark for administering anesthetic during childbirth.
17. Name a common site of femur fractures, especially in the elderly.
18. What correlation exists between the length of the femur and body height?
19. Name the central depression of the head of the femur.
20. What structures form the knee joint?
21. What bone forms the connecting link between the foot and the leg?
22. What bone contributes to the knee and ankle joints?
23. Name the largest tarsal bone. Where is it located?
24. What structure on the tibia contributes to the ankle joint?
25. What structure on the fibula contributes to the ankle joint?

MODULE 5 The Skeletal System

HEADS UP!
*With the study of the joints, or articulations, we can tie together the concepts that you have learned in the **Skeletal System** and integrate them with those you will learn in the next module—the **Muscular System**. With this in mind, you will want to refer back to this section of the **Skeletal System** while you are working on the **Muscular System**.*

SELECT ANIMATION
Synovial Joint — PLAY

After viewing the animation, answer the following questions:

1. A joint is _____.

2. What is the most common type of joint?

3. How many types of synovial joints are found in the body? What distinguishes each one from the others?

4. Which type of synovial joint allows the most mobility? Give an example.

5. Which type of synovial joint allows the least mobility? Give an example.

6. A typical synovial joint is characterized by _____.

7. Describe the joint capsule.

8. What is synovial fluid? Where is it produced?

9. What tissue covers the surface of adjoining bones? What is its function?

10. What is the function of the meniscus of the knee joint?

11. What is the function of bursae?

SELECT ANIMATION
Joint Movements: Temporomandibular Joint (TMJ) — PLAY

After viewing the animation, answer the following questions:

1. Describe the temporomandibular joint.

2. Describe depression of a body part.

3. Describe depression of the mandible.

4. Describe elevation of a body part.

5. Describe elevation of the mandible.

6. Describe protrusion of the mandible. What is protraction?

7. Describe retraction of the mandible.

SELECT ANIMATION
Joints/Osteoarthritis/Rheumatoid Arthritis — PLAY

After viewing the animation, answer the following questions:

1. What are the functions of the bones of the skeletal system?

2. Define a joint.

3. List the three features of joints.

4. List the three major categories of joints.

5. Which category of joints is affected by arthritis?

6. Describe the mobility associated with most of these joints.

7. List four examples of synovial joints.

8. Describe the structures in the joint cavity of a synovial joint.

9. Describe the structure that seals the joint cavity.

10. What is the function of synovial fluid? How does it accomplish this function?

11. How does synovial fluid allow a joint to move freely?

12. Define arthritis.

13. List the two most common types of arthritis.

14. What is the most common type of arthritis? What is another name for this malady?

15. What causes the most common type of arthritis?

16. What are the symptoms of the most common type of arthritis?

17. What is the second most common type of arthritis?

18. What is the exact cause of this arthritis?

19. How is rheumatoid arthritis characterized?

20. In rheumatoid arthritis, which immune cells infiltrate the tissue of the joint?

21. What chemicals do these cells release?

22. What affect do these chemicals have on the joint tissue?

23. How is rheumatoid arthritis like osteoarthritis?

24. How is rheumatoid arthritis unlike osteoarthritis?

25. What eventually happens to the articular cartilage in a joint with rheumatoid arthritis?

26. What then occurs to the joint?

27. What happens to the joint over time?

28. What is this condition called?

29. List the treatments for osteoarthritis.

30. List the treatments for rheumatoid arthritis.

31. What are biologics?

32. What is sometimes an option in severe joint damage? How does this help?

EXERCISE 5.74:
Temporomandibular Joint

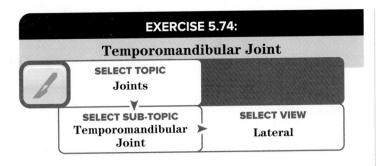

Click **LAYER 1** *in the* **LAYER CONTROLS** *window, and you will see the following image:*

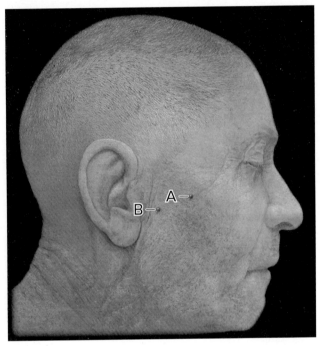

©McGraw-Hill Education

Mouse-over the pins on the screen to find the information necessary to identify the following structures:

A. _____

B. _____

Click **LAYER 2** *in the* **LAYER CONTROLS** *window, and you will see the following image:*

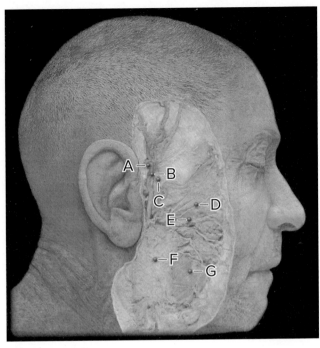

©McGraw-Hill Education

Mouse-over the pins on the screen to find the information necessary to identify the following structures:

A. _____

B. _____

C. _____

D. _____

E. _____

F. _____

G. _____

Click **LAYER 3** *in the* **LAYER CONTROLS** *window, and you will see the following image:*

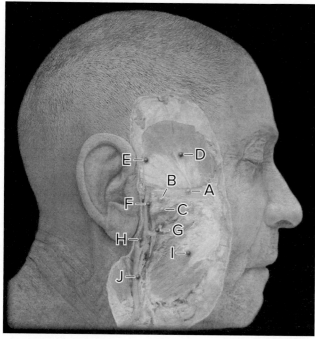

©McGraw-Hill Education

Mouse-over the pins on the screen to find the information necessary to identify the following structures:

A. _____
B. _____
C. _____

Nonskeletal System Structures (blue pins)

D. _____
E. _____
F. _____
G. _____
H. _____
I. _____
J. _____

Click **LAYER 4** *in the* **LAYER CONTROLS** *window, and you will see the following image:*

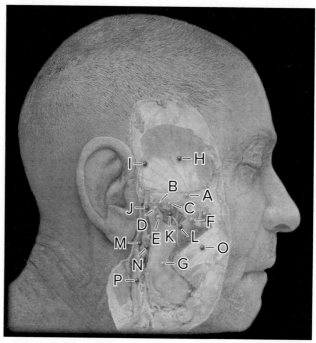

©McGraw-Hill Education

Mouse-over the pins on the screen to find the information necessary to identify the following structures:

A. _____
B. _____
C. _____
D. _____
E. _____
F. _____
G. _____

Nonskeletal System Structures (blue pins)

H. _____
I. _____
J. _____
K. _____
L. _____
M. _____
N. _____
O. _____
P. _____

Click **LAYER 5** *in the* **LAYER CONTROLS** *window, and you will see the following image:*

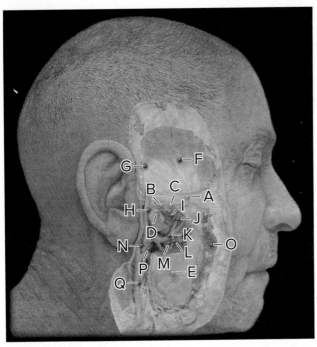

©McGraw-Hill Education

Mouse-over the pins on the screen to find the information necessary to identify the following structures:

A. _____
B. _____
C. _____
D. _____
E. _____

Nonskeletal System Structures (blue pins)

F. _____
G. _____
H. _____
I. _____
J. _____
K. _____
L. _____
M. _____
N. _____
O. _____
P. _____
Q. _____

Click **LAYER 6** *in the* **LAYER CONTROLS** *window, and you will see the following image:*

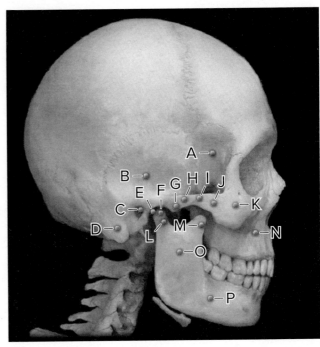

©McGraw-Hill Education

Mouse-over the pins on the screen to find the information necessary to identify the following structures:

A. _____
B. _____
C. _____
D. _____
E. _____
F. _____
G. _____
H. _____
I. _____
J. _____
K. _____
L. _____
M. _____
N. _____
O. _____
P. _____

CHECK POINT

Temporomandibular Joint

1. Name the structure that encloses the temporomandibular joint.
2. What is the function of the synovial membrane?
3. What is the function of the lateral ligament of the temporomandibular joint?
4. What is the cause of "jaw clicking"?
5. What is the location of the articular disk of the temporomandibular joint.

| SELECT ANIMATION Joint Movements: Glenohumeral Joint | PLAY |

After viewing the animation, answer the following questions:

1. Describe the glenohumeral joint.

2. Describe flexion of a joint.

3. Describe flexion of the glenohumeral joint. Give an example of this flexion.

4. Describe extension of a joint.

5. Describe extension of the glenohumeral joint. Give an example of this extension.

6. Describe abduction of a joint.

7. Describe abduction of the glenohumeral joint. Give an example of this abduction.

8. Describe adduction of a joint.

9. Describe adduction of the glenohumeral joint. Give an example of this adduction.

10. Describe medial rotation of a joint. Which joints are capable of medial rotation?

11. Describe medial rotation of the glenohumeral joint.

12. Describe lateral rotation of a joint. Which joints are capable of medial rotation?

13. Describe lateral rotation of the glenohumeral joint.

14. Describe circumduction of a joint.

15. Describe circumduction of the glenohumeral joint. Give an example of this circumduction.

EXERCISE 5.75:
Glenohumeral (Shoulder) Joint

| SELECT TOPIC Glenohumeral (Shoulder) Joint | ▶ | SELECT VIEW Anterior |

*Click **LAYER 1** in the **LAYER CONTROLS** window, and you will see the following image:*

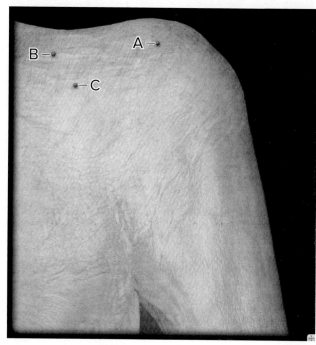

©McGraw-Hill Education

Mouse-over the pins on the screen to find the information necessary to identify the following structures:

A. _____

B. _____

C. _____

246 MODULE 5 The Skeletal System

Click **LAYER 2** *in the* **LAYER CONTROLS** *window, and you will see the following image:*

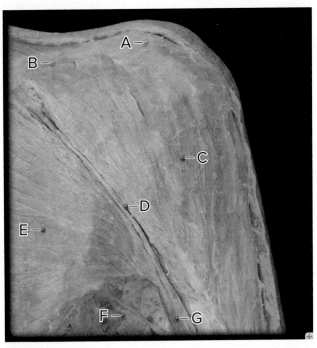

©McGraw-Hill Education

Mouse-over the pins on the screen to find the information necessary to identify the following structures:

A. _____
B. _____

Nonskeletal System Structures (blue pins)

C. _____
D. _____
E. _____
F. _____
G. _____

Click **LAYER 3** *in the* **LAYER CONTROLS** *window, and you will see the following image:*

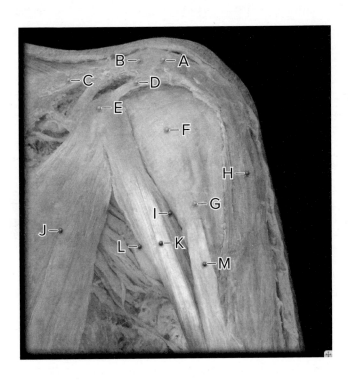

©McGraw-Hill Education

Mouse-over the pins on the screen to find the information necessary to identify the following structures:

A. _____
B. _____
C. _____
D. _____
E. _____
F. _____
G. _____

Nonskeletal System Structures (blue pins)

H. _____
I. _____
J. _____
K. _____
L. _____
M. _____

Click **LAYER 4** *in the* **LAYER CONTROLS** *window, and you will see the following image:*

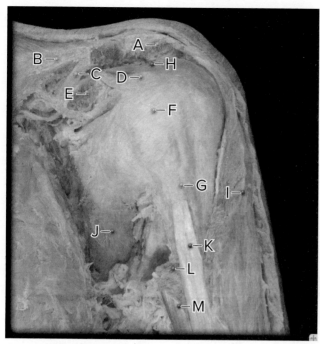

©McGraw-Hill Education

Mouse-over the pins on the screen to find the information necessary to identify the following structures:

A. _____
B. _____
C. _____
D. _____
E. _____
F. _____
G. _____

Nonskeletal System Structures (blue pins)

H. _____
I. _____
J. _____
K. _____
L. _____
M. _____

Click **LAYER 5** *in the* **LAYER CONTROLS** *window, and you will see the following image:*

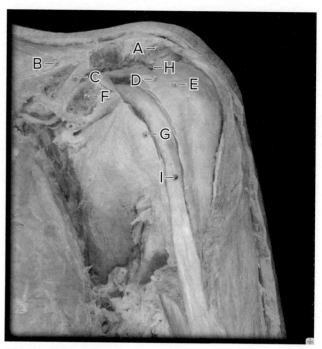

©McGraw-Hill Education

Mouse-over the pins on the screen to find the information necessary to identify the following structures:

A. _____
B. _____
C. _____
D. _____
E. _____
F. _____
G. _____

Nonskeletal System Structures (blue pins)

H. _____
I. _____

248 MODULE 5 The Skeletal System

Click **LAYER 6** *in the* **LAYER CONTROLS** *window, and you will see the following image:*

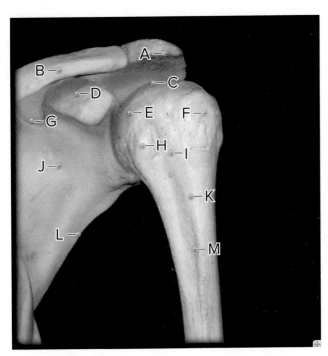

©McGraw-Hill Education

Mouse-over the pins on the screen to find the information necessary to identify the following structures:

A. _____
B. _____
C. _____
D. _____
E. _____
F. _____
G. _____
H. _____
I. _____
J. _____
K. _____
L. _____
M. _____

CHECK POINT

Glenohumeral (Shoulder) Joint

1. Name the structure enclosing the acromioclavicular joint.
2. What is the function of the coracohumeral ligament?
3. What is the function of the coracoclavicular ligament?
4. What is the result of a torn coracoclavicular ligament?
5. Name the tough, fibrous envelope of the glenohumeral joint.

| SELECT ANIMATION Joint Movements: Elbow Joint | PLAY |

After viewing the animation, answer the following questions:

1. Describe the elbow joint.

2. Describe flexion of a joint.

3. Describe flexion the elbow joint. Give an example of this flexion.

4. Describe extension of a joint.

5. Describe extension of the elbow joint. Give an example of this extension.

EXERCISE 5.76:
Elbow Joint

| SELECT TOPIC Elbow Joint ▶ | SELECT VIEW Anterior |

Click **LAYER 1** *in the* **LAYER CONTROLS** *window, and you will see the following image:*

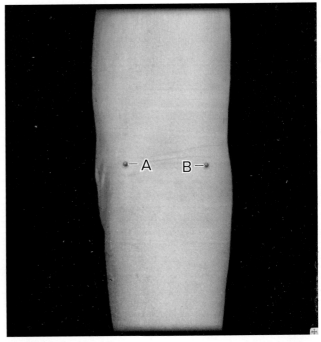

©McGraw-Hill Education

Mouse-over the pins on the screen to find the information necessary to identify the following structures:

A. _____
B. _____

Click **LAYER 2** *in the* **LAYER CONTROLS** *window, and you will see the following image:*

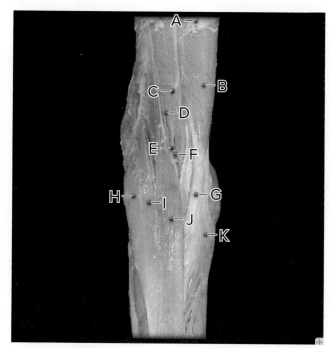

©McGraw-Hill Education

Mouse-over the pins on the screen to find the information necessary to identify the following structures:

A. _____
B. _____
C. _____
D. _____
E. _____
F. _____
G. _____
H. _____
I. _____
J. _____
K. _____

Click **LAYER 3** *in the* **LAYER CONTROLS** *window, and you will see the following image:*

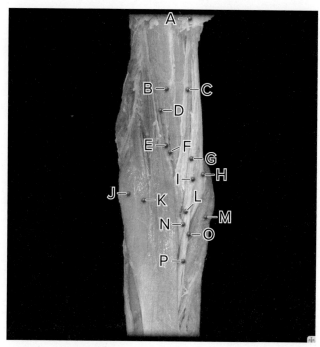

©McGraw-Hill Education

Mouse-over the pins on the screen to find the information necessary to identify the following structures:

A. _____
B. _____
C. _____
D. _____
E. _____
F. _____
G. _____
H. _____
I. _____
J. _____
K. _____
L. _____
M. _____
N. _____
O. _____
P. _____

250 MODULE 5 The Skeletal System

Click **LAYER 4** *in the* **LAYER CONTROLS** *window, and you will see the following image:*

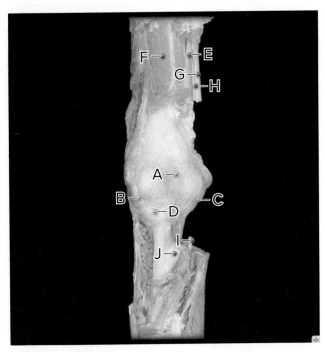

©McGraw-Hill Education

Mouse-over the pins on the screen to find the information necessary to identify the following structures:

A. _____

B. _____

C. _____

D. _____

Nonskeletal System Structure (blue pins)

E. _____

F. _____

G. _____

H. _____

I. _____

J. _____

Click **LAYER 5** *in the* **LAYER CONTROLS** *window, and you will see the following image:*

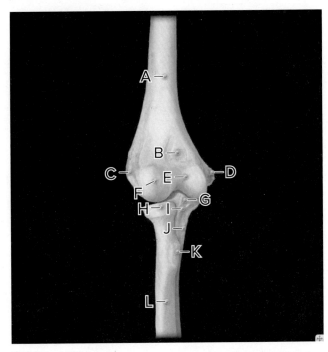

©McGraw-Hill Education

Mouse-over the pins on the screen to find the information necessary to identify the following structures:

A. _____

B. _____

C. _____

D. _____

E. _____

F. _____

G. _____

H. _____

I. _____

J. _____

K. _____

L. _____

CHECK POINT

Elbow Joint

1. Name the structure that encloses the elbow joint.
2. What is the function of the radial collateral ligament?
3. What is the function of the ulnar collateral ligament?
4. What is the function of the anular ligament of the radius?
5. Name the bones associated with the elbow joint.

MODULE 5 The Skeletal System 251

SELECT ANIMATION	
Joint Movements: Knee Joint	PLAY

After viewing the animation, answer the following questions:

1. Describe the knee joint.

2. Describe flexion of a joint.

3. Describe flexion of the knee joint. Give an example of this flexion.

4. Describe extension of a joint.

5. Describe extension of the knee joint. Give an example of this extension.

EXERCISE 5.77:
Knee Joint—Anterior View

SELECT TOPIC	SELECT VIEW
Knee Joint ▸	Anterior

Click **LAYER 1** *in the* **LAYER CONTROLS** *window, and you will see the following image:*

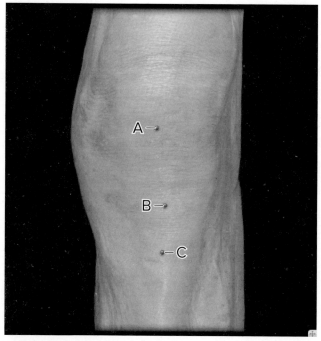

©McGraw-Hill Education

Mouse-over the pins on the screen to find the information necessary to identify the following structures:

A. _____

B. _____

C. _____

Click **LAYER 2** *in the* **LAYER CONTROLS** *window, and you will see the following image:*

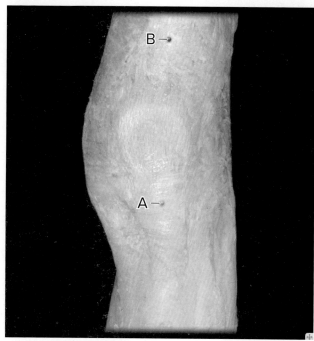

©McGraw-Hill Education

Mouse-over the pins on the screen to find the information necessary to identify the following structure:

A. _____

Nonskeletal System Structure (blue pin)

B. _____

252 MODULE 5 The Skeletal System

Click **LAYER 3** *in the* **LAYER CONTROLS** *window, and you will see the following image:*

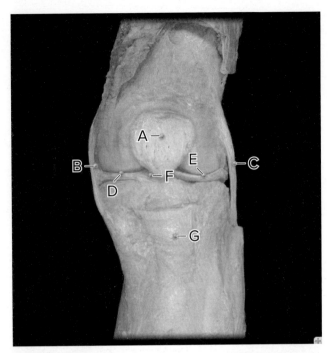

©McGraw-Hill Education

Mouse-over the pins on the screen to find the information necessary to identify the following structures:

A. _____
B. _____
C. _____
D. _____
E. _____
F. _____
G. _____

Click **LAYER 4** *in the* **LAYER CONTROLS** *window, and you will see the following image:*

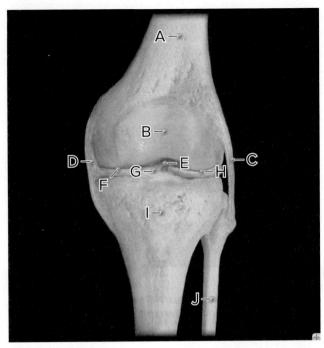

©McGraw-Hill Education

Mouse-over the pins on the screen to find the information necessary to identify the following structures:

A. _____
B. _____
C. _____
D. _____
E. _____
F. _____
G. _____
H. _____
I. _____
J. _____

Click **LAYER 5** *in the* **LAYER CONTROLS** *window, and you will see the following image:*

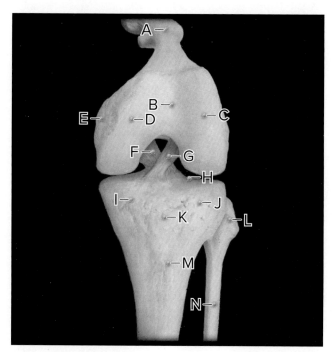

©McGraw-Hill Education

Mouse-over the pins on the screen to find the information necessary to identify the following structures:

A. _____
B. _____
C. _____
D. _____
E. _____
F. _____
G. _____
H. _____
I. _____
J. _____
K. _____
L. _____
M. _____
N. _____

EXERCISE 5.78:
Knee Joint—Posterior View

SELECT TOPIC	SELECT VIEW
Knee Joint	Posterior

Click **LAYER 1** *in the* **LAYER CONTROLS** *window, and you will see the following image:*

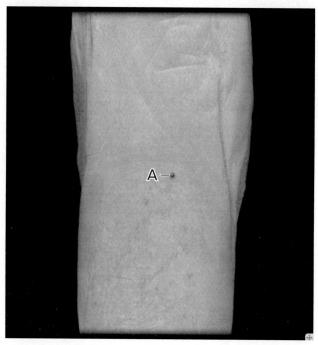

©McGraw-Hill Education

Mouse-over the pin on the screen to find the information necessary to identify the following structure:

A. _____

Click **LAYER 2** *in the* **LAYER CONTROLS** *window, and you will see the following image:*

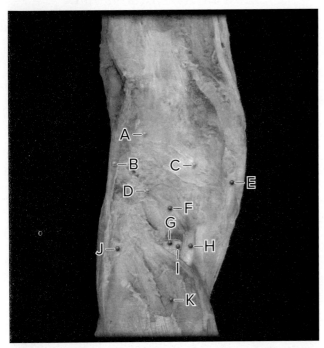

©McGraw-Hill Education

Mouse-over the pins on the screen to find the information necessary to identify the following structures:

A. _____
B. _____
C. _____
D. _____

Nonskeletal System Structures (blue pins)

E. _____
F. _____
G. _____
H. _____
I. _____
J. _____
K. _____

Click **LAYER 3** *in the* **LAYER CONTROLS** *window, and you will see the following image:*

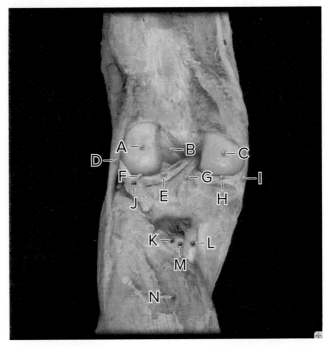

©McGraw-Hill Education

Mouse-over the pins on the screen to find the information necessary to identify the following structures:

A. _____
B. _____
C. _____
D. _____
E. _____
F. _____
G. _____
H. _____
I. _____

Nonskeletal System Structures (blue pins)

J. _____
K. _____
L. _____
M. _____
N. _____

Click **LAYER 4** *in the* **LAYER CONTROLS** *window, and you will see the following image:*

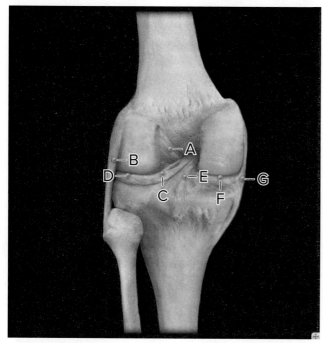

©McGraw-Hill Education

Mouse-over the pins on the screen to find the information necessary to identify the following structures:

A. _____
B. _____
C. _____
D. _____
E. _____
F. _____
G. _____

Click **LAYER 5** *in the* **LAYER CONTROLS** *window, and you will see the following image:*

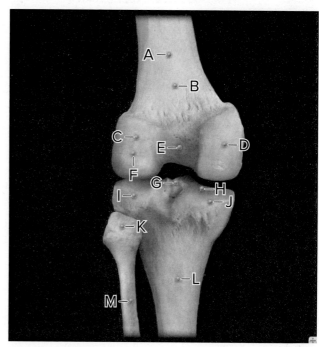

©McGraw-Hill Education

Mouse-over the pins on the screen to find the information necessary to identify the following structures:

A. _____
B. _____
C. _____
D. _____
E. _____
F. _____
G. _____
H. _____
I. _____
J. _____
K. _____
L. _____
M. _____

CHECK POINT

Knee Joint

1. Name the structures that make up the "unhappy triad."
2. Why are these structures referred to as the "unhappy triad"?
3. Name a ligament absent in approximately 40 percent of knees.
4. What is the function of the anterior cruciate ligament?
5. What is the function of the posterior cruciate ligament?

256 MODULE 5 The Skeletal System

| SELECT ANIMATION |
| Joint Movements: |
| Tibiofibulotalar Joint (Ankle) | PLAY |

After viewing the animation, answer the following questions:

1. Describe the tibiofibulotalar joint.

2. Describe dorsiflexion. Dorsiflexion describes a movement unique to _____.

3. Describe dorsiflexion of the tibiofibulotalar joint. Give an example of this dorsiflexion.

4. Describe plantar flexion. Plantar flexion describes a movement unique to _____.

5. Describe plantar flexion of the tibiofibulotalar joint. Give an example of this dorsiflexion.

| SELECT ANIMATION |
| Joint Movements: |
| Intertarsal Joints | PLAY |

After viewing the animation, answer the following questions:

1. Describe intertarsal joints.

2. Describe inversion. Inversion describes a movement unique to _____.

3. Describe eversion. Eversion describes a movement unique to _____.

| SELECT ANIMATION |
| Joint Movements: |
| Interphalangeal (IP) Joints of Toes | PLAY |

After viewing the animation, answer the following questions:

1. How are the interphalangeal joints of the toes formed?

2. How does the interphalangeal joint of the big toe compare to the interphalangeal joints of fingers 2–5?

3. Describe flexion of a joint.

4. Describe flexion of the interphalangeal joints of the toes.

5. Describe extension of a joint.

6. Describe extension of the interphalangeal joints of the toes.

CHECK POINT

Joint Movements

1. A movement that raises a body part is _____.
2. A movement of a body part away from the main axis of the body or structure is _____.
3. A movement at the ankle so that the dorsum of the foot is elevated is _____.
4. A movement that decreases the angle between two bones at a joint is _____.
5. A movement that increases the angle between two bones at a joint is _____.

EXERCISE 5.79:

Ankle Joint—Lateral View

| SELECT TOPIC | SELECT VIEW |
| Ankle Joint | Lateral |

Click **LAYER 1** *in the* **LAYER CONTROLS** *window, and you will see the following image:*

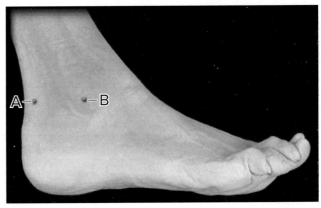

©McGraw-Hill Education

Mouse-over the pins on the screen to find the information necessary to identify the following structures:

A. _____

B. _____

Click **LAYER 2** *in the* **LAYER CONTROLS** *window, and you will see the following image:*

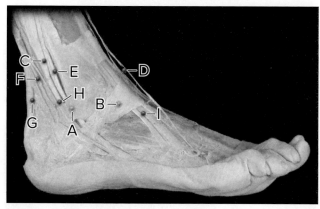

©McGraw-Hill Education

Mouse-over the pins on the screen to find the information necessary to identify the following structures:

A. _____

B. _____

Nonskeletal System Structures (blue pins)

C. _____

D. _____

E. _____

F. _____

G. _____

H. _____

I. _____

Click **LAYER 3** *in the* **LAYER CONTROLS** *window, and you will see the following image:*

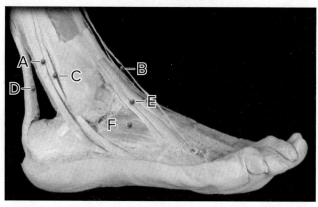

©McGraw-Hill Education

Mouse-over the pins on the screen to find the information necessary to identify the following structures:

A. _____

B. _____

C. _____

D. _____

E. _____

F. _____

Click **LAYER 4** *in the* **LAYER CONTROLS** *window, and you will see the following image:*

©McGraw-Hill Education

Mouse-over the pins on the screen to find the information necessary to identify the following structures:

A. _____

B. _____

Nonskeletal System Structures (blue pins)

C. _____

D. _____

E. _____

F. _____

G. _____

Click **LAYER 5** *in the* **LAYER CONTROLS** *window, and you will see the following image:*

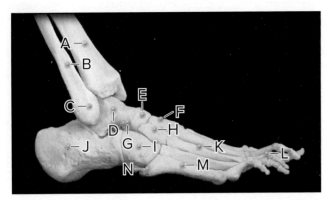

©McGraw-Hill Education

Mouse-over the pins on the screen to find the information necessary to identify the following structures:

A. _____
B. _____
C. _____
D. _____
E. _____
F. _____
G. _____
H. _____
I. _____
J. _____
K. _____
L. _____
M. _____
N. _____

EXERCISE 5.80:
Ankle Joint—Medial View

SELECT TOPIC	SELECT VIEW
Ankle Joint	Medial

Click **LAYER 1** *in the* **LAYER CONTROLS** *window, and you will see the following image:*

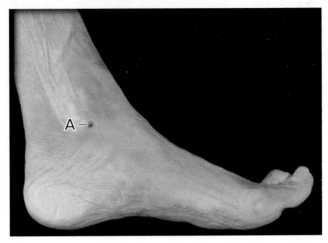

©McGraw-Hill Education

Mouse-over the pin on the screen to find the information necessary to identify the following structure:

A. _____

Click **LAYER 2** *in the* **LAYER CONTROLS** *window, and you will see the following image:*

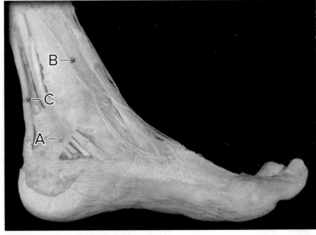

©McGraw-Hill Education

Mouse-over the pin on the screen to find the information necessary to identify the following structure:

A. _____

Nonskeletal System Structures (blue pins)

B. _____
C. _____

MODULE 5 The Skeletal System 259

Click **LAYER 3** *in the* **LAYER CONTROLS** *window, and you will see the following image:*

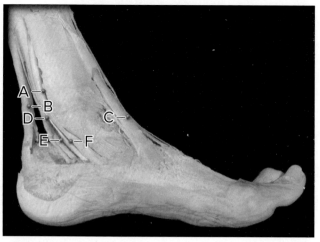

©McGraw-Hill Education

Mouse-over the pins on the screen to find the information necessary to identify the following structures:

A. _____
B. _____
C. _____
D. _____
E. _____
F. _____

Click **LAYER 4** *in the* **LAYER CONTROLS** *window, and you will see the following image:*

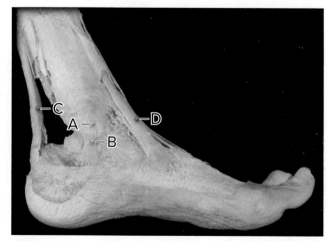

©McGraw-Hill Education

Mouse-over the pins on the screen to find the information necessary to identify the following structures:

A. _____
B. _____

Nonskeletal System Structures (blue pins)

C. _____
D. _____

Click **LAYER 5** *in the* **LAYER CONTROLS** *window, and you will see the following image:*

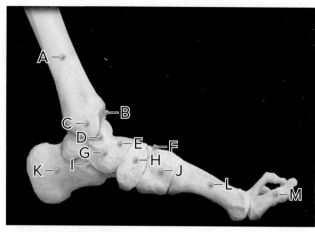

©McGraw-Hill Education

Mouse-over the pins on the screen to find the information necessary to identify the following structures:

A. _____
B. _____
C. _____
D. _____
E. _____
F. _____
G. _____
H. _____
I. _____
J. _____
K. _____
L. _____
M. _____

CHECK POINT

Ankle Joint

1. Name the strongest tendon of the body. Why is it considered the strongest?
2. What is the extensor retinaculum of the foot? What is its function?
3. What is the function of the lateral ligament of the ankle?
4. What is the flexor retinaculum of the foot? What is its function?
5. What is the function of the medial (deltoid) ligament of the ankle?

Self-Quiz

Take this opportunity to check your progress by taking the **QUIZ**. See the **Introduction Module** for a reminder on how to access the **QUIZ** for this Study Area.

IN REVIEW

What Have I Learned?

The following questions cover the material that you have just learned: joints. Apply what you have learned in answering these questions on a separate piece of paper.

1. Name the structure that encloses the temporomandibular joint.
2. What is the function of the synovial membrane?
3. What is the cause of "jaw clicking"?
4. Name the structure enclosing the acromioclavicular joint.
5. What is the result of a torn coracoclavicular ligament?
6. Name the tough, fibrous envelope of the glenohumeral joint.
7. Name the structure that encloses the elbow joint.
8. Name the bones associated with the elbow joint.
9. Name the structures that make up the "unhappy triad."
10. Name a ligament absent in approximately 40 percent of knees.
11. What is the function of the anterior cruciate ligament?
12. What is the function of the posterior cruciate ligament?
13. Name the strongest tendon of the body. Why is it considered the strongest?
14. What is the extensor retinaculum of the foot? What is its function?

EXERCISE 5.81a:
Elastic Cartilage—Histology

SELECT TOPIC	SELECT VIEW
Elastic Cartilage	LM: Medium Magnification

Click the **TAGS ON/OFF** button, and you will see the following image:

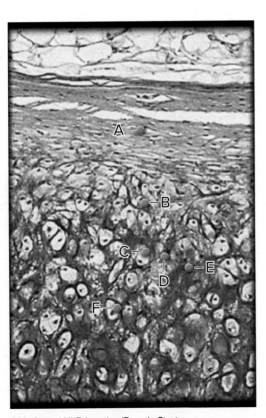

©McGraw-Hill Education/Dennis Strete

Mouse-over the pins on the screen to find the information necessary to identify the following structures:

A. _____
B. _____
C. _____
D. _____
E. _____
F. _____

EXERCISE 5.81b:
Elastic Cartilage—Histology

SELECT TOPIC	SELECT VIEW
Elastic Cartilage	LM: High Magnification

Click the **TAGS ON/OFF** button, and you will see the following image:

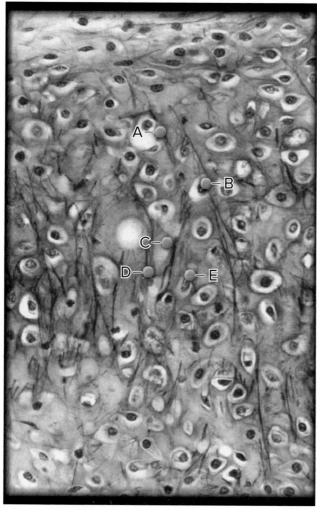

©Victor Eroschenko

Mouse-over the pins on the screen to find the information necessary to identify the following structures:

A. _____
B. _____
C. _____
D. _____
E. _____

EXERCISE 5.82a:
Fibrocartilage—Histology

SELECT TOPIC	SELECT VIEW
Fibrocartilage	LM: Low Magnification

Click the **TAGS ON/OFF** button, and you will see the following image:

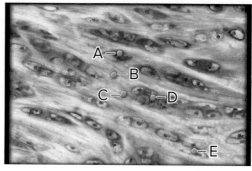

©Victor Eroschenko

Mouse-over the pins on the screen to find the information necessary to identify the following structures:

A. _____
B. _____
C. _____
D. _____
E. _____

EXERCISE 5.82b:
Fibrocartilage—Histology

SELECT TOPIC	SELECT VIEW
Fibrocartilage	LM: Medium Magnification

Click the **TAGS ON/OFF** button, and you will see the following image:

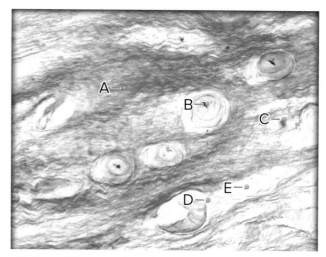

©McGraw-Hill Education/Steve Sullivan

262   MODULE 5   The Skeletal System

Mouse-over the pins on the screen to find the information necessary to identify the following structures:

A. _____
B. _____
C. _____
D. _____
E. _____

EXERCISE 5.83a:
Hyaline Cartilage—Histology

SELECT TOPIC: Hyaline Cartilage ▸ SELECT VIEW: LM: Medium Magnification

*Click the **TAGS ON/OFF** button, and you will see the following image:*

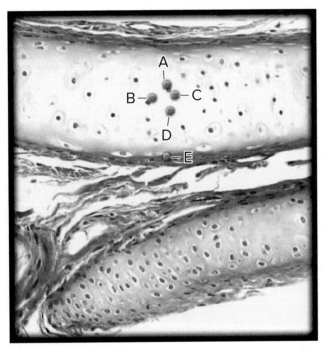

©McGraw-Hill Education/Dennis Strete

Mouse-over the pins on the screen to find the information necessary to identify the following structures:

A. _____
B. _____
C. _____
D. _____
E. _____

EXERCISE 5.83b:
Hyaline Cartilage—Histology

SELECT TOPIC: Hyaline Cartilage ▸ SELECT VIEW: LM: High Magnification

*Click the **TAGS ON/OFF** button, and you will see the following image:*

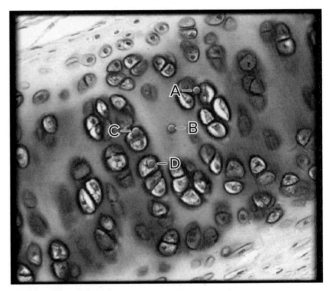

©Victor Eroschenko

Mouse-over the pins on the screen to find the information necessary to identify the following structures:

A. _____
B. _____
C. _____
D. _____

SELECT ANIMATION: Bone Cells and Bone Formation (3D)   PLAY

After viewing the animation, answer the following questions:

1. Bone consists of a dense _____ layer and a spongy _____ structure.

2. Compare the structure of compact bone with that of cancellous bone.

3. _____ and _____ are the _____ or bone cells that make up bone tissue.

4. What is remodeling, and how does it affect bone tissue?

5. What is the function of osteoclasts?

6. What is the function of osteoblasts?

7. In normal bone, what two processes initiate bone remodeling?

8. Osteoblasts emit a _____ that transforms immature _____ into mature _____.

9. Describe the process of resorption.

10. Then _____ release chemicals that stimulate immature _____ to mature and release _____.

11. What is the result of the release of osteoprotegerin?

12. What is the response to the termination of bone resorption?

13. Mature osteoblasts deposit _____, a _____ that contains _____ such as _____ and _____ and a strong flexible protein called _____.

14. What is the function of calcium and phosphorus in bone formation?

15. What structure is formed to complete the bone formation process?

16. In both men and women before midlife, bone _____ and bone _____ are balanced.

17. What influences bone remodeling?

18. What hormone suppresses osteoclasts? How does this help maintain bone strength?

19. What occurs after midlife to affect osteoclast activity?

20. What is the result of increased resorption?

21. What is osteoporosis?

22. What effect does this have on the trabeculae in cancellous bone?

23. What effect does this have on compact bone?

24. What is the overall effect on bone tissue?

25. What are the treatments for osteoporosis and what are their effects?

26. What activity adds stress to bone and promotes bone remodeling?

264 MODULE 5 The Skeletal System

EXERCISE 5.84a:
Compact Bone—Histology

SELECT TOPIC: Compact Bone ▸ SELECT VIEW: LM: Low Magnification

Click the **TAGS ON/OFF** button, and you will see the following image:

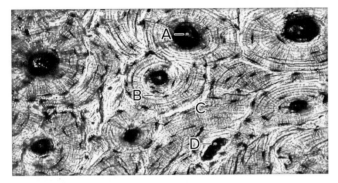

©McGraw-Hill Education/Dennis Strete

Mouse-over the pins on the screen to find the information necessary to identify the following structures:

A. _____
B. _____
C. _____
D. _____

EXERCISE 5.84b:
Compact Bone—Histology

SELECT TOPIC: Compact Bone ▸ SELECT VIEW: LM: High Magnification

Click the **TAGS ON/OFF** button, and you will see the following image:

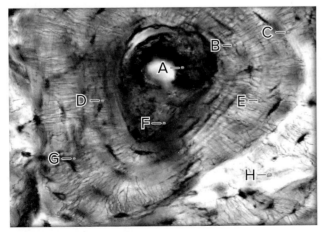

©McGraw-Hill Education/Al Telser

Mouse-over the pins on the screen to find the information necessary to identify the following structures:

A. _____
B. _____
C. _____
D. _____
E. _____
F. _____
G. _____
H. _____

EXERCISE 5.85a:
Spongy Bone—Histology

SELECT TOPIC: Spongy Bone ▸ SELECT VIEW: LM: Low Magnification

Click the **TURN ON/OFF** button, and you will see the following image:

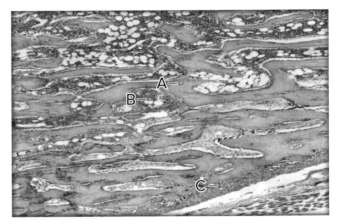

©McGraw-Hill Education/Steve Sullivan

Mouse-over the pins on the screen to find the information necessary to identify the following structures:

A. _____
B. _____
C. _____

EXERCISE 5.85b:
Spongy Bone—Histology

SELECT TOPIC	SELECT VIEW
Spongy Bone	LM: Medium Magnification

Click the **TAGS ON/OFF** button, and you will see the following image:

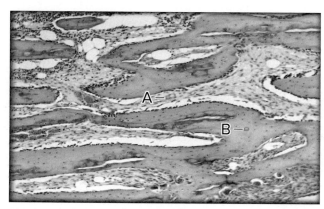

©McGraw-Hill Education/Steve Sullivan

Mouse-over the pins on the screen to find the information necessary to identify the following structures:

A. _____

B. _____

EXERCISE 5.85c:
Spongy Bone—Histology

SELECT TOPIC	SELECT VIEW
Spongy Bone	SEM: Low Magnification

Click the **TAGS ON/OFF** button, and you will see the following image:

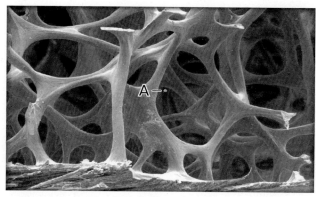

©Steve Gschmeissner/Science Source

Mouse-over the pin on the screen to find the information necessary to identify the following structure:

A. _____

Self-Quiz
Take this opportunity to check your progress by taking the **QUIZ**. See the **Introduction Module** for a reminder on how to access the **QUIZ** for this Study Area.

IN REVIEW

What Have I Learned?

The following questions cover the material that you have just learned: histology. Apply what you have learned in answering these questions on a separate piece of paper.

1. Name the outer layer of most cartilage.

2. Give five examples of elastic cartilage.

3. What are lacunae? What are contained within the lacunae?

4. Give four examples of fibrocartilage.

5. What is the function of the chondrogenic layer of the perichondrium?

6. Describe an osteon.

7. Describe a perforating canal.

8. Interstitial lamellae result from _____.

9. What component of bone tissue is essential to bone nutrition, growth, and repair?

10. What are trabeculae?

266 MODULE 5 The Skeletal System

EXERCISE 5.86:
Coloring Exercise

Look up the bones of the skull and then color them in with colored pens or pencils. Write the names of important bones or bone features in the blank label lines.

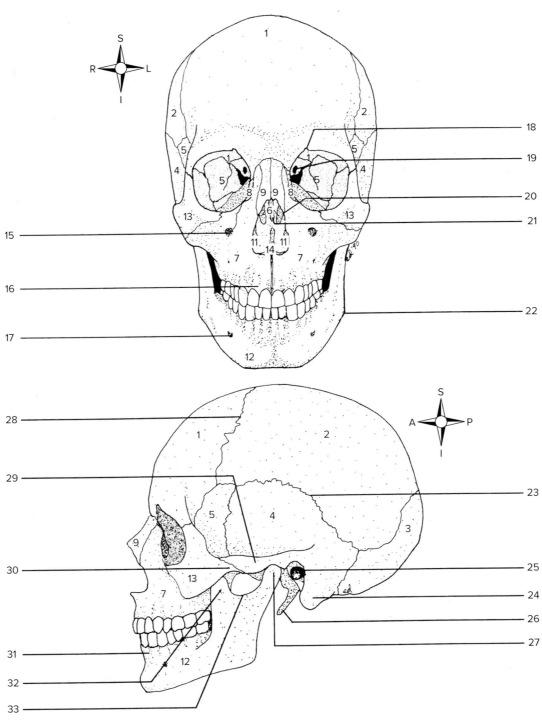

©McGraw-Hill Education

MODULE 5   The Skeletal System   **267**

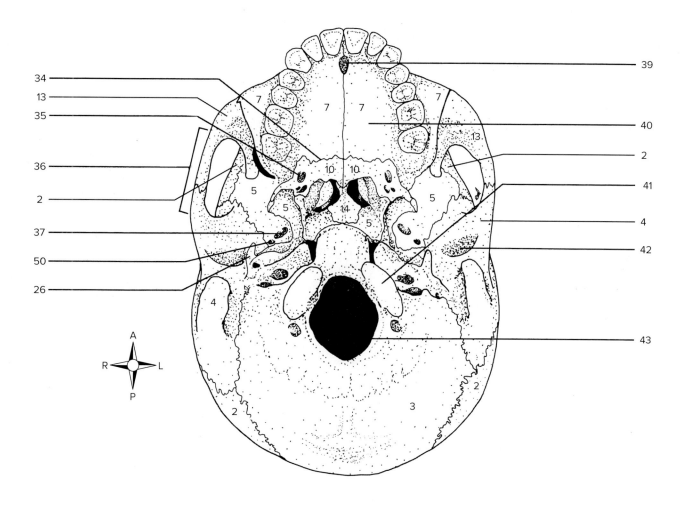

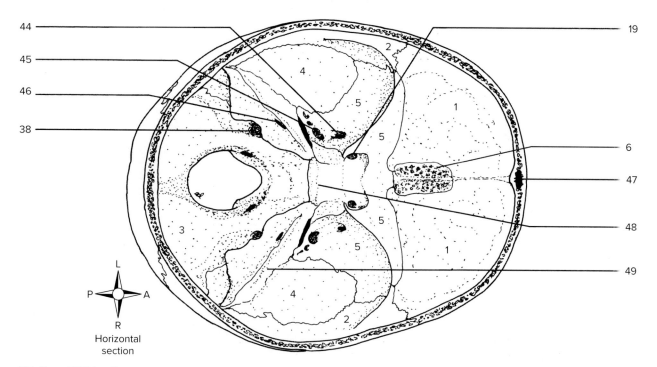

Horizontal section

©McGraw-Hill Education

MODULE 6

The Muscular System

Overview: Muscular System

Without the skeletal muscular system, the bones that we learned in the previous module would be incapable of movement. Our bodies are equipped with some 600 skeletal muscles not only to put those 206 bones into motion, but also to generate as much as 85 percent of our body heat, maintain our posture, control the openings involved with the entrance and exit of materials, and to express our emotions and thoughts through movements of our facial muscles.

Three important structural terms to understand as you begin your study of the skeletal muscular system are a muscle's **origin, insertion,** and **belly.** The ends of most muscles are attached to separate bones. This assures that each muscle or its tendon will span at least one joint. When a muscle contracts, one bone acts as an anchor and remains relatively stationary while the second bone will move. As we have previously stated, the end of the muscle attached to the relatively stationary bone is called the **origin,** while the end of the muscle attached to the freely moving bone is called the **insertion.** Again, one way to remember the difference is to think of your birthplace, your **origin.** No matter where you may move throughout your life, your **origin** remains the same—*it does not move!* You may **insert** yourself at several locations throughout your life—away to college, a job in a different town, and so on. These require *moving!*

Also, many muscles are narrow at their ends, their origin and insertion, and thick in the middle. This thicker middle region is called the muscle's **belly.**

From the **HOME** *screen, click the drop-down box on the* **MODULE** *menu.*

From the systems listed, click on **Muscular.**

Animations: Anatomy and Physiology

Before we begin our study of the muscular system, it is important that we first have an understanding of how the muscles function—not only their anatomy, but also their physiology. The animations presented here are designed to help you understand the anatomy and physiology of muscles and muscle contractions. After we lay this groundwork, we will be better prepared to build on this foundation by learning their names, origins, and insertions.

The first animation will introduce you to the structure of muscle fibers, and then, the remaining animations, will walk you through the process of muscle contraction. The final animation will provide a 3D overview of muscle structure and function to help clarify the information in the previous animations.

Strategically placed among these animations are three interactive animations that allow you to initiate the action of specific events in muscle physiology. This hands-on approach to learning will help you better understand the concepts presented. These interactive animations are designed for you to complete in a sequence of steps, so be sure to interact with them at every opportunity presented.

If after viewing these animations the process is not clear to you, repeat the series of animations in sequence until you understand how this important process unfolds. You can pause the animations and repeat sections as needed to answer the questions following each animation.

SELECT ANIMATION
Skeletal Muscle
PLAY

After viewing the animation, answer the following questions:

1. _____ _____ is responsible for voluntary movement of the human body.

2. What is the structure of these muscles?

3. What is the epimysium?

4. At the ends of the muscle, the epimysium is continuous with the _____ and _____ of the bone.

5. What is the perimysium?

6. What is a fascicle?

7. What is the endomysium?

8. Describe a muscle fiber.

9. What is unique about the nuclei of a muscle fiber? Where are they located?

10. What structures dominate the interior of the muscle fiber?

11. What are the two types of protein filaments that compose the myofibril?

12. Name the orderly contractile unit of the muscle fiber.

13. The shortened sarcomeres result in _____ _____, and ultimately _____ of the _____.

SELECT ANIMATION
Neuromuscular Junction
PLAY

After viewing the animation, answer the following questions:

1. What sequence of events occurs when an action potential arrives at the presynaptic terminal?

2. What sequence of events occurs when the calcium ions enter the presynaptic terminal?

3. What is the function of acetylcholine?

4. The movement of _____ ions into the muscle cell results in . . .

5. Once threshold has been reached, . . .

6. What happens to the acetylcholine after the generation of an action potential on the muscle cell membrane?

7. What happens to the acetylcholine breakdown products?

This Interactive Animation will walk you through the structure and function of the Neuromuscular Junction. Be sure to interact with the animation to glean all the information available to you.

Open the **Module 6–Muscular System ANIMATION LIST.**

Under the second heading of **Skeletal Muscular Contraction (Interactive)**, select **1. The Neuromuscular Junction** and you will see this image.

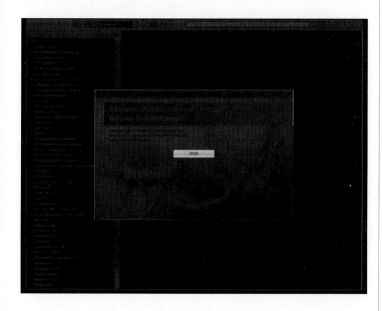

Select the **BEGIN** button in the middle, and you will see this screen and the narration will begin.

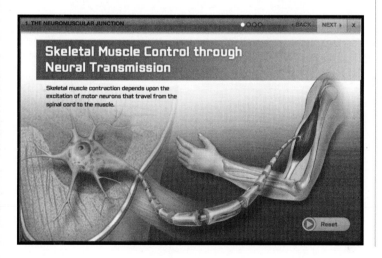

Observe as the animation demonstrates the pathway of an action potential from the soma to the axon hillock, and then down the axon to the neuromuscular junction.

Select the **NEXT** button flashing at the top-right. You will see this screen and the narration will resume.

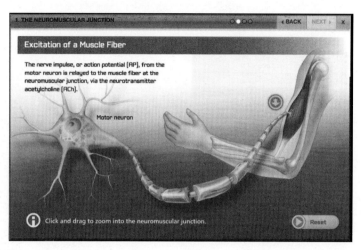

After listening to the narration, click on and drag the blue arrow downward to zoom into the neuromuscular junction.

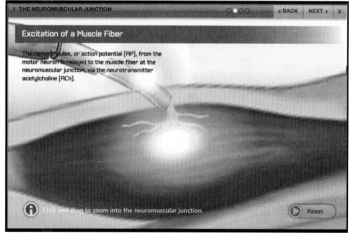

If you would like to interact with this image again, click the blue **RESET** button in the bottom-right corner.

When finished with this image, click **NEXT**. You will then see this screen and the narration will continue.

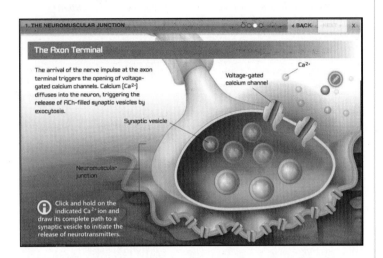

As indicated on the image, click and hold the indicated Ca^{2+} ion and draw its path to a synaptic vesicle.

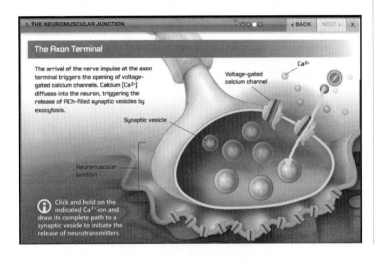

When finished with the interaction, click **NEXT**. You will then see this screen and the narration will continue.

Be sure to click and drag the ACh to their receptor sites.

When finished with the interactions, click the **X** in the top-right corner to close the animation.

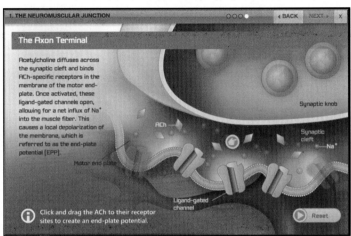

After reviewing the animation, answer the following questions:

1. Skeletal muscle contraction depends upon the _____ of _____ _____ that travel from the _____ _____ to the _____.

2. What is another name for a nerve impulse?

3. Where is this impulse relayed?

4. What neurotransmitter is employed by the motor neuron?

5. What is triggered by the arrival of the nerve impulse at the axon terminal?

6. _____ diffuses into the _____, triggering the release of ACh-filled _____ _____ by _____.

7. What does the ACh accomplish as it diffuses across the synaptic cleft? What then flows into the muscle fiber?

8. Describe the end-plate potential [EPP].

SELECT ANIMATION
Sliding Filament — PLAY

After viewing the animation, answer the following questions:

1. Describe the muscle myofilaments in a relaxed muscle.

2. What interaction do these myofilaments have during contraction?

3. What is the result?

4. Describe the muscle myofilaments in a fully contracted muscle.

SELECT ANIMATION
Excitation-Contraction Coupling — PLAY

After viewing the animation, answer the following questions:

1. Action potentials are propagated over which structure of the skeletal muscle fiber?

2. How does the action potential arrive into the interior of the muscle fiber?

3. What does the entry of the action potential cause to occur inside the muscle fiber?

4. This causes _____ ions to diffuse from the _____ _____ into the _____.

5. Where is the tropomyosin located? What is it covering?

6. Describe the molecules attached to the tropomyosin.

7. What events occur when the calcium ions bind to the troponin molecule?

8. How are cross-bridges formed?

This Interactive Animation will walk you through the steps of Excitation-contraction Coupling. Be sure to interact with the animation to glean all the information available to you.

Open the **Module 6–Muscular System ANIMATION LIST.**

Under the second heading of **Skeletal Muscular Contraction (Interactive),** *select 2.* **Excitation-contraction Coupling** *and you will see this image:*

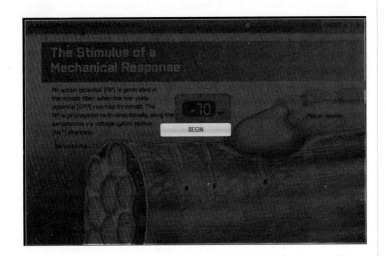

Select the **BEGIN** *button in the middle, and interact with the four images contained in this animation.*

When finished with the interactions, click the **X** *in the top-right corner to close the animation.*

After reviewing the animation, answer the following questions:

1. When is an action potential generated in the muscle fiber?

2. How is the action potential propagated?

3. What structures ensure that all regions of the myofibril are capable of responding?

4. How do these structures respond?

5. What is a triad?

6. Where do triads occur?

7. In what manner do these structures allow the propagation of the action potential?

8. At the triad, what receptors are in the T-tubule and what receptors are they coupled with in the sarcoplasmic reticulum membrane?

9. What series of changes occur to these receptors upon the arrival of the action potential?

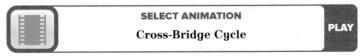

SELECT ANIMATION
Cross-Bridge Cycle
PLAY

After viewing the animation, answer the following questions:

1. During contraction of a muscle, _____ _____ bind to _____. This moves _____ out of the way and uncovers _____ _____ for _____ on the _____ _____.

2. What molecules are attached to the myosin head from the previous cycle of movement?

3. How are cross-bridges formed? What is released in the process?

4. How is the myosin head moved? What does this cause? What is released from the myosin head when this occurs?

5. How is the bond between the actin and myosin filaments broken? How is energy released? What happens to this energy?

6. After the myosin head returns to its upright position, what occurs if calcium ions are still present?

This Interactive Animation will walk you through the steps of cross-bridge Cycling. Be sure to interact with the animation to glean all the information available to you.

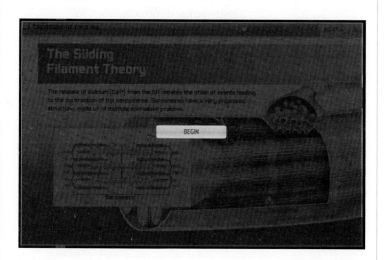

Open the **Module 6–Muscular System ANIMATION LIST.**

Under the second heading of **Skeletal Muscular Contraction (Interactive),** *select* **3. Cross-bridge Cycling** *and you will see this image:*

Select the **BEGIN** *button in the middle, and interact with the seven images contained in this animation.*

When finished with the interactions, click the **X** *in the top-right corner to close the animation.*

After reviewing the animation, answer the following questions:

1. What initiates the chain of events leading up to the contraction of the sarcomere?

2. Describe the structure of the sarcomere.

3. Describe the Z discs.

4. Describe the M line.

5. Describe the structure of the thin filaments.

6. Describe the structure of the thick filaments.

7. Describe the structure of myosin molecules.

8. What are the two forms of the myosin heads?

9. Describe how and why the myosin heads become energized.

10. Describe the detailed structure of the thin filaments.

11. What sequence of events is initiated by calcium ions?

12. Describe cross-bridge formation.

13. What interaction occurs between the myosin head and the actin molecule once cross-bridge formation occurs?

14. What causes cross-bridge detachment?

15. What then returns the myosin head to its energized state?

16. What must occur for muscle contraction to continue?

17. What then is the result of each successive neuronal stimulus?

18. What ultimately produces a unified contraction of the whole muscle?

SELECT ANIMATION
Skeletal Muscle Contraction (3D) PLAY

After viewing the animation, answer the following questions:

1. Skeletal muscles are _____.

2. How are skeletal muscles controlled?

3. Skeletal muscles are made up of _____, which are bundles of muscle fibers.

4. These bundles of muscle fibers are surrounded by _____, _____ _____, and _____.

5. Motor nerve _____ supply one or more _____ _____ _____, constituting a _____ _____.

6. Inside a muscle fiber, thread like structures called _____ are organized into contractile units or _____, reflecting the _____ characteristic of _____ muscle.

7. Each sarcomere is made up of thick and thin _____, strands of protein called _____ and _____.

8. Which protein forms thin filaments?

9. The protruding heads of _____ _____ form _____ _____ that bind to _____ _____ and ____, an energy transport molecule.

10. Describe the structure of thin filaments.

11. At rest, _____ molecules bind _____ and _____, two _____ proteins that inhibit _____.

12. Where does skeletal muscle contraction begin?

13. What are T-tubules?

14. T-tubules connect to a membranous network called the _____ _____ that store _____ and transfer _____ via voltage-gated _____ into the myofibrils _____ to regulate muscle _____.

15. What is the first step of muscle contraction?

16. What occurs during this step?

17. What is the second step of muscle contraction?

18. What occurs during this step?

19. Along the myofilaments, _____ binds _____ and separates from _____.

20. What is the third step of muscle contraction?

21. What occurs during this step?

22. What is the fourth step of muscle contraction?

23. What occurs during this step?

Muscular System: Head and Neck

EXERCISE 6.1:
Skeletal Muscle—Head and Neck—Anterior View

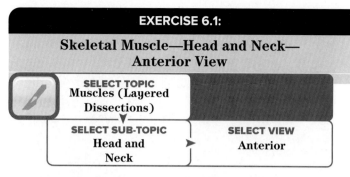

Click **LAYER 1** *in the* **LAYER CONTROLS** *window, and you will see the following image:*

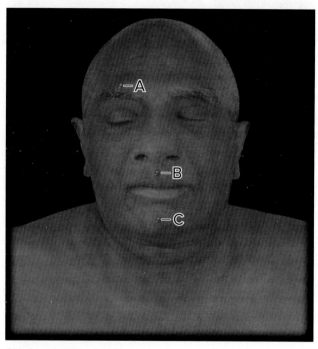

©McGraw-Hill Education

Mouse-over the pins on the screen to find the information necessary to identify the following nonmuscular system structures:

A. _____

B. _____

C. _____

CHECK POINT

Head and Neck, Anterior View

1. Name the ridge superior to each orbit on the anterior side.
2. What bone is that ridge part of?
3. What is the name of the shallow midline groove of the upper lip?

Click **LAYER 2** *in the* **LAYER CONTROLS** *window, and you will see the following image:*

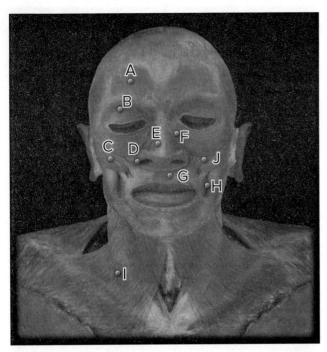

©McGraw-Hill Education

Click **LAYER 3** *in the* **LAYER CONTROLS** *window, and you will see the following image:*

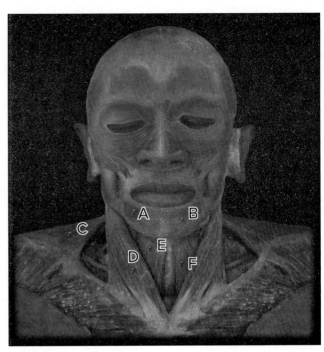

©McGraw-Hill Education

Mouse-over the pins on the screen to find the information necessary to identify the following structures:

A. _____
B. _____
C. _____
D. _____
E. _____
F. _____
G. _____
H. _____
I. _____
J. _____

Mouse-over the pins on the screen to find the information necessary to identify the following structures:

A. _____
B. _____
C. _____
D. _____
E. _____
F. _____

CHECK POINT

Head and Neck, Anterior View, *continued*

7. Name two muscles whose insertion is the hyoid bone.
8. Name the muscle responsible for depression of the angle of the mouth to grimace.
9. Name the muscle responsible for depression of the lower lip while pouting.

CHECK POINT

Head and Neck, Anterior View, *continued*

4. Name the muscle that closes the eye when winking or blinking.
5. Name the muscle responsible for compression of the cheek as in inflating a balloon or playing a wind instrument.
6. Name the muscle that closes and protrudes the lips.

Click **LAYER 4** *in the* **LAYER CONTROLS** *window, and you will see the following image:*

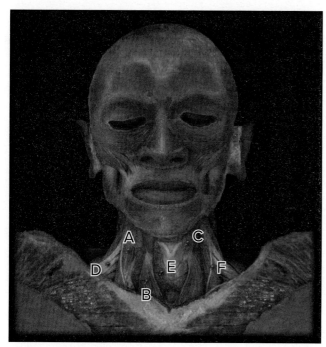

©McGraw-Hill Education

Click **LAYER 5** *in the* **LAYER CONTROLS** *window, and you will see the following image:*

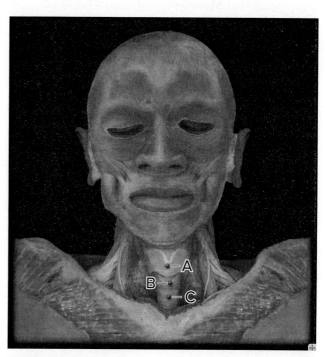

©McGraw-Hill Education

Mouse-over the pins on the screen to find the information necessary to identify the following structures:

A. _____
B. _____
C. _____

Nonmuscular System Structures (blue pins)

D. _____
E. _____
F. _____

Mouse-over the pins on the screen to find the information necessary to identify the following nonmuscular system structures:

A. _____
B. _____
C. _____

Animations: Muscle Actions

SELECT ANIMATION
Buccinator Muscle
PLAY

After viewing this animation, select and view these additional animations:

- Frontalis muscle
- Levator labii superioris alaeque nasi muscle
- Orbicularis oculi muscle
- Orbicularis oris muscle
- Trapezius muscle

CHECK POINT

Head and Neck, Anterior View, *continued*

10. The roots of the brachial plexus are located between the anterior and middle _____ muscles.
11. Name a muscle responsible for elevation of the larynx and depression of the hyoid bone.
12. Name the four infrahyoid muscles.

EXERCISE 6.2:
Skeletal Muscle—Head and Neck—Lateral View

SELECT TOPIC	SELECT VIEW
Head and Neck	Lateral

Click **LAYER 1** *in the* **LAYER CONTROLS** *window, and you will see the following image:*

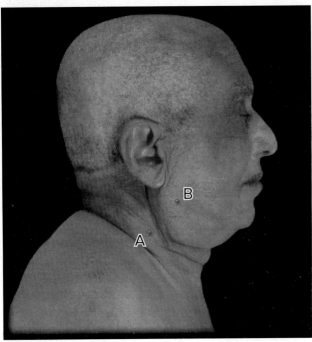

©McGraw-Hill Education

Mouse-over the pins on the screen to find the information necessary to identify the following structure:

A. _____

Nonmuscular System Structure (blue pin)

B. _____

CHECK POINT
Head and Neck, Lateral View

1. What location on the mandible provides an attachment site for the masseter muscle?
2. What other muscle attaches at this point?

Click **LAYER 2** *in the* **LAYER CONTROLS** *window, and you will see the following image:*

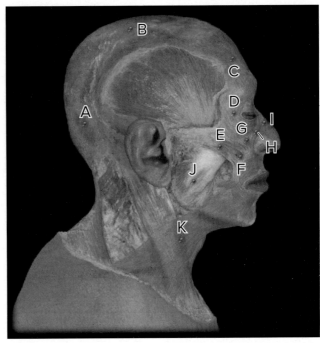

©McGraw-Hill Education

Mouse-over the pins on the screen to find the information necessary to identify the following structures:

A. _____
B. _____
C. _____
D. _____
E. _____
F. _____
G. _____
H. _____
I. _____
J. _____
K. _____

CHECK POINT
Head and Neck, Lateral View, *continued*

3. Name the muscle responsible for elevation of the upper lip in a sneer.
4. Name the two muscles responsible for elevation of the upper lip in a smile.
5. Name the muscle that elevates and creases the skin of the neck as well as depresses the lower lip and the angle of the mouth.

Click **LAYER 3** *in the* **LAYER CONTROLS** *window, and you will see the following image:*

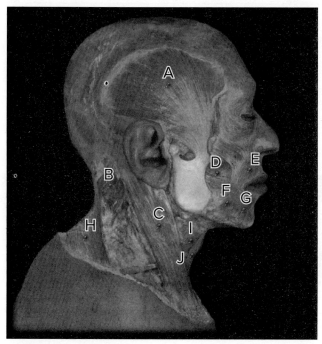

©McGraw-Hill Education

Click **LAYER 4** *in the* **LAYER CONTROLS** *window, and you will see the following image:*

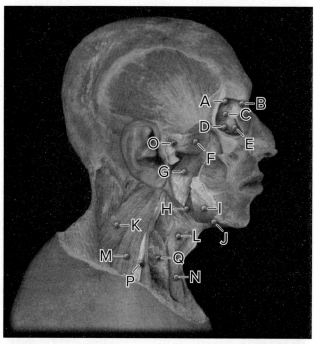

©McGraw-Hill Education

Mouse-over the pins on the screen to find the information necessary to identify the following structures:

A. _____
B. _____
C. _____
D. _____
E. _____
F. _____
G. _____
H. _____
I. _____
J. _____

CHECK POINT

Head and Neck, Lateral View, *continued*

6. Name a muscle with two bellies (superior and inferior) joined by an intermediate tendon.
7. What is the "kissing muscle"?

Mouse-over the pins on the screen to find the information necessary to identify the following structures:

A. _____
B. _____
C. _____
D. _____
E. _____
F. _____
G. _____
H. _____
I. _____
J. _____
K. _____
L. _____
M. _____
N. _____

Nonmuscular System Structures (blue pins)

O. _____
P. _____
Q. _____

CHECK POINT

Head and Neck, Lateral View, *continued*

8. Name a muscle responsible for the protrusion of the mandible.
9. Name a muscle responsible for the elevation of the scapula, as in shrugging the shoulders.
10. Name the muscle involved in abduction of the eyeball.

Click **LAYER 5** in the **LAYER CONTROLS** *window, and you will see the following image:*

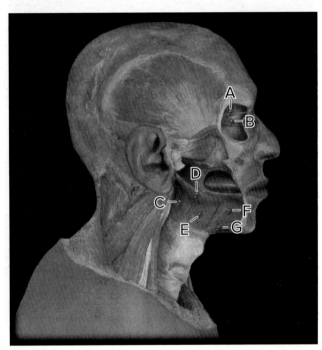

©McGraw-Hill Education

Mouse-over the pins on the screen to find the information necessary to identify the following structures:

A. _____
B. _____
C. _____
D. _____
E. _____
F. _____
G. _____

CHECK POINT

Head and Neck, Lateral View, *continued*

11. Name the muscle involved with adduction of the eyeball.
12. Name the muscle whose tendon passes through a trochlea.
13. Which muscle allows you to stick out your tongue?

| SELECT ANIMATION |
| Buccinator Muscle | PLAY |

After viewing this animation, select and view these additional animations:

- Frontalis muscle
- Levator labii superioris alaeque nasi muscle
- Masseter muscle
- Platysma muscle
- Temporalis muscle
- Trapezius muscle

EXERCISE 6.3:

Skeletal Muscles—Head and Neck—Midsagittal View

| SELECT TOPIC | SELECT VIEW |
| Head and Neck | Midsagittal |

Click **LAYER 1** in the **LAYER CONTROLS** *window, and you will see the following image:*

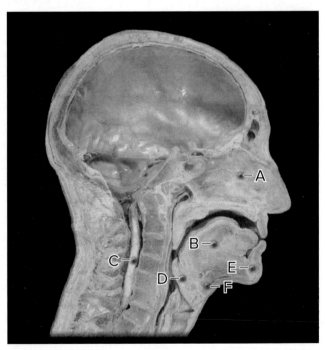

©McGraw-Hill Education

Mouse-over the pins on the screen to find the information necessary to identify the following nonmuscular system structures:

A. _____
B. _____
C. _____
D. _____
E. _____
F. _____

CHECK POINT

Head and Neck, Midsagittal View

1. Name the muscular structure that separates the oropharynx from the nasopharynx.
2. Name a muscle that blends with the musculature of the tongue.

Click **LAYER 2** *in the* **LAYER CONTROLS** *window, and you will see the following image:*

Click **LAYER 3** *in the* **LAYER CONTROLS** *window, and you will see the following image:*

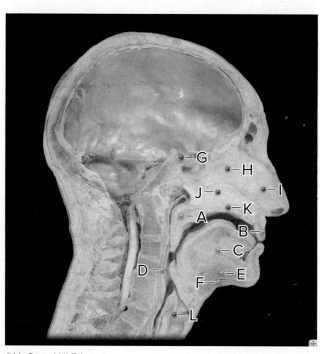
©McGraw-Hill Education

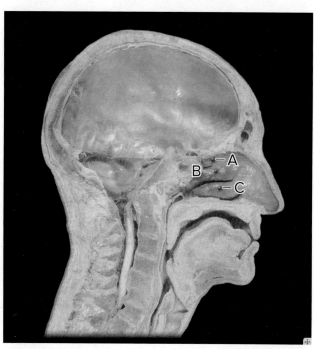
©McGraw-Hill Education

Mouse-over the pins on the screen to find the information necessary to identify the following structures:

A. _____
B. _____
C. _____
D. _____
E. _____
F. _____

Nonmuscular System Structures (blue pins)

G. _____
H. _____
I. _____
J. _____
K. _____
L. _____

Mouse-over the pins on the screen to find the information necessary to identify the following nonmuscular system structures:

A. _____
B. _____
C. _____

284 MODULE 6 The Muscular System

EXERCISE 6.4:
Skeletal Muscle—Head and Neck—Posterior View

SELECT TOPIC	SELECT VIEW
Head and Neck	▶ Posterior

Click **LAYER 1** *in the* **LAYER CONTROLS** *window, and you will see the following image:*

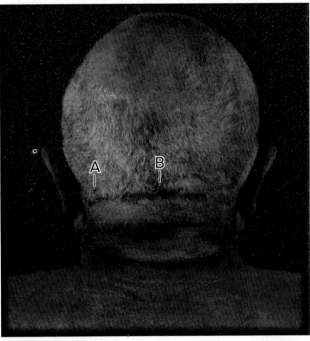

©McGraw-Hill Education

Mouse-over the pins on the screen to find the information necessary to identify the following nonmuscular system structures:

A. _____

B. _____

CHECK POINT
Head and Neck, Posterior View

1. Name the three muscles that attach to the mastoid process.
2. What is the attachment point to the skull for the nuchal ligament?

Click **LAYER 2** *in the* **LAYER CONTROLS** *window, and you will see the following image:*

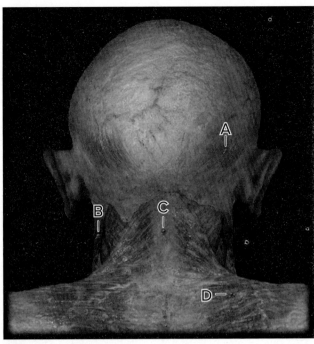

©McGraw-Hill Education

Mouse-over the pins on the screen to find the information necessary to identify the following structures:

A. _____

B. _____

C. _____

D. _____

CHECK POINT
Head and Neck, Posterior View, *continued*

3. Name two muscles attached to the nuchal ligament.
4. Name the two origins and the one insertion for the sternocleidomastoid muscle.
5. Name the large superficial muscle located from the posterior neck to the shoulders and the posterior midline.

Click **LAYER 3** *in the* **LAYER CONTROLS** *window, and you will see the following image:*

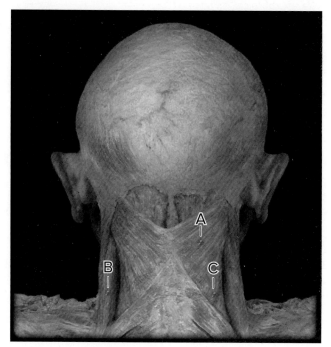

©McGraw-Hill Education

Mouse-over the pins on the screen to find the information necessary to identify the following structures:

A. _____

B. _____

C. _____

Click **LAYER 4** *in the* **LAYER CONTROLS** *window, and you will see the following image:*

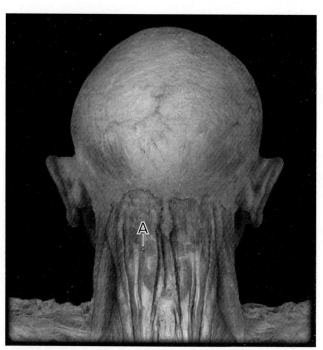

©McGraw-Hill Education

Mouse-over the pin on the screen to find the information necessary to identify the following structure:

A. _____

Click **LAYER 6** *in the* **LAYER CONTROLS** *window, and you will see the following image:*

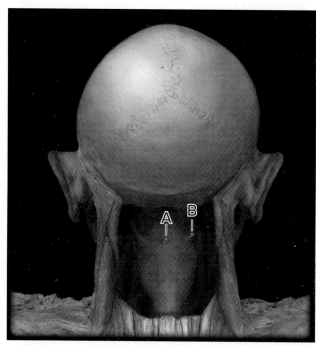

©McGraw-Hill Education

Mouse-over the pins on the screen to find the information necessary to identify the following structures:

A. _____

B. _____

CHECK POINT

Head and Neck, Posterior View, *continued*

6. Name a muscle responsible for elevation of the pharynx during swallowing.

Self-Quiz

Take this opportunity to check your progress by taking the **QUIZ**. See the **Introduction Module** for a reminder on how to access the **QUIZ** for this Study Area.

IN REVIEW

What Have I Learned?

The following questions cover the material that you have just learned, the muscles of the head and neck. Apply what you have learned to answer these questions on a separate piece of paper.

1. Name a muscle responsible for elevation of the larynx.

2. Name the muscle that flares the nostrils.

3. The scapula is elevated by which muscles?

4. When this muscle contracts, the head rotates so that the face turns downward and to the opposite side.

5. Name three muscles responsible for closing the mouth.

6. Name three muscles responsible for depression of the hyoid bone.

7. What muscle is responsible for flexion of the head to look downward?

8. Name the group of muscles responsible for the peristaltic waves of swallowing.

9. Name three muscles involved in moving the tongue.

10. Name the muscle involved in elevating the eyebrow and creasing the skin of the forehead.

11. Name a muscle responsible for depression of the larynx.

12. There is a muscle complex that lies deep to the scalp from the forehead to the posterior skull. What is the name of that complex and the two muscles that it consists of?

13. List all of the muscles involved with eye movement, and describe the movement involved with each muscle.

14. Name the anatomical structure commonly called the chin.

Muscular System: Trunk, Shoulder Girdle, and Upper Limb

EXERCISE 6.5:
Skeletal Muscle—Thorax—Anterior View

SELECT TOPIC	SELECT VIEW
Thorax	Anterior

Click **LAYER 1** *in the* **LAYER CONTROLS** *window, and you will see the following image:*

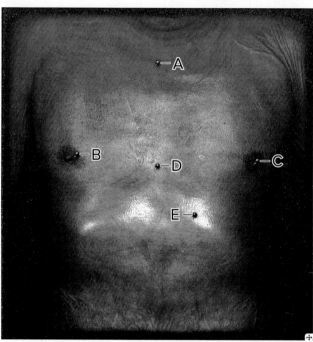

©McGraw-Hill Education

Mouse-over the pins on the screen to find the information necessary to identify the following nonmuscular system structures:

A. surface projection of jugular notch of sternum
B. areola
C. nipple
D. surface projection of xiphoid process
E. surface projection of costal margin

CHECK POINT
Thorax, Anterior View

1. What is the name for the superficially visible inferior border of costal cartilages 7–10?
2. What structures attach to this location?
3. What are the two names for the shallow notch in the superficially visible superior border of the manubrium?

Click **LAYER 2** *in the* **LAYER CONTROLS** *window, and you will see the following image:*

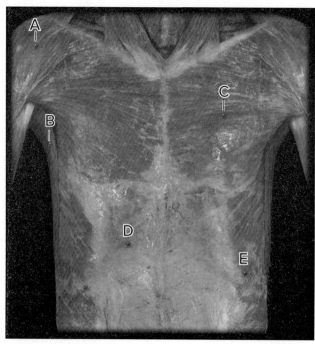

©McGraw-Hill Education

Mouse-over the pins on the screen to find the information necessary to identify the following structures:

A. deltoid m.
B. latissimus dorsi m.
C. pectoralis major m.
D. anterior rectus sheath
E. external abdominal oblique m.

CHECK POINT
Thorax, Anterior View, *continued*

4. Name the muscle involved with adduction, extension, and medial rotation of the arm.
5. Name the muscle involved with abduction, flexion, extension, and lateral and medial rotation of the arm.
6. What is the name for the fibrous compartment enclosing the rectus abdominis muscle?
7. What is an aponeurosis?

Click **LAYER 3** *in the* **LAYER CONTROLS** *window, and you will see the following image:*

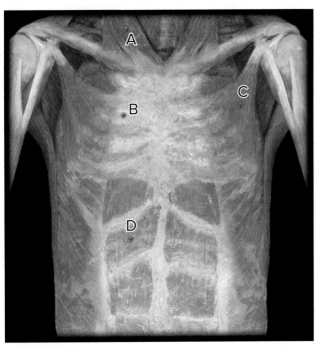

©McGraw-Hill Education

Mouse-over the pins on the screen to find the information necessary to identify the following structures:

A. sternocleidomastoid
B. external intercostal & membrane
C. pectoralis minor
D. rectus abdominis

CHECK POINT

Thorax, Anterior View, *continued*

8. Name the muscle that consists of three to four bellies, separated by tendinous intersections.
9. Name the muscle with its origin at the medial clavicle and the manubrium of the sternum and its insertion at the mastoid process.
10. Name the muscle that stabilizes the scapula and is involved in its lateral rotation.

Click **LAYER 4** *in the* **LAYER CONTROLS** *window, and you will see the following image:*

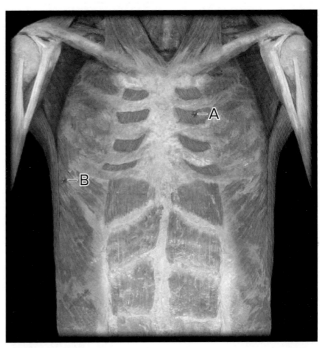

©McGraw-Hill Education

Mouse-over the pins on the screen to find the information necessary to identify the following structures:

A. internal intercostal m
B. serratus anterior

Click **LAYER 5** *in the* **LAYER CONTROLS** *window, and you will see the following image:*

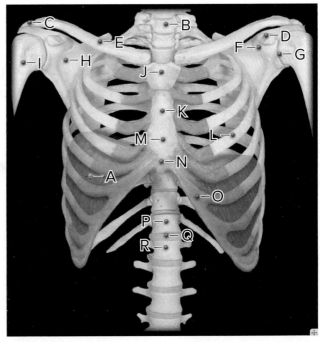

©McGraw-Hill Education

Mouse-over the pins on the screen to find the information necessary to identify the following structure:

A. diaphragm

Nonmuscular System Structures (blue pins)

B. cervical vertebra
C. acromion
D. coracoid process
E. clavicle
F. suprascapular notch
G. glenoid cavity
H. scapula
I. humerus
J. manubrium
K. sternum
L. ribs 1-12
M. body of sternum
N. xiphoid process
O. costal cartilages
P. thoracic vertebra
Q. intervertebral disc
R. lumbar vertebra

 SELECT ANIMATION — **Deltoid Muscle** PLAY

After viewing this animation, select and view these additional animations:

- External abdominal oblique muscle
- Latissimus dorsi muscle
- Pectoralis major muscle
- Rectus abdominus muscle
- Serratus anterior muscle

Self-Quiz

Take this opportunity to check your progress by taking the **QUIZ**. See the **Introduction Module** for a reminder on how to access the **QUIZ** for this Study Area.

IN REVIEW

What Have I Learned?

The following questions cover the material that you have just learned, the muscles of the thorax. Apply what you have learned to answer these questions on a separate piece of paper.

1. Name the structure formed by the tendons of three abdominal muscles.

2. Name the three primary muscles of respiration.

3. Name the muscle responsible for the adduction, extension, and medial rotation of the humerus.

4. Name the two muscles that stabilize the scapula.

5. Name the muscle that is the site of intramuscular injections of the arm.

EXERCISE 6.6:
Skeletal Muscle—Abdomen—Anterior View

SELECT TOPIC	SELECT VIEW
Abdomen	Anterior

Click **LAYER 1** *in the* **LAYER CONTROLS** *window, and you will see the following image:*

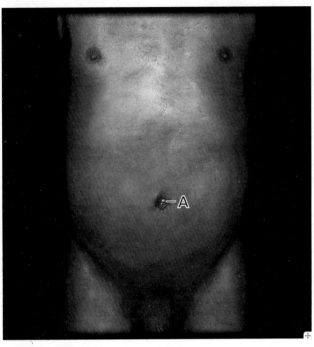

©McGraw-Hill Education

Mouse-over the pin on the screen to find the information necessary to identify the following nonmuscular system structure:

A. _umbilicus_

CHECK POINT
Abdomen, Anterior View

1. Describe umbilicus variability.
2. What is the umbilicus a landmark for?
3. Where is it located on lean individuals?

Click **LAYER 2** *in the* **LAYER CONTROLS** *window, and you will see the following image:*

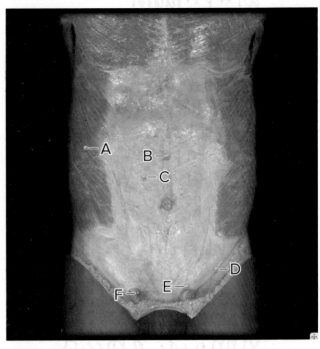

©McGraw-Hill Education

Mouse-over the pins on the screen to find the information necessary to identify the following structures:

A. _external abdominal oblique_
B. _linea alba_
C. _anterior rectus sheath_
D. _inguinal ligament_
E. _superficial inguinal ring_

Nonmuscular System Structure (blue pin)

F. _spermatic cord_

CHECK POINT
Abdomen, Anterior View, *continued*

4. Name the common site for male inguinal hernias.
5. Opening the abdominal wall by incision through the _____ avoids cutting muscle fibers.
6. What abdominal muscle has its fibers running at right angles to the internal abdominal oblique?

MODULE 6 The Muscular System 291

Click **LAYER 3** *in the* **LAYER CONTROLS** *window, and you will see the following image:*

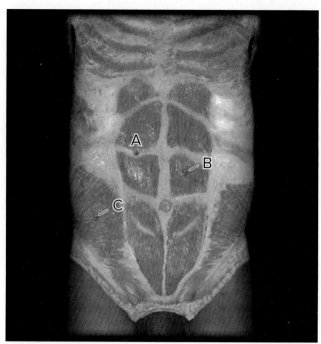

©McGraw-Hill Education

Mouse-over the pins on the screen to find the information necessary to identify the following structures:

A. ten dinous intersections of rectus abdominus
C. internal abdominal oblique
B. rectus abdominis

CHECK POINT

Abdomen, Anterior View, *continued*

7. Name the structures that subdivide the rectus abdominis muscle into three to four bellies.
8. What abdominal muscle has its fibers running at right angles to the external abdominal oblique?
9. Name the abdominal muscles in this view important in straining and abdominal breathing.

Click **LAYER 4** *in the* **LAYER CONTROLS** *window, and you will see the following image:*

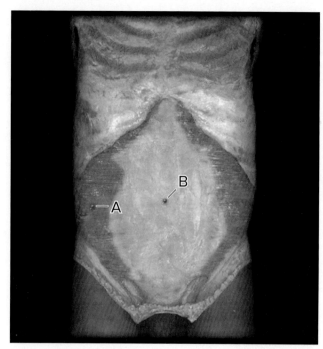

©McGraw-Hill Education

Mouse-over the pins on the screen to find the information necessary to identify the following structures:

A. transverse abdominis
B. posterior rectus sheath

CHECK POINT

Abdomen, Anterior View, *continued*

10. Name the abdominal muscle whose fibers run in a transverse plane.
11. What is the anatomical term for "flat tendons"?
12. What two structures come together to form the posterior rectus sheath?

MODULE 6 The Muscular System

Click **LAYER 5** *in the* **LAYER CONTROLS** *window, and you will see the following image:*

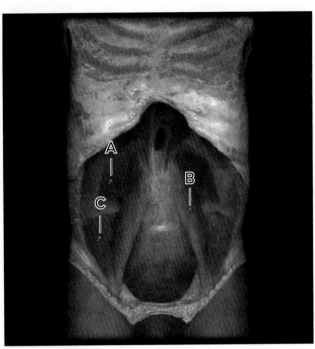

©McGraw-Hill Education

Mouse-over the pins on the screen to find the information necessary to identify the following structures:

A. _Quadratus lumborum_
B. _psoas major_
C. _Iliacus_

CHECK POINT

Abdomen, Anterior View, *continued*

13. Name a muscle of the posterior abdominal wall involved in respiration.

SELECT ANIMATION
External Abdominal Oblique Muscle   PLAY

After viewing this animation, select and view these additional animations:

- Iliacus muscle
- Rectus abdominis muscle

Self-Quiz
Take this opportunity to check your progress by taking the **QUIZ**. See the **Introduction Module** for a reminder on how to access the **QUIZ** for this Study Area.

IN REVIEW

What Have I Learned?

The following questions cover the material that you have just learned, the muscles of the thorax. Apply what you have learned to answer these questions on a separate piece of paper.

1. Name the abdominal wall muscles responsible for abdominal breathing.

2. What is the term for a "seam" where two structures meet?

3. Two individual muscles of the abdomen unite to form a single muscle, the most powerful flexor of the hip. Name those two individual muscles and the muscle they unite to form.

4. Two pairs of abdominal wall muscles have their structures running at right angles to each other. What are those two pairs of muscles?

5. Name the abdominal wall muscles important in straining, such as while lifting.

EXERCISE 6.7:
Skeletal Muscle—Pelvis—Superior View

SELECT TOPIC: Pelvis ▶ **SELECT VIEW**: Superior

Click **LAYER 1** in the **LAYER CONTROLS** window, and you will see the following image:

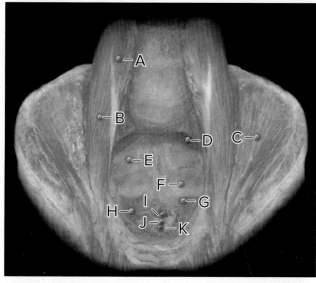

©McGraw-Hill Education

Mouse-over the pins on the screen to find the information necessary to identify the following structures:

A. psoas minor
B. psoas major
C. iliacus
D. piriformis
E. ischiococcygenus
F. pelvic diaphragm
G. levator ani
H. tendinous arch of levator ani

Nonmuscular System Structures (blue pins)

I. rectum
J. prostatic part of urethra
K. prostate

Self-Quiz
Take this opportunity to check your progress by taking the **QUIZ**. See the **Introduction Module** for a reminder on how to access the **QUIZ** for this Study Area.

IN REVIEW

What Have I Learned?

The following questions cover the material that you have just learned, the muscles of the pelvis. Apply what you have learned to answer these questions on a separate piece of paper.

1. Name the pelvic muscle involved with lateral rotation of the femur and that exits the pelvis through the greater sciatic foramen.

2. Name the two muscles that make up the pelvic diaphragm. What are their functions?

3. Name the structure that serves as the origin for part of the levator ani muscle.

EXERCISE 6.8:
Skeletal Muscle—Back—Posterior View

SELECT TOPIC	SELECT VIEW
Back	Posterior

Click **LAYER 1** *in the* **LAYER CONTROLS** *window, and you will see the following image:*

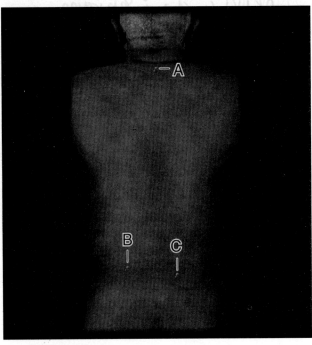

©McGraw-Hill Education

Mouse-over the pins on the screen to find the information necessary to identify the following nonmuscular system structures:

A. _vertebra prominens_
B. _surface projection of illiac crest_
C. _"   "   "   posterior superior illiac spine_

CHECK POINT

Back, Posterior View

1. Name the landmark for intramuscular injections of the hip.
2. The shallow skin depression (dimple) in the lower back marks what point?
3. What is the name for the prominent surface projection produced by the spinous process of vertebra C7?

Click **LAYER 2** *in the* **LAYER CONTROLS** *window, and you will see the following image:*

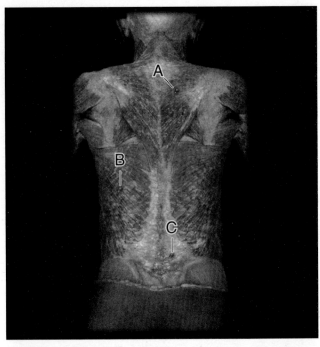

©McGraw-Hill Education

Mouse-over the pins on the screen to find the information necessary to identify the following structures:

A. _trapezius_
B. _latissimus dorsi_
C. _thoracolumbar fascia_

CHECK POINT

Back, Posterior View, *continued*

4. Name the superficial "kite-shaped" muscle of the back that spans from the nuchal line of the occipital bone to vertebra T12.
5. Name the deep fascia whose attached structures include the latissimus dorsi muscle.
6. Name the superficial muscle whose name describes its location as it spans from the back to the side of the body.

Click **LAYER 3** *in the* **LAYER CONTROLS** *window, and you will see the following image:*

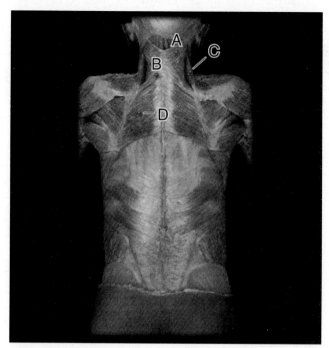

©McGraw-Hill Education

Mouse-over the pins on the screen to find the information necessary to identify the following structures:

A. _____splenius capitis_____
B. _____rhomboid minor_____
C. _____levator scapulae_____
D. _____rhomboid major_____

CHECK POINT

Back, Posterior View, *continued*

7. Name two muscles involved in the retraction and elevation of the scapula.
8. Name a muscle that allows the shrugging of the shoulders.

Click **LAYER 4** *in the* **LAYER CONTROLS** *window, and you will see the following image:*

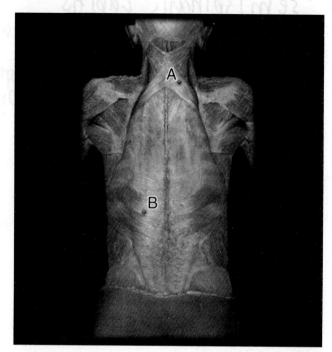

©McGraw-Hill Education

Mouse-over the pins on the screen to find the information necessary to identify the following structures:

A. _____serratus posterior superior_____
B. _____serratus posterior inferior m_____

Click **LAYER 5** *in the* **LAYER CONTROLS** *window, and you will see the following image:*

©McGraw-Hill Education

Mouse-over the pins on the screen to find the information necessary to identify the following structures:

A. semispinalis capitis
B. spinalis part of erector spinae
C. erector spinae
D. iliocostalis part of erector spinae
E. longissimus part of erector spinae

CHECK POINT

Back, Posterior View, *continued*

9. Name the muscle known as the "antigravity muscle."
10. This muscle consists of three separate muscles. What are they?

SELECT ANIMATION
Erector Spinae Muscle (extension)   PLAY

After viewing this animation, select and view these additional animations:

- Erector spinae muscle (lateral flexion)
- Infraspinatus muscle
- Latissimus dorsi muscle
- Rhomboid major and minor muscles
- Subscapularis muscle
- Supraspinatus muscle
- Trapezius muscle

Self-Quiz
Take this opportunity to check your progress by taking the **QUIZ**. See the **Introduction Module** for a reminder on how to access the **QUIZ** for this Study Area.

EXERCISE 6.9:
Skeletal Muscle—Shoulder and Arm—Anterior View

SELECT TOPIC Shoulder and Arm ▶ **SELECT VIEW** Anterior

Click **LAYER 1** *in the* **LAYER CONTROLS** *window, and you will see the following image:*

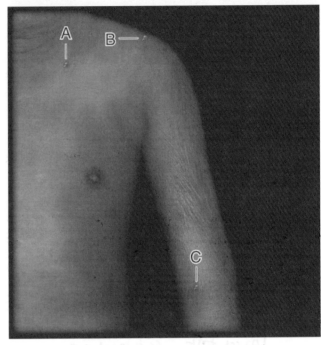

©McGraw-Hill Education

Mouse-over the pins on the screen to find the information necessary to identify the following nonmuscular system structures:

A. surface projection of clavicle
B. surface projection of acromion
C. surface projection of cubital fossa

CHECK POINT

Shoulder and Arm, Anterior View

1. Name the structure referred to as the collar bone.
2. Name the structure that is the flattened, lateral part of the scapular spine.
3. What is the name for the triangular concavity of the anterior elbow?

Click **LAYER 2** *in the* **LAYER CONTROLS** *window, and you will see the following image:*

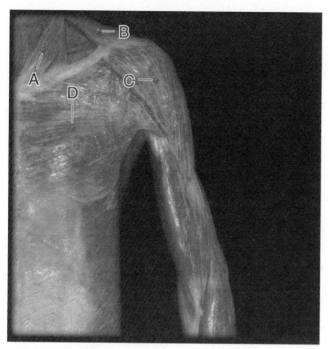

©McGraw-Hill Education

Mouse-over the pins on the screen to find the information necessary to identify the following structures:

A. sternocleidomastoid
B. trapezius
C. deltoid
D. pectoralis major

CHECK POINT

Shoulder and Arm, Anterior View, *continued*

4. Name the superficial muscle of the chest.
5. Name the muscle that contributes to the roundness of the shoulder.
6. Name the mostly posterior muscle that has its insertion at the clavicle and scapula.

Click **LAYER 3** *in the* **LAYER CONTROLS** *window, and you will see the following image:*

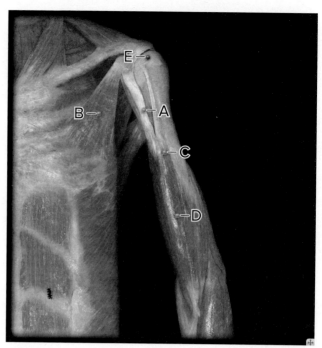

©McGraw-Hill Education

Mouse-over the pins on the screen to find the information necessary to identify the following structures:

A. short head of biceps brachii
B. pectoralis minor
C. long head of biceps brachii
D. biceps brachii

Nonmuscular System Structure (blue pin)

E. fibrous capsule of glenohumeral joint

CHECK POINT

Shoulder and Arm, Anterior View, *continued*

7. Name the muscle of the arm that has two heads.
8. Name the tough fibrous envelope that surrounds the joint where the arm attaches to the pectoral girdle.
9. Name the two muscles referred to as the pecs.

298 MODULE 6 The Muscular System

Click **LAYER 4** *in the* **LAYER CONTROLS** *window, and you will see the following image:*

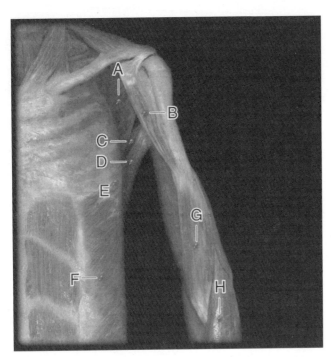
©McGraw-Hill Education

Mouse-over the pins on the screen to find the information necessary to identify the following structures:

A. sub scapularis
B. coracobrachialis
C. teres major
D. latissimus dorsi
E. serratus anterior
F. external abdominal oblique
G. brachialis
H. brachioradialis

CHECK POINT

Shoulder and Arm, Anterior View, *continued*

10. Name the four rotator cuff muscles.
11. What is the function of the rotator cuff muscles?
12. Name the muscle deep to the biceps brachii.

Click **LAYER 5** *in the* **LAYER CONTROLS** *window, and you will see the following image:*

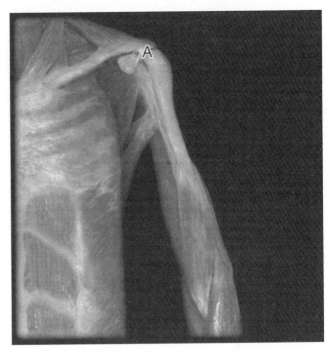
©McGraw-Hill Education

Mouse-over the pin on the screen to find the information necessary to identify the following nonmuscular system structure:

A. coraco-acromial ligament

 SELECT ANIMATION — **Biceps Brachii Muscle** — PLAY

After viewing this animation, select and view these additional animations:

- Brachialis muscle
- Deltoid muscle
- External abdominal oblique muscle
- Latissimus dorsi muscle
- Pectoralis major muscle
- Serratus anterior muscle
- Subscapularis muscle
- Teres major muscle
- Trapezius muscle
- Triceps brachii muscle

EXERCISE 6.10:
Skeletal Muscle—Shoulder and Arm—Posterior View

SELECT TOPIC: Shoulder and Arm ▸ **SELECT VIEW**: Posterior

Click **LAYER 1** *in the* **LAYER CONTROLS** *window, and you will see the following image:*

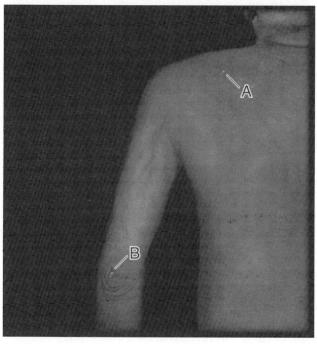

©McGraw-Hill Education

Mouse-over the pins on the screen to find the information necessary to identify the following nonmuscular system structures:

A. surface projection of spine of scapula
B. surface projection of olecranon

CHECK POINT

Shoulder and Arm, Posterior View

1. What is the name for the point of the elbow?
2. What specific structure of what bone constitutes the point of the elbow?
3. Name the prominent ridge on the posterior surface of the scapula.

Click **LAYER 2** *in the* **LAYER CONTROLS** *window, and you will see the following image:*

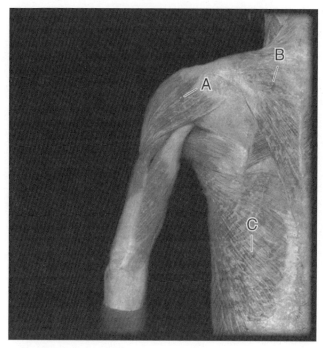

©McGraw-Hill Education

Mouse-over the pins on the screen to find the information necessary to identify the following structures:

A. deltoid
B. trapezius
C. lattissimus dorsi

CHECK POINT

Shoulder and Arm, Posterior View, *continued*

4. Name the triangle-shaped muscle of the shoulder.
5. Name the large lateral muscle responsible for adduction, extension, and medial rotation of the arm.
6. Name the muscle responsible for the elevation, medial rotation, adduction, and depression of the scapula.

Click **LAYER 3** *in the* **LAYER CONTROLS** *window, and you will see the following image:*

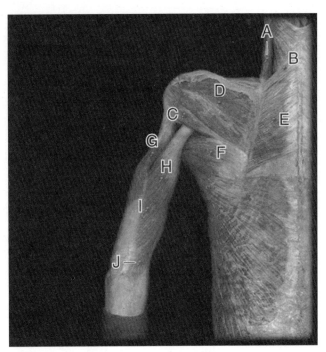

©McGraw-Hill Education

Mouse-over the pins on the screen to find the information necessary to identify the following structures:

A. levator scapulae
B. rhomboid minor
C. teres minor
D. infraspinatus
E. rhomboid major
F. teres major
G. lateral head of triceps brachii
H. long head of triceps brachii
I. triceps brachii
J. medial head of triceps brachii

CHECK POINT

Shoulder and Arm, Posterior View, *continued*

7. Name the muscle of the arm with three heads.
8. Name the muscle found in the infraspinous fossa of the scapula.
9. Name two muscles with their insertions on the medial border of the scapula.

Click **LAYER 4** *in the* **LAYER CONTROLS** *window, and you will see the following image:*

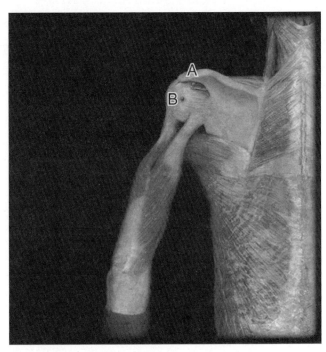

©McGraw-Hill Education

Mouse-over the pin on the screen to find the information necessary to identify the following structure:

A. supraspinatus

Nonmuscular System Structure (blue pin)

B. fibrous capsule of glenohumeral joint

CHECK POINT

Shoulder and Arm, Posterior View, *continued*

10. Name the muscle located in the supraspinous fossa of the scapula.
11. Name a muscle that holds the head of the humerus in the glenoid cavity.

Self-Quiz

Take this opportunity to check your progress by taking the **QUIZ**. See the **Introduction Module** for a reminder on how to access the **QUIZ** for this Study Area.

IN REVIEW

What Have I Learned?

The following questions cover the material that you have just learned, the muscles of the shoulder and arm. Apply what you have learned to answer these questions on a separate piece of paper.

1. Name the landmark for intramuscular injections of the shoulder.

2. Name the superficial "kite-shaped" muscle of the back that spans from the nuchal line of the occipital bone to vertebra T12.

3. Name the superficial muscle whose name describes its location as it spans from the back to the side of the body.

4. Name two muscles involved in the retraction and elevation of the scapula.

5. Name a muscle that allows the shrugging of the shoulders.

6. Name the muscle known as the antigravity muscle. What three muscles combine to form this muscle?

7. Name the superficial muscle of the chest.

8. Name the muscle of the arm that has two heads.

9. Name the muscle of the arm that has three heads.

10. Name the four rotator cuff muscles. What is their function?

11. Name the large lateral muscle responsible for adduction, extension, and medial rotation of the humerus.

12. Name the muscle responsible for the elevation, medial rotation, adduction, and depression of the scapula.

13. Name a muscle that holds the head of the humerus in the glenoid cavity.

EXERCISE 6.11:

Skeletal Muscles—Forearm and Hand—Anterior View

SELECT TOPIC: Forearm and Hand ▶ SELECT VIEW: Anterior

Click **LAYER 1** in the **LAYER CONTROLS** window, and you will see the following image:

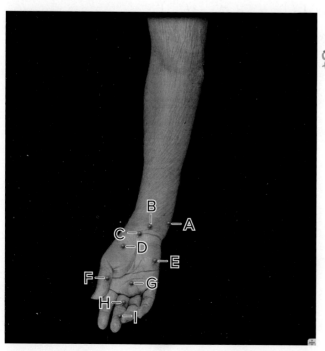

©McGraw-Hill Education

Mouse-over the pins on the screen to find the information necessary to identify the following nonmuscular system structures:

A. surface proj of styloid process of ulna
B. proximal wrist crease
C. distal wrist crease
D. thenar eminence
E. hypothenar eminence
F. S.P. of interphalangeal joint of thumb
G. S.P. of metacarpalphalangeal joint
H. S.P. of proximal interphalangeal joint of finger
I. S.P of distal " "

Click **LAYER 2** *in the* **LAYER CONTROLS** *window, and you will see the following image:*

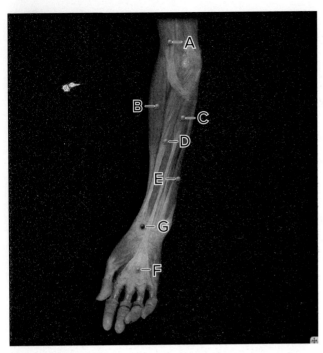

©McGraw-Hill Education

Mouse-over the pins on the screen to find the information necessary to identify the following structures:

A. biceps brachii
B. brachioradialis
C. palmaris longus
D. flexor carpi radialis
E. flexor carpi ulnaris
F. palmar aponeurosis

Nonmuscular System Structure (blue pin)

G. flexor retinaculum of hand

CHECK POINT

Forearm and Hand, Anterior View

1. Name three muscles that flex the wrist/hand.
2. Name two muscles that flex the forearm.
3. Name the structure that forms the carpal tunnel.

Click **LAYER 3** *in the* **LAYER CONTROLS** *window, and you will see the following image:*

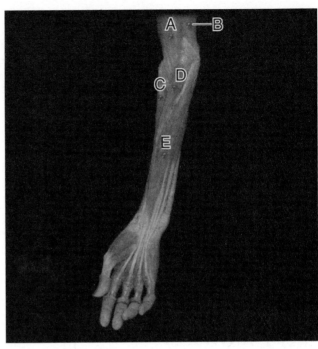

©McGraw-Hill Education

Mouse-over the pins on the screen to find the information necessary to identify the following structures:

A. brachialis
B. triceps brachii
C. supinator
D. pronator teres
E. flexor digitorum superficialis

CHECK POINT

Forearm and Hand, Anterior View, *continued*

4. Name a muscle involved in the pronation of the forearm.
5. Name a muscle involved in the supination of the forearm.
6. Name a muscle involved in the extension of the forearm.

Click **LAYER 4** *in the* **LAYER CONTROLS** *window, and you will see the following image:*

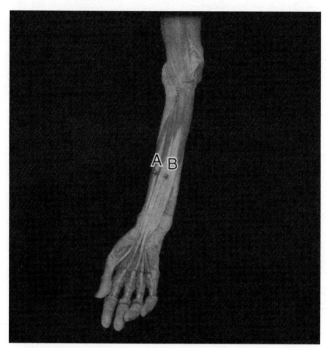

©McGraw-Hill Education

Mouse-over the pins on the screen to find the information necessary to identify the following structures:

A. flexor pollicus longus
B. flexor digitorum profundus

CHECK POINT

Forearm and Hand, Anterior View, *continued*

7. Name the only muscle that flexes the distal interphalangeal joint.
8. Name the only muscle that flexes the distal phalanx of the thumb.
9. What is the anatomical name for the thumb?

Click **LAYER 5** *in the* **LAYER CONTROLS** *window, and you will see the following image:*

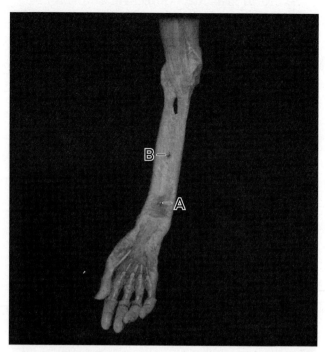

©McGraw-Hill Education

Mouse-over the pin on the screen to find the information necessary to identify the following structure:

A. pronator quadratus

Nonmuscular System Structure (blue pin)

B. introsseous membrane of forearm

CHECK POINT

Forearm and Hand, Anterior View, *continued*

10. Name the thick sheet of connective tissue between the ulna and the radius.
11. What is its function?

EXERCISE 6.12:
Skeletal Muscles—Forearm and Hand—Posterior View

SELECT TOPIC: Forearm and Hand ▸ **SELECT VIEW**: Posterior

Click **LAYER 1** *in the* **LAYER CONTROLS** *window, and you will see the following image:*

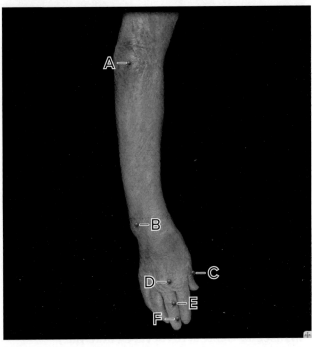

©McGraw-Hill Education

Mouse-over the pins on the screen to find the information necessary to identify the following nonmuscular structures:

A. S.P. of olecranon
B. S.P. of styloid process of ulna
C. S.P. of interphalangeal joint of thumb
D. S.P. of metacarpalphalgeal joint
E. S.P. of prox interphalg. joint of finger
F. " distal " " "

Click **LAYER 2** *in the* **LAYER CONTROLS** *window, and you will see the following image:*

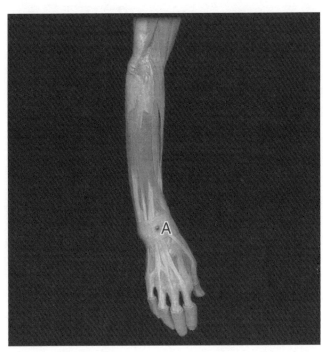

©McGraw-Hill Education

Mouse-over the pin on the screen to find the information necessary to identify the following nonmuscular system structure:

A. extensor retinaculum of hand

CHECK POINT
Forearm and Hand, Posterior View

What is the relationship between the retinaculum and the extensor tendons of the forearm?

Click **LAYER 3** *in the* **LAYER CONTROLS** *window, and you will see the following image:*

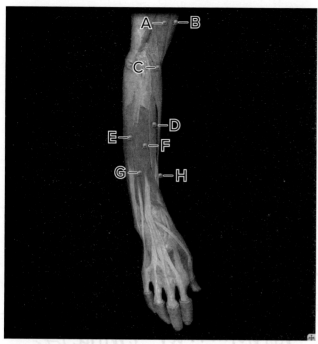

©McGraw-Hill Education

Click **LAYER 4** *in the* **LAYER CONTROLS** *window, and you will see the following image:*

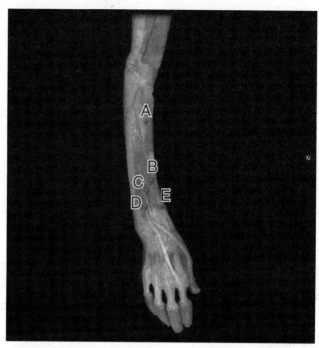

©McGraw-Hill Education

Mouse-over the pins on the screen to find the information necessary to identify the following structures:

A. brachialis
B. biceps brachii
C. brachioradialis
D. extensor carpi radialis brevis
E. extensor carpi ulnaris
F. extensor of digitorum
G. extensor digiti minimi
H. extensor carpi radialis longus

Mouse-over the pins on the screen to find the information necessary to identify the following structures:

A. supinator
B. abductor pollicis longus
C. extensor pollicis longus
D. extensor indicis
E. extensor pollicis brevis

CHECK POINT

Forearm and Hand, Posterior View, *continued*

5. Name a muscle that both extends and abducts the thumb.
6. Name two muscles that only extend the thumb.
7. Name a muscle that extends the index finger.

CHECK POINT

Forearm and Hand, Posterior View, *continued*

2. Name three muscles that extend the hand.
3. Name a muscle that assists in both pronation and supination of the forearm.
4. Name a muscle that extends the fifth finger.

Self-Quiz

Take this opportunity to check your progress by taking the **QUIZ**. See the **Introduction Module** for a reminder on how to access the **QUIZ** for this Study Area.

IN REVIEW

What Have I Learned?

The following questions cover the material that you have just learned, the muscles of the forearm and hand. Apply what you have learned to answer these questions on a separate piece of paper.

1. Name three muscles that flex the wrist.
2. Name two muscles that flex the forearm.
3. Name the structure that forms the carpal tunnel.
4. Name a muscle involved in the pronation of the forearm.
5. Name a muscle involved in the supination of the forearm.
6. What is the anatomical name for the thumb?
7. Name the thick sheet of connective tissue between the ulna and the radius. What is its function?
8. Name three muscles that extend the hand.
9. Name a muscle that both extends and abducts the thumb.

EXERCISE 6.13:
Skeletal Muscles—Wrist and Hand—Anterior View

SELECT TOPIC	SELECT VIEW
Wrist and Hand	Anterior

Click **LAYER 1** *in the* **LAYER CONTROLS** *window, and you will see the following image:*

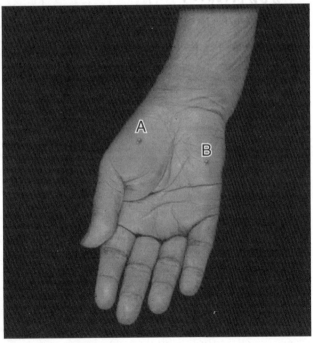

©McGraw-Hill Education

Mouse-over the pins on the screen to find the information necessary to identify the following nonmuscular system structures:

A. thenar eminence
B. hypothenar eminence

CHECK POINT
Wrist and Hand, Anterior View

1. Name the thick, fleshy eminence at the base of the first digit.
2. Name the thick, fleshy eminence at the base of the fifth digit.

Click **LAYER 2** *in the* **LAYER CONTROLS** *window, and you will see the following image:*

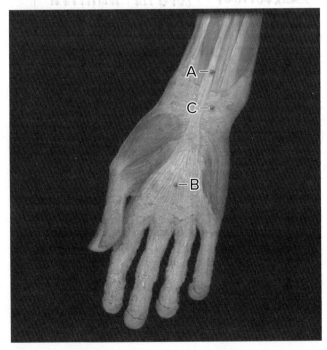

©McGraw-Hill Education

Mouse-over the pins on the screen to find the information necessary to identify the following structures:

A. tendon of palmaris longus
B. palmar aponeurosis

Nonmuscular System Structure (blue pin)

C. flexor retinaculum of hand

CHECK POINT

Wrist and Hand, Anterior View, *continued*

3. Name a muscle often missing on one or both forearms.

Click **LAYER 3** *in the* **LAYER CONTROLS** *window, and you will see the following image:*

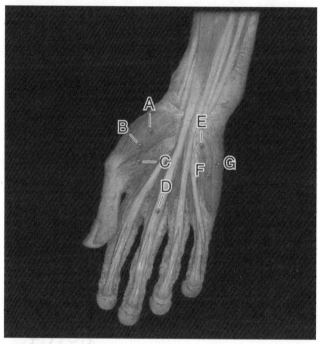

Mouse-over the pins on the screen to find the information necessary to identify the following structures:

A. abductor pollicis brevis
B. thenar
C. flexor pollicis brevis
D. flexor digitorum superficialis
E. hypothenar
F. flexor digiti minimi brevis m. of hand
G. abductor digiti minimi m. of hand

CHECK POINT

Wrist and Hand, Anterior View, *continued*

4. Name the three thenar muscles.
5. Name the three hypothenar muscles.
6. Which digits do each of the above six muscles act upon?

Click **LAYER 4** *in the* **LAYER CONTROLS** *window, and you will see the following image:*

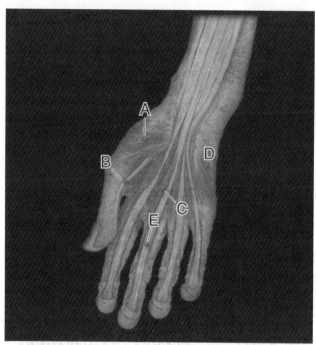

Mouse-over the pins on the screen to find the information necessary to identify the following structures:

A. opponens pollicis
B. flexor pollicis longus
C. lumbrical mm of hand
D. opponens digiti minimi m
E. tendons of flexor digitorum profundus

CHECK POINT

Wrist and Hand, Anterior View, *continued*

7. Name the muscles that both flex the metacarpophalangeal joint and extend the interphalangeal joints.
8. Name the muscle that allows the fifth finger to touch the tip of the first finger.
9. Name the muscle that allows the tip of the first finger to touch the tips of the other fingers.

Click **LAYER 5** *in the* **LAYER CONTROLS** *window, and you will see the following image:*

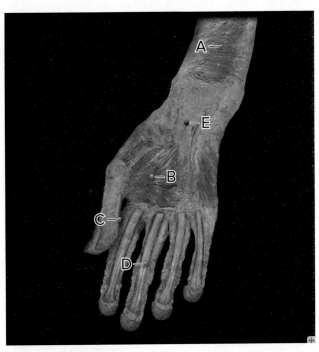

©McGraw-Hill Education

Mouse-over the pins on the screen to find the information necessary to identify the following structures:

A. pronator quadratus
B. adductor pollicis
C. tendon of flexor pollicis longus
D. tendons of flexor digitorum profundus

Nonmuscular System Structure (blue pin)

E. carpal tunnel

CHECK POINT

Wrist and Hand, Anterior View, *continued*

10. Name the only muscle that flexes the distal phalanx of the first digit.
11. Name the only muscle that flexes the distal interphalangeal joint of digits II to V.
12. Name the distal pronator of the forearm.

EXERCISE 6.14:
Skeletal Muscles—Wrist and Hand—Posterior View

| SELECT TOPIC | SELECT VIEW |
| Wrist and Hand | Posterior |

Click **LAYER 1** *in the* **LAYER CONTROLS** *window, and you will see the following image:*

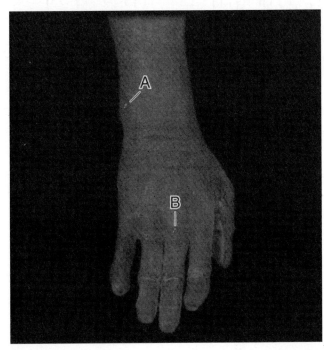

©McGraw-Hill Education

Mouse-over the pins on the screen to find the information necessary to identify the following nonmuscular system structures:

A. S.P. of styloid process of ulna
B. S.P. of metacarpophalangeal joint

CHECK POINT

Wrist and Hand, Posterior View

1. Flexion of which joint makes the knuckles prominent?
2. What structures are visible as the knuckles?

Click **LAYER 2** *in the* **LAYER CONTROLS** *window, and you will see the following image:*

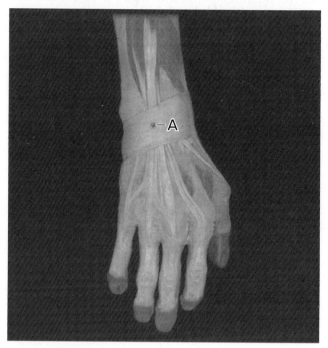

©McGraw-Hill Education

Mouse-over the pin on the screen to find the information necessary to identify the following nonmuscular system structure:

A. extensor retinculum of hand

Click **LAYER 3** *in the* **LAYER CONTROLS** *window, and you will see the following image:*

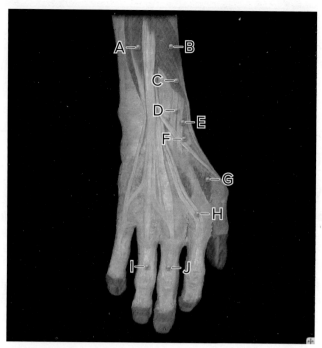

©McGraw-Hill Education

Mouse-over the pins on the screen to find the information necessary to identify the following structures:

A. extensor digiti minimi
B. abductor pollicis longus
C. extensor pollicis brevis
D. tendon of extensor carpi radialis brevis
E. "   "   " longus
F. extensor pollicis longus
G. first dorsal introsseous m. of hand
H. extensor indicis
I. extensor expansion
J. extensor digitorum

CHECK POINT

Wrist and Hand, Posterior View, *continued*

4. Name a muscle that extends the index finger.
5. Name a muscle that extends fingers II through V.
6. Name a muscle that extends the thumb.
7. Name a muscle that extends the little finger.

SELECT ANIMATION — **Extensor Digitorum Muscle** — **PLAY**

After viewing this animation, select and view this additional animation:

- **Flexor digitorum superficialis and profundus muscles**

Self-Quiz

Take this opportunity to check your progress by taking the **QUIZ**. See the **Introduction Module** for a reminder on how to access the **QUIZ** for this Study Area.

IN REVIEW

What Have I Learned?

The following questions cover the material that you have just learned, the muscles of the wrist and hand. Apply what you have learned to answer these questions on a separate piece of paper.

1. Name the thick, fleshy eminence at the base of the first digit.
2. Name the thick, fleshy eminence at the base of the fifth digit.
3. Name a muscle often missing on one or both forearms.
4. Name the only muscle that flexes the distal phalanx of the first digit.
5. Flexion of which joint makes the knuckles prominent?
6. What structures are visible as the knuckles?

EXERCISE 6.15:

Skeletal Muscles—Hip and Thigh—Anterior View

SELECT TOPIC Hip and Thigh ▶ **SELECT VIEW** Anterior

Click **LAYER 1** in the **LAYER CONTROLS** window, and you will see the following image:

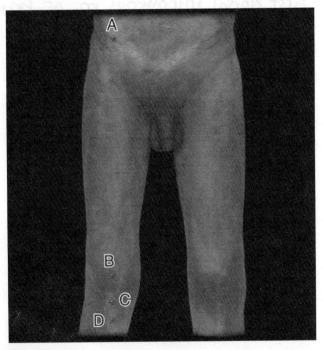

©McGraw-Hill Education

Mouse-over the pins on the screen to find the information necessary to identify the following nonmuscular system structures:

A. S.P. of anterior superior iliac spine
B. S.P. of patella
C. S.P. of patellar ligament
D. S.P. of tibial tuberosity

CHECK POINT

Hip and Thigh, Anterior View

1. Name the superficially visible anterior subcutaneous end of the iliac crest.
2. Name the point of attachment for the quadriceps femoris muscles by way of the patellar ligament.
3. Name the ligament that connects the patella to the tuberosity of the tibia.

MODULE 6 The Muscular System 311

Click **LAYER 2** in the **LAYER CONTROLS** window, and you will see the following image:

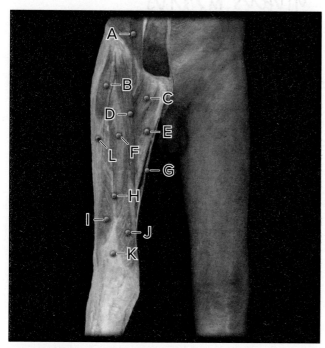

©McGraw-Hill Education

Mouse-over the pins on the screen to find the information necessary to identify the following structures:

A. iliopsoas m
B. tensor fasciae latae
C. pectineus
D. sartorius
E. adductor longus
F. rectus femoris
G. gracilis
H. quadriceps femoris
I. vastus lateralis
J. vastus medialis
K. tendon of quadriceps femoris

Nonmuscular System Structure (blue pin)

L. iliotibial tract

CHECK POINT

Hip and Thigh, Anterior View, *continued*

4. Name the muscle whose origin is the anterior superior iliac spine of the ilium and whose insertion is the proximal medial shaft of the tibia.
5. Name the four muscles of the quadriceps femoris.
6. Name the most powerful flexor of the thigh.

Click **LAYER 3** in the **LAYER CONTROLS** window, and you will see the following image:

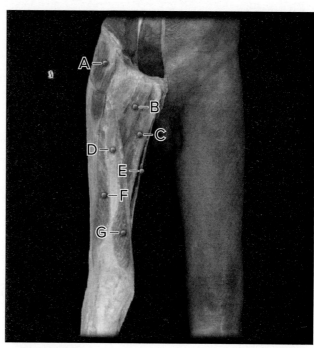

©McGraw-Hill Education

Mouse-over the pins on the screen to find the information necessary to identify the following structures:

A. gluteus medius
B. pectineus
C. adductor longus
D. vastus intermedius
E. gracilis
F. vastus lateralis
G. vastus medialis

CHECK POINT

Hip and Thigh, Anterior View, *continued*

7. Name the muscle of the thigh that is weak in humans and used in muscle transplants.
8. Name the muscle often involved in a "pulled groin."

312　MODULE 6　The Muscular System

Click **LAYER 4** *in the* **LAYER CONTROLS** *window, and you will see the following image:*

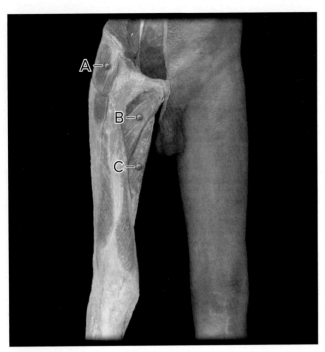

©McGraw-Hill Education

Mouse-over the pins on the screen to find the information necessary to identify the following structures:

A. gluteus medius
B. adductor brevis
C. adductor magnus

Click **LAYER 5** *in the* **LAYER CONTROLS** *window, and you will see the following image:*

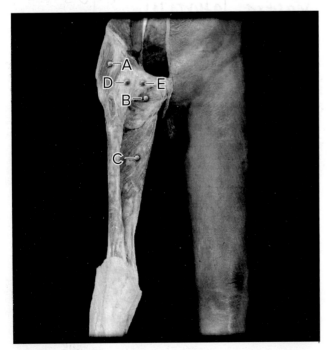

©McGraw-Hill Education

Mouse-over the pins on the screen to find the information necessary to identify the following structures:

A. gluteus medius
B. obturator externus
C. adductor magnus

Nonmuscular System Structures (blue pins)

D. iliofemoral ligament
E. pubofemoral ligament

CHECK POINT

Hip and Thigh, Anterior View, *continued*

9. Name the strongest ligament around the hip joint.
10. Name the ligament that resists excessive abduction of the hip.
11. Name the ligament that resists hyperextension of the hip joint.

EXERCISE 6.16:

Skeletal Muscles—Hip and Thigh—Posterior View

SELECT TOPIC	SELECT VIEW
Hip and Thigh	Posterior

Click **LAYER 1** *in the* **LAYER CONTROLS** *window, and you will see the following image:*

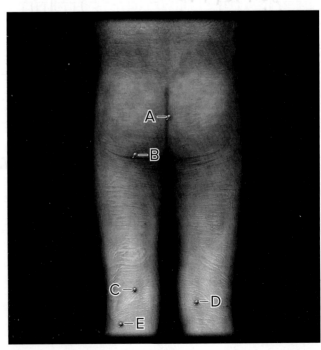

©McGraw-Hill Education

Mouse-over the pins on the screen to find the information necessary to identify the following nonmuscular system structures:

A. natal cleft
B. gluteal fold
C. surface proj of popliteal fossa
D. S.P. of hamstring (proj) tendons
E. S.P. of head of fibula

CHECK POINT

Hip and Thigh, Posterior View

1. Name the muscle whose tendon is the lateral hamstring.
2. Name the muscles whose tendons are the medial hamstring.
3. Name the structure that provides attachment for the fibular collateral ligament of the knee and the biceps femoris muscle.
4. What is the natal cleft? What is its function?
5. What are the gluteal folds? What do they represent?

*Click **LAYER 2** in the **LAYER CONTROLS** window, and you will see the following image:*

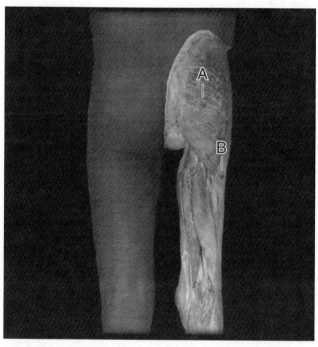

©McGraw-Hill Education

Mouse-over the pins on the screen to find the information necessary to identify the following structures:

A. gluteus maximus
B. iliotibial tract

CHECK POINT

Hip and Thigh, Posterior View, *continued*

6. Name a muscle of the posterior thigh not important in walking.
7. Name a muscle of the posterior thigh important for powerful extension of the femur as in running, climbing stairs, and rising from the seated position.
8. Name the structure that provides attachment for the tensor fascia latae and gluteus maximus muscles.

*Click **LAYER 3** in the **LAYER CONTROLS** window, and you will see the following image:*

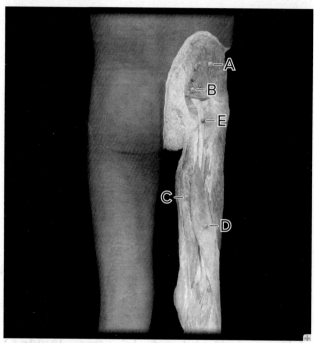

©McGraw-Hill Education

Mouse-over the pins on the screen to find the information necessary to identify the following structures:

A. gluteus medius
B. piriformis
C. semitendonosus
D. long head of biceps femoris

Nonmuscular System Structure (blue pin)

E. sciatic

CHECK POINT

Hip and Thigh, Posterior View, *continued*

9. Name the two muscles that allow the non-weight-bearing limb to swing forward during walking.
10. Name the two heads of the biceps femoris.
11. Name the largest nerve in the body.

Click **LAYER 4** *in the* **LAYER CONTROLS** *window, and you will see the following image:*

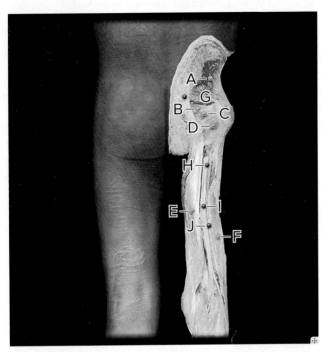

©McGraw-Hill Education

Mouse-over the pins on the screen to find the information necessary to identify the following structures:

A. gluteus minimus
B. obturator internus
C. superior & inferior gemellus mm.
D. Quadratus femoris
E. Semimembranosus
F. Short head of biceps femoris

Nonmuscular System Structures (blue pins)

G. Sacrotuberous ligament
H. sciatic
I. tibial
J. common fibular

CHECK POINT

Hip and Thigh, Posterior View, *continued*

12. Name the structure that is an important anchor of the sacrum to the hip bone.
13. Name the two components of the sciatic nerve.

Click **LAYER 5** *in the* **LAYER CONTROLS** *window, and you will see the following image:*

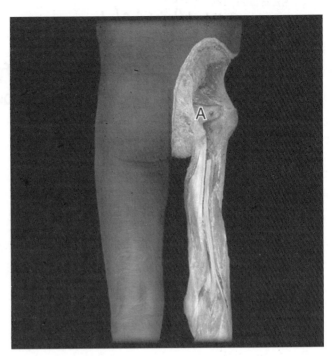

©McGraw-Hill Education

Mouse-over the pin on the screen to find the information necessary to identify the following nonmuscular system structure:

A. Ischiofemoral ligament

CHECK POINT

Hip and Thigh, Posterior View, *continued*

14. Name the thick fibrous band fused to the posterior surface of the hip joint capsule.
15. Name the ligament that resists hyperflexion of the hip.

SELECT ANIMATION
Adductor Magnus Muscle

PLAY

After viewing this animation, select and view these additional animations:

– **Short head of biceps femoris muscle**
– **Gluteus maximus muscle**

Continued on next page . . .

Additional animations continued . . .

- **Gluteus medius muscle**
- **Hamstring muscles**
- **Quadriceps femoris muscle**
- **Sartorius muscle**

Self-Quiz

Take this opportunity to check your progress by taking the **QUIZ**. See the **Introduction Module** for a reminder on how to access the **QUIZ** for this Study Area.

IN REVIEW

What Have I Learned?

The following questions cover the material that you have just learned, the muscles of the hip and thigh. Apply what you have learned to answer these questions on a separate piece of paper.

1. Name the four muscles of the quadriceps femoris.

2. Name the most powerful flexor of the hip joint.

3. Name the muscle of the thigh weak in humans and used in muscle transplants.

4. Name the muscle often involved in a pulled groin.

5. Name the strongest ligament around the hip joint.

6. Name a muscle of the posterior thigh important for powerful extension of the femur as in running, climbing stairs, and rising from the seated position.

7. Name the two muscles that allow the non-weight-bearing limb to swing forward during walking.

8. Name the two heads of the biceps femoris.

9. Name the thick fibrous band fused to the posterior surface of the hip joint capsule.

EXERCISE 6.17:

Skeletal Muscles—Leg and Foot—Anterior View

SELECT TOPIC	SELECT VIEW
Leg and Foot	Anterior

Click **LAYER 1** in the **LAYER CONTROLS** window, and you will see the following image:

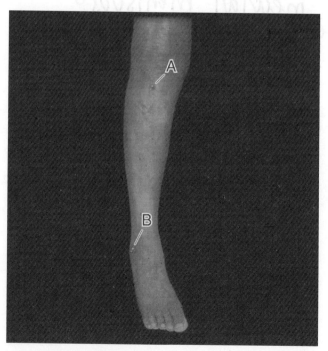

©McGraw-Hill Education

Mouse-over the pins on the screen to find the information necessary to identify the following nonmuscular system structures:

A. S.P. of tibial tuberosity
B. S.P. of lateral malleovlus

CHECK POINT

Leg and Foot, Anterior View

1. Name the bony elevation of the anterior proximal tibia.
2. Name the lateral subcutaneous projection that contributes to the ankle joint.

Click **LAYER 2** *in the* **LAYER CONTROLS** *window, and you will see the following image:*

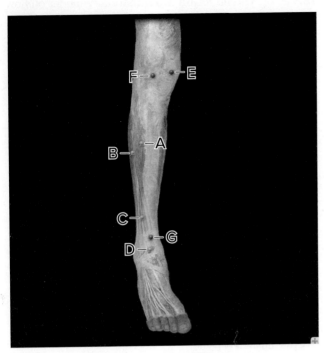

©McGraw-Hill Education

Mouse-over the pins on the screen to find the information necessary to identify the following structures:

A. tibialis anterior
B. fibularis longus and brevis
C. extensor digitorum longus
D. extensor hallucis longus

Nonmuscular System Structures (blue pins)

E. fibrous capsule of knee joint
F. patellar ligament
G. extensor retinaculum of ankle

CHECK POINT

Leg and Foot, Anterior View, *continued*

3. What is the anatomical term for the first toe?
4. The tendons of which muscles are subcutaneous on the dorsum of the foot?
5. Name the structure that serves to bind in place the tendons from the anterior compartment of the leg as they cross the ankle joint.

Click **LAYER 3** *in the* **LAYER CONTROLS** *window, and you will see the following image:*

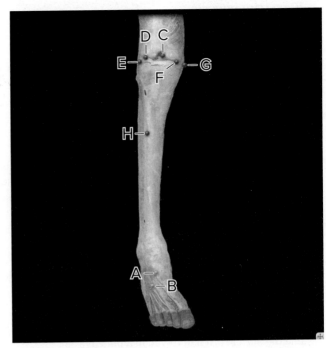

©McGraw-Hill Education

Mouse-over the pins on the screen to find the information necessary to identify the following structures:

A. extensor hallucis brevis
B. extensor digitorum brevis

Nonmuscular System Structures (blue pins)

C. anterior cruciate ligament
D. lateral meniscus
E. fibular collateral ligament
F. medial meniscus
G. tibial collateral ligament
H. interosseous membrane of leg

CHECK POINT

Leg and Foot, Anterior View, *continued*

6. Name the thick sheet of connective tissue between the tibia and fibula.
7. What is the function of this sheet of connective tissue?
8. Name the structures referred to as the unhappy triad.

EXERCISE 6.18:

Skeletal Muscles—Leg and Foot—Posterior View

SELECT TOPIC	SELECT VIEW
Leg and Foot	Posterior

Click **LAYER 1** *in the* **LAYER CONTROLS** *window, and you will see the following image:*

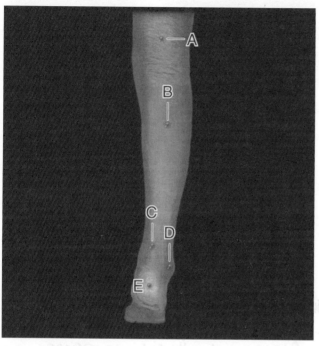

©McGraw-Hill Education

Mouse-over the pins on the screen to find the information necessary to identify the following nonmuscular system structures:

A. S.p. of popliteal fossa
B. calf prominence
C. calcaneal tendon
D. S.p. of lateral malleus
E. S.p. of calcaneus

CHECK POINT

Leg and Foot, Posterior View

1. Name the strongest tendon in the body.
2. Give an example of this tendon's strength from the **STRUCTURE INFORMATION** window.
3. Name the tendon also known as the "Achilles" tendon.

Click **LAYER 2** *in the* **LAYER CONTROLS** *window, and you will see the following image:*

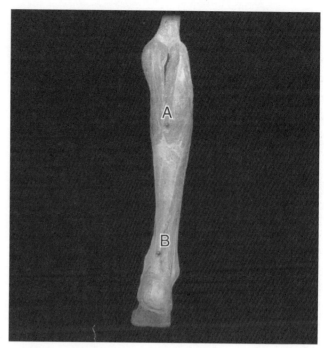

©McGraw-Hill Education

Mouse-over the pins on the screen to find the information necessary to identify the following structures:

A. gastrocnemius
B. calcaneal tendon

CHECK POINT

Leg and Foot, Posterior View, *continued*

4. The tendons of which two muscles contribute to the calcaneal tendon?
5. Name the calf muscle that consists of a medial and a lateral belly.
6. Name the superficial calf muscle.

Click **LAYER 3** *in the* **LAYER CONTROLS** *window, and you will see the following image:*

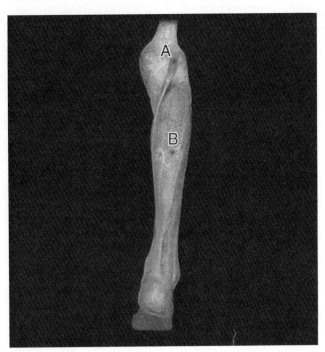

©McGraw-Hill Education

Mouse-over the pins on the screen to find the information necessary to identify the following structures:

A. plantaris
B. soleus

CHECK POINT

Leg and Foot, Posterior View, *continued*

7. Name the calf muscle deep to the gastrocnemius.
8. Name the long thin tendon that is a common source for tendon transplants.

Click **LAYER 4** *in the* **LAYER CONTROLS** *window, and you will see the following image:*

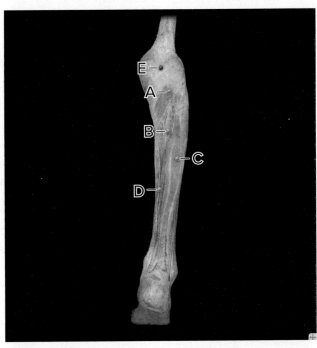

©McGraw-Hill Education

Mouse-over the pins on the screen to find the information necessary to identify the following structures:

E/A. fibrous capsule of knee joint
B. tibialis posterior
C. flexor hallucis longus
D. flexor digitorum longus

Nonmuscular System Structure (blue pin)

A. popliteus

CHECK POINT

Leg and Foot, Posterior View, *continued*

9. Name the muscle that helps "unlock" the knee joint from full extension.
10. Name the two structures that maintain the position of the femur on the tibia in full knee flexion such as squatting.
11. Name the powerful muscle for "push-off" of the foot during walking or running.

MODULE 6 The Muscular System **319**

Click **LAYER 5** *in the* **LAYER CONTROLS** *window, and you will see the following image:*

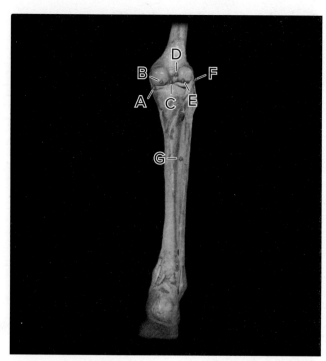

©McGraw-Hill Education

Mouse-over the pins on the screen to find the information necessary to identify the following nonmuscular system structures:

A. tibial collateral ligament
B. medial meniscus
C. posterior cruciate ligament
D. anterior " "
E. lateral meniscus
F. fibular collateral ligament
G. interosseous membrane of leg

CHECK POINT

Leg and Foot, Posterior View, *continued*

12. Name the thinner and weaker of the cruciate ligaments.
13. Name the cartilaginous structure on the tibia that articulates with the medial condyle of the femur.
14. Name the structure that limits rotation between the femur and the tibia.

SELECT ANIMATION
Fibularis Longus and Brevis Muscle PLAY

After viewing this animation, select and view this additional animation:

– **Gastrocnemius muscle**

Self-Quiz
Take this opportunity to check your progress by taking the **QUIZ**. See the **Introduction Module** for a reminder on how to access the **QUIZ** for this Study Area.

IN REVIEW

What Have I Learned?

The following questions cover the material that you have just learned, the muscles of the leg and foot. Apply what you have learned to answer these questions on a separate piece of paper.

1. What is the anatomical term for the first toe?

2. Name the thick sheet of connective tissue between the tibia and fibula. What is its function?

3. Name the structure that serves to bind the tendons from the anterior compartment of the leg in place as they cross the ankle joint.

4. Name the strongest tendon in the body.

5. The tendons of which two muscles contribute to the calcaneal tendon?

6. Name the long thin tendon that is a common source for tendon transplants.

7. Name the powerful muscle for "push-off" of the foot during walking or running.

8. Name the cruciate ligaments. Which of the two is thinner and weaker?

EXERCISE 6.19:

Skeletal Muscles—Foot—Plantar View

SELECT TOPIC	SELECT VIEW
Foot	Plantar

Click **LAYER 1** *in the* **LAYER CONTROLS** *window, and you will see the following image:*

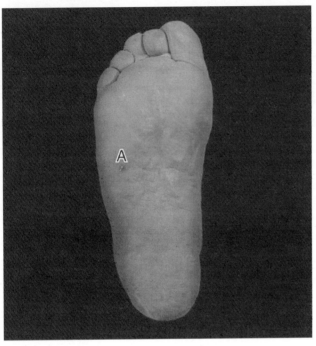

©McGraw-Hill Education

Mouse-over the pin on the screen to find the information necessary to identify the following nonmuscular system structure:

A. _____

CHECK POINT

Foot, Plantar View

1. Name the structures that serve as contact points of the foot for weight-bearing.

Click **LAYER 2** *in the* **LAYER CONTROLS** *window, and you will see the following image:*

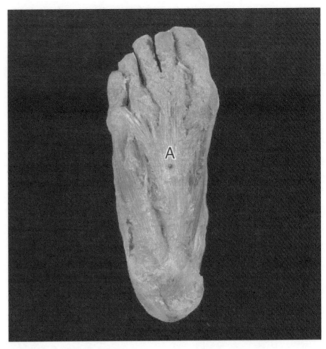

©McGraw-Hill Education

Mouse-over the pin on the screen to find the information necessary to identify the following structure:

A. _____

CHECK POINT

Foot, Plantar View, *continued*

2. Name the structure that protects the muscles, vessels, and nerves of plantar foot.

Click **LAYER 3** *in the* **LAYER CONTROLS** *window, and you will see the following image:*

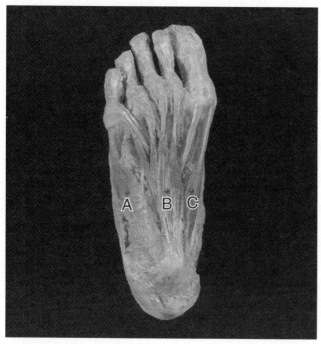

©McGraw-Hill Education

Mouse-over the pins on the screen to find the information necessary to identify the following structures:

A. _____

B. _____

C. _____

CHECK POINT

Foot, Plantar View, *continued*

3. What is the anatomical term for the first toe?
4. Name the muscle responsible for flexion of toes II to V.
5. Name the muscle that supports the medial longitudinal arch of the foot during weight-bearing.

Click **LAYER 4** *in the* **LAYER CONTROLS** *window, and you will see the following image:*

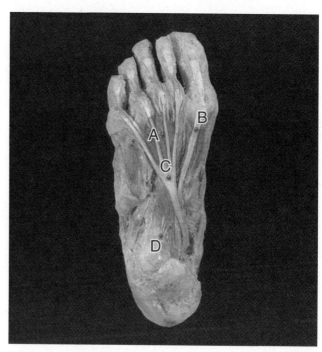

©McGraw-Hill Education

Mouse-over the pins on the screen to find the information necessary to identify the following structures:

A. _____

B. _____

C. _____

D. _____

CHECK POINT

Foot, Plantar View, *continued*

6. Name two muscles located on the posterior leg whose tendons run along the plantar foot.
7. Name a muscle of the foot that uses the tendons of another muscle to produce toe flexion.

Click **LAYER 5** *in the* **LAYER CONTROLS** *window, and you will see the following image:*

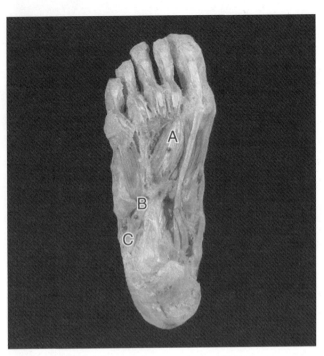

©McGraw-Hill Education

Mouse-over the pins on the screen to find the information necessary to identify the following structures:

A. _____

B. _____

C. _____

CHECK POINT

Foot, Plantar View, *continued*

8. Name a muscle that resists separation ("spreading") of the metatarsals during weight-bearing.
9. Name two muscles of the lateral leg whose tendons run along the plantar foot.
10. Name a muscle responsible for adduction of the first toe.

Self-Quiz

Take this opportunity to check your progress by taking the **QUIZ**. See the **Introduction Module** for a reminder on how to access the **QUIZ** for this Study Area.

IN REVIEW

What Have I Learned?

The following questions cover the material that you have just learned, the muscles of the plantar foot. Apply what you have learned to answer these questions on a separate piece of paper.

1. Name the structures that serve as contact points of the foot for weight-bearing.
2. Name the structure that protects the muscles, vessels, and nerves of plantar foot.
3. Name the muscle that supports the medial longitudinal arch of the foot during weight-bearing.
4. Name a muscle that resists separation ("spreading") of the metatarsals during weight-bearing.

Muscle Attachments (on Skeleton)

We will now concentrate on how and where the skeletal muscles attach to the skeleton. After all, without this attachment between the two, we would be unable to accomplish any movement. Don't be intimidated by this section of *Anatomy and Physiology | Revealed®*. It is very basic physics. Think of your computer mouse (assuming you are using one). You place your hand on the mouse, and by moving your arm, the mouse moves in the same direction. Likewise, when the insertion end of a muscle is attached to a bone, when the muscle moves (contracts), the bone it is attached to moves in the direction of contraction. Remember that the bone located at the *origin* of the muscle will not be moving, but will instead function as an anchor for the muscle during contraction. The bone at the *insertion* end of the muscle will be moving, as the contracting muscle exerts a force on it.

As we view the images in *Anatomy and Physiology | Revealed®*, remember that the origins of muscles are shaded in red and the insertions are shaded in blue. A box will be visible on the screen showing the overall region being viewed, with a key to remind you which color represents which end of the muscle.

MODULE 6 The Muscular System **323**

EXERCISE 6.20:
Skull—Anterior View

SELECT TOPIC
Muscle Attachments (on skeleton)

SELECT SUB-TOPIC **SELECT VIEW**
Skull ▶ Anterior

Click **LAYER 1** *in the* **LAYER CONTROLS** *window, and you will see the following image:*

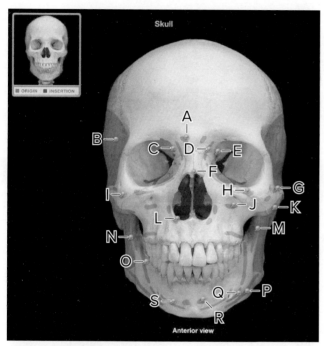

©McGraw-Hill Education

Mouse-over the pins on the screen to find the information necessary to identify the following attachments:

A. _____
B. _____
C. _____
D. _____
E. _____
F. _____
G. _____
H. _____
I. _____
J. _____
K. _____
L. _____
M. _____
N. _____
O. _____
P. _____
Q. _____
R. _____
S. _____

EXERCISE 6.21:
Skull—Lateral View

SELECT TOPIC **SELECT VIEW**
Skull ▶ Lateral

Click **LAYER 1** *in the* **LAYER CONTROLS** *window, and you will see the following image:*

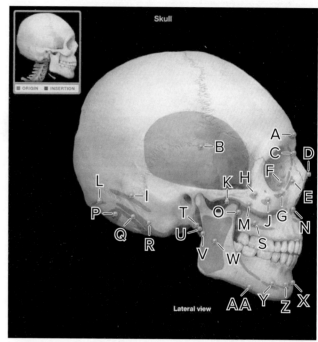

©McGraw-Hill Education

Mouse-over the pins on the screen to find the information necessary to identify the following attachments:

A. _____
B. _____
C. _____
D. _____
E. _____
F. _____
G. _____
H. _____
I. _____
J. _____
K. _____
L. _____

324 MODULE 6 The Muscular System

M. _____
N. _____
O. _____
P. _____
Q. _____
R. _____
S. _____
T. _____
U. _____
V. _____
W. _____
X. _____
Y. _____
Z. _____
AA. _____

Mouse-over the pins on the screen to find the information necessary to identify the following attachments:

A. _____
B. _____
C. _____
D. _____
E. _____
F. _____
G. _____
H. _____
I. _____
J. _____
K. _____
L. _____

EXERCISE 6.22:
Skull—Midsagittal View

SELECT TOPIC	SELECT VIEW
Skull	Midsagittal

Click **LAYER 1** *in the* **LAYER CONTROLS** *window, and you will see the following image:*

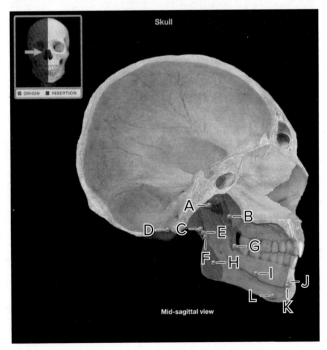

©McGraw-Hill Education

EXERCISE 6.23:
Skull—Posterior View

SELECT TOPIC	SELECT VIEW
Skull	Posterior

Click **LAYER 1** *in the* **LAYER CONTROLS** *window, and you will see the following image:*

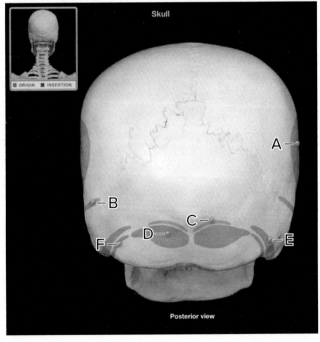

©McGraw-Hill Education

MODULE 6   The Muscular System   325

Mouse-over the pins on the screen to find the information necessary to identify the following attachments:

A. _____
B. _____
C. _____
D. _____
E. _____
F. _____
J. _____
K. _____
L. _____
M. _____
N. _____
O. _____
P. _____
Q. _____

EXERCISE 6.24:
Skull—Inferior View

SELECT TOPIC	SELECT VIEW
Skull	Inferior

Click **LAYER 1** *in the* **LAYER CONTROLS** *window, and you will see the following image:*

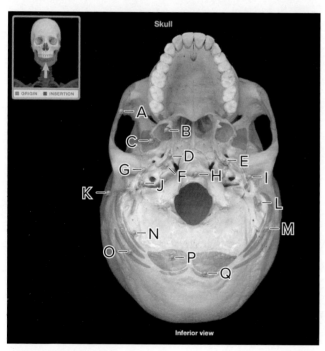

©McGraw-Hill Education

Mouse-over the pins on the screen to find the information necessary to identify the following attachments:

A. _____
B. _____
C. _____
D. _____
E. _____
F. _____
G. _____
H. _____
I. _____

EXERCISE 6.25:
Orbit—Anterolateral View

SELECT TOPIC	SELECT VIEW
Orbit	Anterolateral

Click **LAYER 1** *in the* **LAYER CONTROLS** *window, and you will see the following image:*

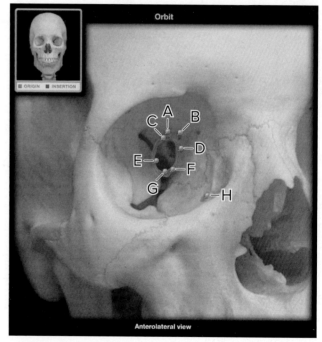

©McGraw-Hill Education

Mouse-over the pins on the screen to find the information necessary to identify the following attachments:

A. _____
B. _____
C. _____
D. _____
E. _____
F. _____
G. _____
H. _____

326 MODULE 6 The Muscular System

EXERCISE 6.26:	EXERCISE 6.27:
Hyoid Bone—Anterior and Lateral View	**Clavicle**

SELECT TOPIC	SELECT VIEW
Hyoid Bone	Anterior and Lateral

SELECT TOPIC	SELECT VIEW
Clavicle	Superior-Inferior

Click **LAYER 1** *in the* **LAYER CONTROLS** *window, and you will see the following image:*

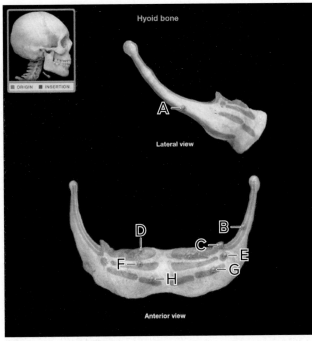

©McGraw-Hill Education

Click **LAYER 1** *in the* **LAYER CONTROLS** *window, and you will see the following image:*

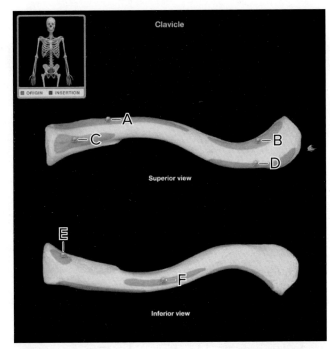

©McGraw-Hill Education

Mouse-over the pins on the screen to find the information necessary to identify the following attachments:

A. _____
B. _____
C. _____
D. _____
E. _____
F. _____
G. _____
H. _____

Mouse-over the pins on the screen to find the information necessary to identify the following attachments:

A. _____
B. _____
C. _____
D. _____
E. _____
F. _____

Self-Quiz
Take this opportunity to check your progress by taking the **QUIZ**. See the **Introduction Module** for a reminder on how to access the **QUIZ** for this Study Area.

EXERCISE 6.28:
Scapula

SELECT TOPIC: Scapula
SELECT VIEW: Anterior, Posterior, and Lateral

Click **LAYER 1** in the **LAYER CONTROLS** window, and you will see the following image:

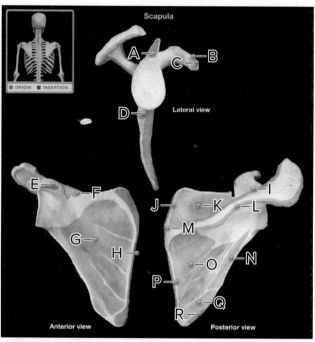

©McGraw-Hill Education

Mouse-over the pins on the screen to find the information necessary to identify the following attachments:

A. _____
B. _____
C. _____
D. _____
E. _____
F. _____
G. _____
H. _____
I. _____
J. _____
K. _____
L. _____
M. _____
N. _____
O. _____
P. _____
Q. _____
R. _____

EXERCISE 6.29:
Humerus

SELECT TOPIC: Humerus
SELECT VIEW: Anterior and Posterior

Click **LAYER 1** in the **LAYER CONTROLS** window, and you will see the following image:

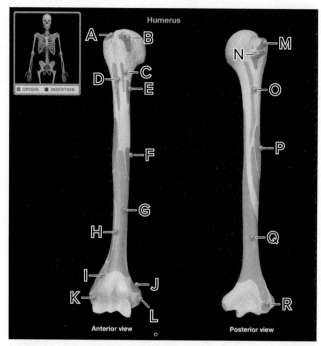

©McGraw-Hill Education

Mouse-over the pins on the screen to find the information necessary to identify the following attachments:

A. _____
B. _____
C. _____
D. _____
E. _____
F. _____
G. _____
H. _____
I. _____
J. _____
K. _____
L. _____
M. _____
N. _____
O. _____
P. _____
Q. _____
R. _____

328 MODULE 6 The Muscular System

Self-Quiz
Take this opportunity to check your progress by taking the **QUIZ**. See the **Introduction Module** for a reminder on how to access the **QUIZ** for this Study Area.

EXERCISE 6.30:
Radius and Ulna

SELECT TOPIC: Radius and Ulna ▶ SELECT VIEW: Anterior and Posterior

Click **LAYER 1** *in the* **LAYER CONTROLS** *window, and you will see the following image:*

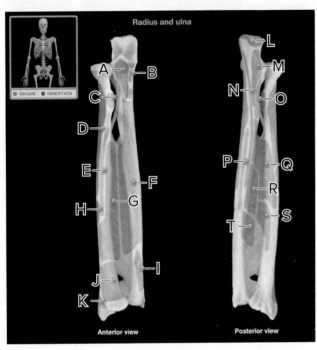

©McGraw-Hill Education

Mouse-over the pins on the screen to find the information necessary to identify the following attachments:

A. _____
B. _____
C. _____
D. _____
E. _____
F. _____
G. _____
H. _____
I. _____
J. _____
K. _____
L. _____
M. _____
N. _____
O. _____
P. _____
Q. _____
R. _____
S. _____
T. _____

EXERCISE 6.31:
Wrist and Hand

SELECT TOPIC: Wrist and Hand ▶ SELECT VIEW: Anterior and Posterior

Click **LAYER 1** *in the* **LAYER CONTROLS** *window, and you will see the following image:*

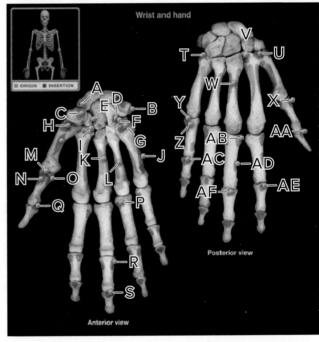

©McGraw-Hill Education

Mouse-over the pins on the screen to find the information necessary to identify the following attachments:

A. _____
B. _____
C. _____
D. _____
E. _____

F. _____
G. _____
H. _____
I. _____
J. _____
K. _____
L. _____
M. _____
N. _____
O. _____
P. _____
Q. _____
R. _____
S. _____
T. _____
U. _____
V. _____
W. _____
X. _____
Y. _____
Z. _____
AA. _____
AB. _____
AC. _____
AD. _____
AE. _____
AF. _____

Self-Quiz
Take this opportunity to check your progress by taking the **QUIZ**. See the **Introduction Module** for a reminder on how to access the **QUIZ** for this Study Area.

EXERCISE 6.32:
Thoracic Cage

SELECT TOPIC	SELECT VIEW
Thoracic Cage	Anterior

Click **LAYER 1** *in the* **LAYER CONTROLS** *window, and you will see the following image:*

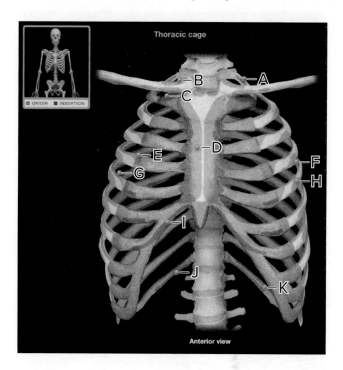

Mouse-over the pins on the screen to find the information necessary to identify the following attachments:

A. _____
B. _____
C. _____
D. _____
E. _____
F. _____
G. _____
H. _____
I. _____
J. _____
K. _____

Self-Quiz
Take this opportunity to check your progress by taking the **QUIZ**. See the **Introduction Module** for a reminder on how to access the **QUIZ** for this Study Area.

330 MODULE 6 The Muscular System

EXERCISE 6.33:
Vertebral Column

SELECT TOPIC: Vertebral Column ▸ **SELECT VIEW**: Anterior, Posterior, and Lateral

Click **LAYER 1** in the **LAYER CONTROLS** window, and you will see the following image:

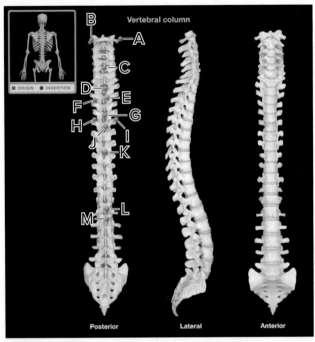

©McGraw-Hill Education

Mouse-over the pins on the screen to find the information necessary to identify the following attachments:

A. _____

B. _____

C. _____

D. _____

E. _____

F. _____

G. _____

H. _____

I. _____

J. _____

K. _____

L. _____

M. _____

EXERCISE 6.34:
Pelvic Girdle—Female

SELECT TOPIC: Pelvic Girdle—Female ▸ **SELECT VIEW**: Anterior

Click **LAYER 1** in the **LAYER CONTROLS** window, and you will see the following image:

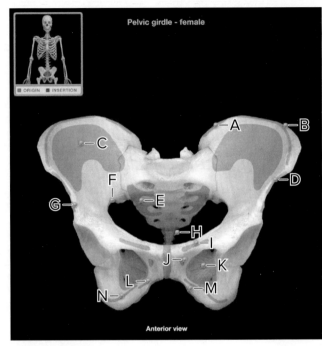

©McGraw-Hill Education

Mouse-over the pins on the screen to find the information necessary to identify the following attachments:

A. _____

B. _____

C. _____

D. _____

E. _____

F. _____

G. _____

H. _____

I. _____

J. _____

K. _____

L. _____

M. _____

N. _____

Self-Quiz
Take this opportunity to check your progress by taking the **QUIZ**. See the **Introduction Module** for a reminder on how to access the **QUIZ** for this Study Area.

EXERCISE 6.35:
Pelvic Girdle—Male

SELECT TOPIC Pelvic Girdle—Male ▸ **SELECT VIEW** Anterior

Click **LAYER 1** *in the* **LAYER CONTROLS** *window, and you will see the following image:*

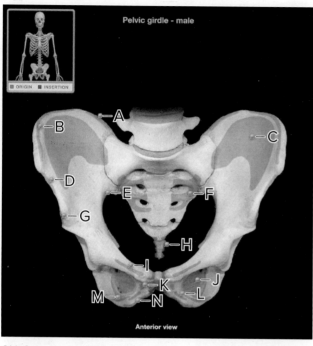

©McGraw-Hill Education

Mouse-over the pins on the screen to find the information necessary to identify the following attachments:

A. _____
B. _____
C. _____
D. _____
E. _____
F. _____
G. _____
H. _____
I. _____
J. _____
K. _____
L. _____
M. _____
N. _____

EXERCISE 6.36:
Sacrum and Coccyx

SELECT TOPIC Sacrum and Coccyx ▸ **SELECT VIEW** Anterior and Posterior

Click **LAYER 1** *in the* **LAYER CONTROLS** *window, and you will see the following image:*

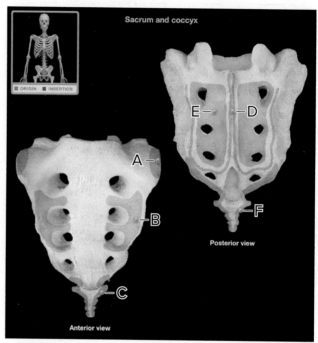

©McGraw-Hill Education

Mouse-over the pins on the screen to find the information necessary to identify the following attachments:

A. _____
B. _____
C. _____
D. _____
E. _____
F. _____

Self-Quiz

Take this opportunity to check your progress by taking the **QUIZ**. See the **Introduction Module** for a reminder on how to access the **QUIZ** for this Study Area.

332 MODULE 6 The Muscular System

EXERCISE 6.37:
Hip Bone

SELECT TOPIC: Hip Bone **SELECT VIEW:** Medial and Lateral

Click **LAYER 1** *in the* **LAYER CONTROLS** *window, and you will see the following image:*

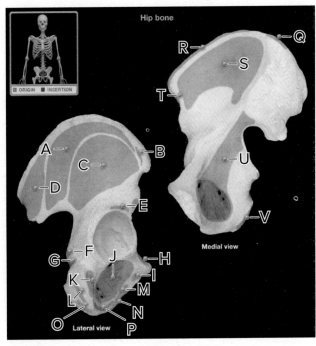

©McGraw-Hill Education

Mouse-over the pins on the screen to find the information necessary to identify the following attachments:

A. _____
B. _____
C. _____
D. _____
E. _____
F. _____
G. _____
H. _____
I. _____
J. _____
K. _____
L. _____
M. _____
N. _____
O. _____
P. _____
Q. _____
R. _____

S. _____
T. _____
U. _____
V. _____

EXERCISE 6.38:
Femur

SELECT TOPIC: Femur **SELECT VIEW:** Anterior and Posterior

Click **LAYER 1** *in the* **LAYER CONTROLS** *window, and you will see the following image:*

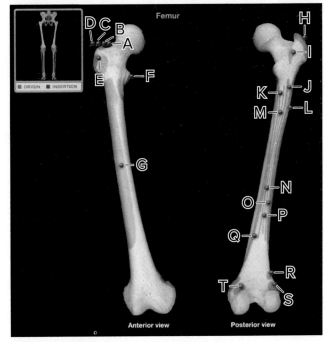

©McGraw-Hill Education

Mouse-over the pins on the screen to find the information necessary to identify the following attachments:

A. _____
B. _____
C. _____
D. _____
E. _____
F. _____
G. _____
H. _____
I. _____
J. _____
K. _____
L. _____
M. _____

N. _____
O. _____
P. _____
Q. _____
R. _____
S. _____
T. _____

G. _____
H. _____
I. _____
J. _____
K. _____
L. _____
M. _____
N. _____
O. _____
P. _____
Q. _____
R. _____

Self-Quiz

Take this opportunity to check your progress by taking the **QUIZ**. See the **Introduction Module** for a reminder on how to access the **QUIZ** for this Study Area.

EXERCISE 6.39:
Tibia and Fibula

SELECT TOPIC	SELECT VIEW
Tibia and Fibula ▶	Anterior and Posterior

Click **LAYER 1** in the **LAYER CONTROLS** *window, and you will see the following image:*

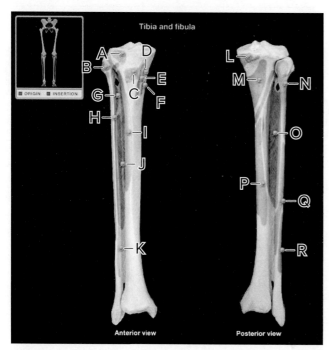

©McGraw-Hill Education

Mouse-over the pins on the screen to find the information necessary to identify the following attachments:

A. _____
B. _____
C. _____
D. _____
E. _____
F. _____

EXERCISE 6.40:
Ankle and Foot

SELECT TOPIC	SELECT VIEW
Ankle and Foot ▶	Superior and Inferior

Click **LAYER 1** in the **LAYER CONTROLS** *window, and you will see the following image:*

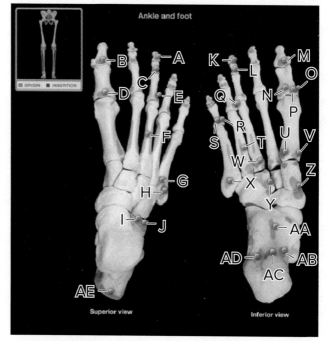

©McGraw-Hill Education

Mouse-over the pins on the screen to find the information necessary to identify the following attachments:

A. _____
B. _____
C. _____

334 MODULE 6 The Muscular System

D. _____
E. _____
F. _____
G. _____
H. _____
I. _____
J. _____
K. _____
L. _____
M. _____
N. _____
O. _____
P. _____
Q. _____
R. _____
S. _____
T. _____
U. _____
V. _____
W. _____
X. _____
Y. _____
Z. _____
AA. _____
AB. _____
AC. _____
AD. _____
AE. _____

Self-Quiz
Take this opportunity to check your progress by taking the **QUIZ**. See the **Introduction Module** for a reminder on how to access the **QUIZ** for this Study Area.

EXERCISE 6.41a:
Skeletal Muscle—Histology

SELECT TOPIC	SELECT VIEW
Skeletal Muscle	LM: Medium Magnification

Click the **TAGS ON/OFF** button, and you will see the following image:

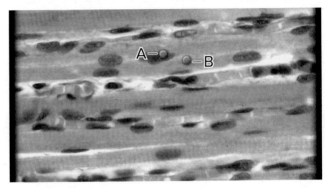

©McGraw-Hill Education/Al Telser

Mouse-over the pins on the screen to find the information necessary to identify the following structures:

A. _____

B. _____

EXERCISE 6.41b:
Skeletal Muscle—Histology

SELECT TOPIC	SELECT VIEW
Skeletal Muscle	LM: High Magnification

Click the **TAGS ON/OFF** button, and you will see the following image:

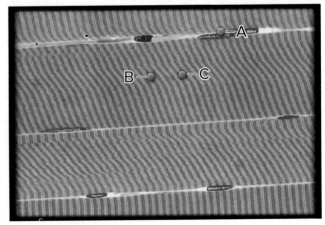

©Victor Eroschenko

Mouse-over the pins on the screen to find the information necessary to identify the following structures:

A. _____

B. _____

C. _____

EXERCISE 6.42:
Skeletal Muscle (Striations)—Histology

SELECT TOPIC: Skeletal Muscle (striations) ▶ **SELECT VIEW**: LM: High Magnification

*Click the **TAGS ON/OFF** button, and you will see the following image:*

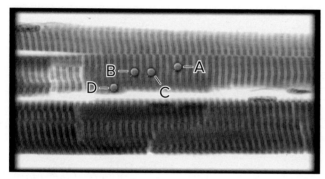

©McGraw-Hill Education/Al Telser

Mouse-over the pins on the screen to find the information necessary to identify the following structures:

A. _____
B. _____
C. _____
D. _____

EXERCISE 6.43:
Sarcomere—Histology

SELECT TOPIC: Sarcomere ▶ **SELECT VIEW**: TEM: High Magnification

*Click the **TAGS ON/OFF** button, and you will see the following image:*

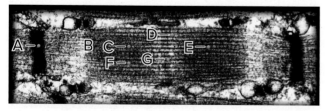

©McGraw-Hill Education

Mouse-over the pins on the screen to find the information necessary to identify the following structures:

A. _____
B. _____

C. _____
D. _____
E. _____
F. _____
G. _____

CHECK POINT

Skeletal Muscle, Histology

1. Describe the A bands of skeletal muscle fibers.
2. Describe the I bands of skeletal muscle fibers.
3. How many nuclei are found in each skeletal muscle fiber? Where are they located?

EXERCISE 6.44a:
Neuromuscular Junction—Histology

SELECT TOPIC: Neuromuscular Junction ▶ **SELECT VIEW**: LM: High Magnification

*Click the **TAGS ON/OFF** button, and you will see the following image:*

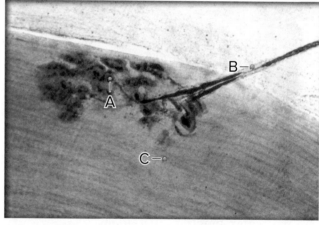

©Biophoto/Science Source

Mouse-over the pins on the screen to find the information necessary to identify the following structures:

A. _____
B. _____
C. _____

EXERCISE 6.44b:

Neuromuscular Junction—Histology

  SELECT TOPIC: Neuromuscular junction ▸ SELECT VIEW: TEM: High Magnification

Click the **TAGS ON/OFF** *button, and you will see the following image:*

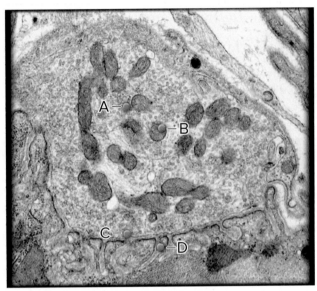

©EM Research Services, Newcastle University

Mouse-over the pins on the screen to find the information necessary to identify the following structures:

A. _____
B. _____
C. _____
D. _____

CHECK POINT

Neuromuscular Junction, Histology

1. Describe the motor end plate. Where is it located?
2. What neurotransmitter is found at the neuromuscular junction?
3. Describe the axon of the motor neuron. What is its function?

EXERCISE 6.45a:

Smooth Muscle—Histology

  SELECT TOPIC: Smooth Muscle ▸ SELECT VIEW: LM: Low Magnification

Click the **TAGS ON/OFF** *button and you will see the following image:*

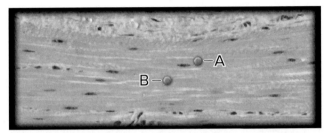

©McGraw-Hill Education/Al Telser

Mouse-over the pins on the screen to find the information necessary to identify the following structures:

A. _____
B. _____

EXERCISE 6.45b:

Smooth Muscle—Histology

  SELECT TOPIC: Smooth Muscle ▸ SELECT VIEW: LM: High Magnification

Click the **TAGS ON/OFF** *button, and you will see the following image:*

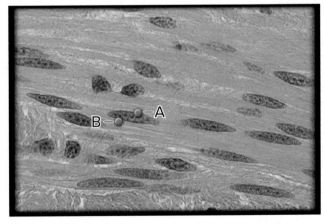

©McGraw-Hill Education/Dennis Strete

Mouse-over the pins on the screen to find the information necessary to identify the following structures:

A. _____
B. _____

CHECK POINT

Smooth Muscle, Histology

1. Describe the smooth muscle fibers.
2. What is another name for smooth muscle?
3. Where is this tissue found?

EXERCISE 6.46:

Cardiac Muscle—Histology

| SELECT TOPIC | SELECT VIEW |
| Cardiac Muscle | LM: High Magnification |

Click the **TAGS ON/OFF** button, and you will see the following image:

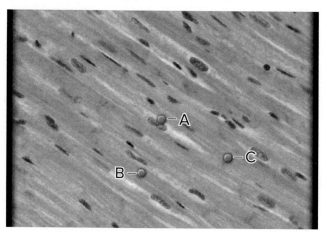

©McGraw-Hill Education/Al Telser

Mouse-over the pins on the screen to find the information necessary to identify the following structures:

A. _____
B. _____
C. _____

EXERCISE 6.47:

Cardiac Muscle (Intercalated Disc)— Histology

| SELECT TOPIC | SELECT VIEW |
| Cardiac Muscle (intercalated disc) | LM: High Magnification |

Click the **TAGS ON/OFF** button, and you will see the following image:

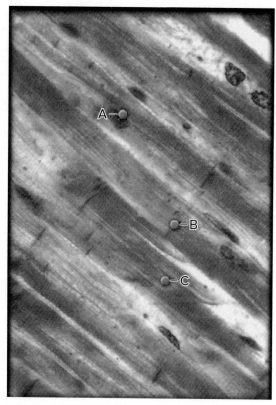

©McGraw-Hill Education/Al Telser

Mouse-over the pins on the screen to find the information necessary to identify the following structures:

A. _____
B. _____
C. _____

CHECK POINT

Cardiac Muscle, Histology

1. State the functions of the intercalated discs.
2. What are desmosomes?
3. How many nuclei are found in each cardiac muscle fiber?

Self-Quiz

Take this opportunity to check your progress by taking the **QUIZ**. See the **Introduction Module** for a reminder on how to access the **QUIZ** for this Study Area.

IN REVIEW

What Have I Learned?

The following questions cover the material you have just learned, the histology of muscle tissue. Apply what you have learned to answer the following questions on a separate piece of paper.

1. State the functions of the intercalated discs.

2. What are desmosomes?

3. How many nuclei are found in each cardiac muscle fiber?

4. Describe the A bands of skeletal muscle fibers.

5. How did the A bands gain their designation?

6. Describe the I bands of skeletal muscle fibers.

7. How did the I bands gain their designation?

8. How many nuclei are found in each skeletal muscle fiber? Where are they located?

9. Describe the smooth muscle fibers.

10. What is another name for smooth muscle?

11. Where is this tissue found?

12. Describe the motor end plate. Where is it located?

13. What neurotransmitter is found at the neuromuscular junction?

14. What is the sole-plate?

15. Describe the axon of the motor neuron. What is its function?

EXERCISE 6.48:
Coloring Exercise

Look up the muscles of the upper extremity and then color them in with colored pens or pencils. Fill in the blank labels with the appropriate names.

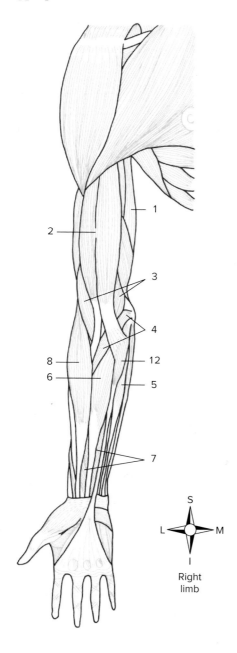

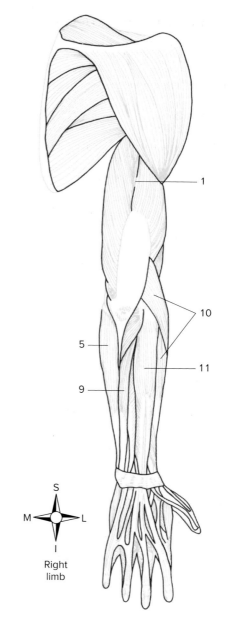

- [ ] 1. _____
- [ ] 2. _____
- [ ] 3. _____
- [ ] 4. _____
- [ ] 5. _____
- [ ] 6. _____

- [ ] 7. _____
- [ ] 8. _____
- [ ] 9. _____
- [ ] 10. _____
- [ ] 11. _____
- [ ] 12. _____

MODULE 7

The Nervous System

Overview: Nervous System

The nervous system is the master controlling and communicating system of your body. It is divided into two major divisions. The **central nervous system** (CNS) consists of the brain and spinal cord, while the **peripheral nervous system** (PNS) is comprised of nerve pathways leading to and from the CNS.

The brain in the adult is one of our largest organs. The average weight for the adult human brain is 1.4 kg, or 3 pounds. It consists of roughly 100 billion neurons and 900 billion glial cells. This means that there are one-fourth as many neurons and 2½ times as many total cells in your brain as there are stars in our Milky Way galaxy! Think about that the next time you are out at night, looking at the stars.

We will begin by exploring the structure and function of neurons, which are the basic functional units of the nervous system. We will then apply what we have learned as we investigate the central nervous system, and then continue with the peripheral nervous system and the special senses.

From the **HOME** *screen, click the drop-down box on the* **MODULE** *menu.*

From the systems listed, click on **Nervous**.

HEADS UP!

Be sure to click all **TAG** *buttons in the* **DISSECTION** *Study Area of* **Anatomy & Physiology | Revealed**®, *even if it is not necessary for completion of the assignments. This will allow you to have a better understanding of the locations of all of the structures that you* do *need to know.*

Neuron Physiology (Interactive)

SELECT ANIMATION
An Introduction to Neuron Physiology — PLAY

This Interactive Animation will walk you through the physiology of a nerve impulse. Be sure to interact with the animation to glean all the information available to you.

Open the **Module 7—Nervous System ANIMATION LIST.**

Under the second heading of **Neuron physiology (Interactive)**, *select* **1. An Introduction to Neuron Physiology** *and you will see the image to the right:*

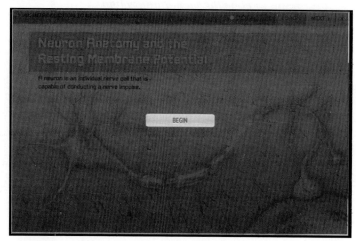
©McGraw-Hill Education

342 MODULE 7 The Nervous System

Select the **BEGIN** button in the middle, and you will see the image to the right:

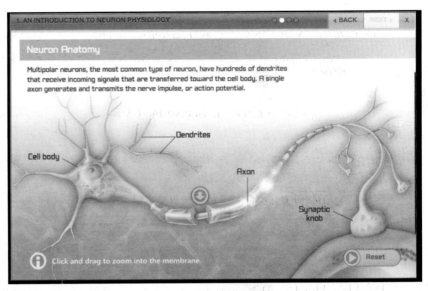
©McGraw-Hill Education

The narration and animation will begin immediately.

Be sure to click and drag the **blue arrow** downward to view the image to the right:

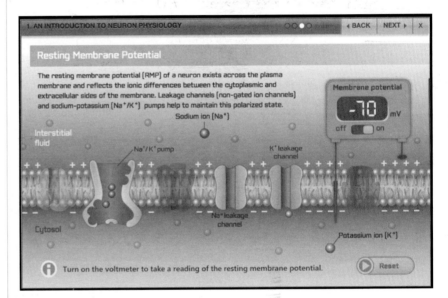
©McGraw-Hill Education

Be sure to slide the voltage meter switch over to **on** to view the membrane voltage.

You can click **Reset** to begin the animation again, or click **NEXT** to go to the next slide:

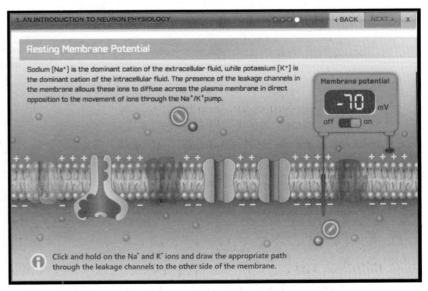
©McGraw-Hill Education

Be sure to move the Na⁺ and K⁺ ions through their leakage channels in and out of the neuron membrane.

*When finished with the interactions with all four images, click the **X** in the top-right corner to close the animation.*

After reviewing the animation, answer the following questions:

1. What is a neuron?

   an individual nerve cell that is capable of conducting a nerve impulse

2. Describe the structure AND function of a multipolar neuron.

   hundreds of dendrites that recieve incoming signals that are transfered toward the cell body. A single axon generates and transmits the nerve impulse, or AP

3. Where does the resting membrane potential exist? What does it reflect?

   exists across the plasma membrane. Reflects the ionic differences between cytoplasmic & extracellular sides of the membrane

4. What structures help to maintain the neuron's polarized state?

   leakage channels (non-gated ion channels) and sodium-potassium pumps help maintain polarized state

5. What is the dominant cation of the extracellular fluid?

   sodium

6. What is the dominant cation of the intracellular fluid?

   potassium

7. What allows the sodium and potassium ions to diffuse across the plasma membrane?

   presence of the leakage channel in membrane

8. What is the relationship between this movement and the movement of ions through the sodium-potassium pump?

   ions diffuse across plasma membrane in direct opposition to the movement of ions through the Na/K pump

SELECT ANIMATION — **The Receptive Segment** — PLAY

This Interactive Animation will walk you through the steps of the neurotransmission of graded potentials. Be sure to interact with the animation to glean all the information available to you.

*Open the **Module 7—Nervous System ANIMATION LIST**.*

*Under the second heading of **Neuron physiology (Interactive)**, select **2. The receptive segment** and you will see the image to the right:*

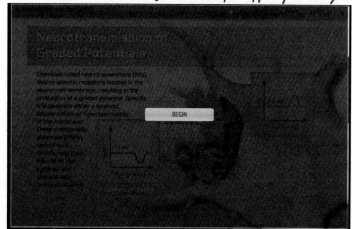

©McGraw-Hill Education

*Select the **BEGIN** button in the middle, and interact with the five images contained in this animation.*

When finished with the interactions, click the X in the top-right corner to close the animation.

After reviewing the animation, answer the following questions:

1. What are neurotransmitters and what is their function?

   chemicals - bind to specific receptors located in neuron cell membrane, resulting in production of a graded potential

2. What two types of responses are generated by specific neurotransmitters?

   localized depolorization hyperpolarization

3. How do these postsynaptic potentials spread? What affects do time and distance have on these potentials?

   mutil directionaly from the site of the synapse and degrade with time and distance

4. Describe an excitatory postsynaptic potential.

occurs when a local region of the cell membrane is depolarized

5. What causes excitatory postsynaptic potentials to occur?

in response to the opening of neurotic ligand-gated ion channels in membrane that result in net influx of Na into cell

6. What is the result of excitatory postsynaptic potentials?

creates a localized region of depolarization on the membrane

7. What causes inhibitory postsynaptic potentials to occur?

when a local region of the cell membrane is hyperpolarized

8. What is triggered by inhibitory postsynaptic potentials?

The opening of either ligand gated K or Cl channels.

9. What determines which outcome will occur?

depending on the specific neurotransmitter

10. What are the results of these outcomes?

efflux of K or influx of Cl, respectively, both of which would result in localized hyperpolarization of the membrane

This Interactive Animation will walk you through the steps of summation at the axon hillock. Be sure to interact with the animation to glean all the information available to you.

Open the **Module 7—Nervous System ANIMATION LIST.**

Under the second heading of **Neuron physiology (Interactive),** *select* **3. The initial segment** *and you will see the image to the right:*

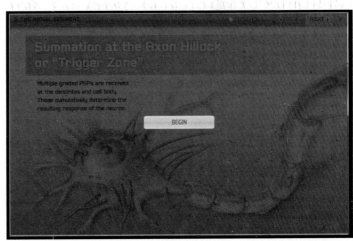
©McGraw-Hill Education

Select the **BEGIN** *button in the middle, and interact with the five images contained in this animation.*

When finished with the interactions, click the **X** *in the top-right corner to close the animation.*

After reviewing the animation, answer the following questions:

1. What are received at the dendrites and cell body of the neuron?

multiple graded PSPs

2. What is their cumulative effect?

the SE of all graded potentials determines if the resulting net depolarization of the axon hillock reaches threshold

3. What determines if the net depolarization of the axon hillock reaches threshold?

determine the resulting response of the neuron

4. What voltage is typically reached at threshold?

−55 mV

5. What is the result if threshold is met?

then an action potential is generated

MODULE 7   The Nervous System   345

6. Describe temporal summation.
   occurs as multiple signals are sent from the same presynaptic terminal in a rapid fire fashion

7. What results in increasing strength of the developing graded potential?
   The reduced time between signals results in strength

8. Describe special summation.
   related to the proximity of the multiple postsynaptic potentials to each other, & their relative distance to axon hillock

9. What changes occur within the axon hillock once threshold has been reached?
   voltage gated Na channels trigger zone open

10. What is the result of these changes?
    an influx of Na ions

11. What does the influx of Na^+ ions cause?
    depolarization

12. What does this begin?
    propagation of an AP

SELECT ANIMATION
The Conductive Segment   PLAY

This Interactive Animation will walk you through the steps of propagation of an action potential. Be sure to interact with the animation to glean all the information available to you.

Open the **Module 7—Nervous System ANIMATION LIST.**

Under the second heading of **Neuron physiology (Interactive)**, select **4. The conductive segment** and you will see the image to the right:

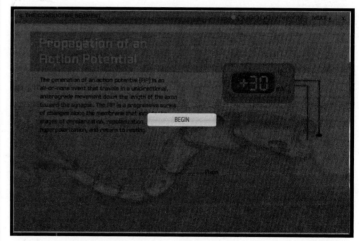
©McGraw-Hill Education

Select the **BEGIN** button in the middle, and interact with the seven images contained in this animation.

When finished with the interactions, click the **X** in the top-right corner to close the animation.

After reviewing the animation, answer the following questions:

1. The generation of an action potential is an __all__ __or__ __none__ event that travels in a __unidirectional__ __antrograde__ movement down the length of the __axon__ toward the __synapse__.

2. The action potential is a progressive series of changes along the __membrane__ that includes the stages of __depolarization__, __repolarization__, __hyperpolarization__ and return to __resting__.

3. Describe depolarization of the membrane, including details of voltage changes and their causes.
   involves a positive change in membrane potential as V-gated Na channels in plasma membrane of axon open, resulting in influx of Na

4. Describe repolarization of the membrane, including details of voltage changes and their causes.

5. Describe hyperpolarization of the membrane, including details of voltage changes and their causes.

6. Describe the reestablishment of resting membrane potential.

MODULE 7 The Nervous System

7. Describe the absolute refractory period. What cannot occur during this time? How does this ensure the unidirectional movement of the signal?

8. In what order does each exposed region of the axon go through this process? How does this affect each adjacent region of the axon?

9. Describe the differences in action potential movement between unmyelinated and myelinated axons.

SELECT ANIMATION — **The Transmissive Segment** — PLAY

This Interactive Animation will walk you through the action potential's flow from the synaptic cleft to the postsynaptic membrane. Be sure to interact with the animation to glean all the information available to you.

Open the **Module 7—Nervous System ANIMATION LIST.**

Under the second heading of **Neuron physiology (Interactive)**, select **5. The transmissive segment** and you will see the image to the right:

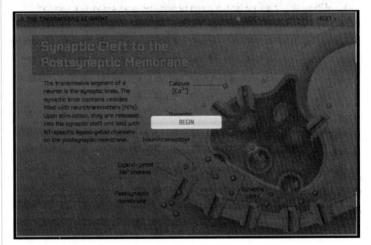
©McGraw-Hill Education

Select the **BEGIN** button in the middle, and interact with the five images contained in this animation.

When finished with the interactions, click the **X** in the top-right corner to close the animation.

After reviewing the animation, answer the following questions:

1. What structure constitutes the transmissive segment of the neuron?

   synaptic knob

2. What structures are contained in the transmissive segment of the neuron?

   vesicles filled with neurotransmitters

3. What happens to these structures upon stimulation? What is the result?

   They are released into the synaptic cleft and bind with NT-specific ligand gated channels

4. Describe the process by which neurotransmitters are released from the synaptic knob, including the trigger for their release.

   pic

5. What is the purpose for the diffusion of neurotransmitters across the synaptic cleft?

   bind to NT-specific receptors present on the postsynaptic membrane

6. What is triggered by this binding? What two options does this cause?

   The opening of ligand gated ion channels, causing depolarization or hyperpolarization of postsynaptic membrane

7. What happens to the neurotransmitters after they accomplish their purpose? What group of chemicals carry out this task?

   either broken down or taken back up into synaptic knob. enzymes within extracellular fluid

8. What specifically is involved in the reuptake of acetylcholine? enzyme
acetylcholinesterase breaks down ACh into acetic acid & choline

9. What drives neural communication between cells? changes in membrane permeability to ions

10. What input do neurons receive? from multiple types of stimuli

11. What are the primary reasons for the summation of graded potentials and the unidirectional propagation of action potentials? changes in Na & K membrane perability

12. What is the ultimate result of these processes? realse of NT's

6. How many of each ion is transported across the membrane by the sodium-potassium pump? 3 Na 2 K

7. Even in a __resting__ neuron, there is the __potential__ for the charge difference to create an __electrical__ __current__. This is called a __resting__ __membrane__ __potential__.

8. Describe a local membrane potential. when an electrical current flows through a dendrite

9. What sequence of events occurs on a dendrite when it detects a stimulus?
 - sodium channel in plasma mem opens & lets sodium into neuron
 - influx of (+) ions reverse charge of mem - dep
 - repolarize - release ka out of neuron

10. What affect does this have across a particular section of the membrane? reverses charge

11. What term describes this process? depolarization

SELECT ANIMATION
Nerve Impulse (3D)   PLAY

After viewing the animation, answer the following questions:

1. List and describe the structures that comprise a typical neuron. cell body
axon hillock - trigger zone that releases impulse
dendrites - extention
axon - elongated fiber

2. What is the function of the axon hillock and what does it determine? maintains an excitation limit or threshold which determines wheater or not a neuron will generate a nerve impulse

3. Describe a nerve impulse. What response does it cause? an electrical signal conducted by a neuron causing a response in another neuron or target cell

4. Describe a neuron's membrane at rest. How do the charges differ from one side of the membrane to the other? its membrane is polarized b/c there are more positive ions outside cell and more (-) ions inside the cell, which creates a charge difference across mem

 Na K

5. What is the function of the sodium-potassium pumps? What mechanism do they use?
across membrane to maintain this charge difference

12. Describe the process of repolarization. K channels open and release K out of neuron

13. What is the function of the nearby sodium-potassium pumps? transports excess sodium out and brings K+ in

14. What is the result of the sodium-potassium pump's actions? restores resting membrane potential

15. What is produced by the flow of reversing charges along a dendrite's membrane? produces wavlike electrical current forwards neurons trigger zone

16. What direction does this flow travel?

348 MODULE 7 The Nervous System

17. What is required for an action potential or nerve impulse to occur?
    if strength of current meets or exceeds the threshold at the trigger zone

18. Describe the function of the trigger zone during a nerve impulse. sends an electrical signal down axon toward space between neurons

19. Describe a synapse.

20. What are the two possible destinations for a nerve impulse?
    synapse
    target cell membrane

SELECT ANIMATION — Action Potential Generation [PLAY]

After viewing the animation, answer the following questions:

1. When the cell membrane is at its resting membrane potential, in what position are the voltage-gated sodium and potassium channels?
   V-G Na closed and inactivation gates open

2. depolarization is initiated by a stimulus that makes the membrane potential more positive

3. What affect does this have on the voltage-gated sodium channels? What occurs at threshold?
   causes them to open
   threshold — Na channels open

4. What causes depolarization? What happens to the voltage-gated potassium channels?
   sodium ions diffuse across membrane
   VGPC also begin to open, but slower

5. Therefore, depolarization occurs because _____.
   more sodium ions diffuse into the cell than potassium diffuse out of it

6. What causes the diffusion of sodium ions to decrease? What happens to the potassium ions?
   p) c

7. What causes the membrane potential to become slightly more negative than the resting value?
   K channels remain open any K ions continue to diffuse out of cell

8. What causes reestablishment of resting membrane potential? active transport of sodium and potassium ions

SELECT ANIMATION — Action Potential Propagation [PLAY]

After viewing the animation, answer the following questions:

1. An action potential is propagated _____.
   in one direction along the axon

2. During an action potential, the inside of the cell membrane _____.
   becomes positive w/ respect to the outside

3. What happens to the membrane immediately adjacent to the action potential?
   AP generates local currents that depolarize it

4. What occurs when depolarization caused by the local currents reaches threshold? a new AP is produced adjacent to the original one

5. Why does action potential propagation occur in only one direction? b/c recently depolarized area of the membrane is in absolute refractory period & cannot generate an AP

SELECT ANIMATION — Chemical Synapse [PLAY]

After viewing the animation, answer the following questions:

1. What is caused by action potentials arriving at the pre-synaptic terminal?
   V-G calcium ion channels to open

2. What occurs when calcium ions diffuse into the cell?
   cause synaptic vesicles to release Ach

3. What do the acetylcholine molecules do? What does this cause?
   p/c

4. If the membrane potential reaches threshold level, _____.
   an AP will be produced

extra efflux of K ions cause

EXERCISE 7.1:
Multipolar Neuron—Golgi Stain— Histology

SELECT TOPIC	SELECT VIEW
Multipolar Neuron— Golgi stain	LM: Low Magnification

Click the **TAGS ON/OFF** *button, and you will see the following image:*

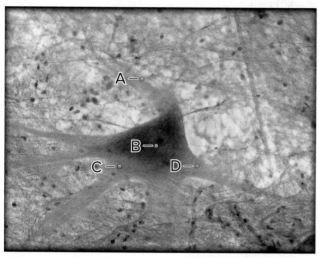

©SPL/Science Source

Mouse-over the pins on the screen to find the information necessary to identify the following structures:

A. _____

B. _____

C. _____

D. _____

CHECK POINT
Multipolar Neuron—Golgi Stain, Histology

1. Name the two types of cell processes.
2. Which of the cell processes convey efferent nerve impulses?
3. Which of the cell processes convey afferent nerve impulses?

EXERCISE 7.2:
Axon Hillock—Histology

SELECT TOPIC	SELECT VIEW
Axon Hillock	LM: High Magnification

Click the **TAGS ON/OFF** *button, and you will see the following image:*

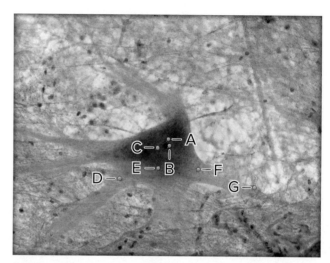

©McGraw-Hill Education/Al Telser

Mouse-over the pins on the screen to find the information necessary to identify the following structures:

A. _____

B. _____

C. _____

D. _____

E. _____

F. _____

G. _____

CHECK POINT
Axon Hillock, Histology

1. Describe the axon hillock.
2. Name the largest membrane-bound organelle of the neuron.
3. What structures are found within this organelle?

350 MODULE 7 The Nervous System

EXERCISE 7.3:
Unmyelinated Axon—Histology

SELECT TOPIC	SELECT VIEW
Axon	TEM: Low Magnification

Click the **TAGS ON/OFF** *button, and you will see the following image:*

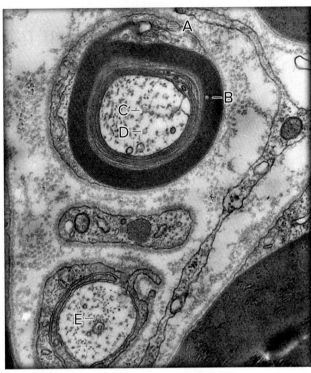

©McGraw-Hill Education/Al Telser

Mouse-over the pins on the screen to find the information necessary to identify the following structures:

A. _____

B. _____

C. _____

D. _____

E. _____

CHECK POINT
Unmyelinated Axon, Histology

1. Name the structure that surrounds the myelinated axon.
2. Name the myelinating cells for the CNS.
3. Name the myelinating cells for the PNS.

EXERCISE 7.4:
Synapse—Histology

SELECT TOPIC	SELECT VIEW
Synapse	TEM: Low Magnification

Click the **TAGS ON/OFF** *button, and you will see the following image:*

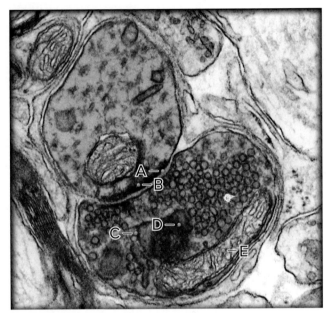

©McGraw-Hill Education/Dr. Dennis Emery, Iowa State University

Mouse-over the pins on the screen to find the information necessary to identify the following structures:

A. _____

B. _____

C. _____

D. _____

E. _____

CHECK POINT
Synapse, Histology

1. Describe the synapse.
2. Where is it located?
3. What is its function?

EXERCISE 7.5a:
Neuromuscular Junction—Histology

 SELECT TOPIC: Neuromuscular Junction ▶ SELECT VIEW: LM: High Magnification

Click the **TAGS ON/OFF** button, and you will see the following image:

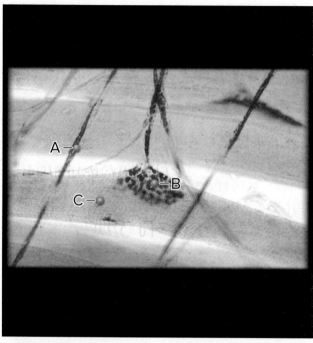

©SPL/Science Source

Mouse-over the pins on the screen to find the information necessary to identify the following structures:

A. _____

B. _____

C. _____

CHECK POINT
Neuromuscular Junction, Histology

1. Define neuromuscular junction.
2. Describe the neuromuscular junction.
3. What neurotransmitter is released at the neuromuscular junction?

EXERCISE 7.5b:
Nervous System—Neuromuscular Junction—Histology

 SELECT TOPIC: Neuromuscular Junction ▶ SELECT VIEW: TEM: High Magnification

Click the **TAGS ON/OFF** button, and you will see the following image:

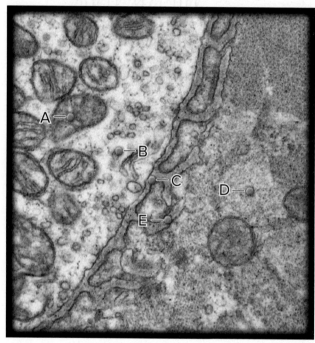

©Dr. Thomas Caceci, Virginia-Maryland Regional College of Veterinary Medicine

Mouse-over the pins on the screen to find the information necessary to identify the following structures:

A. _____

B. _____

C. _____

D. _____

E. _____

CHECK POINT
Neuromuscular Junction, Histology, *continued*

4. Describe the motor end plate. Where is it located?
5. What is the sole-plate?
6. Name the neurotransmitter at the neuromuscular junction.
7. Describe the primary and secondary synaptic clefts. Where are they located?

352 MODULE 7 The Nervous System

EXERCISE 7.6a:
Schwann Cell—Histology

SELECT TOPIC	SELECT VIEW
Schwann Cell	LM: Low Magnification

Click the **TAGS ON/OFF** button, and you will see the following image:

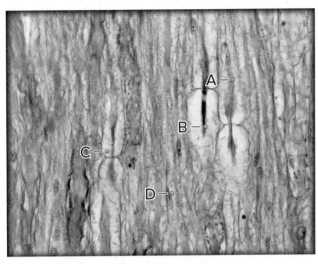

©EM Research Services, Newcastle University

Mouse-over the pins on the screen to find the information necessary to identify the following structures:

A. _____

B. _____

C. _____

D. _____

CHECK POINT

Schwann Cell, Histology

1. What is the name for the cleft between the inter-nodes of the myelin sheath?
2. What is saltatory conduction?
3. What is lacking between myelin layers?

EXERCISE 7.6b:
Nervous System—Schwann Cell—Histology

SELECT TOPIC	SELECT VIEW
Schwann Cell	TEM: Medium Magnification

Click the **TAGS ON/OFF** button, and you will see the following image:

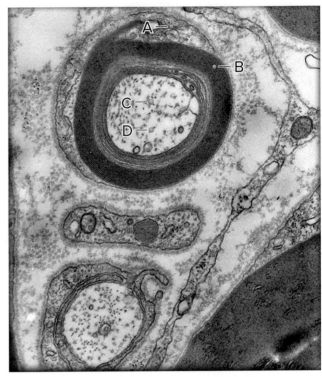

©Biophoto Assoc./Science Source

Mouse-over the pins on the screen to find the information necessary to identify the following structures:

A. _____

B. _____

C. _____

D. _____

CHECK POINT

Schwann Cell, Histology, *continued*

4. Describe salutatory conduction.
5. Describe the node of Ranvier. Where is it located? What is its function?
6. Name the cell responsible for the formation of the myelin sheath in the CNS. In the PNS. How do they compare?

HEADS UP!
Take this time to review the following animations we covered earlier in this module:
- *Action potential generation*
- *Action potential propagation*
- *Chemical synapse*

Self-Quiz
Take this opportunity to check your progress by taking the **QUIZ**. See the **Introduction Module** for a reminder on how to access the **QUIZ** for this Study Area.

IN REVIEW

What Have I Learned?

The following questions cover the material that you have just learned, histology. Apply what you have learned to answer these questions on a separate piece of paper.

1. Name the two types of cell processes.

2. Which of the cell processes convey efferent nerve impulses?

3. Which of the cell processes convey afferent nerve impulses?

4. What are Nissl bodies?

5. What structures of the soma support a neuron's structure and function?

6. Which cell processes are not myelinated?

7. Which specific cell processes are usually myelinated?

8. Which cell processes arise from the axon hillock?

9. Which cell processes are usually without branches near the cell of origin?

10. Which cell processes receive information from other neurons?

11. Which cell processes convey information to other neurons or effectors?

12. Name a type of neuron composed of a soma, multiple dendrites, and a single axon.

13. Describe the axon hillock.

14. Name the largest membrane-bound organelle of the neuron.

15. What is located within this organelle?

16. Name the distinct round, dark-staining organelle in the nucleus of the neuron. What is its function?

17. Name the structure that surrounds the myelinated neuron.

18. Name the myelinating cells for the CNS.

19. Name the myelinating cells for the PNS.

20. At what speeds do nerve impulses travel along unmyelinated neurons?

21. At what speeds do nerve impulses travel along comparable myelinated neurons?

22. Where do unmyelinated axons rest?

23. Describe the myelin sheath.

24. What is cytosol?

25. Describe the synapse.

26. Where is it located?

27. What is its function?

28. Describe the synaptic cleft.

29. Where is it located?

30. What is its function?

31. Describe synaptic vesicles.

32. Where are they located?

33. What is their function?

34. Define a neurotransmitter. Give two examples of neurotransmitters.

35. What is the function of the mitochondria?

36. Describe the presynaptic terminals.

37. Where are they located?

38. What is their function?

39. Define neuromuscular junction.

40. Describe the neuromuscular junction.

41. What neurotransmitter is released at the neuromuscular junction?

42. Describe the axon of a motor neuron.

43. What is its function?

44. What is the name for the cleft between the internodes of the myelin sheath?

45. What is salutatory conduction?

46. What is lacking between myelin layers?

SELECT ANIMATION
Divisions of Brain
PLAY

After viewing the animation, answer the following questions:

1. Name the four regions of the brain.

2. Describe the cerebrum.

3. How many cerebral hemispheres exist?

4. Name the five lobes of the cerebral hemispheres.

5. The cerebrum is considered _____ responsible for _____.

6. The cerebellum is composed of _____.

7. The primary function of the cerebellum is _____.

8. The diencephalon is composed of three structures. What are they, and what are their functions?

9. Name and describe the three divisions of the brainstem.

10. Name the functions of each division of the brainstem.

11. Where do the cerebellar hemispheres attach to the brainstem?

12. The medulla oblongata extends from _____.

13. The medulla oblongata is continuous with _____.

The Brain

HEADS UP!

A note about **ROSTRAL** vs **CAUDAL** when studying the brain.

We normally use a specific set of anatomical terms when describing the location of structures within the body. We will often use terms such as anterior and posterior or superior and inferior to refer to a structure's location within the body itself, or its relationship to the structures that surround it. When describing the brain, we are faced with a unique situation. Our brain does not align along a vertical axis as our body does—from superior to inferior—nor does it align in a horizontal axis as does the body of a dog or a cat. It actually aligns along both axes! The "anterior" portion aligns horizontally while the "posterior" portion aligns vertically. Therefore, when we are describing the portion of the brain in the proximity of the frontal bone, we use the term **rostral**. Conversely, when we describe the portions of the brain in the proximity of the foramen magnum of the occipital bone, we use the term **caudal**. This distinction may be seen in **Exercise 7.1**, where the **LAYER CONTROLS** window lists **ROSTRAL** along the slider for **LAYER 1** and **CAUDAL** is listed under **LAYER 6**. As you go through the sequence of dissections in this exercise, note how the plane of dissection changes as you progress from rostral to caudal.

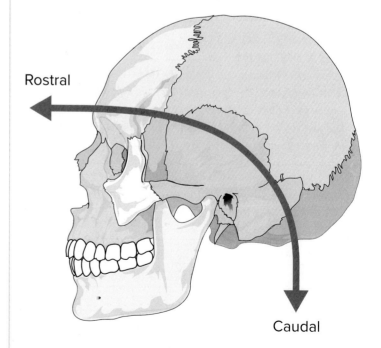

356 MODULE 7 The Nervous System

EXERCISE 7.7:
Nervous System—Brain—Coronal View

SELECT TOPIC	SELECT VIEW
Brain	Coronal

Click **LAYER 1** *in the* **LAYER CONTROLS** *window, and you will see the following image:*

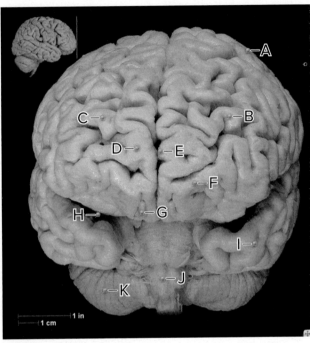

©McGraw-Hill Education

HEADS UP!
There are two images visible in the **IMAGE AREA**, *the large image with the identification pins and a smaller image in the top-left corner. This second, smaller image is to show you the plane of section, which allows you to keep the larger image in perspective.*

Mouse-over the pins on the screen to find the information necessary to identify the following structures:

A. _____
B. _____
C. _____
D. _____
E. _____
F. _____
G. _____
H. _____
I. _____
J. _____
K. _____

CHECK POINT
Brain, Coronal View

1. Name the deep grooves that separate the temporal lobes from the frontal and parietal lobes.
2. Name the deep groove that separates the right and left cerebral hemispheres.
3. Name the site of synapse for the olfactory neurons after they pass through the cribriform plate.

Click **LAYER 2** *in the* **LAYER CONTROLS** *window, and you will see the following image:*

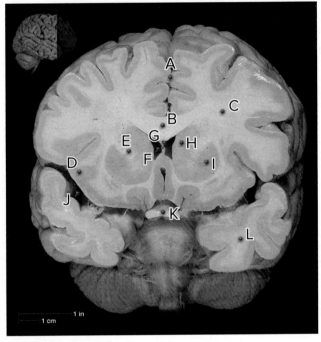

©McGraw-Hill Education

Mouse-over the pins on the screen to find the information necessary to identify the following structures:

A. _____
B. _____
C. _____
D. _____
E. _____
F. _____
G. _____
H. _____
I. _____
J. _____
K. _____
L. _____

CHECK POINT

Brain, Coronal View, *continued*

4. What is the term for unmyelinated nervous tissue?
5. What is the term for the collection of myelinated axons in the brain?
6. What is the large myelinated fiber tract that connects the right and left cerebral hemispheres?

Click **LAYER 3** *in the* **LAYER CONTROLS** *window, and you will see the following image:*

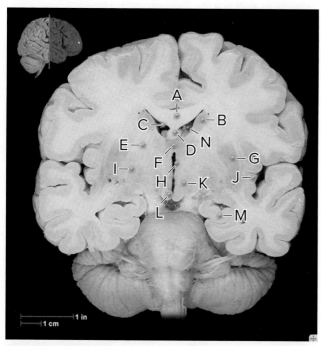

©McGraw-Hill Education

Mouse-over the pins on the screen to find the information necessary to identify the following structures:

A. _____
B. _____
C. _____
D. _____
E. _____
F. _____
G. _____
H. _____
I. _____
J. _____
K. _____
L. _____
M. _____

Nonnervous System Structure (blue pin)

N. _____

CHECK POINT

Brain, Coronal View, *continued*

7. Name the paired, rounded projections involved in regulation of autonomic functions, emotional behavior, and memory.
8. Name the structure located in the cerebral ventricles that is the site of the production of cerebrospinal fluid (CSF).
9. Name the structure that is primarily for the relay of sensory information to the cerebral cortex.

Click **LAYER 4** *in the* **LAYER CONTROLS** *window, and you will see the following image:*

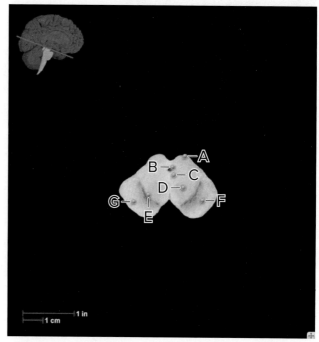

©McGraw-Hill Education

Mouse-over the pins on the screen to find the information necessary to identify the following structures:

A. _____
B. _____
C. _____

358 MODULE 7 The Nervous System

D. _____
E. _____
F. _____
G. _____

CHECK POINT
Brain, Coronal View, *continued*

10. Name the structure that coordinates orienting movements of the eyes and head.
11. Name the narrow midline channel between the third and fourth ventricles.
12. Name the structure involved with suppression and modulation of pain.

CHECK POINT
Brain, Coronal View, *continued*

13. Name the structure that controls voluntary movement.
14. Name the major afferent pathway for information from the motor cortex to the cerebellum.
15. Name the cerebrospinal fluid-filled pyramidal cavity that is continuous with the cerebral aqueduct and the central canal of the spinal cord.

Click **LAYER 5** *in the* **LAYER CONTROLS** *window, and you will see the following image:*

Click **LAYER 6** *in the* **LAYER CONTROLS** *window, and you will see the following image:*

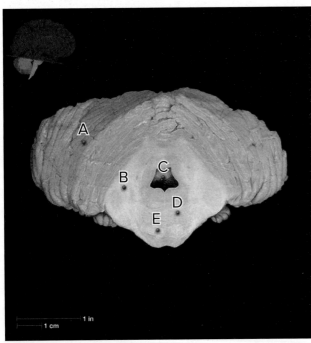

©McGraw-Hill Education

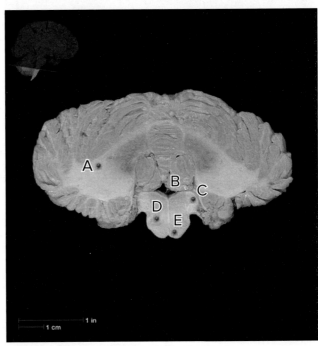

©McGraw-Hill Education

Mouse-over the pins on the screen to find the information necessary to identify the following structures:

A. _____
B. _____
C. _____
D. _____
E. _____

Mouse-over the pins on the screen to find the information necessary to identify the following structures:

A. _____
B. _____
C. _____
D. _____
E. _____

CHECK POINT

Brain, Coronal View, *continued*

16. Name the structure that processes and sends information to the cerebellum from many CNS nuclei and skeletal muscle proprioceptors.
17. Name the structure of the medulla oblongata that controls voluntary movement.
18. Name the structure that carries information about muscle performance from the spinal cord to the cerebellum.

SELECT ANIMATION
Meninges
PLAY

After viewing the animation, answer the following questions:

1. List the three meninges in order from superficial to deep.

2. Name the two layers of the most superficial of the meninges.

3. Name the structures formed where these two layers split.

4. Name the space located between the middle and deepest meninges. What fills this space?

Self-Quiz
Take this opportunity to check your progress by taking the **QUIZ**. See the **Introduction Module** for a reminder on how to access the **QUIZ** for this Study Area.

EXERCISE 7.8:
Nervous System—Brain—Lateral View

SELECT TOPIC	SELECT VIEW
Brain	Lateral

Click **LAYER 1** *in the* **LAYER CONTROLS** *window, and you will see the following image:*

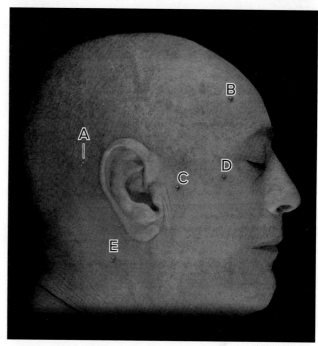

©McGraw-Hill Education

Mouse-over the pins on the screen to find the information necessary to identify the following structures:

A. _____
B. _____
C. _____
D. _____
E. _____

CHECK POINT

Brain, Lateral View

1. What is a dermatome?
2. The skin of the superior face is innervated by which nerve?
3. The external ear is innervated by which nerve?

Click **LAYER 3** *in the* **LAYER CONTROLS** *window, and you will see the following image:*

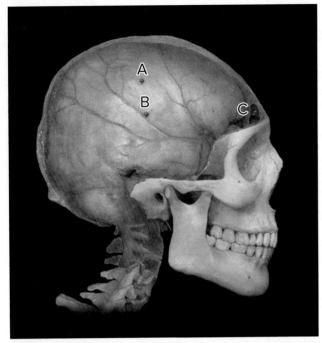

©McGraw-Hill Education

Mouse-over the pin on the screen to find the information necessary to identify the following structure:

A. _____

Nonnervous System Structures (blue pins)

B. _____

C. _____

CHECK POINT

Brain, Lateral View, *continued*

4. Name the paired mucous membrane-lined cavities within the frontal bone.
5. Name the most external of the meninges.
6. Name an artery that courses between the dura mater and the cranium.

Click **LAYER 4** *in the* **LAYER CONTROLS** *window, and you will see the following image:*

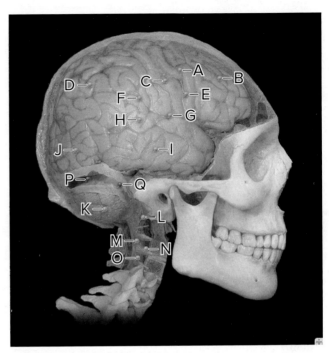

©McGraw-Hill Education

Mouse-over the pins on the screen to find the information necessary to identify the following structures:

A. _____
B. _____
C. _____
D. _____
E. _____
F. _____
G. _____
H. _____
I. _____
J. _____
K. _____
L. _____
M. _____
N. _____
O. _____

Nonnervous System Structures (blue pins)

P. _____
Q. _____

MODULE 7 The Nervous System 361

CHECK POINT

Brain, Lateral View, *continued*

7. Name the distinct fold at the posterior border of the frontal lobe that controls voluntary movement.
8. Name the distinct fold at the anterior border of the parietal lobe that receives somatosensory information from the body.
9. Name the groove that forms the boundary between the frontal and parietal lobes.

CHECK POINT

Brain, Lateral View, *continued*

10. Name the large, crescent-shaped fold of the dura mater that separates the two cerebral hemispheres.
11. Name the structure that contains arachnoid granulations. What is the function of these granulations?
12. The confluence of the sinuses is the meeting point for four different sinuses. What are they?

Click **LAYER 5** *in the* **LAYER CONTROLS** *window, and you will see the following image:*

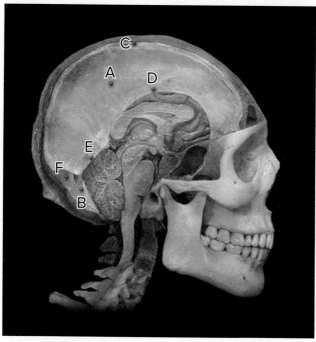

©McGraw-Hill Education

Mouse-over the pins on the screen to find the information necessary to identify the following structures:

A. _____

B. _____

Nonnervous System Structures (blue pins)

C. _____

D. _____

E. _____

F. _____

Click **LAYER 6** *in the* **LAYER CONTROLS** *window, and you will see the following image:*

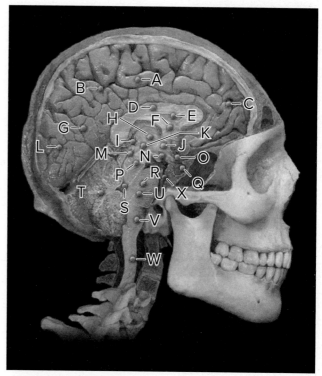

©McGraw-Hill Education

Mouse-over the pins on the screen to find the information necessary to identify the following structures:

A. _____

B. _____

C. _____

D. _____

E. _____

362 MODULE 7 The Nervous System

F. _____
G. _____
H. _____
I. _____
J. _____
K. _____
L. _____
M. _____
N. _____
O. _____
P. _____
Q. _____
R. _____
S. _____
T. _____
U. _____
V. _____
W. _____

Nonnervous System Structure (blue pin)

X. _____

CHECK POINT

Brain, Lateral View, *continued*

13. Name the groove that separates the parietal and occipital lobes of the brain.
14. Name the pea-sized endocrine gland attached to the roof of the third ventricle. What hormone does it secrete?
15. Name the narrow cerebrospinal fluid-filled channel between the third and fourth ventricles.

Self-Quiz
Take this opportunity to check your progress by taking the **QUIZ**. See the **Introduction Module** for a reminder on how to access the **QUIZ** for this Study Area.

EXERCISE 7.9:
Nervous System—Brain—Superior View

SELECT TOPIC	SELECT VIEW
Brain	Superior

Click **LAYER 1** *in the* **LAYER CONTROLS** *window, and you will see the following image:*

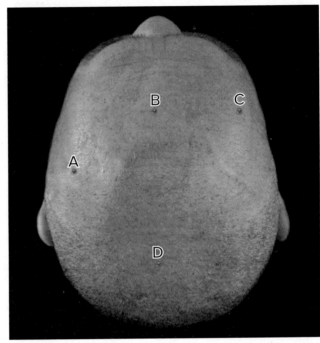

©McGraw-Hill Education

Mouse-over the pins on the screen to find the information necessary to identify the following structures:

A. _____
B. _____
C. _____
D. _____

CHECK POINT

Brain, Superior View

1. Name the spinal nerve that innervates the posterior scalp.
2. Name the nerve that innervates the skin over the temple and the anterior portion of the external ear.
3. Name the nerve that innervates the skin over the anterior scalp and forehead.

Click **LAYER 2** *in the* **LAYER CONTROLS** *window, and you will see the following image:*

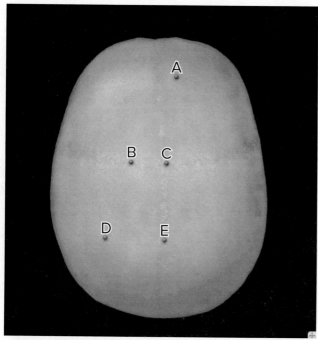

©McGraw-Hill Education

Mouse-over the pins on the screen to find the information necessary to identify the following structures:

A. _____
B. _____
C. _____
D. _____
E. _____

Click **LAYER 3** *in the* **LAYER CONTROLS** *window, and you will see the following image:*

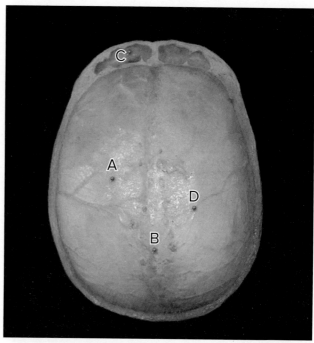

©McGraw-Hill Education

Mouse-over the pins on the screen to find the information necessary to identify the following structures:

A. _____
B. _____

Nonnervous System Structures (blue pins)

C. _____
D. _____

CHECK POINT

Brain, Superior View, *continued*

4. Name the structure that allows the return of cerebrospinal fluid to the venous circulation.
5. Name an artery that courses between the dura mater and the cranium.

364 MODULE 7 The Nervous System

Click **LAYER 4** *in the* **LAYER CONTROLS** *window, and you will see the following image:*

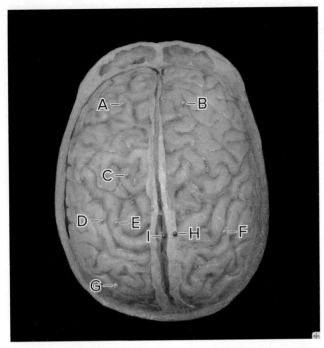

©McGraw-Hill Education

Mouse-over the pins on the screen to find the information necessary to identify the following structures:

A. _____
B. _____
C. _____
D. _____
E. _____
F. _____
G. _____
H. _____

Nonnervous System Structure (blue pin)

I. _____

CHECK POINT

Brain, Superior View, *continued*

6. Name an unpaired dural venous sinus that terminates at the confluence of sinuses.
7. Name the structure that is also called the primary motor cortex.
8. Name the structure that is also called the primary somatosensory cortex.

Click **LAYER 5** *in the* **LAYER CONTROLS** *window, and you will see the following image:*

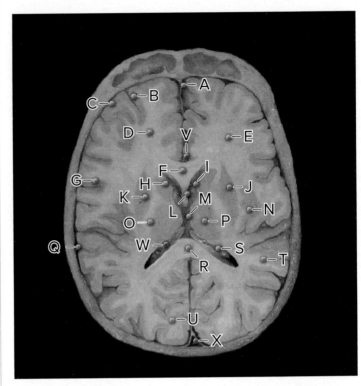

©McGraw-Hill Education

Mouse-over the pins on the screen to find the information necessary to identify the following structures:

A. _____
B. _____
C. _____
D. _____
E. _____
F. _____
G. _____
H. _____
I. _____
J. _____
K. _____
L. _____
M. _____
N. _____
O. _____
P. _____
Q. _____
R. _____
S. _____
T. _____
U. _____

Nonnervous System Structures (blue pins)

V. _____

W. _____

X. _____

CHECK POINT

Brain, Superior View, *continued*

9. The cerebral ventricles are lined with tufts of capillaries covered by specialized ependymal cells. What are these tufts called?
10. Both the superior sagittal sinus and the inferior sagittal sinus are located in the margins of the _____.
11. Name the midline cavity that separates the right and left halves of the diencephalon.

Click **LAYER 6** *in the* **LAYER CONTROLS** *window, and you will see the following image:*

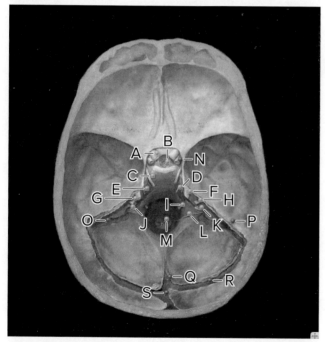

©McGraw-Hill Education

Mouse-over the pins on the screen to find the information necessary to identify the following structures:

A. _____

B. _____

C. _____

D. _____

E. _____

F. _____

G. _____

H. _____

I. _____

J. _____

K. _____

L. _____

M. _____

Nonnervous System Structures (blue pins)

N. _____

O. _____

P. _____

Q. _____

R. _____

S. _____

CHECK POINT

Brain, Superior View, *continued*

12. Name the point of attachment of the pituitary gland to the hypothalamus.
13. Name the S-shaped groove on the inner aspect of the temporal bone.
14. Name the cranial nerve that controls the superior oblique muscle.

Self-Quiz

Take this opportunity to check your progress by taking the **QUIZ**. See the **Introduction Module** for a reminder on how to access the **QUIZ** for this Study Area.

SELECT ANIMATION
Brain Ventricles Fly-through (3D) — PLAY

After viewing the animation, answer the following questions:

1. Name the four ventricles of the brain.

2. Describe the lateral ventricles. Where are they located?

3. Describe the third ventricle. How does it connect to the lateral ventricles?

4. Describe the cerebral aqueduct.

5. Describe the fourth ventricle. Where is it located?

6. Where is CSF located? How and where is it produced?

7. Describe the circulation of CSF.

SELECT ANIMATION
CSF Flow — PLAY

After viewing the animation, answer the following questions:

1. In what brain structures would you expect to find (CSF)?

2. Where is this CSF produced?

3. What structure produces the CSF?

4. Beginning in the lateral ventricles, trace the flow of CSF.

5. What are arachnoid granulations?

SELECT ANIMATION
Dural Sinus Blood Flow — PLAY

After viewing the animation, answer the following questions:

1. What are the dural venous sinuses?

2. Where are they located?

3. Name the two dural sinuses located along the midline.

4. Name the three sinuses that unite at the confluence of sinuses.

5. What vessels do the sigmoid sinuses become?

EXERCISE 7.10:

Nervous System—Brain—Inferior View

SELECT TOPIC: **Brain** ▸ SELECT VIEW: **Inferior**

Click **LAYER 1** *in the* **LAYER CONTROLS** *window, and you will see the following image:*

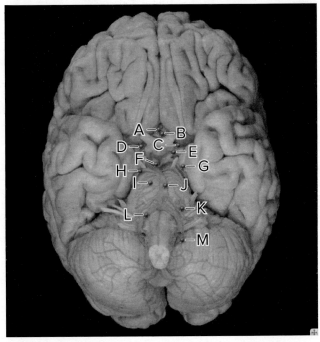

©McGraw-Hill Education

Mouse-over the pins on the screen to find the information necessary to identify the following nonnervous system structures:

A. _____
B. _____
C. _____
D. _____
E. _____
F. _____
G. _____
H. _____
I. _____
J. _____
K. _____
L. _____
M. _____

Mouse-over the pins on the screen to find the information necessary to identify the following structures:

A. _____
B. _____
C. _____
D. _____
E. _____
F. _____
G. _____
H. _____
I. _____
J. _____
K. _____
L. _____
M. _____
N. _____
O. _____
P. _____
Q. _____
R. _____
S. _____
T. _____
U. _____
V. _____
W. _____
X. _____
Y. _____

CHECK POINT

Brain, Inferior View

1. Name the circular anastomosis on the ventral surface of the brain which also referred to as the "Circle of Willis."
2. Name the artery that passes through the transverse foramina of the cervical vertebrae.
3. Name the unpaired midline artery that ascends on the anterior surface of the pons.

Click **LAYER 2** *in the* **LAYER CONTROLS** *window, and you will see the following image:*

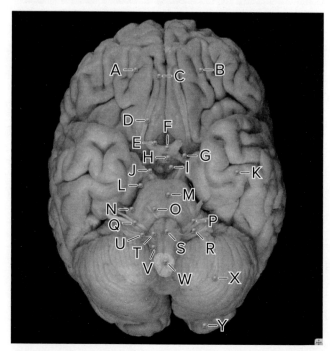

©McGraw-Hill Education

CHECK POINT

Brain, Inferior View, *continued*

4. Name the crossing white-matter tract between the optic nerve and the optic tracts.
5. Name the brain structure whose name means bridge.
6. Name the most caudal portion of the brain. What are its functions?

Self-Quiz

Take this opportunity to check your progress by taking the **QUIZ**. See the **Introduction Module** for a reminder on how to access the **QUIZ** for this Study Area.

EXERCISE 7.11:
Nervous System—Brain—Inferior View (close-up)

SELECT TOPIC	SELECT VIEW
Brain	Inferior (close-up)

Click **LAYER 1** *in the* **LAYER CONTROLS** *window, and you will see the following image:*

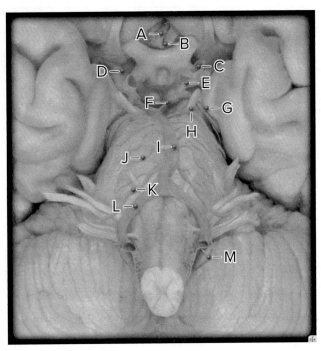

©McGraw-Hill Education

Mouse-over the pins on the screen to find the information necessary to identify the following nonnervous system structures:

A. _____
B. _____
C. _____
D. _____
E. _____
F. _____
G. _____
H. _____
I. _____
J. _____
K. _____
L. _____
M. _____

CHECK POINT
Brain, Inferior View (close-up)

1. Name the artery whose significant branches include the ophthalmic, anterior cerebral, and middle cerebral arteries.
2. Name the artery whose major branches include the pontine and superior cerebellar arteries.
3. Name the artery whose numerous branches course laterally across the surface of the pons.

Click **LAYER 2** *in the* **LAYER CONTROLS** *window, and you will see the following image:*

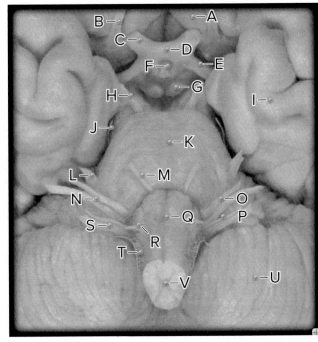

©McGraw-Hill Education

Mouse-over the pins on the screen to find the information necessary to identify the following structures:

A. _____
B. _____
C. _____
D. _____
E. _____
F. _____
G. _____
H. _____
I. _____
J. _____

K. _____
L. _____
M. _____
N. _____
O. _____
P. _____
Q. _____
R. _____
S. _____
T. _____
U. _____
V. _____

CHECK POINT

Brain, Inferior View (close-up), *continued*

4. Name the paired, small, rounded projections of the hypothalamus involved in the regulation of autonomic functions, emotional behavior, and memory.
5. Name the "funnel-shaped" extension of the floor of the third ventricle that continues as the pituitary stalk.
6. Name the structure that gives rise to eight pairs of spinal nerves and contains the cervical enlargement.

Self-Quiz
Take this opportunity to check your progress by taking the **QUIZ.** See the **Introduction Module** for a reminder on how to access the **QUIZ** for this Study Area.

EXERCISE 7.12a:
Imaging—Brain

SELECT TOPIC	SELECT VIEW
Brain	MRI—Axial 1

Click the **TAGS ON/OFF** *button, and you will see the following image:*

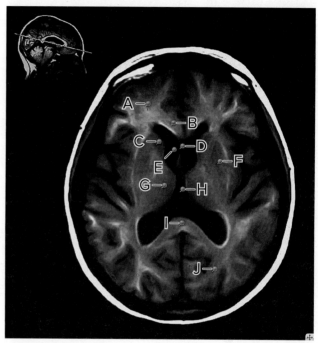

©McGraw-Hill Education

Mouse-over the pins on the screen to find the information necessary to identify the following structures:

A. _____
B. _____
C. _____
D. _____
E. _____
F. _____
G. _____
H. _____
I. _____
J. _____

370 MODULE 7 The Nervous System

EXERCISE 7.12b:
Imaging—Brain

SELECT TOPIC	SELECT VIEW
Brain	MRI—Axial 2

Click the **TAGS ON/OFF** *button, and you will see the following image:*

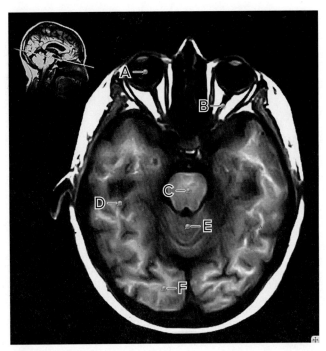

©McGraw-Hill Education

Mouse-over the pins on the screen to find the information necessary to identify the following structures:

A. _____
B. _____
C. _____
D. _____
E. _____
F. _____

EXERCISE 7.12c:
Imaging—Brain

SELECT TOPIC	SELECT VIEW
Brain	Sagittal

Click the **TAGS ON/OFF** *button, and you will see the following image:*

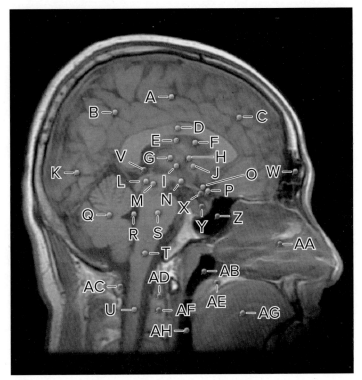

©McGraw-Hill Education

Mouse-over the pins on the screen to find the information necessary to identify the following structures:

A. _____
B. _____
C. _____
D. _____
E. _____
F. _____
G. _____
H. _____
I. _____
J. _____
K. _____
L. _____
M. _____
N. _____
O. _____
P. _____
Q. _____

R. _____
S. _____
T. _____
U. _____

Nonnervous System Structures (blue pins)

V. _____
W. _____
X. _____
Y. _____
Z. _____
AA. _____
AB. _____
AC. _____
AD. _____
AE. _____
AF. _____
AG. _____
AH. _____

EXERCISE 7.12d:
Imaging—Brain

SELECT TOPIC: **Brain** ▸ SELECT VIEW: **MRI—Coronal**

Click the **TAGS ON/OFF** button, and you will see the following image:

Mouse-over the pins on the screen to find the information necessary to identify the following structures:

A. _____
B. _____
C. _____
D. _____
E. _____
F. _____
G. _____
H. _____
I. _____
J. _____
K. _____
L. _____
M. _____
N. _____
O. _____
P. _____
Q. _____

Nonnervous System Structures (blue pins)

R. _____
S. _____
T. _____

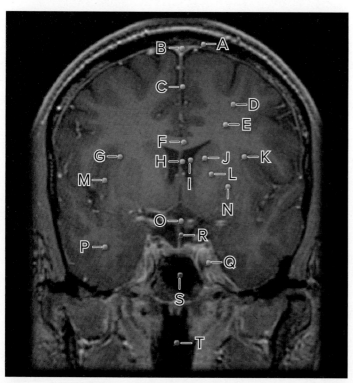

©McGraw-Hill Education

IN REVIEW

What Have I Learned?

The following questions cover the material that you have just learned, the brain. Apply what you have learned to answer these questions on a separate piece of paper.

1. Name the division of the brain that includes the mid-brain, pons, and medulla oblongata.

2. Name the division of the brain that coordinates complex movements and smooths muscle contractions.

3. The primary hearing and smell areas are located in which lobe of the brain?

4. Memory is located in which lobe of the brain?

5. Speech perception and recognition areas are located in which hemisphere of which lobe of the brain?

6. What is the term that refers to the superficial gray matter of the cerebrum?

7. The reception of general sensory information from the body occurs in which lobe of the brain?

8. Tactile object recognition occurs in which lobe of the brain?

9. Language and verbatim repetition of terms occurs in which lobe of which hemisphere of the brain?

10. Name the structure that forms the floor of the longitudinal fissure.

11. Name the paired cavities containing cerebrospinal fluid in the brain.

12. Where is the falx cerebri located?

13. Name the crossing white-matter tracts between the optic nerves and the optic tracts.

14. Name two structures responsible for planning and execution of movement, muscle tone, and posture.

15. Name the slitlike, fluid-filled cavity that separates the right and left halves of the diencephalon.

16. Name the site of emotional behavior.

17. Name the structure that regulates body temperature, eating, and drinking.

18. Name the structure that controls the autonomic nervous system.

19. Name the lobe that is the primary visual area.

20. Name the lobe that controls voluntary motor activity.

21. Name the lobe that is the site of higher mental processing.

22. Name the structure continuous with the brain at the foramen magnum.

Cranial Nerves

EXERCISE 7.13:
Nervous System—Cranial Nerves—Inferior Brain (CN I-XII)

SELECT TOPIC: **Cranial Nerves** ▸ SELECT VIEW: **Inferior Brain (CN I-XII)**

Click **LAYER 1** in the **LAYER CONTROLS** window, and you will see the image to the right:

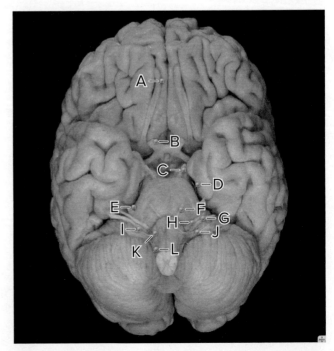

©McGraw-Hill Education

Mouse-over the pins on the screen to find the information necessary to identify the following structures:

A. _____
B. _____
C. _____
D. _____
E. _____
F. _____
G. _____
H. _____
I. _____
J. _____
K. _____
L. _____

EXERCISE 7.14:

Nervous System—Cranial Nerves—CN I Olfactory

SELECT TOPIC	SELECT VIEW
Cranial Nerves	CN I Olfactory

Click **LAYER 3** in the **LAYER CONTROLS** window, and you will see the following image:

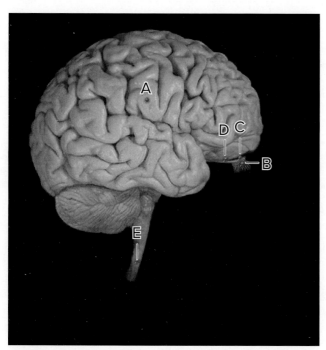

©McGraw-Hill Education

Mouse-over the pins on the screen to find the information necessary to identify the following structures:

A. _____
B. _____
C. _____
D. _____
E. _____

Click **LAYER 4** in the **LAYER CONTROLS** window, and you will see the following image:

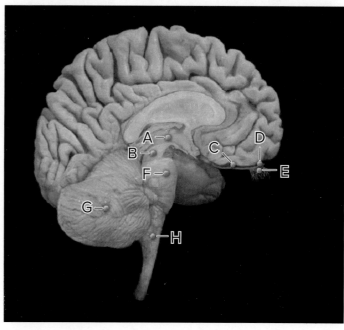

©McGraw-Hill Education

Mouse-over the pins on the screen to find the information necessary to identify the following structures:

A. _____
B. _____
C. _____
D. _____
E. _____
F. _____
G. _____
H. _____

EXERCISE 7.15:
Nervous System—Cranial Nerves—CN II Optic

SELECT TOPIC	SELECT VIEW
Cranial Nerves	CN II Optic

Click **LAYER 3** *in the* **LAYER CONTROLS** *window, and you will see the following image:*

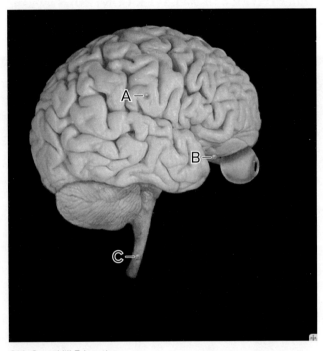

©McGraw-Hill Education

Mouse-over the pins on the screen to find the information necessary to identify the following structures:

A. _____
B. _____
C. _____

Click **LAYER 4** *in the* **LAYER CONTROLS** *window, and you will see the following image:*

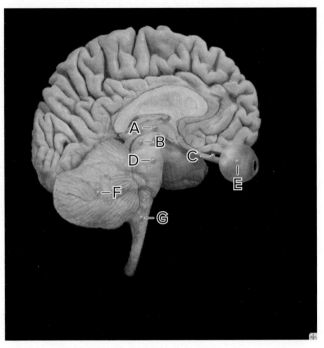

©McGraw-Hill Education

Mouse-over the pins on the screen to find the information necessary to identify the following structures:

A. _____
B. _____
C. _____
D. _____
E. _____
F. _____
G. _____

SELECT ANIMATION	
Vision	PLAY

After viewing the animation, answer the following questions:

1. What are the three structures involved in vision?

2. The optic nerve runs between which two structures?

3. Describe the two parts of the retina.

4. Describe the light pathway from the eye to the brain.

5. Name the structure where the optic nerves converge.

6. Images are perceived in which lobe of the brain?

EXERCISE 7.16:
Nervous System—Cranial Nerves—CN III Oculomotor

SELECT TOPIC	SELECT VIEW
Cranial Nerves	CN III Oculomotor

Click **LAYER 3** *in the* **LAYER CONTROLS** *window, and you will see the following image:*

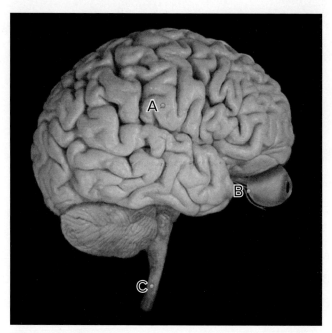

©McGraw-Hill Education

Mouse-over the pins on the screen to find the information necessary to identify the following structures:

A. _____
B. _____
C. _____

Click **LAYER 4** *in the* **LAYER CONTROLS** *window, and you will see the following image:*

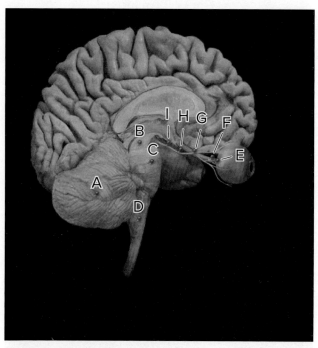

©McGraw-Hill Education

Mouse-over the pins on the screen to find the information necessary to identify the following structures:

A. _____
B. _____
C. _____
D. _____
E. _____
F. _____
G. _____
H. _____
I. _____

376 MODULE 7 The Nervous System

EXERCISE 7.17:
Nervous System—Cranial Nerves—CN IV Trochlear

SELECT TOPIC	SELECT VIEW
Cranial Nerves	CN IV Trochlear

Click **LAYER 4** *in the* **LAYER CONTROLS** *window, and you will see the following image:*

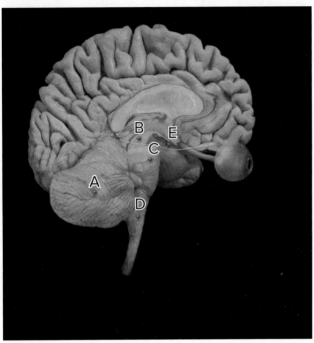

©McGraw-Hill Education

Mouse-over the pins on the screen to find the information necessary to identify the following structures:

A. _____
B. _____
C. _____
D. _____
E. _____

EXERCISE 7.18:
Nervous System—Cranial Nerves—CN V Trigeminal

SELECT TOPIC	SELECT VIEW
Cranial Nerves	CN V Trigeminal

Click **LAYER 3** *in the* **LAYER CONTROLS** *window, and you will see the following image:*

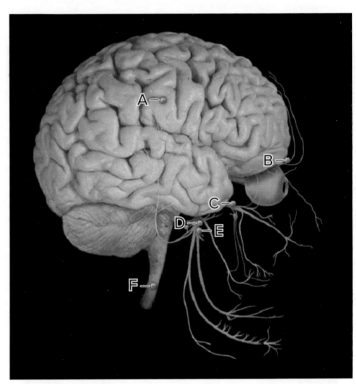

©McGraw-Hill Education

Mouse-over the pins on the screen to find the information necessary to identify the following structures:

A. _____
B. _____
C. _____
D. _____
E. _____
F. _____

Click **LAYER 4** *in the* **LAYER CONTROLS** *window, and you will see the following image:*

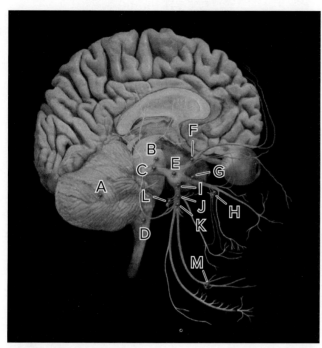

©McGraw-Hill Education

Mouse-over the pins on the screen to find the information necessary to identify the following structures:

A. _____
B. _____
C. _____
D. _____
E. _____
F. _____
G. _____
H. _____
I. _____
J. _____
K. _____
L. _____
M. _____

EXERCISE 7.19:
Nervous System—Cranial Nerves— CN VI Abducens

SELECT TOPIC	SELECT VIEW
Cranial Nerves	CN VI Abducens

Click **LAYER 3** *in the* **LAYER CONTROLS** *window, and you will see the following image:*

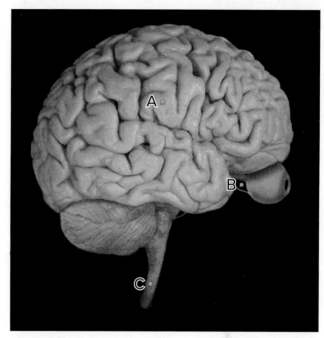

©McGraw-Hill Education

Mouse-over the pins on the screen to find the information necessary to identify the following structures:

A. _____
B. _____
C. _____

Click **LAYER 4** *in the* **LAYER CONTROLS** *window, and you will see the following image:*

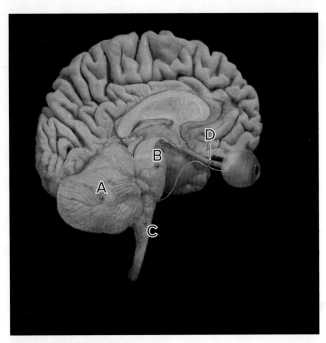

©McGraw-Hill Education

Mouse-over the pins on the screen to find the information necessary to identify the following structures:

A. _____

B. _____

C. _____

D. _____

EXERCISE 7.20:
Nervous System—Cranial Nerves— CN VII Facial

SELECT TOPIC	SELECT VIEW
Cranial Nerves	CN VII Facial

Click **LAYER 3** *in the* **LAYER CONTROLS** *window, and you will see the following image:*

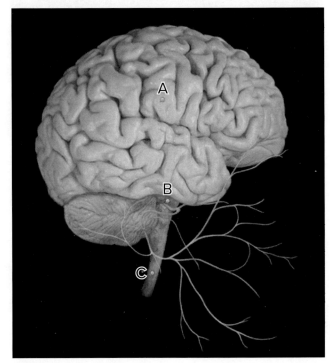

©McGraw-Hill Education

Mouse-over the pins on the screen to find the information necessary to identify the following structures:

A. _____

B. _____

C. _____

Click **LAYER 4** *in the* **LAYER CONTROLS** *window, and you will see the following image:*

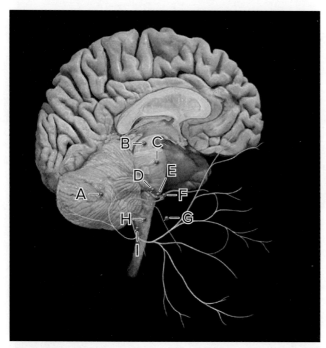

©McGraw-Hill Education

Mouse-over the pins on the screen to find the information necessary to identify the following structures:

A. _____
B. _____
C. _____
D. _____
E. _____
F. _____
G. _____
H. _____
I. _____

EXERCISE 7.21:
Nervous System—Cranial Nerves—CN VIII Vestibulocochlear

SELECT TOPIC	SELECT VIEW
Cranial Nerves	CN VIII Vestibulocochlear

Click **LAYER 3** *in the* **LAYER CONTROLS** *window, and you will see the following image:*

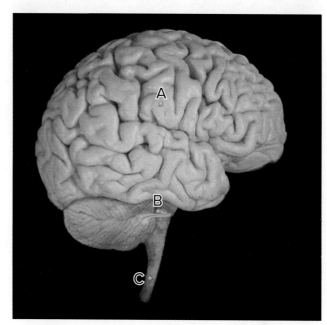

©McGraw-Hill Education

Mouse-over the pins on the screen to find the information necessary to identify the following structures:

A. _____
B. _____
C. _____

Click **LAYER 4** *in the* **LAYER CONTROLS** *window, and you will see the following image:*

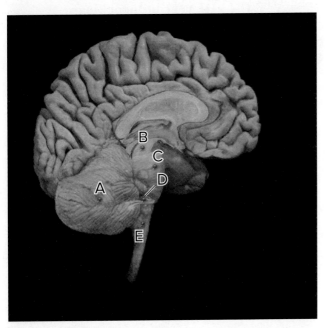

©McGraw-Hill Education

Mouse-over the pins on the screen to find the information necessary to identify the following structures:

A. _____

B. _____

C. _____

D. _____

E. _____

After viewing the animation, answer the following questions:

1. What structure do sound waves strike and cause to vibrate?

2. The vibrations are transferred to the three bones of the middle ear. From lateral to medial, what are those bones?

3. What structure transfers this vibration to the oval window?

4. The vibrations are then transferred to a fluid-filled chamber of the inner ear. What is that fluid, and what is the name of that chamber?

5. What is the difference in location for the detection of high-pitched and low-pitched sounds?

EXERCISE 7.22:

Nervous System—Cranial Nerves— CN IX Glossopharyngeal

 SELECT TOPIC: Cranial Nerves ▸ SELECT VIEW: CN IX Glossopharyngeal

Click **LAYER 3** *in the* **LAYER CONTROLS** *window, and you will see the following image:*

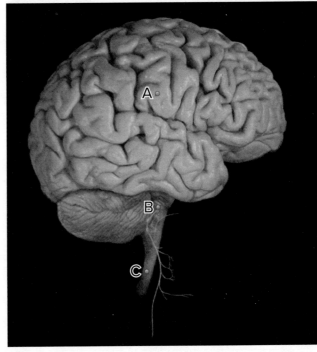

©McGraw-Hill Education

Mouse-over the pins on the screen to find the information necessary to identify the following structures:

A. _____

B. _____

C. _____

Click **LAYER 4** *in the* **LAYER CONTROLS** *window, and you will see the following image:*

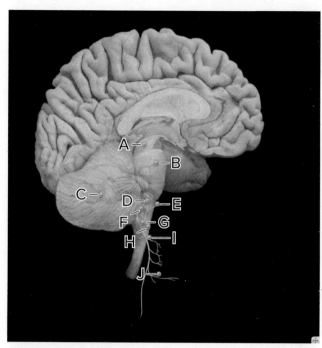

©McGraw-Hill Education

Mouse-over the pins on the screen to find the information necessary to identify the following structures:

A. _____
B. _____
C. _____
D. _____
E. _____
F. _____
G. _____
H. _____
I. _____
J. _____

EXERCISE 7.23:
Nervous System—Cranial Nerves—CN X Vagus

SELECT TOPIC	SELECT VIEW
Cranial Nerves	CN X Vagus

Click **LAYER 3** *in the* **LAYER CONTROLS** *window, and you will see the following image:*

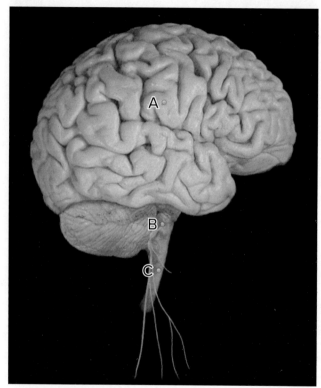

©McGraw-Hill Education

Mouse-over the pins on the screen to find the information necessary to identify the following structures:

A. _____
B. _____
C. _____

Click **LAYER 4** *in the* **LAYER CONTROLS** *window, and you will see the following image:*

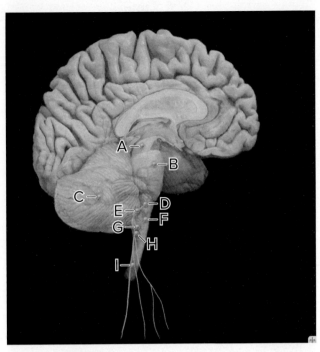

©McGraw-Hill Education

Mouse-over the pins on the screen to find the information necessary to identify the following structures:

A. _____
B. _____
C. _____
D. _____
E. _____
F. _____
G. _____
H. _____
I. _____

EXERCISE 7.24:
Cranial Nerves, CN XI Accessory

SELECT TOPIC	SELECT VIEW
Cranial Nerves	CN XI Accessory

Click **LAYER 3** *in the* **LAYER CONTROLS** *window, and you will see the following image:*

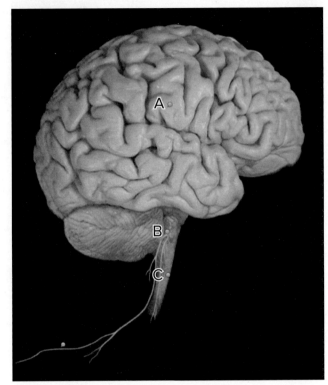

©McGraw-Hill Education

Mouse-over the pins on the screen to find the information necessary to identify the following structures:

A. _____
B. _____
C. _____

Click **LAYER 4** *in the* **LAYER CONTROLS** *window, and you will see the following image:*

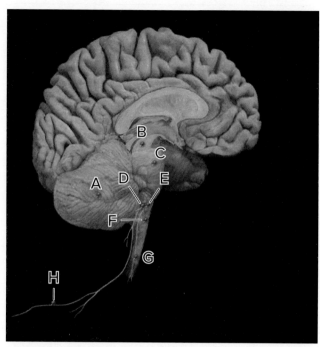

©McGraw-Hill Education

Mouse-over the pins on the screen to find the information necessary to identify the following structures:

A. _____
B. _____
C. _____
D. _____
E. _____
F. _____
G. _____
H. _____

EXERCISE 7.25:
Nervous System—Cranial Nerves—CN XII Hypoglossal

SELECT TOPIC	SELECT VIEW
Cranial Nerves	CN XII Hypoglossal

Click **LAYER 3** *in the* **LAYER CONTROLS** *window, and you will see the following image:*

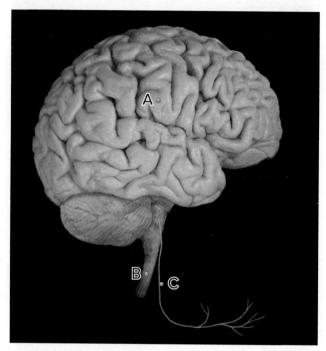

©McGraw-Hill Education

Mouse-over the pins on the screen to find the information necessary to identify the following structures:

A. _____
B. _____
C. _____

Click **LAYER 4** *in the* **LAYER CONTROLS** *window, and you will see the following image:*

Mouse-over the pins on the screen to find the information necessary to identify the following structures:

A. _____

B. _____

C. _____

D. _____

E. _____

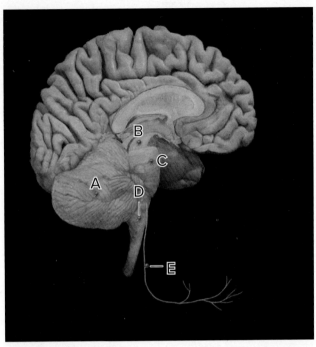

©McGraw-Hill Education

IN REVIEW

What Have I Learned?

The following questions cover the material that you have just learned, the cranial nerves. Apply what you have learned to answer these questions on a separate piece of paper.

1. Which cranial nerve is composed of the ophthalmic, the maxillary, and the mandibular nerves?

2. Name the cranial nerve responsible for the intrinsic and extrinsic muscles of the tongue.

3. Which cranial nerve has sensory fibers that monitor blood pressure at the carotid sinus?

4. Which cranial nerve is responsible for the constriction of the pupillary sphincter and accommodation of the lens for near vision?

5. Which cranial nerve is responsible for taste from the anterior two-thirds of the tongue and the muscles of facial expression?

6. Which cranial nerve has olfactory neurons on the mucosa of the anterosuperior nasal cavity?

7. Which cranial nerve controls the lateral rectus muscle of the eye?

8. Which cranial nerve is involved with hearing and balance?

9. Which cranial nerve is responsible for vision?

10. Which cranial nerve controls the extraocular muscle of the superior oblique?

11. Which cranial nerve controls all but one of the muscles of the palate, the pharynx, and the intrinsic muscles of the larynx?

12. Which cranial nerve is the only one that extends beyond the head and neck?

SELECT ANIMATION
Reflex Arc
PLAY

After viewing the animation, answer the following questions:

1. List the two components of a simple reflex arc.

2. How is the reflex initiated? Where are these located?

3. Stimulation of a sensory receptor results in _____.

4. Where do these axons end?

5. Where do these neurons synapse with a motor neuron?

6. Where does the motor neuron carry the efferent impulse?

7. The reflex functions to _____.

8. The response takes place before _____.

Spinal Cord

EXERCISE 7.26:
Nervous System—Spinal Cord—Overview

SELECT TOPIC	SELECT VIEW
Spinal Cord	Overview

Click **LAYER 1** *in the* **LAYER CONTROLS** *window, and you will see the following image:*

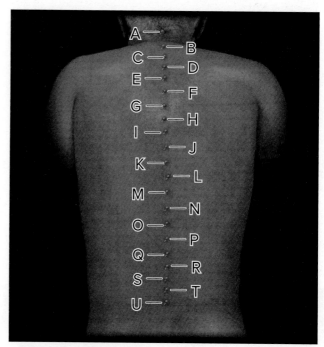

©McGraw-Hill Education

Mouse-over the pins on the screen to find the information necessary to identify the following structures:

A. _____
B. _____
C. _____
D. _____
E. _____
F. _____
G. _____
H. _____
I. _____
J. _____

386 MODULE 7 The Nervous System

K. _____
L. _____
M. _____
N. _____
O. _____
P. _____
Q. _____
R. _____
S. _____
T. _____
U. _____

Nonnervous System Structures (blue pins)

C. _____
D. _____
E. _____
F. _____
G. _____
H. _____
I. _____
J. _____
K. _____
L. _____
M. _____

CHECK POINT

Spinal Cord, Overview

1. Describe a ventral ramus. Where are ventral rami located?
2. List the general sensation and motor regions for the subcostal nerve.

CHECK POINT

Spinal Cord, Overview, *continued*

3. What is a dermatome?
4. Which dermatome includes the skin of the foot, including the middle three toes?
5. Which dermatome includes the skin over the knee and on the medial foot, including the great toe?

Click **LAYER 2** *in the* **LAYER CONTROLS** *window, and you will see the following image:*

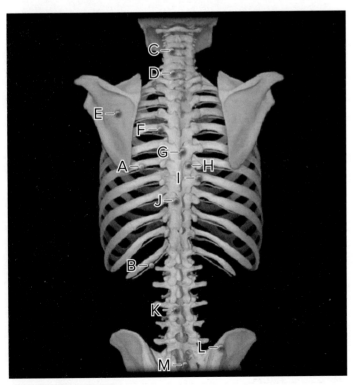

©McGraw-Hill Education

Click **LAYER 3** *in the* **LAYER CONTROLS** *window, and you will see the following image:*

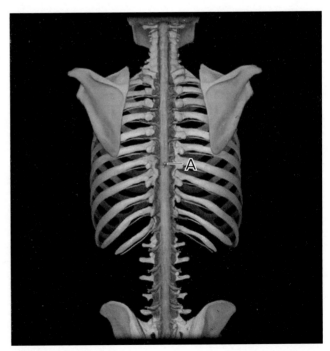

©McGraw-Hill Education

Mouse-over the pins on the screen to find the information necessary to identify the following structures:

A. _____
B. _____

Mouse-over the pin on the screen to find the information necessary to identify the following nonnervous system structure:

A. _____

CHECK POINT

Spinal Cord, Overview, *continued*

6. Name the outermost tough connective tissue that surrounds the brain and spinal cord.

Click **LAYER 4** *in the* **LAYER CONTROLS** *window, and you will see the following image:*

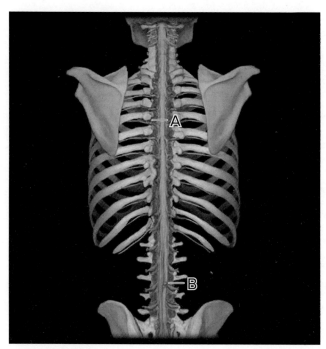

©McGraw-Hill Education

Mouse-over the pins on the screen to find the information necessary to identify the following structures:

A. _____

B. _____

CHECK POINT

Spinal Cord, Overview, *continued*

7. Name the large bundle of dorsal and ventral roots for spinal nerves below L2.
8. Name the structure that contains the sensory ganglion for each dorsal root.
9. Name the structure whose Latin name means horse tail.

Click **LAYER 5** *in the* **LAYER CONTROLS** *window, and you will see the following image:*

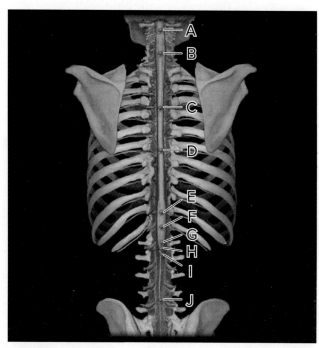

©McGraw-Hill Education

Mouse-over the pins on the screen to find the information necessary to identify the following structures:

A. _____

B. _____

C. _____

D. _____

E. _____

F. _____

G. _____

H. _____

I. _____

J. _____

CHECK POINT

Spinal Cord, Overview, *continued*

10. What is the cervical enlargement?
11. Name the structure that contains neurons for lower-limb innervation.
12. Name the tapered inferior end of the spinal cord.

EXERCISE 7.27:

Nervous System—Spinal Cord—Typical Spinal Nerve

SELECT TOPIC	SELECT VIEW
Spinal Cord	▶ Typical Spinal Nerve

Click **LAYER 1** *in the* **LAYER CONTROLS** *window, and you will see the following image:*

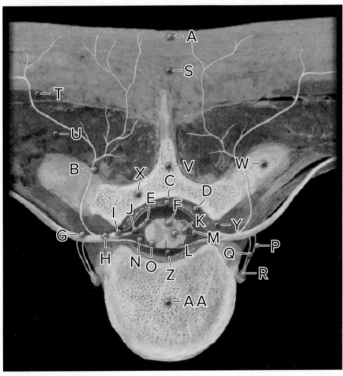

©McGraw-Hill Education

Mouse-over the pins on the screen to find the information necessary to identify the following structures:

A. _____
B. _____
C. _____
D. _____
E. _____
F. _____
G. _____
H. _____
I. _____
J. _____
K. _____
L. _____
M. _____
N. _____
O. _____
P. _____
Q. _____
R. _____

Nonnervous System Structures (blue pins)

S. _____
T. _____
U. _____
V. _____
W. _____
X. _____
Y. _____
Z. _____
AA. _____

CHECK POINT

Spinal Cord, Typical Spinal Nerve

1. In what respect is the spinal cord structurally opposite to the brain?
2. What structure makes up the core of the spinal cord?
3. Name the structure that consists of ascending and descending bundles of myelinated axons.

MODULE 7 The Nervous System 389

SELECT ANIMATION
Typical Spinal Nerve   PLAY

After viewing the animation, answer the following questions:

1. How many spinal nerves exist for each vertebral level?

2. What structures connect each spinal nerve to the spinal cord?

3. What structure is made up of bundles of nerve fibers carrying sensory information from the skin to the spinal cord?

4. Where are the cell bodies (somas) of these sensory nerve fibers located?

5. Name the structure consisting of bundles of motor (efferent) fibers carrying impulses away from the spinal cord to the skeletal muscles.

6. Where are the cell bodies (somas) associated with these nerve fibers located?

7. What structures unite to form the spinal nerve?

8. The spinal nerves exit the vertebral column through what structure?

9. What is a mixed nerve?

10. Name the two branches that form from each spinal nerve.

11. What structures innervate the muscles and skin of the back?

12. What structures innervate the muscles and skin of the lateral and ventral trunk and the limbs?

EXERCISE 7.28:
Nervous System—Spinal Cord—Cervical Region

SELECT TOPIC Spinal Cord ▶ **SELECT VIEW** Cervical Region

Click **LAYER 1** *in the* **LAYER CONTROLS** *window, and you will see the following image:*

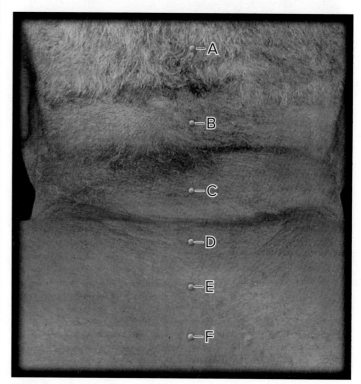

©McGraw-Hill Education

Mouse-over the pins on the screen to find the information necessary to identify the following structures:

A. _____

B. _____

C. _____

D. _____

E. _____

F. _____

390 MODULE 7 The Nervous System

Click **LAYER 3** *in the* **LAYER CONTROLS** *window, and you will see the following image:*

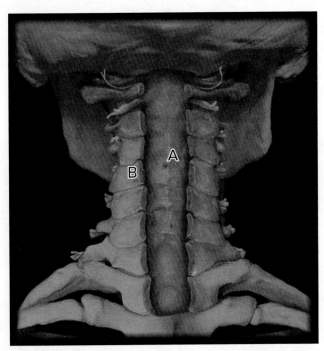
©McGraw-Hill Education

Mouse-over the pin on the screen to find the information necessary to identify the following structure:

A. _____

Nonnervous System Structure (blue pin)

B. _____

Click **LAYER 4** *in the* **LAYER CONTROLS** *window, and you will see the following image:*

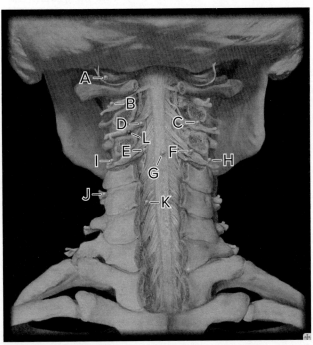
©McGraw-Hill Education

Mouse-over the pins on the screen to find the information necessary to identify the following structures:

A. _____
B. _____
C. _____
D. _____
E. _____
F. _____
G. _____
H. _____
I. _____
J. _____
K. _____

Nonnervous System Structure (blue pin)

L. _____

CHECK POINT

Spinal Cord, Cervical Region

1. Name the series of small nerves branching from the dorsal length of the spinal cord.
2. Name the afferent (sensory) limb of each spinal nerve.
3. Name the two components of each spinal nerve.

MODULE 7 The Nervous System 391

EXERCISE 7.29:
Nervous System—Spinal Cord—Thoracic Region

SELECT TOPIC	SELECT VIEW
Spinal Cord	Thoracic Region

Click **LAYER 1** *in the* **LAYER CONTROLS** *window, and you will see the following image:*

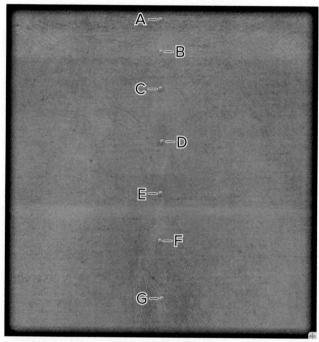

©McGraw-Hill Education

Mouse-over the pins on the screen to find the information necessary to identify the following structures:

A. _____
B. _____
C. _____
D. _____
E. _____
F. _____
G. _____

Click **LAYER 2** *in the* **LAYER CONTROLS** *window, and you will see the following image:*

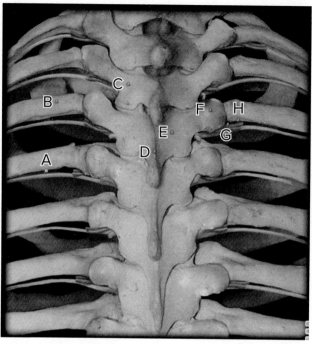

©McGraw-Hill Education

Mouse-over the pins on the screen to find the information necessary to identify the following structures:

A. _____
B. _____
C. _____
D. _____
E. _____
F. _____
G. _____
H. _____

Click **LAYER 3** *in the* **LAYER CONTROLS** *window, and you will see the following image:*

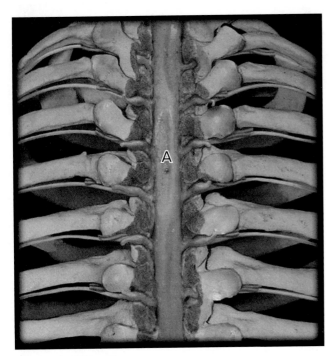

©McGraw-Hill Education

Mouse-over the pin on the screen to find the information necessary to identify the following structure:

A. _____

Click **LAYER 4** *in the* **LAYER CONTROLS** *window, and you will see the following image:*

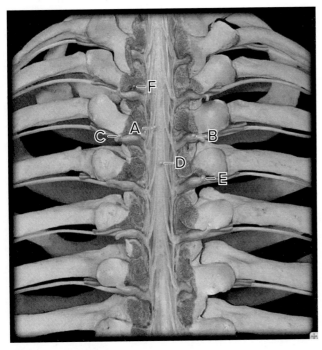

©McGraw-Hill Education

Mouse-over the pins on the screen to find the information necessary to identify the following structures:

A. _____
B. _____
C. _____
D. _____
E. _____

Nonnervous System Structure (blue pin)

F. _____

EXERCISE 7.30:
Nervous System—Spinal Cord— Lumbar Region

SELECT TOPIC	SELECT VIEW
Spinal Cord	Lumbar Region

Click **LAYER 1** *in the* **LAYER CONTROLS** *window, and you will see the following image:*

©McGraw-Hill Education

Mouse-over the pins on the screen to find the information necessary to identify the following structures:

A. _____
B. _____
C. _____
D. _____
E. _____
F. _____
G. _____

Click **LAYER 3** *in the* **LAYER CONTROLS** *window, and you will see the following image:*

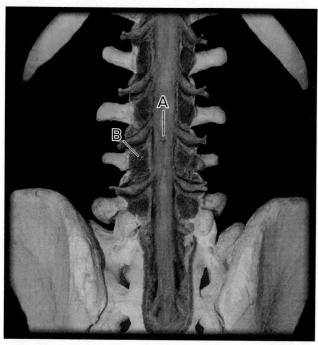

©McGraw-Hill Education

Mouse-over the pin on the screen to find the information necessary to identify the following structure:

A. _____

Nonnervous System Structure (blue pin)

B. _____

Click **LAYER 4** *in the* **LAYER CONTROLS** *window, and you will see the following image:*

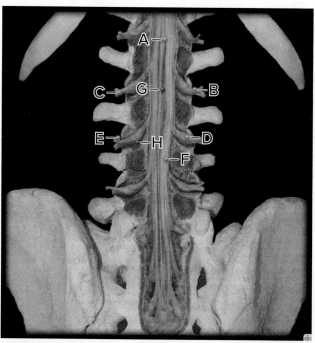

©McGraw-Hill Education

Mouse-over the pins on the screen to find the information necessary to identify the following structures:

A. _____
B. _____
C. _____
D. _____
E. _____
F. _____
G. _____

Nonnervous System Structure (blue pin)

H. _____

CHECK POINT

Spinal Cord, Lumbar Region

Name the structure of the pia mater caudal to the spinal cord.

Self-Quiz
Take this opportunity to check your progress by taking the **QUIZ**. See the **Introduction Module** for a reminder on how to access the **QUIZ** for this Study Area.

IN REVIEW

What Have I Learned?

The following questions cover the material that you have just learned, the spinal cord. Apply what you have learned to answer these questions on a separate piece of paper.

1. Name the structure that consists of a filament of pia mater that forms at the conus medullaris, passes through the sacral hiatus, and attaches to the coccyx.

2. In adults, where does the spinal cord end?

3. Name the structures that contribute to the lumbosacral enlargement.

4. Name the bony structure through which each spinal nerve passes.

5. Name the structure that is the sensory ganglion of each dorsal root.

6. Which terminal branch of the spinal nerves contains motor innervation to muscles of the suboccipital triangle?

7. What do the dorsal and ventral roots at the same spinal cord level unite to form?

Peripheral Nerves

EXERCISE 7.31:
Nervous System—Peripheral Nerves—Cervical Plexus

SELECT TOPIC	SELECT VIEW
Peripheral Nerves	Cervical Plexus

Click **LAYER 1** *in the* **LAYER CONTROLS** *window, and you will see the following image:*

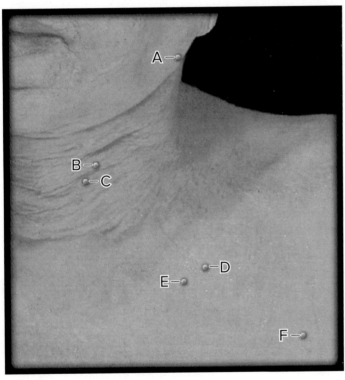

©McGraw-Hill Education

Mouse-over the pins on the screen to find the information necessary to identify the following structures:

A. _____
B. _____
C. _____
D. _____
E. _____
F. _____

Click **LAYER 2** *in the* **LAYER CONTROLS** *window, and you will see the following image:*

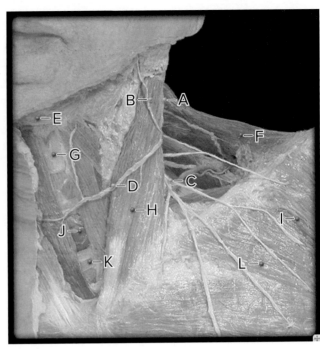

©McGraw-Hill Education

Mouse-over the pins on the screen to find the information necessary to identify the following structures:

A. _____
B. _____
C. _____
D. _____

Nonnervous System Structures (blue pins)

E. _____
F. _____
G. _____
H. _____
I. _____
J. _____
K. _____
L. _____

CHECK POINT
Peripheral Nerves, Cervical Plexus

1. Name the four sensory branches of the cervical plexus.
2. Which of these four innervates the skin over the anterior and lateral neck?
3. Which of these branches innervates the lateral scalp and posterior auricle of the ear?

MODULE 7   The Nervous System   395

Click **LAYER 3** *in the* **LAYER CONTROLS** *window, and you will see the following image:*

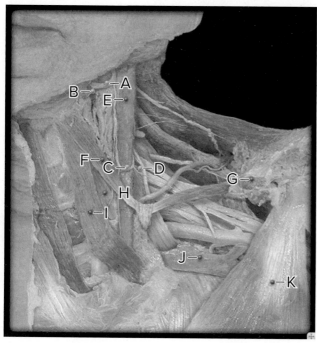
©McGraw-Hill Education

Mouse-over the pins on the screen to find the information necessary to identify the following structures:

A. _____
B. _____
C. _____
D. _____

Nonnervous System Structures (blue pins)

E. _____
F. _____
G. _____
H. _____
I. _____
J. _____
K. _____

CHECK POINT

Peripheral Nerves, Cervical Plexus, *continued*

4. Which nerve innervates the genioglossus, hyoglossus, styloglossus, and intrinsic muscles of the tongue?
5. Which nerve fibers "hitchhike" on this nerve?
6. Which nerve's Latin name means neck loop?

Click **LAYER 4** *in the* **LAYER CONTROLS** *window, and you will see the following image:*

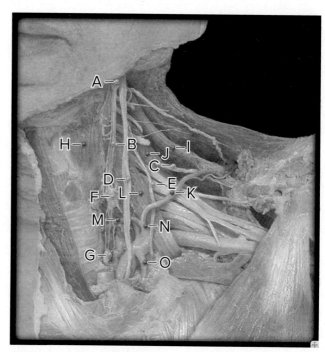
©McGraw-Hill Education

Mouse-over the pins on the screen to find the information necessary to identify the following structures:

A. _____
B. _____
C. _____
D. _____
E. _____
F. _____
G. _____

Nonnervous System Structures (blue pins)

H. _____
I. _____
J. _____
K. _____
L. _____
M. _____
N. _____
O. _____

CHECK POINT

Peripheral Nerves, Cervical Plexus, *continued*

7. Name the three sympathetic ganglia in the neck area.
8. Which of these distributes all postganglionic sympathetic nerve fibers to the head?

CHECK POINT

Peripheral Nerves, Cervical Plexus, *continued*

9. Name the structure that consists of nerves distributed to the upper limb.
10. What are the ventral rami of spinal nerves C1–C4 referred to as?

Click **LAYER 5** *in the* **LAYER CONTROLS** *window, and you will see the following image:*

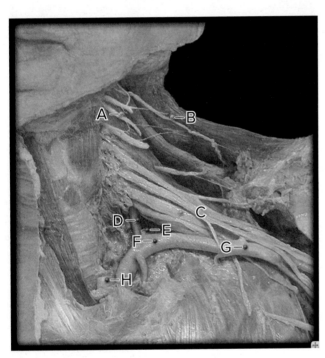

©McGraw-Hill Education

Mouse-over the pins on the screen to find the information necessary to identify the following structures:

A. _____
B. _____
C. _____

Nonnervous System Structures (blue pins)

D. _____
E. _____
F. _____
G. _____
H. _____

Self-Quiz

Take this opportunity to check your progress by taking the **QUIZ**. See the **Introduction Module** for a reminder on how to access the **QUIZ** for this Study Area.

EXERCISE 7.32:
Nervous System—Peripheral Nerves—Brachial Plexus

SELECT TOPIC	SELECT VIEW
Peripheral Nerves	Brachial Plexus

Click **LAYER 1** *in the* **LAYER CONTROLS** *window, and you will see the following image:*

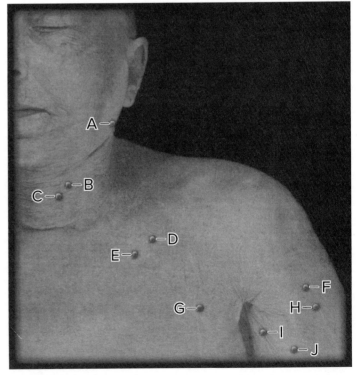

©McGraw-Hill Education

Mouse-over the pins on the screen to find the information necessary to identify the following structures:

A. _____
B. _____
C. _____

MODULE 7 The Nervous System 397

D. _____
E. _____
F. _____
G. _____
H. _____
I. _____
J. _____

CHECK POINT

Peripheral Nerves, Brachial Plexus

1. Name the nerve that supplies cutaneous innervation to the skin of and around the auricle of the ear.

Nonnervous System Structures (blue pins)

G. _____
H. _____
I. _____
J. _____
K. _____
L. _____
M. _____
N. _____
O. _____
P. _____

Click **LAYER 4** *in the* **LAYER CONTROLS** *window, and you will see the following image:*

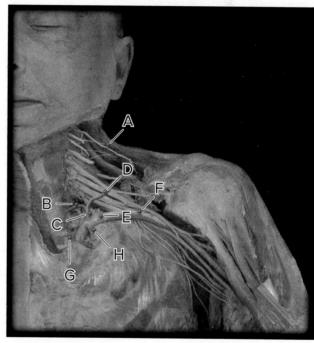

©McGraw-Hill Education

Mouse-over the pin on the screen to find the information necessary to identify the following structure:

A. _____

Nonnervous System Structures (blue pins)

B. _____
C. _____
D. _____
E. _____
F. _____
G. _____
H. _____

Click **LAYER 3** *in the* **LAYER CONTROLS** *window, and you will see the following image:*

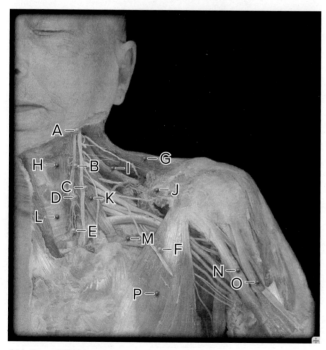

©McGraw-Hill Education

Mouse-over the pins on the screen to find the information necessary to identify the following structures:

A. _____
B. _____
C. _____
D. _____
E. _____
F. _____

398 MODULE 7 The Nervous System

Click **LAYER 5** *in the* **LAYER CONTROLS** *window, and you will see the following image:*

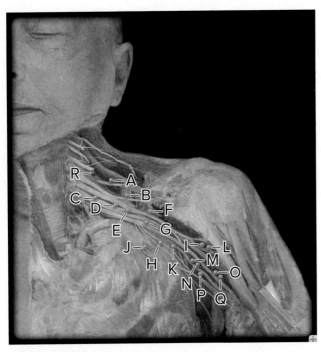

©McGraw-Hill Education

Mouse-over the pins on the screen to find the information necessary to identify the following structures:

A. _____
B. _____
C. _____
D. _____
E. _____
F. _____
G. _____
H. _____
I. _____
J. _____
K. _____
L. _____
M. _____
N. _____
O. _____
P. _____
Q. _____

Nonnervous System Structure (blue pin)

R. _____

CHECK POINT

Peripheral Nerves, Brachial Plexus, *continued*

2. Which nerve is stimulated when you strike your "funny bone"?
3. Name a nerve that innervates the pectoralis major and minor muscles.
4. Which brachial plexus roots contribute to the long thoracic nerve?

Click **LAYER 6** *in the* **LAYER CONTROLS** *window, and you will see the following image:*

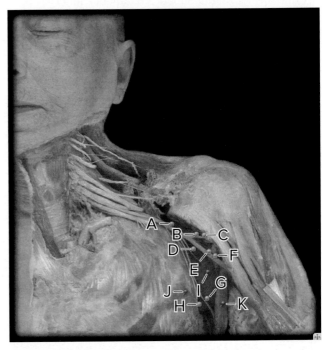

©McGraw-Hill Education

Mouse-over the pins on the screen to find the information necessary to identify the following structures:

A. _____
B. _____
C. _____
D. _____
E. _____
F. _____
G. _____
H. _____

Nonnervous System Structures (blue pins)

I. _____
J. _____
K. _____

Self-Quiz

Take this opportunity to check your progress by taking the **QUIZ**. See the **Introduction Module** for a reminder on how to access the **QUIZ** for this Study Area.

EXERCISE 7.33:

Nervous System—Peripheral Nerves—Upper Limb—Anterior View

SELECT TOPIC	SELECT VIEW
Peripheral Nerves	Upper Limb—Anterior

Click **LAYER 1** *in the* **LAYER CONTROLS** *window, and you will see the following image:*

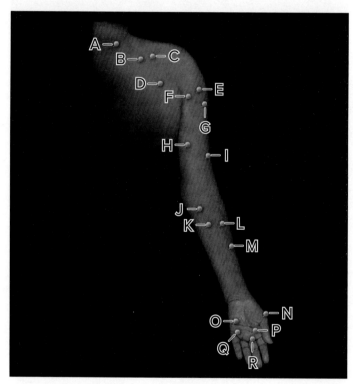

©McGraw-Hill Education

Mouse-over the pins on the screen to find the information necessary to identify the following structures:

A. _____
B. _____
C. _____
D. _____
E. _____
F. _____
G. _____
H. _____
I. _____
J. _____
K. _____
L. _____
M. _____
N. _____
O. _____
P. _____
Q. _____
R. _____

CHECK POINT

Peripheral Nerves, Upper Limb, Anterior View

1. Which nerve supplies the cutaneous innervation to the skin over the medial arm and forearm?
2. Which nerve supplies the cutaneous innervation to the skin over finger V?
3. Name two nerves responsible for the cutaneous innervation of the shoulder.

Click **LAYER 3** *in the* **LAYER CONTROLS** *window, and you will see the following image:*

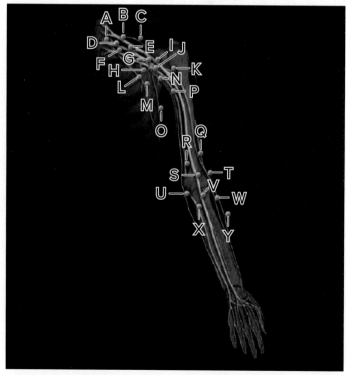

©McGraw-Hill Education

400 MODULE 7 The Nervous System

Mouse-over the pins on the screen to find the information necessary to identify the following structures:

A. _____
B. _____
C. _____
D. _____
E. _____
F. _____
G. _____
H. _____
I. _____
J. _____
K. _____
L. _____
M. _____
N. _____
O. _____
P. _____
Q. _____
R. _____
S. _____
T. _____
U. _____
V. _____
W. _____
X. _____
Y. _____

CHECK POINT

Peripheral Nerves, Upper Limb, Anterior View, *continued*

4. Name the nerve that innervates the joints of the hand.
5. Which division of the brachial plexus gives rise to nerves that distribute to the anterior aspect of the forearm?
6. Name the nerve responsible for the innervation of the glenohumeral joint.

Self-Quiz
Take this opportunity to check your progress by taking the **QUIZ**. See the **Introduction Module** for a reminder on how to access the **QUIZ** for this Study Area.

EXERCISE 7.34:
Nervous System—Peripheral Nerves—Upper Limb—Posterior View

SELECT TOPIC	SELECT VIEW
Peripheral Nerves	Upper Limb—Posterior

Click **LAYER 1** *in the* **LAYER CONTROLS** *window, and you will see the following image:*

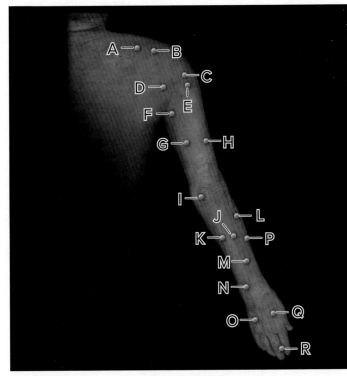

©McGraw-Hill Education

Mouse-over the pins on the screen to find the information necessary to identify the following structures:

A. _____
B. _____
C. _____
D. _____
E. _____
F. _____
G. _____
H. _____
I. _____
J. _____
K. _____
L. _____
M. _____
N. _____
O. _____

P. _____
Q. _____
R. _____

R. _____
S. _____
T. _____
U. _____
V. _____

Click **LAYER 3** *in the* **LAYER CONTROLS** *window, and you will see the following image:*

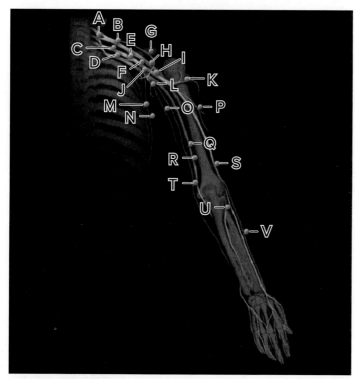

©McGraw-Hill Education

CHECK POINT

Peripheral Nerves, Upper Limb, Posterior View

1. Name a nerve that innervates the serratus anterior muscle.
2. Name a nerve that innervates the supraspinatus and infraspinatus muscles.
3. Name a nerve that innervates the subscapularis and teres major muscles.

Self-Quiz

Take this opportunity to check your progress by taking the **QUIZ**. See the **Introduction Module** for a reminder on how to access the **QUIZ** for this Study Area.

EXERCISE 7.35:

Nervous System—Peripheral Nerves— Shoulder and Arm—Anterior View

 SELECT TOPIC Peripheral Nerves ▶ **SELECT VIEW** Shoulder and Arm— Anterior

Click **LAYER 1** *in the* **LAYER CONTROLS** *window, and you will see the following image:*

Mouse-over the pins on the screen to find the information necessary to identify the following structures:

A. _____
B. _____
C. _____
D. _____
E. _____
F. _____
G. _____
H. _____
I. _____
J. _____
K. _____
L. _____
M. _____
N. _____
O. _____
P. _____
Q. _____

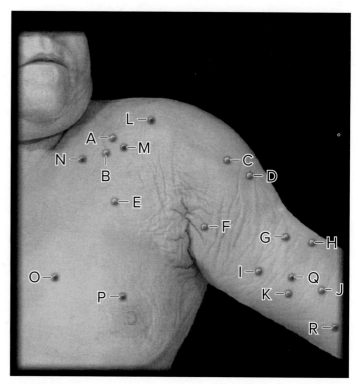

©McGraw-Hill Education

MODULE 7 The Nervous System **401**

402 MODULE 7 The Nervous System

Mouse-over the pins on the screen to find the information necessary to identify the following structures:

A. _____
B. _____
C. _____
D. _____
E. _____
F. _____
G. _____
H. _____
I. _____
J. _____
K. _____

Nonnervous System Structures (blue pins)

L. _____
M. _____
N. _____
O. _____
P. _____
Q. _____
R. _____

*Click **LAYER 3** in the **LAYER CONTROLS** window, and you will see the following image:*

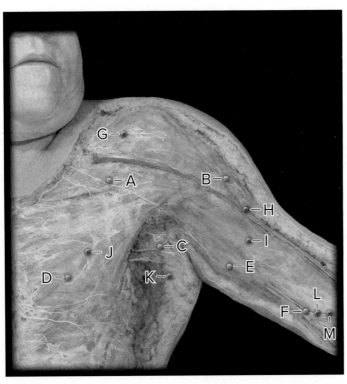

©McGraw-Hill Education

Mouse-over the pins on the screen to find the information necessary to identify the following structures:

A. _____
B. _____
C. _____
D. _____
E. _____
F. _____

Nonnervous System Structures (blue pins)

G. _____
H. _____
I. _____
J. _____
K. _____
L. _____
M. _____

*Click **LAYER 4** in the **LAYER CONTROLS** window, and you will see the following image:*

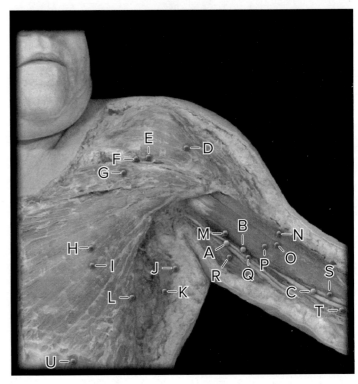

©McGraw-Hill Education

Mouse-over the pins on the screen to find the information necessary to identify the following structures:

A. _____
B. _____
C. _____

Nonnervous System Structures (blue pins)

D. _____
E. _____
F. _____
G. _____
H. _____
I. _____
J. _____
K. _____
L. _____
M. _____
N. _____
O. _____
P. _____
Q. _____
R. _____
S. _____
T. _____
U. _____

Click **LAYER 5** *in the* **LAYER CONTROLS** *window, and you will see the following image:*

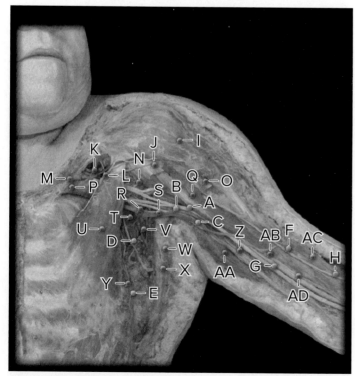

Mouse-over the pins on the screen to find the information necessary to identify the following structures:

A. _____
B. _____
C. _____
D. _____
E. _____
F. _____
G. _____
H. _____

Nonnervous System Structures (blue pins)

I. _____
J. _____
K. _____
L. _____
M. _____
N. _____
O. _____
P. _____
Q. _____
R. _____
S. _____
T. _____
U. _____
V. _____
W. _____
X. _____
Y. _____
Z. _____
AA. _____
AB. _____
AC. _____
AD. _____

404 MODULE 7 The Nervous System

Click **LAYER 6** in the **LAYER CONTROLS** window, and you will see the following image:

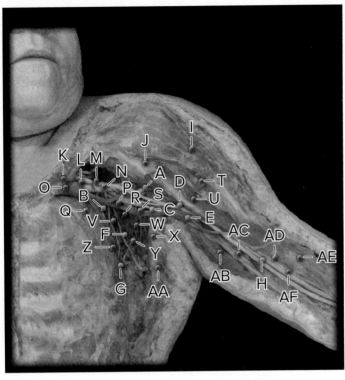

©McGraw-Hill Education

Mouse-over the pins on the screen to find the information necessary to identify the following structures:

A. _____
B. _____
C. _____
D. _____
E. _____
F. _____
G. _____
H. _____

Nonnervous System Structures (blue pins)

I. _____
J. _____
K. _____
L. _____
M. _____
N. _____
O. _____
P. _____
Q. _____

R. _____
S. _____
T. _____
U. _____
V. _____
W. _____
X. _____
Y. _____
Z. _____
AA. _____
AB. _____
AC. _____
AD. _____
AE. _____
AF. _____

EXERCISE 7.36:

Nervous System—Peripheral Nerves—Shoulder and Arm—Posterior View

 SELECT TOPIC
Peripheral Nerves

SELECT VIEW
Shoulder and Arm—Posterior

Click **LAYER 1** in the **LAYER CONTROLS** window, and you will see the following image:

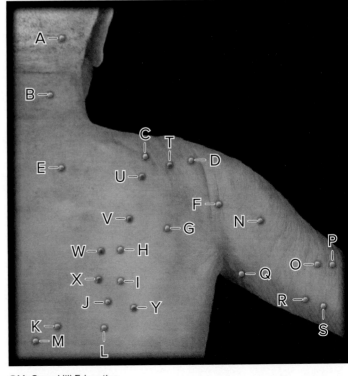

©McGraw-Hill Education

Mouse-over the pins on the screen to find the information necessary to identify the following structures:

A. _____
B. _____
C. _____
D. _____
E. _____
F. _____
G. _____
H. _____
I. _____
J. _____
K. _____
L. _____
M. _____
N. _____
O. _____
P. _____
Q. _____
R. _____
S. _____

Nonnervous System Structures (blue pins)

T. _____
U. _____
V. _____
W. _____
X. _____
Y. _____

Click **LAYER 3** *in the* **LAYER CONTROLS** *window, and you will see the following image:*

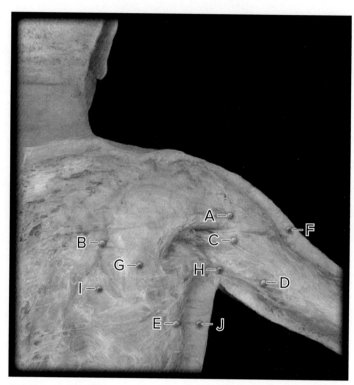

©McGraw-Hill Education

Mouse-over the pins on the screen to find the information necessary to identify the following structures:

A. _____
B. _____
C. _____
D. _____
E. _____

Nonnervous System Structures (blue pins)

F. _____
G. _____
H. _____
I. _____
J. _____

406 MODULE 7 The Nervous System

Click **LAYER 5** *in the* **LAYER CONTROLS** *window, and you will see the following image:*

©McGraw-Hill Education

Mouse-over the pins on the screen to find the information necessary to identify the following structures:

A. _____
B. _____
C. _____
D. _____

Nonnervous System Structures (blue pins)

E. _____
F. _____
G. _____
H. _____
I. _____
J. _____
K. _____
L. _____
M. _____
N. _____
O. _____
P. _____
Q. _____
R. _____
S. _____
T. _____
U. _____
V. _____
W. _____
X. _____
Y. _____
Z. _____
AA. _____
AB. _____
AC. _____
AD. _____

Click **LAYER 6** *in the* **LAYER CONTROLS** *window, and you will see the following image:*

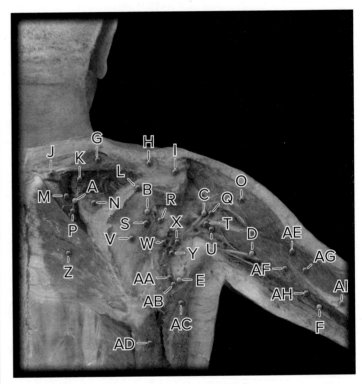

©McGraw-Hill Education

Mouse-over the pins on the screen to find the information necessary to identify the following structures:

A. _____
B. _____
C. _____
D. _____
E. _____
F. _____

Nonnervous System Structures (blue pins)

G. _____
H. _____
I. _____
J. _____
K. _____
L. _____
M. _____
N. _____
O. _____
P. _____
Q. _____
R. _____
S. _____
T. _____
U. _____
V. _____
W. _____
X. _____
Y. _____
Z. _____
AA. _____
AB. _____
AC. _____
AD. _____
AE. _____
AF. _____
AG. _____
AH. _____
AI. _____

CHECK POINT

Nervous System—Peripheral Nerves, Shoulder and Arm

1. What roots join to form the median nerve?
2. Describe the roots of the brachial plexus.
3. Name the two roots of the accessory nerve. Where do these roots unite?

EXERCISE 7.37:

Nervous System—Peripheral Nerves—Arm, Forearm, and Hand—Anterior View

SELECT TOPIC	SELECT VIEW
Peripheral Nerves	Arm, Forearm, and Hand—Anterior

*Click **LAYER 1** in the **LAYER CONTROLS** window, and you will see the following image:*

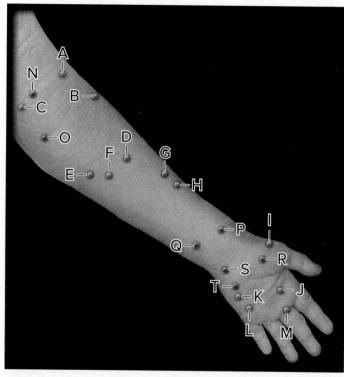

©McGraw-Hill Education

Mouse-over the pins on the screen to find the information necessary to identify the following structures:

A. _____
B. _____
C. _____
D. _____
E. _____
F. _____
G. _____
H. _____
I. _____
J. _____
K. _____
L. _____
M. _____

MODULE 7 The Nervous System

Nonnervous System Structures (blue pins)

N. _____
O. _____
P. _____
Q. _____
R. _____
S. _____
T. _____

Click **LAYER 3** *in the* **LAYER CONTROLS** *window, and you will see the following image:*

L. _____
M. _____
N. _____

Click **LAYER 4** *in the* **LAYER CONTROLS** *window, and you will see the following image:*

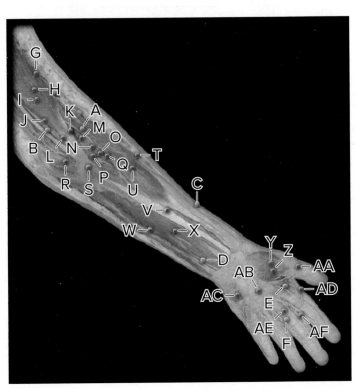
©McGraw-Hill Education

Mouse-over the pins on the screen to find the information necessary to identify the following structures:

A. _____
B. _____
C. _____
D. _____
E. _____

Nonnervous System Structures (blue pins)

F. _____
G. _____
H. _____
I. _____
J. _____
K. _____

Mouse-over the pins on the screen to find the information necessary to identify the following structures:

A. _____
B. _____
C. _____
D. _____
E. _____
F. _____

Nonnervous System Structures (blue pins)

G. _____
H. _____
I. _____
J. _____
K. _____
L. _____
M. _____
N. _____
O. _____

MODULE 7 The Nervous System 409

P. _____
Q. _____
R. _____
S. _____
T. _____
U. _____
V. _____
W. _____
X. _____
Y. _____
Z. _____
AA. _____
AB. _____
AC. _____
AD. _____
AE. _____
AF. _____

Click **LAYER 5** *in the* **LAYER CONTROLS** *window, and you will see the following image:*

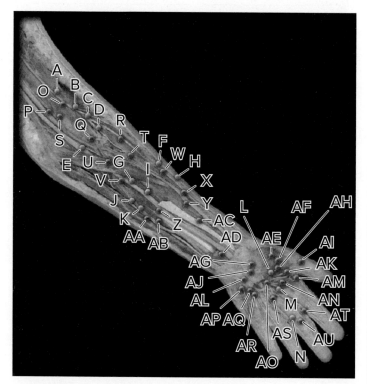
©McGraw-Hill Education

Mouse-over the pins on the screen to find the information necessary to identify the following structures:

A. _____
B. _____

C. _____
D. _____
E. _____
F. _____
G. _____
H. _____
I. _____
J. _____
K. _____
L. _____
M. _____
N. _____

Nonnervous System Structures (blue pins)

O. _____
P. _____
Q. _____
R. _____
S. _____
T. _____
U. _____
V. _____
W. _____
X. _____
Y. _____
Z. _____
AA. _____
AB. _____
AC. _____
AD. _____
AE. _____
AF. _____
AG. _____
AH. _____
AI. _____
AJ. _____
AK. _____
AL. _____
AM. _____
AN. _____
AO. _____

410 MODULE 7 The Nervous System

AP. _____
AQ. _____
AR. _____
AS. _____
AT. _____
AU. _____

Click **LAYER 6** *in the* **LAYER CONTROLS** *window, and you will see the following image:*

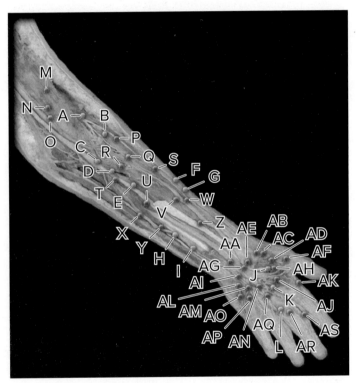

©McGraw-Hill Education

Mouse-over the pins on the screen to find the information necessary to identify the following structures:

A. _____
B. _____
C. _____
D. _____
E. _____
F. _____
G. _____
H. _____
I. _____
J. _____
K. _____
L. _____

Nonnervous System Structures (blue pins)

M. _____
N. _____
O. _____
P. _____
Q. _____
R. _____
S. _____
T. _____
U. _____
V. _____
W. _____
X. _____
Y. _____
Z. _____
AA. _____
AB. _____
AC. _____
AD. _____
AE. _____
AF. _____
AG. _____
AH. _____
AI. _____
AJ. _____
AK. _____
AL. _____
AM. _____
AN. _____
AO. _____
AP. _____
AQ. _____
AR. _____
AS. _____

MODULE 7 The Nervous System 411

EXERCISE 7.38:
Nervous System—Peripheral Nerves—Arm, Forearm, and Hand—Posterior View

SELECT TOPIC	SELECT VIEW
Peripheral Nerves	Arm, Forearm, and Hand—Posterior

Click **LAYER 1** in the **LAYER CONTROLS** window, and you will see the following image:

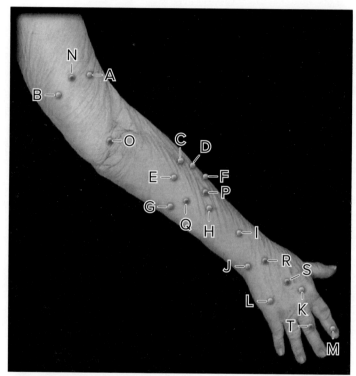

©McGraw-Hill Education

Mouse-over the pins on the screen to find the information necessary to identify the following structures:

A. _____
B. _____
C. _____
D. _____
E. _____
F. _____
G. _____
H. _____
I. _____
J. _____
K. _____
L. _____
M. _____

Nonnervous System Structures (blue pins)

N. _____
O. _____
P. _____
Q. _____
R. _____
S. _____
T. _____

Click **LAYER 3** in the **LAYER CONTROLS** window, and you will see the following image:

©McGraw-Hill Education

Mouse-over the pins on the screen to find the information necessary to identify the following structures:

A. _____
B. _____
C. _____
D. _____
E. _____
F. _____
G. _____
H. _____
I. _____

Nonnervous System Structures (blue pins)

J. _____
K. _____
L. _____

412 MODULE 7 The Nervous System

M. _____
N. _____
O. _____
P. _____
Q. _____

Click **LAYER 4** *in the* **LAYER CONTROLS** *window, and you will see the following image:*

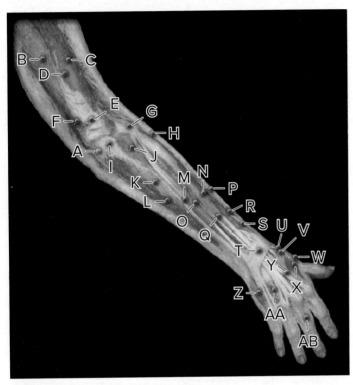

©McGraw-Hill Education

Mouse-over the pin on the screen to find the information necessary to identify the following structure:

A. _____

Nonnervous System Structures (blue pins)

B. _____
C. _____
D. _____
E. _____
F. _____
G. _____
H. _____
I. _____
J. _____
K. _____
L. _____
M. _____

N. _____
O. _____
P. _____
Q. _____
R. _____
S. _____
T. _____
U. _____
V. _____
W. _____
X. _____
Y. _____
Z. _____
AA. _____
AB. _____

Click **LAYER 5** *in the* **LAYER CONTROLS** *window, and you will see the following image:*

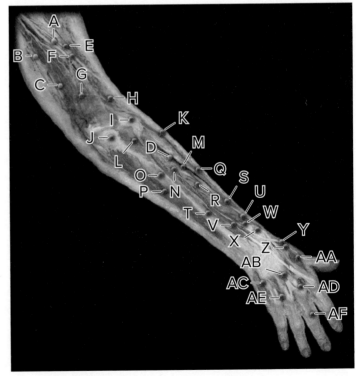

©McGraw-Hill Education

Mouse-over the pins on the screen to find the information necessary to identify the following structures:

A. _____
B. _____
C. _____
D. _____

Nonnervous System Structures (blue pins)

E. _____
F. _____
G. _____
H. _____
I. _____
J. _____
K. _____
L. _____
M. _____
N. _____
O. _____
P. _____
Q. _____
R. _____
S. _____
T. _____
U. _____
V. _____
W. _____
X. _____
Y. _____
Z. _____
AA. _____
AB. _____
AC. _____
AD. _____
AE. _____
AF. _____

Click **LAYER 6** *in the* **LAYER CONTROLS** *window, and you will see the following image:*

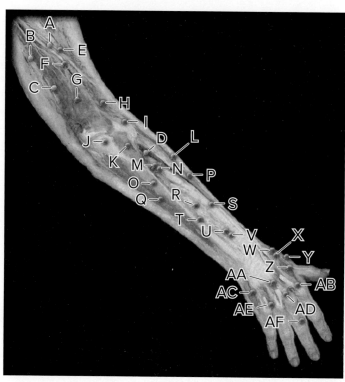

©McGraw-Hill Education

Mouse-over the pins on the screen to find the information necessary to identify the following structures:

A. _____
B. _____
C. _____
D. _____

Nonnervous System Structures (blue pins)

E. _____
F. _____
G. _____
H. _____
I. _____
J. _____
K. _____
L. _____
M. _____
N. _____
O. _____
P. _____
Q. _____
R. _____
S. _____
T. _____

414 MODULE 7 The Nervous System

U. _____
V. _____
W. _____
X. _____
Y. _____
Z. _____
AA. _____
AB. _____
AC. _____
AD. _____
AE. _____
AF. _____

CHECK POINT

Peripheral Nerves, Arm, Forearm, and Hand

1. Name the nerve that innervates the axilla, medial arm, anterior forearm, and lateral palmar hand. What is its composition?
2. A blow to which nerve is said to tingle your "funny bone."
3. Name a nerve with motor composition that innervates all extensor muscles in the arm and forearm.

EXERCISE 7.39:

Nervous System—Peripheral Nerves—Hand—Palmar View

SELECT TOPIC: Peripheral Nerves ▸ SELECT VIEW: Hand—Palmar

Click **LAYER 1** in the **LAYER CONTROLS** window, and you will see the following image:

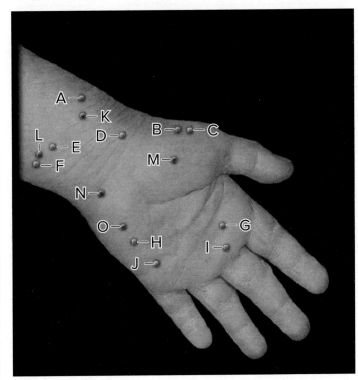

©McGraw-Hill Education

Mouse-over the pins on the screen to find the information necessary to identify the following structures:

A. _____
B. _____
C. _____
D. _____
E. _____
F. _____
G. _____
H. _____
I. _____
J. _____

Nonnervous System Structures (blue pins)

K. _____
L. _____
M. _____

N. _____
O. _____

Click **LAYER 3** *in the* **LAYER CONTROLS** *window, and you will see the following image:*

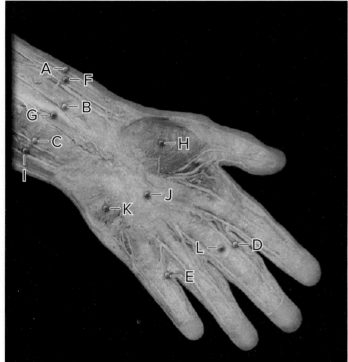

©McGraw-Hill Education

Mouse-over the pins on the screen to find the information necessary to identify the following structures:

A. _____
B. _____
C. _____
D. _____
E. _____

Nonnervous System Structures (blue pins)

F. _____
G. _____
H. _____
I. _____
J. _____
K. _____
L. _____

Click **LAYER 4** *in the* **LAYER CONTROLS** *window, and you will see the following image:*

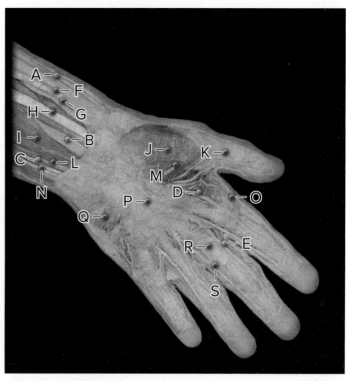

©McGraw-Hill Education

Mouse-over the pins on the screen to find the information necessary to identify the following structures:

A. _____
B. _____
C. _____
D. _____
E. _____

Nonnervous System Structures (blue pins)

F. _____
G. _____
H. _____
I. _____
J. _____
K. _____
L. _____
M. _____
N. _____
O. _____
P. _____
Q. _____
R. _____
S. _____

Click **LAYER 5** *in the* **LAYER CONTROLS** *window, and you will see the following image:*

Mouse-over the pins on the screen to find the information necessary to identify the following structures:

A. _____
B. _____
C. _____
D. _____
E. _____
F. _____
G. _____
H. _____
I. _____

Nonnervous System Structures (blue pins)

J. _____
K. _____
L. _____
M. _____
N. _____
O. _____
P. _____
Q. _____
R. _____
S. _____
T. _____
U. _____
V. _____
W. _____
X. _____
Y. _____
Z. _____
AA. _____
AB. _____
AC. _____
AD. _____
AE. _____
AF. _____
AG. _____

Click **LAYER 6** *in the* **LAYER CONTROLS** *window, and you will see the following image:*

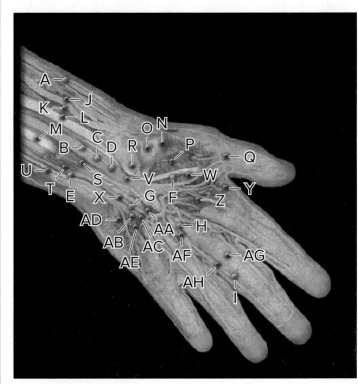

©McGraw-Hill Education

Mouse-over the pins on the screen to find the information necessary to identify the following structures:

A. _____
B. _____
C. _____
D. _____

E. _____
F. _____
G. _____
H. _____
I. _____

Nonnervous System Structures (blue pins)

J. _____
K. _____
L. _____
M. _____
N. _____
O. _____
P. _____
Q. _____
R. _____
S. _____
T. _____
U. _____
V. _____
W. _____
X. _____
Y. _____
Z. _____
AA. _____
AB. _____
AC. _____
AD. _____
AE. _____
AF. _____
AG. _____
AH. _____

EXERCISE 7.40:
Nervous System—Peripheral Nerves—Hand—Dorsum View

SELECT TOPIC	SELECT VIEW
Peripheral Nerves	Hand—Dorsum

Click **LAYER 1** in the **LAYER CONTROLS** window, and you will see the following image:

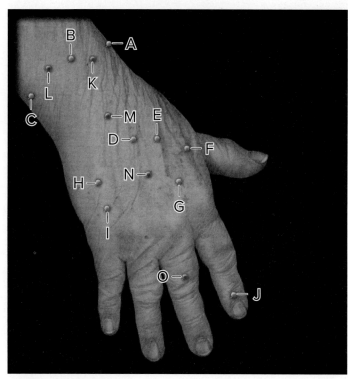

©McGraw-Hill Education

Mouse-over the pins on the screen to find the information necessary to identify the following structures:

A. _____
B. _____
C. _____
D. _____
E. _____
F. _____
G. _____
H. _____
I. _____
J. _____

Nonnervous System Structures (blue pins)

K. _____
L. _____

M. _____
N. _____
O. _____

Click **LAYER 3** *in the* **LAYER CONTROLS** *window, and you will see the following image:*

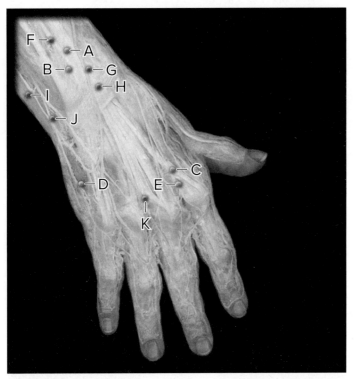

©McGraw-Hill Education

Mouse-over the pins on the screen to find the information necessary to identify the following structures:

A. _____
B. _____
C. _____
D. _____
E. _____

Nonnervous System Structures (blue pins)

F. _____
G. _____
H. _____
I. _____
J. _____
K. _____

Click **LAYER 4** *in the* **LAYER CONTROLS** *window, and you will see the following image:*

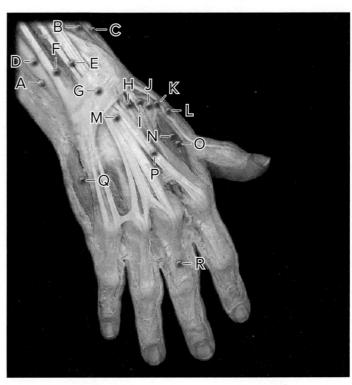

©McGraw-Hill Education

Mouse-over the pin on the screen to find the information necessary to identify the following structure:

A. _____

Nonnervous System Structures (blue pins)

B. _____
C. _____
D. _____
E. _____
F. _____
G. _____
H. _____
I. _____
J. _____
K. _____
L. _____
M. _____
N. _____
O. _____
P. _____
Q. _____
R. _____

Click **LAYER 6** *in the* **LAYER CONTROLS** *window, and you will see the following image:*

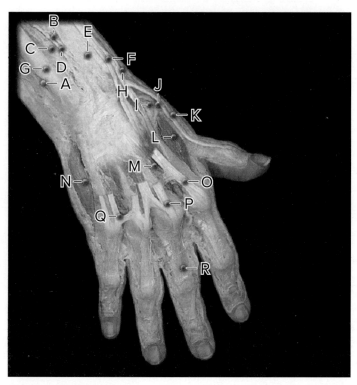

©McGraw-Hill Education

Mouse-over the pin on the screen to find the information necessary to identify the following structure:

A. _____

Nonnervous System Structures (blue pins)

B. _____
C. _____
D. _____
E. _____
F. _____
G. _____
H. _____
I. _____
J. _____
K. _____
L. _____
M. _____
N. _____
O. _____
P. _____
Q. _____
R. _____

CHECK POINT

Peripheral Nerves, Hand

1. Name a nerve that arises from the median nerve distal to the carpal tunnel.
2. Name a general sensation nerve of the skin and joints of the medial and palmar aspects of the fingers.
3. Name a motor nerve that innervates most of the small muscles of the hand.

EXERCISE 7.41:

Nervous System—Peripheral Nerves—Trunk

SELECT TOPIC	SELECT VIEW
Peripheral Nerves	Trunk

Click **LAYER 1** *in the* **LAYER CONTROLS** *window, and you will see the following image:*

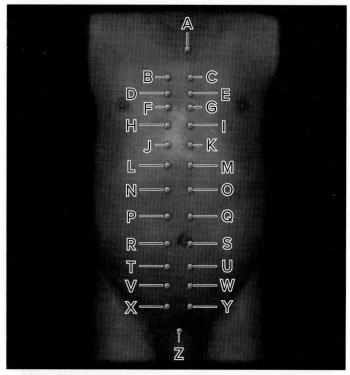

©McGraw-Hill Education

Mouse-over the pins on the screen to find the information necessary to identify the following structures:

A. _____
B. _____
C. _____
D. _____
E. _____
F. _____

G. _____
H. _____
I. _____
J. _____
K. _____
L. _____
M. _____
N. _____
O. _____
P. _____
Q. _____
R. _____
S. _____
T. _____
U. _____
V. _____
W. _____
X. _____
Y. _____
Z. _____

Click **LAYER 3** *in the* **LAYER CONTROLS** *window, and you will see the following image:*

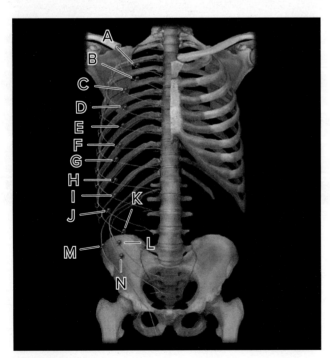
©McGraw-Hill Education

Mouse-over the pins on the screen to find the information necessary to identify the following structures:

A. _____
B. _____
C. _____
D. _____
E. _____
F. _____
G. _____
H. _____
I. _____
J. _____
K. _____
L. _____
M. _____
N. _____

CHECK POINT

Peripheral Nerves, Trunk

1. Name five nerves that innervate the skin of the abdomen.
2. Which group of nerves are located within the intercostal space between the innermost and the internal intercostal muscles?
3. Name five of these nerves.

Self-Quiz

Take this opportunity to check your progress by taking the **QUIZ**. See the **Introduction Module** for a reminder on how to access the **QUIZ** for this Study Area.

MODULE 7 The Nervous System 421

EXERCISE 7.42:
Nervous System—Peripheral Nerves—Lumbosacral Plexus

SELECT TOPIC	SELECT VIEW
Peripheral Nerves	Lumbosacral Plexus

Click **LAYER 1** in the **LAYER CONTROLS** window, and you will see the following image:

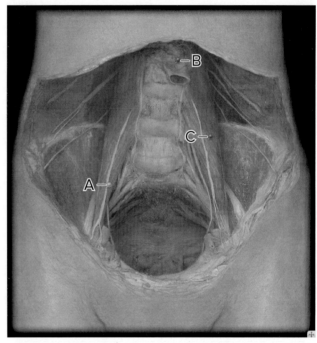

©McGraw-Hill Education

Mouse-over the pins on the screen to find the information necessary to identify the following structures:

A. _____
B. _____
C. _____
D. _____
E. _____
F. _____
G. _____
H. _____
I. _____
J. _____
K. _____
L. _____
M. _____
N. _____
O. _____
P. _____
Q. _____
R. _____

Click **LAYER 2** in the **LAYER CONTROLS** window, and you will see the following image:

©McGraw-Hill Education

Mouse-over the pin on the screen to find the information necessary to identify the following structure:

A. _____

Nonnervous System Structures (blue pins)

B. _____
C. _____

Click **LAYER 3** *in the* **LAYER CONTROLS** *window, and you will see the following image:*

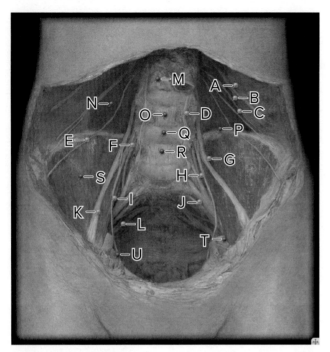

©McGraw-Hill Education

Mouse-over the pins on the screen to find the information necessary to identify the following structures:

A. _____
B. _____
C. _____
D. _____
E. _____
F. _____
G. _____
H. _____
I. _____
J. _____
K. _____
L. _____

Nonnervous System Structures (blue pins)

M. _____
N. _____
O. _____
P. _____
Q. _____
R. _____
S. _____
T. _____
U. _____

CHECK POINT

Peripheral Nerves, Lumbosacral Plexus

1. Nerves derived from what structure distribute to the lower anterior abdominal wall, spermatic cord, thigh, medial leg and foot, and the sacral plexus?
2. Name a nerve that supplies sensory innervation to the lateral thigh.
3. Name the seven terminal branches of the lumbar plexus.

Self-Quiz

Take this opportunity to check your progress by taking the **QUIZ**. See the **Introduction Module** for a reminder on how to access the **QUIZ** for this Study Area.

EXERCISE 7.43:

Nervous System—Peripheral Nerves— Lower Limb—Anterior View

SELECT TOPIC: **Peripheral Nerves** ▶ SELECT VIEW: **Lower Limb—Anterior**

Click **LAYER 1** *in the* **LAYER CONTROLS** *window, and you will see the following image:*

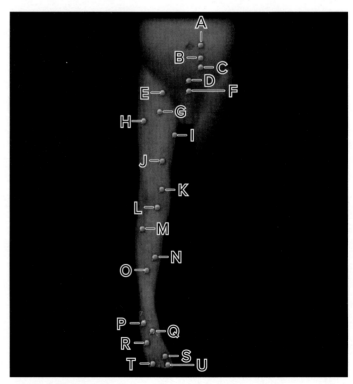

©McGraw-Hill Education

Mouse-over the pins on the screen to find the information necessary to identify the following structures:

A. _____
B. _____
C. _____
D. _____
E. _____
F. _____
G. _____
H. _____
I. _____
J. _____
K. _____
L. _____
M. _____
N. _____
O. _____
P. _____
Q. _____
R. _____
S. _____
T. _____
U. _____

CHECK POINT

Peripheral Nerves, Lower Limb—Anterior View

1. Name the nerve providing cutaneous innervation to the lateral thigh.
2. Name the nerve providing cutaneous innervation to the anterior thigh, medial leg, and the medial margin of the foot.
3. Name the nerve providing cutaneous innervation to the medial leg and medial margin of the foot.

Click LAYER 2 in the LAYER CONTROLS window, and you will see the following image:

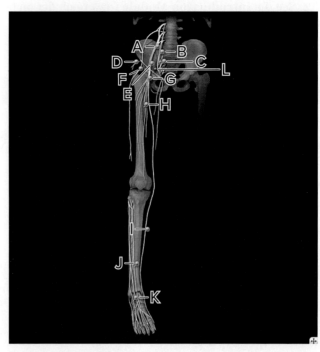

©McGraw-Hill Education

Mouse-over the pins on the screen to find the information necessary to identify the following structures:

A. _____
B. _____
C. _____
D. _____
E. _____
F. _____
G. _____
H. _____
I. _____
J. _____
K. _____

Nonnervous System Structure (blue pin)

L. _____

CHECK POINT

Peripheral Nerves, Lower Limb—Anterior View, *continued*

4. Name the branch of the femoral nerve that provides sensory innervation to the medial leg and the medial margin of the foot.
5. Name the nerve providing motor innervation to the gluteus medius, gluteus minimus, and the tensor fascia latae muscles.
6. Name the nerve providing motor and sensory innervation to the muscles of the lateral leg.

424 MODULE 7 The Nervous System

Self-Quiz

Take this opportunity to check your progress by taking the **QUIZ**. See the **Introduction Module** for a reminder on how to access the **QUIZ** for this Study Area.

EXERCISE 7.44:
Nervous System—Peripheral Nerves—Lower Limb—Posterior View

SELECT TOPIC: Peripheral Nerves ▶ SELECT VIEW: Lower Limb—Posterior

Click **LAYER 1** *in the* **LAYER CONTROLS** *window, and you will see the following image:*

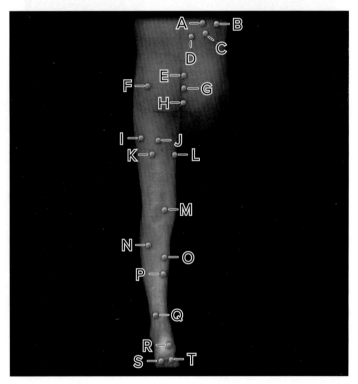

©McGraw-Hill Education

Mouse-over the pins on the screen to find the information necessary to identify the following structures:

A. _____
B. _____
C. _____
D. _____
E. _____
F. _____
G. _____
H. _____
I. _____
J. _____
K. _____
L. _____
M. _____
N. _____
O. _____
P. _____
Q. _____
R. _____
S. _____
T. _____

CHECK POINT

Peripheral Nerves, Lower Limb—Posterior View

1. Name the nerve providing cutaneous innervation to the posterior distal and lateral proximal leg and the lateral margin of the foot.
2. Name the two nerves providing cutaneous innervation to the sole of the foot.
3. Name the nerve providing innervation to the skin of the medial thigh.

Click **LAYER 2** *in the* **LAYER CONTROLS** *window, and you will see the following image:*

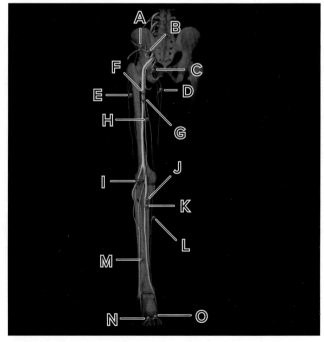

©McGraw-Hill Education

Mouse-over the pins on the screen to find the information necessary to identify the following structures:

A. _____
B. _____

C. _____
D. _____
E. _____
F. _____
G. _____
H. _____
I. _____
J. _____
K. _____
L. _____
M. _____
N. _____
O. _____

CHECK POINT

Peripheral Nerves, Lower Limb—Posterior View, *continued*

4. Name the nerve supplying motor innervation to the anterior and lateral leg muscles and the muscles of the dorsum of the foot.
5. Name the nerve supplying motor innervation to the adductor muscles of the medial thigh.
6. Name the nerve supplying sensory innervation to the medial leg and the medial margin of the foot.

EXERCISE 7.45:

Nervous System—Peripheral Nerves— Hip and Thigh—Anterior View

SELECT TOPIC	SELECT VIEW
Peripheral Nerves	Hip and Thigh—Anterior

Click **LAYER 1** *in the* **LAYER CONTROLS** *window, and you will see the following image:*

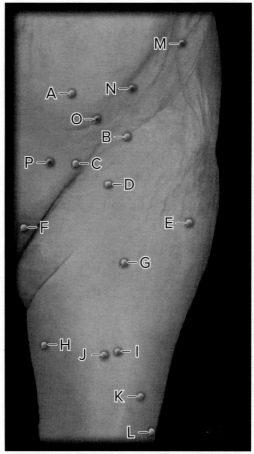

©McGraw-Hill Education

Mouse-over the pins on the screen to find the information necessary to identify the following structures:

A. _____
B. _____
C. _____
D. _____
E. _____
F. _____
G. _____
H. _____
I. _____
J. _____
K. _____
L. _____

426 MODULE 7 The Nervous System

Nonnervous System Structures (blue pins)

M. _____

N. _____

O. _____

P. _____

Click **LAYER 2** *in the* **LAYER CONTROLS** *window, and you will see the following image:*

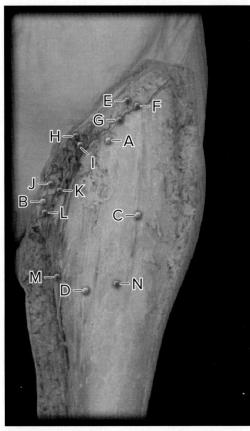

©McGraw-Hill Education

Mouse-over the pins on the screen to find the information necessary to identify the following structures:

A. _____

B. _____

C. _____

D. _____

Nonnervous System Structures (blue pins)

E. _____

F. _____

G. _____

H. _____

I. _____

J. _____

K. _____

L. _____

M. _____

N. _____

Click **LAYER 3** *in the* **LAYER CONTROLS** *window, and you will see the following image:*

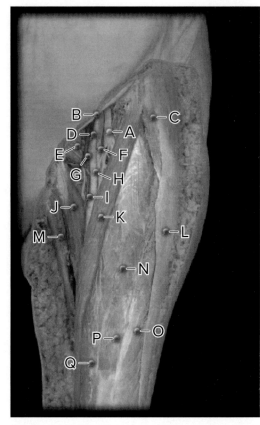

©McGraw-Hill Education

Mouse-over the pin on the screen to find the information necessary to identify the following structure:

A. _____

Nonnervous System Structures (blue pins)

B. _____

C. _____

D. _____

E. _____

F. _____

G. _____

H. _____

I. _____

J. _____

K. _____
L. _____
M. _____
N. _____
O. _____
P. _____
Q. _____

Click **LAYER 4** *in the* **LAYER CONTROLS** *window, and you will see the following image:*

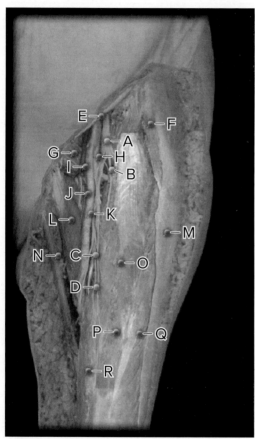

©McGraw-Hill Education

Mouse-over the pins on the screen to find the information necessary to identify the following structures:

A. _____
B. _____
C. _____
D. _____

Nonnervous System Structures (blue pins)

E. _____
F. _____
G. _____
H. _____

I. _____
J. _____
K. _____
L. _____
M. _____
N. _____
O. _____
P. _____
Q. _____
R. _____

Click **LAYER 5** *in the* **LAYER CONTROLS** *window, and you will see the following image:*

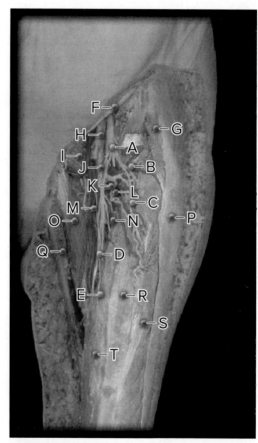

©McGraw-Hill Education

Mouse-over the pins on the screen to find the information necessary to identify the following structures:

A. _____
B. _____
C. _____
D. _____
E. _____

Nonnervous System Structures (blue pins)

F. _____
G. _____
H. _____
I. _____
J. _____
K. _____
L. _____
M. _____
N. _____
O. _____
P. _____
Q. _____
R. _____
S. _____
T. _____

Click **LAYER 6** *in the* **LAYER CONTROLS** *window, and you will see the following image:*

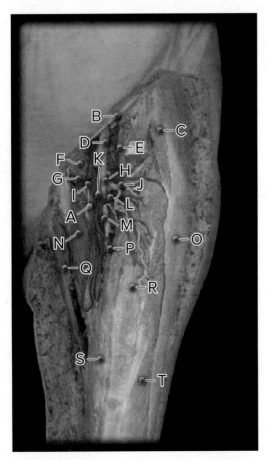

©McGraw-Hill Education

Mouse-over the pin on the screen to find the information necessary to identify the following structure:

A. _____

Nonnervous System Structures (blue pins)

B. _____
C. _____
D. _____
E. _____
F. _____
G. _____
H. _____
I. _____
J. _____
K. _____
L. _____
M. _____
N. _____
O. _____
P. _____
Q. _____
R. _____
S. _____
T. _____

EXERCISE 7.46:
Nervous System—Peripheral Nerves—Hip and Thigh—Posterior View

SELECT TOPIC: Peripheral Nerves ▸ **SELECT VIEW**: Hip and Thigh—Posterior

Click **LAYER 1** *in the* **LAYER CONTROLS** *window, and you will see the following image:*

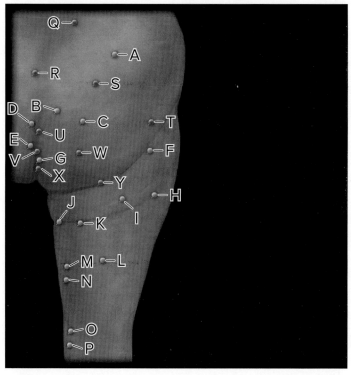

©McGraw-Hill Education

Mouse-over the pins on the screen to find the information necessary to identify the following structures:

A. _____
B. _____
C. _____
D. _____
E. _____
F. _____
G. _____
H. _____
I. _____
J. _____
K. _____
L. _____
M. _____
N. _____
O. _____
P. _____

Nonnervous System Structures (blue pins)

Q. _____
R. _____
S. _____
T. _____
U. _____
V. _____
W. _____
X. _____
Y. _____

Click **LAYER 2** *in the* **LAYER CONTROLS** *window, and you will see the following image:*

©McGraw-Hill Education

Mouse-over the pins on the screen to find the information necessary to identify the following structures:

A. _____
B. _____
C. _____
D. _____

Nonnervous System Structures (blue pins)

E. _____
F. _____

430 MODULE 7 The Nervous System

Click **LAYER 3** *in the* **LAYER CONTROLS** *window, and you will see the following image:*

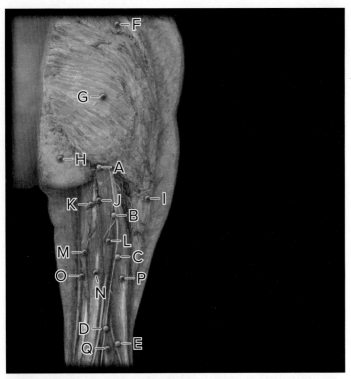

©McGraw-Hill Education

Mouse-over the pins on the screen to find the information necessary to identify the following structures:

A. _____
B. _____
C. _____
D. _____
E. _____

Nonnervous System Structures (blue pins)

F. _____
G. _____
H. _____
I. _____
J. _____
K. _____
L. _____
M. _____
N. _____
O. _____
P. _____
Q. _____

Click **LAYER 4** *in the* **LAYER CONTROLS** *window, and you will see the following image:*

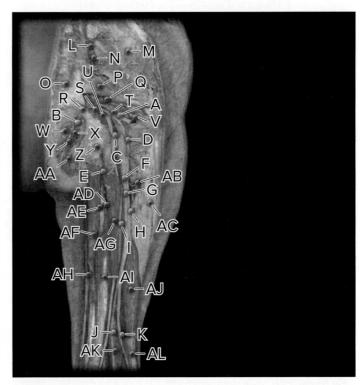

©McGraw-Hill Education

Mouse-over the pins on the screen to find the information necessary to identify the following structures:

A. _____
B. _____
C. _____
D. _____
E. _____
F. _____
G. _____
H. _____
I. _____
J. _____
K. _____

Nonnervous System Structures (blue pins)

L. _____
M. _____
N. _____
O. _____
P. _____
Q. _____
R. _____

S. _____
T. _____
U. _____
V. _____
W. _____
X. _____
Y. _____
Z. _____
AA. _____
AB. _____
AC. _____
AD. _____
AE. _____
AF. _____
AG. _____
AH. _____
AI. _____
AJ. _____
AK. _____
AL. _____

Click **LAYER 5** *in the* **LAYER CONTROLS** *window, and you will see the following image:*

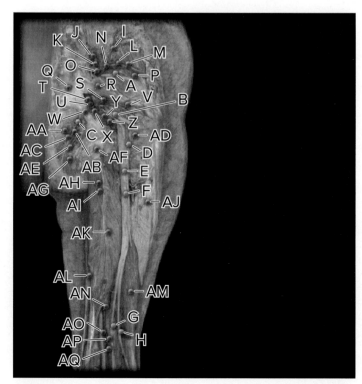

©McGraw-Hill Education

Mouse-over the pins on the screen to find the information necessary to identify the following structures:

A. _____
B. _____
C. _____
D. _____
E. _____
F. _____
G. _____
H. _____

Nonnervous System Structures (blue pins)

I. _____
J. _____
K. _____
L. _____
M. _____
N. _____
O. _____
P. _____
Q. _____
R. _____
S. _____
T. _____
U. _____
V. _____
W. _____
X. _____
Y. _____
Z. _____
AA. _____
AB. _____
AC. _____
AD. _____
AE. _____
AF. _____
AG. _____
AH. _____
AI. _____
AJ. _____
AK. _____

432 MODULE 7 The Nervous System

AL. _____
AM. _____
AN. _____
AO. _____
AP. _____
AQ. _____

Click **LAYER 6** *in the* **LAYER CONTROLS** *window, and you will see the following image:*

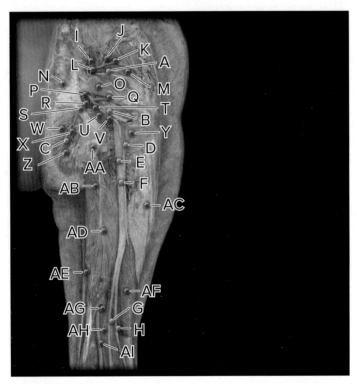

©McGraw-Hill Education

Mouse-over the pins on the screen to find the information necessary to identify the following structures:

A. _____
B. _____
C. _____
D. _____
E. _____
F. _____
G. _____
H. _____

Nonnervous System Structures (blue pins)

I. _____
J. _____
K. _____

L. _____
M. _____
N. _____
O. _____
P. _____
Q. _____
R. _____
S. _____
T. _____
U. _____
V. _____
W. _____
X. _____
Y. _____
Z. _____
AA. _____
AB. _____
AC. _____
AD. _____
AE. _____
AF. _____
AG. _____
AH. _____
AI. _____

CHECK POINT

Nervous System—Peripheral Nerves, Hip and Thigh

1. Name the largest nerve of the body. What two nerves join to form it?
2. Name the nerve that enters the thigh posterior to the inguinal ligament and ends in the proximal thigh as numerous named branches?
3. What conditions might contribute to sciatica?

EXERCISE 7.47:
Nervous System—Peripheral Nerves— Knee—Anterior View

SELECT TOPIC	SELECT VIEW
Peripheral Nerves	Knee—Anterior

Click **LAYER 1** *in the* **LAYER CONTROLS** *window, and you will see the following image:*

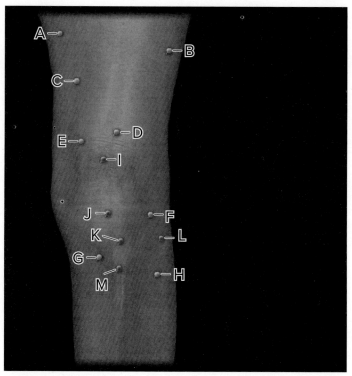

©McGraw-Hill Education

Mouse-over the pins on the screen to find the information necessary to identify the following structures:

A. _____
B. _____
C. _____
D. _____
E. _____
F. _____
G. _____
H. _____

Nonnervous System Structures (blue pins)

I. _____
J. _____
K. _____
L. _____
M. _____

Click **LAYER 2** *in the* **LAYER CONTROLS** *window, and you will see the following image:*

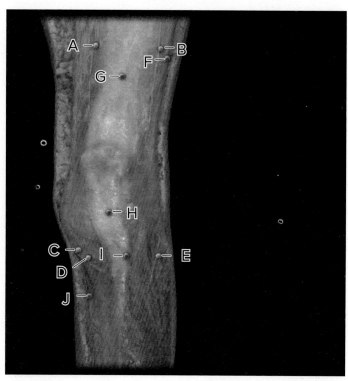

©McGraw-Hill Education

Mouse-over the pins on the screen to find the information necessary to identify the following structures:

A. _____
B. _____
C. _____
D. _____
E. _____

Nonnervous System Structures (blue pins)

F. _____
G. _____
H. _____
I. _____
J. _____

Click **LAYER 4** *in the* **LAYER CONTROLS** *window, and you will see the following image:*

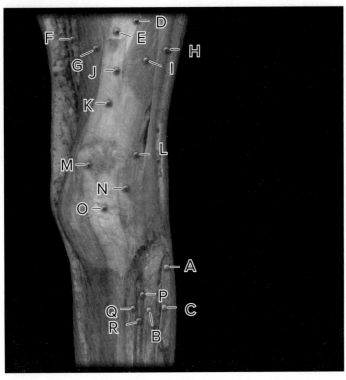

©McGraw-Hill Education

Mouse-over the pins on the screen to find the information necessary to identify the following structures:

A. _____
B. _____
C. _____

Nonnervous System Structures (blue pins)

D. _____
E. _____
F. _____
G. _____
H. _____
I. _____
J. _____
K. _____
L. _____
M. _____
N. _____
O. _____
P. _____
Q. _____
R. _____

EXERCISE 7.48:

Nervous System—Peripheral Nerves— Knee—Posterior View

SELECT TOPIC	SELECT VIEW
Peripheral Nerves	Knee—Posterior

Click **LAYER 1** *in the* **LAYER CONTROLS** *window, and you will see the following image:*

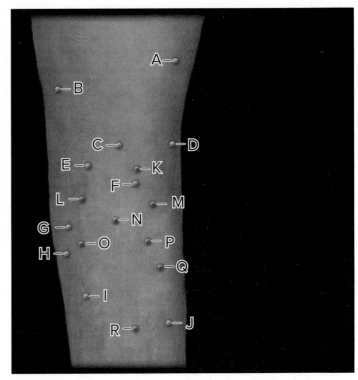

©McGraw-Hill Education

Mouse-over the pins on the screen to find the information necessary to identify the following structures:

A. _____
B. _____
C. _____
D. _____
E. _____
F. _____
G. _____
H. _____
I. _____
J. _____

Nonnervous System Structures (blue pins)

K. _____
L. _____
M. _____

N. _____
O. _____
P. _____
Q. _____
R. _____

Click **LAYER 2** *in the* **LAYER CONTROLS** *window, and you will see the following image:*

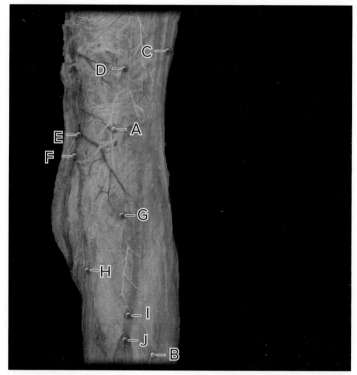

©McGraw-Hill Education

Mouse-over the pins on the screen to find the information necessary to identify the following structures:

A. _____
B. _____

Nonnervous System Structures (blue pins)

C. _____
D. _____
E. _____
F. _____
G. _____
H. _____
I. _____
J. _____

Click **LAYER 3** *in the* **LAYER CONTROLS** *window, and you will see the following image:*

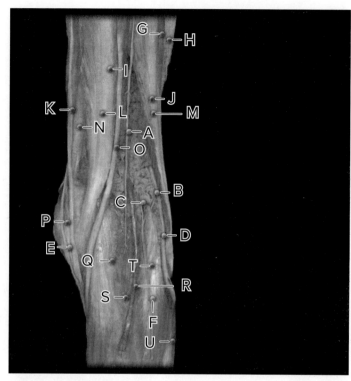

©McGraw-Hill Education

Mouse-over the pins on the screen to find the information necessary to identify the following structures:

A. _____
B. _____
C. _____
D. _____
E. _____
F. _____

Nonnervous System Structures (blue pins)

G. _____
H. _____
I. _____
J. _____
K. _____
L. _____
M. _____
N. _____
O. _____
P. _____
Q. _____
R. _____

S. _____
T. _____
U. _____

Q. _____
R. _____
S. _____
T. _____
U. _____
V. _____
W. _____
X. _____

Click **LAYER 4** *in the* **LAYER CONTROLS** *window, and you will see the following image:*

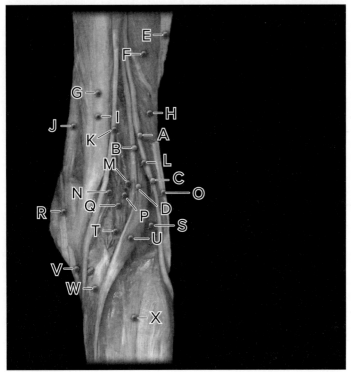

©McGraw-Hill Education

Click **LAYER 5** *in the* **LAYER CONTROLS** *window, and you will see the following image:*

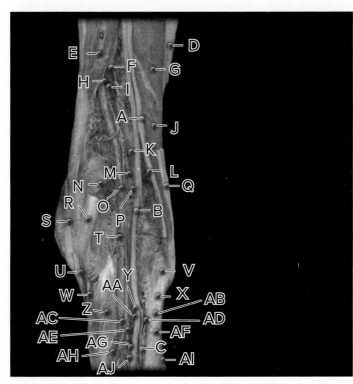

©McGraw-Hill Education

Mouse-over the pins on the screen to find the information necessary to identify the following structures:

A. _____
B. _____
C. _____
D. _____

Nonnervous System Structures (blue pins)

E. _____
F. _____
G. _____
H. _____
I. _____
J. _____
K. _____
L. _____
M. _____
N. _____
O. _____
P. _____

Mouse-over the pins on the screen to find the information necessary to identify the following structures:

A. _____
B. _____
C. _____

Nonnervous System Structures (blue pins)

D. _____
E. _____
F. _____
G. _____
H. _____
I. _____
J. _____
K. _____
L. _____

M. _____
N. _____
O. _____
P. _____
Q. _____
R. _____
S. _____
T. _____
U. _____
V. _____
W. _____
X. _____
Y. _____
Z. _____
AA. _____
AB. _____
AC. _____
AD. _____
AE. _____
AF. _____
AG. _____
AH. _____
AI. _____
AJ. _____

Mouse-over the pins on the screen to find the information necessary to identify the following structures:

A. _____
B. _____
C. _____

Nonnervous System Structures (blue pins)

D. _____
E. _____
F. _____
G. _____
H. _____
I. _____
J. _____
K. _____
L. _____
M. _____
N. _____
O. _____
P. _____
Q. _____
R. _____
S. _____
T. _____
U. _____
V. _____
W. _____
X. _____
Y. _____
Z. _____
AA. _____
AB. _____
AC. _____
AD. _____

Click **LAYER 6** *in the* **LAYER CONTROLS** *window, and you will see the following image:*

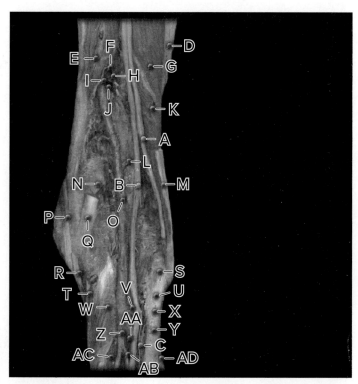

©McGraw-Hill Education

CHECK POINT

Peripheral Nerves, Knee

1. Name the branch of the femoral nerve responsible for general sensation of the skin of the medial leg and medial margin of the foot.
2. Name a nerve that lies on the interosseous membrane.
3. What nerve is vulnerable to lateral knee trauma? Why?

438 MODULE 7 The Nervous System

EXERCISE 7.49:
Nervous System—Peripheral Nerves—Leg and Foot—Anterior View

SELECT TOPIC	SELECT VIEW
Peripheral Nerves	Leg and Foot—Anterior

Click **LAYER 1** *in the* **LAYER CONTROLS** *window, and you will see the following image:*

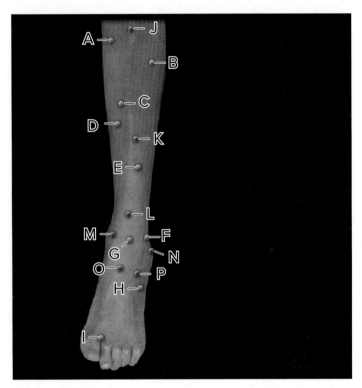

©McGraw-Hill Education

Mouse-over the pins on the screen to find the information necessary to identify the following structures:

A. _____
B. _____
C. _____
D. _____
E. _____
F. _____
G. _____
H. _____
I. _____

Nonnervous System Structures (blue pins)

J. _____
K. _____
L. _____
M. _____
N. _____
O. _____
P. _____

Click **LAYER 2** *in the* **LAYER CONTROLS** *window, and you will see the following image:*

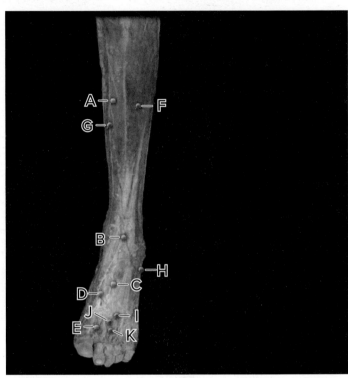

©McGraw-Hill Education

Mouse-over the pins on the screen to find the information necessary to identify the following structures:

A. _____
B. _____
C. _____
D. _____
E. _____

Nonnervous System Structures (blue pins)

F. _____
G. _____
H. _____
I. _____
J. _____
K. _____

MODULE 7   The Nervous System   **439**

Click **LAYER 4** *in the* **LAYER CONTROLS** *window, and you will see the following image:*

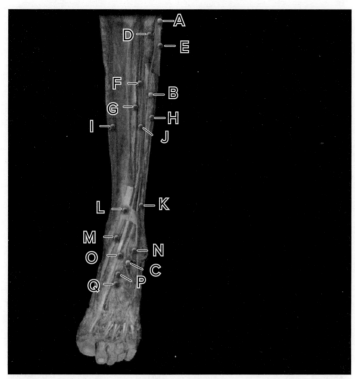

©McGraw-Hill Education

Mouse-over the pins on the screen to find the information necessary to identify the following structures:

A. _____

B. _____

C. _____

Nonnervous System Structures (blue pins)

D. _____

E. _____

F. _____

G. _____

H. _____

I. _____

J. _____

K. _____

L. _____

M. _____

N. _____

O. _____

P. _____

Q. _____

Click **LAYER 5** *in the* **LAYER CONTROLS** *window, and you will see the following image:*

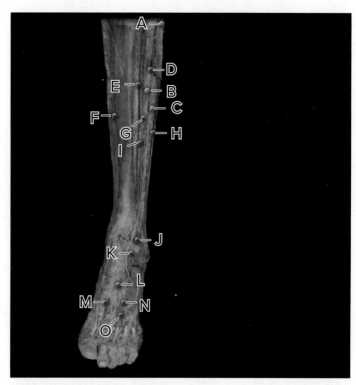

©McGraw-Hill Education

Mouse-over the pins on the screen to find the information necessary to identify the following structures:

A. _____

B. _____

C. _____

Nonnervous System Structures (blue pins)

D. _____

E. _____

F. _____

G. _____

H. _____

I. _____

J. _____

K. _____

L. _____

M. _____

N. _____

O. _____

440 MODULE 7 The Nervous System

EXERCISE 7.50:
Nervous System—Peripheral Nerves—Leg and Foot—Posterior View

SELECT TOPIC	SELECT VIEW
Peripheral Nerves	Leg and Foot—Posterior

Click **LAYER 1** *in the* **LAYER CONTROLS** *window, and you will see the following image:*

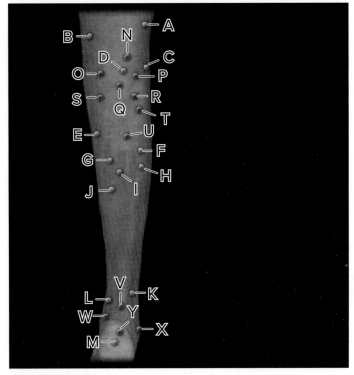

©McGraw-Hill Education

Mouse-over the pins on the screen to find the information necessary to identify the following structures:

A. _____
B. _____
C. _____
D. _____
E. _____
F. _____
G. _____
H. _____
I. _____
J. _____
K. _____
L. _____
M. _____

Nonnervous System Structures (blue pins)

N. _____
O. _____
P. _____
Q. _____
R. _____
S. _____
T. _____
U. _____
V. _____
W. _____
X. _____
Y. _____

Click **LAYER 2** *in the* **LAYER CONTROLS** *window, and you will see the following image:*

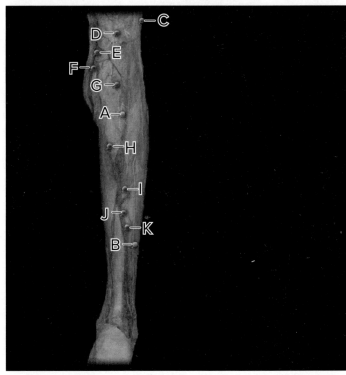

©McGraw-Hill Education

Mouse-over the pins on the screen to find the information necessary to identify the following structures:

A. _____
B. _____

Nonnervous System Structures (blue pins)

C. _____
D. _____
E. _____
F. _____
G. _____
H. _____
I. _____

J. _____
K. _____

Click **LAYER 3** *in the* **LAYER CONTROLS** *window, and you will see the following image:*

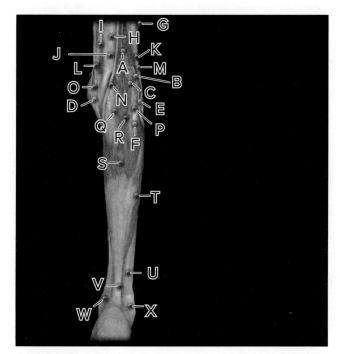

©McGraw-Hill Education

Mouse-over the pins on the screen to find the information necessary to identify the following structures:

A. _____
B. _____
C. _____
D. _____
E. _____
F. _____

Nonnervous System Structures (blue pins)

G. _____
H. _____
I. _____
J. _____
K. _____
L. _____
M. _____
N. _____
O. _____
P. _____
Q. _____
R. _____

S. _____
T. _____
U. _____
V. _____
W. _____
X. _____

Click **LAYER 4** *in the* **LAYER CONTROLS** *window, and you will see the following image:*

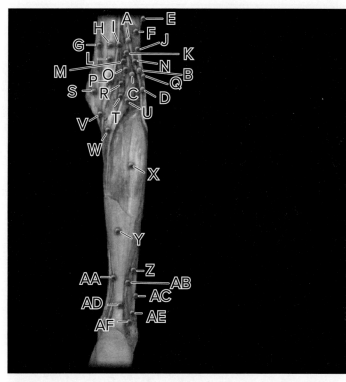

©McGraw-Hill Education

Mouse-over the pins on the screen to find the information necessary to identify the following structures:

A. _____
B. _____
C. _____
D. _____

Nonnervous System Structures (blue pins)

E. _____
F. _____
G. _____
H. _____
I. _____
J. _____
K. _____
L. _____
M. _____

N. _____
O. _____
P. _____
Q. _____
R. _____
S. _____
T. _____
U. _____
V. _____
W. _____
X. _____
Y. _____
Z. _____
AA. _____
AB. _____
AC. _____
AD. _____
AE. _____
AF. _____

Click **LAYER 5** *in the* **LAYER CONTROLS** *window, and you will see the following image:*

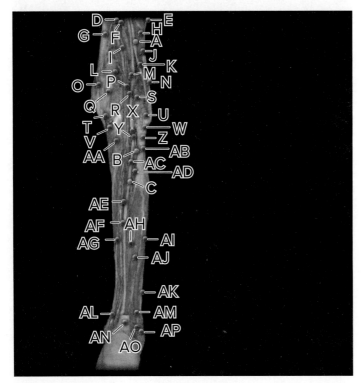

©McGraw-Hill Education

Mouse-over the pins on the screen to find the information necessary to identify the following structures:

A. _____
B. _____
C. _____

Nonnervous System Structures (blue pins)

D. _____
E. _____
F. _____
G. _____
H. _____
I. _____
J. _____
K. _____
L. _____
M. _____
N. _____
O. _____
P. _____
Q. _____
R. _____
S. _____
T. _____
U. _____
V. _____
W. _____
X. _____
Y. _____
Z. _____
AA. _____
AB. _____
AC. _____
AD. _____
AE. _____
AF. _____
AG. _____
AH. _____
AI. _____

AJ. _____
AK. _____
AL. _____
AM. _____
AN. _____
AO. _____
AP. _____

Click **LAYER 6** *in the* **LAYER CONTROLS** *window, and you will see the following image:*

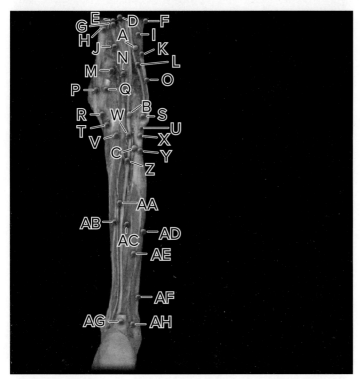

©McGraw-Hill Education

Mouse-over the pins on the screen to find the information necessary to identify the following structures:

A. _____
B. _____
C. _____

Nonnervous System Structures (blue pins)

D. _____
E. _____
F. _____
G. _____
H. _____

I. _____
J. _____
K. _____
L. _____
M. _____
N. _____
O. _____
P. _____
Q. _____
R. _____
S. _____
T. _____
U. _____
V. _____
W. _____
X. _____
Y. _____
Z. _____
AA. _____
AB. _____
AC. _____
AD. _____
AE. _____
AF. _____
AG. _____
AH. _____

CHECK POINT

Peripheral Nerves, Leg and Foot

1. Name a nerve with motor composition to the muscles of the lateral leg and sensory composition from the skin of the anterior distal leg and dorsum of the foot.
2. Name a nerve with sensory composition from the skin on adjacent sides of toes 1 and 2.
3. Name a nerve that divides at the medial ankle into medial and lateral plantar nerves.

EXERCISE 7.51:
Nervous System—Peripheral Nerves—Foot—Dorsum View

SELECT TOPIC: Peripheral Nerves **SELECT VIEW**: Foot—Dorsum

Click **LAYER 1** *in the* **LAYER CONTROLS** *window, and you will see the following image:*

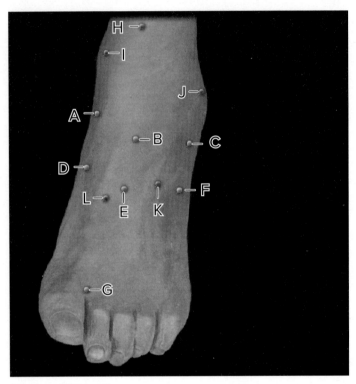

©McGraw-Hill Education

Mouse-over the pins on the screen to find the information necessary to identify the following structures:

A. _____
B. _____
C. _____
D. _____
E. _____
F. _____
G. _____

Nonnervous System Structures (blue pins)

H. _____
I. _____
J. _____
K. _____
L. _____

Click **LAYER 2** *in the* **LAYER CONTROLS** *window, and you will see the following image:*

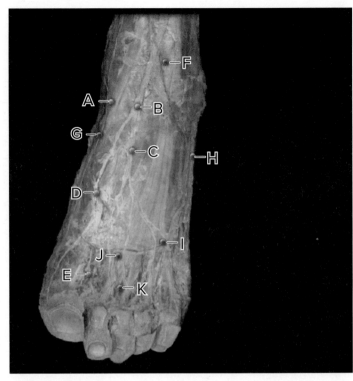

©McGraw-Hill Education

Mouse-over the pins on the screen to find the information necessary to identify the following structures:

A. _____
B. _____
C. _____
D. _____
E. _____

Nonnervous System Structures (blue pins)

F. _____
G. _____
H. _____
I. _____
J. _____
K. _____

Click **LAYER 3** *in the* **LAYER CONTROLS** *window, and you will see the following image:*

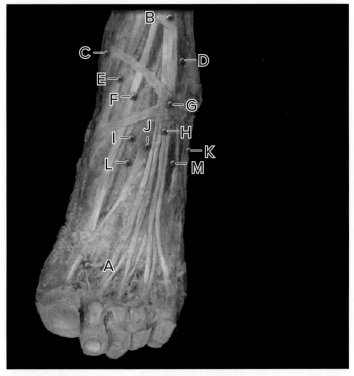

©McGraw-Hill Education

Mouse-over the pin on the screen to find the information necessary to identify the following structure:

A. _____

Nonnervous System Structures (blue pins)

B. _____
C. _____
D. _____
E. _____
F. _____
G. _____
H. _____
I. _____
J. _____
K. _____
L. _____
M. _____

Click **LAYER 4** *in the* **LAYER CONTROLS** *window, and you will see the following image:*

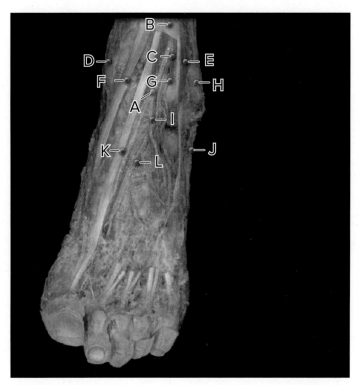

©McGraw-Hill Education

Mouse-over the pin on the screen to find the information necessary to identify the following structure:

A. _____

Nonnervous System Structures (blue pins)

B. _____
C. _____
D. _____
E. _____
F. _____
G. _____
H. _____
I. _____
J. _____
K. _____
L. _____

446 MODULE 7 The Nervous System

Click **LAYER 5** *in the* **LAYER CONTROLS** *window, and you will see the following image:*

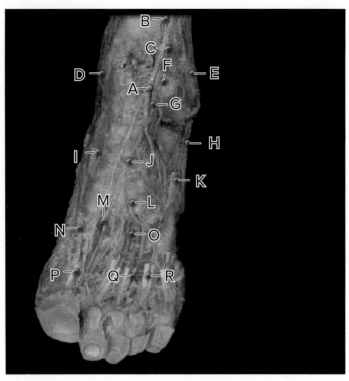

©McGraw-Hill Education

Mouse-over the pin on the screen to find the information necessary to identify the following structure:

A. _____

Nonnervous System Structures (blue pins)

B. _____
C. _____
D. _____
E. _____
F. _____
G. _____
H. _____
I. _____
J. _____
K. _____
L. _____
M. _____
N. _____
O. _____
P. _____
Q. _____
R. _____

EXERCISE 7.52:
Nervous System—Peripheral Nerves— Foot—Plantar View

SELECT TOPIC	SELECT VIEW
Peripheral Nerves	Foot—Plantar

Click **LAYER 1** *in the* **LAYER CONTROLS** *window, and you will see the following image:*

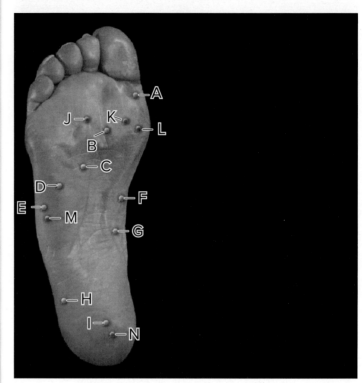

©McGraw-Hill Education

Mouse-over the pins on the screen to find the information necessary to identify the following structures:

A. _____
B. _____
C. _____
D. _____
E. _____
F. _____
G. _____
H. _____
I. _____

Nonnervous System Structures (blue pins)

J. _____
K. _____
L. _____
M. _____
N. _____

Click **LAYER 2** *in the* **LAYER CONTROLS** *window, and you will see the following image:*

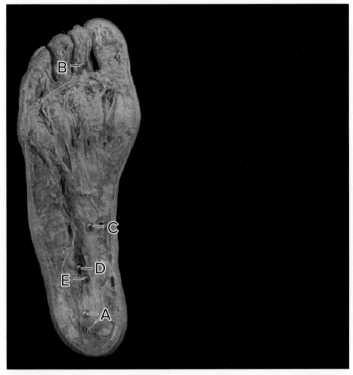

©McGraw-Hill Education

Mouse-over the pin on the screen to find the information necessary to identify the following structure:

A. _____

Nonnervous System Structures (blue pins)

B. _____
C. _____
D. _____
E. _____

Click **LAYER 3** *in the* **LAYER CONTROLS** *window, and you will see the following image:*

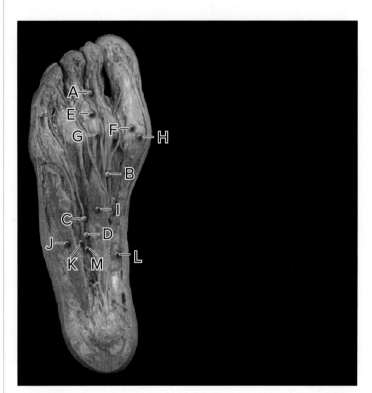

©McGraw-Hill Education

Mouse-over the pins on the screen to find the information necessary to identify the following structures:

A. _____
B. _____
C. _____
D. _____

Nonnervous System Structures (blue pins)

E. _____
F. _____
G. _____
H. _____
I. _____
J. _____
K. _____
L. _____
M. _____

Click **LAYER 4** *in the* **LAYER CONTROLS** *window, and you will see the following image:*

Mouse-over the pins on the screen to find the information necessary to identify the following structures:

A. _____
B. _____
C. _____
D. _____
E. _____

Nonnervous System Structures (blue pins)

F. _____
G. _____
H. _____
I. _____
J. _____
K. _____
L. _____
M. _____
N. _____
O. _____
P. _____
Q. _____
R. _____
S. _____
T. _____

Click **LAYER 5** *in the* **LAYER CONTROLS** *window, and you will see the following image:*

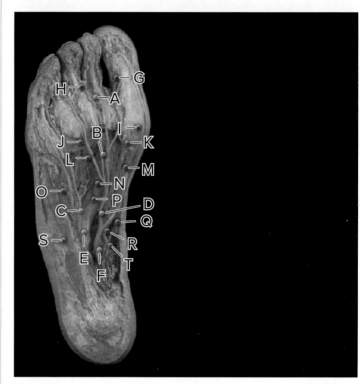

Mouse-over the pins on the screen to find the information necessary to identify the following structures:

A. _____
B. _____
C. _____
D. _____
E. _____
F. _____

Nonnervous System Structures (blue pins)

G. _____
H. _____
I. _____
J. _____
K. _____
L. _____
M. _____
N. _____
O. _____
P. _____

MODULE 7 The Nervous System 449

Q. _____
R. _____
S. _____
T. _____

O. _____
P. _____

Click **LAYER 6** *in the* **LAYER CONTROLS** *window, and you will see the following image:*

CHECK POINT

Peripheral Nerves, Foot

1. Name a nerve with motor composition to the muscles of the lateral leg and sensory composition from the skin of the anterior distal leg and the dorsum of the foot.
2. Name the motor nerves of the intrinsic foot muscles.
3. Name a nerve with general sensation composition from the skin and joints of the toes.

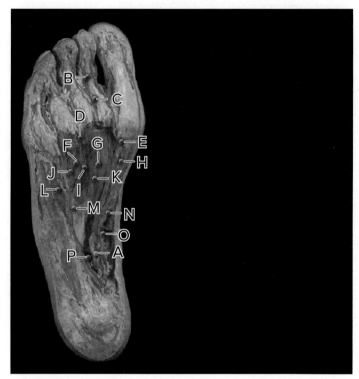

©McGraw-Hill Education

Mouse-over the pin on the screen to find the information necessary to identify the following structure:

A. _____

Nonnervous System Structures (blue pins)

B. _____
C. _____
D. _____
E. _____
F. _____
G. _____
H. _____
I. _____
J. _____
K. _____
L. _____
M. _____
N. _____

Self-Quiz

Take this opportunity to check your progress by taking the **QUIZ**. See the **Introduction Module** for a reminder on how to access the **QUIZ** for this Study Area.

Autonomic Nervous System

EXERCISE 7.53:
Nervous System—Sympathetic (ANS)—Overview

SELECT TOPIC: Sympathetic (ANS)   SELECT VIEW: Overview

Click **LAYER 1** *in the* **LAYER CONTROLS** *window, and you will see the following image:*

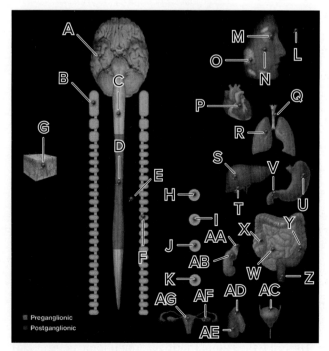

©McGraw-Hill Education

450 MODULE 7 The Nervous System

Mouse-over the pins on the screen to find the information necessary to identify the following structures:

A. _____
B. _____
C. _____
D. _____
E. _____
F. _____
G. _____
H. _____
I. _____
J. _____
K. _____
L. _____
M. _____
N. _____
O. _____
P. _____
Q. _____
R. _____
S. _____
T. _____
U. _____
V. _____
W. _____
X. _____
Y. _____
Z. _____
AA. _____
AB. _____
AC. _____
AD. _____
AE. _____
AF. _____
AG. _____

EXERCISE 7.54:
Nervous System—Sympathetic (ANS)—Thoracic Region

SELECT TOPIC	SELECT VIEW
Sympathetic (ANS)	Thoracic Region

After clicking through **LAYER 2** *and* **LAYER 3** *to gain perspective for this location, click* **LAYER 4** *in the* **LAYER CONTROLS** *window, and you will see the following image:*

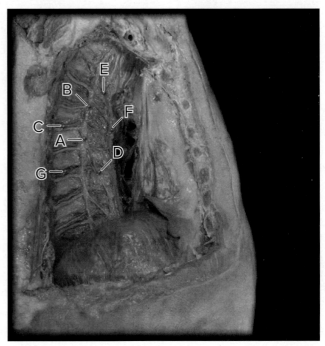

©McGraw-Hill Education

Mouse-over the pins on the screen to find the information necessary to identify the following structures:

A. _____
B. _____
C. _____
D. _____

Nonnervous System Structures (blue pins)

E. _____
F. _____
G. _____

CHECK POINT

Sympathetic (ANS), Overview

1. Name the structure also known as the sympathetic chain.
2. Name the sympathetic ganglion that distributes postganglionic neuronal processes to the stomach, duodenum, and spleen.
3. Describe the sympathetic postganglionic neuron and pathway of the adrenal medulla.

CHECK POINT

Sympathetic (ANS), Thoracic Region

1. Name the location of the postganglionic sympathetic neuronal cell bodies (somas).
2. What structures give the sympathetic trunk a beaded appearance?
3. Name the three nerves that contain preganglionic sympathetic nerve fibers that enter the abdomen to synapse in prevertebral ganglia.

MODULE 7 The Nervous System 451

Self-Quiz

Take this opportunity to check your progress by taking the **QUIZ**. See the **Introduction Module** for a reminder on how to access the **QUIZ** for this Study Area.

EXERCISE 7.55:
Nervous System—Parasympathetic (ANS)—Overview

SELECT TOPIC: Parasympathetic (ANS) ▶ **SELECT VIEW**: Overview

Click **LAYER 1** in the **LAYER CONTROLS** window, and you will see the following image:

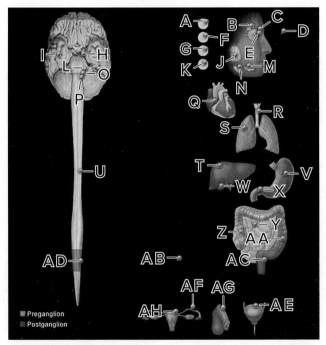

©McGraw-Hill Education

Mouse-over the pins on the screen to find the information necessary to identify the following structures:

A. _____
B. _____
C. _____
D. _____
E. _____
F. _____
G. _____
H. _____
I. _____
J. _____
K. _____
L. _____
M. _____
N. _____
O. _____
P. _____
Q. _____
R. _____
S. _____
T. _____
U. _____
V. _____
W. _____
X. _____
Y. _____
Z. _____
AA. _____
AB. _____
AC. _____
AD. _____
AE. _____
AF. _____
AG. _____
AH. _____

CHECK POINT

Parasympathetic (ANS), Overview

1. Where do the parasympathetic fibers terminate in the heart?
2. Name the nerves that distribute to the pelvic viscera via blood vessels.
3. The postganglionic fibers from which ganglion are distributed along branches of the auriculotemporal nerve to the parotid gland?

EXERCISE 7.56:
Nervous System—Parasympathetic (ANS)—Inferior Brain

SELECT TOPIC: Parasympathetic (ANS) ▶ **SELECT VIEW**: Inferior Brain

Click **LAYER 1** in the **LAYER CONTROLS** window, and you will see the following image:

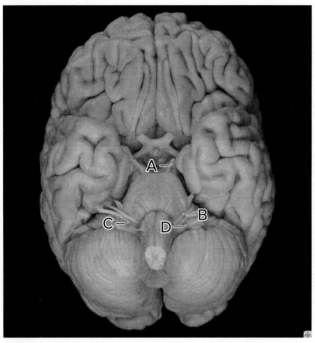

©McGraw-Hill Education

Mouse-over the pins on the screen to find the information necessary to identify the following structures:

A. _____

B. _____

C. _____

D. _____

CHECK POINT

Parasympathetic (ANS), Inferior Brain

1. Which cranial nerve has its postganglionic parasympathetic cell bodies (somas) located in the ciliary ganglia?

IN REVIEW

What Have I Learned?

The following questions cover the material that you have just learned, the peripheral nerves and autonomic nervous system. Apply what you have learned to answer these questions on a separate piece of paper.

1. What structure is also known as the sympathetic chain ganglia?

2. What nerve carries parasympathetic impulses to the smooth muscle of the thoracic and abdominal viscera?

3. Which nerve supplies all motor innervation to the diaphragm?

4. Name the sympathetic ganglion that distributes postganglionic neuronal processes to the kidneys and gonads.

5. Which division of the ANS has its terminal ganglia near or in the wall of the innervated organ?

6. Which division of the ANS has its terminal ganglia near the spinal cord?

7. Name a network of nerves with a branch that passes through the spermatic cord.

The Senses

EXERCISE 7.57:
Nervous System—Taste—Inferior Brain

SELECT TOPIC	SELECT VIEW
Taste	Inferior Brain

Click **LAYER 1** *in the* **LAYER CONTROLS** *window, and you will see the following image:*

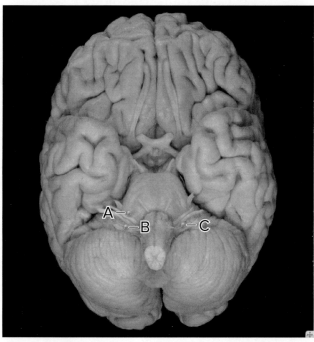

©McGraw-Hill Education

Mouse-over the pins on the screen to find the information necessary to identify the following structures:

A. _____
B. _____
C. _____

EXERCISE 7.58:
Nervous System—Taste—Tongue—Superior View

SELECT TOPIC	SELECT VIEW
Taste	Tongue—Superior

Click **LAYER 2** *in the* **LAYER CONTROLS** *window, and you will see the following image:*

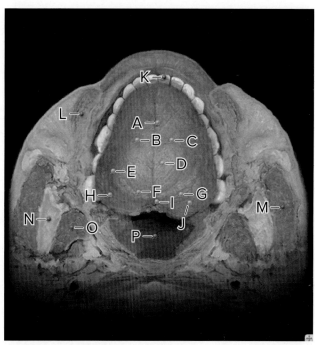

©McGraw-Hill Education

Mouse-over the pins on the screen to find the information necessary to identify the following structures:

A. _____
B. _____
C. _____
D. _____
E. _____
F. _____
G. _____
H. _____
I. _____
J. _____

Nonnervous System Structures (blue pins)

K. _____
L. _____
M. _____
N. _____
O. _____
P. _____

CHECK POINT

Taste

1. Name the shallow median longitudinal groove on the anterior part of the tongue.
2. Name the nerve that innervates the posterior tongue, pharynx, and middle ear.
3. Name the tonsil associated with the tongue.

CHECK POINT

Taste, Histology—Vallate Papilla

1. Name the nerve that receives special sensory information from the taste buds.
2. In what structure are the taste buds located?
3. Name the five primary taste sensations.

EXERCISE 7.59:

Nervous System—Taste—Vallate Papilla—Histology

SELECT TOPIC	SELECT VIEW
Taste—Vallate Papilla	LM: High Magnification

Click the **TAGS ON/OFF** button, and you will see the following image:

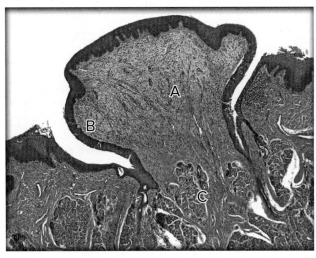

©McGraw-Hill Education/Al Telser

Mouse-over the pins on the screen to find the information necessary to identify the following structures:

A. _____

B. _____

C. _____

EXERCISE 7.60:

Nervous System—Taste— Taste Bud—Histology

SELECT TOPIC	SELECT VIEW
Taste—Taste Bud	LM: High Magnification

Click the **TAGS ON/OFF** button, and you will see the following image:

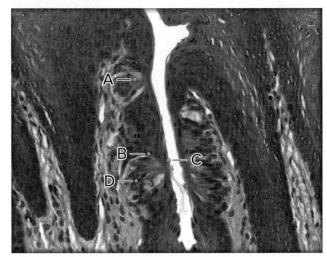

©McGraw-Hill Education/Al Telser

Mouse-over the pins on the screen to find the information necessary to identify the following structures:

A. _____

B. _____

C. _____

D. _____

CHECK POINT

Taste—Taste Bud, Histology

1. Name the structures that project into the taste pore.
2. What type of cells are taste cells?
3. What is the site of taste reception?

MODULE 7   The Nervous System   455

EXERCISE 7.61:
Nervous System—Smell—Inferior Brain

SELECT TOPIC	SELECT VIEW
Smell	Inferior Brain

Click **LAYER 1** *in the* **LAYER CONTROLS** *window, and you will see the following image:*

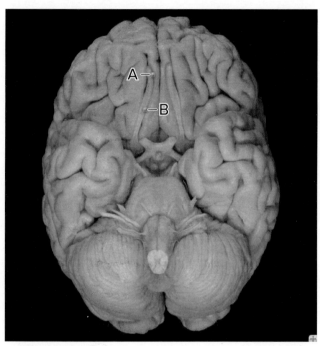

©McGraw-Hill Education

Mouse-over the pins on the screen to find the information necessary to identify the following structures:

A. _____

B. _____

CHECK POINT

Smell, Inferior Brain

1. Name the type of neurons found in the mucous membranes of the nasal cavities.
2. What bony structure do these neurons pass through to synapse with the olfactory nerve?
3. With what specific portion of the olfactory neuron do these neurons synapse?

EXERCISE 7.62:
Nervous System—Smell—Nasal Cavity—Lateral View

SELECT TOPIC	SELECT VIEW
Smell	Nasal Cavity—Lateral

Click **LAYER 2** *in the* **LAYER CONTROLS** *window, and you will see the following image:*

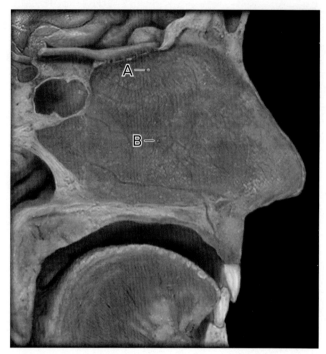

©McGraw-Hill Education

Mouse-over the pin on the screen to find the information necessary to identify the following structure:

A. _____

Nonnervous System Structure (blue pin)

B. _____

Click **LAYER 3** *in the* **LAYER CONTROLS** *window, and you will see the following image:*

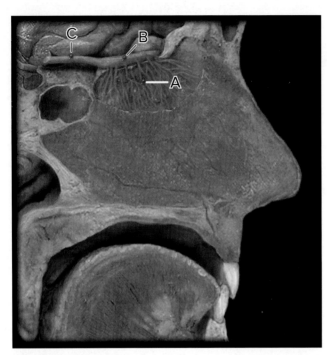

©McGraw-Hill Education

Mouse-over the pins on the screen to find the information necessary to identify the following structures:

A. _____

B. _____

C. _____

CHECK POINT

Smell, Nasal Cavity, Lateral View

1. Where in the brain does the olfactory nerve connect?
2. What structure detects odorants in the nasal cavities?
3. Name the expanded anterior end of the olfactory tract.

EXERCISE 7.63:

Nervous System—Smell—Olfactory Mucosa—Histology

SELECT TOPIC	SELECT VIEW
Smell—Olfactory Mucosa	LM: High Magnification

Click the **TAGS ON/OFF** *button, and you will see the following image:*

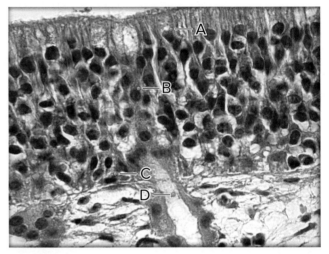

©McGraw-Hill Education/Steve Sullivan

Mouse-over the pins on the screen to find the information necessary to identify the following structures:

A. _____

B. _____

C. _____

D. _____

CHECK POINT

Smell, Histology—Olfactory Mucosa

1. Name the structures that produce seromucus in the nasal cavities.
2. What is the function of the seromucus?

EXERCISE 7.64:
Nervous System—Hearing/Balance— Inferior Brain

SELECT TOPIC: Hearing/Balance ▸ **SELECT VIEW**: Inferior Brain

Click **LAYER 1** in the **LAYER CONTROLS** window, and you will see the following image:

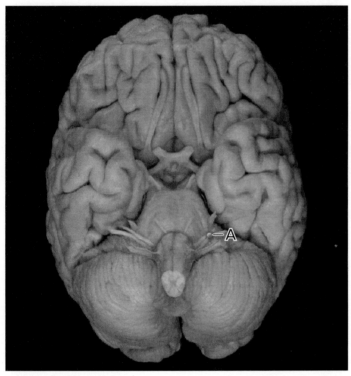

©McGraw-Hill Education

Mouse-over the pin on the screen to find the information necessary to identify the following structure:

A. _____

CHECK POINT

Hearing/Balance, Inferior Brain

1. What two special senses are provided by the vestibulocochlear nerve?
2. Which branches of the vestibulocochlear nerve function for each of the two special senses?
3. What is the CNS connection for the vestibulocochlear nerve?

EXERCISE 7.65:
Nervous System—Hearing/Balance— Ear—Anterior View

SELECT TOPIC: Hearing/Balance ▸ **SELECT VIEW**: Ear—Anterior

Click **LAYER 1** in the **LAYER CONTROLS** window, and you will see the following image:

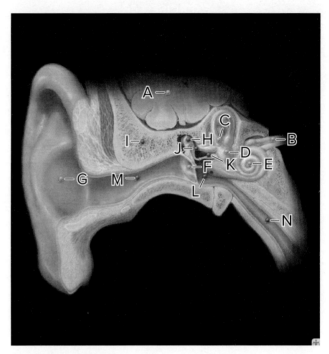

©McGraw-Hill Education

Mouse-over the pins on the screen to find the information necessary to identify the following structures:

A. _____
B. _____
C. _____
D. _____
E. _____
F. _____
G. _____

Nonnervous System Structures (blue pins)

H. _____
I. _____
J. _____
K. _____
L. _____
M. _____
N. _____

CHECK POINT

Hearing/Balance, Ear, Anterior View

1. Name the organ of hearing.
2. Name the structure of the ear that senses linear movement.
3. Name the organ of equilibrium.

EXERCISE 7.66:

Imaging—Tympanic Membrane

SELECT TOPIC: Tympanic Membrane ▶ SELECT VIEW: Otoscopic Photograph—Lateral

Click the **TAGS ON/OFF** button, and you will see the following image:

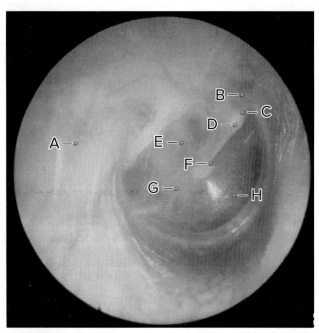

©SPL/Science Source

Mouse-over the pins on the screen to find the information necessary to identify the following structures:

A. _____
B. _____
C. _____
D. _____
E. _____
F. _____
G. _____
H. _____

EXERCISE 7.67:

Imaging—Tympanic Membrane with Otitis Media

SELECT TOPIC: Tympanic Membrane with Otitis Media ▶ SELECT VIEW: Otoscopic Photograph—Lateral

Click the **TAGS ON/OFF** button, and you will see the following image:

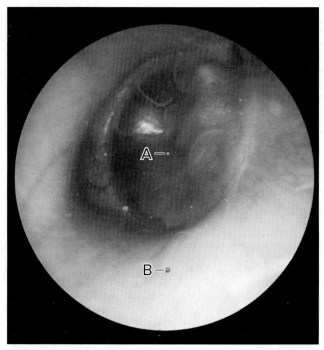

©McGraw-Hill Education

Mouse-over the pins on the screen to find the information necessary to identify the following structures:

A. _____
B. _____

MODULE 7 The Nervous System 459

EXERCISE 7.68a:
Nervous System—Hearing/Balance—Cochlea—Histology

SELECT TOPIC	SELECT VIEW
Hearing/balance, Cochlea	LM: Low Magnification

*Click the **TAGS ON/OFF** button, and you will see the following image:*

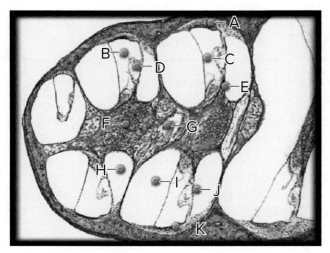

©Victor Eroschenko

Mouse-over the pins on the screen to find the information necessary to identify the following structures:

A. _____
B. _____
C. _____
D. _____
E. _____
F. _____
G. _____
H. _____
I. _____
J. _____
K. _____

EXERCISE 7.68b:
Nervous System—Hearing/Balance—Cochlea—Histology

SELECT TOPIC	SELECT VIEW
Hearing/Balance—Cochlea	LM: High Magnification

*Click the **TAGS ON/OFF** button, and you will see the following image:*

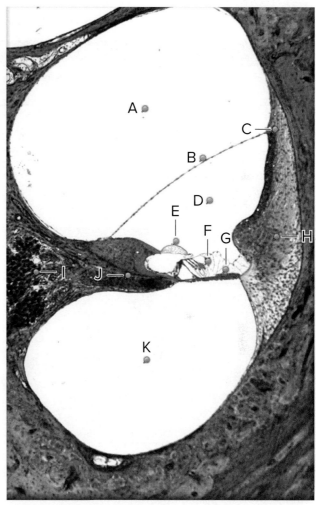

©Biophoto Assoc./Science Source

Mouse-over the pins on the screen to find the information necessary to identify the following structures:

A. _____
B. _____
C. _____
D. _____
E. _____
F. _____
G. _____
H. _____
I. _____
J. _____
K. _____

CHECK POINT

Hearing/Balance, Histology—Cochlea

1. Name the three fluid-filled chambers of the cochlea, from superior to inferior.
2. What specific structure of the cochlea is the site of conversion of sound vibrations into electrochemical signals?
3. Name the structure that supports this structure.

CHECK POINT

Hearing/Balance, Histology—Spiral Organ

1. Name the structure with the function of transmitting vibrations to embedded stereocilia of sensory cells.
2. What specific structures function as the receptors for hearing?
3. Describe the function of stereocilia.

EXERCISE 7.69:

Nervous System—Hearing/Balance, Spiral Organ—Histology

SELECT TOPIC	SELECT VIEW
Hearing/Balance—Spiral Organ ▶	Low Magnification

Click the **TAGS ON/OFF** button, and you will see the following image:

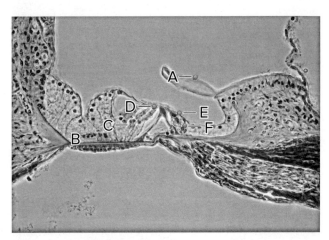

©J. Carrillo-Farg/Science Source

Mouse-over the pins on the screen to find the information necessary to identify the following structures:

A. _____

B. _____

C. _____

D. _____

E. _____

F. _____

 SELECT ANIMATION

Hearing PLAY

After viewing the animation, answer the following questions:

1. What structure do sound waves strike and cause to vibrate?

2. The vibrations are transferred to the three bones of the middle ear. From lateral to medial, what are those bones?

3. What structure transfers this vibration to the oval window?

4. The vibrations are then transferred to a fluid-filled chamber of the inner ear. What is that fluid, and what is the name of that chamber?

5. What is the difference in location for the detection of high-pitched and low-pitched sounds?

MODULE 7 The Nervous System 461

EXERCISE 7.70:
Nervous System—Vision—Inferior Brain

SELECT TOPIC	SELECT VIEW
Vision	Inferior Brain

Click **LAYER 1** *in the* **LAYER CONTROLS** *window, and you will see the following image:*

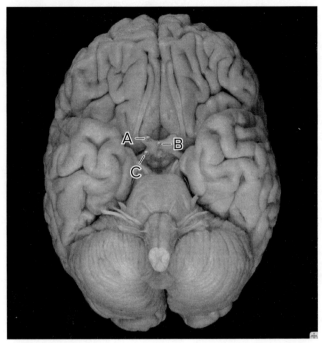

©McGraw-Hill Education

Mouse-over the pins on the screen to find the information necessary to identify the following structures:

A. _____

B. _____

C. _____

CHECK POINT

Vision, Inferior Brain

1. Name the cranial nerve of vision.
2. What is this nerve's origin?
3. Where does this nerve terminate?

EXERCISE 7.71:
Nervous System—Vision— Orbit—Lateral View

SELECT TOPIC	SELECT VIEW
Vision	Orbit—Lateral

Click **LAYER 2** *to gain perspective on this structure, and then click* **LAYER 3** *in the* **LAYER CONTROLS** *window, and you will see the following image:*

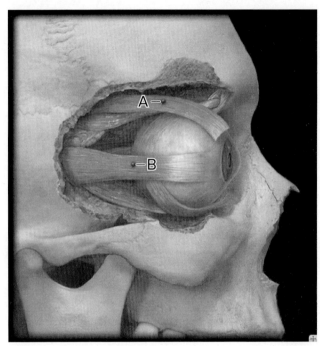

©McGraw-Hill Education

Mouse-over the pins on the screen to find the information necessary to identify the following nonnervous system structures:

A. _____

B. _____

462 MODULE 7 The Nervous System

Click **LAYER 4** *in the* **LAYER CONTROLS** *window, and you will see the following image:*

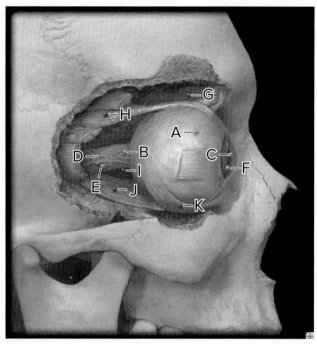

©McGraw-Hill Education

Mouse-over the pins on the screen to find the information necessary to identify the following structures:

A. _____
B. _____
C. _____
D. _____
E. _____
F. _____

Nonnervous System Structures (blue pins)

G. _____
H. _____
I. _____
J. _____
K. _____

CHECK POINT

Vision, Orbit, Lateral View

1. Name and describe the colored part of the eye.
2. What is its function?
3. Describe the short ciliary nerves.

Click **LAYER 5** *in the* **LAYER CONTROLS** *window, and you will see the following image:*

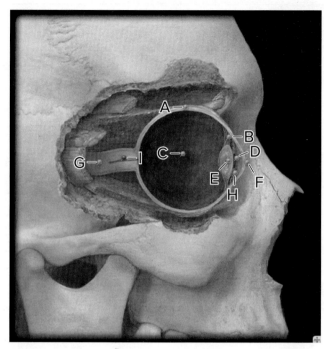

©McGraw-Hill Education

Mouse-over the pins on the screen to find the information necessary to identify the following structures:

A. _____
B. _____
C. _____
D. _____
E. _____
F. _____
G. _____
H. _____

Nonnervous System Structure (blue pin)

I. _____

CHECK POINT

Vision, Orbit, Lateral View, *continued*

4. Name the nerve responsible for the special sensation of vision.
5. Name the outer layer of the posterior five-sixths of the eye.
6. Name the outer layer of the anterior one-sixth of the eye.

MODULE 7 The Nervous System 463

EXERCISE 7.72:
Nervous System—Vision—Eye—Lateral View

SELECT TOPIC	SELECT VIEW
Vision	Eye—Lateral

Click **LAYER 1** in the **LAYER CONTROLS** window, and you will see the following image:

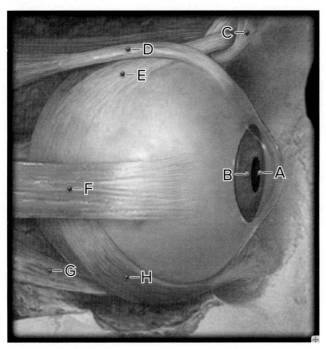

©McGraw-Hill Education

Mouse-over the pins on the screen to find the information necessary to identify the following structures:

A. _____
B. _____

Nonnervous System Structures (blue pins)

C. _____
D. _____
E. _____
F. _____
G. _____
H. _____

CHECK POINT
Vision, Eye, Lateral View

1. Name the muscle responsible for abduction of the eyeball.
2. Name the muscle responsible for depression and medial rotation of the eyeball.

Click **LAYER 2** in the **LAYER CONTROLS** window, and you will see the following image:

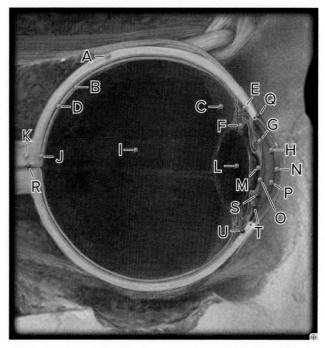

©McGraw-Hill Education

Mouse-over the pins on the screen to find the information necessary to identify the following structures:

A. _____
B. _____
C. _____
D. _____
E. _____
F. _____
G. _____
H. _____
I. _____
J. _____
K. _____
L. _____
M. _____
N. _____
O. _____
P. _____

Nonnervous System Structures (blue pins)

Q. _____
R. _____
S. _____
T. _____
U. _____

464 MODULE 7 The Nervous System

CHECK POINT

Vision, Eye, Lateral View, *continued*

3. Name the structure referred to as the "white" of the eye.
4. Name the artery that, when blocked, may cause blindness.
5. Name the internal structure of the eye that focuses incoming light.

IN REVIEW

What Have I Learned?

The following questions cover the material that you have just learned, vision, the orbit, and the eye. Apply what you have learned to answer the following questions on a separate piece of paper.

1. Name and describe the colored part of the eye.
2. What is its function?
3. Name the opening in the middle of this structure.
4. Describe the short ciliary nerves.
5. Describe the ciliary ganglion.
6. Name the muscle that allows looking upward and outward.
7. Name the muscle that allows looking downward and medially.
8. Name the nerve responsible for the special sensation of vision.
9. Name the outer layer of the posterior five-sixths of the eye.
10. Describe the tissue of this structure.
11. Name the outer layer of the anterior one-sixth of the eye.
12. Describe the tissue of this structure.

EXERCISE 7.73:
Nervous System—Vision—Retina—Histology

SELECT TOPIC	SELECT VIEW
Vision—Retina	LM: High Magnification

Click the **TAGS ON/OFF** button, and you will see this image:

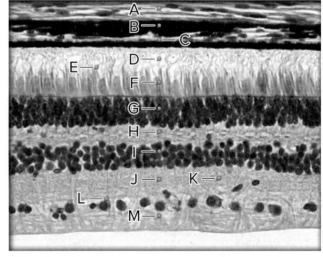

©Victor Eroschenko

Mouse-over the pins on the screen to find the information necessary to identify the following structures:

A. _____
B. _____
C. _____
D. _____
E. _____
F. _____
G. _____
H. _____
I. _____
J. _____
K. _____
L. _____
M. _____

Mouse-over the pins on the screen to find the information necessary to identify the following structures:

A. _____
B. _____
C. _____

CHECK POINT

Vision—Retina

1. Which eye layer prevents light reflection within the eye?
2. What pigment is located in this layer?
3. Which organelles are abundant in the rod and cone inner segments? What are their functions?
4. Name the structures that transmit signals from ganglion cells to the optic nerve?
5. What is the function of the rods? What visual pigments do they contain?
6. What is the function of the cones? What visual pigments do they contain?

EXERCISE 7.74:

Nervous System—Vision— Retinal Rods and Cones—Histology

SELECT TOPIC	SELECT VIEW
Vision—Retinal Rods and Cones	SEM: High Magnification

Click the **TAGS ON/OFF** *button, and you will see the following image:*

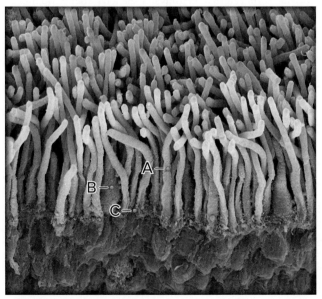

©Science Photo Library RF

EXERCISE 7.75:

Imaging—Retina

SELECT TOPIC	SELECT VIEW
Retina	Ophthalmoscopic Photograph—Anterior

Click the **TAGS ON/OFF** *button, and you will see the following image:*

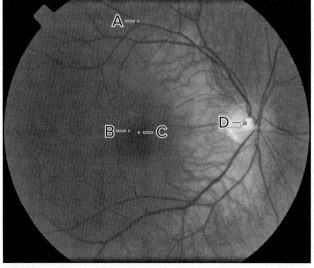

©SPL/Science Source

Mouse-over the pins on the screen to find the information necessary to identify the following structures:

A. _____
B. _____
C. _____
D. _____

CHECK POINT

Imaging—Retina

1. Identify the eye's "blind spot." What causes this phenomenon?
2. Describe the macula lutea. Where is it located?
3. Describe the fovea centralis. Where is it located?
4. Describe the foveola. Where is it located?

IN REVIEW

What Have I Learned?

The following questions cover the material that you have just learned, the senses. Apply what you have learned to answer these questions on a separate piece of paper.

1. Name the cranial nerves responsible for taste. Name the structures innervated by each of these nerves.
2. Name the ovoid collections of lymphoid tissue covered with mucous membrane on the posterior aspect of the tongue.
3. Name the three subdivisions of the pharynx.
4. Name the three types of cells found in taste buds.
5. What is another name for olfactory neurons?
6. Name the cranial nerve responsible for hearing and balance.
7. Name the organ of hearing.
8. Name the two organs of balance.
9. Name the bony canal of the external ear.
10. What is the anatomical name for the eardrum?
11. Name the three smallest bones in the body from lateral to medial.
12. What is the function of these three bones?
13. What is the name for the passage between the tympanic cavity and nasopharynx?
14. What is the function of this passage?
15. The auditory ossicles articulate with what structure of the inner ear?
16. What are the functions of the three fluid-filled chambers of the cochlea?
17. Where does the lens of the eye focus the incoming light?
18. What term refers to changes of the lens' shape when focusing?
19. Name the structure that prevents light scatter in the eye.

We conclude the Nervous System module by exploring five practical clinical applications of the information you have learned in this module. Thoughtfully complete the following exercises, knowing that you will possibly encounter these scenarios in either your personal life or in your professional career.

Clinical Application

 SELECT ANIMATION Neuron/Alzheimer's Disease (3D) **PLAY**

After viewing the animation, answer the following questions:

1. The brain's cognitive functions include _____, which is the process of _____ _____ and _____.
2. Where do memories form?
3. What structures compose this portion of the brain?
4. How does each neuron receive chemical messages from other neurons?
5. What happens to the chemical message?

6. What is the message called when changed from a chemical to an electrical message?

7. What are neurotransmitters? Where are neurotransmitters released? Where do they bind?

8. What is the function of the neurotransmitters once they bind to the receptors?

9. _____ _____ is a gradual degenerative brain condition in which _____ in the memory and other cognitive areas of the brain _____ _____ and _____.

10. This results in _____ _____ _____ first, then difficulties _____ and _____, and eventually failing basic functions like _____.

11. In healthy neurons, what process occurs to help with normal cell function?

12. In Alzheimer's, abnormal _____ processing produces _____ that include a stick _____ called _____ _____.

13. _____ _____ accumulates in the _____ _____, forming clumps called _____ _____.

14. These clumps _____ the _____ and _____ connections between _____.

15. Inside healthy neurons, _____ and _____ proteins form an _____ _____ that carries _____ and _____ within the cell.

16. In Alzheimer's, what are the results of chemical changes inside the neuron?

17. Describe neurofibrillary tangles.

18. What is the result of the neurofibrillary tangles?

19. What causes neural connections to diminish?

20. What then results as the cells become malnourished?

21. What is the cure for Alzheimer's?

22. Describe the current treatments for Alzheimer's.

SELECT ANIMATION
Spinal Cord Anatomy and Injury (3D) PLAY

After viewing the animation, answer the following questions:

1. Where does the spinal cord begin? Where does it exit the skull? How far does it extend?

2. List the regions of the spinal cord.

3. The spinal cord gives rise to pairs of _____ _____ that supply _____ and _____ function to the body.

4. Describe the cauda equine and its location.

5. The spinal cord is composed of _____ _____ and _____ _____.

6. What structures are contained in gray matter?

7. What structures are contained in white matter?

8. Describe the pathway of sensory nerve impulses.

9. Describe the pathway of motor nerve impulses.

10. What structures merge to form each spinal nerve?

11. How does the American Spinal Injury Association classify spinal cord injuries?

12. Describe a complete or level A spinal cord injury.

13. Describe dermatomes.

14. Describe the sensory and motor results of a C4 to C5 level A spinal cord injury.

15. Describe tetraplegia or quadriplegia.

16. Structures with no motor function may include the _____, requiring _____ _____ _____.

17. Describe acute treatment for a C4 to C5 level A spinal cord injury.

18. What are used to reduce inflammation in a C4 to C5 level A spinal cord injury?

19. What are the results of physical and occupational therapy?

SELECT ANIMATION
Concussion (3D)
PLAY

After viewing the animation, answer the following questions:

1. Describe a concussion.

2. Describe the physical characteristics of the brain that require physical protection from the environment.

3. What structures provide a hard external shield from outside impacts.

4. Describe the brain's environment inside the cranium.

5. What is the function of cerebrospinal fluid?

6. What can cause the brain to rotate and strike the inside of the skull.

7. What are the results of this injury?

8. What are the results of the impact force?

9. What are the results of diffuse axonal shearing?

10. List the physical symptoms of concussion.

11. List the cognitive and emotional symptoms of concussion.

12. What determines the treatment for concussion?

13. List the treatments for mild concussion.

14. What two items should be avoided with mild concussion.

15. What is required for severe concussions?

16. In many cases, _____ are used to _____ _____ and _____ in the brain.

17. What evaluations are used to rule out internal damage or hemorrhages?

SELECT ANIMATION
Stroke: Cerebrovascular Accident (3D) PLAY

After viewing the animation, answer the following questions:

1. List the two major pairs of arteries supplying blood to the brain.

2. What do the branches of these arteries supply to the brain?

3. Describe a stroke or CVA.

4. What is ischemia?

5. What are the results of ischemia?

6. List results that may be caused by a stroke.

7. What is a Transient Ischemic Attack, or TIA?

8. What causes a TIA?

9. List the effects a TIA may cause.

10. What can a TIA signal?

11. What is the most common type of stroke?

12. What are the two classifications of ischemic stroke?

13. Describe a thrombotic ischemic stroke.

14. Describe an embolic ischemic stroke.

15. What is the other type of stroke?

16. Describe this type of stroke.

17. Describe intracerebral hemorrhagic strokes.

18. Describe subarachnoid hemorrhagic strokes.

19. List two causes of hemorrhagic stroke.

20. Describe acute thrombotic stroke care.

21. What is the time limit for this treatment?

22. Describe long-term stroke treatment.

23. What do these treatments prevent?

24. How are lost or impaired skills recovered?

25. How do patients re-learn motor function and daily activities?

26. What therapies help the patient express themselves and cope with new challenges.

SELECT ANIMATION
Hearing/Hearing Loss (3D) — PLAY

After viewing the animation, answer the following questions:

1. Describe the process of hearing up to the cochlear duct.

2. Describe in detail the structures that constitute the organ of Corti.

3. Describe in detail how these structures create a nerve impulse in the nerve fibers.

4. Nerve fibers throughout the _____ combine to form the _____ branch of the _____ or _____ nerve which transmits the sound nerve impulse to the _____.

5. What causes sensorineural hearing loss?

6. What can cause damage to the auditory nerve pathway?

7. What is another common cause of hearing loss?

8. Describe Meniere's disease.

9. List diagnostic tests for hearing loss.

10. Depending on the cause and severity of _____ _____ _____, sensorineural hearing loss is treated with devices such as _____ _____ or _____ _____.

EXERCISE 7.76:
Coloring Exercise

Identify these features of the ventral surface of the brain and the cranial nerves. Then, using colored pens or pencils, color in the figure.

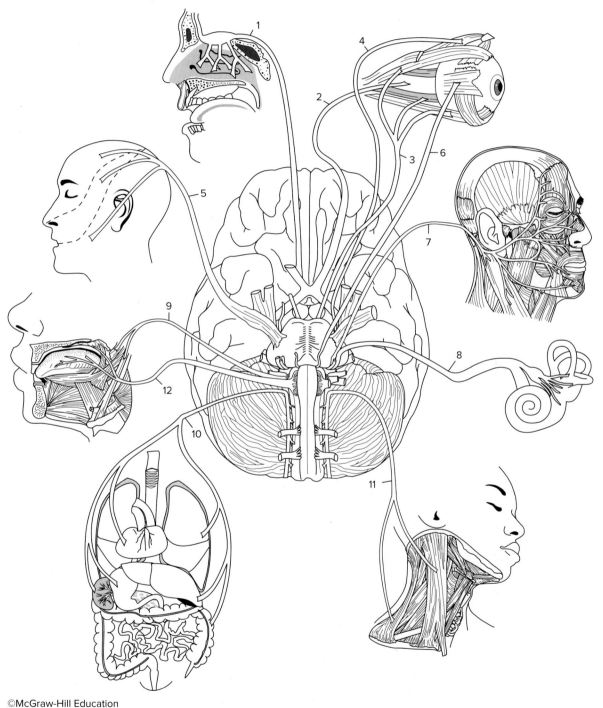

©McGraw-Hill Education

- ☐ 1. _____
- ☐ 2. _____
- ☐ 3. _____
- ☐ 4. _____
- ☐ 5. _____
- ☐ 6. _____
- ☐ 7. _____
- ☐ 8. _____
- ☐ 9. _____
- ☐ 10. _____
- ☐ 11. _____
- ☐ 12. _____

472　MODULE 7　The Nervous System

EXERCISE 7.77:
Coloring Exercise

Identify the parts of the eye and accessory structures. Then color them with colored pencils.

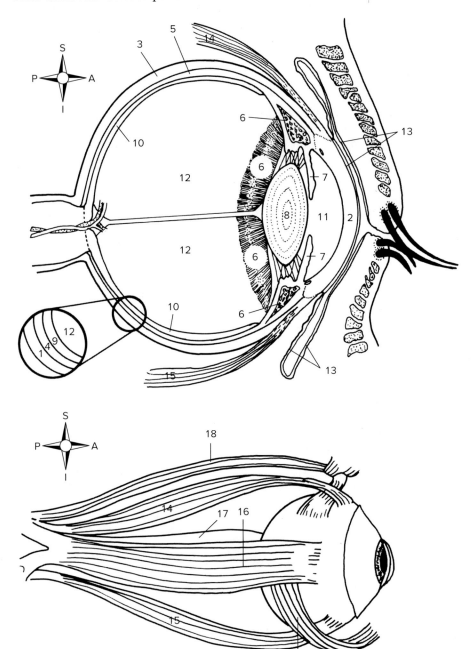

☐ 1. _____
☐ 2. _____
☐ 3. _____
☐ 4. _____
☐ 5. _____
☐ 6. _____
☐ 7. _____
☐ 8. _____
☐ 9. _____
☐ 10. _____
☐ 11. _____
☐ 12. _____
☐ 13. _____
☐ 14. _____
☐ 15. _____
☐ 16. _____
☐ 17. _____
☐ 18. _____
☐ 19. _____

©McGraw-Hill Education

MODULE 8

The Endocrine System

Overview: The Endocrine System

The endocrine system is one of the two organ systems involved with internal communication. It often works hand-in-hand with the other—the nervous system. An example of this complement between the two systems occurs between the autonomic nervous system and the suprarenal (adrenal) gland of the endocrine system during the fight-or-flight response.

We will revisit the function of the endocrine system when we discuss the reproductive system in the last module. The body orchestrates the dynamic events of puberty and reproduction with intercellular messengers called hormones. Hormones regulate many other bodily functions, including but not limited to digestion, blood glucose levels, mood, and growth. Your textbook has an excellent discussion of this topic.

We will begin with a discussion of what has been termed the master gland because of its regulation of other endocrine glands and conclude with the endocrine system's involvement with reproduction.

From the **HOME** *screen, click the drop-down box on the* **MODULE** *menu.*

From the systems listed, click on **Endocrine**.

SELECT ANIMATION
Hypothalamus and Pituitary Gland   PLAY

After viewing the animation, answer the following questions:

1. The hypothalamus is sometimes referred to as the _____ _____ _____. Why?

2. Where in the brain is the hypothalamus located?

3. Describe the structure of the hypothalamus.

4. What is the infundibulum? What is its function?

5. Where is the pituitary gland located? How is it divided?

6. What is another name for the anterior pituitary? How is it connected to the hypothalamus?

7. What travels along this pathway? What is their function?

8. What is another name for the posterior pituitary? How is it connected to the hypothalamus?

9. What travels along this pathway? How are they transported? What is their destination?

10. Name the two classes of hypothalamic hormones that regulate the anterior pituitary. How do they reach the anterior pituitary? What is their function?

11. How do anterior pituitary hormones arrive at their target tissues?

12. Describe an example of these hormones and their function.

13. Name the hormones produced by the posterior pituitary. What is the source of posterior pituitary hormones?

14. Name two posterior pituitary hormones. How do they arrive at the posterior pituitary?

15. Name the structures that store the posterior pituitary hormones. What causes their release? Where are they released?

16. Name the functions of each posterior pituitary hormone.

Self-Quiz
Take this opportunity to check your progress by taking the **QUIZ**. See the **Introduction Module** for a reminder on how to access the **QUIZ** for this Study Area.

SELECT ANIMATION — Hormonal Communication — **PLAY**

After viewing the animation, answer the following questions:

1. In general, how does hormonal communication begin? What reaction then occurs?

2. How are hormones transported to target cells?

3. What occurs when the hormones arrive at their target cells?

4. What then triggers changes in the target cells?

Self-Quiz
Take this opportunity to check your progress by taking the **QUIZ**. See the **Introduction Module** for a reminder on how to access the **QUIZ** for this Study Area.

SELECT ANIMATION — Intracellular Receptor Model — **PLAY**

After viewing the animation, answer the following questions:

1. Describe aldosterone, the hormone used in the animation.

2. What does aldosterone bind with in the cytoplasm of the cell?

3. Where does the aldosterone-receptor complex go, and where does it bind?

4. This binding stimulates the synthesis of what molecule? What is the function of this molecule?

5. Where does this mRNA molecule go, and what does it do?

6. What is directed by this binding? What response is produced?

Self-Quiz
Take this opportunity to check your progress by taking the **QUIZ**. See the **Introduction Module** for a reminder on how to access the **QUIZ** for this Study Area.

SELECT ANIMATION — Receptors and G Proteins — **PLAY**

After viewing the animation, answer the following questions:

1. What is located on the membrane-bound receptor on the outside of the cell?

2. What is a ligand?

3. To what does the portion of the membrane-bound receptor on the inside of the cell bind?

4. What are the three subunits of this protein? What is attached to the alpha subunit?

5. What changes occur in the G protein when the ligand binds to the receptor site? What changes occur to the alpha subunit?

6. What now occurs with the activated alpha subunit? How long can this step be repeated?

7. What occurs when the ligand separates from the receptor site?

8. How is the alpha subunit inactivated?

9. What occurs with the G protein subunits after this inactivation?

SELECT ANIMATION
Second Messengers: The cAMP & Ca^{2+} Pathways
PLAY

After viewing the animation, answer the following questions:

1. How do second messengers affect changes inside of cells?

2. The signal molecule is referred to as the _____ messenger.

3. What molecule is activated by the binding of the signal molecule to its receptor?

4. Typically, the G protein activates . . .

5. Some second messenger systems involve activating . . .

6. The other type of second messenger system involves _____ ions.

7. Whether triggering phosphorylation or the release of Ca^{2+}, second messenger systems . . .

Self-Quiz
Take this opportunity to check your progress by taking the **QUIZ**. See the **Introduction Module** for a reminder on how to access the **QUIZ** for this Study Area.

The Hypothalamus, Pituitary, and Pineal Glands

EXERCISE 8.1:
Endocrine System—Hypothalamus/Pituitary/Pineal—Lateral View

 | **SELECT TOPIC** Hypothalamus/Pituitary/Pineal ▶ **SELECT VIEW** Lateral

After clicking **LAYER 2** *to orient yourself to our location, click* **LAYER 3** *in the* **LAYER CONTROLS** *window, and you will see the following image:*

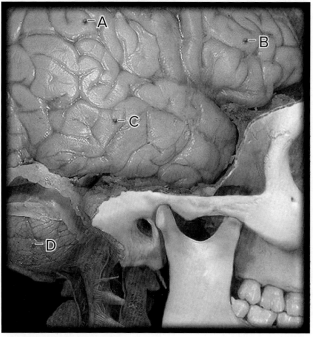

©McGraw-Hill Education

Mouse-over the pins on the screen to find the information necessary to identify the following nonendocrine system structures:

A. _____

B. _____

C. _____

D. _____

476 MODULE 8 The Endocrine System

Click **LAYER 4** *in the* **LAYER CONTROLS** *window, and you will see the following image:*

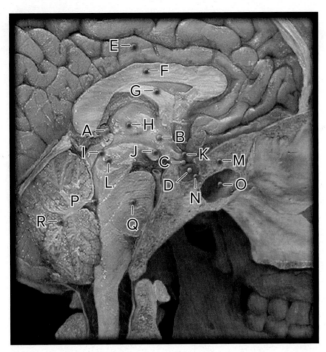

©McGraw-Hill Education

Mouse-over the pins on the screen to find the information necessary to identify the following structures:

A. _____

B. _____

C. _____

D. _____

Nonendocrine System Structures (blue pins)

E. _____

F. _____

G. _____

H. _____

I. _____

J. _____

K. _____

L. _____

M. _____

N. _____

O. _____

P. _____

Q. _____

R. _____

CHECK POINT

Hypothalamus/Pituitary/Pineal Glands, Lateral View

1. Describe the structure of the hypothalamus.
2. Name the functions of the hypothalamus.
3. Describe the size and location of the pineal gland. What are its known functions?
4. Describe the location and structure of the pituitary gland.
5. List the functions of the hypothalamus.

EXERCISE 8.2:

Imaging—Hypothalamus and Pituitary Gland

 SELECT TOPIC: **Hypothalamus and Pituitary Gland** ▶ SELECT VIEW: **Sagittal**

Click the **TAGS ON/OFF** *button, and you will see the following image:*

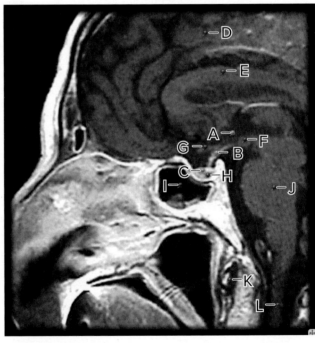

©McGraw-Hill Education

Mouse-over the pins on the screen to find the information necessary to identify the following structures:

A. _____

B. _____

C. _____

Nonendocrine System Structures (blue pins)

D. _____

E. _____

F. _____

G. _____

H. _____
I. _____
J. _____
K. _____
L. _____

The Pituitary Gland

EXERCISE 8.3:
Endocrine System—Pituitary—Histology

SELECT TOPIC: **Pituitary** ▸ SELECT VIEW: **LM: Low Magnification**

Click the **TAGS ON/OFF** button, and you will see the following image:

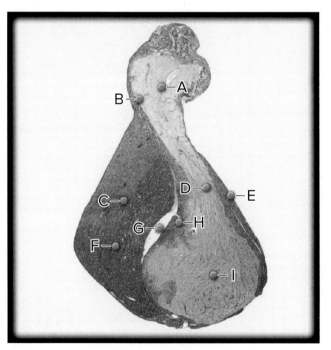

©Michael Ross/Photo Researchers, Inc.

Mouse-over the pins on the screen to find the information necessary to identify the following structures:

A. _____
B. _____
C. _____
D. _____
E. _____
F. _____
G. _____
H. _____
I. _____

EXERCISE 8.4:
Endocrine System—Anterior Pituitary—Histology

SELECT TOPIC: **Anterior Pituitary** ▸ SELECT VIEW: **LM: High Magnification**

Click the **TAGS ON/OFF** button, and you will see the following image:

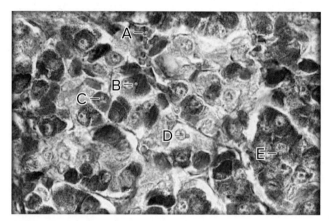

©Victor Eroschenko

Mouse-over the pins on the screen to find the information necessary to identify the following structures:

A. _____
B. _____
C. _____
D. _____
E. _____

CHECK POINT

Anterior Pituitary, Histology

1. What are chromophils? Name the two subtypes of chromophils.
2. Which chromophils secrete somatotropin and prolactin? What is somatotropin?
3. Name the functions of the basophils.

478 MODULE 8 The Endocrine System

EXERCISE 8.5:
Endocrine System—Posterior Pituitary—Histology

SELECT TOPIC	SELECT VIEW
Posterior Pituitary	LM: Medium Magnification

Click the **TAGS ON/OFF** *button, and you will see the following image:*

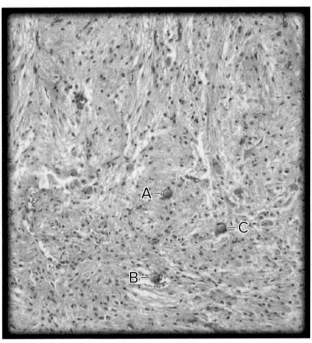

©Lutz Slomianka

Mouse-over the pins on the screen to find the information necessary to identify the following structures:

A. _____

B. _____

C. _____

EXERCISE 8.6:
Endocrine System—Posterior Pituitary—Histology

SELECT TOPIC	SELECT VIEW
Posterior Pituitary	LM: High Magnification

Click the **TAGS ON/OFF** *button, and you will see the following image:*

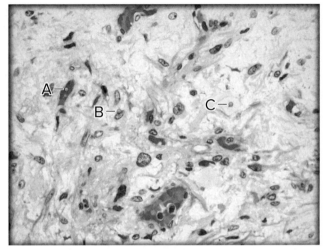

©McGraw-Hill Education/Steve Sullivan

Mouse-over the pins on the screen to find the information necessary to identify the following structures:

A. _____

B. _____

C. _____

CHECK POINT

Posterior Pituitary, Histology

1. What is another name for antidiuretic hormone?
2. What are herring bodies? What hormones are associated with them?
3. What is the function of the small blood vessels of the posterior pituitary?

The Thyroid Gland

SELECT ANIMATION
Thyroid Gland — PLAY

After viewing the animation, answer the following questions:

1. Name the largest endocrine gland. Describe its location and structure.

2. What causes the thyroid's reddish color? Name the blood vessels that distribute to and drain the thyroid gland. What other function do the veins have?

3. The thyroid gland is composed of spherical structures called _____ _____. Describe the wall of these structures. What do they surround?

4. Describe what is located in the lumen of the follicles.

5. What are located in the loose connective tissue between the follicles? Which cells secrete hormones?

6. Name the hormone secreted and its function. What is its antagonist?

7. Name the two different hormones referred to as thyroid hormone.

8. Name the hormone that maintains TH synthesis and secretion. Where is this hormone secreted?

9. What is contained in thyroglobulin?

10. Name the molecules that cross the follicular cell from the blood in the first phase of TH production.

11. What is the destination of these molecules? What are they converted to?

12. What two molecules combine in the lumen of the follicles? What is formed by this combination? How long can it be stored in the follicle lumen?

13. In the second phase of TH production, what becomes of the combined molecules formed in phase one?

14. Name the primary effect of TH. Describe its importance in children.

15. What is hyperthyroidism? What can result from this condition?

16. What is hypothyroidism? What can result from this condition?

Clinical Application

SELECT ANIMATION
Thyroid Function/ Hyperthyroidism (3D) — PLAY

After viewing the animation, answer the following questions:

1. Which of your body's endocrine glands is the largest?

2. Where is it located?

3. List a function of the thyroid.

4. What is the function of this hormone?

5. How does thyroid hormone affect specific organs to increase metabolic rate?

6. Thyroid hormone interacts with the _____ and the _____ _____ in a _____ _____ _____ to affect metabolic rate.

7. What effect does low levels of thyroid hormone in the blood have on the hypothalamus?

8. Thyrotropin releasing hormone then . . .

9. Thyroid stimulating hormone then . . .

10. What are the two hormones secreted as thyroid hormone?

11. Where are the effects of T3 and T4 manifested?

12. What organ detects the elevated levels of thyroxine? What is its response?

13. What is the result of this negative feedback loop?

14. What is hyperthyroidism?

15. What is the most common cause of hyperthyroidism?

16. Describe Grave's Disease.

17. How does Grave's Disease stimulate the thyroid gland to continue producing thyroid hormone?

18. What are the chronic manifestations of Grave's Disease?

19. What is thyroid hypertrophy or goiter? What causes a goiter in this example?

20. List the three common treatments of hyperthyroidism and how they treat the cause of hyperthyroidism.

Self-Quiz
Take this opportunity to check your progress by taking the **QUIZ**. See the **Introduction Module** for a reminder on how to access the **QUIZ** for this Study Area.

EXERCISE 8.7:
Endocrine System—Thyroid Gland—Anterior View

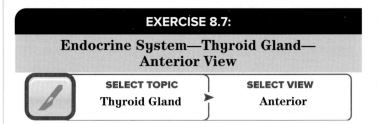

Click **LAYER 1** *in the* **LAYER CONTROLS** *window, and you will see the following image:*

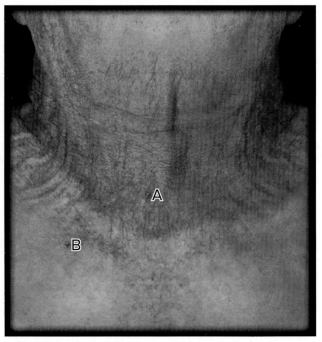

©McGraw-Hill Education

Mouse-over the pins on the screen to find the information necessary to identify the following structures:

A. _____

B. _____

*Click **LAYER 2** in the **LAYER CONTROLS** window, and you will see the following image:*

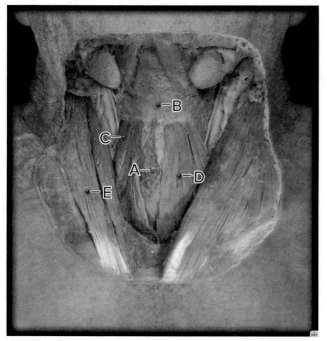

©McGraw-Hill Education

Mouse-over the pins on the screen to find the information necessary to identify the following structures:

A. _____

Nonendocrine System Structures (blue pins)

B. _____

C. _____

D. _____

E. _____

*Click **LAYER 3** in the **LAYER CONTROLS** window, and you will see the following image:*

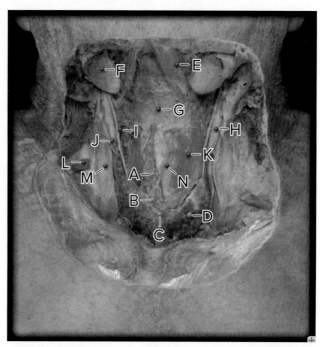

©McGraw-Hill Education

Mouse-over the pins on the screen to find the information necessary to identify the following structures:

A. _____

B. _____

C. _____

D. _____

Nonendocrine System Structures (blue pins)

E. _____

F. _____

G. _____

H. _____

I. _____

J. _____

K. _____

L. _____

M. _____

N. _____

CHECK POINT

Thyroid Gland, Anterior View

1. Describe the structure of the thyroid gland.
2. What is the function of the thyroid gland?
3. What is the name for an enlarged thyroid gland? What is the cause of this condition?

EXERCISE 8.8a:
Endocrine System—Thyroid Gland—Histology

SELECT TOPIC: Thyroid Gland ▸ **SELECT VIEW**: LM: Medium Magnification

Click the **TAGS ON/OFF** button, and you will see the following image:

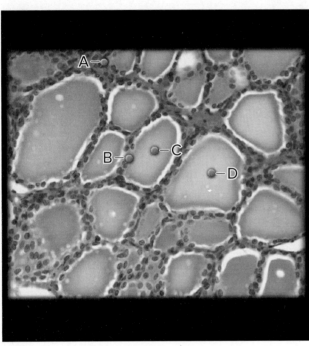

©McGraw-Hill Education/Al Telser

Mouse-over the pins on the screen to find the information necessary to identify the following structures:

A. _____

B. _____

C. _____

D. _____

EXERCISE 8.8b:
Endocrine System—Thyroid Gland—Histology

SELECT TOPIC: Thyroid Gland ▸ **SELECT VIEW**: LM: High Magnification

Click the **TAGS ON/OFF** button, and you will see the following image:

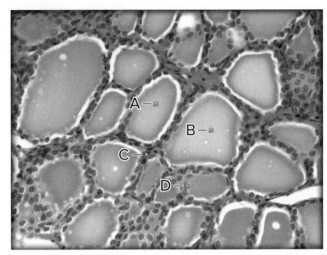

©McGraw-Hill Education/Al Telser

Mouse-over the pins on the screen to find the information necessary to identify the following structures:

A. _____

B. _____

C. _____

D. _____

CHECK POINT

Thyroid Gland, Histology

1. Name the thyroid cells that produce calcitonin. What is the function of calcitonin?
2. Describe the structure and function of the thyroid follicle.
3. What is follicular colloid? How long of a supply can be stored?

The Parathyroid Glands

 SELECT ANIMATION — Parathyroid Glands — **PLAY**

After viewing the animation, answer the following questions:

1. Describe the location of the parathyroid glands. How many are there? What arteries supply these glands?

2. Name the two types of parathyroid gland cells. What are their functions?

3. What causes the release of PTH?

4. How does PTH raise blood calcium levels?

5. What are the symptoms that may result from hyperparathyroidism?

6. What is the most common cause of hypoparathyroidism?

7. What are the symptoms of hypoparathyroidism?

Self-Quiz
Take this opportunity to check your progress by taking the **QUIZ**. See the **Introduction Module** for a reminder on how to access the **QUIZ** for this Study Area.

EXERCISE 8.9a:
Endocrine System—Parathyroid Gland—Histology

 SELECT TOPIC Parathyroid Gland ▶ **SELECT VIEW** LM: Low Magnification

Click the **TAGS ON/OFF** *button, and you will see the following image:*

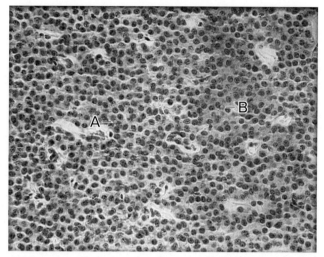

©McGraw-Hill Education/Steve Sullivan

Mouse-over the pins on the screen to find the information necessary to identify the following structures:

A. _____

B. _____

484 MODULE 8 The Endocrine System

EXERCISE 8.9b:
Endocrine System—Parathyroid Gland—Histology

SELECT TOPIC
Parathyroid Gland
▶
SELECT VIEW
LM: Medium Magnification

Click the **TAGS ON/OFF** *button, and you will see the following image:*

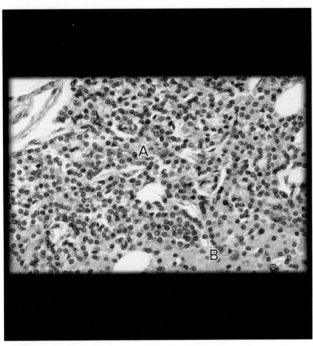

©Victor Eroschenko

Mouse-over the pins on the screen to find the information necessary to identify the following structures:

A. _____

B. _____

EXERCISE 8.9c:
Endocrine System—Parathyroid Gland—Histology

SELECT TOPIC
Parathyroid Gland
▶
SELECT VIEW
LM: High Magnification

Click the **TAGS ON/OFF** *button, and you will see the following image:*

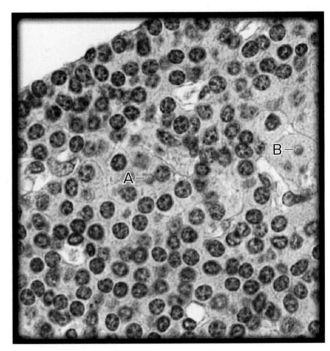

©Victor Eroschenko

Mouse-over the pins on the screen to find the information necessary to identify the following structures:

A. _____

B. _____

CHECK POINT

Parathyroid Gland, Histology

1. Name the most numerous functional cell type in the parathyroid gland. What hormone do they secrete?
2. What is the function of this hormone? What hormone is an antagonist to this one?
3. Name the parathyroid cells that increase in number with age.

The Pancreas

SELECT ANIMATION — Pancreas — PLAY

After viewing the animation, answer the following questions:

1. The organs of the endocrine system secrete _____ directly into the _____ to . . .

2. Describe the location of the pancreas. Is it endocrine, exocrine, or both?

3. The primary cells of the pancreas are . . . What do these cells secrete, and what is the function of these secretions?

4. Where are the endocrine cells of the pancreas located? What are they called?

5. Name the four types of cells located in the pancreatic islets. What type of substances do these cells release?

6. Which cells are activated by declining blood glucose levels? What hormone do they release? What action does this hormone stimulate?

7. Which cells are activated by increasing blood glucose levels? What hormone do they release? What action does this hormone stimulate?

8. Which cells release somatostatin? What actions does this hormone initiate?

9. Which cells secrete pancreatic polypeptide? What is the function of this secretion?

10. What do these pancreatic hormones together provide for?

Self-Quiz
Take this opportunity to check your progress by taking the **QUIZ**. See the **Introduction Module** for a reminder on how to access the **QUIZ** for this Study Area.

Clinical Application

SELECT ANIMATION — Type 1 Diabetes (3D) — PLAY

After viewing the animation, answer the following questions:

1. What hormone is secreted as one of the functions of the pancreas?

2. Describe in detail the cells that produce this hormone and where they are located.

3. After consumption of a meal, increasing amounts of _____ trigger _____ cells in the _____ to secrete the appropriate amount of _____ _____, which travels through the _____ _____ to _____ _____ where it promotes the transport of _____ into the _____.

4. Why is glucose needed inside the cells? What affect does this have on cellular processes?

5. What are the special requirements of certain tissues to allow the entry of glucose into their cells?

6. How does insulin function to fulfill this requirement?

7. As cells take up glucose, what happens to the blood glucose level?

8. What is type 1 diabetes? What is the result?

9. What is the connection between type 1 diabetes, lymphocytes, and antibodies?

10. What does the lack of sufficient insulin production prevent at the cellular level, and what is the result?

11. What is this condition called?

12. How are blood glucose levels affected?

13. The _____ filter out the excess glucose, which is lost in _____, resulting in _____ or large quantities of _____ in the _____.

14. List the common symptoms of hyperglycemia in type 1 diabetes.

15. What is the result of the continued insulin deficiency? What alternative sources of energy are used?

16. What are the acidic breakdown byproducts of fat catabolism? Where do they accumulate? What is the resulting condition?

17. What is diabetic ketoacidosis?

18. Type 1 diabetes can cause _____ _____ _____.

19. What are the long-term complications?

20. What treatment is available for type 2 diabetes?

21. How are localized tissue damage and absorption problems prevented?

22. What are the two delivery methods?

23. What is the result of this treatment and what does it facilitate?

24. What are other benefits of insulin injection therapy? What do they restore?

25. What monitoring must type 1 diabetes patients do frequently?

26. What are optimum blood glucose levels while fasting and two hours after starting a meal?

27. What other periodic tests should patients use to monitor blood glucose levels?

28. What do these tests measure?

29. What is glycated hemoglobin and how is it created?

30. How do plasma glucose levels affect the formation of glycated hemoglobin?

31. What is the desired hemoglobin A1C level for people with diabetes?

32. How do hemoglobin A1C levels affect the risk of complications from diabetes?

33. What other actions can patients take to monitor their glucose levels more closely?

34. How may patients prevent the occurrence of the complications of diabetes?

SELECT ANIMATION
Type 2 Diabetes (3D) — PLAY

After viewing the animation, answer the following questions:

1. One of the _____ functions of the pancreas is to secrete a _____ called _____ into the _____.

2. What is the function of insulin in the bloodstream?

3. Describe type 2 diabetes.

4. How is type 2 diabetes different from type 1 diabetes?

5. What causes insulin resistance?

6. What affect does a defect in insulin receptors have on the cells?

7. What is the result?

8. What is hyperglycemia?

9. What stimulatory affect does hyperglycemia have in the pancreas?

10. What causes the reduction in insulin production?

11. List the classic symptoms of type 2 diabetes.

12. List the symptoms of type 2 diabetes that appear over time.

13. What are the alternative energy sources in type 2 diabetes?

14. What are ketone bodies? Where do they accumulate?

15. What is the condition caused by this accumulation?

16. What is diabetic ketoacidosis?

17. Describe hypoglycemia. What causes hypoglycemia?

18. What is the effect of an excessive dose of insulin or oral hypoglycemic medication?

19. What is the brain's primary source of energy?

20. What blood glucose levels affect the brain first?

21. What effect does this have on the neurons?

22. What are the symptoms presented?

23. What happens if the glucose levels continue to drop?

24. What conditions can be caused by poorly controlled type 2 diabetes? What are the results?

25. What oral hypoglycemic medications are prescribed to treat type 2 diabetes?

26. What are the results from these medications?

27. What are the two nondrug therapies for the control of type 2 diabetes?

28. What are ideal fasting blood glucose levels for type 2 diabetes patients?

29. What are ideal blood glucose levels for type 2 diabetes patients two hours after starting a meal?

30. When are insulin injections indicated for type 2 diabetes patients?

31. Medication should be continued with the use of _____ _____ therapy options.

EXERCISE 8.10:
Endocrine System—Pancreas—Anterior View

SELECT TOPIC	SELECT VIEW
Pancreas	Anterior

*After clicking **LAYERS 1** through **3** to orient yourself to our location, click **LAYER 4** in the **LAYER CONTROLS** window, and you will see the following image:*

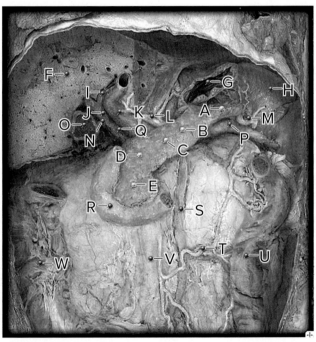

©McGraw-Hill Education

Mouse-over the pins on the screen to find the information necessary to identify the following structures:

A. _____

B. _____

C. _____

D. _____

E. _____

Nonendocrine System Structures (blue pins)

F. _____
G. _____
H. _____
I. _____
J. _____
K. _____
L. _____
M. _____
N. _____
O. _____
P. _____
Q. _____
R. _____
S. _____
T. _____
U. _____
V. _____
W. _____

CHECK POINT

Pancreas, Anterior View

1. What is the exocrine function of the pancreas?
2. What is the endocrine function of the pancreas?
3. Name two events that can result in diabetes mellitus.

EXERCISE 8.11a:
Endocrine System—Pancreas—Histology

SELECT TOPIC: Pancreas ▶ SELECT VIEW: LM: Low Magnification

Click the **TAGS ON/OFF** *button, and you will see the following image:*

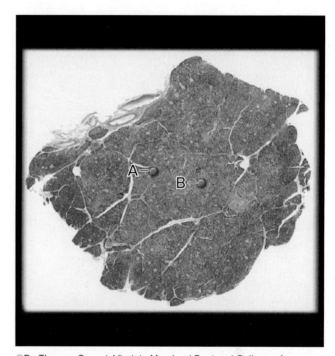

©Dr. Thomas Caceci, Virginia-Maryland Regional College of Veterinary Medicine

Mouse-over the pins on the screen to find the information necessary to identify the following structures:

A. _____
B. _____

EXERCISE 8.11b:
Endocrine System—Pancreas—Histology

SELECT TOPIC	SELECT VIEW
Pancreas	LM: Medium Magnification

Click the **TAGS ON/OFF** button, and you will see the following image:

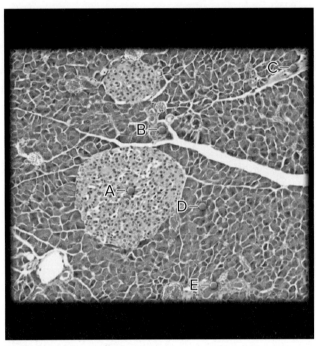

©McGraw-Hill Education/Al Telser

Mouse-over the pins on the screen to find the information necessary to identify the following structures:

A. _____
B. _____
C. _____
D. _____
E. _____

EXERCISE 8.12:
Endocrine System—Pancreas Alpha Cell—Histology

SELECT TOPIC	SELECT VIEW
Pancreas Alpha Cell	LM: High Magnification

Click the **TAGS ON/OFF** button, and you will see the following image:

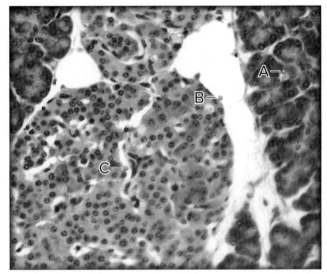

©McGraw-Hill Education/Steve Sullivan

Mouse-over the pins on the screen to find the information necessary to identify the following structures:

A. _____
B. _____
C. _____

MODULE 8 The Endocrine System 491

EXERCISE 8.13:
Endocrine System—Pancreas Beta Cell—Histology

SELECT TOPIC
Pancreas Beta Cell

SELECT VIEW
LM: High Magnification

Click the **TAGS ON/OFF** *button, and you will see the following image:*

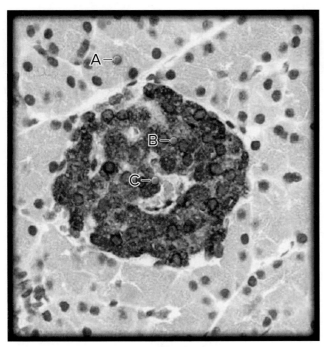

©Dr. Thomas Caceci, Virginia-Maryland Regional College of Veterinary Medicine

Mouse-over the pins on the screen to find the information necessary to identify the following structures:

A. _____

B. _____

C. _____

EXERCISE 8.14:
Endocrine System—Pancreas (Endocrine)—Histology

SELECT TOPIC
Pancreas (Endocrine)

SELECT VIEW
LM: High Magnification

Click the **TAGS ON/OFF** *button, and you will see the following image:*

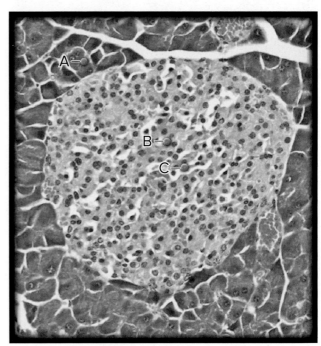

©McGraw-Hill Education/Al Telser

Mouse-over the pins on the screen to find the information necessary to identify the following structures:

A. _____

B. _____

C. _____

CHECK POINT

Pancreas (Endocrine), Histology

1. Name the clusters of cells that form the endocrine portion of the pancreas.
2. Name the hormones produced by these cells.
3. Where are these hormones secreted?

492 MODULE 8 The Endocrine System

EXERCISE 8.15:
Imaging—Pancreas

SELECT TOPIC: Pancreas → SELECT VIEW: Axial

Click the **TAGS ON/OFF** button, and you will see the following image:

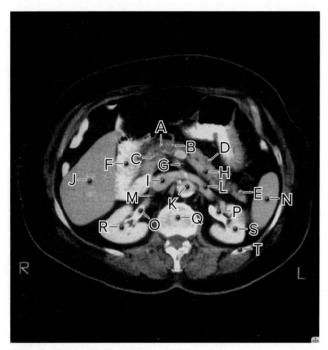

©McGraw-Hill Education

Mouse-over the pins on the screen to find the information necessary to identify the following structures:

A. _____
B. _____
C. _____
D. _____
E. _____

Nonendocrine System Structures (blue pins)

F. _____
G. _____
H. _____
I. _____
J. _____
K. _____
L. _____
M. _____
N. _____

O. _____
P. _____
Q. _____
R. _____
S. _____
T. _____

The Suprarenal (Adrenal) Gland

SELECT ANIMATION: Suprarenal (Adrenal) Gland — PLAY

After viewing the animation, answer the following questions:

1. The organs of the endocrine system secrete _____ directly into the _____ to . . .

2. Describe the location of the suprarenal glands. Where do they receive their blood supply? Name the single vein that drains them.

3. The suprarenal glands are composed of what two layers?

4. What are corticosteroids? Where are they synthesized?

5. What are mineralocorticoids? Where are they synthesized?

6. What is the principal mineralocorticoid? What is its function?

7. What are glucocorticoids? Where are they synthesized?

8. What are the two most common glucocorticoids?

9. What are the adrenal sex hormones? Where are they synthesized?

10. Name the inner core of the suprarenal glands. What cells are located in large numbers there?

11. What hormones are produced in this inner core area? What stimulates their release? What is their function?

Self-Quiz
Take this opportunity to check your progress by taking the **QUIZ**. See the **Introduction Module** for a reminder on how to access the **QUIZ** for this Study Area.

EXERCISE 8.16:
Endocrine System—Suprarenal (Adrenal) Gland—Anterior View

SELECT TOPIC	SELECT VIEW
Suprarenal Gland	Anterior

Click **LAYER 1** *in the* **LAYER CONTROLS** *window, and you will see the following image:*

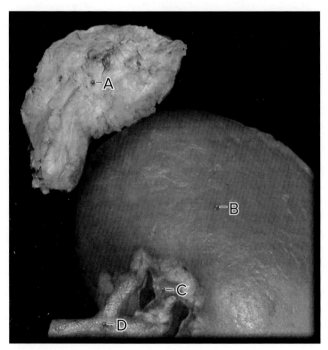

©McGraw-Hill Education

Mouse-over the pins on the screen to find the information necessary to identify the following structure:

A. _____

Nonendocrine System Structures (blue pins)

B. _____
C. _____
D. _____

CHECK POINT

Suprarenal (Adrenal) Gland, Anterior View

1. Name the endocrine gland located superior to the kidney.
2. Name the two parts of this gland.
3. What are the functions of these two parts?

Click **LAYER 2** *in the* **LAYER CONTROLS** *window, and you will see the following image:*

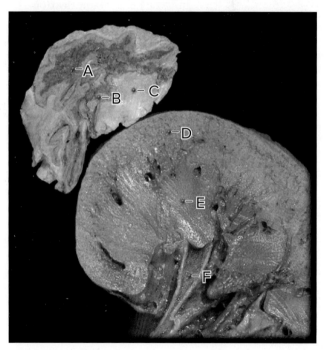

©McGraw-Hill Education

Mouse-over the pins on the screen to find the information necessary to identify the following structures:

A. _____
B. _____
C. _____

Nonendocrine System Structures (blue pins)

D. _____
E. _____
F. _____

494 MODULE 8 The Endocrine System

CHECK POINT

Suprarenal (Adrenal) Gland, Anterior View, *continued*

4. Name the connective tissue that surrounds the suprarenal gland.
5. Name the lipid-rich outer part of the suprarenal gland. What is the function of each of the three regions of this part?
6. Name the reddish-brown core of the suprarenal gland. What is its function?

EXERCISE 8.17a:
Endocrine System—Suprarenal Gland— Histology

SELECT TOPIC	SELECT VIEW
Suprarenal Gland	LM: Low Magnification

Click the **TAGS ON/OFF** button, and you will see the following image:

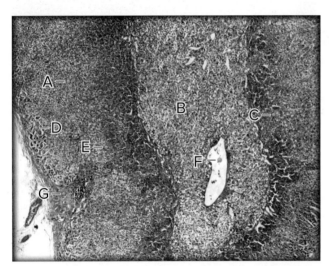

©McGraw-Hill Education/Steve Sullivan

Mouse-over the pins on the screen to find the information necessary to identify the following structures:

A. _____
B. _____
C. _____
D. _____
E. _____
F. _____
G. _____

EXERCISE 8.17b:
Endocrine System—Suprarenal Gland— Histology

SELECT TOPIC	SELECT VIEW
Suprarenal Gland	LM: Medium Magnification

Click the **TAGS ON/OFF** button, and you will see the following image:

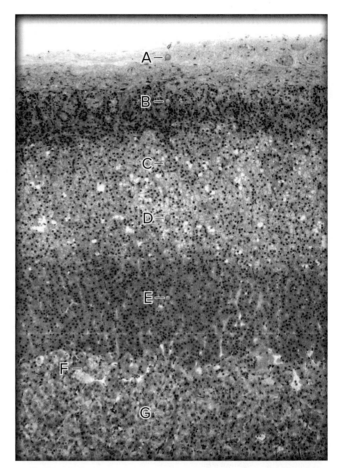

©Victor Eroschenko

Mouse-over the pins on the screen to find the information necessary to identify the following structures:

A. _____
B. _____
C. _____
D. _____
E. _____
F. _____
G. _____

MODULE 8 The Endocrine System 495

EXERCISE 8.18:

Endocrine System—Suprarenal Gland, Zona Glomerulosa, and Fasciculata—Histology

SELECT TOPIC	SELECT VIEW
Suprarenal Gland, Zona Glomerulosa, and Fasciculata	LM: High Magnification

Click the **TAGS ON/OFF** *button, and you will see the following image:*

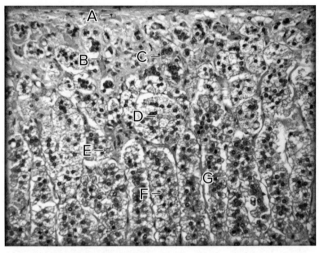

©McGraw-Hill Education/Greg Reeder

Mouse-over the pins on the screen to find the information necessary to identify the following structures:

A. _____
B. _____
C. _____
D. _____
E. _____
F. _____
G. _____

EXERCISE 8.19:

Endocrine System—Suprarenal Gland, Zona Reticularis—Histology

SELECT TOPIC	SELECT VIEW
Suprarenal Gland Zona Reticularis	LM: High Magnification

Click the **TAGS ON/OFF** *button, and you will see the following image:*

©Dr. Thomas Caceci, Virginia-Maryland Regional College of Veterinary Medicine

Mouse-over the pins on the screen to find the information necessary to identify the following structures:

A. _____
B. _____
C. _____
D. _____

CHECK POINT

Suprarenal (Adrenal) Gland, Histology

1. Name the five layers of the suprarenal gland from superficial to deep.
2. Name the thick fibroblastic outer coat of the suprarenal gland. What is its function?
3. Name the layer of the suprarenal gland responsible for sex hormone production.

The Ovary

EXERCISE 8.20:
Endocrine System—Ovary—Superior View

SELECT TOPIC	SELECT VIEW
Ovary	Superior

*After clicking **LAYER 1** to orient yourself to our location, click **LAYER 2** in the **LAYER CONTROLS** window, and you will see the following image:*

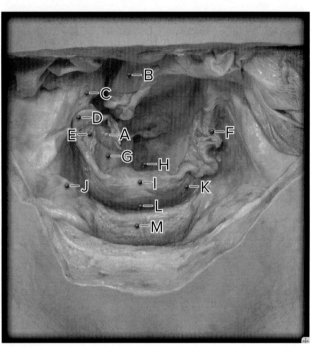

©McGraw-Hill Education

Mouse-over the pins on the screen to find the information necessary to identify the following structure:

A. _____

Nonendocrine System Structures (blue pins)

B. _____
C. _____
D. _____
E. _____
F. _____
G. _____
H. _____
I. _____
J. _____
K. _____
L. _____
M. _____

CHECK POINT

Ovary, Superior View

1. Describe the structure of the female gonads.
2. Name the hormones produced by the female gonads.
3. What change occurs in these structures after menopause?

The Testis

EXERCISE 8.21:
Endocrine System—Testis and Spermatic Cord—Anterior View

SELECT TOPIC	SELECT VIEW
Testis and Spermatic Cord	Anterior

*After clicking **LAYERS 1** through **3** to orient yourself to our location, click **LAYER 4** in the **LAYER CONTROLS** window, and you will see the following image:*

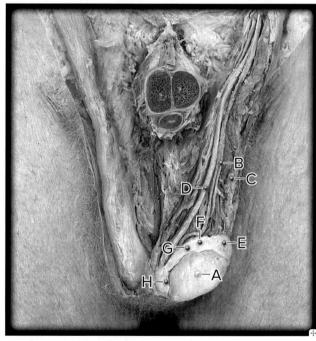

©McGraw-Hill Education

Mouse-over the pins on the screen to find the information necessary to identify the following structure:

A. _____

Nonendocrine System Structures (blue pins)

B. _____
C. _____
D. _____

E. _____
F. _____
G. _____
H. _____

CHECK POINT

Testis and Spermatic Cord, Anterior View

1. Describe the structure of the male gonads.
2. Name the hormones produced by the male gonads.
3. What is the function of testosterone?

Click **LAYER 5** *in the* **LAYER CONTROLS** *window, and you will see the following image:*

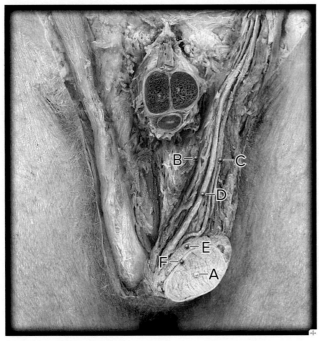

©McGraw-Hill Education

Mouse-over the pins on the screen to find the information necessary to identify the following structure:

A. _____

Nonendocrine System Structures (blue pins)

B. _____
C. _____
D. _____
E. _____
F. _____

EXERCISE 8.22:

Endocrine System—Testis and Spermatic Cord (Isolated)—Lateral View

 SELECT TOPIC Testis and Spermatic Cord (Isolated) ▶ **SELECT VIEW** Lateral

Click **LAYER 1** *in the* **LAYER CONTROLS** *window, and you will see the following image:*

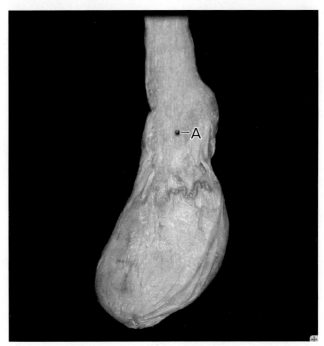

©McGraw-Hill Education

Mouse-over the pin on the screen to find the information necessary to identify the following nonendocrine system structure:

A. _____

498 MODULE 8 The Endocrine System

Click **LAYER 2** *in the* **LAYER CONTROLS** *window, and you will see the following image:*

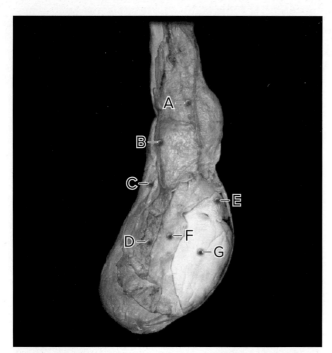

©McGraw-Hill Education

Mouse-over the pins on the screen to find the information necessary to identify the following nonendocrine system structures:

A. _____

B. _____

C. _____

D. _____

E. _____

F. _____

G. _____

Click **LAYER 3** *in the* **LAYER CONTROLS** *window, and you will see the following image:*

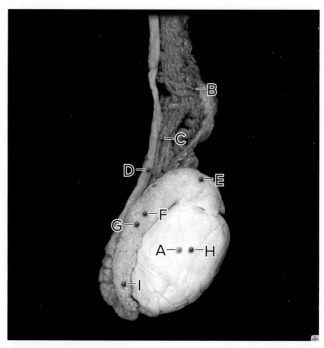

©McGraw-Hill Education

Mouse-over the pins on the screen to find the information necessary to identify the following structure:

A. _____

Nonendocrine System Structures (blue pins)

B. _____

C. _____

D. _____

E. _____

F. _____

G. _____

H. _____

I. _____

Click **LAYER 4** *in the* **LAYER CONTROLS** *window, and you will see the following image:*

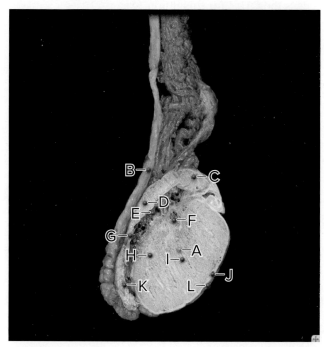

©McGraw-Hill Education

Mouse-over the pins on the screen to find the information necessary to identify the following structure:

A. _____

Nonendocrine System Structures (blue pins)

B. _____
C. _____
D. _____
E. _____
F. _____
G. _____
H. _____
I. _____
J. _____
K. _____
L. _____

The Seminiferous Tubule

EXERCISE 8.23:
Endocrine System—Interstitial (Leydig) Cell of Seminiferous Tubule—Histology

SELECT TOPIC
Interstitial (Leydig) Cell of Seminiferous Tubule ▶

SELECT VIEW
High Magnification

Click the **TAGS ON/OFF** *button, and you will see the following image:*

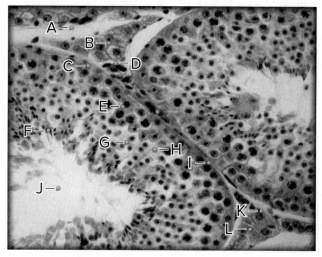

©McGraw-Hill Education/Greg Reeder

Mouse-over the pins on the screen to find the information necessary to identify the following structures:

A. _____
B. _____
C. _____
D. _____
E. _____
F. _____
G. _____
H. _____
I. _____
J. _____
K. _____
L. _____

CHECK POINT

Seminiferous Tubule, Histology

1. Name the structures responsible for testosterone production.
2. What is another name for these cells?
3. Name the structures that these cells surround. What correlation do they have with these structures?

Self-Quiz

Take this opportunity to check your progress by taking the **QUIZ**. See the **Introduction Module** for a reminder on how to access the **QUIZ** for this Study Area.

IN REVIEW

What Have I Learned?

The following questions cover the material that you have just learned, the endocrine system. Apply what you have learned to answer these questions on a separate piece of paper.

1. The pituitary gland rests in what structure of which bone?
2. Name the fossa that surrounds the pituitary gland.
3. Describe the structure and function of the mammillary body.
4. Name the cell that represents a chromophil depleted of hormone.
5. List the hormones released by the anterior pituitary gland.
6. List the hormones released by the posterior pituitary gland.
7. Name the function of the following hormones: T_3 and T_4 and calcitonin. (This may require a little research on your part.)
8. Which division of the autonomic nervous system is responsible for the fight-or-flight response? What two hormones regulate this response? What is the origin of these hormones?
9. Name two disorders of suprarenal cortex hormone secretion.
10. Name the layer of the suprarenal gland responsible for mineralocorticoid production.
11. Name the layer of the suprarenal gland responsible for glucocorticoid production.
12. In the chart below, list the hormones produced by the suprarenal gland, where they are produced, and their function.

HORMONE	WHERE PRODUCED	FUNCTION

MODULE 9

The Cardiovascular System

Overview: Cardiovascular System

Your heart and circulatory system accomplish amazing feats! Think about these accomplishments:

- Your heart, roughly the size of your fist, beats over 100,000 times per day. This equals approximately 36 million times per year.
- Your body contains an average volume of 5 liters of blood, the equivalent volume of which circulates through your heart, and thus your body, once every minute.
- It has been estimated that your heart pumps the equivalent volume of 1,900 gallons (7,200 liters) of blood through approximately 60,000–100,000 miles (96,560–63,730 kilometers) of blood vessels every day. That's 2½ to 3 times around the earth at the equator—every day.
- If you live 70 years, your heart will beat some 2.5 billion times. In this 70-year lifetime, it will pump roughly 1 million barrels of blood, enough to fill more than three super tankers.

Pretty impressive accomplishments, wouldn't you say? Let's begin our exploration of the wonders of this amazing cardiovascular system.

We'll begin with an overview of the entire system, followed by an in depth study of blood, without which we wouldn't have a cardiovascular system! We will then focus on the structures of the cardiovascular system in detail region by region, beginning with the heart. Let's get started!

From the **HOME** *screen, click the drop-down box on the* **MODULE** *menu.*

From the systems listed, click **Cardiovascular**.

SELECT ANIMATION
Cardiovascular System Overview
PLAY

After viewing the animation, answer the following questions:

1. The cardiovascular consists of what three structures?

2. The heart distributes and receives blood through which structures?

3. Define an artery.

4. Gases and nutrients are exchanged with the tissues through which blood vessels?

5. Define a vein.

Self-Quiz

Take this opportunity to check your progress by taking the **QUIZ**. See the **Introduction Module** for a reminder on how to access the **QUIZ** for this Study Area.

Blood

You may have heard the saying that "the life of the body is in the blood," and this is true. Without the blood circulating through our blood vessels, our cells would quickly die from lack of oxygen. But this is just one aspect of the life provided by the blood.

Your circulatory system can be compared to a major superhighway, where freight is transported from one location to another. In a similar manner, oxygen is transported by your red blood cells from your lungs to your cells, nutrients are shipped from your digestive system and liver to every cell of your body, and wastes are shipped from your cells to your kidneys for removal. And the white blood cells of your immune system are policing your tissues and traveling on this superhighway as well.

So, as we begin by looking at blood, the fundamental component of the cardiovascular system, keep this superhighway concept in mind and see just how "the life of the body is in the blood."

SELECT ANIMATION
Hemopoiesis
PLAY

After viewing the animation, answer the following questions:

1. Hemopoiesis is the process of _____ _____, which occurs primarily . . .

2. Hemopoiesis begins with undifferentiated cells called _____.

3. These _____ _____ give rise to . . .

4. What types of hormones influence the differentiation of the blood stem cells?

5. What are the two lines of cells that differentiate from the hemocytoblast?

6. Which groups of cells arise from each of these lines?

7. Which cells are produced by the lymphoid cell line?

8. Erythropoiesis produces . . .

9. List the steps for red blood cell (erythrocyte) production.

10. Thrombopoiesis produces . . .

11. What is the name for the committed progenitor cell in thrombopoiesis?

12. In response to the hormone _____, the _____ differentiates into a _____.

13. Platelets are formed when . . .

14. Leukopoiesis is the . . .

15. The myeloid cell line gives rise to . . .

16. The lymphoid cell line produces _____.

17. Eosinophilic myelocytes differentiate into _____.

18. Basophilic myelocytes differentiate into _____.

19. Neutrophilic myelocytes develop into _____.

20. Monocytes are derived from . . .

21. Lymphocytes are derived from . . .

22. The two types of lymphocytes that develop are . . .

Self-Quiz
Take this opportunity to check your progress by taking the **QUIZ**. See the **Introduction Module** for a reminder on how to access the **QUIZ** for this Study Area.

SELECT ANIMATION
Hemoglobin Breakdown
PLAY

After viewing the animation, answer the following questions:

1. What happens to the hemoglobin released by the rupture of old red blood cells?

2. The globin chains . . .

3. What is released from the heme?

4. The remaining structure of the heme goes through a two-step process, being converted to the following two products sequentially _____ _____.

5. What plasma protein transports the iron?

6. Where is iron transported for storage (two locations)?

7. Where is iron transported to make new hemoglobin?

8. What plasma protein transports free bilirubin?

9. Where is it transported?

10. Liver cells make _____ _____, excreted as part of _____ into the _____ _____.

11. _____ _____ convert _____ into _____, which contribute to the . . .

12. Some of the _____ _____ are absorbed into the blood and excreted from the _____ in the _____. (This product is urochrome, which gives urine its yellow color.)

Self-Quiz
Take this opportunity to check your progress by taking the **QUIZ**. See the **Introduction Module** for a reminder on how to access the **QUIZ** for this Study Area.

IN REVIEW

What Have I Learned?

The following questions cover the material that you have just learned, hemopoiesis and hemoglobin breakdown. Apply what you have learned to answer these questions on a separate piece of paper.

1. Hemopoiesis is the process of _____ , which occurs primarily _____ .

2. What types of hormones influence the differentiation of the blood stem cells?

3. Erythropoiesis produces _____ .

4. Describe the production of platelets.

5. Name the two types of lymphocytes. Where does each type mature?

6. What is hemopoiesis?

7. Name the oxygen-carrying molecule in red blood cells.

8. What two components of this molecule are released by hemopoiesis?

9. Describe how these components are recycled.

10. What products of hemoglobin breakdown end up in the feces? In the urine?

EXERCISE 9.1:

Cardiovascular System—Blood (Peripheral Smear)—Histology

SELECT TOPIC	SELECT VIEW
Blood (Peripheral Smear) ▶	LM: Medium Magnification

Click the **TAGS ON/OFF** button, and you will see the following image:

Mouse-over the pins on the screen to find the information necessary to identify the following structures:

A. _____

B. _____

C. _____

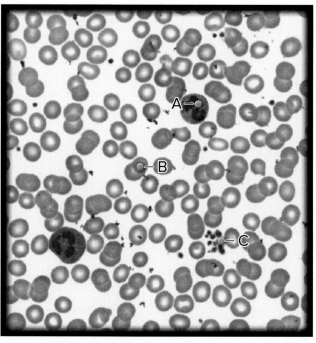

©McGraw-Hill Education/Al Telser

504 MODULE 9 The Cardiovascular System

EXERCISE 9.2a:
Cardiovascular System—Erythrocyte—Histology

SELECT TOPIC: Erythrocyte → SELECT VIEW: LM: Low Magnification

Click the **TAGS ON/OFF** *button, and you will see the following image:*

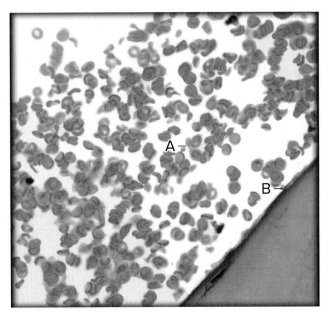

©Science Photo Library/Alamy Stock Photo

Mouse-over the pins on the screen to find the information necessary to identify the following structures:

A. _____

B. _____

EXERCISE 9.2b:
Cardiovascular System—Erythrocyte—Histology

SELECT TOPIC: Erythrocyte → SELECT VIEW: TEM: High Magnification

Click the **TAGS ON/OFF** *button, and you will see the following image:*

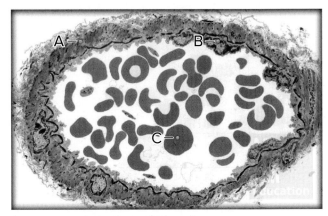

©Science Photo Library/Alamy Stock Photo

Mouse-over the pins on the screen to find the information necessary to identify the following structures:

A. _____

B. _____

C. _____

MODULE 9   The Cardiovascular System   505

EXERCISE 9.2c:
Cardiovascular System—Erythrocyte—Histology

SELECT TOPIC	SELECT VIEW
Erythrocyte	SEM: High Magnification

Click the **TAGS ON/OFF** *button, and you will see the following image:*

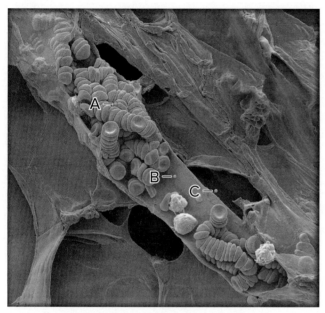

©Science Photo Library/Alamy Stock Photo

Mouse-over the pins on the screen to find the information necessary to identify the following structures:

A. _____

B. _____

C. _____

EXERCISE 9.3:
Cardiovascular System—Neutrophil—Histology

SELECT TOPIC	SELECT VIEW
Neutrophil	LM: High Magnification

Click the **TAGS ON/OFF** *button, and you will see the following image:*

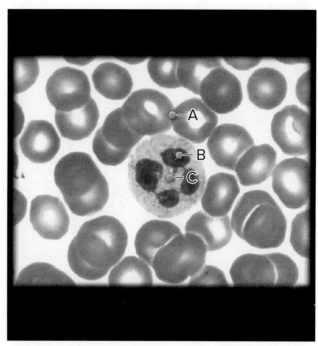

©McGraw-Hill Education/Al Telser

Mouse-over the pins on the screen to find the information necessary to identify the following structures:

A. _____

B. _____

C. _____

EXERCISE 9.4:
Cardiovascular System—Eosinophil—Histology

SELECT TOPIC	SELECT VIEW
Eosinophil	LM: High Magnification

Click the **TAGS ON/OFF** *button, and you will see the following image:*

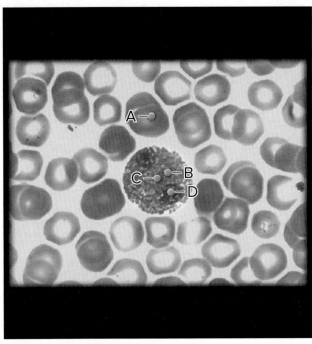

©McGraw-Hill Education/Al Telser

Mouse-over the pins on the screen to find the information necessary to identify the following structures:

A. _____

B. _____

C. _____

D. _____

EXERCISE 9.5:
Cardiovascular System—Basophil—Histology

SELECT TOPIC	SELECT VIEW
Basophil	LM: High Magnification

Click the **TAGS ON/OFF** *button, and you will see the following image:*

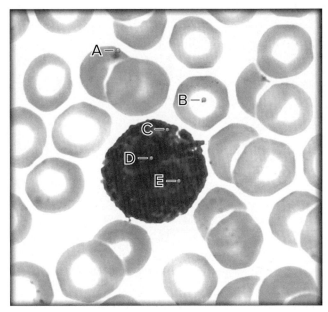

©McGraw-Hill Education/Al Telser

Mouse-over the pins on the screen to find the information necessary to identify the following structures:

A. _____

B. _____

C. _____

D. _____

E. _____

MODULE 9   The Cardiovascular System   507

EXERCISE 9.6:
Cardiovascular System—Lymphocyte—Histology

SELECT TOPIC	SELECT VIEW
Lymphocyte	LM: High Magnification

Click the **TAGS ON/OFF** *button, and you will see the following image:*

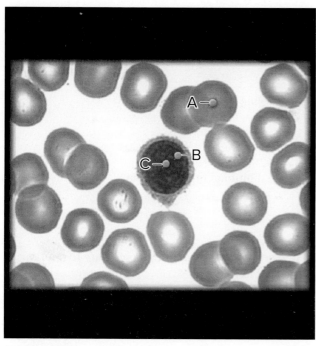

©McGraw-Hill Education/Al Telser

Mouse-over the pins on the screen to find the information necessary to identify the following structures:

A. _____

B. _____

C. _____

EXERCISE 9.7:
Cardiovascular System—Monocyte—Histology

SELECT TOPIC	SELECT VIEW
Monocyte	LM: High Magnification

Click the **TAGS ON/OFF** *button, and you will see the following image:*

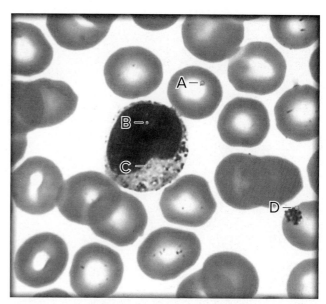

©McGraw-Hill Education/Al Telser

Mouse-over the pins on the screen to find the information necessary to identify the following structures:

A. _____

B. _____

C. _____

D. _____

EXERCISE 9.8:
Cardiovascular System—Megakaryocyte—Histology

SELECT TOPIC: Megakaryocyte ▶ **SELECT VIEW**: LM: High Magnification

Click the **TAGS ON/OFF** *button, and you will see the following image:*

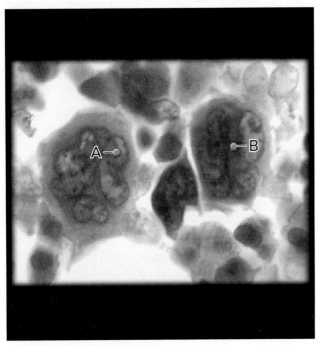

©McGraw-Hill Education/Al Telser

Mouse-over the pins on the screen to find the information necessary to identify the following structures:

A. _____

B. _____

Clinical Applications

SELECT ANIMATION: Fluid and Electrolyte Imbalances (3D) **PLAY**

After viewing the animation, answer the following questions:

1. The body's fluids are a mix of . . .

2. What are solutes?

3. The majority of solutes in the extracellular fluid are electrically charged particles called _____.

4. _____ is one of the most abundant _____.

5. What is homeostasis? What is its effect on electrolytes?

6. What is serum sodium? What does it indicate?

7. Normal serum sodium is maintained at _____ to _____ milliequivalents per liter.

8. Define hyponatremia?

9. Define dilutional hyponatremia?

10. How does profuse sweating contribute to dilutional hyponatremia?

11. In response, _____ is stimulated and _____ intake increases—sometimes in _____ amounts.

12. How do these processes lead to dilutional hyponatremia?

13. How does chronic kidney disease contribute to dilutional hyponatremia?

14. In an attempt to restore water and electrolyte balance, _____ _____ moves into _____.

15. What is the result?

16. Describe treatments for hyponatremia.

17. What is a hypertonic saline solution? What is its affect?

18. Describe the function of antidiuretic hormone inhibitor medication.

SELECT ANIMATION
Inflammation (3D) — PLAY

After viewing the animation, answer the following questions:

1. Define inflammation. Include its symptoms.

2. Immediately after injury, inflammation begins with . . .

3. This results in . . .

4. What is the response of local cells? What stimulates this response?

5. List the vasoactive chemicals and the responses that they initiate.

6. What effect do these chemicals have on endothelial cells in small blood vessels?

7. This increased _____ _____ allows . . .

8. Next, during a multistage process called _____, circulating immune cells called _____ move . . .

9. What is the function of the neutrophils?

10. How does chemotaxis begin?

11. Next, in a process called _____, neutrophils . . .

12. The neutrophils migrate to the _____ _____ by following a _____ _____.

13. Upon arrival . . .

14. Define phagocytosis.

15. After what processes does tissue repair begin?

16. The process of tissue repair begins when . . .

17. What is the function of the collagen produced?

18. What is the function of anti-inflammatory drugs? How do they accomplish this?

19. What are the most common drugs for inflammation?

20. Two examples of these are . . .

21. They contain an enzyme called _____ or _____ that _____ the production of _____ and several other _____ chemicals.

22. NSAIDS reduce . . .

SELECT ANIMATION
Acidosis (3D) — PLAY

After viewing the animation, answer the following questions:

1. Define acid-base balance.

2. Define pH.

3. What is expressed as a pH value?

4. What is the range of normal blood pH?

5. Metabolic processes constantly release _____ which freely release _____ ions, resulting in . . .

6. What is the body's chemical *and* physiological response?

7. When does respiratory acidosis occur?

8. If respiration cannot keep pace with _____ _____ production, _____ _____ builds up in the _____.

9. Excess _____ _____ combines with _____ to produce _____ _____.

10. _____ _____ disassociates into _____ and _____ ions.

11. What lowers pH and causes acidosis?

12. Metabolic acidosis occurs when . . .

13. List the conditions causing metabolic acidosis.

14. This results in . . .

15. What is the first step in treatment for acidosis?

16. What may be used to treat metabolic acidosis?

17. What may be used to treat respiratory acidosis?

SELECT ANIMATION
Alkalosis (3D) — PLAY

After viewing the animation, answer the following questions:

1. Define acid-base balance.

2. Define pH.

3. What is expressed as a pH value?

4. What is the range of normal blood pH?

5. Metabolic processes constantly release _____ which freely release _____ ions, resulting in . . .

6. What is the body's chemical *and* physiological response?

7. When does respiratory alkalosis occur?

8. What causes a carbon dioxide deficit?

9. What is the result of a carbon dioxide deficit?

10. Less _____ _____ dissociates into fewer free _____ ions.

11. What causes alkalosis?

12. Metabolic alkalosis occurs when . . .

13. List the conditions causing metabolic alkalosis.

14. This results in . . .

15. Discuss the primary treatment for respiratory alkalosis.

16. Discuss the treatment for metabolic alkalosis.

17. When does the treatment of alkalosis include anti-emetic drugs? How do they help?

SELECT ANIMATION
Blood Flow Through Heart — PLAY

After viewing the animation, answer the following questions:

1. Name the four chambers of the heart. What are the differences between them?

2. Name the two major vessels that deliver oxygen-poor blood to the heart.

3. What areas of the body do these two vessels drain?

4. What is the route of venous blood from the heart?

5. Trace the route of blood as it flows through the heart. Be sure to list *all* structures involved.

6. Name the only arteries to carry oxygen-poor blood.

7. Name the only veins to carry oxygen-rich blood.

Self-Quiz
Take this opportunity to check your progress by taking the **QUIZ**. See the **Introduction Module** for a reminder on how to access the **QUIZ** for this Study Area.

SELECT ANIMATION
Heart Fly-through — PLAY

After viewing the animation, answer the following questions:

1. List the organs in the thoracic cavity.

2. How does the heart function?

3. How is one-way flow through the heart maintained?

4. Which chambers make up the right side of the heart?

5. Is the blood in the right side of the heart oxygen-rich or oxygen-poor?

6. Where does the right ventricle send the blood? Why?

7. Which chambers make up the left side of the heart?

8. Is the blood in the left side of the heart oxygen rich or oxygen poor?

9. Where does the left side of the heart direct this blood?

10. How does venous blood from the systemic circulation return to the heart?

11. What is the function of the coronary sinus?

12. What structures make the right atrium distinct?

13. Where is the tricuspid valve located?

14. What is the function of the chordae tendineae and their papillary muscles?

15. Contraction of the right ventricle causes _____.

16. How does the pulmonary trunk divide?

17. How does blood return to the heart? What chamber receives this blood?

18. How is the left atrium similar to the right atrium?

19. What is the function of the bicuspid or mitral valve?

20. Contraction of the left ventricle _____.

21. How is the heart able to support the delivery of blood to the entire body?

HEADS UP!

The heart in the following exercise have been removed from the cadaver, and thus will not have all of the surrounding structures visible for reference. The subsequent section of the blood vessels of the thorax will cover the anatomy surrounding the heart in detail.

Elastic Artery—The Aorta

EXERCISES 9.9a:

Cardiovascular System—Elastic Artery (Aorta)—Histology

SELECT TOPIC	SELECT VIEW
Elastic Artery (Aorta)	LM: Medium Magnification

*Click the **TAGS ON/OFF** button, and you will see the following image:*

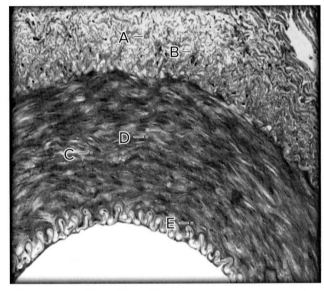

©Science Photo Library/Alamy Stock Photo

Mouse-over the pins on the screen to find the information necessary to identify the following structures:

A. _____

B. _____

C. _____

D. _____

E. _____

CHECK POINT

Elastic Artery (Aorta), Histology

1. What is the name for the outermost layer of the vessel wall?
2. What is the name for the middle layer of the vessel wall?
3. What is the name for the innermost layer of the vessel wall?

EXERCISE 9.9b:
Cardiovascular System—Elastic Artery (Aorta)—Histology

SELECT TOPIC	SELECT VIEW
Elastic Artery (Aorta)	LM: High Magnification

Click the **TAGS ON/OFF** button, and you will see the following image:

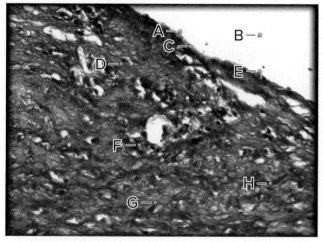

©McGraw-Hill Education/Steve Sullivan

Mouse-over the pins on the screen to find the information necessary to identify the following structures:

A. _____
B. _____
C. _____
D. _____
E. _____
F. _____
G. _____
H. _____

IN REVIEW

What Have I Learned?

The following questions cover the material that you have just learned, the histology of an elastic anterior. Apply what you have learned to answer these questions on a separate piece of paper.

1. In arteries, which tunic or layer is the thickest?

2. In veins, which tunic or layer is the thickest?

3. What is the function of the tunica externa of elastic arteries?

4. What is the function of the tunica media of elastic arteries?

5. What is the function of the tunica intima of elastic arteries?

6. Which of the three tunics contains the endothelium? Did you know that the endothelium of the blood vessels is continuous with the endocardium of the heart?

7. Which structures in the walls of medium and large arteries provide the elastic properties of those blood vessel walls?

8. Name the small blood vessels in the tunica externa.

9. How are the other two tunics supplied with oxygen and nutrients?

Large Vein—The Inferior Vena Cava

EXERCISE 9.10:
Cardiovascular System—Large Vein (Inferior Vena Cava)—Histology

SELECT TOPIC	SELECT VIEW
Large Vein (Inferior Vena Cava) ▶	LM: High Magnification

Click the **TAGS ON/OFF** button, and you will see the following image:

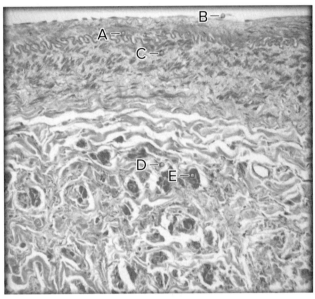

©McGraw-Hill Education/Greg Reeder

Mouse-over the pins on the screen to find the information necessary to identify the following structures:

A. _____

B. _____

C. _____

D. _____

E. _____

CHECK POINT

Large Vein (Inferior Vena Cava), Histology

1. What is the name for the outermost layer of the vessel wall?
2. What is the name for the middle layer of the vessel wall?
3. What is the name for the innermost layer of the vessel wall?

IN REVIEW

What Have I Learned?

The following questions cover the material that you have just learned, the histology of a large vein. Apply what you have learned to answer these questions on a separate piece of paper.

1. Where are the smooth muscle bundles located in veins?

2. What is their function?

3. How does the tunica intima of the inferior vena cava compare with that of the aorta?

4. How does the tunica media of the inferior vena cava compare with that of the aorta?

5. How does the tunica externa of the inferior vena cava compare with that of the aorta?

Muscular Artery and Medium-sized Vein

EXERCISE 9.11:
Cardiovascular System—Muscular Artery and Medium-sized Vein—Histology

SELECT TOPIC	SELECT VIEW
Muscular Artery and Medium-sized Vein ▶	LM: High Magnification

Click the **TAGS ON/OFF** button, and you will see the following image:

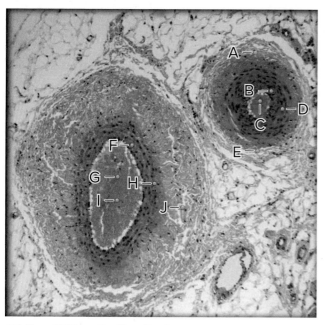

©McGraw-Hill Education/Greg Reeder

Mouse-over the pins on the screen to find the information necessary to identify the following structures:

A. _____
B. _____
C. _____
D. _____
E. _____
F. _____
G. _____
H. _____
I. _____
J. _____

CHECK POINT

Muscular Artery and Medium-sized Vein, Histology

1. Where are muscular arteries located?
2. Where are medium-sized veins located?
3. What is the function of muscular arteries?

IN REVIEW

What Have I Learned?

The following questions cover the material that you have just learned, the histology of muscular arteries and medium-sized veins. Apply what you have learned to answer these questions on a separate piece of paper.

1. What tissue predominates the tunica media of a muscular artery?
2. Give three examples of muscular arteries.
3. What is the function of medium-sized veins?
4. What tissues predominate the tunica media of these veins?
5. Give two examples of medium-sized veins.

Arteriole and Venule

EXERCISE 9.12:
Cardiovascular System—Arteriole and Venule—Histology

SELECT TOPIC	SELECT VIEW
Arteriole and Venule	LM: High Magnification

Click the **TAGS ON/OFF** button, and you will see the following image:

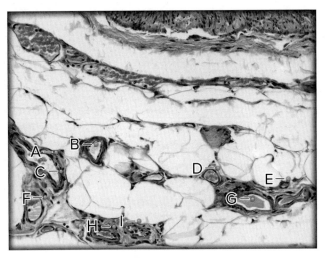

©McGraw-Hill Education/Al Telser

Mouse-over the pins on the screen to find the information necessary to identify the following structures:

A. _____
B. _____
C. _____
D. _____
E. _____
F. _____
G. _____
H. _____
I. _____

CHECK POINT

Arteriole and Venule, Histology

1. What is the primary point for control of blood flow to organs?
2. Where are these vessels located?
3. Where are venules located?

IN REVIEW

What Have I Learned?

The following questions cover the material that you have just learned, the histology of arterioles and venules. Apply what you have learned to answer these questions on a separate piece of paper.

1. What is the diameter of an arteriole?
2. Describe the tunica media of an arteriole.
3. What is the function of an arteriole?
4. What is the diameter of a venule?
5. What is the function of venules?

Neurovascular Bundle

EXERCISE 9.13:
Cardiovascular System—Neurovascular Bundle—Histology

SELECT TOPIC	SELECT VIEW
Neurovascular Bundle ▶	LM: High Magnification

Click the **TAGS ON/OFF** *button, and you will see the following image:*

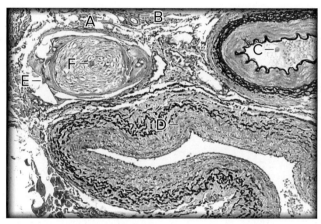

©McGraw-Hill Education/Dennis Strete

Mouse-over the pins on the screen to find the information necessary to identify the following structures:

A. _____

B. _____

C. _____

D. _____

Noncardiovascular System Structures (blue pins)

E. _____

F. _____

CHECK POINT
Neurovascular Bundle, Histology

1. Where are peripheral nerves located?
2. Describe peripheral nerves.
3. Give at least two examples of peripheral nerves.

IN REVIEW

What Have I Learned?

The following questions cover the material that you have just learned, the histology of a neurovascular bundle. Apply what you have learned to answer these questions on a separate piece of paper.

1. Describe adipose connective tissue.

2. What is its function?

3. Name five structures, other than connective tissue, found in neurovascular bundles.

518 MODULE 9 The Cardiovascular System

SELECT ANIMATION: Capillary Exchange — PLAY

After viewing the animation, answer the following questions:

1. Describe the structure of capillaries.

2. What is the function of capillaries? What term is used to describe this function?

3. What specific characteristics of capillaries make them optimal for their function?

4. Fluid moves out of the capillaries at the _____ end.

5. What happens to most of that fluid?

6. Name the forces that drive fluids and their dissolved contents into and out of the capillaries.

7. What is net filtration pressure? What does net filtration pressure regulate?

8. What force favors movement of fluid from the blood into the interstitial spaces? Where does this occur?

9. What force causes a shift from filtration to reabsorption in the capillary? Where does this occur?

10. What percent of the fluid that leaves the blood capillary at its arterial end will reenter the capillary at its venous end?

11. What happens to the remaining fluid that does not reenter the blood capillary?

12. Using the information from this animation, fill in the blank spaces in the Fluid Exchange Summary table:

Fluid Exchange Summary

ARTERIAL END		VENOUS END
_____	Net hydrostatic pressure	_____
_____	Oncotic pressure	_____
_____	Net filtration pressure	_____

Self-Quiz
Take this opportunity to check your progress by taking the **QUIZ**. See the **Introduction Module** for a reminder on how to access the **QUIZ** for this Study Area.

The Heart

EXERCISE 9.14:
Cardiovascular System—Heart—Internal Features—Anterior View

SELECT TOPIC	SELECT VIEW
Heart	Internal Features—Anterior

Click **LAYER 1** *in the* **LAYER CONTROLS** *window, and you will see the following image:*

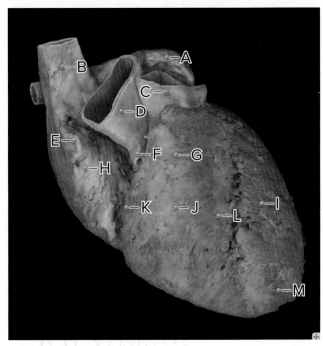

©McGraw-Hill Education

HEADS UP!
When you click on the pins for blood vessels of the cardiovascular system, the vessels will be highlighted red for arteries and blue for veins. The heart muscle will also be highlighted in red when the corresponding pins are clicked. Nerves will be highlighted in yellow.

Mouse-over the pins on the screen to find the information necessary to identify the following structures:

A. left auricle
B. superior vena cava
C. pulmonary trunk
D. ascending aorta
E. right auricle
F. right coronary a.
G. heart

H. right atrium
I. left ventricle
J. right ventricle
K. coronary sulcus
L. anterior interventricular sulcus
M. apex of heart

CHECK POINT

Heart, Internal Features, Anterior View

1. Name the major blood vessel between the heart and the aortic arch.
2. What are its two branches?
3. Name the branches of the right coronary artery.

Click **LAYER 2** in the **LAYER CONTROLS** window, and you will see the following image:

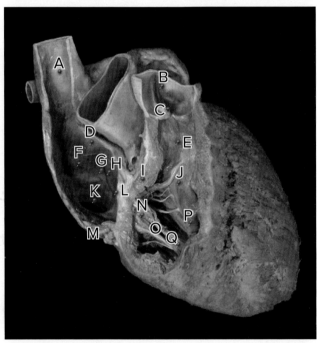

©McGraw-Hill Education

Mouse-over the pins on the screen to find the information necessary to identify the following structures:

A. superior vena cava
B. pulmonary trunk
C. pulmonary valve
D. crista terminalis
E. conus arteriosus
F. right atrium
G. fossa ovalis
H. limbus fossa ovalis
I. myocardium of right ventricle
J. right ventricle
K. opening of inferior vena cava
L. opening of coronary sinus
M. pectinate mm
N. right atrioventricular valve
O. chordae tendineae
P. septomarginal trabecula
Q. papillary mm

CHECK POINT

Heart, Internal Features, Anterior View, *continued*

4. What are the two terms for the valve between the right atrium and right ventricle?
5. Name the structure that prevents reflux of blood into the right ventricle.
6. Name the two branches of the pulmonary trunk.

Click **LAYER 3** *in the* **LAYER CONTROLS** *window, and you will see the following image:*

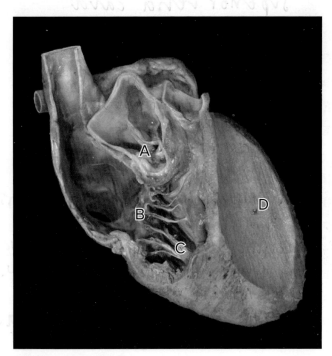

©McGraw-Hill Education

Mouse-over the pins on the screen to find the information necessary to identify the following structures:

A. aortic valve
B. right atroventricular valve
C. chordae tendineae
D. myocardium of left ventricle

CHECK POINT

Heart, Internal Features, Anterior View, *continued*

7. Name the structure that prevents reflux of blood into the left ventricle.
8. Name the muscle layer of the heart wall.
9. Where is that muscle layer thinner and thicker?

Click **LAYER 4** *in the* **LAYER CONTROLS** *window, and you will see the following image:*

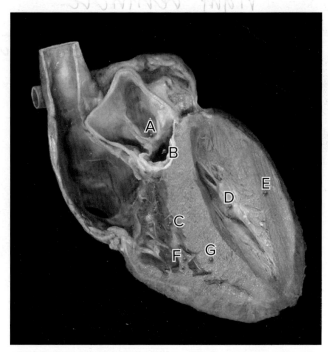

©McGraw-Hill Education

Mouse-over the pins on the screen to find the information necessary to identify the following structures:

A. aortic valve
B. aortic vestibule
C. right ventricle
D. left ventricle
E. myocardium of left ventricle
F. trabeculae carneae
G. myocardium of right ventricle

CHECK POINT

Heart, Internal Features, Anterior View, *continued*

10. What heart chamber is responsible for pumping oxygen-rich blood to the body (except the lungs)?
11. What heart chamber is responsible for pumping oxygen-poor blood to the lungs?
12. Name the irregular, muscular elevations on the internal surface of both ventricles.

Click **LAYER 5** *in the* **LAYER CONTROLS** *window, and you will see the following image:*

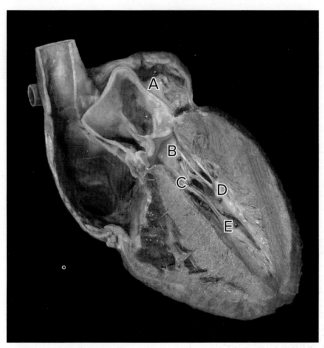

©McGraw-Hill Education

Mouse-over the pins on the screen to find the information necessary to identify the following structures:

A. left atrium
B. left atrioventricular valve
C. chordae tendineae
D. left ventricle
E. papillary mm

CHECK POINT

Heart, Internal Features, Anterior View, *continued*

13. Name the fibrous strands that attach the free edges of the atrioventricular valve cusps to the papillary muscles.
14. Name the heart chamber that receives oxygen-rich blood from the lungs.
15. Name the blood vessels that bring this blood from the lungs to the heart. How many are there?

Click **LAYER 6** *in the* **LAYER CONTROLS** *window, and you will see the following image:*

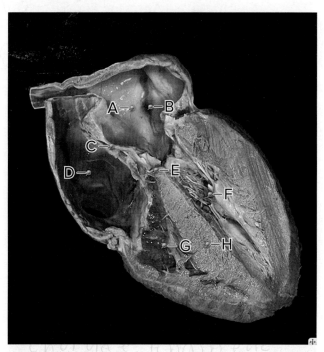

©McGraw-Hill Education

Mouse-over the pins on the screen to find the information necessary to identify the following structures:

A. left atrium
B. opening of pulmonary
C. interatrial septum
D. right atrium
E. membranous interventricular septum
F. left ventricle
G. right ventricle
H. muscular interventricular septum

CHECK POINT

Heart, Internal Features, Anterior View, *continued*

16. Name the structure that separates the right and left atria.
17. Name the structure that separates the right and left ventricles.
18. What is the name for the superior membranous part of the structure that separates the right and left ventricles?

EXERCISE 9.15:

Cardiovascular System—Heart—Vasculature—Anterior View

SELECT TOPIC: **Heart** ▸ SELECT VIEW: **Vasculature—Anterior**

Click **LAYER 1** *in the* **LAYER CONTROLS** *window, and you will see the following image:*

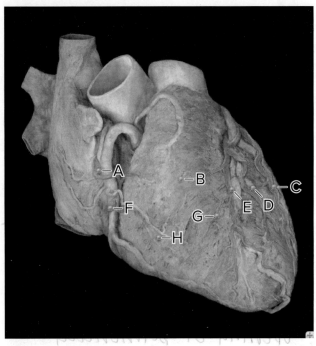

©McGraw-Hill Education

Mouse-over the pins on the screen to find the information necessary to identify the following structures:

A. _____

B. _____

C. _____

D. _____

E. _____

F. _____

G. _____

H. _____

CHECK POINT

Heart, Vasculature, Anterior View

1. Name the vein that ascends the anterior interventricular sulcus.
2. Name the vein that ascends across the anterior surface of the left ventricle.
3. Name the numerous veins that course across the anterior surface of the right ventricle.

Click **LAYER 2** *in the* **LAYER CONTROLS** *window, and you will see the following image:*

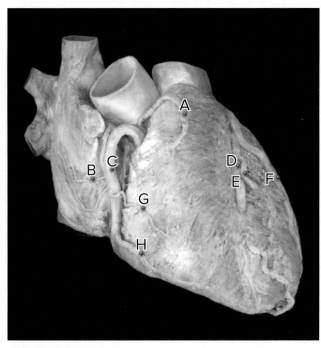

©McGraw-Hill Education

Mouse-over the pins on the screen to find the information necessary to identify the following structures:

A. _____

B. _____

C. _____

D. _____

E. _____

F. _____

G. _____

H. _____

CHECK POINT

Heart, Vasculature, Anterior View, *continued*

4. Name the artery that lies in the right coronary sulcus.
5. What are the branches of this artery?
6. Name the artery that descends along the margin of the left ventricle.

Click **LAYER 3** *in the* **LAYER CONTROLS** *window, and you will see the following image:*

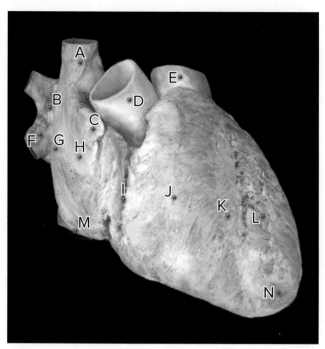

©McGraw-Hill Education

Mouse-over the pins on the screen to find the information necessary to identify the following structures:

A. _____
B. _____
C. _____
D. _____
E. _____
F. _____
G. _____
H. _____
I. _____
J. _____
K. _____
L. _____
M. _____
N. _____

CHECK POINT

Heart, Vasculature, Anterior View, *continued*

7. Name the major vein that drains everything inferior to the diaphragm.
8. Name the tributaries of the superior vena cava.
9. Name the inferolateral point of the heart.

EXERCISE 9.16:
Cardiovascular System—Heart—Vasculature—Posterior View

SELECT TOPIC	SELECT VIEW
Heart	Vasculature—Posterior

Click **LAYER 1** *in the* **LAYER CONTROLS** *window, and you will see the following image:*

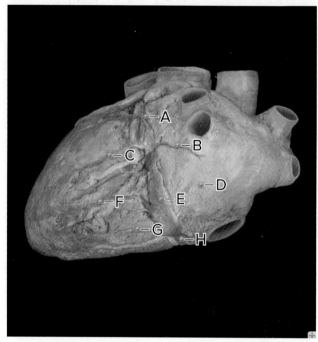

©McGraw-Hill Education

Mouse-over the pins on the screen to find the information necessary to identify the following structures:

A. _____
B. _____
C. _____
D. _____
E. _____
F. _____
G. _____
H. _____

CHECK POINT

Heart, Vasculature, Posterior View

1. Name the structure that receives venous blood from the heart.
2. What are the three major tributaries of this structure?
3. List the structures drained by the great cardiac vein.

Click **LAYER 2** *in the* **LAYER CONTROLS** *window, and you will see the following image:*

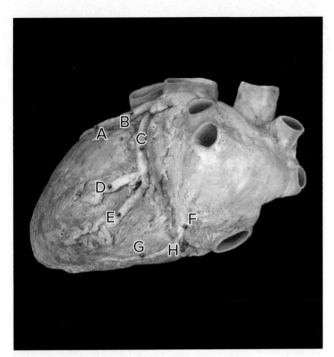

©McGraw-Hill Education

Mouse-over the pins on the screen to find the information necessary to identify the following structures:

A. _____
B. _____
C. _____
D. _____
E. _____
F. _____
G. _____
H. _____

CHECK POINT

Heart, Vasculature, Posterior View, *continued*

4. What is the term that refers to the end-to-end union of blood vessels?
5. Name the arteries located in the posterior interventricular sulcus.
6. Name the artery that descends in the anterior interventricular sulcus.

Click **LAYER 3** *in the* **LAYER CONTROLS** *window, and you will see the following image:*

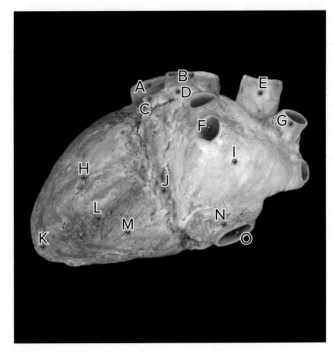

©McGraw-Hill Education

Mouse-over the pins on the screen to find the information necessary to identify the following structures:

A. _____
B. _____
C. _____
D. _____
E. _____
F. _____
G. _____
H. _____
I. _____
J. _____
K. _____
L. _____
M. _____
N. _____
O. _____

MODULE 9 The Cardiovascular System 525

EXERCISE 9.17:
Imaging—Heart—Left Ventricle

SELECT TOPIC	SELECT VIEW
Heart—Left Ventricle	CT: Axial

Click the **TAGS ON/OFF** button, and you will see the following image:

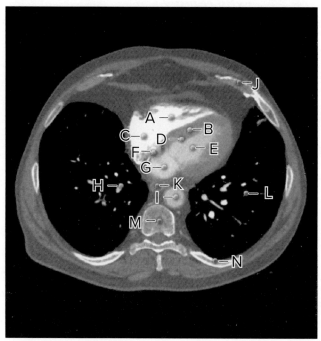

©McGraw-Hill Education

Mouse-over the pins on the screen to find the information necessary to identify the following structures:

A. _____

B. _____

C. _____

D. _____

E. _____

F. _____

G. _____

H. _____

I. _____

Noncardiovascular System Structures (blue pins)

J. _____

K. _____

L. _____

M. _____

N. _____

EXERCISE 9.18:
Imaging—Heart—Right Ventricle

SELECT TOPIC	SELECT VIEW
Heart—Right Ventricle	CT: Axial

Click the **TAGS ON/OFF** button, and you will see the following image:

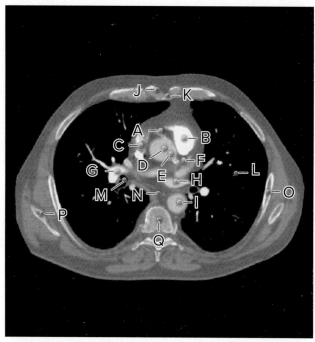

©McGraw-Hill Education

Mouse-over the pins on the screen to find the information necessary to identify the following structures:

A. _____

B. _____

C. _____

D. _____

E. _____

F. _____

G. _____

H. _____

I. _____

Noncardiovascular System Structures (blue pins)

J. _____

K. _____

L. _____

M. _____

N. _____

O. _____

P. _____

Q. _____

526 MODULE 9 The Cardiovascular System

SELECT ANIMATION
Conducting System of Heart PLAY

After viewing the animation, answer the following questions:

1. Where do action potentials associated with heartbeat regulation originate?

2. From their origin, the action potentials travel across the wall of the atrium to the _____ (_____) _____.

3. Name the structure in the interventricular septum where the action potentials pass upon leaving the atria.

4. The structure in question 3 then divides into right and left _____ _____.

5. After passing the apex of the ventricles, the action potentials pass along what structure to the ventricle walls?

6. What causes the ventricular muscle cells to contract in unison, providing a strong contraction?

Self-Quiz
Take this opportunity to check your progress by taking the **QUIZ**. See the **Introduction Module** for a reminder on how to access the **QUIZ** for this Study Area.

Cardiac Muscle

EXERCISE 9.19:

Cardiovascular System—Cardiac Muscle—Histology

 SELECT TOPIC Cardiac Muscle ▶ **SELECT VIEW** LM: High Magnification

Click the **TAGS ON/OFF** *button, and you will see the following image:*

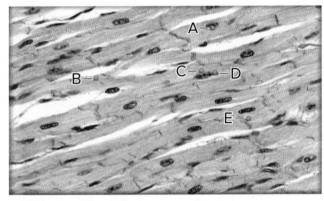

©McGraw-Hill Education

Mouse-over the pins on the screen to find the information necessary to identify the following structures:

A. _____

B. _____

C. _____

D. _____

E. _____

CHECK POINT

Cardiac Muscle, Histology

1. Name three locations where cardiac muscle is found.
2. What is the function of cardiac muscle tissue?
3. This function may be modulated by _____.

IN REVIEW

What Have I Learned?

The following questions cover the material that you have just learned, the histology of cardiac muscle. Apply what you have learned to answer these questions on a separate piece of paper.

1. Unlike skeletal muscle, cardiac muscle fibers _____.

2. Name one way skeletal muscle tissue is similar to cardiac muscle tissue.

3. Describe intercalated disks.

4. What is their function?

5. What is sarcoplasm?

SELECT ANIMATION — Cardiac Cycle with ECG — **PLAY**

After viewing the animation, answer the following questions:

1. A single cardiac cycle is made up of . . .

2. What is the term that refers to the relaxation of a heart chamber?

3. What physically occurs in a heart chamber during relaxation?

4. The term that refers to the contraction of a heart chamber is _____.

5. Atrial depolarization is represented by which wave on an electrocardiogram?

6. What is initiated by atrial depolarization?

7. This phenomenon is represented by what portion of the ECG?

8. Which section of the ECG represents ventricular depolarization?

9. This same section of the ECG in question 8 masks what portion of the cardiac cycle?

10. What is S1? How is it often described?

11. What is represented by the S-T segment of the ECG?

12. The T-wave on the ECG represents _____ _____.

13. What happens in the ventricles at this time?

14. What is S2? How is it often described?

15. What causes the beginning of the next cardiac cycle?

Self-Quiz

Take this opportunity to check your progress by taking the **QUIZ**. See the **Introduction Module** for a reminder on how to access the **QUIZ** for this Study Area.

SELECT ANIMATION — Cardiac Cycle (3D) — **PLAY**

After viewing the animation, answer the following questions:

1. The cardiac cycle is . . .

2. Define systole.

3. Define diastole.

4. How does the cycle begin?

5. Both _____ valves are _____ while the _____ and _____ _____ valves are closed.

6. Blood flows into the _____ _____ through the _____ and _____ _____ _____.

7. Blood flows from the lungs to the _____ _____ through the _____ veins.

8. Then, blood moves from both atria into the _____ through the open _____ valves.

9. During atrial systole . . .

10. The ventricles are still in _____, allowing them to . . .

11. During _____ _____, the ventricles _____.

12. The _____ valves close, preventing . . .

13. The _____ _____ valve opens and the _____ _____ expels blood into the _____ _____ to the _____.

14. Likewise, the _____ _____ valve opens and the . . .

15. After ventricular systole, the cardiac cycle begins again as both the _____ and _____ enter ____ to allow the _____ to fill with blood.

16. Normally, this cycle repeats _____ to _____ times a minute.

Cardiac Cycle (Interactive)

SELECT ANIMATION
Cardiac Structure and Function
PLAY

This Interactive Animation will walk you through the structure and function of the heart. Be sure to interact with the animation to glean all the information available to you.

Open the **Module 9—Cardiovascular System ANIMATION LIST**.

Under the second heading of **Cardiac Cycle (Interactive)**, *select* **1. Cardiac Structure and Function** *and you will see the image to the right:*

©McGraw-Hill Education

Select the **BEGIN** *button in the middle, and the narration and animation will begin immediately.*

Interact with the five images contained in this animation.

You can click **Reset** *to begin the animation again, or click* **NEXT** *to go to the next slide.*

When finished interacting with all five images, click the **X** *in the top-right corner to close the animation.*

After reviewing the animation, answer the following questions:

1. What is another term for a single cardiac cycle?

2. What events are included in a single cardiac cycle?

3. What is involved in a cardiac cycle?

4. What is produced by a cardiac cycle?

5. Distinct histological features of cardiac tissue enable the heart to _____ and _____ beat, _____ _____ and adapt to changes in _____.

6. List these histological features.

7. What structures constitute the electrical conduction system of the heart?

8. What makes these structures unique?

9. List the primary nodal cells.

10. List the other nodal cells.

11. What is the function of these other nodal cells?

12. What important roles do the valves of the cardiac system play?

13. Describe the two categories of valves and their locations.

SELECT ANIMATION
Regulation of Blood Flow through the Heart
PLAY

This Interactive Animation will walk you through the process of the heart's blood flow regulation. Be sure to interact with the animation to glean all the information available to you.

Open the **Module 9—Cardiovascular System ANIMATION LIST.**

Under the second heading of **Cardiac Cycle (Interactive)**, select **2. Regulation of Blood Flow through the Heart** *and you will see the image to the right:*

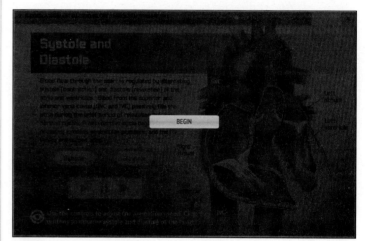

©McGraw-Hill Education

Select the **BEGIN** *button in the middle, and interact with the five images contained in this animation.*

When finished with the interactions, click the **X** *in the top-right corner to close the animation.*

After reviewing the animation, answer the following questions:

1. How is blood flow through the heart regulated?

2. What is systole?

3. What is diastole?

4. What occurs during the brief period of relaxation between cardiac cycles?

5. Describe the process of beginning a new cardiac cycle.

6. _____ percent of blood flow from the atria to the ventricles is passive.

7. _____ percent of blood flow from the atria to the ventricles is actively pumped.

8. What occurs during this phase?

9. What is end diastolic volume?

10. Describe early ventricular systole.

11. Why is this phase called isovolumic contraction?

12. How are the AV valves kept closed during this phase?

13. What event indicates the end of early ventricular systole?

14. What occurs during late ventricular systole?

15. What is stroke volume?

16. What causes the SL valves to close?

17. What is end systolic volume?

18. What is another term for early ventricular diastole?

19. What occurs during early ventricular diastole?

20. What initiates late ventricular diastole?

21. What represents the end of one cardiac cycle?

SELECT ANIMATION
Electrocardiogram (ECG) and the Cardiac Cycle
PLAY

This Interactive Animation will walk you through the correlation of the events of the cardiac cycle with the changes indicated by an electrocardiogram (ECG). Be sure to interact with the animation to glean all the information available to you.

Open the **Module 09—Cardiovascular System ANIMATION LIST**.

Under the second heading of **Cardiac Cycle (Interactive)**, select **3. Electrocardiogram (ECG) and the Cardiac Cycle** *and you will see the image to the right:*

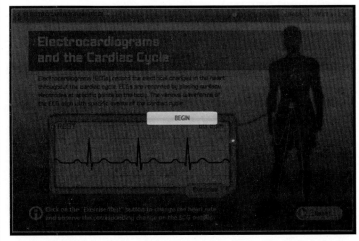

©McGraw-Hill Education

Select the **BEGIN** *button in the middle, and interact with the five images contained in this animation.*

When finished with the interactions, click the **X** *in the top-right corner to close the animation.*

After reviewing the animation, answer the following questions:

1. What information does the electrocardiogram record?

2. How are ECGs recorded?

3. What is indicated by the various waveforms of the ECG?

4. What is represented by the P wave?

5. What produces the coordinated contraction of the right and left atria?

6. Describe the events indicated by the P-Q segment of the ECG.

7. The QRS complex of the ECG represents…

8. How is this initiated?

9. What occurs during the S-T segment of the ECG?

10. What is represented by the T wave? What is it associated with?

11. List the three waveforms of an ECG for one complete cardiac cycle.

12. The P-Q segment separates…

13. The S-T segment separates…

14. What is stimulated by the cardiac conduction system?

15. What is the purpose of the cardiac cycle?

16. What is a record of these cumulative events?

Clinical Applications

SELECT ANIMATION
Blood Pressure/Hypertension (3D)
PLAY

After viewing the animation, answer the following questions:

1. The cardiac cycle begins with _____.

2. What events occur in the heart during this time?

3. _____ completes the cardiac cycle.

4. What events occur in the heart during this time?

5. What is blood pressure?

6. When is blood pressure higher?

7. When is blood pressure lower?

8. Normal systolic blood pressure should be below _____.

9. Normal diastolic blood pressure should be below _____.

10. What is another term for high blood pressure?

11. Define high blood pressure.

12. What risk factors are associated with hypertension?

13. To what do these risk factors all contribute?

14. This increased _____ causes increased _____ _____ the heart must push against.

15. What occurs as the heart works harder?

16. What affect does this have on the blood ejected with each cardiac cycle.

17. How does the brain respond to this?

18. This eventually leads to _____ _____ and _____ to other organs such as the _____ and the _____.

19. What are the treatments for hypertension?

20. Lifestyle changes include . . .

21. How does drug therapy reduce blood pressure?

22. List the anti-hypertensive drugs and their methods of lowering blood pressure.

SELECT ANIMATION
Heart Failure Overview (3D) — PLAY

After viewing the animation, answer the following questions:

1. How does deoxygenated blood flow in the normal heart?

2. What is preload? Where does it move?

3. What causes the blood to flow out of the heart?

4. Where does this blood go? What occurs there?

5. Where does the oxygenated blood move?

6. Where does this blood go *en route* to the systemic circulation?

7. What is afterload?

8. Define heart failure.

9. Define left-sided heart failure.

10. Define pulmonary congestion.

11. How does the heart compensate for the low oxygen levels?

12. What affect does this have on the heart?

13. What continues the cycle of cardiac muscle damage?

14. Define right-sided heart failure.

15. What is the result of right-sided heart failure?

16. What does this cause?

17. What is edema?

18. What effect does this have on the kidneys?

19. What effect does this have on fluid balance and vascular resistance?

20. What is the result?

21. What is the treatment for heart failure?

22. Commonly, therapies for right-sided heart failure are geared toward . . .

23. In heart failure, an increase in _____ _____ places added _____ on the _____ _____.

24. How do diuretic medications help?

25. What effect does this have on the urine?

26. How do ACE inhibitors help?

27. How does digoxin help?

28. To compensate for the decreased cardiac output in heart failure, the sympathetic nervous system . . .

29. How do beta-blockers help?

SELECT ANIMATION
Left-sided Heart Failure (3D) PLAY

After viewing the animation, answer the following questions:

1. Define left-sided heart failure.

2. What is preload?

3. What does this condition cause? What is the result?

4. Define systolic left-sided heart failure.

5. What is afterload?

6. After a left ventricular contraction . . .

7. What does it lead to?

8. Define diastolic left-sided heart failure.

9. What longstanding diseases can thicken and stiffen the ventricular muscle?

10. What is the result?

11. As a result of reduced myocardial contractility . . .

12. What is the result of this hypertrophy of the heart muscle?

13. What occurs in both systolic and diastolic left-sided heart failure? What does this cause?

14. What are crackles and rhonchi?

15. What symptoms occur as the lungs fill with fluid?

16. Define dyspnea.

17. Define orthopnea.

18. Define tachypnea.

19. How does the heart compensate for the left ventricle's weakened state?

20. This in time leads to . . .

21. How does this continue the cycle of cardiac muscle damage?

22. Describe the treatment for heart failure.

23. List the medications commonly prescribed for heart failure and how they work.

SELECT ANIMATION
Right-sided Heart Failure (3D) — PLAY

After viewing the animation, answer the following questions:

1. Define right-sided heart failure.

2. What is the most common cause of right-sided heart failure?

3. This is often the result of . . .

4. What elevates pumping pressure?

5. What does this cause over time?

6. What is cor pulmonale?

7. List the conditions that contribute to right-sided heart failure.

8. What causes the right ventricle wall hypertrophy? What is the result?

9. Ineffective pumping causes . . .

10. This results in . . .

11. What does this force the right ventricle to do?

12. The heart muscle _____ and pumps _____.

13. What is the result?

14. What congests the venous circulation?

15. This venous circulation includes . . .

16. What does the buildup of blood and fluid in the liver's tissues cause?

17. How does reduced cardiac output affect the kidneys?

18. Define nocturia. What does it often reflect?

19. What is caused by retained sodium and water? What does *this* cause?

20. What is the response to venous congestion?

21. What does this cause?

22. What does treatment for heart failure involve?

23. What are medications for right-sided heart failure geared toward?

24. List the medications commonly prescribed for heart failure and how they work.

SELECT ANIMATION
Coronary Artery Disease (3D) — PLAY

After viewing the animation, answer the following questions:

1. Where do the coronary arteries arise?

2. What is their function?

3. What is coronary artery disease?

4. What is cholesterol?

5. Where do cholesterol and triglyceride molecules bind in the bloodstream?

6. Where do these fats then circulate?

7. When do the cholesterol/LDL molecules adhere to the coronary artery walls?

8. What is another name for LDL?

9. What is a fatty streak? What causes this? What does it trigger?

10. _____ enter the lesion and transform into _____.

11. What is their function?

12. What is an atheroma?

13. Define atherosclerosis? What does it cause?

14. What happens to advanced atherosclerotic plaque over time?

15. What is a thrombus? What causes it? What is the result of a thrombus?

16. Define ischemia. What does it lead to?

17. Define infarction.

18. List the treatments for CAD and their effects.

19. What is angioplasty? What are other names for angioplasty?

20. What is an endovascular stent? What is its function?

IN REVIEW

What Have I Learned?

The following questions cover the material that you have just learned, the heart. Apply what you have learned to answer these questions on a separate piece of paper.

1. Name the body areas drained by the superior vena cava.
2. Name the small pouchlike extensions of the atria.
3. Name the vessel that conveys oxygen-poor blood from the right ventricle.
4. Name the three blood vessels that drain into the right atrium.
5. What external structure of the heart marks the position of the junction between the atria and the ventricles?
6. What external structure of the heart marks the position of the interventricular septum?
7. What is the term that refers to the blunt tip of the left ventricle?
8. Name the conical elevations of myocardium in the ventricular walls. What is their function?
9. Name the remnant of the fetal blood shunt from the right atrium to the left atrium.
10. Name the muscle layer of the heart wall.

The Thorax

EXERCISE 9.20:
Cardiovascular System—Thorax—Arteries—Anterior View

SELECT TOPIC	SELECT VIEW
Thorax	Arteries—Anterior

Click **LAYER 1** in the **LAYER CONTROLS** window, and you will see the following image:

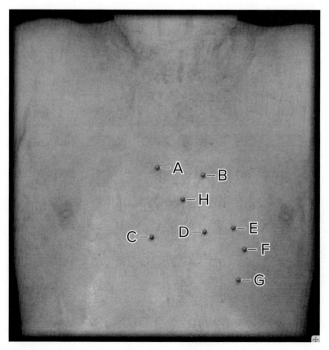

©McGraw-Hill Education

Mouse-over the pins on the screen to find the information necessary to identify the following noncardiovascular system structures:

A. auscultation point for aortic valve
B. auscultation point for pulmonary valve
C. surface projection of right border of heart
D. auscultation point for right atrioventricular
E. surface projection of left border of heart
F. auscultation point for left atrioventricular
G. surface projection of apex of heart

Noncardiovascular System Structure (blue pin)

H. surface projection of sternum

CHECK POINT

Thorax, Arteries, Anterior View

1. Where do heart valve sounds resonate?
2. What structure forms the right border of the heart?
3. What structure forms the apex of the heart?

HEADS UP!

*Click on all of the blue pins in each exercise for **Anatomy & Physiology | Revealed**®. The "What Have I Learned?" questions may cover the information for the blue pins, as well as the green ones.*

CHECK POINT

Thorax, Arteries, Anterior View, *continued*

4. What artery is also called the internal mammary artery?
5. What vein is also called the internal mammary vein?

Click **LAYER 2** *in the* **LAYER CONTROLS** *window, and you will see the following image:*

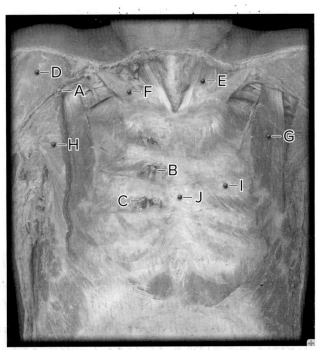
©McGraw-Hill Education

Click **LAYER 3** *in the* **LAYER CONTROLS** *window, and you will see the following image:*

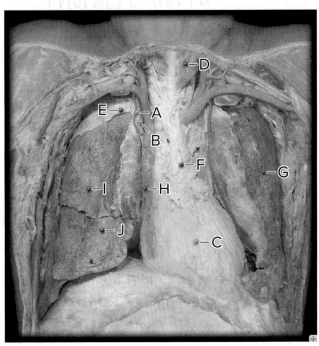
©McGraw-Hill Education

Mouse-over the pins on the screen to find the information necessary to identify the following structures:

A. cephalic v
B. internal thoracic a
C. internal thoracic v

Noncardiovascular System Structures (blue pins)

D. deltoid m
E. sternocleidomastoid m
F. clavicle
G. pectoralis minor m
H. pectoralis major m
I. costal cartilages
J. sternum

Mouse-over the pins on the screen to find the information necessary to identify the following structures:

A. internal thoracic v
B. internal thoracic a
C. fibrous pericardium

Noncardiovascular System Structures (blue pins)

D. thyroid gland
E. parietal pleura
F. remnant of thymus gland
G. left lung
H. phrenic n
I. right lung
J. visceral pleura

MODULE 9 The Cardiovascular System

CHECK POINT
Thorax, Arteries, Anterior View, *continued*

6. Name the artery that descends adjacent to the sternum within the thoracic cavity.

Click **LAYER 4** *in the* **LAYER CONTROLS** *window, and you will see the following image:*

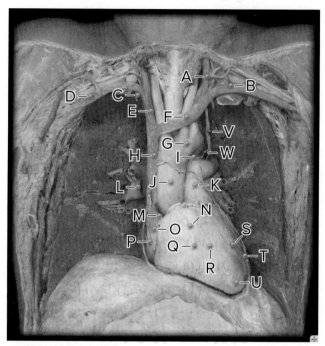

©McGraw-Hill Education

Mouse-over the pins on the screen to find the information necessary to identify the following structures:

A. internal jugular v
B. subclavian v
C. internal thoracic v
D. axillary v
E. right brachiocephalic v
F. left brachiocephalic v
G. arch of aorta
H. superior vena cava
I. ligamentum arteriosum
J. ascending aorta
K. pulmonary trunk
L. pulmonary a
M. right auricle
N. visceral layer of serous pericardium
O. right atrium
P. right border of heart

Q. right ventricle
R. heart
S. left ventricle
T. left border of heart
U. apex of heart

Noncardiovascular System Structures (blue pins)

V. vagus n CN X
W. left recurrent laryngeal n

CHECK POINT
Thorax, Arteries, Anterior View, *continued*

7. Name the large artery that originates at the aortic valve and ascends 5 mm within the pericardium.
8. Name the arched continuation of the artery in question 7.
9. Name the three branches of the artery in question 8.

Click **LAYER 5** *in the* **LAYER CONTROLS** *window, and you will see the following image:*

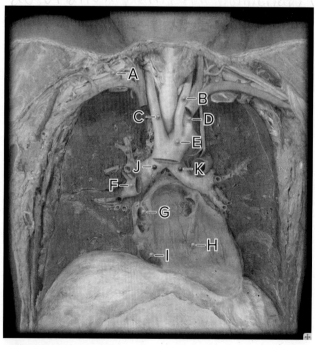

©McGraw-Hill Education

Mouse-over the pins on the screen to find the information necessary to identify the following structures:

A. right subclavian a
B. left common carotid a
C. brachiocephalic trunk
D. left subclavian a

540 MODULE 9 The Cardiovascular System

E. arch of aorta
F. pulmonary a
G. pulmonary vv
H. parietal layer of serous pericard
I. inferior vena cava

Noncardiovascular System Structures (blue pins)

J. right main bronchus
K. left main bronchus

D. axillary a
E. left subclavian a
F. left common carotid a
G. brachiocephalic trunk
H. arch of aorta
I. arch of azygos v
J. posterior intercostal a
K. azygos v
L. posterior intercostal v
M. thoracic aorta

Noncardiovascular System Structures (blue pins)

N. phrenic n
O. anterior scalene m
P. Rib 1
Q. trachea
R. right main bronchus
S. intercostal nn
T. left main bronchus
U. esophagus
V. sympathetic trunk
W. innermost intercostal mm
X. thoracic duct

CHECK POINT

Thorax, Arteries, Anterior View, *continued*

10. Name the blood vessel also known as the innominate artery. What two vessels are found at its terminus?
11. Name the second artery to branch off of the aortic arch. What are its terminal branches?
12. Name the third artery to branch off of the aortic arch. Name the vessel that is the continuation of this artery beginning at the lateral border of rib 1.

CHECK POINT

Thorax, Arteries, Anterior View, *continued*

13. Name the major artery that passes through the axilla.
14. Name the artery that ascends into the neck from the brachiocephalic trunk.
15. What are the two terminal branches of the artery in question 14?

Click **LAYER 6** *in the* **LAYER CONTROLS** *window, and you will see the following image:*

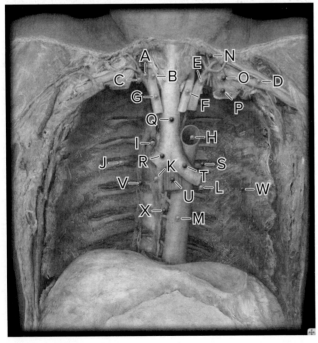

©McGraw-Hill Education

Mouse-over the pins on the screen to find the information necessary to identify the following structures:

A. right subclavian a
B. right common carotid a
C. internal thoracic a

MODULE 9 The Cardiovascular System 541

EXERCISE 9.21a:
Imaging—Aortic Arch

SELECT TOPIC	SELECT VIEW
Aortic Arch	CTA: Anterior

Click the **TAGS ON/OFF** button, and you will see the following image:

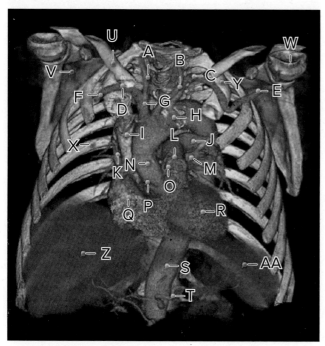

©McGraw-Hill Education

Mouse-over the pins on the screen to find the information necessary to identify the following structures:

A. _____
B. _____
C. _____
D. _____
E. _____
F. _____
G. _____
H. _____
I. _____
J. _____
K. _____
L. _____
M. _____
N. _____
O. _____
P. _____
Q. _____
R. _____
S. _____
T. _____

Noncardiovascular System Structures (blue pins)

U. _____
V. _____
W. _____
X. _____
Y. _____
Z. _____
AA. _____

EXERCISE 9.21b:
Imaging—Aortic Arch

SELECT TOPIC	SELECT VIEW
Aortic Arch	CTA: Oblique

Click the **TAGS ON/OFF** button, and you will see the following image:

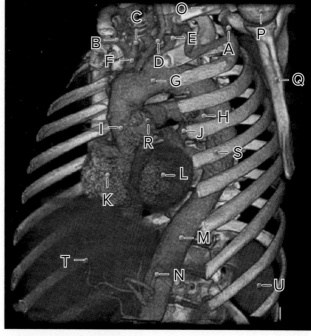

©McGraw-Hill Education

Mouse-over the pins on the screen to find the information necessary to identify the following structures:

A. _____
B. _____
C. _____
D. _____

542 MODULE 9 The Cardiovascular System

E. _____
F. _____
G. _____
H. _____
I. _____
J. _____
K. _____
L. _____
M. _____
N. _____

Noncardiovascular System Structures (blue pins)

O. _____
P. _____
Q. _____
R. _____
S. _____
T. _____
U. _____

EXERCISE 9.22:
Cardiovascular System—Thorax—Veins—Anterior View

SELECT TOPIC	SELECT VIEW
Thorax	Veins—Anterior

HEADS UP!
We are skipping **LAYER 1** for this exercise because it is a repeat of Exercise 9.6. It is advisable to review this layer before you complete this exercise.

Click **LAYER 2** *in the* **LAYER CONTROLS** *window, and you will see the following image:*

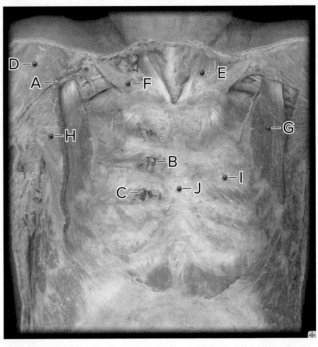

©McGraw-Hill Education

Mouse-over the pins on the screen to find the information necessary to identify the following structures:

A. _____
B. _____
C. _____

Noncardiovascular System Structures (blue pins)

D. _____
E. _____
F. _____
G. _____
H. _____
I. _____
J. _____

CHECK POINT
Thorax, Veins, Anterior View

1. Name the vein that ascends from the dorsum of the hand to the anterolateral forearm and arm and into the deltopectoral triangle.
2. Name the vein also known as the internal mammary vein.

Click **LAYER 3** *in the* **LAYER CONTROLS** *window, and you will see the following image:*

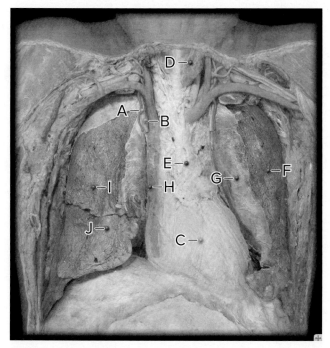

©McGraw-Hill Education

Mouse-over the pins on the screen to find the information necessary to identify the following structures:

A. _____
B. _____
C. _____

Noncardiovascular System Structures (blue pins)

D. _____
E. _____
F. _____
G. _____
H. _____
I. _____
J. _____

CHECK POINT

Thorax, Veins, Anterior View, *continued*

3. Name the areas drained by the internal thoracic vein.

Click **LAYER 4** *in the* **LAYER CONTROLS** *window, and you will see the following image:*

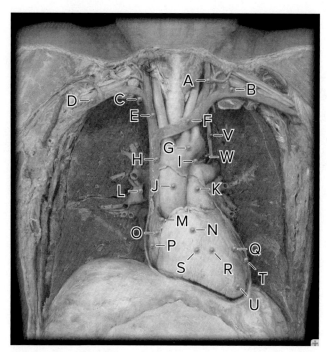

©McGraw-Hill Education

Mouse-over the pins on the screen to find the information necessary to identify the following structures:

A. _____
B. _____
C. _____
D. _____
E. _____
F. _____
G. _____
H. _____
I. _____
J. _____
K. _____
L. _____
M. _____
N. _____
O. _____
P. _____
Q. _____
R. _____
S. _____
T. _____
U. _____

Noncardiovascular System Structures (blue pins)

V. _____

W. _____

Noncardiovascular System Structures (blue pins)

I. _____

J. _____

CHECK POINT

Thorax, Veins, Anterior View, *continued*

4. Name the areas drained by the superior vena cava. Where does it terminate?
5. What two vessels unite to form the superior vena cava?
6. Name the continuation of the sigmoid dural sinus of the cranial cavity that drains the brain, face, and neck.

CHECK POINT

Thorax, Veins, Anterior View, *continued*

7. Name the vessels that carry oxygen-rich blood to the heart.
8. Where do they terminate?
9. Name the drainage of the inferior vena cava. Where does it terminate?

Click **LAYER 5** *in the* **LAYER CONTROLS** *window, and you will see the following image:*

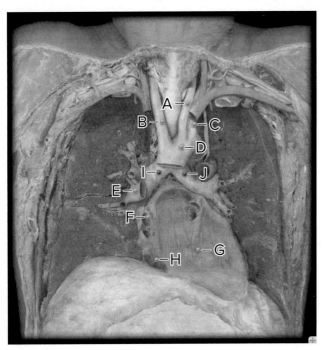

©McGraw-Hill Education

Click **LAYER 6** *in the* **LAYER CONTROLS** *window, and you will see the following image:*

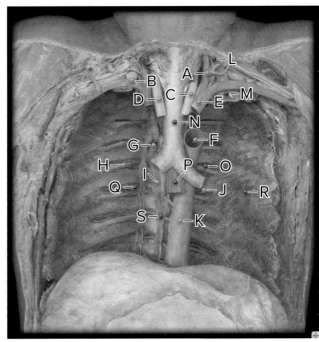

©McGraw-Hill Education

Mouse-over the pins on the screen to find the information necessary to identify the following structures:

A. _____
B. _____
C. _____
D. _____
E. _____
F. _____
G. _____
H. _____

Mouse-over the pins on the screen to find the information necessary to identify the following structures:

A. _____
B. _____
C. _____
D. _____
E. _____
F. _____
G. _____
H. _____
I. _____

J. _____
K. _____

Noncardiovascular System Structures (blue pins)

L. _____
M. _____
N. _____
O. _____
P. _____
Q. _____
R. _____
S. _____

CHECK POINT

Thorax, Veins, Anterior View, *continued*

10. Name the venous system that forms a collateral pathway between the superior and inferior vena cavae.

SELECT ANIMATION
Systemic vs. Pulmonary Circulation — PLAY

After viewing the animation, answer the following questions:

1. Name the two divisions of the cardiovascular system.

2. What are the destinations of these two circuits?

3. In the systemic circulation, where does gas exchange occur?

4. In the pulmonary circulation, where does gas exchange occur?

5. Name the blood vessels that carry oxygen-rich blood to the heart. How many are there? Where do they terminate?

Self-Quiz
Take this opportunity to check your progress by taking the **QUIZ**. See the **Introduction Module** for a reminder on how to access the **QUIZ** for this Study Area.

SELECT ANIMATION
Pulmonary and Systemic Circulation (3D) — PLAY

After viewing the animation, answer the following questions:

1. The _____ side of the heart produces _____ circulation.

2. Describe pulmonary circulation.

3. Where does the blood become oxygenated?

4. How does the reoxygenated blood return to the heart?

5. The _____ side of the heart produces _____ circulation.

6. Describe systemic circulation.

EXERCISE 9.23:
Imaging—Thorax

SELECT TOPIC: Thorax ▶ SELECT VIEW: X ray—Posterior–Anterior

Click the **TAGS ON/OFF** button, and you will see the following image:

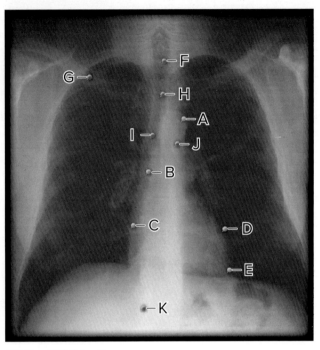

©McGraw-Hill Education

Mouse-over the pins on the screen to find the information necessary to identify the following structures:

A. _____
B. _____
C. _____
D. _____
E. _____

Noncardiovascular System Structures (blue pins)

F. _____
G. _____
H. _____
I. _____
J. _____
K. _____

Self-Quiz
Take this opportunity to check your progress by taking the **QUIZ**. See the **Introduction Module** for a reminder on how to access the **QUIZ** for this Study Area.

IN REVIEW

What Have I Learned?

The following questions cover the material that you have just learned, the blood vessels of the thorax. Apply what you have learned to answer these questions on a separate piece of paper.

1. What is the name for the fibrous sac that encloses the heart?

2. Name the lymphatic organ that is large in children but atrophies during adolescence.

3. Name the bilobed endocrine gland located lateral to the trachea and larynx.

4. How do large arteries supply blood to body structures?

5. Name the large vessel that conveys oxygen-poor blood from the right ventricle of the heart.

6. Name the two branches of the blood vessel mentioned in question 5 that convey oxygen-poor blood to the lungs.

7. Name the blunt tip of the left ventricle.

8. What is the carotid sheath? What structures are found within it?

9. What is the serous pericardium?

10. Name the structure that served to shunt blood in the fetus from the pulmonary artery to the aorta, bypassing the lungs. What is it called in the adult?

11. Name all of the different sections of the aorta, and describe where they are found.

12. Name the two major vessels that return oxygen-poor blood to the heart. What are the drainages for each? Where do they terminate?

The Head and Neck

EXERCISE 9.24:
Cardiovascular System—Head and Neck—Vasculature—Lateral View

SELECT TOPIC	SELECT VIEW
Head and Neck	Vasculature—Lateral

Click **LAYER 1** *in the* **LAYER CONTROLS** *window, and you will see the following image:*

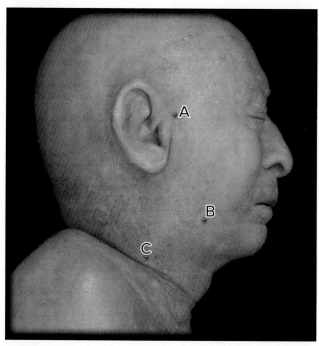

©McGraw-Hill Education

Mouse-over the pins on the screen to find the information necessary to identify the following structures:

A. _____
B. _____
C. _____

CHECK POINT

Head and Neck, Vasculature, Lateral View

1. What blood vessel is responsible for the superficial temporal pulse?
2. What blood vessel is responsible for the facial pulse?
3. Between which two structures of the neck would you find the carotid pulse?

Click **LAYER 2** *in the* **LAYER CONTROLS** *window, and you will see the following image:*

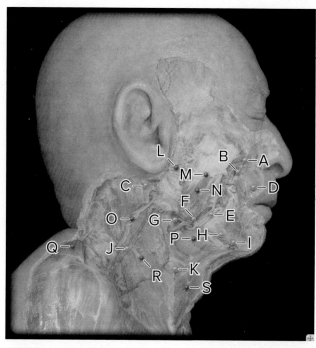

©McGraw-Hill Education

Mouse-over the pins on the screen to find the information necessary to identify the following structures:

A. _____
B. _____
C. _____
D. _____
E. _____
F. _____
G. _____
H. _____
I. _____
J. _____
K. _____

Noncardiovascular System Structures (blue pins)

L. _____
M. _____
N. _____
O. _____
P. _____
Q. _____
R. _____
S. _____

CHECK POINT

Head and Neck, Vasculature, Lateral View, *continued*

4. Name a subcutaneous vein with tributaries that drain the scalp and the face and terminates in the subclavian vein.
5. Name a vein, the size of which is inversely proportional to the vein in question 4.
6. Name a vein that drains the muscles, glands, and mucous membranes of the upper lip.

Noncardiovascular System Structures (blue pins)

M. _____
N. _____
O. _____

CHECK POINT

Head and Neck, Vasculature, Lateral View, *continued*

7. Name the large vein that drains the cranial cavity, including the brain, as well as the face and neck.
8. Name a vein that drains the superior molar and premolar teeth and the mucous membrane of the maxillary sinus.
9. Name a vein that drains the deep face and temporal region and descends within the parotid gland.

Click **LAYER 3** *in the* **LAYER CONTROLS** *window, and you will see the following image:*

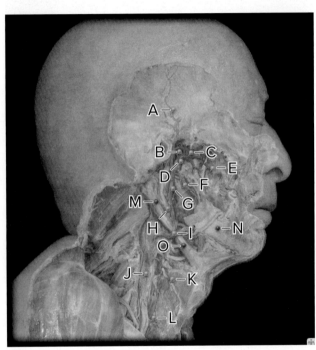

©McGraw-Hill Education

Mouse-over the pins on the screen to find the information necessary to identify the following structures:

A. _____
B. _____
C. _____
D. _____
E. _____
F. _____
G. _____
H. _____
I. _____
J. _____
K. _____
L. _____

Click **LAYER 4** *in the* **LAYER CONTROLS** *window, and you will see the following image:*

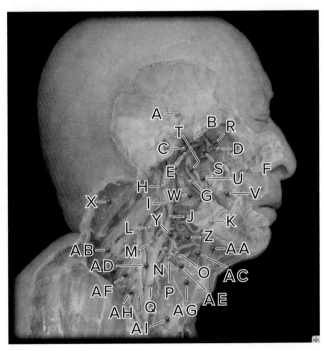

©McGraw-Hill Education

Mouse-over the pins on the screen to find the information necessary to identify the following structures:

A. _____
B. _____
C. _____
D. _____
E. _____
F. _____
G. _____

H. _____
I. _____
J. _____
K. _____
L. _____
M. _____
N. _____
O. _____
P. _____
Q. _____

Noncardiovascular System Structures (blue pins)

R. _____
S. _____
T. _____
U. _____
V. _____
W. _____
X. _____
Y. _____
Z. _____
AA. _____
AB. _____
AC. _____
AD. _____
AE. _____
AF. _____
AG. _____
AH. _____
AI. _____

CHECK POINT

Head and Neck, Vasculature, Lateral View, *continued*

10. Name the major arteries of the neck where the left one originates from the aortic arch and the right one originates from the brachiocephalic trunk.
11. Name the branch of the arteries in question 10 that distributes to the exterior of the head (except the orbit), the face, meninges, and neck structures.
12. Name the branch of the arteries in question 10 that enters the cranial cavity and terminates in the cerebral arterial circle.

Click **LAYER 5** *in the* **LAYER CONTROLS** *window, and you will see the following image:*

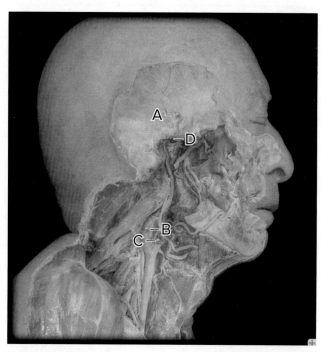
©McGraw-Hill Education

Mouse-over the pins on the screen to find the information necessary to identify the following structures:

A. _____
B. _____
C. _____

Noncardiovascular System Structure (blue pin)

D. _____

CHECK POINT

Head and Neck, Vasculature, Lateral View, *continued*

13. List the structures contained within the carotid sheath.

EXERCISE 9.25:
Imaging—Carotid Artery

SELECT TOPIC: Carotid Artery ▸ **SELECT VIEW**: CTA: Oblique

Click the **TAGS ON/OFF** *button, and you will see the following image:*

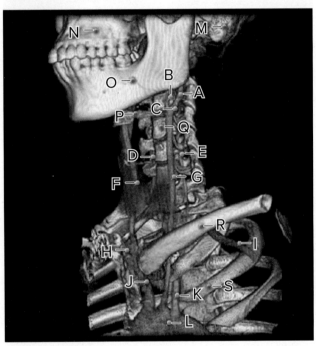

©McGraw-Hill Education

Mouse-over the pins on the screen to find the information necessary to identify the following structures:

A. _____
B. _____
C. _____
D. _____
E. _____
F. _____
G. _____
H. _____
I. _____
J. _____
K. _____
L. _____

Noncardiovascular System Structures (blue pins)

M. _____
N. _____
O. _____
P. _____
Q. _____
R. _____
S. _____

SELECT ANIMATION: Baroreceptor Reflex — **PLAY**

After viewing the animation, answer the following questions:

1. Where are baroreceptors located?

2. What is their function?

3. How do the arteries with baroreceptors, and all arteries for that matter, respond to increased blood pressure?

4. What response do the baroreceptors have to this increased blood pressure?

5. Where are these action potentials conducted?

6. What nerves conduct these impulses?

7. What is the parasympathetic response to and result of this stimulation?

8. What is the sympathetic response and result?

9. What about sympathetic stimulation of the blood vessels?

10. What physical events combine to bring elevated blood pressure back toward normal?

Self-Quiz

Take this opportunity to check your progress by taking the **QUIZ**. See the **Introduction Module** for a reminder on how to access the **QUIZ** for this Study Area.

SELECT ANIMATION — Chemoreceptor Reflex — **PLAY**

After viewing the animation, answer the following questions:

1. Where are chemoreceptors located?

2. What is their function?

3. Where are the impulses from the chemoreceptors conducted?

4. What nerves conduct these impulses?

5. What three events decrease parasympathetic stimulation of the heart?

6. What effect does this decreased stimulation have on the physiology of the heart?

7. What sympathetic response occurs during the three events listed in question 5?

8. What sympathetic response occurs in the blood vessels?

9. If the chemoreceptors are stimulated by decreased blood oxygen, what physical changes occur as a result of the changes in autonomic stimulation?

10. What would you deduce is the effect on blood pressure as a result of the answer to question 9?

Self-Quiz
Take this opportunity to check your progress by taking the **QUIZ**. See the **Introduction Module** for a reminder on how to access the **QUIZ** for this Study Area.

EXERCISE 9.26a:
Imaging—Head and Neck

SELECT TOPIC: Head and Neck ▶ **SELECT VIEW**: Angiogram Arteries—Anterior–Posterior

Click the **TAGS ON/OFF** *button, and you will see the following image:*

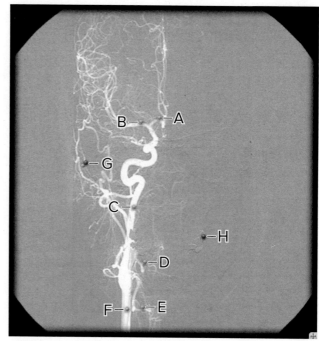

©McGraw-Hill Education

Mouse-over the pins on the screen to find the information necessary to identify the following structures:

A. _____
B. _____
C. _____
D. _____
E. _____
F. _____

Noncardiovascular System Structures (blue pins)

G. _____
H. _____

552 MODULE 9 The Cardiovascular System

EXERCISE 9.26b:
Imaging—Head and Neck

SELECT TOPIC	SELECT VIEW
Head and Neck	Angiogram Arteries—Lateral

Click the **TAGS ON/OFF** *button, and you will see the following image:*

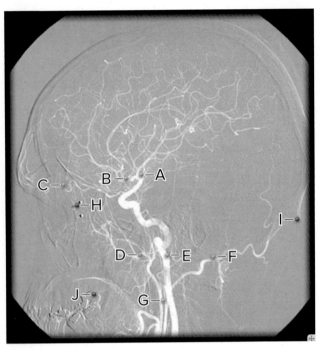

©McGraw-Hill Education

Mouse-over the pins on the screen to find the information necessary to identify the following structures:

A. _____
B. _____
C. _____
D. _____
E. _____
F. _____
G. _____

Noncardiovascular System Structures (blue pins)

H. _____
I. _____
J. _____

EXERCISE 9.26c:
Imaging—Head and Neck

SELECT TOPIC	SELECT VIEW
Head and Neck	Angiogram Veins—Anterior–Posterior

Click the **TAGS ON/OFF** *button, and you will see the following image:*

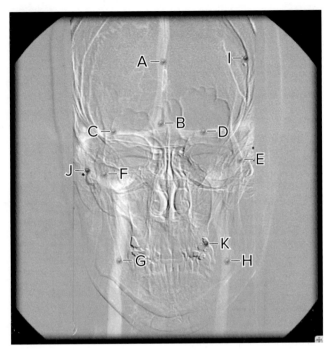

©McGraw-Hill Education

Mouse-over the pins on the screen to find the information necessary to identify the following structures:

A. _____
B. _____
C. _____
D. _____
E. _____
F. _____
G. _____
H. _____

Noncardiovascular System Structures (blue pins)

I. _____
J. _____
K. _____

EXERCISE 9.26d:
Imaging—Head and Neck

SELECT TOPIC	SELECT VIEW
Head and Neck	Angiogram Veins—Lateral

Click the **TAGS ON/OFF** button, and you will see the following image:

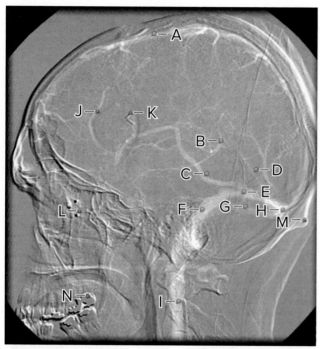

©McGraw-Hill Education

Mouse-over the pins on the screen to find the information necessary to identify the following structures:

A. _____
B. _____
C. _____
D. _____
E. _____
F. _____
G. _____
H. _____
I. _____

Noncardiovascular System Structures (blue pins)

J. _____
K. _____
L. _____
M. _____
N. _____

Self-Quiz
Take this opportunity to check your progress by taking the **QUIZ**. See the **Introduction Module** for a reminder on how to access the **QUIZ** for this Study Area.

IN REVIEW

What Have I Learned?

The following questions cover the material that you have just learned, the blood vessels of the head and neck. Apply what you have learned to answer these questions on a separate piece of paper.

1. Name the gland that produces 25 percent of your saliva.

2. List three structures that pass through this gland.

3. Where does the duct from this gland empty?

4. Name two veins of the head and neck that lack valves.

5. Name a superficial vein that drains the temporal region.

6. List the areas drained by the maxillary veins.

7. What is an anastomosis? Give an example of one in the head and neck region.

The Brain

EXERCISE 9.27:
Cardiovascular System—Brain—Arteries—Inferior View

SELECT TOPIC	SELECT VIEW
Brain	Arteries—Inferior

Click **LAYER 1** *in the* **LAYER CONTROLS** *window, and you will see the following image:*

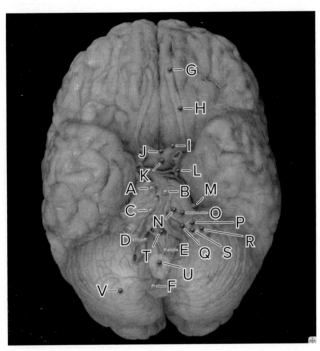

©McGraw-Hill Education

Mouse-over the pins on the screen to find the information necessary to identify the following structures:

A. _____
B. _____
C. _____
D. _____
E. _____
F. _____

Noncardiovascular System Structures (blue pins)

G. _____
H. _____
I. _____
J. _____
K. _____
L. _____
M. _____
N. _____
O. _____
P. _____
Q. _____
R. _____
S. _____
T. _____
U. _____
V. _____

CHECK POINT
Brain, Arteries, Inferior View

1. The central branches of which artery supply the anterior two-thirds of the spinal cord?
2. Name the artery that distributes to the posterior cerebellum and lateral medulla oblongata.
3. Name the artery that distributes to the anterior-inferior cerebellum, pons, and medulla oblongata.

Click **LAYER 2** *in the* **LAYER CONTROLS** *window, and you will see the following image:*

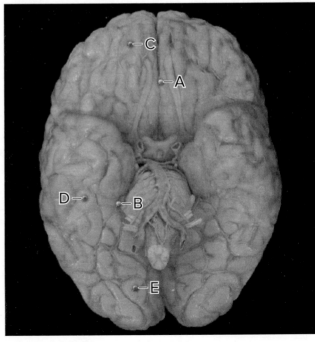

©McGraw-Hill Education

Mouse-over the pins on the screen to find the information necessary to identify the following structures:

A. _____
B. _____

Noncardiovascular System Structures (blue pins)

C. _____

D. _____

E. _____

G. _____

H. _____

Noncardiovascular System Structures (blue pins)

I. _____

J. _____

K. _____

CHECK POINT

Brain, Arteries, Inferior View, *continued*

4. Name the artery that distributes to the medial aspect of the frontal lobes of the cerebral cortex.
5. Name the artery that distributes to the temporal and occipital lobes of the cerebral cortex.

CHECK POINT

Brain, Arteries, Inferior View, *continued*

6. Name the arterial anastomosis of the ventral surface of the brain complete in only 20 percent of individuals.
7. Name the paired arteries that course along each lateral sulcus of the cerebral hemisphere.
8. Name the cerebral lobe not visible from the surface, also known as the isle of Reil.

Click **LAYER 3** *in the* **LAYER CONTROLS** *window, and you will see the following image:*

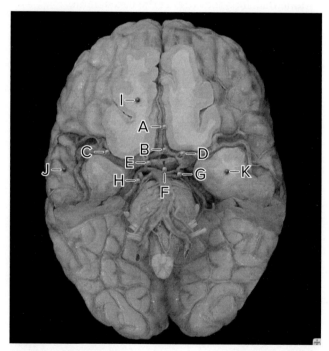

©McGraw-Hill Education

Mouse-over the pins on the screen to find the information necessary to identify the following structures:

A. _____

B. _____

C. _____

D. _____

E. _____

F. _____

EXERCISE 9.28:

Cardiovascular System—Brain—Veins—Lateral View

SELECT TOPIC	SELECT VIEW
Brain	Veins—Lateral

After clicking **LAYER 3** *in the* **LAYER CONTROLS** *window, to review the bones of the skull, click* **LAYER 3** *and you will see the following image:*

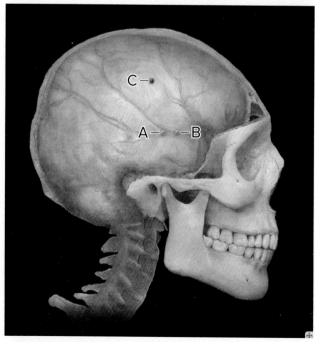

©McGraw-Hill Education

Mouse-over the pins on the screen to find the information necessary to identify the following structures:

A. _____

B. _____

Noncardiovascular System Structure (blue pin)

C. _____

CHECK POINT

Brain, Veins, Lateral View

1. Name the most exterior of the meninges. Where is it located? What is its function?
2. Name the artery whose distribution includes the dura mater, skull, trigeminal, and facial ganglia.
3. Name the vein that drains the dura mater and bones of the anterior and middle cranial fossae.

Click **LAYER 4** *in the* **LAYER CONTROLS** *window, and you will see the following image:*

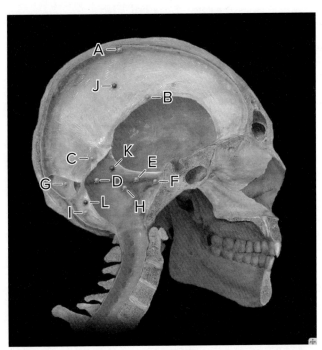

©McGraw-Hill Education

Mouse-over the pins on the screen to find the information necessary to identify the following structures:

A. _____
B. _____
C. _____
D. _____
E. _____

F. _____
G. _____
H. _____
I. _____

Noncardiovascular System Structures (blue pins)

J. _____
K. _____
L. _____

CHECK POINT

Brain, Veins, Lateral View, *continued*

4. What are dural venous sinuses?
5. Name the structure located in the internal occipital protuberance that drains the superior sagittal, straight, and occipital sinuses.
6. Name the sinus that drains the cerebellum.

Click **LAYER 5** *in the* **LAYER CONTROLS** *window, and you will see the following image:*

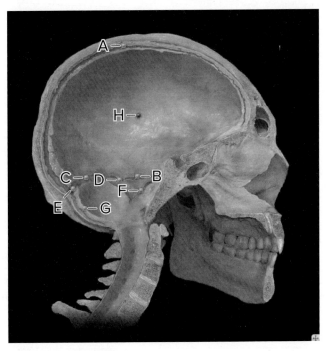

©McGraw-Hill Education

Mouse-over the pins on the screen to find the information necessary to identify the following structures:

A. _____
B. _____
C. _____
D. _____

E. _____
F. _____
G. _____

Noncardiovascular System Structure (blue pin)

H. _____

CHECK POINT

Brain, Veins, Lateral View, *continued*

7. Name the s-shaped continuation of the transverse sinus.
8. Name a paired dural venous sinus that terminates at the jugular foramen.
9. List the areas drained by the transverse sinus.

SELECT ANIMATION
Blood Flow Through Brain — PLAY

After viewing the animation, answer the following questions:

1. Name the three arteries through which blood flows to the brain.

2. Which two arteries supply blood to 80 percent of the cerebrum? These are the terminal branches of which arteries?

3. Which arteries enter the cranial cavity through the foramen magnum? They unite to form which artery?

4. Which arteries supply the occipital and temporal lobes of the cerebrum?

5. Which arteries form an anastomosis at the base of the brain? What is the name of this anastomosis?

6. What is the function of the anastomosis?

7. Blood is drained from the brain through small veins that empty into vessel channels called _____.

8. From where does blood flow to enter the confluence of sinuses?

9. From the confluence of sinuses, where does blood flow before leaving the skull?

10. The blood leaves the skull via the _____.

Self-Quiz

Take this opportunity to check your progress by taking the **QUIZ**. See the **Introduction Module** for a reminder on how to access the **QUIZ** for this Study Area.

SELECT ANIMATION
Stroke: Cerebrovascular Accident (3D) — PLAY

After viewing the animation, answer the following questions:

1. List the two major pairs of arteries supplying blood to the brain.

2. What do the branches of these arteries supply to the brain?

3. Describe a stroke or CVA.

4. What is ischemia?

5. What are the results of ischemia?

6. List results that may be caused by a stroke.

7. What is a Transient Ischemic Attack, or TIA?

8. What causes a TIA?

9. List the effects a TIA may cause.

10. What can a TIA signal?

11. What is the most common type of stroke?

12. What are the two classifications of ischemic stroke?

13. Describe a thrombotic ischemic stroke.

14. Describe an embolic ischemic stroke.

15. What is the other type of stroke?

16. Describe this type of stroke.

17. Describe intracerebral hemorrhagic strokes.

18. Describe subarachnoid hemorrhagic strokes.

19. List two causes of hemorrhagic stroke.

20. Describe acute thrombotic stroke care.

21. What is the time limit for this treatment?

22. Describe long-term stroke treatment.

23. What do these treatments prevent?

24. How are lost or impaired skills recovered?

25. How do patients relearn motor function and daily activities?

26. What therapies help the patient express themselves and cope with new challenges.

IN REVIEW

What Have I Learned?

The following questions cover the material that you have just learned, the blood vessels of the brain. Apply what you have learned to answer these questions on a separate piece of paper.

1. Name the large crescent-shaped fold of dura mater that separates the right and left cerebral hemispheres.

2. Name the small crescent-shaped fold of dura mater that separates the right and left cerebellar hemispheres.

3. Name the dural venous sinus that contains arachnoid granulations. What is the function of these granulations?

4. Name the dural venous sinus that drains the medial cerebral hemispheres.

5. Name the horizontal crescent-shaped fold of the dura mater that separates the cerebral hemispheres and the cerebellum.

The Shoulder

EXERCISE 9.29:
Cardiovascular System—Shoulder— Arteries—Anterior View

SELECT TOPIC	SELECT VIEW
Shoulder	Arteries—Anterior

Click **LAYER 2** *in the* **LAYER CONTROLS** *window, and you will see the following image:*

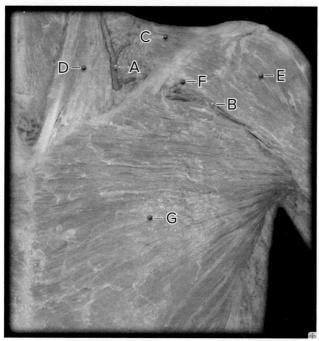

©McGraw-Hill Education

Mouse-over the pins on the screen to find the information necessary to identify the following structures:

A. _____
B. _____

Noncardiovascular System Structures (blue pins)

C. _____
D. _____
E. _____
F. _____
G. _____

CHECK POINT

The Shoulder, Arteries, Anterior View

1. Name the large vein of the neck that drains the face and scalp.
2. What are the major tributaries of this vein?
3. List the structures drained by the cephalic vein.

Click **LAYER 3** *in the* **LAYER CONTROLS** *window, and you will see the following image:*

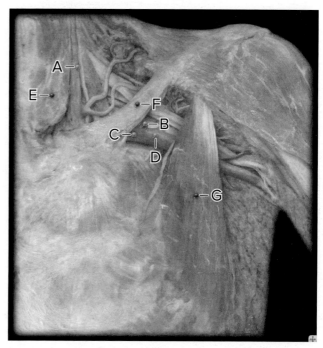

©McGraw-Hill Education

Mouse-over the pins on the screen to find the information necessary to identify the following structures:

A. _____
B. _____
C. _____
D. _____

Noncardiovascular System Structures (blue pins)

E. _____
F. _____
G. _____

CHECK POINT

The Shoulder, Arteries, Anterior View, *continued*

4. Name the large vein within the carotid sheath that is a continuation of the sigmoid dural sinus of the cranial cavity.
5. Which two veins unite to form the brachiocephalic vein?
6. Where does the axillary vein become the subclavian vein?

Click **LAYER 4** *in the* **LAYER CONTROLS** *window, and you will see the following image:*

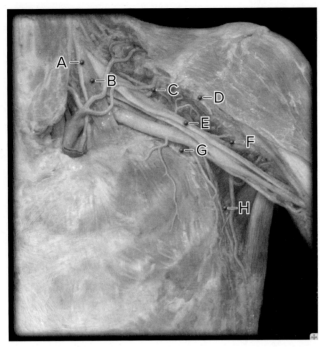

©McGraw-Hill Education

Mouse-over the pins on the screen to find the information necessary to identify the following noncardiovascular system structures:

A. _____
B. _____
C. _____
D. _____
E. _____
F. _____
G. _____
H. _____

Click **LAYER 5** *in the* **LAYER CONTROLS** *window, and you will see the following image:*

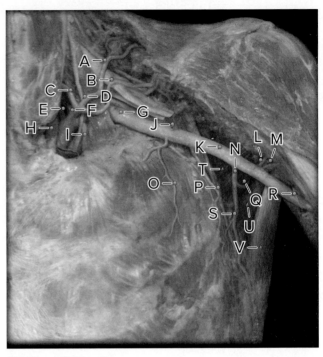

©McGraw-Hill Education

Mouse-over the pins on the screen to find the information necessary to identify the following structures:

A. _____
B. _____
C. _____
D. _____
E. _____
F. _____
G. _____
H. _____
I. _____
J. _____
K. _____
L. _____
M. _____
N. _____
O. _____
P. _____
Q. _____
R. _____
S. _____

Noncardiovascular System Structures (blue pins)

T. _____

U. _____

V. _____

CHECK POINT

The Shoulder, Arteries, Anterior View, *continued*

7. What is the definition of a "trunk"? (This one will require inference from the information available in the **STRUCTURE INFORMATION** window.)
8. Name a trunk that ascends in the lower neck.
9. Where does the subclavian artery become the axial artery? Where does the axillary artery become the brachial artery?

EXERCISE 9.30:

Cardiovascular System—Shoulder—Veins—Anterior View

SELECT TOPIC	SELECT VIEW
Shoulder	Veins—Anterior

Click **LAYER 2** *in the* **LAYER CONTROLS** *window, and you will see the following image:*

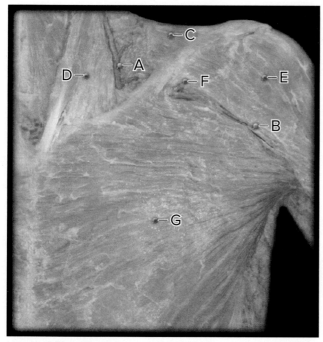

©McGraw-Hill Education

Mouse-over the pins on the screen to find the information necessary to identify the following structures:

A. _____

B. _____

Noncardiovascular System Structures (blue pins)

C. _____

D. _____

E. _____

F. _____

G. _____

Click **LAYER 4** *in the* **LAYER CONTROLS** *window, and you will see the following image:*

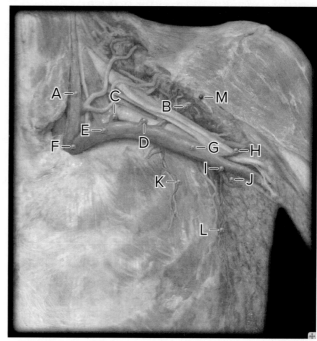

©McGraw-Hill Education

Mouse-over the pins on the screen to find the information necessary to identify the following structures:

A. _____

B. _____

C. _____

D. _____

E. _____

F. _____

G. _____

H. _____

I. _____

J. _____

K. _____

L. _____

Noncardiovascular System Structure (blue pin)

M. _____

562 MODULE 9 The Cardiovascular System

CHECK POINT

Shoulder, Veins, Anterior View

1. List the areas drained by the left brachiocephalic vein. What is the other name for this vein in Latin and in English?
2. Where does the left brachiocephalic vein terminate?
3. List the areas drained by the axillary vein. What are its tributaries?
4. Does the left brachiocephalic vein contain valves?

Click **LAYER 5** in the **LAYER CONTROLS** window, and you will see the following image:

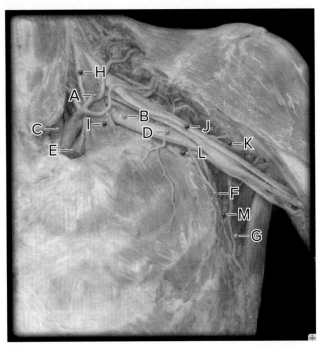

©McGraw-Hill Education

Mouse-over the pins on the screen to find the information necessary to identify the following structures:

A. _____
B. _____
C. _____
D. _____
E. _____
F. _____
G. _____

Noncardiovascular System Structures (blue pins)

H. _____
I. _____
J. _____
K. _____
L. _____
M. _____

Self-Quiz

Take this opportunity to check your progress by taking the **QUIZ**. See the **Introduction Module** for a reminder on how to access the **QUIZ** for this Study Area.

IN REVIEW

What Have I Learned?

The following questions cover the material that you have just learned, the blood vessels of the shoulder. Apply what you have learned to answer these questions on a separate piece of paper.

1. What is the deltopectoral triangle? What is another name for it?
2. Name three arteries that distribute to the muscles of the scapula.
3. Name the paired arteries that ascend through the neck via the transverse foramina of the cervical vertebrae.
4. Where do these paired arteries enter the cranial cavity?

The Shoulder and Arm

EXERCISE 9.31:
Cardiovascular System—Shoulder and Arm—Vasculature—Anterior View

SELECT TOPIC: Shoulder and Arm ▸ SELECT VIEW: Vasculature—Anterior

Click **LAYER 1** *in the* **LAYER CONTROLS** *window, and you will see the following image:*

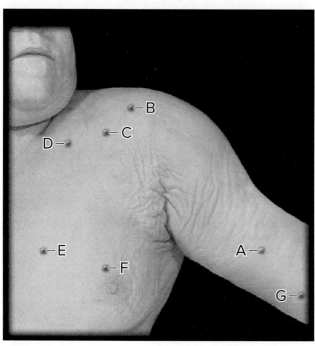

©McGraw-Hill Education

Mouse-over the pin on the screen to find the information necessary to identify the following structure:

A. _____

Noncardiovascular System Structures (blue pins)

B. _____
C. _____
D. _____
E. _____
F. _____
G. _____

Click **LAYER 3** *in the* **LAYER CONTROLS** *window, and you will see the following image:*

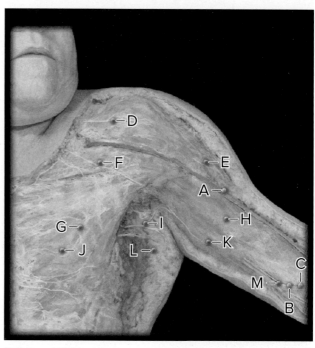

©McGraw-Hill Education

Mouse-over the pins on the screen to find the information necessary to identify the following structures:

A. _____
B. _____
C. _____

Noncardiovascular System Structures (blue pins)

D. _____
E. _____
F. _____
G. _____
H. _____
I. _____
J. _____
K. _____
L. _____
M. _____

564 MODULE 9 The Cardiovascular System

Click **LAYER 4** *in the* **LAYER CONTROLS** *window, and you will see the following image:*

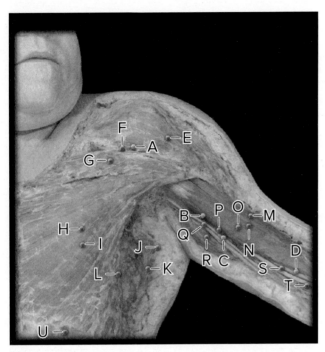

©McGraw-Hill Education

Mouse-over the pins on the screen to find the information necessary to identify the following structures:

A. _____
B. _____
C. _____
D. _____

Noncardiovascular System Structures (blue pins)

E. _____
F. _____
G. _____
H. _____
I. _____
J. _____
K. _____
L. _____
M. _____
N. _____
O. _____
P. _____
Q. _____
R. _____
S. _____
T. _____
U. _____

Click **LAYER 5** *in the* **LAYER CONTROLS** *window, and you will see the following image:*

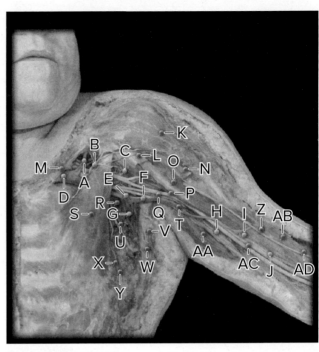

©McGraw-Hill Education

Mouse-over the pins on the screen to find the information necessary to identify the following structures:

A. _____
B. _____
C. _____
D. _____
E. _____
F. _____
G. _____
H. _____
I. _____
J. _____

Noncardiovascular System Structures (blue pins)

K. _____
L. _____
M. _____
N. _____
O. _____
P. _____
Q. _____
R. _____
S. _____
T. _____

U. _____
V. _____
W. _____
X. _____
Y. _____
Z. _____
AA. _____
AB. _____
AC. _____
AD. _____

Click **LAYER 6** *in the* **LAYER CONTROLS** *window, and you will see the following image:*

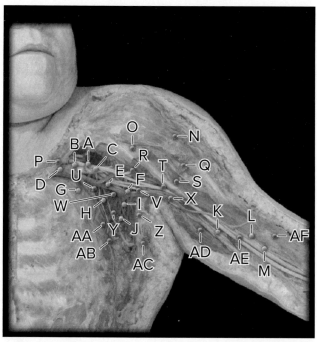

©McGraw-Hill Education

Mouse-over the pins on the screen to find the information necessary to identify the following structures:

A. _____
B. _____
C. _____

D. _____
E. _____
F. _____
G. _____
H. _____
I. _____
J. _____
K. _____
L. _____
M. _____

Noncardiovascular System Structures (blue pins)

N. _____
O. _____
P. _____
Q. _____
R. _____
S. _____
T. _____
U. _____
V. _____
W. _____
X. _____
Y. _____
Z. _____
AA. _____
AB. _____
AC. _____
AD. _____
AE. _____
AF. _____

566 MODULE 9 The Cardiovascular System

EXERCISE 9.32:
Cardiovascular System—Shoulder and Arm—Vasculature—Posterior View

SELECT TOPIC	SELECT VIEW
Shoulder and Arm	Vasculature—Posterior

Click **LAYER 5** in the **LAYER CONTROLS** window, and you will see the following image:

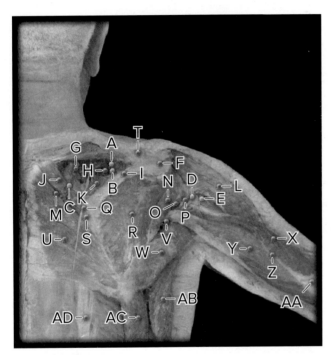

©McGraw-Hill Education

Mouse-over the pins on the screen to find the information necessary to identify the following structures:

A. _____
B. _____
C. _____
D. _____
E. _____

Noncardiovascular System Structures (blue pins)

F. _____
G. _____
H. _____
I. _____
J. _____
K. _____
L. _____
M. _____
N. _____
O. _____
P. _____
Q. _____
R. _____
S. _____
T. _____
U. _____
V. _____
W. _____
X. _____
Y. _____
Z. _____
AA. _____
AB. _____
AC. _____
AD. _____

Click **LAYER 6** in the **LAYER CONTROLS** window, and you will see the following image:

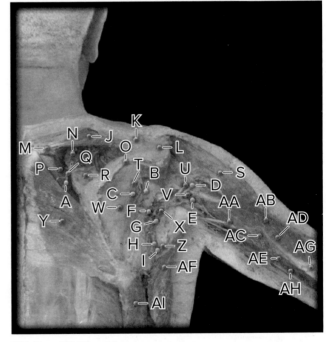

©McGraw-Hill Education

Mouse-over the pins on the screen to find the information necessary to identify the following structures:

A. _____
B. _____
C. _____
D. _____
E. _____
F. _____

G. _____
H. _____
I. _____

Noncardiovascular System Structures (blue pins)

J. _____
K. _____
L. _____
M. _____
N. _____
O. _____
P. _____
Q. _____
R. _____
S. _____
T. _____
U. _____
V. _____
W. _____
X. _____
Y. _____
Z. _____
AA. _____
AB. _____
AC. _____
AD. _____
AE. _____
AF. _____
AG. _____
AH. _____
AI. _____

EXERCISE 9.33:
Cardiovascular System—Shoulder and Arm—Arteries—Anterior View

SELECT TOPIC	SELECT VIEW
Shoulder and Arm	Arteries—Anterior

Click **LAYER 3** in the **LAYER CONTROLS** window, and you will see the following image:

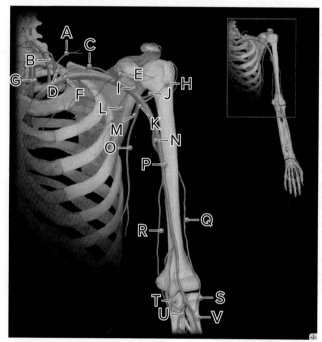

©McGraw-Hill Education

Mouse-over the pins on the screen to find the information necessary to identify the following structures:

A. _____
B. _____
C. _____
D. _____
E. _____
F. _____
G. _____
H. _____
I. _____
J. _____
K. _____
L. _____
M. _____
N. _____
O. _____
P. _____

Q. _____
R. _____
S. _____
T. _____
U. _____
V. _____

CHECK POINT

Shoulder and Arm, Arteries, Anterior View, *continued*

4. Name the major artery of the arm.
5. What two arteries branch from this artery at the anterior elbow?
6. What artery is also known as the *profunda brachii* artery? How does *profunda brachii* translate?

EXERCISE 9.34:

Cardiovascular System—Shoulder and Arm—Veins—Anterior View

SELECT TOPIC	SELECT VIEW
Shoulder and Arm	Veins—Anterior

Click **LAYER 2** in the **LAYER CONTROLS** window, and you will see the following image:

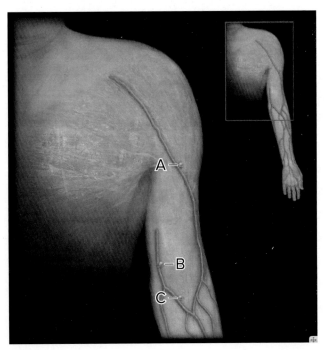

©McGraw-Hill Education

Mouse-over the pins on the screen to find the information necessary to identify the following structures:

A. _____
B. _____
C. _____

CHECK POINT

Shoulder and Arm, Veins, Anterior View

1. Which vein of the arm is frequently used for venipuncture? What is venipuncture?
2. What structures are drained by the basilica vein?
3. What structures are drained by the cephalic vein?

Click **LAYER 4** in the **LAYER CONTROLS** window, and you will see the following image:

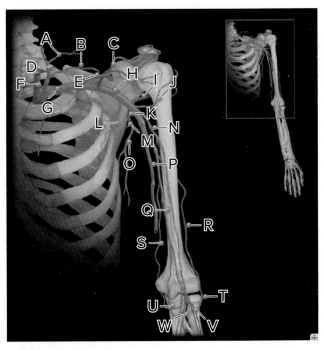

©McGraw-Hill Education

Mouse-over the pins on the screen to find the information necessary to identify the following structures:

A. _____
B. _____
C. _____
D. _____
E. _____

F. _____
G. _____
H. _____
I. _____
J. _____
K. _____
L. _____
M. _____
N. _____
O. _____
P. _____
Q. _____
R. _____
S. _____
T. _____
U. _____

V. _____
W. _____

CHECK POINT

Shoulder and Arm, Veins, Anterior View, *continued*

4. Name the paired veins that ascend along the brachial artery.
5. Name the structures drained by the axillary vein. What are its major tributaries?
6. Name the large vein that terminates in the axillary vein.

Self-Quiz

Take this opportunity to check your progress by taking the **QUIZ**. See the **Introduction Module** for a reminder on how to access the **QUIZ** for this Study Area.

IN REVIEW

What Have I Learned?

The following questions cover the material that you have just learned, the blood vessels of the shoulder and arm. Apply what you have learned to answer these questions on a separate piece of paper.

1. Name the two arteries that arise from the brachiocephalic trunk.

2. Which of the two ascends into the neck?

3. Name the two branches at the terminus of this artery.

4. Name the arteries that form an anastomosis around the elbow.

5. Name two arteries that form an anastomosis around the scapula.

6. Name the structures drained by the external jugular vein. What are its major tributaries?

7. Name the structures drained by the internal jugular vein. What are its major tributaries?

8. Name two locations where venous anastomoses occur in the shoulder and arm.

570 MODULE 9 The Cardiovascular System

EXERCISE 9.35:
Cardiovascular System—Arm, Forearm, and Hand Vasculature—Anterior View

SELECT TOPIC	SELECT VIEW
Arm, Forearm, and Hand	Vasculature—Anterior

Click **LAYER 1** in the **LAYER CONTROLS** window, and you will see the following image:

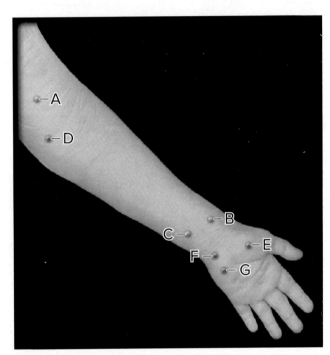

©McGraw-Hill Education

Mouse-over the pins on the screen to find the information necessary to identify the following structures:

A. _____
B. _____
C. _____

Noncardiovascular System Structures (blue pins)

D. _____
E. _____
F. _____
G. _____

Click **LAYER 3** in the **LAYER CONTROLS** window, and you will see the following image:

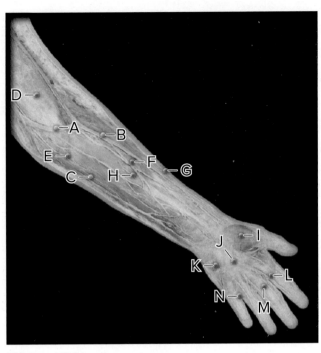

©McGraw-Hill Education

Mouse-over the pins on the screen to find the information necessary to identify the following structures:

A. _____
B. _____
C. _____

Noncardiovascular System Structures (blue pins)

D. _____
E. _____
F. _____
G. _____
H. _____
I. _____
J. _____
K. _____
L. _____
M. _____
N. _____

Click **LAYER 4** *in the* **LAYER CONTROLS** *window, and you will see the following image:*

©McGraw-Hill Education

Mouse-over the pins on the screen to find the information necessary to identify the following structures:

A. _____
B. _____
C. _____
D. _____
E. _____
F. _____
G. _____
H. _____

Noncardiovascular System Structures (blue pins)

I. _____
J. _____
K. _____
L. _____
M. _____
N. _____
O. _____
P. _____
Q. _____
R. _____
S. _____
T. _____
U. _____
V. _____
W. _____
X. _____
Y. _____
Z. _____
AA. _____
AB. _____
AC. _____
AD. _____
AE. _____
AF. _____

Click **LAYER 5** *in the* **LAYER CONTROLS** *window, and you will see the following image:*

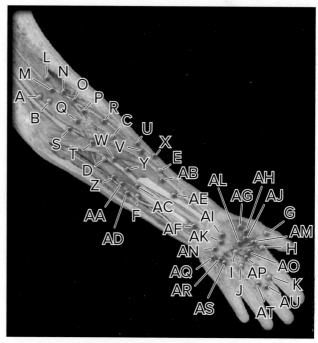

©McGraw-Hill Education

Mouse-over the pins on the screen to find the information necessary to identify the following structures:

A. _____
B. _____
C. _____
D. _____
E. _____
F. _____
G. _____
H. _____
I. _____

572 MODULE 9 The Cardiovascular System

J. _____
K. _____

Noncardiovascular System Structures (blue pins)

L. _____
M. _____
N. _____
O. _____
P. _____
Q. _____
R. _____
S. _____
T. _____
U. _____
V. _____
W. _____
X. _____
Y. _____
Z. _____
AA. _____
AB. _____
AC. _____
AD. _____
AE. _____
AF. _____
AG. _____
AH. _____
AI. _____
AJ. _____
AK. _____
AL. _____
AM. _____
AN. _____
AO. _____
AP. _____
AQ. _____
AR. _____
AS. _____
AT. _____
AU. _____

Click **LAYER 6** *in the* **LAYER CONTROLS** *window, and you will see the following image:*

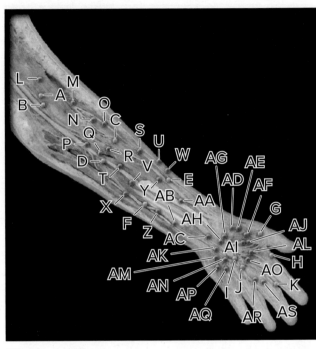

©McGraw-Hill Education

Mouse-over the pins on the screen to find the information necessary to identify the following structures:

A. _____
B. _____
C. _____
D. _____
E. _____
F. _____
G. _____
H. _____
I. _____
J. _____
K. _____

Noncardiovascular System Structures (blue pins)

L. _____
M. _____
N. _____
O. _____
P. _____
Q. _____
R. _____

S. _____
T. _____
U. _____
V. _____
W. _____
X. _____
Y. _____
Z. _____
AA. _____
AB. _____
AC. _____
AD. _____
AE. _____
AF. _____
AG. _____
AH. _____
AI. _____
AJ. _____
AK. _____
AL. _____
AM. _____
AN. _____
AO. _____
AP. _____
AQ. _____
AR. _____
AS. _____

EXERCISE 9.36:
Cardiovascular System—Arm, Forearm, and Hand Vasculature—Posterior View

SELECT TOPIC	SELECT VIEW
Arm, Forearm, and Hand	Vasculature—Posterior

Click **LAYER 3** in the **LAYER CONTROLS** window, and you will see the following image:

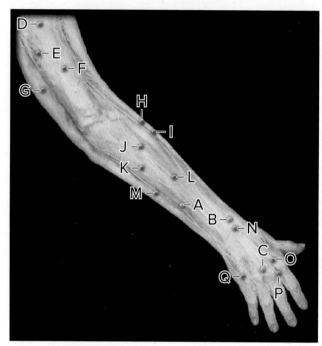

©McGraw-Hill Education

Mouse-over the pins on the screen to find the information necessary to identify the following structures:

A. _____
B. _____
C. _____

Noncardiovascular System Structures (blue pins)

D. _____
E. _____
F. _____
G. _____
H. _____
I. _____
J. _____
K. _____
L. _____
M. _____
N. _____

O. _____
P. _____
Q. _____

Click **LAYER 4** *in the* **LAYER CONTROLS** *window, and you will see the following image:*

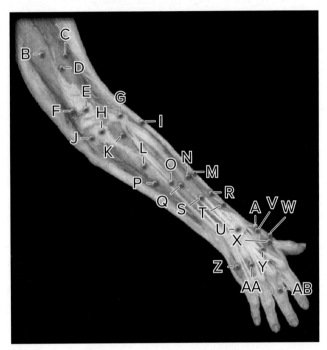

©McGraw-Hill Education

Mouse-over the pin on the screen to find the information necessary to identify the following structure:

A. _____

Noncardiovascular System Structures (blue pins)

B. _____
C. _____
D. _____
E. _____
F. _____
G. _____
H. _____
I. _____
J. _____
K. _____
L. _____
M. _____
N. _____
O. _____
P. _____

Q. _____
R. _____
S. _____
T. _____
U. _____
V. _____
W. _____
X. _____
Y. _____
Z. _____
AA. _____
AB. _____

Click **LAYER 5** *in the* **LAYER CONTROLS** *window, and you will see the following image:*

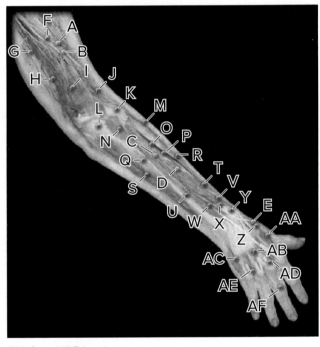

©McGraw-Hill Education

Mouse-over the pins on the screen to find the information necessary to identify the following structures:

A. _____
B. _____
C. _____
D. _____
E. _____

Noncardiovascular System Structures (blue pins)

F. _____
G. _____

H. _____
I. _____
J. _____
K. _____
L. _____
M. _____
N. _____
O. _____
P. _____
Q. _____
R. _____
S. _____
T. _____
U. _____
V. _____
W. _____
X. _____
Y. _____
Z. _____
AA. _____
AB. _____
AC. _____
AD. _____
AE. _____
AF. _____

Click **LAYER 6** *in the* **LAYER CONTROLS** *window, and you will see the following image:*

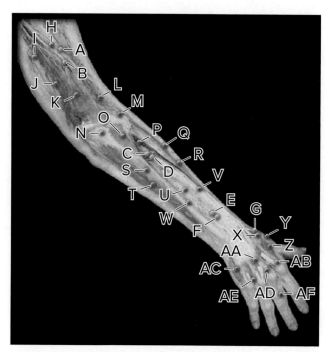

©McGraw-Hill Education

Mouse-over the pins on the screen to find the information necessary to identify the following structures:

A. _____
B. _____
C. _____
D. _____
E. _____
F. _____
G. _____

Noncardiovascular System Structures (blue pins)

H. _____
I. _____
J. _____
K. _____
L. _____
M. _____
N. _____
O. _____
P. _____
Q. _____
R. _____

576 MODULE 9 The Cardiovascular System

S. _____
T. _____
U. _____
V. _____
W. _____
X. _____
Y. _____
Z. _____
AA. _____
AB. _____
AC. _____
AD. _____
AE. _____
AF. _____

The Forearm and Hand

EXERCISE 9.37:

Cardiovascular System—Forearm and Hand—Arteries—Anterior View

  SELECT TOPIC: Forearm and Hand ▸ SELECT VIEW: Arteries—Anterior

Click **LAYER 1** in the **LAYER CONTROLS** window, and you will see the following image:

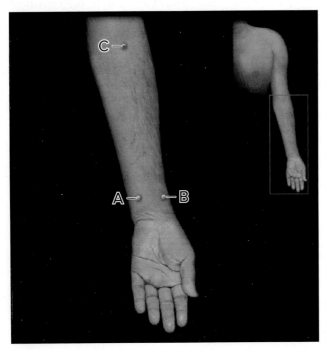

©McGraw-Hill Education

Mouse-over the pins on the screen to find the information necessary to identify the following structures:

A. _____
B. _____

Noncardiovascular System Structure (blue pin)

C. _____

CHECK POINT

Forearm and Hand, Arteries, Anterior View

1. Name the location commonly used to take a pulse at the wrist.
2. Name another location for taking a pulse at the wrist.
3. Name a location for venipuncture.

Click **LAYER 2** in the **LAYER CONTROLS** window, and you will see the following image:

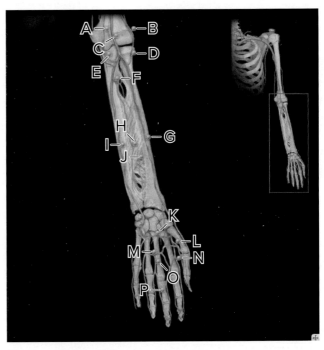

©McGraw-Hill Education

Mouse-over the pins on the screen to find the information necessary to identify the following structures:

A. _____
B. _____
C. _____
D. _____
E. _____
F. _____

G. _____
H. _____
I. _____
J. _____
K. _____
L. _____
M. _____
N. _____
O. _____
P. _____

CHECK POINT

Forearm and Hand, Arteries, Anterior View, *continued*

4. Name the two arteries that form an anastomosis through the superficial and deep palmar arch.
5. Name the artery that descends along the anterior aspect of the interosseous membrane of the forearm.
6. Name the artery commonly used for taking a pulse at the wrist.

EXERCISE 9.38:

Cardiovascular System—Forearm and Hand—Veins—Anterior View

SELECT TOPIC	SELECT VIEW
Forearm and Hand	Veins—Anterior

Click **LAYER 2** *in the* **LAYER CONTROLS** *window, and you will see the following image:*

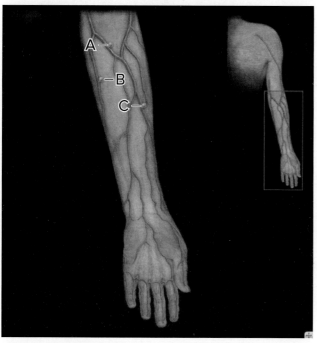

©McGraw-Hill Education

Mouse-over the pins on the screen to find the information necessary to identify the following structures:

A. _____
B. _____
C. _____

CHECK POINT

Forearm and Hand, Veins, Anterior View

1. Name the two veins that extend from the dorsal hand to the axillary vein.
2. Name the vein that has an oblique subcutaneous path between these two veins over the cubital fossa.

Click **LAYER 3** *in the* **LAYER CONTROLS** *window, and you will see the following image:*

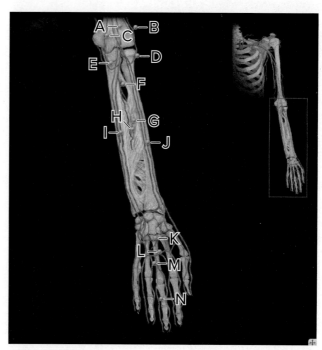

©McGraw-Hill Education

Mouse-over the pins on the screen to find the information necessary to identify the following structures:

A. _____
B. _____
C. _____
D. _____
E. _____
F. _____
G. _____
H. _____

I. _____
J. _____
K. _____
L. _____
M. _____
N. _____

CHECK POINT

Forearm and Hand, Veins, Anterior View, *continued*

3. Name the paired veins that ascend along the lateral aspect of the forearm.
4. Name the paired veins that ascend along the medial aspect of the forearm.
5. Name a vein of the forearm that is not always present.

EXERCISE 9.39:

Imaging—Hand

SELECT TOPIC	SELECT VIEW
Hand	Angiogram Arteries—Anterior–Posterior

Click the **TAGS ON/OFF** button, and you will see the following image:

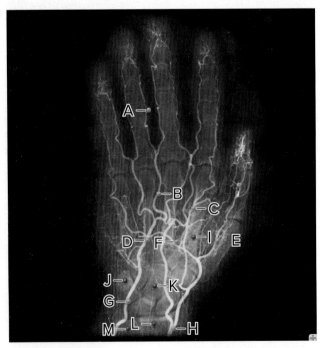

©McGraw-Hill Education

Mouse-over the pins on the screen to find the information necessary to identify the following structures:

A. _____
B. _____
C. _____
D. _____
E. _____
F. _____
G. _____
H. _____

Noncardiovascular System Structures (blue pins)

I. _____
J. _____
K. _____
L. _____
M. _____

EXERCISE 9.40:

Cardiovascular System—Hand Vasculature—Dorsum

SELECT TOPIC	SELECT VIEW
Hand	Vasculature—Dorsum

Click **LAYER 3** in the **LAYER CONTROLS** window, and you will see the following image:

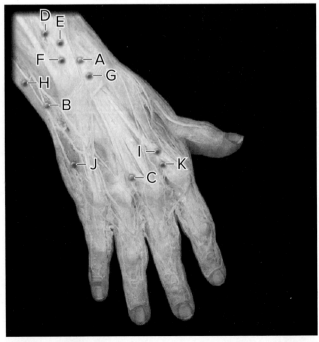

©McGraw-Hill Education

MODULE 9 The Cardiovascular System 579

Mouse-over the pins on the screen to find the information necessary to identify the following structures:

A. _____
B. _____
C. _____

Noncardiovascular System Structures (blue pins)

D. _____
E. _____
F. _____
G. _____
H. _____
I. _____
J. _____
K. _____

Click **LAYER 4** *in the* **LAYER CONTROLS** *window, and you will see the following image:*

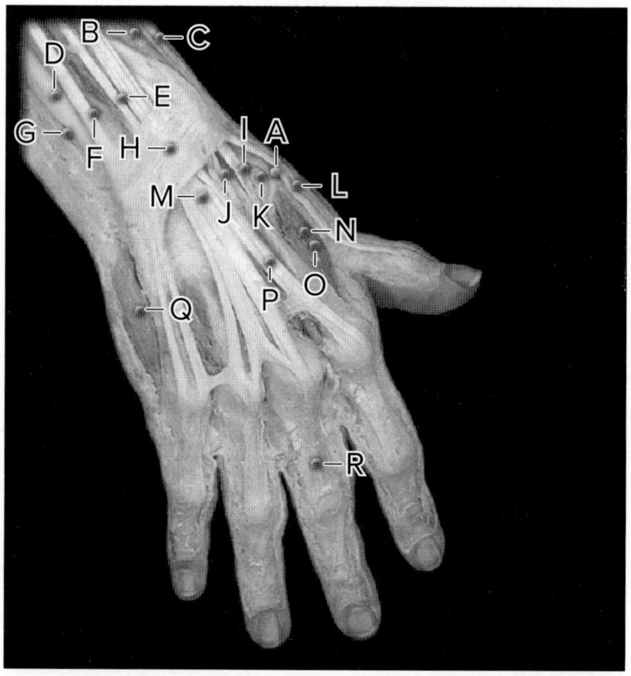

©McGraw-Hill Education

Mouse-over the pin on the screen to find the information necessary to identify the following structure:

A. _____

Noncardiovascular System Structures (blue pins)

B. _____
C. _____
D. _____
E. _____

F. _____
G. _____
H. _____
I. _____
J. _____
K. _____
L. _____
M. _____
N. _____
O. _____
P. _____
Q. _____
R. _____

Click **LAYER 5** *in the* **LAYER CONTROLS** *window, and you will see the following image:*

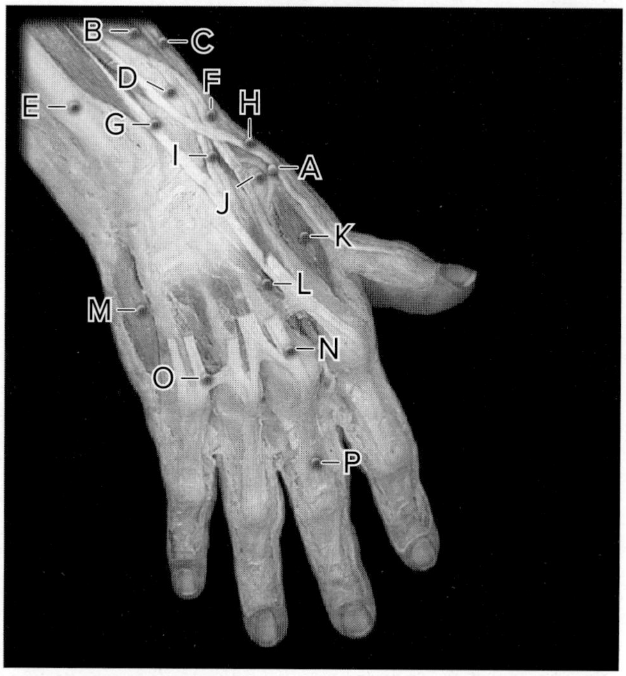

©McGraw-Hill Education

Mouse-over the pin on the screen to find the information necessary to identify the following structure:

A. _____

Noncardiovascular System Structures (blue pins)

B. _____
C. _____
D. _____
E. _____
F. _____

580 MODULE 9 The Cardiovascular System

G. _____
H. _____
I. _____
J. _____
K. _____
L. _____
M. _____
N. _____
O. _____
P. _____

Click **LAYER 6** *in the* **LAYER CONTROLS** *window, and you will see the following image:*

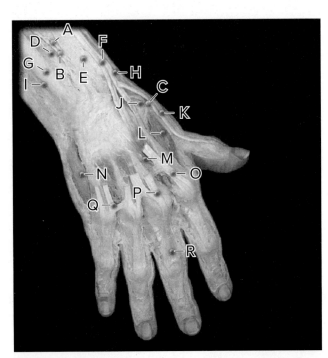

©McGraw-Hill Education

Mouse-over the pins on the screen to find the information necessary to identify the following structures:

A. _____
B. _____
C. _____

Noncardiovascular System Structures (blue pins)

D. _____
E. _____
F. _____
G. _____
H. _____
I. _____

J. _____
K. _____
L. _____
M. _____
N. _____
O. _____
P. _____
Q. _____
R. _____

EXERCISE 9.41:
Cardiovascular System—Hand Vasculature—Palmar

SELECT TOPIC	SELECT VIEW
Hand	Vasculature—Palmar

Click **LAYER 1** *in the* **LAYER CONTROLS** *window, and you will see the following image:*

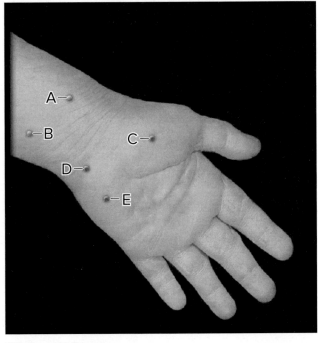

©McGraw-Hill Education

Mouse-over the pins on the screen to find the information necessary to identify the following structures:

A. _____
B. _____

Noncardiovascular System Structures (blue pins)

C. _____
D. _____
E. _____

MODULE 9 The Cardiovascular System 581

Click **LAYER 3** *in the* **LAYER CONTROLS** *window, and you will see the following image:*

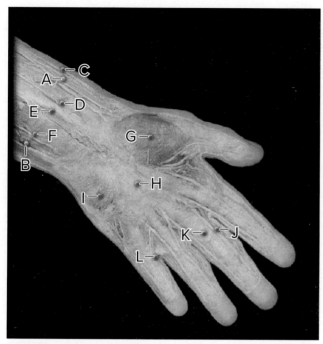

©McGraw-Hill Education

Mouse-over the pins on the screen to find the information necessary to identify the following structures:

A. _____
B. _____

Noncardiovascular System Structures (blue pins)

C. _____
D. _____
E. _____
F. _____
G. _____
H. _____
I. _____
J. _____
K. _____
L. _____

Click **LAYER 4** *in the* **LAYER CONTROLS** *window, and you will see the following image:*

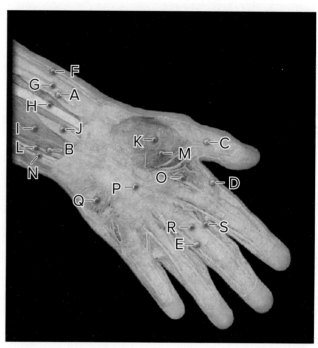

©McGraw-Hill Education

Mouse-over the pins on the screen to find the information necessary to identify the following structures:

A. _____
B. _____
C. _____
D. _____
E. _____

Noncardiovascular System Structures (blue pins)

F. _____
G. _____
H. _____
I. _____
J. _____
K. _____
L. _____
M. _____
N. _____
O. _____
P. _____
Q. _____
R. _____
S. _____

582   MODULE 9   The Cardiovascular System

Click **LAYER 5** *in the* **LAYER CONTROLS** *window, and you will see the following image:*

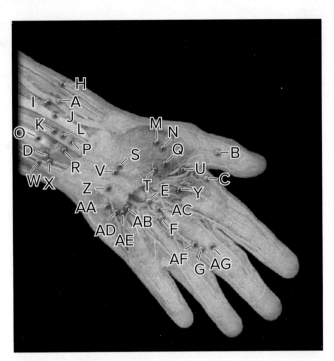

©McGraw-Hill Education

Mouse-over the pins on the screen to find the information necessary to identify the following structures:

A. _____
B. _____
C. _____
D. _____
E. _____
F. _____
G. _____

Noncardiovascular System Structures (blue pins)

H. _____
I. _____
J. _____
K. _____
L. _____
M. _____
N. _____
O. _____
P. _____
Q. _____
R. _____
S. _____
T. _____
U. _____
V. _____
W. _____
X. _____
Y. _____
Z. _____
AA. _____
AB. _____
AC. _____
AD. _____
AE. _____
AF. _____
AG. _____

Click **LAYER 6** *in the* **LAYER CONTROLS** *window, and you will see the following image:*

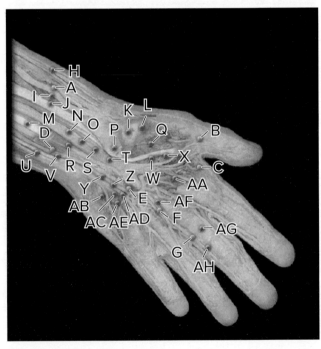

©McGraw-Hill Education

Mouse-over the pins on the screen to find the information necessary to identify the following structures:

A. _____
B. _____
C. _____
D. _____
E. _____
F. _____
G. _____

Noncardiovascular System Structures (blue pins)

H. _____
I. _____
J. _____
K. _____
L. _____
M. _____
N. _____
O. _____
P. _____
Q. _____
R. _____
S. _____
T. _____
U. _____
V. _____
W. _____
X. _____
Y. _____
Z. _____
AA. _____
AB. _____
AC. _____
AD. _____
AE. _____
AF. _____
AG. _____
AH. _____

Self-Quiz

Take this opportunity to check your progress by taking the **QUIZ**. See the **Introduction Module** for a reminder on how to access the **QUIZ** for this Study Area.

IN REVIEW

What Have I Learned?

The following questions cover the material that you have just learned, the blood vessels of the forearm and hand. Apply what you have learned to answer these questions on a separate piece of paper.

1. Describe the pathway of arterial blood flow from the arm through the elbow and lateral forearm to the palm of the hand.

2. Describe the pathway of arterial blood flow from the arm through the elbow and medial forearm to the palm of the hand.

3. Describe the pathway of arterial blood flow from the palm of the hand to the middle finger.

4. Describe the *superficial* pathway of venous blood drainage from the anterior forearm to the elbow.

5. Describe the *deep* pathway of venous blood drainage from the middle finger to the anterior elbow and forearm.

The Abdomen

EXERCISE 9.42:
Cardiovascular System—Abdomen—Celiac Trunk—Anterior View

SELECT TOPIC	SELECT VIEW
Abdomen	Celiac Trunk—Anterior

Click **LAYER 1** *in the* **LAYER CONTROLS** *window, and you will see the following image:*

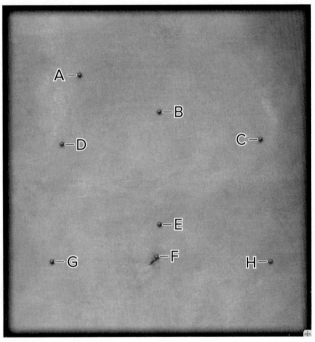

©McGraw-Hill Education

Mouse-over the pins on the screen to find the information necessary to identify the following noncardiovascular system structures:

A. _____
B. _____
C. _____
D. _____
E. _____
F. _____
G. _____
H. _____

CHECK POINT

Abdomen, Celiac Trunk, Anterior View

1. Name the upper median abdominal region.
2. Name the two abdominal regions lateral to question 1.
3. Name the median abdominal region.

Click **LAYER 2** *in the* **LAYER CONTROLS** *window, and you will see the following image:*

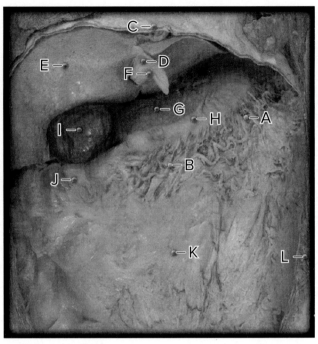

©McGraw-Hill Education

Mouse-over the pins on the screen to find the information necessary to identify the following structures:

A. _____
B. _____

Noncardiovascular System Structures (blue pins)

C. _____
D. _____
E. _____
F. _____
G. _____
H. _____
I. _____
J. _____
K. _____
L. _____

CHECK POINT

Abdomen, Celiac Trunk, Anterior View, *continued*

4. Name the structure that is the remnant of the umbilical vein of the fetus.
5. Name the anastomosis that has distribution to the stomach and the greater omentum.
6. Once again, what is an anastomosis?

Click **LAYER 3** *in the* **LAYER CONTROLS** *window, and you will see the following image:*

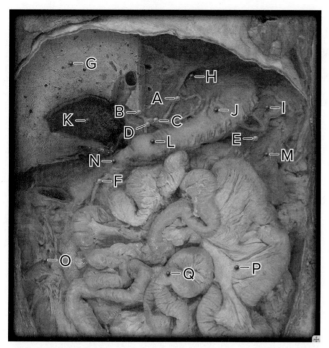

©McGraw-Hill Education

Mouse-over the pins on the screen to find the information necessary to identify the following structures:

A. _____
B. _____
C. _____
D. _____
E. _____
F. _____

Noncardiovascular System Structures (blue pins)

G. _____
H. _____
I. _____
J. _____
K. _____
L. _____
M. _____
N. _____
O. _____
P. _____
Q. _____

CHECK POINT

Abdomen, Celiac Trunk, Anterior View, *continued*

7. Name two arteries that originate at the celiac artery.
8. What is the smallest branch of the celiac artery? What is its distribution?
9. Name the large, highly vascular, accessory digestive organ. What are its four lobes?

Click **LAYER 4** *in the* **LAYER CONTROLS** *window, and you will see the following image:*

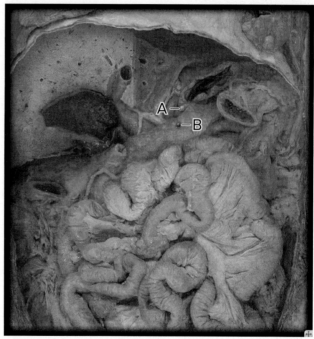

©McGraw-Hill Education

Mouse-over the pin on the screen to find the information necessary to identify the following structure:

A. _____

Noncardiovascular System Structure (blue pin)

B. _____

CHECK POINT

Abdomen, Celiac Trunk, Anterior View, *continued*

10. Name the artery that is the largest branch of the celiac artery.
11. List the branches of this artery.
12. What is meant by serpentine?

586 MODULE 9 The Cardiovascular System

Click **LAYER 5** *in the* **LAYER CONTROLS** *window, and you will see the following image:*

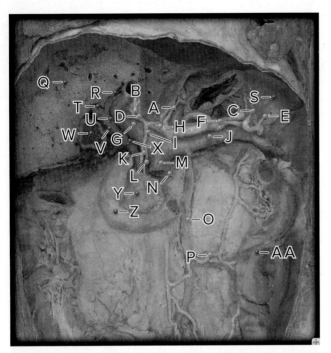

©McGraw-Hill Education

Mouse-over the pins on the screen to find the information necessary to identify the following structures:

A. _____
B. _____
C. _____
D. _____
E. _____
F. _____
G. _____
H. _____
I. _____
J. _____
K. _____
L. _____
M. _____
N. _____
O. _____
P. _____

Noncardiovascular System Structures (blue pins)

Q. _____
R. _____
S. _____
T. _____
U. _____
V. _____
W. _____
X. _____
Y. _____
Z. _____
AA. _____

Click **LAYER 6** *in the* **LAYER CONTROLS** *window, and you will see the following image:*

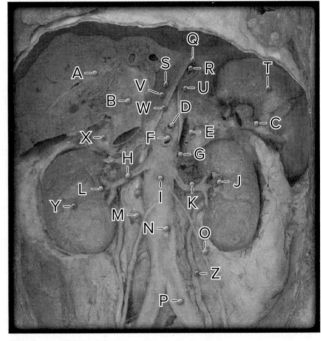

©McGraw-Hill Education

Mouse-over the pins on the screen to find the information necessary to identify the following structures:

A. _____
B. _____
C. _____
D. _____
E. _____
F. _____
G. _____
H. _____
I. _____
J. _____

K. _____
L. _____
M. _____
N. _____
O. _____
P. _____

Noncardiovascular System Structures (blue pins)

Q. _____
R. _____
S. _____
T. _____
U. _____
V. _____
W. _____
X. _____
Y. _____
Z. _____

CHECK POINT

Abdomen, Celiac Trunk, Anterior View, *continued*

13. Name the unpaired anterior artery immediately inferior to the celiac artery.
14. What is the distribution of this artery?
15. Name the arteries that originate at the celiac artery.
16. Name the three arteries that branch off the anterior abdominal aorta, from superior to inferior.
17. Name the large paired arteries that branch off from the abdominal aorta laterally to the kidneys.
18. Name the small paired arteries that branch off from the abdominal aorta inferiolaterally in the vicinity of the kidneys. What is their name in the male? In the female?

EXERCISE 9.43:
Imaging—Abdominal Aorta and Branches

SELECT TOPIC: Abdominal Aorta and Branches ▸ SELECT VIEW: CTA—Anterior

Click the **TAGS ON/OFF** *button, and you will see the following image:*

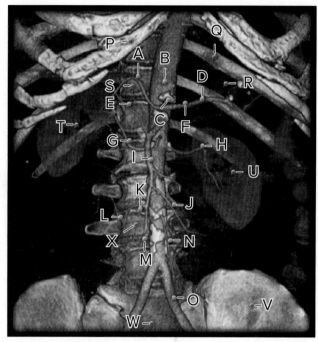

©McGraw-Hill Education

Mouse-over the pins on the screen to find the information necessary to identify the following structures:

A. _____
B. _____
C. _____
D. _____
E. _____
F. _____
G. _____
H. _____
I. _____
J. _____
K. _____
L. _____

M. _____
N. _____
O. _____

Noncardiovascular System Structures (blue pins)

P. _____
Q. _____
R. _____
S. _____
T. _____
U. _____
V. _____
W. _____
X. _____

EXERCISE 9.44:
Imaging—Abdominal Aorta and Iliac Arteries

SELECT TOPIC	SELECT VIEW
Abdominal Aorta and Iliac Arteries ▶	CTA—Anterior

Click the **TAGS ON/OFF** button, and you will see the following image:

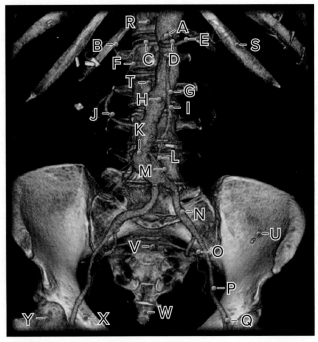

©McGraw-Hill Education

Mouse-over the pins on the screen to find the information necessary to identify the following structures:

A. _____
B. _____
C. _____
D. _____
E. _____
F. _____
G. _____
H. _____
I. _____
J. _____
K. _____
L. _____
M. _____
N. _____
O. _____
P. _____
Q. _____

Noncardiovascular System Structures (blue pins)

R. _____
S. _____
T. _____
U. _____
V. _____
W. _____
X. _____
Y. _____

EXERCISE 9.45a:
Imaging—Aortic Aneurysm

SELECT TOPIC: Aortic Aneurysm **SELECT VIEW:** CTA—Anterior

Click the **TAGS ON/OFF** *button, and you will see the following image:*

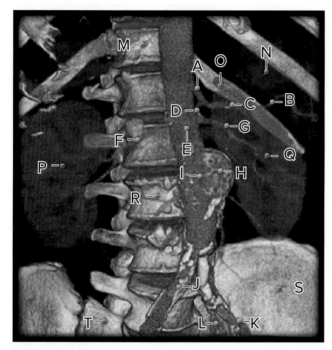

©McGraw-Hill Education

Mouse-over the pins on the screen to find the information necessary to identify the following structures:

A. _____
B. _____
C. _____
D. _____
E. _____
F. _____
G. _____
H. _____
I. _____
J. _____
K. _____
L. _____

Noncardiovascular System Structures (blue pins)

M. _____
N. _____

O. _____
P. _____
Q. _____
R. _____
S. _____
T. _____

EXERCISE 9.45b:
Imaging—Aortic Aneurysm

SELECT TOPIC: Aortic Aneurysm **SELECT VIEW:** CT—Sagittal

Click the **TAGS ON/OFF** *button, and you will see the following image:*

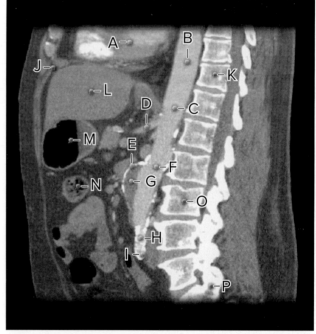

©McGraw-Hill Education

Mouse-over the pins on the screen to find the information necessary to identify the following structures:

A. _____
B. _____
C. _____
D. _____
E. _____
F. _____
G. _____
H. _____
I. _____

590 MODULE 9 The Cardiovascular System

Noncardiovascular System Structures (blue pins)

J. _____
K. _____
L. _____
M. _____
N. _____
O. _____
P. _____

EXERCISE 9.46:
Cardiovascular System—Abdomen—Mesenteric Arteries—Anterior View

SELECT TOPIC	SELECT VIEW
Abdomen	▶ Mesenteric Arteries—Anterior

Click **LAYER 1** in the **LAYER CONTROLS** window, and you will see the following image:

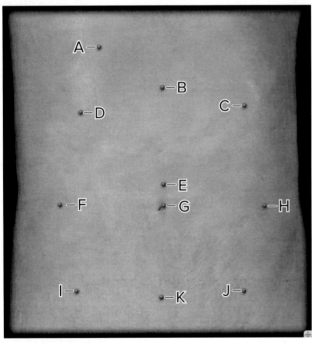

©McGraw-Hill Education

Mouse-over the pins on the screen to find the information necessary to identify the following noncardiovascular system structures:

A. _____
B. _____
C. _____
D. _____
E. _____
F. _____
G. _____
H. _____
I. _____
J. _____
K. _____

CHECK POINT

Abdomen, Mesenteric Arteries, Anterior View

1. Name the lower medial abdominal region.
2. What regions flank this lower medial region?
3. What regions flank the umbilical region?

Click **LAYER 2** in the **LAYER CONTROLS** window, and you will see the following image:

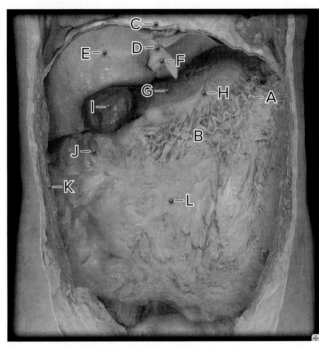

©McGraw-Hill Education

Mouse-over the pins on the screen to find the information necessary to identify the following structures:

A. _____
B. _____

Noncardiovascular System Structures (blue pins)

C. _____
D. _____
E. _____
F. _____
G. _____

H. _____
I. _____
J. _____
K. _____
L. _____

CHECK POINT

Abdomen, Mesenteric Arteries, Anterior View, *continued*

4. Name two arteries that distribute to the stomach and the greater omentum.
5. Describe the greater omentum.
6. Name the largest visceral organ.

Click **LAYER 3** *in the* **LAYER CONTROLS** *window, and you will see the following image:*

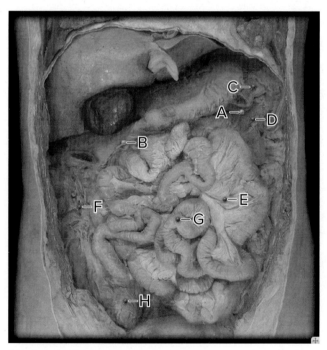

©McGraw-Hill Education

Mouse-over the pins on the screen to find the information necessary to identify the following structures:

A. _____
B. _____

Noncardiovascular System Structures (blue pins)

C. _____
D. _____

E. _____
F. _____
G. _____
H. _____

CHECK POINT

Abdomen, Mesenteric Arteries, Anterior View, *continued*

7. What is the origin of the middle colic artery?
8. What is the origin of the left colic artery?
9. An anastomosis is formed between these two colic arteries and what other artery?

Click **LAYER 4** *in the* **LAYER CONTROLS** *window, and you will see the following image:*

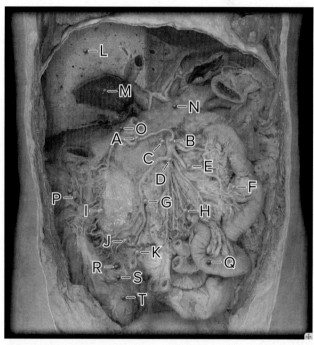

©McGraw-Hill Education

Mouse-over the pins on the screen to find the information necessary to identify the following structures:

A. _____
B. _____
C. _____
D. _____
E. _____
F. _____
G. _____

MODULE 9 The Cardiovascular System

H. _____
I. _____
J. _____
K. _____

Noncardiovascular System Structures (blue pins)

L. _____
M. _____
N. _____
O. _____
P. _____
Q. _____
R. _____
S. _____
T. _____

CHECK POINT

Abdomen, Mesenteric Arteries, Anterior View, *continued*

10. Name the abdominal artery that courses inferiorly to enter the mesentery of the small intestine.
11. What is the termination of this artery?

Click **LAYER 5** *in the* **LAYER CONTROLS** *window, and you will see the following image:*

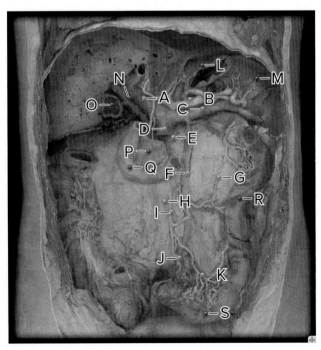

©McGraw-Hill Education

Mouse-over the pins on the screen to find the information necessary to identify the following structures:

A. _____
B. _____
C. _____
D. _____
E. _____
F. _____
G. _____
H. _____
I. _____
J. _____
K. _____

Noncardiovascular System Structures (blue pins)

L. _____
M. _____
N. _____
O. _____
P. _____
Q. _____
R. _____
S. _____

CHECK POINT

Abdomen, Mesenteric Arteries, Anterior View, *continued*

12. What is the distribution of the inferior mesenteric artery?
13. Name the artery with the distribution to the lower part of the descending colon and the sigmoid colon.
14. Name the artery that is the direct continuation of the inferior mesenteric artery and distributes to the rectum.

Click **LAYER 6** *in the* **LAYER CONTROLS** *window, and you will see the following image:*

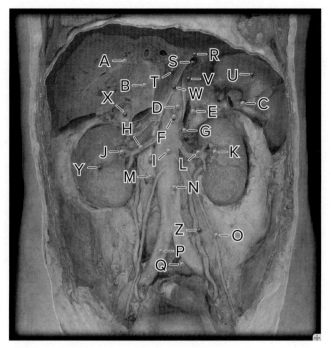

©McGraw-Hill Education

Mouse-over the pins on the screen to find the information necessary to identify the following structures:

A. _____
B. _____
C. _____
D. _____
E. _____
F. _____
G. _____
H. _____
I. _____
J. _____
K. _____
L. _____
M. _____
N. _____
O. _____
P. _____
Q. _____

Noncardiovascular System Structures (blue pins)

R. _____
S. _____
T. _____
U. _____
V. _____
W. _____
X. _____
Y. _____
Z. _____

CHECK POINT

Abdomen, Mesenteric Arteries, Anterior View, *continued*

15. List the drainages of the common iliac veins. Where do they terminate?
16. What is the origin and termination of the common iliac arteries?
17. Name the vein that drains the spleen, pancreas, and the fundus and greater curvature of the stomach.

SELECT ANIMATION
Hepatic Portal System
PLAY

After viewing the animation, answer the following questions:

1. What is the hepatic portal system?

2. What does this do for the liver?

3. The hepatic portal system originates from which organs?

4. Which vein is the largest in the hepatic portal system?

5. Which two veins unite to form it?

6. Which areas are drained by the splenic vein?

7. Which areas are drained by the superior mesenteric vein?

8. Blood from which structures drains directly into the hepatic portal vein?

594 MODULE 9 The Cardiovascular System

9. What happens to the hepatic portal vein upon entering the liver?

10. What are hepatic sinusoids?

11. What occurs there?

12. What occurs in central veins of the liver?

13. What veins are formed by the union of thousands of these central veins?

14. Upon exiting the liver, these veins _____.

Self-Quiz
Take this opportunity to check your progress by taking the **QUIZ**. See the **Introduction Module** for a reminder on how to access the **QUIZ** for this Study Area.

Mouse-over the pins on the screen to find the information necessary to identify the following structures:

A. _____
B. _____

Noncardiovascular System Structures (blue pins)

C. _____
D. _____
E. _____
F. _____
G. _____
H. _____
I. _____
J. _____
K. _____
L. _____

EXERCISE 9.47:
Cardiovascular System—Abdomen—Veins—Anterior View

SELECT TOPIC	SELECT VIEW
Abdomen	Veins—Anterior

After reviewing the structures of **LAYER 1**, *click* **LAYER 2** *in the* **LAYER CONTROLS** *window, and you will see the following image:*

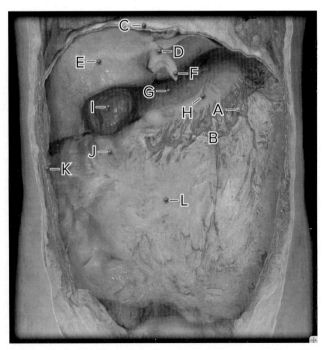

©McGraw-Hill Education

CHECK POINT
Abdomen, Veins, Anterior View

1. Name the veins that drain the stomach and greater omentum.

Click **LAYER 3** *in the* **LAYER CONTROLS** *window, and you will see the following image:*

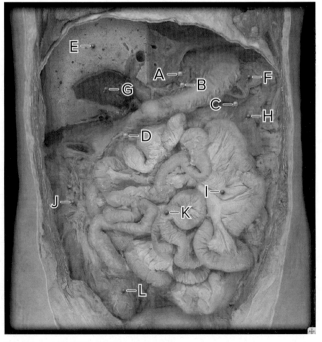

©McGraw-Hill Education

Mouse-over the pins on the screen to find the information necessary to identify the following structures:

A. _____
B. _____
C. _____
D. _____

Noncardiovascular System Structures (blue pins)

E. _____
F. _____
G. _____
H. _____
I. _____
J. _____
K. _____
L. _____

CHECK POINT

Abdomen, Veins, Anterior View, *continued*

2. Name a vein that drains the stomach and the inferior thoracic and abdominal parts of the esophagus.
3. Name a vein that drains only the stomach.
4. Where do these two veins terminate?

Click **LAYER 4** *in the* **LAYER CONTROLS** *window, and you will see the following image:*

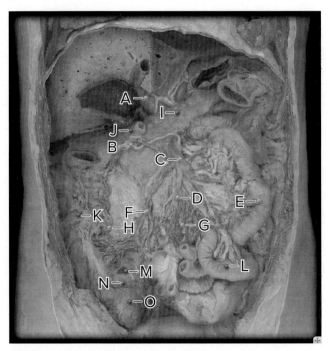

©McGraw-Hill Education

Mouse-over the pins on the screen to find the information necessary to identify the following structures:

A. _____
B. _____
C. _____
D. _____
E. _____
F. _____
G. _____
H. _____

Noncardiovascular System Structures (blue pins)

I. _____
J. _____
K. _____
L. _____
M. _____
N. _____
O. _____

CHECK POINT

Abdomen, Veins, Anterior View, *continued*

5. List the structures drained by the hepatic portal vein.
6. What do marginal venous arcades represent?
7. Where is the termination of the intestinal venous arcades?

Click **LAYER 5** *in the* **LAYER CONTROLS** *window, and you will see the following image:*

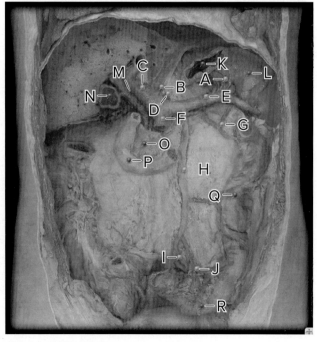

©McGraw-Hill Education

Mouse-over the pins on the screen to find the information necessary to identify the following structures:

A. _____
B. _____
C. _____
D. _____
E. _____
F. _____
G. _____
H. _____
I. _____
J. _____

Noncardiovascular System Structures (blue pins)

K. _____
L. _____
M. _____
N. _____
O. _____
P. _____
Q. _____
R. _____

CHECK POINT

Abdomen, Veins, Anterior View, *continued*

8. Name the vein that ascends from the true pelvis to become the inferior mesenteric vein in the false pelvis.

Click **LAYER 6** *in the* **LAYER CONTROLS** *window, and you will see the following image:*

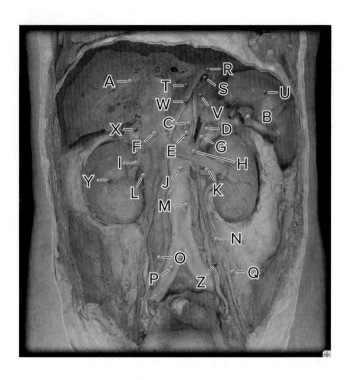

Mouse-over the pins on the screen to find the information necessary to identify the following structures:

A. _____
B. _____
C. _____
D. _____
E. _____
F. _____
G. _____
H. _____
I. _____
J. _____
K. _____
L. _____
M. _____
N. _____
O. _____
P. _____
Q. _____

Noncardiovascular System Structures (blue pins)

R. _____
S. _____
T. _____
U. _____
V. _____
W. _____
X. _____
Y. _____
Z. _____

CHECK POINT

Abdomen, Veins, Anterior View, *continued*

9. What is normally on the right side of the body, the abdominal aorta or the inferior vena cava?
10. Which are normally anterior to the others, the renal arteries or the renal veins?
11. What are the two terminations of the gonadal veins? (Be sure to note which one terminates where.)

Self-Quiz

Take this opportunity to check your progress by taking the **QUIZ**. See the **Introduction Module** for a reminder on how to access the **QUIZ** for this Study Area.

IN REVIEW

What Have I Learned?

The following questions cover the material that you have just learned, blood vessels of the abdomen. Apply what you have learned to answer these questions on a separate piece of paper.

1. Using the following figure, label the nine abdominal regions.

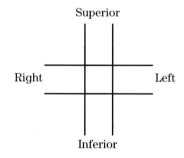

2. What is the umbilicus? (Not the region, the structure.)

3. Name the insertion point for the diaphragm.

4. Name the vein that carries absorbed products of digestion to the liver.

5. What two veins unite to form the hepatic portal vein?

6. List the organs drained by the splenic vein.

7. Name the opening that transmits the aorta from the thoracic to the abdominal cavity.

8. Name the opening that transmits the esophagus from the thoracic to the abdominal cavity.

9. Name the paired arteries at the terminus of the abdominal aorta.

10. Name the vein that drains the diaphragm and the left suprarenal gland.

The Pelvis

EXERCISE 9.48:
Cardiovascular System—Pelvis—Female—Vasculature—Anterior View

SELECT TOPIC	SELECT VIEW
Pelvis—Female	Vasculature—Anterior

Click **LAYER 1** *in the* **LAYER CONTROLS** *window, and you will see the following image:*

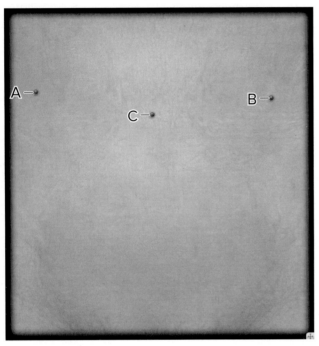

©McGraw-Hill Education

Mouse-over the pins on the screen to find the information necessary to identify the following noncardiovascular system structures:

A. _____
B. _____
C. _____

Click **LAYER 2** *in the* **LAYER CONTROLS** *window, and you will see the following image:*

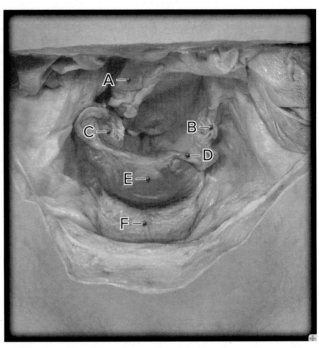

©McGraw-Hill Education

Mouse-over the pins on the screen to find the information necessary to identify the following noncardiovascular system structures:

A. _____
B. _____
C. _____
D. _____
E. _____
F. _____

Click **LAYER 3** *in the* **LAYER CONTROLS** *window, and you will see the following image:*

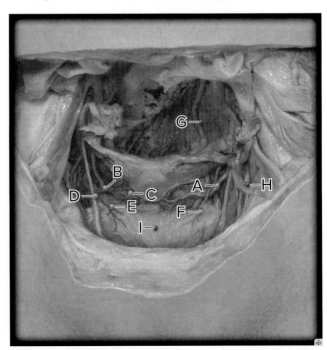

©McGraw-Hill Education

Mouse-over the pins on the screen to find the information necessary to identify the following structures:

A. _____
B. _____
C. _____
D. _____
E. _____
F. _____

Noncardiovascular System Structures (blue pins)

G. _____
H. _____
I. _____

CHECK POINT

Pelvis—Female, Vasculature, Anterior View

1. Name the artery that reflects along the side of the uterus, distributing to the uterus and vagina.
2. Name the vein that drains the uterus.

Click **LAYER 4** *in the* **LAYER CONTROLS** *window, and you will see the following image:*

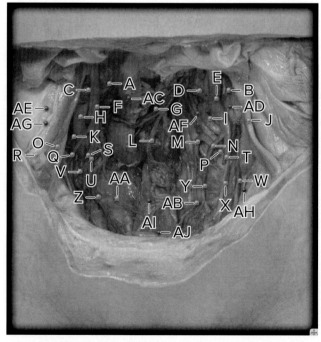

©McGraw-Hill Education

Mouse-over the pins on the screen to find the information necessary to identify the following structures:

A. _____
B. _____
C. _____
D. _____
E. _____
F. _____
G. _____
H. _____
I. _____
J. _____
K. _____
L. _____
M. _____
N. _____
O. _____
P. _____
Q. _____
R. _____
S. _____
T. _____
U. _____
V. _____
W. _____
X. _____
Y. _____
Z. _____
AA. _____
AB. _____

Noncardiovascular System Structures (blue pins)

AC. _____
AD. _____
AE. _____
AF. _____
AG. _____
AH. _____
AI. _____
AJ. _____

CHECK POINT

Pelvis—Female, Vasculature, Anterior View, *continued*

3. Name the vein that drains the pelvis and gluteal region and terminates in the common iliac vein.
4. Name two veins that are tributaries to the vein in question 4.
5. Name the artery that originates at the common iliac artery and continues as the femoral artery.

Self-Quiz

Take this opportunity to check your progress by taking the **QUIZ**. See the **Introduction Module** for a reminder on how to access the **QUIZ** for this Study Area.

EXERCISE 9.49:

Cardiovascular System—Pelvis—Male— Vasculature—Anterior View

SELECT TOPIC	SELECT VIEW
Pelvis—Male	Vasculature—Anterior

Click **LAYER 1** in the **LAYER CONTROLS** *window, and you will see the following image:*

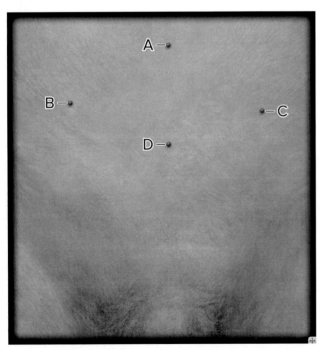

©McGraw-Hill Education

Mouse-over the pins on the screen to find the information necessary to identify the following noncardiovascular system structures:

A. _____

B. _____

C. _____

D. _____

Click **LAYER 2** in the **LAYER CONTROLS** *window, and you will see the following image:*

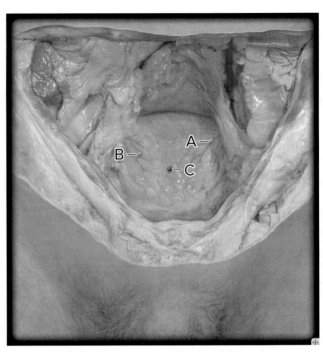

©McGraw-Hill Education

Mouse-over the pins on the screen to find the information necessary to identify the following structures:

A. _____

B. _____

Noncardiovascular System Structure (blue pin)

C. _____

CHECK POINT

Pelvis—Male, Vasculature, Anterior View

1. Name an artery that distributes to the urinary bladder.
2. Name a vein that drains the urinary bladder.
3. The volume of which organ effects the position of the surrounding organs?

Click **LAYER 3** *in the* **LAYER CONTROLS** *window, and you will see the following image:*

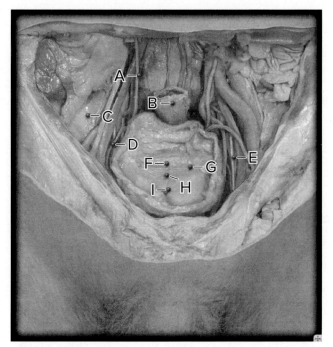

©McGraw-Hill Education

Mouse-over the pins on the screen to find the information necessary to identify the following noncardiovascular system structures:

A. _____
B. _____
C. _____
D. _____
E. _____
F. _____
G. _____
H. _____
I. _____

Click **LAYER 4** *in the* **LAYER CONTROLS** *window, and you will see the following image:*

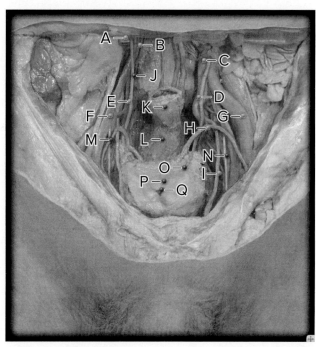

©McGraw-Hill Education

Mouse-over the pins on the screen to find the information necessary to identify the following structures:

A. _____
B. _____
C. _____
D. _____
E. _____
F. _____
G. _____
H. _____
I. _____

Noncardiovascular System Structures (blue pins)

J. _____
K. _____
L. _____
M. _____
N. _____
O. _____
P. _____
Q. _____

602 MODULE 9 The Cardiovascular System

CHECK POINT

Pelvis—Male, Vasculature, Anterior View, *continued*

4. Name an artery that passes medially to reach the superior surface of the urinary bladder.

Click **LAYER 5** *in the* **LAYER CONTROLS** *window, and you will see the following image:*

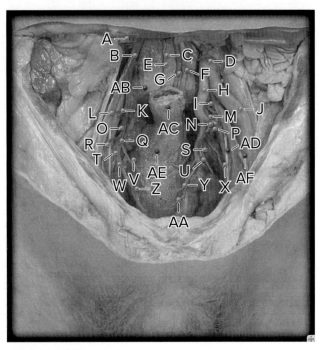

©McGraw-Hill Education

Mouse-over the pins on the screen to find the information necessary to identify the following structures:

A. _____
B. _____
C. _____
D. _____
E. _____
F. _____
G. _____
H. _____
I. _____
J. _____
K. _____
L. _____
M. _____
N. _____
O. _____
P. _____
Q. _____
R. _____
S. _____
T. _____
U. _____
V. _____
W. _____
X. _____
Y. _____
Z. _____
AA. _____

Noncardiovascular System Structures (blue pins)

AB. _____
AC. _____
AD. _____
AE. _____
AF. _____

CHECK POINT

Pelvis—Male, Vasculature, Anterior View, *continued*

5. Name a vein that passes laterally from the inferior surface of the urinary bladder and drains the urinary bladder and the prostate gland. What is this vein known as in the female?
6. List the structures drained by the median sacral vein.

Self-Quiz

Take this opportunity to check your progress by taking the **QUIZ**. See the **Introduction Module** for a reminder on how to access the **QUIZ** for this Study Area.

IN REVIEW

What Have I Learned?

The following questions cover the material that you have just learned, the vasculature of the female and male pelvis. Apply what you have learned to answer these questions on a separate piece of paper.

1. Name the thick-walled, pear-shaped hollow muscular organ that is the site of implantation of the blastocyst.

2. The position of the uterus varies with _____.

3. Name the two parts of the uterus.

4. Name the paired female gonads.

5. What is their function?

6. Name the specific site of fertilization of the egg.

7. Name the artery that descends into the pelvis and passes posteriorly toward the superior margin of the greater sciatic notch.

8. Name an artery that distributes to the anal canal and external genitalia of both sexes.

9. Name an artery whose distribution includes the rectum and vagina in females and the rectum, prostate, and seminal vesicle in males.

10. Name a pelvic vein that drains the muscles and skin of the medial thigh and terminates at the internal iliac vein.

EXERCISE 9.50:

Cardiovascular System—Hip and Thigh—Vasculature—Anterior View

SELECT TOPIC	SELECT VIEW
Hip and Thigh	Vasculature—Anterior

Click **LAYER 2** *in the* **LAYER CONTROLS** *window, and you will see the following image:*

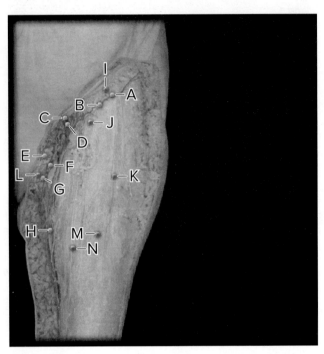

©McGraw-Hill Education

Mouse-over the pins on the screen to find the information necessary to identify the following structures:

A. _____
B. _____
C. _____
D. _____
E. _____
F. _____
G. _____
H. _____

Noncardiovascular System Structures (blue pins)

I. _____
J. _____
K. _____
L. _____
M. _____
N. _____

Click **LAYER 3** *in the* **LAYER CONTROLS** *window, and you will see the following image:*

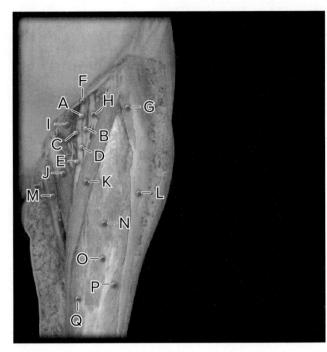

©McGraw-Hill Education

Mouse-over the pins on the screen to find the information necessary to identify the following structures:

A. _____
B. _____
C. _____
D. _____
E. _____

Noncardiovascular System Structures (blue pins)

F. _____
G. _____
H. _____
I. _____
J. _____
K. _____
L. _____
M. _____
N. _____
O. _____
P. _____
Q. _____

MODULE 9 The Cardiovascular System 605

Click **LAYER 4** *in the* **LAYER CONTROLS** *window, and you will see the following image:*

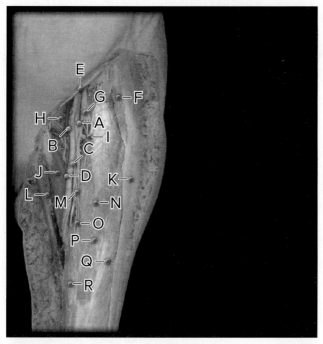

©McGraw-Hill Education

Mouse-over the pins on the screen to find the information necessary to identify the following structures:

A. _____
B. _____
C. _____
D. _____

Noncardiovascular System Structures (blue pins)

E. _____
F. _____
G. _____
H. _____
I. _____
J. _____
K. _____
L. _____
M. _____
N. _____
O. _____
P. _____
Q. _____
R. _____

Click **LAYER 5** *in the* **LAYER CONTROLS** *window, and you will see the following image:*

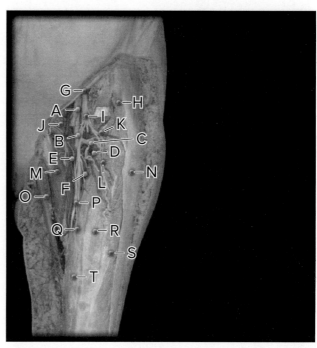

©McGraw-Hill Education

Mouse-over the pins on the screen to find the information necessary to identify the following structures:

A. _____
B. _____
C. _____
D. _____
E. _____
F. _____

Noncardiovascular System Structures (blue pins)

G. _____
H. _____
I. _____
J. _____
K. _____
L. _____
M. _____
N. _____
O. _____
P. _____
Q. _____
R. _____
S. _____
T. _____

Click **LAYER 6** *in the* **LAYER CONTROLS** *window, and you will see the following image:*

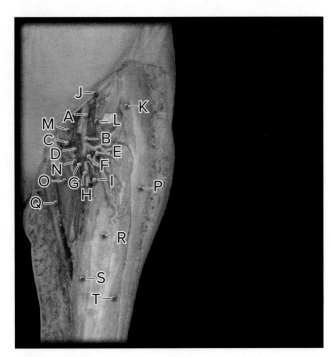

©McGraw-Hill Education

Mouse-over the pins on the screen to find the information necessary to identify the following structures:

A. _____
B. _____
C. _____
D. _____
E. _____
F. _____
G. _____
H. _____
I. _____

Noncardiovascular System Structures (blue pins)

J. _____
K. _____
L. _____
M. _____
N. _____
O. _____
P. _____
Q. _____
R. _____
S. _____
T. _____

EXERCISE 9.51:

Cardiovascular System—Hip and Thigh—Vasculature—Posterior View

 SELECT TOPIC: **Hip and Thigh** ▶ SELECT VIEW: **Vasculature—Posterior**

Click **LAYER 2** *in the* **LAYER CONTROLS** *window, and you will see the following image:*

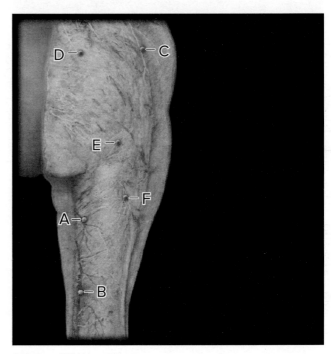

©McGraw-Hill Education

Mouse-over the pins on the screen to find the information necessary to identify the following structures:

A. _____
B. _____

Noncardiovascular System Structures (blue pins)

C. _____
D. _____
E. _____
F. _____

MODULE 9 The Cardiovascular System 607

Click **LAYER 3** *in the* **LAYER CONTROLS** *window, and you will see the following image:*

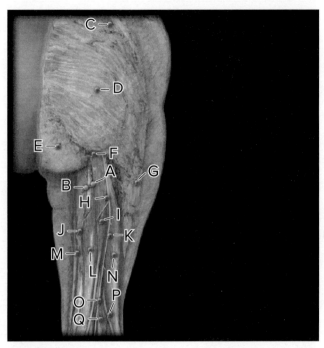

©McGraw-Hill Education

Mouse-over the pins on the screen to find the information necessary to identify the following structures:

A. _____
B. _____

Noncardiovascular System Structures (blue pins)

C. _____
D. _____
E. _____
F. _____
G. _____
H. _____
I. _____
J. _____
K. _____
L. _____
M. _____
N. _____
O. _____
P. _____
Q. _____

Click **LAYER 4** *in the* **LAYER CONTROLS** *window, and you will see the following image:*

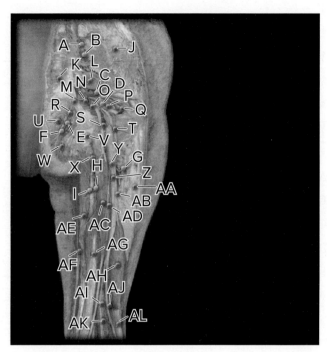

©McGraw-Hill Education

Mouse-over the pins on the screen to find the information necessary to identify the following structures:

A. _____
B. _____
C. _____
D. _____
E. _____
F. _____
G. _____
H. _____
I. _____

Noncardiovascular System Structures (blue pins)

J. _____
K. _____
L. _____
M. _____
N. _____
O. _____
P. _____
Q. _____

R. _____
S. _____
T. _____
U. _____
V. _____
W. _____
X. _____
Y. _____
Z. _____
AA. _____
AB. _____
AC. _____
AD. _____
AE. _____
AF. _____
AG. _____
AH. _____
AI. _____
AJ. _____
AK. _____
AL. _____

Click **LAYER 5** in the **LAYER CONTROLS** window, and you will see the following image:

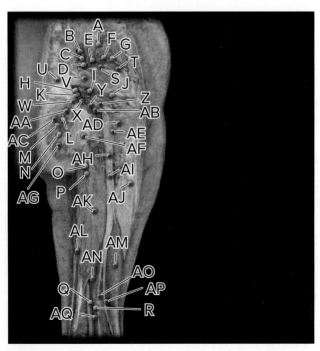

Mouse-over the pins on the screen to find the information necessary to identify the following structures:

A. _____
B. _____
C. _____
D. _____
E. _____
F. _____
G. _____
H. _____
I. _____
J. _____
K. _____
L. _____
M. _____
N. _____
O. _____
P. _____
Q. _____
R. _____

Noncardiovascular System Structures (blue pins)

S. _____
T. _____
U. _____
V. _____
W. _____
X. _____
Y. _____
Z. _____
AA. _____
AB. _____
AC. _____
AD. _____
AE. _____
AF. _____
AG. _____
AH. _____
AI. _____

MODULE 9 The Cardiovascular System 609

AJ. _____
AK. _____
AL. _____
AM. _____
AN. _____
AO. _____
AP. _____
AQ. _____

Click **LAYER 6** *in the* **LAYER CONTROLS** *window, and you will see the following image:*

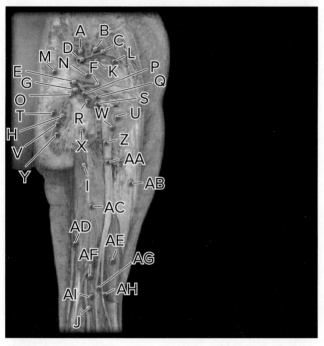

©McGraw-Hill Education

Mouse-over the pins on the screen to find the information necessary to identify the following structures:

A. _____
B. _____
C. _____
D. _____
E. _____
F. _____
G. _____
H. _____
I. _____
J. _____

Noncardiovascular System Structures (blue pins)

K. _____
L. _____
M. _____
N. _____
O. _____
P. _____
Q. _____
R. _____
S. _____
T. _____
U. _____
V. _____
W. _____
X. _____
Y. _____
Z. _____
AA. _____
AB. _____
AC. _____
AD. _____
AE. _____
AF. _____
AG. _____
AH. _____
AI. _____

EXERCISE 9.52:

Cardiovascular System—Hip and Thigh—Arteries—Anterior View

SELECT TOPIC	SELECT VIEW
Hip and Thigh	Arteries—Anterior

Click **LAYER 1** *in the* **LAYER CONTROLS** *window, and you will see the following image:*

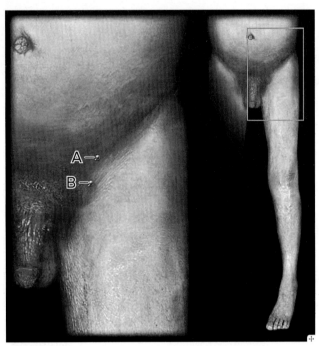

©McGraw-Hill Education

Mouse-over the pins on the screen to find the information necessary to identify the following structures:

A. _____
B. _____

CHECK POINT

Hip and Thigh, Arteries, Anterior View

1. Name the location for checking the pulse at the groin.
2. What emergency function is applicable at this location?
3. What structures are contained within the femoral triangle?

Click **LAYER 2** *in the* **LAYER CONTROLS** *window, and you will see the following image:*

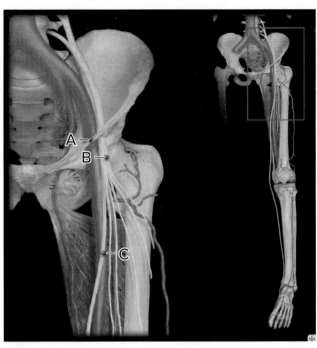

©McGraw-Hill Education

Mouse-over the pins on the screen to find the information necessary to identify the following noncardiovascular system structures:

A. _____
B. _____
C. _____

CHECK POINT

Hip and Thigh, Arteries, Anterior View, *continued*

4. What structure forms the floor of the inguinal canal?
5. Name the nerve and its branches that enter the thigh posterior to the inguinal ligament.
6. Which branch is the largest?

Click **LAYER 3** *in the* **LAYER CONTROLS** *window, and you will see the following image:*

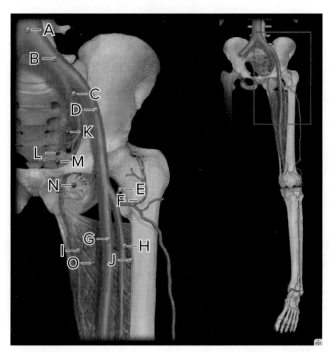

©McGraw-Hill Education

Mouse-over the pins on the screen to find the information necessary to identify the following structures:

A. _____
B. _____
C. _____
D. _____
E. _____
F. _____
G. _____
H. _____
I. _____
J. _____

Noncardiovascular System Structures (blue pins)

K. _____
L. _____
M. _____
N. _____
O. _____

CHECK POINT

Hip and Thigh, Arteries, Anterior View, *continued*

7. List the sequence of the continuous major arteries from the abdominal aorta through the thigh.
8. List the branches of the femoral artery.
9. List the distribution of the obturator artery.

EXERCISE 9.53:

Cardiovascular System—Hip and Thigh—Arteries—Posterior View

SELECT TOPIC	SELECT VIEW
Hip and Thigh	Arteries—Posterior

After viewing the structures in **LAYER 1**, *click* **LAYER 2** *in the* **LAYER CONTROLS** *window, and you will see the following image:*

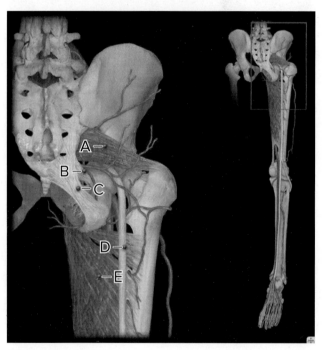

©McGraw-Hill Education

Mouse-over the pins on the screen to find the information necessary to identify the following noncardiovascular system structures:

A. _____
B. _____
C. _____
D. _____
E. _____

612 MODULE 9 The Cardiovascular System

CHECK POINT

Hip and Thigh, Arteries, Posterior View

1. Name the largest nerve in the body.
2. Name the two nerves that form that nerve.
3. List the sensory and motor innervations of this nerve.

Click **LAYER 3** *in the* **LAYER CONTROLS** *window, and you will see the following image:*

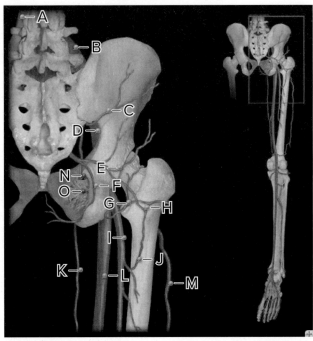

©McGraw-Hill Education

Mouse-over the pins on the screen to find the information necessary to identify the following structures:

A. _____
B. _____
C. _____
D. _____
E. _____
F. _____
G. _____
H. _____
I. _____
J. _____
K. _____
L. _____
M. _____

Noncardiovascular System Structures (blue pins)

N. _____
O. _____

CHECK POINT

Hip and Thigh, Arteries, Posterior View, *continued*

4. Name the largest branch of the internal iliac artery. What is its distribution?
5. List the distribution of the abdominal aorta.
6. What artery is usually crossed by the ureter and gonadal vessels at its origin?

EXERCISE 9.54:

Cardiovascular System—Hip and Thigh—Veins—Anterior View

SELECT TOPIC	SELECT VIEW
Hip and Thigh	Veins—Anterior

After viewing the structures in **LAYER 1,** *click* **LAYER 2** *in the* **LAYER CONTROLS** *window, and you will see the following image:*

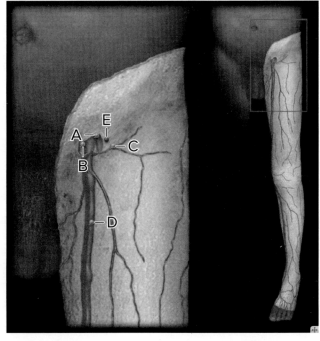

©McGraw-Hill Education

Mouse-over the pins on the screen to find the information necessary to identify the following structures:

A. _____
B. _____
C. _____
D. _____

Noncardiovascular System Structure (blue pin)

E. _____

CHECK POINT

Hip and Thigh, Veins, Anterior View

1. Name the common source of vessel tissue for coronary bypass surgery.
2. Defective valves in this vein may cause _____.
3. What is another name for this vein?

Click **LAYER 4** *in the* **LAYER CONTROLS** *window, and you will see the following image:*

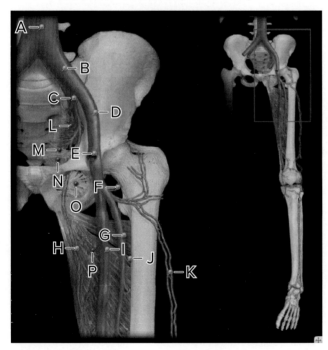

©McGraw-Hill Education

Mouse-over the pins on the screen to find the information necessary to identify the following structures:

A. _____
B. _____
C. _____
D. _____
E. _____
F. _____
G. _____

H. _____
I. _____
J. _____
K. _____

Noncardiovascular System Structures (blue pins)

L. _____
M. _____
N. _____
O. _____
P. _____

CHECK POINT

Hip and Thigh, Veins, Anterior View, *continued*

4. Name a vein of the thigh that has the popliteal vein as a tributary.
5. What is the drainage for this vein?
6. List the sequence of this vein and those that continue with it superiorly to the heart.

EXERCISE 9.55:

Cardiovascular System—Hip and Thigh—Veins—Posterior View

SELECT TOPIC	SELECT VIEW
Hip and Thigh	Veins—Posterior

After viewing the structures in **LAYER 1,** *click* **LAYER 2** *in the* **LAYER CONTROLS** *window, and you will see the following image:*

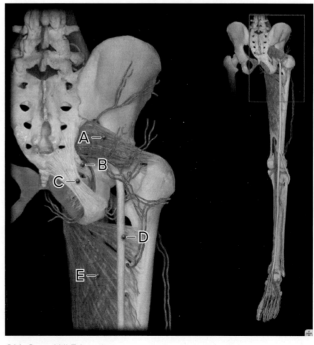

©McGraw-Hill Education

Mouse-over the pins on the screen to find the information necessary to identify the following noncardiovascular system structures:

A. _____
B. _____
C. _____
D. _____
E. _____

*Click **LAYER 3** in the **LAYER CONTROLS** window, and you will see the following image:*

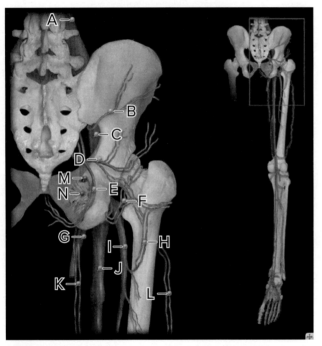

©McGraw-Hill Education

Mouse-over the pins on the screen to find the information necessary to identify the following structures:

A. _____
B. _____
C. _____
D. _____
E. _____
F. _____
G. _____
H. _____
I. _____
J. _____
K. _____
L. _____

Noncardiovascular System Structures (blue pins)

M. _____
N. _____

CHECK POINT

Hip and Thigh, Veins, Posterior View

1. Name the medial vein of the leg that spans from the anterior medial malleolus to the femoral vein just inferior to the inguinal ligament.
2. List the drainages of the deep vein of the thigh.

Self-Quiz

Take this opportunity to check your progress by taking the **QUIZ**. See the **Introduction Module** for a reminder on how to access the **QUIZ** for this Study Area.

IN REVIEW

What Have I Learned?

The following questions cover the material that you have just learned, the blood vessels of the hip and thigh. Apply what you have learned to answer these questions on a separate piece of paper.

1. Name the four arteries that contribute to the cruciate anastomosis of the hip joint.

2. Name the three veins that contribute to the anastomosis of the hip joint.

3. What is the largest vein in the body? Did you know that it is also the thickest *vessel* in your body? Which do you suppose is the longest vein in your body?

4. Name the triangular region of the groin that contains the femoral nerve, artery, and vein.

5. Name the opening in the deep fascia through which the great saphenous vein passes.

6. Name the subcutaneous vein that runs parallel to the inguinal ligament.

7. Name the deep vertical space that separates the left and right buttocks.

8. What structure represents the superior limit of the posterior thigh?

The Knee

EXERCISE 9.56:
Cardiovascular System—Knee Vasculature—Anterior View

SELECT TOPIC	SELECT VIEW
Knee	Vasculature—Anterior

Click **LAYER 2** in the **LAYER CONTROLS** window, and you will see the following image:

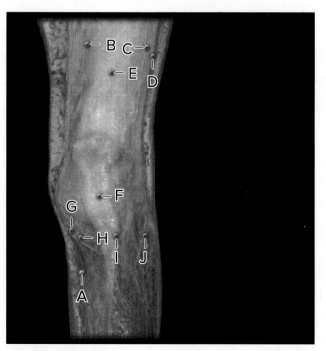

©McGraw-Hill Education

Mouse-over the pin on the screen to find the information necessary to identify the following structure:

A. _____

Noncardiovascular System Structures (blue pins)

B. _____
C. _____
D. _____
E. _____
F. _____
G. _____
H. _____
I. _____
J. _____

616 MODULE 9 The Cardiovascular System

Click **LAYER 3** *in the* **LAYER CONTROLS** *window, and you will see the following image:*

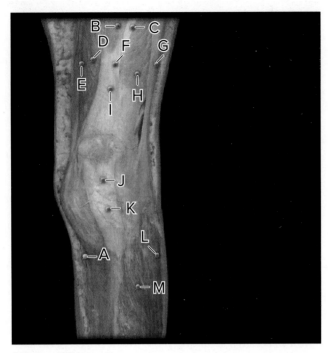

©McGraw-Hill Education

Mouse-over the pin on the screen to find the information necessary to identify the following structure:

A. _____

Noncardiovascular System Structures (blue pins)

B. _____
C. _____
D. _____
E. _____
F. _____
G. _____
H. _____
I. _____
J. _____
K. _____
L. _____
M. _____

Click **LAYER 4** *in the* **LAYER CONTROLS** *window, and you will see the following image:*

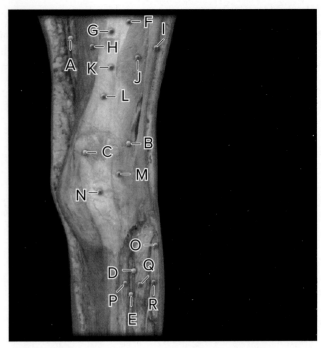

©McGraw-Hill Education

Mouse-over the pins on the screen to find the information necessary to identify the following structures:

A. _____
B. _____
C. _____
D. _____
E. _____

Noncardiovascular System Structures (blue pins)

F. _____
G. _____
H. _____
I. _____
J. _____
K. _____
L. _____
M. _____
N. _____
O. _____
P. _____
Q. _____
R. _____

EXERCISE 9.57:
Cardiovascular System—Knee Vasculature—Posterior View

SELECT TOPIC: Knee **SELECT VIEW:** Vasculature—Posterior

Click **LAYER 2** *in the* **LAYER CONTROLS** *window, and you will see the following image:*

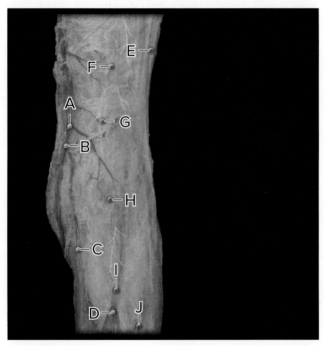

©McGraw-Hill Education

Mouse-over the pins on the screen to find the information necessary to identify the following structures:

A. _____
B. _____
C. _____
D. _____

Noncardiovascular System Structures (blue pins)

E. _____
F. _____
G. _____
H. _____
I. _____
J. _____

Click **LAYER 3** *in the* **LAYER CONTROLS** *window, and you will see the following image:*

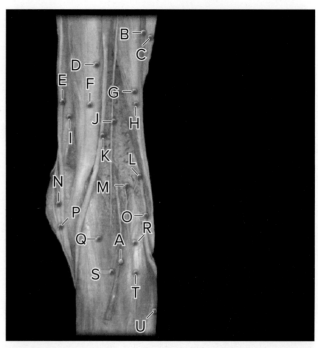

©McGraw-Hill Education

Mouse-over the pin on the screen to find the information necessary to identify the following structure:

A. _____

Noncardiovascular System Structures (blue pins)

B. _____
C. _____
D. _____
E. _____
F. _____
G. _____
H. _____
I. _____
J. _____
K. _____
L. _____
M. _____
N. _____
O. _____
P. _____
Q. _____
R. _____
S. _____
T. _____
U. _____

Click **LAYER 4** *in the* **LAYER CONTROLS** *window, and you will see the following image:*

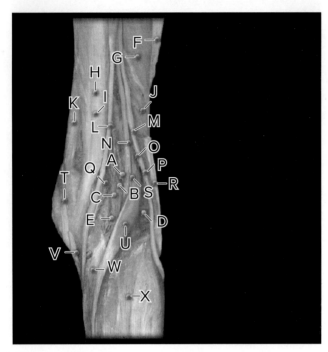

©McGraw-Hill Education

Mouse-over the pins on the screen to find the information necessary to identify the following structures:

A. _____
B. _____
C. _____
D. _____
E. _____

Noncardiovascular System Structures (blue pins)

F. _____
G. _____
H. _____
I. _____
J. _____
K. _____
L. _____
M. _____
N. _____
O. _____
P. _____
Q. _____
R. _____
S. _____
T. _____
U. _____
V. _____
W. _____
X. _____

Click **LAYER 5** *in the* **LAYER CONTROLS** *window, and you will see the following image:*

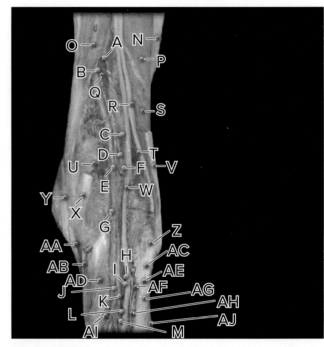

©McGraw-Hill Education

Mouse-over the pins on the screen to find the information necessary to identify the following structures:

A. _____
B. _____
C. _____
D. _____
E. _____
F. _____
G. _____
H. _____
I. _____
J. _____
K. _____
L. _____
M. _____

Noncardiovascular System Structures (blue pins)

N. _____
O. _____

P. _____
Q. _____
R. _____
S. _____
T. _____
U. _____
V. _____
W. _____
X. _____
Y. _____
Z. _____
AA. _____
AB. _____
AC. _____
AD. _____
AE. _____
AF. _____
AG. _____
AH. _____
AI. _____
AJ. _____

Click **LAYER 6** *in the* **LAYER CONTROLS** *window, and you will see the following image:*

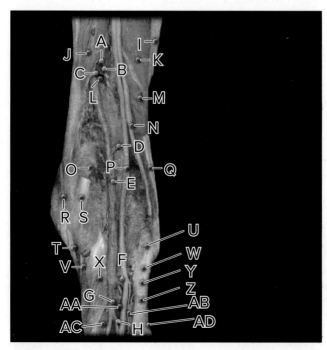

©McGraw-Hill Education

Mouse-over the pins on the screen to find the information necessary to identify the following structures:

A. _____
B. _____
C. _____
D. _____
E. _____
F. _____
G. _____
H. _____

Noncardiovascular System Structures (blue pins)

I. _____
J. _____
K. _____
L. _____
M. _____
N. _____
O. _____
P. _____
Q. _____
R. _____
S. _____
T. _____
U. _____
V. _____
W. _____
X. _____
Y. _____
Z. _____
AA. _____
AB. _____
AC. _____
AD. _____

EXERCISE 9.58:

Cardiovascular System—Knee—Arteries—Anterior View

SELECT TOPIC	SELECT VIEW
Knee	Arteries—Anterior

Click **LAYER 1** *in the* **LAYER CONTROLS** *window, and you will see the following image:*

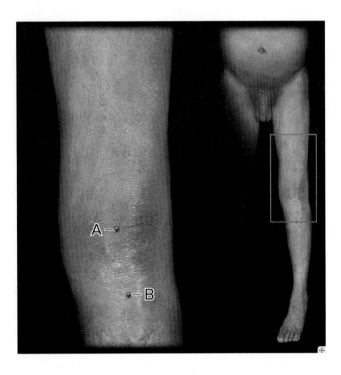

Mouse-over the pins on the screen to find the information necessary to identify the following noncardiovascular system structures:

A. _____

B. _____

CHECK POINT

Knee, Arteries, Anterior View

1. Name the sesamoid bone embedded in the tendon of the quadriceps femoris muscles.
2. Name the bony elevation on the proximal shaft of the tibia.

Click **LAYER 2** *in the* **LAYER CONTROLS** *window, and you will see the following image:*

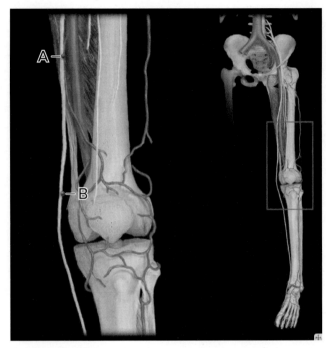

©McGraw-Hill Education

Mouse-over the pins on the screen to find the information necessary to identify the following noncardiovascular system structures:

A. _____

B. _____

Click **LAYER 3** *in the* **LAYER CONTROLS** *window, and you will see the following image:*

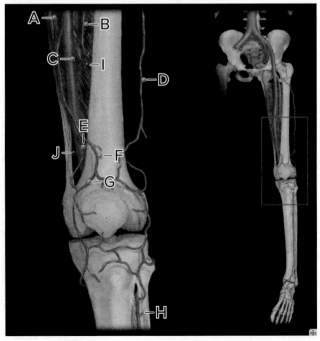

©McGraw-Hill Education

Mouse-over the pins on the screen to find the information necessary to identify the following structures:

A. _____
B. _____
C. _____
D. _____
E. _____
F. _____
G. _____
H. _____

Noncardiovascular System Structures (blue pins)

I. _____
J. _____

CHECK POINT

Knee, Arteries, Anterior View, *continued*

3. List the areas distributed by the femoral artery.
4. Name a deep artery of the thigh that distributes to the hip joint and the quadriceps femoris muscle.
5. Name an artery that distributes to the muscles and skin of the medial thigh.

EXERCISE 9.59:

Cardiovascular System—Knee—Arteries—Posterior View

SELECT TOPIC	SELECT VIEW
Knee	Arteries—Posterior

Click **LAYER 1** *in the* **LAYER CONTROLS** *window, and you will see the following image:*

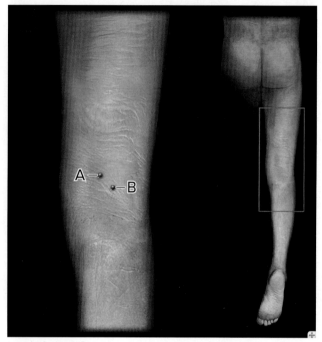

©McGraw-Hill Education

Mouse-over the pin on the screen to find the information necessary to identify the following structure:

A. _____

Noncardiovascular System Structure (blue pin)

B. _____

CHECK POINT

Knees, Arteries, Posterior View

1. Name the diamond-shaped area of the posterior knee.
2. Name the artery that provides a pulse in this area.
3. What special requirements must be met to appreciate this pulse?

Click **LAYER 2** *in the* **LAYER CONTROLS** *window, and you will see the following image:*

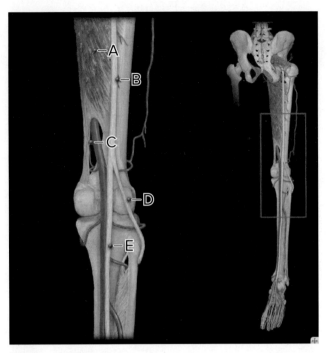

©McGraw-Hill Education

Mouse-over the pins on the screen to find the information necessary to identify the following noncardiovascular system structures:

A. _____
B. _____
C. _____
D. _____
E. _____

CHECK POINT

Knee, Arteries, Posterior View, *continued*

4. Name the large nerve of the posterior hip and thigh. Name the two nerves that merge to form this nerve.
5. What effect will a herniated intervertebral disk in the lower lumbar region have along the distribution of this nerve?
6. What is the adductor hiatus?

Click **LAYER 3** *in the* **LAYER CONTROLS** *window, and you will see the following image:*

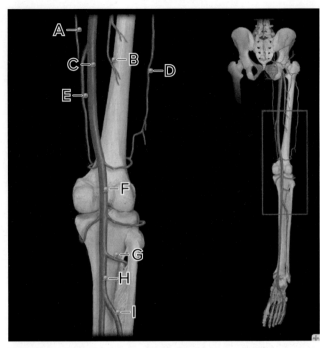

©McGraw-Hill Education

Mouse-over the pins on the screen to find the information necessary to identify the following structures:

A. _____
B. _____
C. _____
D. _____
E. _____
F. _____
G. _____
H. _____
I. _____

CHECK POINT

Knee, Arteries, Posterior View, *continued*

7. What artery does the femoral artery become in the vicinity of the knee?
8. Where exactly does this name change occur?
9. Name an artery that distributes to the anterior (extensor) compartment of the leg.

EXERCISE 9.60:
Cardiovascular System—Knee—Veins—Anterior View

SELECT TOPIC	SELECT VIEW
Knee	Veins—Anterior

Click **LAYER 2** *in the* **LAYER CONTROLS** *window, and you will see the following image:*

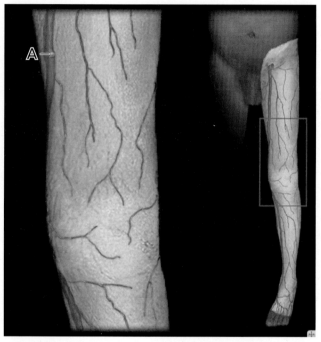

©McGraw-Hill Education

Mouse-over the pin on the screen to find the information necessary to identify the following structure:

A. _____

CHECK POINT
Knee, Veins, Anterior View

1. Name the large subcutaneous vein of the median leg and thigh.
2. What is its drainage?

Click **LAYER 3** *in the* **LAYER CONTROLS** *window, and you will see the following image:*

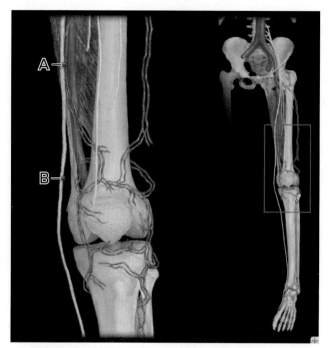

©McGraw-Hill Education

Mouse-over the pins on the screen to find the information necessary to identify the following noncardiovascular system structures:

A. _____
B. _____

Click **LAYER 4** *in the* **LAYER CONTROLS** *window, and you will see the following image:*

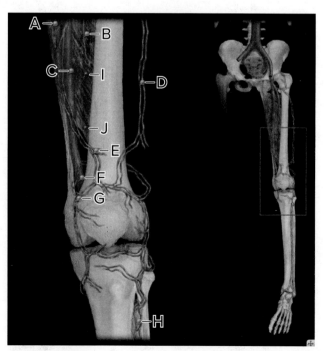

©McGraw-Hill Education

Mouse-over the pins on the screen to find the information necessary to identify the following structures:

A. _____
B. _____
C. _____
D. _____
E. _____
F. _____
G. _____
H. _____

Noncardiovascular System Structures (blue pins)

I. _____
J. _____

CHECK POINT

Knee, Veins, Anterior View, *continued*

3. List the structures drained by the femoral vein.
4. Name the veins that drain the anterior (extensor) compartment of the leg.
5. Name the venous anastomosis that forms a network around the knee. What veins contribute to this anastomosis?

EXERCISE 9.61:

Cardiovascular System—Knee—Veins—Posterior View

SELECT TOPIC	SELECT VIEW
Knee	Veins—Posterior

Click **LAYER 2** *in the* **LAYER CONTROLS** *window, and you will see the following image:*

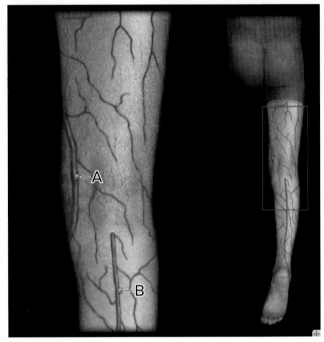

©McGraw-Hill Education

Mouse-over the pins on the screen to find the information necessary to identify the following structures:

A. _____
B. _____

CHECK POINT

Knee, Veins, Posterior View

1. Name the vein providing drainage for the dorsum of the foot and subcutaneous posterior leg.
2. Name the vein providing drainage for the dorsum of the foot and subcutaneous medial leg and thigh.

Click **LAYER 4** *in the* **LAYER CONTROLS** *window, and you will see the following image:*

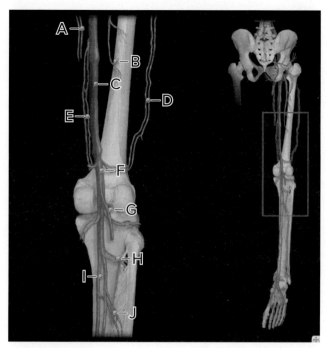

©McGraw-Hill Education

Mouse-over the pins on the screen to find the information necessary to identify the following structures:

A. _____
B. _____
C. _____
D. _____
E. _____
F. _____
G. _____
H. _____
I. _____
J. _____

CHECK POINT

Knee, Veins, Posterior View, *continued*

3. Name the vein that drains the knee joint and the surrounding structures.
4. Name the paired veins that drain the posterior muscles of the leg.
5. List the drainage of the fibular veins.

EXERCISE 9.62:
Imaging—Knee

SELECT TOPIC	SELECT VIEW
Knee	Angiogram Arteries—Anterior–Posterior

Click the **TAGS ON/OFF** *button, and you will see the following image:*

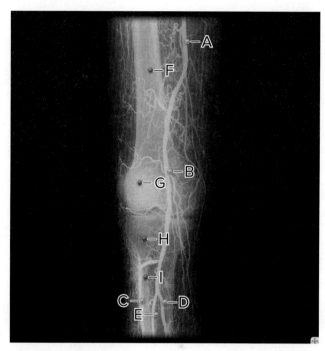

©McGraw-Hill Education

Mouse-over the pins on the screen to find the information necessary to identify the following structures:

A. _____
B. _____
C. _____
D. _____
E. _____

Noncardiovascular System Structures (blue pins)

F. _____
G. _____
H. _____
I. _____

Self-Quiz
Take this opportunity to check your progress by taking the **QUIZ**. See the **Introduction Module** for a reminder on how to access the **QUIZ** for this Study Area.

IN REVIEW

What Have I Learned?

The following questions cover the material that you have just learned, the blood vessels of the knee. Apply what you have learned to answer these questions on a separate piece of paper.

1. Name two blood vessels that change their names at the posterior knee.

2. Where exactly does this name change occur?

3. What new names do these vessels acquire?

4. The terminal branches of which artery supplies everything inferior to the knee?

5. Which artery is the primary blood supply for the posterior leg muscles?

6. List the distribution of the fibular artery.

7. Name the artery with the distribution to the knee joint and the structures around it.

8. Many arteries of the limbs are accompanied by _____.

9. What is the meaning of the Latin word *genu*?

10. Name the paired veins that drain the quadriceps femoris muscles and the hip joint.

The Leg and Foot

EXERCISE 9.63:

Cardiovascular System—Leg and Foot Vasculature—Anterior View

SELECT TOPIC	SELECT VIEW
Leg and Foot	Vasculature—Anterior

Click **LAYER 2** in the **LAYER CONTROLS** window, and you will see the following image:

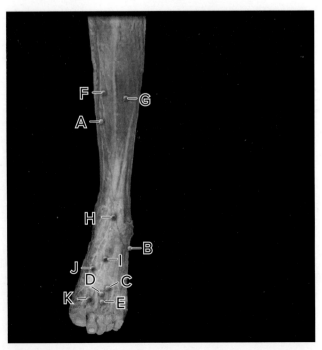

©McGraw-Hill Education

Mouse-over the pins on the screen to find the information necessary to identify the following structures:

A. _____
B. _____
C. _____
D. _____
E. _____

Noncardiovascular System Structures (blue pins)

F. _____
G. _____
H. _____
I. _____
J. _____
K. _____

Click **LAYER 3** *in the* **LAYER CONTROLS** *window, and you will see the following image:*

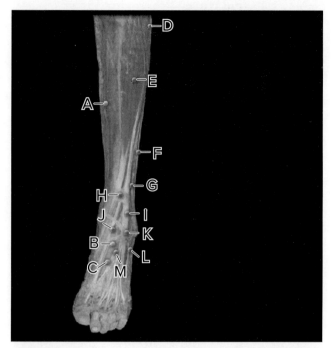

©McGraw-Hill Education

Mouse-over the pins on the screen to find the information necessary to identify the following structures:

A. _____
B. _____
C. _____

Noncardiovascular System Structures (blue pins)

D. _____
E. _____
F. _____
G. _____
H. _____
I. _____
J. _____
K. _____
L. _____
M. _____

Click **LAYER 4** *in the* **LAYER CONTROLS** *window, and you will see the following image:*

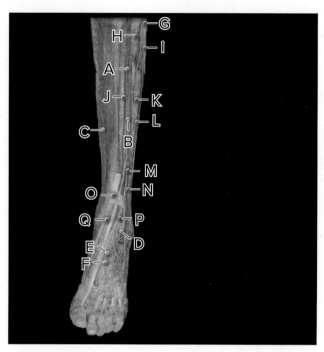

©McGraw-Hill Education

Mouse-over the pins on the screen to find the information necessary to identify the following structures:

A. _____
B. _____
C. _____
D. _____
E. _____
F. _____

Noncardiovascular System Structures (blue pins)

G. _____
H. _____
I. _____
J. _____
K. _____
L. _____
M. _____
N. _____
O. _____
P. _____
Q. _____

628 MODULE 9 The Cardiovascular System

Click **LAYER 5** *in the* **LAYER CONTROLS** *window, and you will see the following image:*

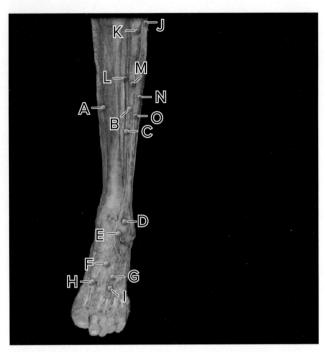

©McGraw-Hill Education

Mouse-over the pins on the screen to find the information necessary to identify the following structures:

A. _____
B. _____
C. _____
D. _____
E. _____
F. _____
G. _____
H. _____
I. _____

Noncardiovascular System Structures (blue pins)

J. _____
K. _____
L. _____
M. _____
N. _____
O. _____

EXERCISE 9.64:
Cardiovascular System—Leg and Foot—Vasculature—Posterior View

SELECT TOPIC	SELECT VIEW
Leg and Foot	Vasculature—Posterior

Click **LAYER 2** *in the* **LAYER CONTROLS** *window, and you will see the following image:*

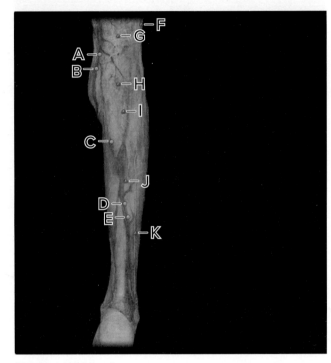

©McGraw-Hill Education

Mouse-over the pin on the screen to find the information necessary to identify the following structure:

A. _____
B. _____
C. _____
D. _____
E. _____
F. _____
G. _____
H. _____
I. _____
J. _____
K. _____

Click **LAYER 3** *in the* **LAYER CONTROLS** *window and you will see the following image:*

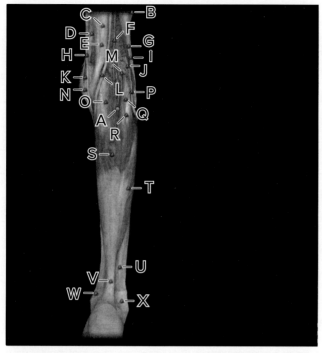

©McGraw-Hill Education

Mouse-over the pin on the screen to find the information necessary to identify the following structure:

A. _____

Noncardiovascular System Structures (blue pins)

B. _____
C. _____
D. _____
E. _____
F. _____
G. _____
H. _____
I. _____
J. _____
K. _____
L. _____
M. _____
N. _____
O. _____
P. _____
Q. _____
R. _____
S. _____
T. _____
U. _____
V. _____
W. _____
X. _____

Click **LAYER 4** *in the* **LAYER CONTROLS** *window, and you will see the following image:*

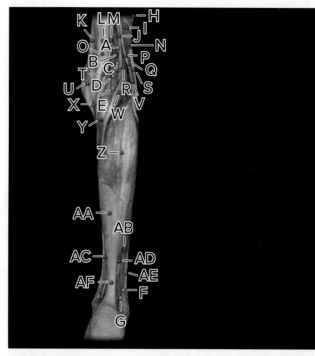

©McGraw-Hill Education

Mouse-over the pins on the screen to find the information necessary to identify the following structures:

A. _____
B. _____
C. _____
D. _____
E. _____
F. _____
G. _____

Noncardiovascular System Structures (blue pins)

H. _____
I. _____
J. _____
K. _____
L. _____
M. _____
N. _____

630 MODULE 9 The Cardiovascular System

O. _____
P. _____
Q. _____
R. _____
S. _____
T. _____
U. _____
V. _____
W. _____
X. _____
Y. _____
Z. _____
AA. _____
AB. _____
AC. _____
AD. _____
AE. _____
AF. _____

Click **LAYER 5** *in the* **LAYER CONTROLS** *window, and you will see the following image:*

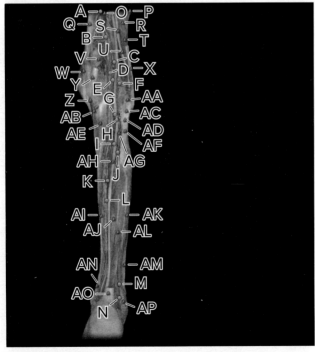

©McGraw-Hill Education

Mouse-over the pins on the screen to find the information necessary to identify the following structures:

A. _____
B. _____

C. _____
D. _____
E. _____
F. _____
G. _____
H. _____
I. _____
J. _____
K. _____
L. _____
M. _____
N. _____

Noncardiovascular System Structures (blue pins)

O. _____
P. _____
Q. _____
R. _____
S. _____
T. _____
U. _____
V. _____
W. _____
X. _____
Y. _____
Z. _____
AA. _____
AB. _____
AC. _____
AD. _____
AE. _____
AF. _____
AG. _____
AH. _____
AI. _____
AJ. _____
AK. _____
AL. _____
AM. _____
AN. _____
AO. _____
AP. _____

Click **LAYER 6** *in the* **LAYER CONTROLS** *window, and you will see the following image:*

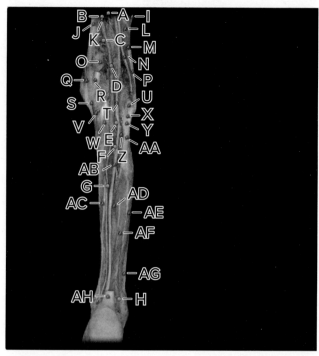

©McGraw-Hill Education

Mouse-over the pins on the screen to find the information necessary to identify the following structures:

A. _____
B. _____
C. _____
D. _____
E. _____
F. _____
G. _____
H. _____

Noncardiovascular System Structures (blue pins)

I. _____
J. _____
K. _____
L. _____
M. _____
N. _____
O. _____
P. _____
Q. _____
R. _____
S. _____
T. _____
U. _____
V. _____
W. _____
X. _____
Y. _____
Z. _____
AA. _____
AB. _____
AC. _____
AD. _____
AE. _____
AF. _____
AG. _____
AH. _____

EXERCISE 9.65:

Cardiovascular System—Leg and Foot—Arteries—Anterior View

 SELECT TOPIC: **Leg and Foot** ▸ SELECT VIEW: **Arteries—Anterior**

Click **LAYER 1** *in the* **LAYER CONTROLS** *window, and you will see the following image:*

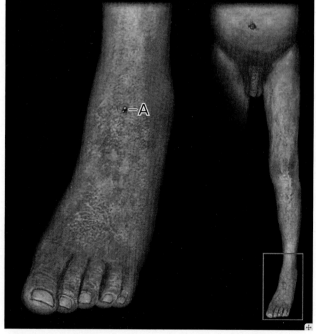

©McGraw-Hill Education

Mouse-over the pin on the screen to find the information necessary to identify the following structure:

A. _____

CHECK POINT

Leg and Foot, Arteries, Anterior View

1. Name the artery used for a pedal pulse.
2. Where is this pulse located?
3. What is this pulse used to assess?

Click **LAYER 2** *in the* **LAYER CONTROLS** *window, and you will see the following image:*

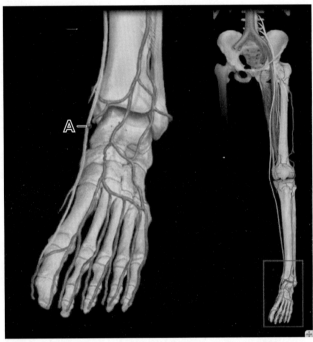

©McGraw-Hill Education

Mouse-over the pin on the screen to find the information necessary to identify the following noncardiovascular system structure:

A. _____

CHECK POINT

Leg and Foot Arteries, Anterior View, *continued*

4. Name the distal branch of the femoral nerve.
5. Where is it located?
6. Where is the general sensory innervation associated with this nerve?

Click **LAYER 3** *in the* **LAYER CONTROLS** *window, and you will see the following image:*

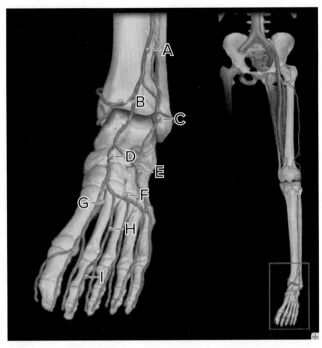

©McGraw-Hill Education

Mouse-over the pins on the screen to find the information necessary to identify the following structures:

A. _____
B. _____
C. _____
D. _____
E. _____
F. _____
G. _____
H. _____
I. _____

CHECK POINT

Leg and Foot, Arteries, Anterior View, *continued*

7. Name the terminal branch of the popliteal artery that continues on the dorsum of the foot.
8. What is the name for this artery on the dorsum of the foot?
9. Name the arteries that distribute to the toes and their joints.

EXERCISE 9.66:
Cardiovascular System—Leg and Foot—Arteries—Posterior View

SELECT TOPIC	SELECT VIEW
Leg and Foot	Arteries—Posterior

Click **LAYER 2** *in the* **LAYER CONTROLS** *window, and you will see the following image:*

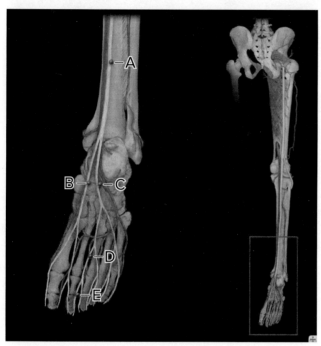

©McGraw-Hill Education

Mouse-over the pins on the screen to find the information necessary to identify the following noncardiovascular system structures:

A. _____
B. _____
C. _____
D. _____
E. _____

CHECK POINT

Leg and Foot, Arteries, Posterior View

1. Name the nerve that is both motor and general sensory and innervates the muscles of the posterior leg.
2. Name the lateral terminal branch of this nerve. Where is its motor and general sensory innervation?
3. Name the medial terminal branch of this nerve. Where is its motor and general sensory innervation?

Click **LAYER 3** *in the* **LAYER CONTROLS** *window, and you will see the following image:*

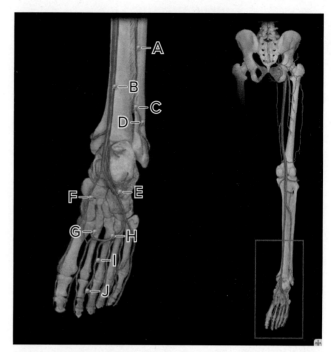

©McGraw-Hill Education

Mouse-over the pins on the screen to find the information necessary to identify the following structures:

A. _____
B. _____
C. _____
D. _____
E. _____
F. _____
G. _____
H. _____
I. _____
J. _____

CHECK POINT

Leg and Foot, Arteries, Posterior View, *continued*

4. Name the arterial vessel that arches across the plantar surface at the bases of metatarsals II–IV.
5. Name the artery that distributes to the muscles and joints of the medial plantar foot.
6. Name the artery that distributes to the muscles and joints of the lateral plantar foot.

EXERCISE 9.67:

Cardiovascular System—Leg and Foot—Veins—Anterior View

SELECT TOPIC	SELECT VIEW
Leg and Foot	Veins—Anterior

Click **LAYER 2** *in the* **LAYER CONTROLS** *window, and you will see the following image:*

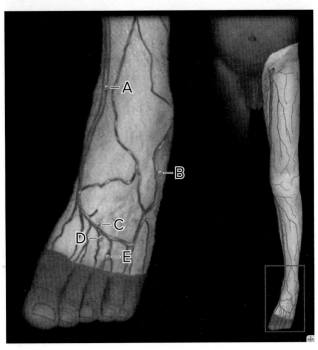

©McGraw-Hill Education

Mouse-over the pins on the screen to find the information necessary to identify the following structures:

A. _____
B. _____
C. _____
D. _____
E. _____

CHECK POINT

Leg and Foot, Veins, Anterior View

1. Name the venous vessel that curves medial to lateral over the proximal ends of the metatarsals.
2. Name the veins that drain the toes, plus the distal dorsum and joints of the foot.
3. Name the veins that course along the medial and lateral sides of the digits and drain the toes.

Click **LAYER 4** *in the* **LAYER CONTROLS** *window, and you will see the following image:*

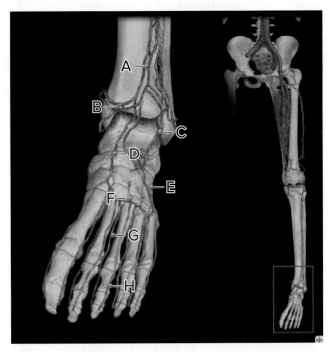

©McGraw-Hill Education

Mouse-over the pins on the screen to find the information necessary to identify the following structures:

A. _____
B. _____
C. _____
D. _____
E. _____
F. _____
G. _____
H. _____

CHECK POINT

Leg and Foot, Veins, Anterior View, *continued*

4. Name the vein that drains the lateral tarsal bones.
5. Name the vein that drains the dorsum of the foot, including the toes.
6. Name the veins that drain the toes and metatarsal region of the foot.

EXERCISE 9.68:
Cardiovascular System—Leg and Foot—Veins—Posterior View

SELECT TOPIC	SELECT VIEW
Leg and Foot	Veins—Posterior

Click **LAYER 2** *in the* **LAYER CONTROLS** *window, and you will see the following image:*

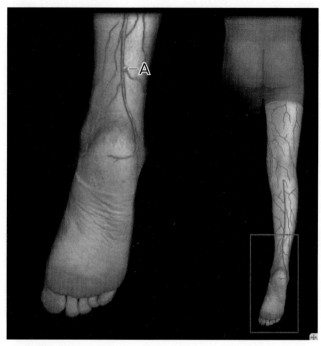

©McGraw-Hill Education

Mouse-over the pin on the screen to find the information necessary to identify the following structure:

A. _____

CHECK POINT

Leg and Foot, Veins, Posterior View

1. List the drainage and the course of the small saphenous vein.

Click **LAYER 4** *in the* **LAYER CONTROLS** *window, and you will see the following image:*

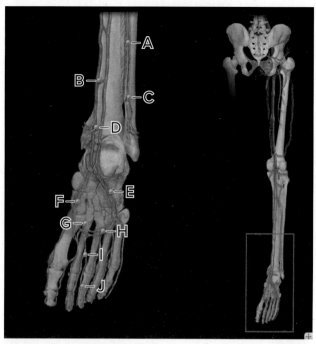

©McGraw-Hill Education

Mouse-over the pins on the screen to find the information necessary to identify the following structures:

A. _____
B. _____
C. _____
D. _____
E. _____
F. _____
G. _____
H. _____
I. _____
J. _____

CHECK POINT

Leg and Foot, Veins, Posterior View, *continued*

1. Name the venous vessels that arch across the plantar surface at the bases of metatarsals II–IV.
2. List the drainages of the medial plantar veins.
3. List the drainages of the lateral plantar veins.

636 MODULE 9 The Cardiovascular System

EXERCISE 9.69:
Cardiovascular System—Foot Vasculature—Dorsum

SELECT TOPIC	SELECT VIEW
Foot	Vasculature—Dorsum

Click **LAYER 2** *in the* **LAYER CONTROLS** *window, and you will see the following image:*

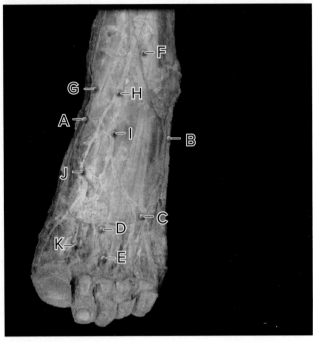

©McGraw-Hill Education

Mouse-over the pins on the screen to find the information necessary to identify the following structures:

A. _____
B. _____
C. _____
D. _____
E. _____

Noncardiovascular System Structures (blue pins)

F. _____
G. _____
H. _____
I. _____
J. _____
K. _____

Click **LAYER 3** *in the* **LAYER CONTROLS** *window, and you will see the following image:*

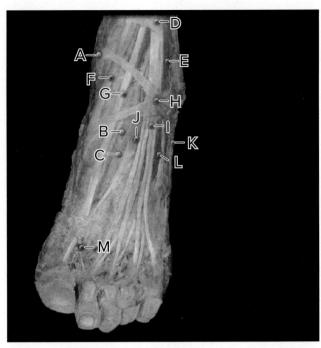

©McGraw-Hill Education

Mouse-over the pins on the screen to find the information necessary to identify the following structures:

A. _____
B. _____
C. _____

Noncardiovascular System Structures (blue pins)

D. _____
E. _____
F. _____
G. _____
H. _____
I. _____
J. _____
K. _____
L. _____
M. _____

Click **LAYER 4** *in the* **LAYER CONTROLS** *window, and you will see the following image:*

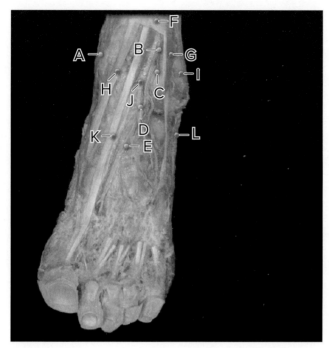

©McGraw-Hill Education

Mouse-over the pins on the screen to find the information necessary to identify the following structures:

A. _____
B. _____
C. _____
D. _____
E. _____

Noncardiovascular System Structures (blue pins)

F. _____
G. _____
H. _____
I. _____
J. _____
K. _____
L. _____

Click **LAYER 5** *in the* **LAYER CONTROLS** *window, and you will see the following image:*

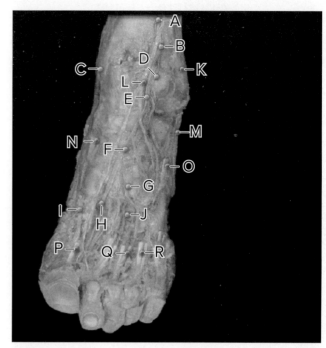

©McGraw-Hill Education

Mouse-over the pins on the screen to find the information necessary to identify the following structures:

A. _____
B. _____
C. _____
D. _____
E. _____
F. _____
G. _____
H. _____
I. _____
J. _____

Noncardiovascular System Structures (blue pins)

K. _____
L. _____
M. _____
N. _____
O. _____
P. _____
Q. _____
R. _____

EXERCISE 9.70:

Cardiovascular System—Foot Vasculature—Plantar

SELECT TOPIC	SELECT VIEW
Foot	Vasculature—Plantar

Click **LAYER 2** *in the* **LAYER CONTROLS** *window, and you will see the following image:*

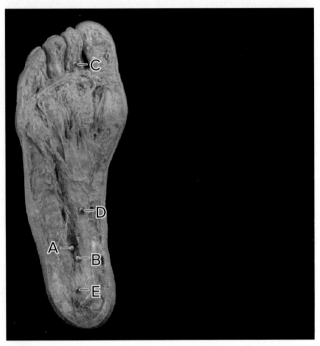

©McGraw-Hill Education

Mouse-over the pins on the screen to find the information necessary to identify the following structures:

A. _____

B. _____

Noncardiovascular System Structures (blue pins)

C. _____

D. _____

E. _____

Click **LAYER 3** *in the* **LAYER CONTROLS** *window, and you will see the following image:*

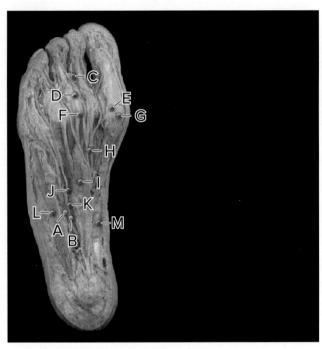

©McGraw-Hill Education

Mouse-over the pins on the screen to find the information necessary to identify the following structures:

A. _____

B. _____

Noncardiovascular System Structures (blue pins)

C. _____

D. _____

E. _____

F. _____

G. _____

H. _____

I. _____

J. _____

K. _____

L. _____

M. _____

Click **LAYER 4** *in the* **LAYER CONTROLS** *window, and you will see the following image:*

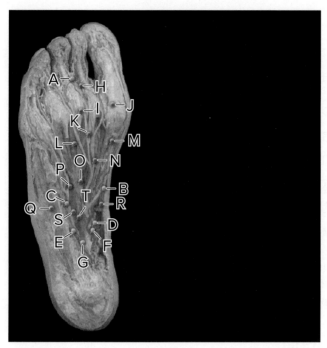

©McGraw-Hill Education

Mouse-over the pins on the screen to find the information necessary to identify the following structures:

A. _____
B. _____
C. _____
D. _____
E. _____
F. _____
G. _____

Noncardiovascular System Structures (blue pins)

H. _____
I. _____
J. _____
K. _____
L. _____
M. _____
N. _____
O. _____
P. _____
Q. _____
R. _____
S. _____
T. _____

Click **LAYER 5** *in the* **LAYER CONTROLS** *window, and you will see the following image:*

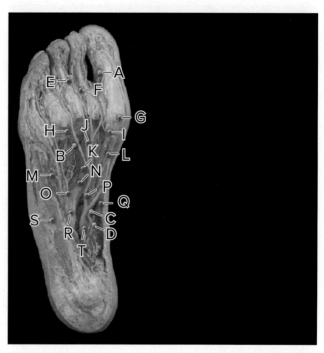

©McGraw-Hill Education

Mouse-over the pins on the screen to find the information necessary to identify the following structures:

A. _____
B. _____
C. _____
D. _____

Noncardiovascular System Structures (blue pins)

E. _____
F. _____
G. _____
H. _____
I. _____
J. _____
K. _____
L. _____
M. _____
N. _____
O. _____
P. _____
Q. _____
R. _____
S. _____
T. _____

Click **LAYER 6** *in the* **LAYER CONTROLS** *window, and you will see the following image:*

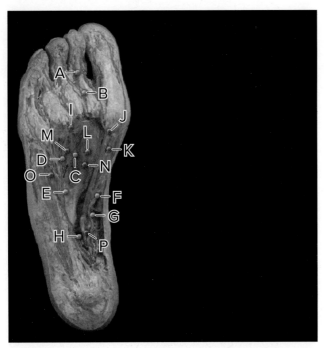

©McGraw-Hill Education

Mouse-over the pins on the screen to find the information necessary to identify the following structures:

A. _____

B. _____

C. _____

D. _____

E. _____

F. _____

G. _____

H. _____

Noncardiovascular System Structures (blue pins)

I. _____

J. _____

K. _____

L. _____

M. _____

N. _____

O. _____

P. _____

Self-Quiz
Take this opportunity to check your progress by taking the **QUIZ**. See the **Introduction Module** for a reminder on how to access the **QUIZ** for this Study Area.

IN REVIEW

What Have I Learned?

The following questions cover the material that you have just learned, the blood vessels of the leg and foot. Apply what you have learned to answer these questions on a separate piece of paper.

1. Name the arteries that pass distally along the metatarsals and distribute to the dorsum of the foot and its joints and toes.

2. Name the artery that distributes to the tarsal bones of the lateral foot.

3. Name the artery used for pedal pulse.

4. Name the artery that distributes to the ankle joint and the dorsum of the foot.

5. Name the artery that distributes to the lateral and medial aspects of the ankle joint.

6. Name the nerve located along the plantar aspect of the metatarsals supplying motor innervation to the intrinsic foot muscles. Where is its general sensory innervation?

7. Name the nerve that is located on the medial and lateral sides of the digits and that supplies general sensory innervation to the skin and joints of the toes.

8. List the distribution of the plantar metatarsal arteries.

9. Name the arteries that course along the medial and lateral sides of the digits, distributing to the toes and their joints.

10. Name the vein that drains the posterior leg muscles as it ascends from the ankle through the posterior leg.

EXERCISE 9.71:
Coloring Exercise

Identify the structures of the heart and then color them with colored pens or pencils.

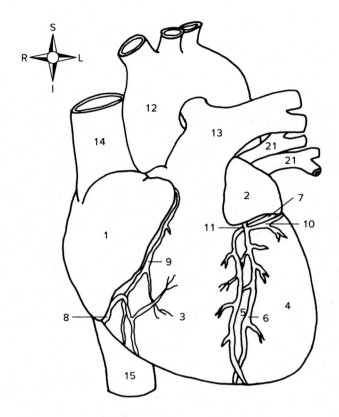

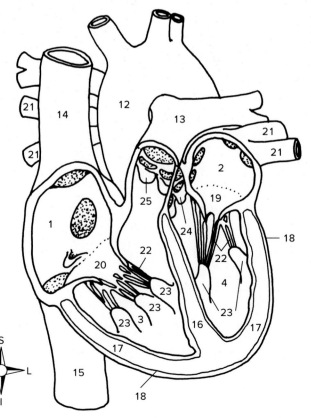

☐ 1. _____
☐ 2. _____
☐ 3. _____
☐ 4. _____
☐ 5. _____
☐ 6. _____
☐ 7. _____
☐ 8. _____
☐ 9. _____
☐ 10. _____
☐ 11. _____
☐ 12. _____
☐ 13. _____
☐ 14. _____
☐ 15. _____
☐ 16. _____
☐ 17. _____
☐ 18. _____
☐ 19. _____
☐ 20. _____
☐ 21. _____
☐ 22. _____
☐ 23. _____
☐ 24. _____
☐ 25. _____

642 MODULE 9 The Cardiovascular System

EXERCISE 9.72:
Labeling Exercise

Major arteries of the body. Write the names of the arteries indicated on the label lines.

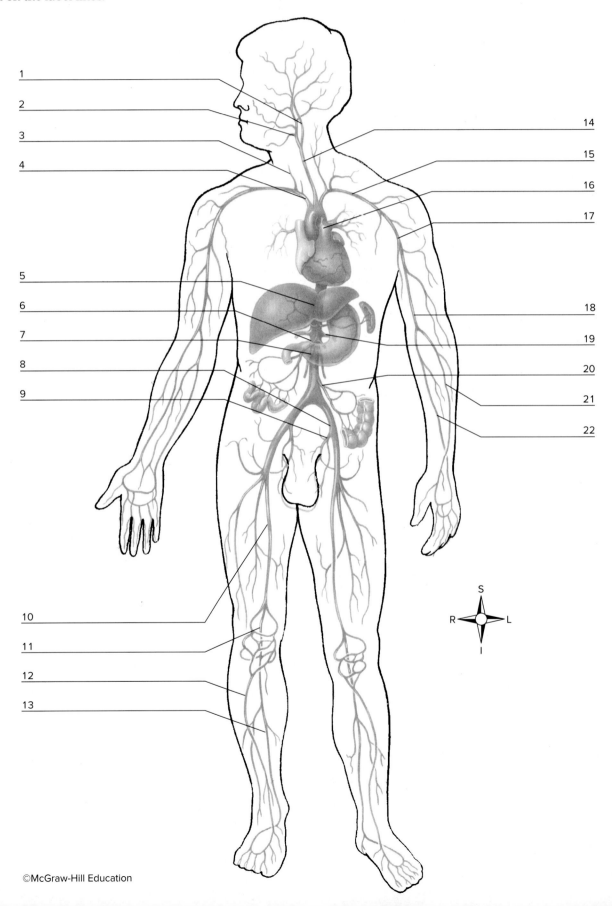

MODULE 9 The Cardiovascular System **643**

EXERCISE 9.73:
Labeling Exercise

Major veins of the body. Write the names of the veins indicated on the label lines.

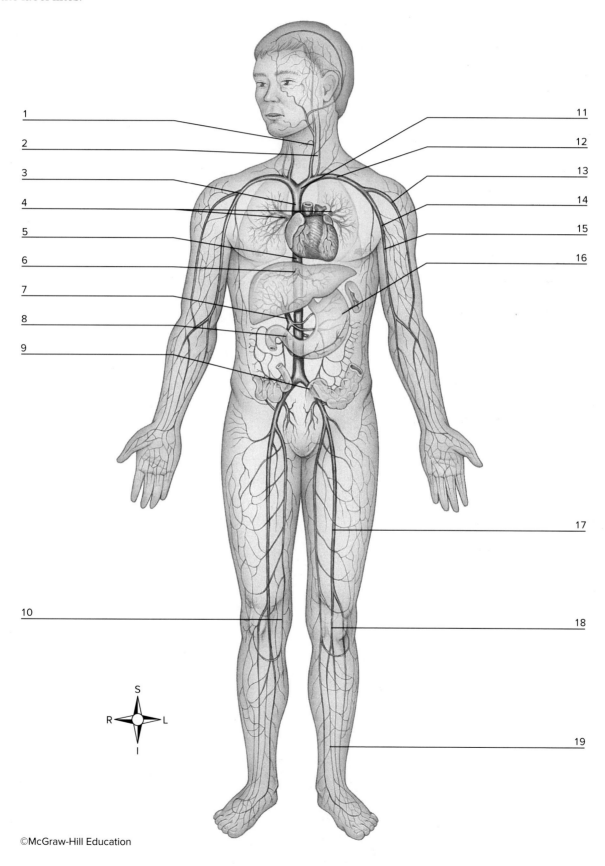

©McGraw-Hill Education

MODULE **10**

The Lymphatic System

Overview: Lymphatic System

You are probably not familiar with the lymphatic system, yet it is vital to your survival. It's not one of those systems that you think about every day—unless you are sick! What does this system do for us? Plenty! It recovers fluid lost from the blood capillaries, is intricately involved with the immune system, and absorbs dietary lipids through lacteals located in the small intestine.

Let's begin our exploration of this mostly unknown organ system by taking an overview of the entire lymphatic system, and then looking at the specific structures in more detail.

From the **HOME** *screen, click the drop-down box on the* **MODULE** *menu.*

From the systems listed, click on **Lymphatic.**

SELECT ANIMATION
Lymphatic System Overview
PLAY

After viewing the animation, answer the following questions:

1. What is the lymphatic system?

2. Name the fluid involved in this system.

3. As a system, what is its function?

4. Name the fluid that seeps from the blood capillaries throughout your body.

5. What percentage of this fluid becomes lymph?

6. What are lymphatic capillaries?

7. What do they form when they converge?

8. What do these vessels form when they merge?

9. Where do these vessels drain?

10. From where does the right lymphatic duct receive lymph?

11. Where does the right lymphatic duct empty?

12. From where does the thoracic duct receive lymph?

13. Where does it empty?

14. Which lymphatic duct is larger, the right lymphatic duct or the thoracic duct?

15. Lymphatic tissues include . . .

16. What are lymphatic nodules?

17. What do they contain?

18. Where are clusters of lymphatic nodules associated?

19. Name the large groups of lymphatic nodules found in the walls of the nasal and oral cavities.

20. _____ tonsils are located in the nasopharynx. When they are inflamed, they are known as _____.

21. Where are the palatine tonsils located? The lingual tonsils?

22. Name the lymphatic organs. What is their structure?

23. What are lymph nodes? Where are prominent clusters of lymph nodes located?

24. What are the primary functions of lymph nodes?

25. What is the basic structure of a lymph node?

26. Describe the passage of lymph into, through, and out of the lymph node.

27. Where is the thymus located? What is its function?

28. When is the thymus most active?

29. What becomes of the thymus, beginning at adolescence?

30. Name the body's largest lymph organ. Where is it located?

31. What is its function?

32. How does it act like a lymph node?

Self-Quiz
Take this opportunity to check your progress by taking the QUIZ. See the Introduction Module for a reminder on how to access the QUIZ for this Study Area.

SELECT ANIMATION
Lymphatic Vascular Network (3D) — PLAY

After viewing the animation, answer the following questions:

1. _____ _____, also called _____ carry a watery fluid, known as _____, from _____ _____, filter _____ through packets of _____ tissue, called _____ _____, and return it to the _____ through a series of larger vessels.

2. Blood capillaries filter _____ from _____ into the _____.

3. The fluid in combination with _____, _____, and _____, comprises _____ _____ in the extracellular space.

4. _____ reabsorb most of the _____ and _____ _____ like _____ _____ and _____ into the _____ blood stream.

5. _____ _____ drain the excess extracellular materials to help maintain _____ _____ and _____.

6. Loose _____ _____ in the _____ _____ allow _____ _____ and _____ to _____, joining _____ _____.

7. _____ _____ converge into _____ _____.

8. _____ _____ occur along these vessels.

9. Inside a lymph node, _____ connect the _____, composed of _____ _____, and the _____, dotted with _____, _____, and _____ _____ _____.

10. As the lymph trickles through the _____, _____ _____ and other _____ cells, such as _____ _____ cells, encounter the _____ cells, which either _____ them or _____ them until an _____ _____ can be mounted to target the infection outside the _____.

11. The _____ lymph, consisting mainly of _____, _____, and _____, exits the _____ via an _____ _____ _____.

12. The _____ _____ _____ push the _____ to the next _____, which opens and allows lymph to move forward, while preventing _____ _____.

13. The _____ _____ combine into _____ _____.

14. Each _____ drains _____ from a _____ _____ _____ and empties into one of two _____ _____.

15. The _____ _____ _____, which drains lymph from the _____ side of the _____, the _____ _____, and the _____ side of the _____, to the _____ _____ _____.

16. The _____ _____, which delivers _____ from the _____ of the _____ to the _____ _____ _____.

17. After passing into the _____ _____, _____ returns to the _____.

Lymph Nodes

SELECT ANIMATION — Antigen Processing — PLAY

After viewing the animation, answer the following questions:

1. What is an antigen?

2. After antigens are produced, where are they transported?

3. What molecules combine with the antigens there? This combination is then transported to the _____ _____ and from there to the _____ _____.

4. What then happens to foreign antigens? To self-antigens?

5. When an antigen originates from outside of the cell, how do the particles enter the cell?

6. What happens to the particles inside the cell?

7. The vesicle containing the foreign particles . . .

8. What then happens to the MHC class II/antigen complex?

SELECT ANIMATION — Cytotoxic T Cells — PLAY

After viewing the animation, answer the following questions:

1. When a virus infects a cell, what does it produce?

2. What happens to some of these proteins?

3. What molecule complexes with these fragments?

4. Where are they displayed?

5. How do the cytotoxic T cells interact with the virus-infected cells?

6. What substances are released by the cytotoxic T cells?

7. The release of these substances results in . . .

8. What is the result for self-proteins?

9. What then becomes of the cytotoxic T cells?

SELECT ANIMATION — Helper T Cells — PLAY

After viewing the animation, answer the following questions:

1. Proteins (antigens) require the cooperation of helper T cells for what purpose?

2. These antigens are therefore said to be _____ _____.

3. What does an antigen presenting cell do in the presence of the antigen?

4. The antigen is then moved . . .

5. How does the helper T cell become activated?

6. What is the activated T cell capable of doing?

7. The antigen reacts with an _____ on the surface of the B cell and is then . . .

8. How does the B cell interact with the activated T cell?

9. The helper T cell produces _____, which stimulate the B cell to . . .

SELECT ANIMATION — IgE-Mediated Hypersensitivity — PLAY

After viewing the animation, answer the following questions:

1. Another name for an allergic reaction is _____.

2. This is mediated by _____.

3. How does sensitization occur?

4. Which tissues are rich in B cells committed to IgE production?

5. IgE-producing cells are more abundant in . . .

6. What do the helper T cells produce, and what is the effect on B cells?

7. Where and how do IgE molecules attach?

8. What are mast cells?

9. When an antigen-sensitive person is exposed a second time to the antigen, where does the antigen bind?

10. What is required to trigger a response?

11. Within seconds, what chemicals are released from the mast cells? What do they trigger?

12. What are some of those symptoms?

SELECT ANIMATION
Immune Response
PLAY

After viewing the animation, answer the following questions:

1. How does activation of the immune response typically begin?

2. Which cells encounter the pathogen and ingest it?

3. What steps occur next?

4. What are antigen-presenting cells?

5. What cell interacts with the antigen-presenting macrophage? What does this cell do?

6. What chemical does the macrophage release during interaction? What is the result?

7. What does the interleukin-2 cause to occur?

8. What are the two paths of the immune response from this point?

9. What response do normal body cells display when infected?

10. Why does the body make millions of different types of cytotoxic T cells?

11. What interaction do cytotoxic T cells have when they recognize the antigen displayed on the surface of infected cells? How does this destroy the pathogen?

12. How are B cells similar to cytotoxic T cells?

13. How are B cells activated? What do they differentiate into?

14. What is the function of plasma cells?

15. What reaction occurs between the antibodies and the antigens?

16. How are pathogens marked for destruction? How are they destroyed?

17. What are memory B cells? How long do they survive?

18. How does a "secondary immune response" compare to an "initial immune response"? How does this compare to vaccination?

Clinical Applications

SELECT ANIMATION: Inflammation (3D) [PLAY]

After viewing the animation, answer the following questions:

1. Define inflammation. Include its symptoms.

2. Immediately after injury, inflammation begins with...

3. This results in...

4. What is the response of local cells? What stimulates this response?

5. List the vasoactive chemicals and the responses that they initiate.

6. What effect do these chemicals have on endothelial cells in small blood vessels?

7. This increased _____ _____ allows...

8. Next, during a multistage process called _____, circulating immune cells called _____ move...

9. What is the function of the neutrophils?

10. How does chemotaxis begin?

11. Next, in a process called _____, neutrophils...

12. The neutrophils migrate to the _____ _____ by following a _____ _____.

13. Upon arrival...

14. Define phagocytosis.

15. After what processes does tissue repair begin?

16. The process of tissue repair begins when...

17. What is the function of the collagen produced?

18. What is the function of anti-inflammatory drugs? How do they accomplish this?

19. What are the most common drugs for inflammation?

20. Two examples of these are...

21. They contain an enzyme called _____ or _____ that _____ the production of _____ and several other _____ chemicals.

22. NSAIDS reduce...

SELECT ANIMATION
Hypersensitivity of Immune System — PLAY

After viewing the animation, answer the following questions:

1. Our body protects us from many _____ _____, such as _____, foreign _____, and _____.

2. The immune system is a _____ system of _____ _____ _____ _____ and _____ that recognize _____ from _____ _____.

3. Define an antigen.

4. Define antibodies. What is their function?

5. Sometimes, our immune system responds _____ to _____, causing _____ and _____ _____.

6. This is called an _____ _____.

7. _____ causing an _____ _____ are _____.

8. List the four major types of hypersensitivity.

9. Which is the most common type of hypersensitivity?

10. Which allergies are included in the most common type?

11. What occurs after initial exposure to this allergen?

12. These antibodies are bound to...

13. Immunoglobulin E . . .

14. This triggers . . .

15. List the inflammatory substances released. Which cells release these substances?

16. What can occur if Type One Hypersensitivity is severe and left untreated?

17. What is anaphylaxis?

18. Anaphylaxis is characterized by . . .

19. Describe treatment for Type One Hypersensitivity.

20. Describe treatment of severe reactions and anaphylaxis.

21. If the hypersensitivity is in response to a necessary drug treatment . . .

22. In some cases . . .

652 MODULE 10 The Lymphatic System

EXERCISE 10.1:
Lymphatic System—B Lymphocyte—Histology

SELECT TOPIC	SELECT VIEW
B Lymphocyte	TEM: High Magnification

Click the **TAGS ON/OFF** *button, and you will see the following image:*

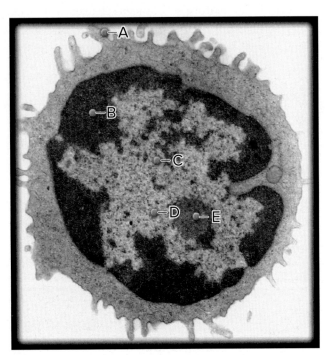

©Don W. Fawcett/Science Source

Mouse-over the pins on the screen to find the information necessary to identify the following structures:

A. _____
B. _____
C. _____
D. _____
E. _____

EXERCISE 10.2:
Lymphatic System—Plasma Cell—Histology

SELECT TOPIC	SELECT VIEW
Plasma Cell	TEM: High Magnification

Click the **TAGS ON/OFF** *button, and you will see the following image:*

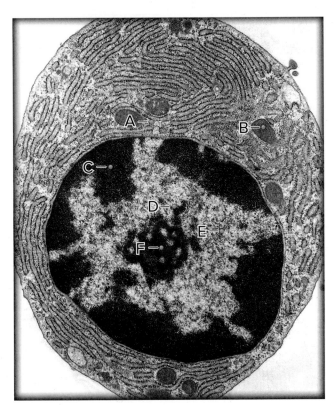

©Don W. Fawcett/Science Source

Mouse-over the pins on the screen to find the information necessary to identify the following structures:

A. _____
B. _____
C. _____
D. _____
E. _____
F. _____

MODULE 10 The Lymphatic System 653

EXERCISE 10.3:
Lymphatic System—Antigen-presenting Cells—Histology

SELECT TOPIC	SELECT VIEW
Antigen-presenting Cells	SEM: High Magnification

Click the **TAGS ON/OFF** button, and you will see the following image:

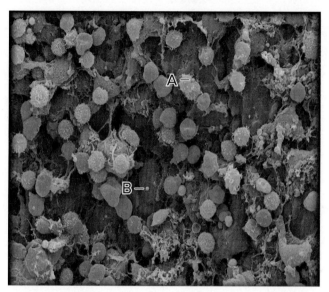

©SPL/Science Source

Mouse-over the pins on the screen to find the information necessary to identify the following structures:

A. _____

B. _____

EXERCISE 10.4:
Lymphatic System—Macrophage—Histology

SELECT TOPIC	SELECT VIEW
Macrophage	SEM: High Magnification

Click the **TAGS ON/OFF** button, and you will see the following image:

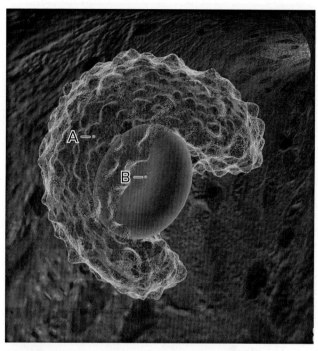

©HankGrebe/Purestock/Superstock

Mouse-over the pins on the screen to find the information necessary to identify the following structures:

A. _____

B. _____

Self-Quiz
Take this opportunity to check your progress by taking the **QUIZ**. See the **Introduction Module** for a reminder on how to access the **QUIZ** for this Study Area.

654 MODULE 10 The Lymphatic System

EXERCISE 10.5:
Lymphatic System—Lymph Node—Histology

SELECT TOPIC	SELECT VIEW
Lymph Node	LM: Low Magnification

Click the **TAGS ON/OFF** button, and you will see the following image:

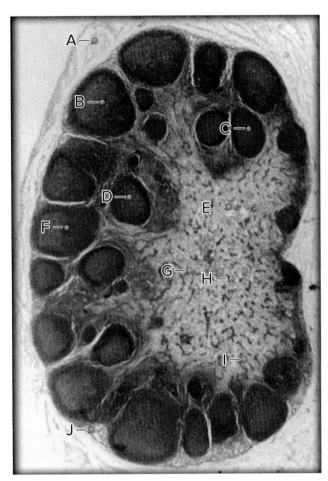

©McGraw-Hill Education/Greg Reeder

Mouse-over the pins on the screen to find the information necessary to identify the following structures:

A. _____
B. _____
C. _____
D. _____
E. _____
F. _____
G. _____
H. _____
I. _____
J. _____

CHECK POINT

Lymph Node, Histology (Low Magnification)

1. Name the dense irregular connective tissue that covers the outer surface of a lymph node.
2. Name the outer zone of the lymph node.
3. What is its function?

EXERCISE 10.6:
Lymphatic System—Lymph Node—Histology

SELECT TOPIC	SELECT VIEW
Lymph Node	LM: Medium Magnification

Click the **TAGS ON/OFF** button, and you will see the following image:

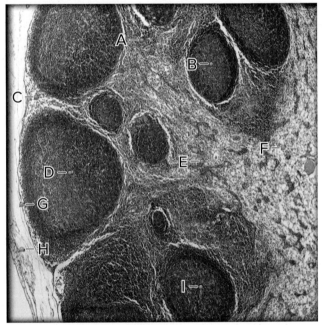

©McGraw-Hill Education/Greg Reeder

Mouse-over the pins on the screen to find the information necessary to identify the following structures:

A. _____
B. _____
C. _____
D. _____
E. _____
F. _____
G. _____
H. _____
I. _____

MODULE 10   The Lymphatic System   655

EXERCISE 10.7a:
Lymphatic System—Lymph Node (Lymphoid Nodule)—Histology

SELECT TOPIC	SELECT VIEW
Lymph Node (Lymphoid Nodule)	LM: Medium Magnification

Click the **TAGS ON/OFF** *button, and you will see the following image:*

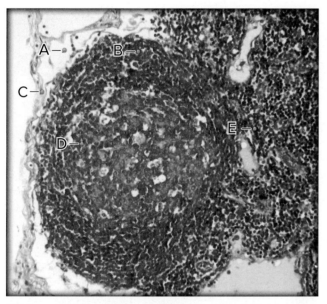

©McGraw-Hill Education/Greg Reeder

Mouse-over the pins on the screen to find the information necessary to identify the following structures:

A. _____
B. _____
C. _____
D. _____
E. _____

EXERCISE 10.7b:
Lymphatic System—Lymph Node (Lymphoid Nodule)—Histology

SELECT TOPIC	SELECT VIEW
Lymph Node (Lymphoid Nodule)	LM: High Magnification

Click the **TAGS ON/OFF** *button, and you will see the following image:*

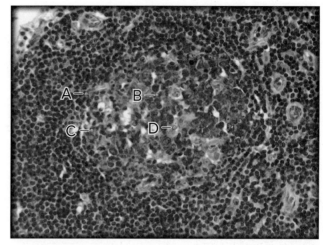

©McGraw-Hill Education/Greg Reeder

Mouse-over the pins on the screen to find the information necessary to identify the following structures:

A. _____
B. _____
C. _____
D. _____

EXERCISE 10.8:
Lymphatic System—Lymph Node (Medullary Sinus)—Histology

SELECT TOPIC	SELECT VIEW
Lymph Node (Medullary Sinus)	LM: High Magnification

Click the **TAGS ON/OFF** *button, and you will see the following image:*

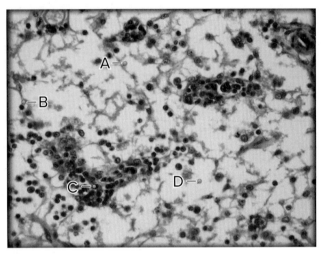

©McGraw-Hill Education/Greg Reeder

Mouse-over the pins on the screen to find the information necessary to identify the following structures:

A. _____

B. _____

C. _____

D. _____

CHECK POINT

Lymph Node, Histology

1. Name three structures that are part of a channel system that allows lymph to filter through the node.
2. Name the structure that provides the main structural support for the lymph node.
3. Name the lymph-filled space between the medullary cords.

EXERCISE 10.9:
Imaging—Pelvis—Lymphangiogram—Anterior–Posterior

SELECT TOPIC	SELECT VIEW
Pelvis	Lymphangiogram—Anterior–Posterior

Click the **TAGS ON/OFF** *button, and you will see the following image:*

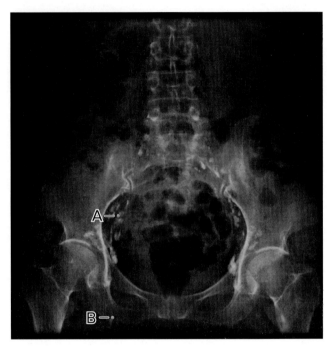

©McGraw-Hill Education

Mouse-over the pins on the screen to find the information necessary to identify the following structures:

A. _____

B. _____

IN REVIEW

What Have I Learned?

The following questions cover the material that you have just learned, the histology of a lymph node. Apply what you have learned to answer these questions on a separate piece of paper.

1. Name the inner zone of the lymph node.

2. What is its function?

3. What percent of the lymphocytes remain in the medulla while the rest leave the node via the lymph?

4. Name the location for memory B lymphocyte and plasma cell formation.

5. Name the site of B lymphocyte localization.

6. Name the structure of the lymph node that forms in response to antigenic challenge. What is its function?

7. Name the site of antibody production. What cells are responsible for this production?

EXERCISE 10.10:
Lymphatic System—Tonsils

SELECT TOPIC	SELECT VIEW
Tonsils	Lateral

Click **LAYER 2** in the **LAYER CONTROLS** window, and you will see the following image:

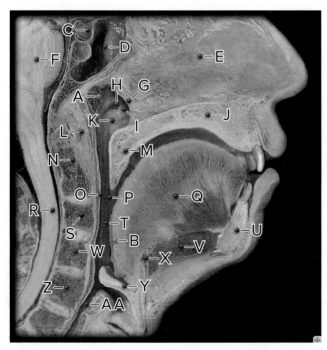

©McGraw-Hill Education

Mouse-over the pins on the screen to find the information necessary to identify the following structures:

A. _____

B. _____

Nonlymphatic System Structures (blue pins)

C. _____
D. _____
E. _____
F. _____
G. _____
H. _____
I. _____
J. _____
K. _____
L. _____
M. _____
N. _____
O. _____
P. _____
Q. _____
R. _____
S. _____
T. _____
U. _____
V. _____
W. _____
X. _____
Y. _____
Z. _____
AA. _____

CHECK POINT

Tonsils

1. Describe the pharyngeal tonsil.
2. Where is it located?
3. What is it known as when infected or inflamed?

Click **LAYER 3** in the **LAYER CONTROLS** window, and you will see the following image:

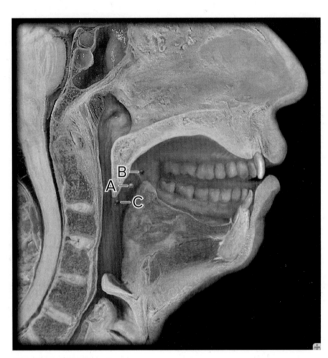
©McGraw-Hill Education

Mouse-over the pin on the screen to find the information necessary to identify the following structure:

A. _____

Nonlymphatic System Structures (blue pins)

B. _____

C. _____

CHECK POINT

Tonsils, *continued*

4. Describe the palatine tonsil.
5. Where is it located?
6. "Tonsils" most commonly refers to which tonsils?

IN REVIEW

What Have I Learned?

The following questions cover the material that you have just learned, the axillary lymph nodes and tonsils. Apply what you have learned to answer these questions on a separate piece of paper.

1. Describe the axillary lymph nodes.

2. Where are they located?

3. What is their function?

4. Describe the pharyngeal tonsil.

5. Where is it located?

6. What is it known as when infected or inflamed?

7. Describe the lingual tonsil.

8. Where is it located?

9. Describe the palatine tonsil.

10. Where is it located?

11. Which tonsils are visible through the open mouth?

12. "Tonsils" most commonly refers to which tonsils?

13. Which tonsil is the largest during childhood? What occurs to this tonsil by middle age?

Palatine Tonsil

EXERCISE 10.11a:
Lymphatic System—Palatine Tonsil—Histology

SELECT TOPIC	SELECT VIEW
Palatine Tonsil	LM: Low Magnification

Click the **TAGS ON/OFF** button, and you will see the following image:

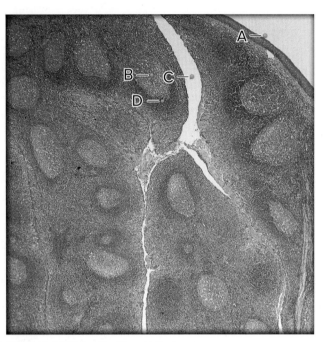

©McGraw-Hill Education/Dennis Strete

Mouse-over the pins on the screen to find the information necessary to identify the following structures:

A. _____

B. _____

C. _____

D. _____

CHECK POINT

Palatine Tonsil, Histology (Low Magnification)

1. Name the superficial structure of the palatine tonsils.
2. What is its function?
3. What is the function of the germinal center?

EXERCISE 10.11b:
Lymphatic System—Palatine Tonsil—Histology

SELECT TOPIC	SELECT VIEW
Palatine Tonsil	LM: High Magnification

Click the **TAGS ON/OFF** button, and you will see the following image:

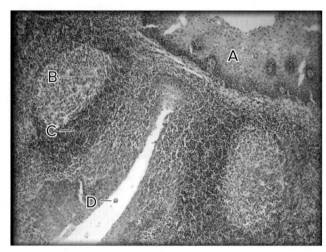

©McGraw-Hill Education/Greg Reeder

Mouse-over the pins on the screen to find the information necessary to identify the following structures:

A. _____

B. _____

C. _____

D. _____

IN REVIEW

What Have I Learned?

The following questions cover the material that you have just learned, the histology of a palatine tonsil. Apply what you have learned to answer these questions on a separate piece of paper.

1. What is the function of the lymphatic nodules?

2. Name the structures that contain discarded epithelial cells, dead white blood cells, bacteria, and debris from the oral cavity.

3. What is the function of these structures? How many are found on each tonsil?

4. Name the structure of the palatine tonsils continuous with the epithelium of the oral cavity and oropharynx.

Peyer's Patch

EXERCISE 10.12:
Lymphatic System—Peyer's Patch—Histology

SELECT TOPIC	SELECT VIEW
Peyer's Patch	LM: Medium Magnification

Click the **TAGS ON/OFF** button, and you will see the following image:

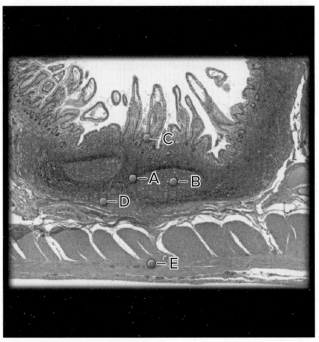

©McGraw-Hill Education/Al Telser

Mouse-over the pins on the screen to find the information necessary to identify the following structures:

A. _____
B. _____
C. _____
D. _____
E. _____

CHECK POINT

Peyer's Patch, Histology

1. Specifically, where are the Peyer's patches found?
2. What is their function?

IN REVIEW

What Have I Learned?

The following questions cover the material that you have just learned, the histology of a Peyer's patch. Apply what you have learned to answer these questions on a separate piece of paper.

1. Where are Peyer's patches located in the body?

2. What is the function of the lymphatic nodules? Why do they form?

3. What is the function of the ileum?

The Thorax

EXERCISE 10.13:
Lymphatic System—Thorax—Anterior View

SELECT TOPIC	SELECT VIEW
Thorax	Anterior

After clicking **LAYERS 2** and **3** to orient yourself to our location, click **LAYER 4** in the **LAYER CONTROLS** window, and you will see the following image:

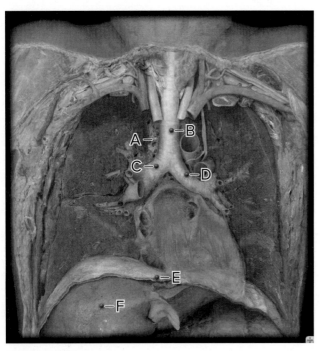

©McGraw-Hill Education

Mouse-over the pin on the screen to find the information necessary to identify the following structure:

A. _____

Nonlymphatic System Structures (blue pins)

B. _____

C. _____

D. _____

E. _____

F. _____

CHECK POINT

Thorax, Anterior View

1. What are lymph nodes?
2. The mediastinal lymph nodes are clusters found along what structures?
3. Where are lymph nodes typically found?

Click **LAYER 5** *in the* **LAYER CONTROLS** *window, and you will see the following image:*

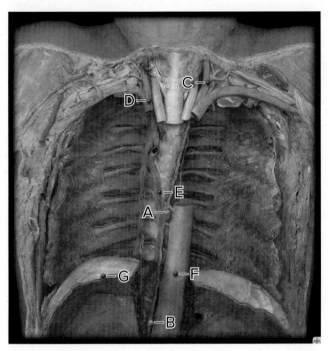

©McGraw-Hill Education

Mouse-over the pins on the screen to find the information necessary to identify the following structures:

A. _____

B. _____

Nonlymphatic System Structures (blue pins)

C. _____

D. _____

E. _____

F. _____

G. _____

CHECK POINT

Thorax, Anterior View, *continued*

4. Name the small irregular-shaped lymph sac found in the abdomen.
5. From where does it receive lymph?
6. This structure in question 4 forms the origin of which lymph duct?
7. List the areas drained by the thoracic duct.

IN REVIEW

What Have I Learned?

The following questions cover the material that you have just learned, the lymphatic system of the thorax. Apply what you have learned to answer these questions on a separate piece of paper.

1. List the areas drained by the thoracic duct.

2. Lymph nodes are _____.

3. Lymph nodes are typically found in _____.

4. Where are these clusters typically found?

5. Name the structure that forms the origin of the thoracic duct.

664 MODULE 10 The Lymphatic System

EXERCISE 10.14:
Lymphatic System—Breast and Axillary Nodes—Female

SELECT TOPIC	SELECT VIEW
Breast and Axillary Nodes—Female ▶	Anterior

After clicking **LAYERS 1** *through* **3** *to orient yourself to our location, click* **LAYER 4** *in the* **LAYER CONTROLS** *window, and you will see the following image:*

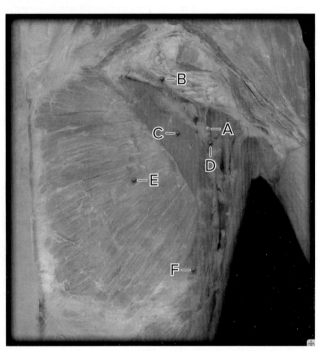

©McGraw-Hill Education

Mouse-over the pin on the screen to find the information necessary to identify the following structure:

A. _____

Nonlymphatic System Structures (blue pins)

B. _____
C. _____
D. _____
E. _____
F. _____

CHECK POINT
Breast and Axillary Nodes—Female

1. Describe the axillary lymph nodes.
2. Where are they located?
3. What is their function?

The Spleen

EXERCISE 10.15:
Lymphatic System—Spleen—Anterior View

SELECT TOPIC	SELECT VIEW
Spleen ▶	Anterior

Click **LAYER 1** *in the* **LAYER CONTROLS** *window, and you will see the following image:*

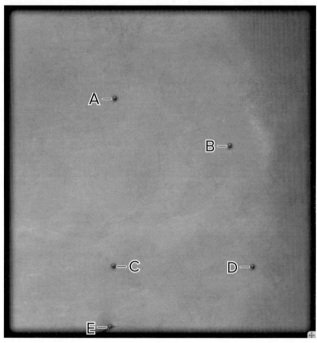

©McGraw-Hill Education

Mouse-over the pins on the screen to find the information necessary to identify the following nonlymphatic system structures:

A. _____
B. _____
C. _____
D. _____
E. _____

Click **LAYER 2** *in the* **LAYER CONTROLS** *window, and you will see the following image:*

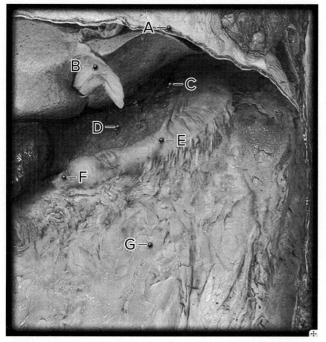

©McGraw-Hill Education

Mouse-over the pins on the screen to find the information necessary to identify the following nonlymphatic system structures:

A. _____
B. _____
C. _____
D. _____
E. _____
F. _____
G. _____

Click **LAYER 3** *in the* **LAYER CONTROLS** *window, and you will see the following image:*

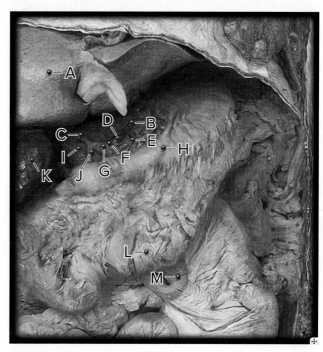

©McGraw-Hill Education

Mouse-over the pins on the screen to find the information necessary to identify the following nonlymphatic system structures:

A. _____
B. _____
C. _____
D. _____
E. _____
F. _____
G. _____
H. _____
I. _____
J. _____
K. _____
L. _____
M. _____

MODULE 10 The Lymphatic System

Click **LAYER 4** *in the* **LAYER CONTROLS** *window, and you will see the following image:*

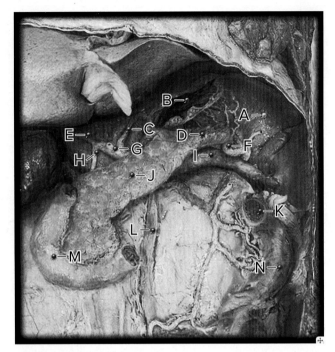

©McGraw-Hill Education

Mouse-over the pin on the screen to find the information necessary to identify the following structure:

A. _____

Nonlymphatic System Structures (blue pins)

B. _____
C. _____
D. _____
E. _____
F. _____
G. _____
H. _____
I. _____
J. _____
K. _____
L. _____
M. _____
N. _____

CHECK POINT

Spleen, Anterior View

1. Name the location of the spleen.
2. What structure of the respiratory system does it contact?
3. Describe the spleen.

Click **LAYER 5** *in the* **LAYER CONTROLS** *window, and you will see the following image:*

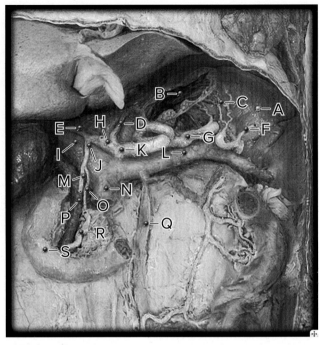

©McGraw-Hill Education

Mouse-over the pin on the screen to find the information necessary to identify the following structure:

A. _____

Nonlymphatic System Structures (blue pins)

B. _____
C. _____
D. _____
E. _____
F. _____
G. _____
H. _____
I. _____
J. _____
K. _____
L. _____
M. _____
N. _____
O. _____
P. _____
Q. _____
R. _____
S. _____

MODULE 10   The Lymphatic System   667

CHECK POINT

Spleen, Anterior View, *continued*

Name four functions of the spleen:
4.
5.
6.
7.

Click **LAYER 6** *in the* **LAYER CONTROLS** *window, and you will see the following image:*

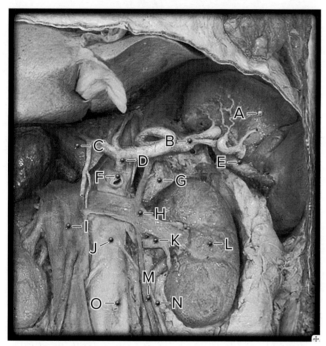

©McGraw-Hill Education

Mouse-over the pin on the screen to find the information necessary to identify the following structure:

A. _____

Nonlymphatic System Structures (blue pins)

B. _____
C. _____
D. _____
E. _____
F. _____
G. _____
H. _____
I. _____
J. _____
K. _____
L. _____

M. _____
N. _____
O. _____

CHECK POINT

Spleen, Anterior View, *continued*

8. Name the structure that protects the spleen posteriorly.

EXERCISE 10.16:

Lymphatic System—Spleen—Histology

SELECT TOPIC	SELECT VIEW
Spleen	LM: Medium Magnification

Click the **TAGS ON/OFF** *button, and you will see the following image:*

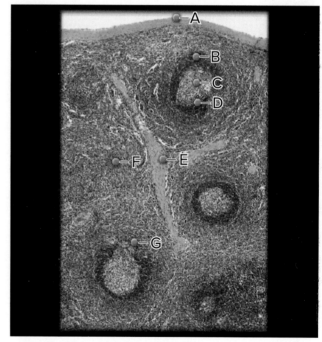

©McGraw-Hill Education/Al Telser

Mouse-over the pins on the screen to find the information necessary to identify the following structures:

A. _____
B. _____
C. _____
D. _____
E. _____
F. _____
G. _____

668 MODULE 10 The Lymphatic System

CHECK POINT
Spleen, Histology

1. Name the outer covering of the spleen. This structure consists of what tissues?
2. What structure provides the main structural support for the spleen?
3. Name the functions of the red pulp.

EXERCISE 10.17:
Lymphatic System—Spleen, Lymphoid Nodule—Histology

SELECT TOPIC	SELECT VIEW
Spleen Lymphoid Nodule	LM: High Magnification

Click the **TAGS ON/OFF** button, and you will see the following image:

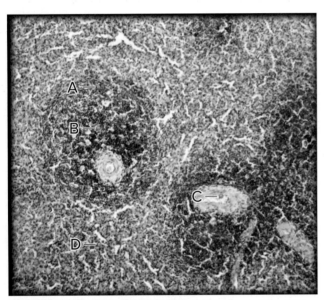

©McGraw-Hill Education/Greg Reeder

Mouse-over the pins on the screen to find the information necessary to identify the following structures:

A. _____
B. _____
C. _____
D. _____

EXERCISE 10.18:
Lymphatic System—Spleen (Unstimulated White Pulp)

SELECT TOPIC	SELECT VIEW
Spleen (Unstimulated White Pulp)	LM: Medium Magnification

Click the **TAGS ON/OFF** button, and you will see the following image:

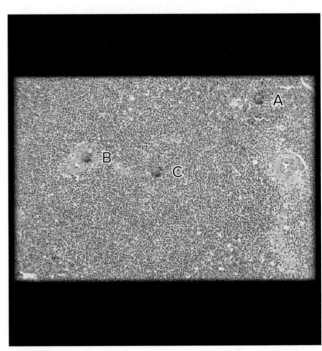

©McGraw-Hill Education/Dennis Strete

Mouse-over the pins on the screen to find the information necessary to identify the following structures:

A. _____
B. _____
C. _____

CHECK POINT
Spleen, Histology

1. Name the source of arterial blood to the white pulp.

IN REVIEW

What Have I Learned?

The following questions cover the material that you have just learned, the spleen. Apply what you have learned to answer these questions on a separate piece of paper.

1. Name the largest lymphatic organ.

2. Name the location of this organ.

3. What structure of the respiratory system does it contact?

4. Describe the largest lymphatic organ.

5. Name four functions of this organ:

6. Name the structure that protects the spleen posteriorly.

7. What is the function of the white pulp? What percent of the total mass of the spleen is white pulp?

8. Name the four structures that make up the white pulp.

9. Name the location for memory B lymphocyte and plasma cell formation. How are they affected by age?

10. What tissue/structure makes up the bulk of the spleen?

11. This tissue/structure is composed of _____.

The Thymus

EXERCISE 10.19:
Lymphatic System—Thymus—Adult—Anterior View

SELECT TOPIC	SELECT VIEW
Thymus—Adult	Anterior

Click **LAYER 1** in the **LAYER CONTROLS** window, and you will see the following image:

Mouse-over the pins on the screen to find the information necessary to identify the following nonlymphatic system structures:

A. _____

B. _____

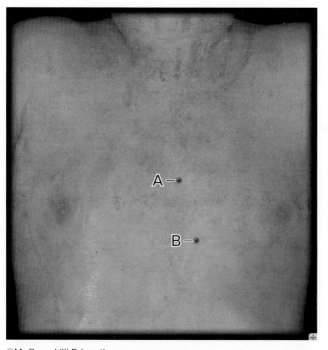

©McGraw-Hill Education

Click **LAYER 2** *in the* **LAYER CONTROLS** *window, and you will see the following image:*

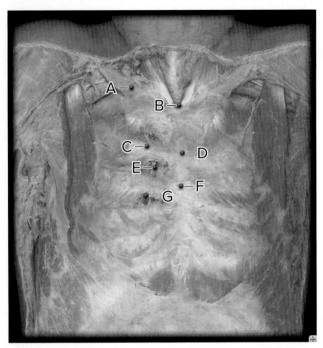

©McGraw-Hill Education

Mouse-over the pins on the screen to find the information necessary to identify the following nonlymphatic system structures:

A. _____
B. _____
C. _____
D. _____
E. _____
F. _____
G. _____

Click **LAYER 3** *in the* **LAYER CONTROLS** *window and you will see the following image:*

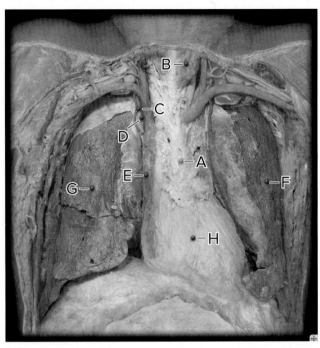

©McGraw-Hill Education

Mouse-over the pin on the screen to find the information necessary to identify the following structure:

A. _____

Nonlymphatic System Structures (blue pins)

B. _____
C. _____
D. _____
E. _____
F. _____
G. _____
H. _____

CHECK POINT

Thymus—Adult, Anterior View

1. Where is the thymus located?
2. Name two functions of the adult thymus.

EXERCISE 10.20:
Lymphatic System—Thymus—Fetus—Anterior View

SELECT TOPIC	SELECT VIEW
Thymus—Fetus	Anterior

Click **LAYER 2** in the **LAYER CONTROLS** *window, and you will see the following image:*

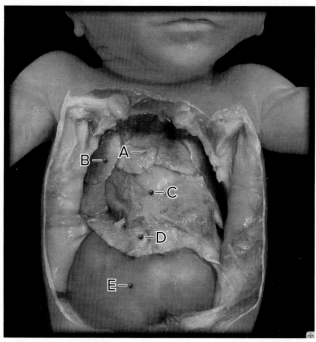

©McGraw-Hill Education

Mouse-over the pin on the screen to find the information necessary to identify the following structure:

A. _____

Nonlymphatic System Structures (blue pins)

B. _____
C. _____
D. _____
E. _____

CHECK POINT

Thymus—Fetus, Anterior View

1. What is the function of thymopoietin and thymosins?
2. What function of the thymus occurs primarily in young individuals?
3. What occurs to the thymus during adolescence?

EXERCISE 10.21a:
Lymphatic System—Fetal Thymus—Histology

SELECT TOPIC	SELECT VIEW
Fetal Thymus	LM: Low Magnification

Click the **TAGS ON/OFF** button, *and you will see the following image:*

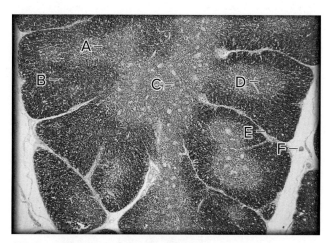

©McGraw-Hill Education/Greg Reeder

Mouse-over the pins on the screen to find the information necessary to identify the following structures:

A. _____
B. _____
C. _____
D. _____
E. _____
F. _____

CHECK POINT

Thymus, Histology (Low Magnification)

1. Where is the cortex of the thymus located? What is its function?
2. Where is the medulla of the thymus located? What is its function?
3. What is the main subdivision of the thymus?

EXERCISE 10.21b:
Lymphatic System—Fetal Thymus—Histology

SELECT TOPIC	SELECT VIEW
Fetal Thymus	LM: Medium Magnification

Click the **TAGS ON/OFF** *button, and you will see the following image:*

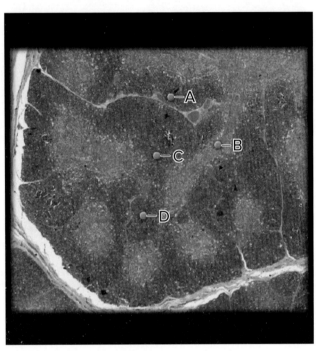

©McGraw-Hill Education/Al Telser

Mouse-over the pins on the screen to find the information necessary to identify the following structures:

A. _____

B. _____

C. _____

D. _____

CHECK POINT

Thymus, Histology (Medium Magnification)

1. Name the structures of the thymus that increase in number with age.
2. What is their function?
3. Where in the thymus are these structures located?

Self-Quiz
Take this opportunity to check your progress by taking the **QUIZ**. See the **Introduction Module** for a reminder on how to access the **QUIZ** for this Study Area.

EXERCISE 10.22:
Lymphatic System—Thymus (Hassall's Corpuscle)—Histology

SELECT TOPIC	SELECT VIEW
Thymus (Hassall's Corpuscle)	LM: High Magnification

Click the **TAGS ON/OFF** *button, and you will see the following image:*

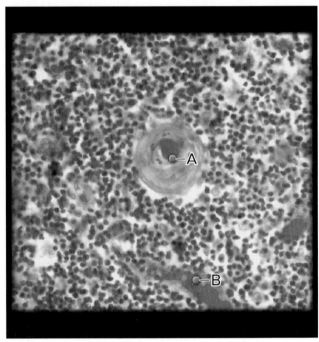

©McGraw-Hill Education/Al Telser

Mouse-over the pins on the screen to find the information necessary to identify the following structures:

A. _____

B. _____

Clinical Application

SELECT ANIMATION: HIV (3D) — PLAY

After viewing the animation, answer the following questions:

1. HIV—or _____ _____ _____—is the _____ that eventually causes _____—or _____ _____ _____.

2. Over time, how does HIV infection affect the body? What is the result?

3. How is HIV transmitted?

4. Which cells defend the body from infection?

5. Which cells are part of a direct immune response against infected cells and tumor cells.

6. Which are the certain cells that are targeted by HIV during HIV infection?

7. What is the function of these targeted cells?

8. Where does HIV replicate? What is the result?

9. How does the body become defenseless? Why?

10. When a CD4 count drops below _____, a person is considered to have _____.

11. This can take as long as _____ _____ from the time a person is infected with HIV.

12. What is caused by this low CD4 count?

13. Patients can die from . . .

14. Is there a cure of vaccine for HIV infection or AIDS?

15. Describe the treatment for HIV infection.

16. _____ _____, such as _____ _____ _____ _____—or _____—combines _____—_____ medications in a daily _____.

17. How does HAART attack HIV?

18. One drug in the cocktail—a _____ _____—blocks HIVE from binding to the _____.

19. A second drug—_____ _____ _____—blocks it from _____.

20. And a third drug—a _____ _____—prevents _____ from _____ a new _____.

21. Though it is not possible to completely eliminate HIV from the body, _____ _____ _____ and significantly reduces _____ from _____—_____ _____.

22. Besides HAART, what other treatment options exist for AIDS? Give examples.

23. Why are blood tests performed regularly?

IN REVIEW

What Have I Learned?

The following questions cover the material that you have just learned, the histology of the thymus. Apply what you have learned to answer these questions on a separate piece of paper.

1. What is the thymus? Where is it located? What is its function?

2. How does it change from birth to adulthood?

3. Name the subdivisions of the thymic lobe. What is its function?

4. What is the function of the septa/trabeculae?

MODULE 11

The Respiratory System

Overview: The Respiratory System

Take a deep breath. Now exhale. What has just occurred? You have taken in much-needed oxygen and breathed out toxic carbon dioxide. This is foundational to all of your body processes as you maintain homeostasis. Your cells require oxygen for metabolic reactions and produce carbon dioxide as a waste product of that metabolism. Without the respiratory system, metabolism, and thus homeostasis, would be impossible.

But, the function of the respiratory system doesn't stop there. It is also involved in regulating the body's acid–base balance, blood gases, and other homeostatic controls of the circulatory system. Like so many of our organ systems, the respiratory system doesn't exist in a vacuum, but works in coordination with the circulatory and urinary systems.

We will begin with an overview of the respiratory system. Next, we will take a regional view of the respiratory structures, followed by a detailed look at these structures.

From the **Home** *screen, click the drop-down box on the* **MODULE** *menu.*

From the systems listed, click on **Respiratory**.

SELECT ANIMATION
Respiratory System Overview
PLAY

After viewing the animation, answer the following questions:

1. What are the two functions of the respiratory system?

2. Name the structures of the upper respiratory tract.

3. Name the structures of the lower respiratory tract.

4. What effect does the nasal cavity have on inhaled air?

5. Name the structure shared by both the respiratory and digestive systems.

6. Name the structure that contains the vestibular and vocal folds. What protective function do these folds have?

7. The vocal folds are also known as the _____ _____ _____. Why?

8. Name the organ that maintains an open passageway to and from the lungs. What particular structure helps to keep this passageway open?

9. This passageway divides into two _____ _____, which upon entering the lungs, continue to divide into smaller _____ _____ _____ until they ultimately divide into _____ _____.

10. Each _____ _____ divides repetitively to form _____ _____, _____ _____, and _____ _____.

11. What structure serves as the site for gas exchange? These rounded structures are surrounded by _____ _____.

12. How do oxygen and carbon dioxide move between the blood and the alveoli?

676 MODULE 11 The Respiratory System

Self-Quiz

Take this opportunity to check your progress by taking the **QUIZ**. See the **Introduction Module** for a reminder on how to access the **QUIZ** for this Study Area.

[SELECT ANIMATION — Respiration (3D) — PLAY]

After viewing the animation, answer the following questions:

1. The respiratory system regulates _____ and _____ _____ levels within _____ _____.

2. What is included in respiration?

3. Inhalation allows . . . (be complete in your answer)

4. What are alveoli? What occurs in the alveoli?

5. Alveoli move freely when . . .

6. What are capillaries? Where are they located?

7. During gas exchange, _____ enters and _____ exits the _____ via the _____ _____.

8. Once _____ molecules move from the _____ into the _____, they _____ into the _____ and enter the _____ _____ _____ or _____.

9. _____ contain millions of soluble _____, called _____.

10. _____ contains four _____ _____, each capable of binding to one _____ of _____.

11. Once one _____ of _____ binds to one of the _____, the other _____ bind _____ more readily.

12. _____ and bound _____ flows through the _____ _____ _____ to _____ within _____.

13. Upon arrival, _____ _____ _____ of the _____ promotes _____ _____.

14. _____ _____ within cells produces _____ _____ _____ as a _____ _____.

15. _____ _____ exits the _____ and _____ and is converted into _____ within the _____.

16. Converting _____ _____ to _____ releases _____ _____ that decrease _____ _____ for _____, freeing the _____ to be delivered to _____ _____.

17. After delivering _____ to the _____, the _____ _____ rich _____ returns to the _____ through the _____ _____ and then to the _____ _____.

18. Inside each _____ the _____ conversion is _____, recreating _____ _____, which _____ across the _____ into the _____ of the _____, and it is excreted out of the body.

MODULE 11 The Respiratory System 677

The Upper Respiratory System

EXERCISE 11.1:
Respiratory System—Upper Respiratory—Lateral View

SELECT TOPIC	SELECT VIEW
Upper Respiratory ▶	Lateral

Click **LAYER 1** in the **LAYER CONTROLS** window, and you will see the following image:

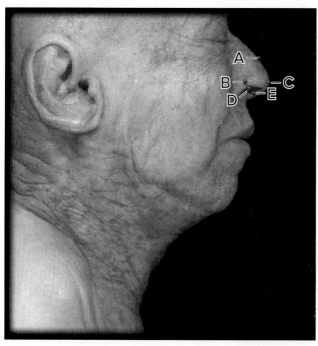

©McGraw-Hill Education

Click **LAYER 2** in the **LAYER CONTROLS** window, and you will see the following image:

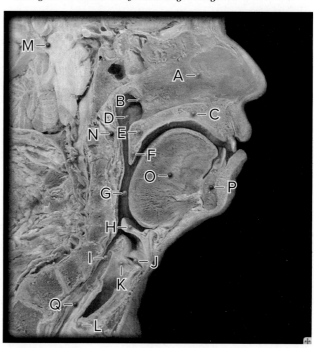

©McGraw-Hill Education

Mouse-over the pins on the screen to find the information necessary to identify the following structures:

A. nasal septum with mucosa
B. choana
C. hard palate
D. nasopharynx
E. soft palate
F. uvula
G. oropharynx
H. epiglottis
I. laryngopharynx
J. vocal fold
K. larynx
L. trachea

Nonrespiratory System Structures (blue pins)

M. brain
N. atlas
O. tongue
P. mandible
Q. esophagus

Mouse-over the pins on the screen to find the information necessary to identify the following structures:

A. dorsum of nose
B. ala of nose
C. apex of nose
D. external naris
E. nasal septum

CHECK POINT

Upper Respiratory, Lateral View

1. Name the external opening of the nose.
2. Name the structure that separates both of the openings of the nose.
3. Name the midline dorsal aspect of the nose.

MODULE 11 The Respiratory System

CHECK POINT

Upper Respiratory, Lateral View, *continued*

4. Name the paired posterior nasal openings.
5. Name two structures contained in the nasopharynx.
6. Name the structure that closes over the laryngeal inlet when swallowing.

CHECK POINT

Upper Respiratory, Lateral View, *continued*

7. Name the expanded area inside of the nares. What does the Latin term mean?
8. The mucosa of what structure has both respiratory and olfactory parts?
9. Name the two shelflike projections of bone covered with mucosa and part of the ethmoid bone.

Click **LAYER 3** in the **LAYER CONTROLS** *window, and you will see the following image:*

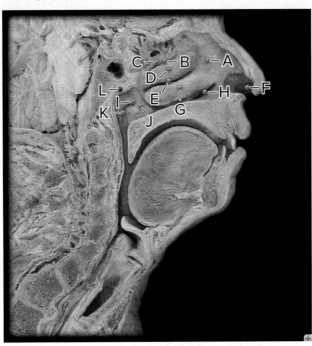

©McGraw-Hill Education

Click **LAYER 4** in the **LAYER CONTROLS** *window, and you will see the following image:*

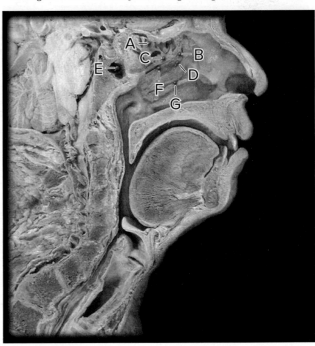

©McGraw-Hill Education

Mouse-over the pins on the screen to find the information necessary to identify the following structures:

A. nasal cavity
B. superior nasal meatus
C. superior nasal concha
D. middle nasal concha
E. middle nasal meatus
F. nasal vestibule
G. inferior nasal concha
H. inferior nasal meatus
I. opening of auditory tube
J. torus tubarius
K. salpingopharyngeal fold

Nonrespiratory System Structure (blue pin)

L. pharyngeal tonsil

Mouse-over the pins on the screen to find the information necessary to identify the following structures:

A. ethmoidal cells
B. ethmoidal bulla
C. superior nasal concha
D. semilunar hiatus
E. sphenoidal sinus
F. middle nasal concha
G. inferior nasal concha

CHECK POINT

Upper Respiratory, Lateral View, *continued*

10. Name the paranasal sinuses.
11. Name the narrow curved gap that contains the openings of the frontal and maxillary sinuses, plus the anterior ethmoid air cells.
12. Name the mucous membrane-lined cavity in the body of the sphenoid bone. Where does it drain?

EXERCISE 11.2:
Respiratory System—Nasal Cavity—Coronal View

SELECT TOPIC	SELECT VIEW
Nasal Cavity	Coronal

Click **LAYER 1** in the **LAYER CONTROLS** window, and you will see the following image:

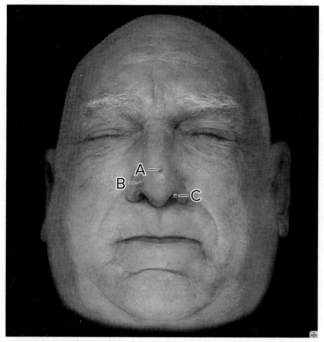

©McGraw-Hill Education

Mouse-over the pins on the screen to find the information necessary to identify the following structures:

A. _____
B. _____
C. _____

CHECK POINT
Nasal Cavity, Coronal View

1. Name the lateral wall of the nostril.
2. What tissue composes this structure?
3. Define ala.

Click **LAYER 2** in the **LAYER CONTROLS** window, and you will see the following image:

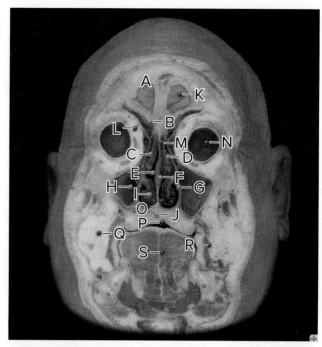

©McGraw-Hill Education

Mouse-over the pins on the screen to find the information necessary to identify the following structures:

A. nasal septum
B. choana
C. hard palate
D. nasopharynx
E. _____
F. _____
G. _____
H. _____
I. _____
J. _____

Nonrespiratory System Structures (blue pins)

K. _____
L. _____
M. _____
N. _____
O. _____
P. _____
Q. _____
R. _____
S. _____

CHECK POINT

Nasal Cavity, Coronal View, *continued*

4. Name a structure commonly deviated from the midline, impacting airflow.
5. Name the structure containing the opening of the nasolacrimal duct.
6. Name the locations where air is warmed, filtered, and humidified.

Click **LAYER 3** *in the* **LAYER CONTROLS** *window, and you will see the following image:*

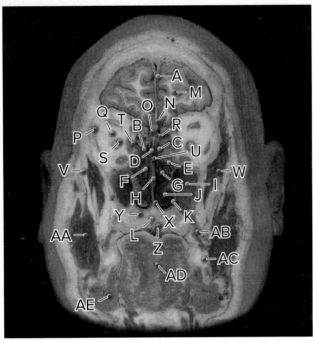

©McGraw-Hill Education

Mouse-over the pins on the screen to find the information necessary to identify the following structures:

A. _____
B. _____
C. _____
D. _____
E. _____
F. _____
G. _____
H. _____
I. _____
J. _____
K. _____
L. _____

Nonrespiratory System Structures (blue pins)

M. _____
N. _____
O. _____
P. _____
Q. _____
R. _____
S. _____
T. _____
U. _____
V. _____
W. _____
X. _____
Y. _____
Z. _____
AA. _____
AB. _____
AC. _____
AD. _____
AE. _____

CHECK POINT

Nasal Cavity, Coronal View, *continued*

7. Name the largest paranasal sinus.
8. This sinus is a common site of _____.
9. Where is this sinus located?

Click **LAYER 4** *in the* **LAYER CONTROLS** *window, and you will see the following image:*

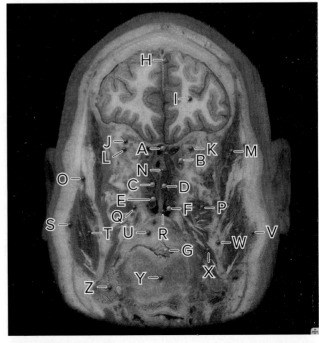

©McGraw-Hill Education

Mouse-over the pins on the screen to find the information necessary to identify the following structures:

A. _____
B. _____
C. _____
D. _____
E. _____
F. _____
G. _____

Nonrespiratory System Structures (blue pins)

H. _____
I. _____
J. _____
K. _____
L. _____
M. _____
N. _____
O. _____
P. _____
Q. _____
R. _____
S. _____
T. _____
U. _____
V. _____
W. _____
X. _____
Y. _____
Z. _____

CHECK POINT

Nasal Cavity, Coronal View, *continued*

10. Name a shelflike projection of the ethmoid bone.
11. What is its function?
12. Name a shelflike projection that is not a part of, but does articulate with, the ethmoid bone.

Click **LAYER 5** *in the* **LAYER CONTROLS** *window, and you will see the following image:*

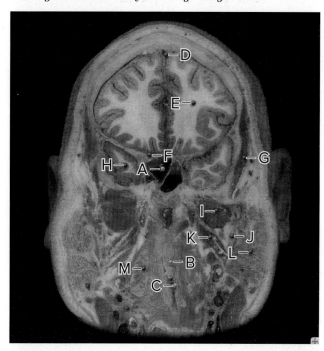

©McGraw-Hill Education

Mouse-over the pins on the screen to find the information necessary to identify the following structures:

A. _____
B. _____
C. _____

Nonrespiratory System Structures (blue pins)

D. _____
E. _____
F. _____
G. _____
H. _____
I. _____
J. _____
K. _____
L. _____
M. _____

CHECK POINT

Nasal Cavity, Coronal View, *continued*

13. Describe the uvula.
14. Where is it located?
15. Name the subdivision of the pharynx that is part of both the respiratory and digestive tracts.

682　MODULE 11　The Respiratory System

EXERCISE 11.3a:
Imaging—Upper Respiratory System

SELECT TOPIC: Upper Respiratory ▸ **SELECT VIEW:** MRI: Coronal

Click the **TAGS ON/OFF** button, and you will see the following image:

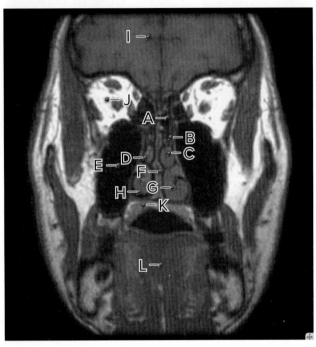

©McGraw-Hill Education

Mouse-over the pins on the screen to find the information necessary to identify the following nonrespiratory system structures:

A. _____
B. _____
C. _____
D. _____
E. _____
F. _____
G. _____
H. _____
I. _____
J. _____
K. _____
L. _____

EXERCISE 11.3b:
Imaging—Upper Respiratory System

SELECT TOPIC: Upper Respiratory ▸ **SELECT VIEW:** MRI: Sagittal

Click the **TAGS ON/OFF** button, and you will see the following image:

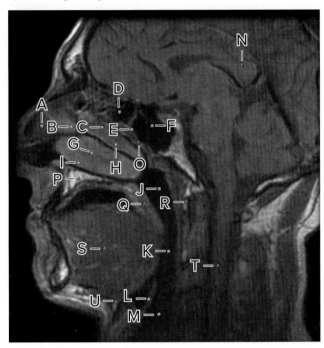

©McGraw-Hill Education

Mouse-over the pins on the screen to find the information necessary to identify the following structures:

A. _____
B. _____
C. _____
D. _____
E. _____
F. _____
G. _____
H. _____
I. _____
J. _____
K. _____
L. _____
M. _____

Nonrespiratory System Structures (blue pins)

N. _____

O. _____

P. _____

Q. _____

R. _____

S. _____

T. _____

U. _____

Self-Quiz

Take this opportunity to check your progress by taking the **QUIZ**. See the **Introduction Module** for a reminder on how to access the **QUIZ** for this Study Area.

IN REVIEW

What Have I Learned?

The following questions cover the material that you have just learned, the upper respiratory system. Apply what you have learned to answer these questions on a separate piece of paper.

1. What is the Latin term for the vestibule? What does this word mean?

2. Name the three midline structures that make up the nasal septum.

3. Name the largest paranasal sinus.

4. Name the portion of the throat posterior to the oral cavity. What lymphatic organs are located there?

5. Name the structure of the soft palate that elevates during swallowing to prevent food from entering the nasopharynx.

6. Name a structure commonly deviated from the midline, impacting airflow.

7. Name the shelflike projection of mucosa-covered bone that is a separate bone.

8. Name the narrow space that contains the opening of the nasolacrimal duct.

9. What is the threefold purpose for increasing the surface area of the nasal cavity with conchae and meatus?

10. Name the collection of lymphatic tissue in the nasopharynx.

11. Name the structure of the nasopharynx that represents the anterior end of the narrow duct between the middle ear and the pharynx. Name the structure of the nasopharynx that surrounds it.

The Lower Respiratory System

EXERCISE 11.4:
Respiratory System—Lower Respiratory—Anterior View

SELECT TOPIC: Lower Respiratory ▸ SELECT VIEW: Anterior

Click **LAYER 1** *in the* **LAYER CONTROLS** *window, and you will see the following image:*

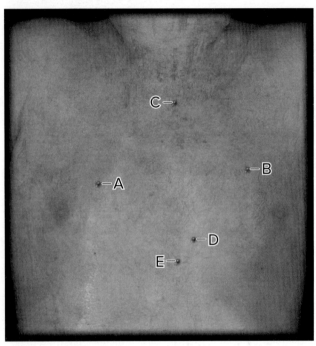

©McGraw-Hill Education

Mouse-over the pins on the screen to find the information necessary to identify the following nonrespiratory system structures:

A. surface projection of right lung
B. surface projection of left lung

Nonrespiratory System Structures (blue pins)

C. surface projection of jugular notch of sternum
D. surface projection of heart
E. surface projection of xiphoid process

Click **LAYER 2** *in the* **LAYER CONTROLS** *window, and you will see the following image:*

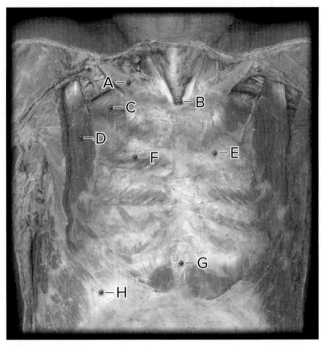

©McGraw-Hill Education

Mouse-over the pins on the screen to find the information necessary to identify the following nonrespiratory system structures:

A. Clavicle
B. Jugular notch of sternum
C. external intercostal m. & membrane
D. pectoralis minor m.
E. costal cartilage
F. internal intercostal m.
G. xiphoid process
H. external abdominal oblique m.

Click **LAYER 3** *in the* **LAYER CONTROLS** *window, and you will see the following image:*

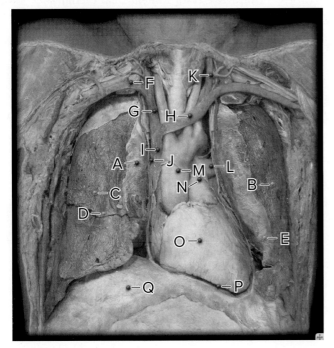

Mouse-over the pins on the screen to find the information necessary to identify the following structures:

A. parietal pleura
B. left lung
C. right lung
D. horizontal fissure of right lung
E. cardiac notch

Nonrespiratory System Structures (blue pins)

F. subclavian v.
G. right brachiocephalic v.
H. left brachiocephalic v.
I. superior vena cava
J. phrenic n.
K. internal jugular v.
L. left pulmonary a.
M. aorta
N. pulmonary trunk
O. heart
P. parietal layer of serous pericardium
Q. diaphragm

CHECK POINT

Lower Respiratory, Anterior View

1. How many lobes are in the left lung? What are they?
2. How many lobes are in the right lung? What are they?
3. Why does the left lung have one fewer lobe than the right lung?

Click **LAYER 4** *in the* **LAYER CONTROLS** *window, and you will see the following image:*

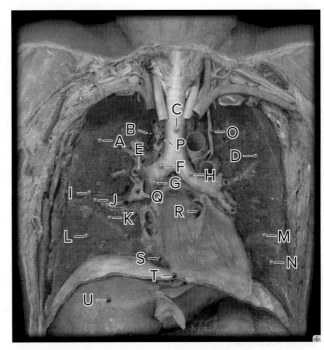

Mouse-over the pins on the screen to find the information necessary to identify the following structures:

A. upper lobe of right lung
B. visceral pleura
C. trachea
D. upper lobe of left lung
E. segmental bronchus and branches
F. right main bronchus
G. lobar bronchus
H. left main bronchus
I. horizontal fissure of right lung
J. middle lobe of right lung
K. oblique fissure of right lung
L. lower lobe of right lung
M. oblique fissure of left lung
N. lower lobe of left lung

Nonrespiratory System Structures (blue pins)

O. vagus n. (CN X)
P. mediastinal lymph nodes
Q. pulmonary a
R. pulmonary vv
S. inferior vena cava
T. diaphragm
U. liver

CHECK POINT

Lower Respiratory, Anterior View, *continued*

4. Name the structures formed by the bifurcation of the trachea. Which of the two is where foreign bodies that enter the trachea tend to pass?
5. Name the branches emanating from the main bronchi. How many are found in each lung?
6. Name the branches emanating from the bronchi in question 5.

Click **LAYER 5** *in the* **LAYER CONTROLS** *window, and you will see the following image:*

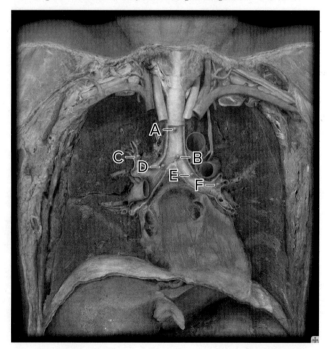

©McGraw-Hill Education

Mouse-over the pins on the screen to find the information necessary to identify the following structures:

A. trachea
B. carina
C. segmental bronchus & branches
D. right main bronchus
E. left main bronchus
F. lobar bronchus

SELECT ANIMATION — **Thoracic Cavity Dimensional Changes** — PLAY

After viewing the animation, answer the following questions:

1. Another term for breathing is _____ _____.

2. Both _____ and _____ result from changes in _____ of the _____ _____.

3. What structures drive these changes?

4. During inspiration, the thoracic cavity increases in _____. Why?

5. What regulates the length of the thoracic cavity?

6. How is the length of the thoracic cavity increased during inspiration?

7. What changes occur with the diaphragm and the thoracic cavity during expiration?

8. What regulates the depth and width of the thoracic cavity? How?

9. How does the elevation of the ribs affect the thoracic cavity width? This motion is similar to . . .

10. What effect does this elevation of the ribs have on the sternum? What effect does this have on the thoracic cavity depth?

11. How does the movement of the sternum and ribs facilitate inspiration?

Self-Quiz

Take this opportunity to check your progress by taking the **QUIZ**. See the **Introduction Module** for a reminder on how to access the **QUIZ** for this Study Area.

MODULE 11 The Respiratory System 687

SELECT ANIMATION
Partial Pressure PLAY

After viewing the animation, answer the following questions:

1. What is the partial pressure of oxygen in fresh air entering the lungs?

2. What effect does moisture in the lungs have on this number?

3. What is the partial pressure of carbon dioxide in fresh air entering the lungs?

4. What effect does carbon dioxide delivered to the lungs from the blood have on this number?

5. Describe the direction of diffusion of oxygen and carbon dioxide in the alveoli.

6. This occurs because of . . .

7. This occurs at the _____ ends of the _____ _____.

8. What occurs as a result of diffusion at the venous ends of the pulmonary capillaries?

9. With no differences in partial pressure, . . .

10. How do oxygen and carbon dioxide diffuse into/out of the tissue capillaries?

11. This occurs because of . . .

12. What occurs at the venous ends of the tissue capillaries?

13. The blood now carries the _____ and _____ _____ to the _____.

14. In the body, all of these exchanges occur _____.

Self-Quiz
Take this opportunity to check your progress by taking the **QUIZ**. See the **Introduction Module** for a reminder on how to access the **QUIZ** for this Study Area.

SELECT ANIMATION
Alveolar Pressure Changes PLAY

After viewing the animation, answer the following questions:

1. At the end of expiration, . . .

2. Therefore, . . .

3. Inspiration begins with _____ of the _____ to . . .

4. This results in . . .

5. The increased alveolar volume causes a _____ in _____ _____ below _____ _____ _____ and _____ flows _____ _____ _____.

6. At the end of inspiration, . . .

7. Airflow into the lungs causes . . .

8. The pressure becomes equal, so . . .

9. During expiration, the _____ of the _____ _____ as the _____ _____, and the _____ and the _____ _____.

10. This results in a _____ in _____ _____ and an _____ in _____ _____.

11. The _____ _____ _____ is now _____ than _____ _____ _____ _____, so air flows _____ of the lungs.

12. Air continues to flow out of the lungs until . . .

Self-Quiz
Take this opportunity to check your progress by taking the **QUIZ**. See the **Introduction Module** for a reminder on how to access the **QUIZ** for this Study Area.

688 MODULE 11 The Respiratory System

SELECT ANIMATION
Diffusion Across Respiratory Membrane — PLAY

After viewing the animation, answer the following questions:

1. In the lungs, gas exchange takes place . . .

2. What are alveoli? Where are they located?

3. What is the diameter of an alveolus? How many are in each lung?

4. What are the two types of specialized cells in the wall of the alveoli?

5. The cells that form 90 percent of the alveolar wall are . . .

6. Describe these cells.

7. Name and describe the second type of cells.

8. What do these cells secrete? What is the function of this secretion?

9. _____ _____ form a network around each alveolus.

10. Name the thin structure that separates the capillary blood from the air in the alveolus.

11. What is this thin structure composed of?

12. This structure has a thickness of only _____, which facilitates . . .

13. What causes the diffusion of gases across this membrane? Explain it for both oxygen and carbon dioxide.

Self-Quiz
Take this opportunity to check your progress by taking the **QUIZ**. See the **Introduction Module** for a reminder on how to access the **QUIZ** for this Study Area.

EXERCISE 11.5a:
Imaging—Lower Respiratory System

SELECT TOPIC: Lower Respiratory → **SELECT VIEW**: X ray—Posterior–Anterior

Click the **TAGS ON/OFF** *button, and you will see the following image:*

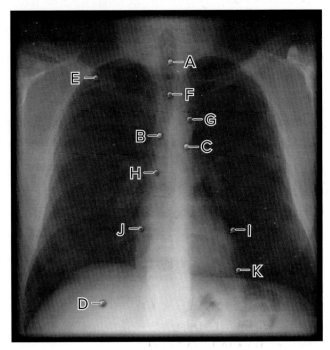

©McGraw-Hill Education

Mouse-over the pins on the screen to find the information necessary to identify the following structures:

A. _____
B. _____
C. _____
D. _____

Nonrespiratory System Structures (blue pins)

E. _____
F. _____
G. _____
H. _____
I. _____
J. _____
K. _____

EXERCISE 11.5b:
Imaging—Lower Respiratory—Pneumonia

SELECT TOPIC	SELECT VIEW
Lower Respiratory—Pneumonia	Bronchogram—Posterior–Anterior

Click the **TAGS ON/OFF** *button, and you will see the following image:*

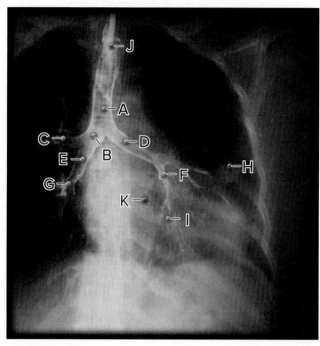

©McGraw-Hill Education

Mouse-over the pins on the screen to find the information necessary to identify the following structures:

A. _____
B. _____
C. _____
D. _____
E. _____
F. _____
G. _____
H. _____
I. _____

Nonrespiratory System Structures (blue pins)

J. _____
K. _____

EXERCISE 11.5c:
Imaging—Lower Respiratory—Pneumonia

SELECT TOPIC	SELECT VIEW
Lower Respiratory—Pneumonia	X ray—Posterior–Anterior

Click the **TAGS ON/OFF** *button, and you will see the following image:*

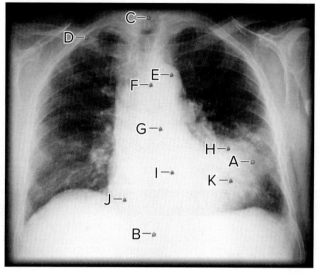

©Shutterstock/Anthony Ricci

Mouse-over the pins on the screen to find the information necessary to identify the following structures:

A. _____
B. _____

Nonrespiratory System Structures (blue pins)

C. _____
D. _____
E. _____
F. _____
G. _____
H. _____
I. _____
J. _____
K. _____

IN REVIEW

What Have I Learned?

The following questions cover the material that you have just learned, the lower respiratory system. Apply what you have learned to answer these questions on a separate piece of paper.

1. Name the serous membrane that covers the lungs.

2. Name the space formed between the visceral and parietal pleura.

3. What is the primary muscle of respiration?

4. What pressure changes occur when the primary muscle contracts?

5. What nerve is responsible for these contractions?

6. The oblique fissure of the left lung separates the _____ and _____ lobes, while in the right lung it separates the . . .

7. The horizontal fissure of the right lung separates the _____ and _____ lobes.

The Trachea

EXERCISE 11.6a:

Respiratory System—Trachea—Histology

SELECT TOPIC	SELECT VIEW
Trachea ▶	LM: Low Magnification

Click the **TAGS ON/OFF** button, and you will see the following image:

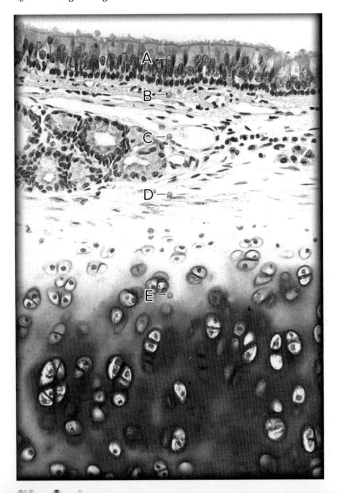

Mouse-over the pins on the screen to find the information necessary to identify the following structures:

A. _____

B. _____

C. _____

D. _____

E. _____

CHECK POINT

Trachea, Histology (Low Magnification)

1. What type of cartilage provides support for the majority of the respiratory tract?
2. Name the fibrocartilage layer that covers the hyaline cartilage.
3. Name the layer of the trachea that contains numerous seromucous glands and is rich in blood and lymph vessels.
4. Name the layer of the trachea that provides support for the epithelial cells. Name the lymphoid elements it may contain.
5. What tissue type makes up the epithelial layer? What is its function?

EXERCISE 11.6b:
Respiratory System—Trachea—Histology

SELECT TOPIC Trachea ▶ **SELECT VIEW** LM: High Magnification

Click the **TAGS ON/OFF** *button, and you will see the following image:*

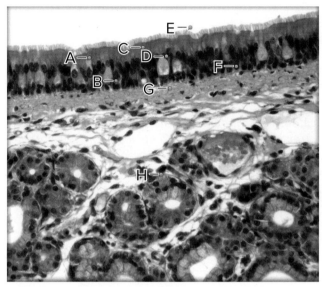

©Science Photo Library/Alamy Stock Photo

Mouse-over the pins on the screen to find the information necessary to identify the following structures:

A. _____
B. _____
C. _____
D. _____
E. _____
F. _____
G. _____
H. _____

EXERCISE 11.7:
Respiratory System—Respiratory Epithelium

SELECT TOPIC Respiratory Epithelium ▶ **SELECT VIEW** LM: High Magnification

Click the **TAGS ON/OFF** *button, and you will see the following image:*

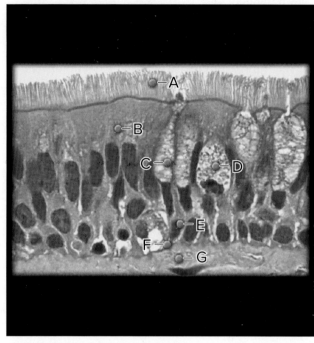

©McGraw-Hill Education/Dennis Strete

Mouse-over the pins on the screen to find the information necessary to identify the following structures:

A. _____
B. _____
C. _____
D. _____
E. _____
F. _____
G. _____

CHECK POINT

Trachea, and Respiratory Epithelium Histology (High Magnification)

1. Name the tissue layer that provides support for most epithelia in the body.
2. Name the stem cell that can produce new epithelial cells.
3. Name the mucous-producing epithelial cells. What is the purpose of the mucus?
4. What is the function of the cilia on the ciliated epithelial cells?

IN REVIEW

What Have I Learned?

The following questions cover the material that you have just learned, the trachea. Apply what you have learned to answer these questions on a separate piece of paper.

1. What type of cartilage provides support for the majority of the respiratory tract?

2. What tissue type makes up the epithelial layer of the trachea? What is its function?

3. Name the tissue layer that attaches the epithelium to the muscularis mucosa.

4. With your knowledge of the function of the cilia in the respiratory tract, explain why cigarette smoking, which paralyzes these cilia, causes the smoker's hacking cough.

The Alveolus and Alveolar Duct

EXERCISE 11.8:
Respiratory System—Alveolar Duct—Histology

SELECT TOPIC: Alveolar Duct ▶ SELECT VIEW: LM: High Magnification

Click the **TAGS ON/OFF** button, and you will see the following image:

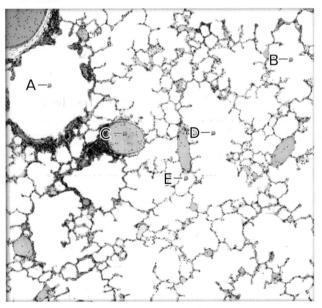

©McGraw-Hill Education/Al Telser

Mouse-over the pins on the screen to find the information necessary to identify the following structures:

A. _____

B. _____

C. _____

D. _____

E. _____

CHECK POINT

Alveolar Duct, Histology

1. Name the structure that provides the respiratory surface for gas exchange.
2. Name the structure that contains two or more alveoli.
3. Name the structure that conducts air to the alveoli.

MODULE 11 The Respiratory System 693

EXERCISE 11.9a:
Respiratory System—Alveolus—Histology

SELECT TOPIC: Alveolus ▶ SELECT VIEW: LM: Low Magnification

Click the **TAGS ON/OFF** button, and you will see the following image:

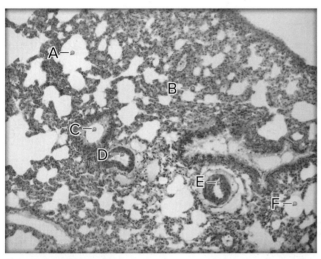

©McGraw-Hill Education/Greg Reeder

Mouse-over the pins on the screen to find the information necessary to identify the following structures:

A. _____
B. _____
C. _____
D. _____
E. _____
F. _____

EXERCISE 11.9b:
Respiratory System—Alveolus—Histology

SELECT TOPIC: Alveolus ▶ SELECT VIEW: LM: High Magnification

Click the **TAGS ON/OFF** button, and you will see the following image:

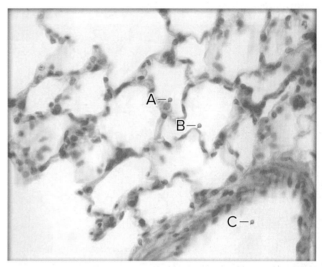

©McGraw-Hill Education/Greg Reeder

Mouse-over the pins on the screen to find the information necessary to identify the following structures:

A. _____
B. _____
C. _____

Self-Quiz
Take this opportunity to check your progress by taking the **QUIZ**. See the **Introduction Module** for a reminder on how to access the **QUIZ** for this Study Area.

IN REVIEW

What Have I Learned?

The following questions cover the material that you have just learned, the alveolus and alveolar duct. Apply what you have learned to answer these questions on a separate piece of paper.

1. What are alveoli? Where are they located?

2. What is the diameter of an alveolus?

3. What are the two types of specialized cells in the wall of the alveoli?

4. What vessels transport deoxygenated blood toward the alveolar capillary plexus?

EXERCISE 11.10:
Respiratory System—Larynx—Anterior View

SELECT TOPIC: **Larynx** ▸ SELECT VIEW: **Anterior**

Click **LAYER 1** in the **LAYER CONTROLS** window, and you will see the following image:

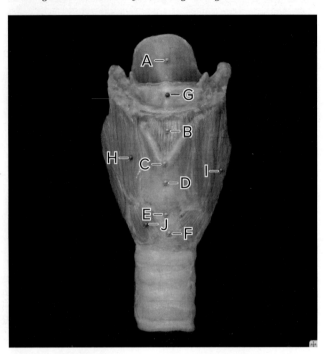

©McGraw-Hill Education

Mouse-over the pins on the screen to find the information necessary to identify the following structures:

A. epiglottis
B. thyrohyoid membrane
C. laryngeal prominence
D. thyroid cartilage
E. cricothyroid ligament
F. cricoid cartilage

Nonrespiratory System Structures (blue pins)

G. hyoid bone
H. thyrohyoid m.
I. inferior pharyngeal constrictor m.
J. cricothyroid m.

CHECK POINT
Larynx, Anterior View

1. Name the cartilaginous structure that closes over the laryngeal inlet when swallowing.
2. Name a muscle that elevates the larynx and depresses the hyoid.
3. Name a muscle that produces waves of contractions during swallowing.

Click **LAYER 2** in the **LAYER CONTROLS** window, and you will see the following image:

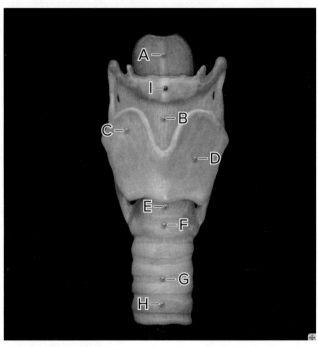

©McGraw-Hill Education

Mouse-over the pins on the screen to find the information necessary to identify the following structures:

A. epiglottis
B. thyrohyoid membrane
C. laryngeal cartilages
D. thyroid cartilage
E. cricothyroid ligament
F. cricoid cartilage
G. trachea
H. tracheal cartilage

Nonrespiratory System Structure (blue pin)

I. hyoid bone

CHECK POINT

Larynx, Anterior View, *continued*

4. Name the bone that does not articulate with any other bone.
5. Name the largest laryngeal cartilage.
6. Name the only cartilage of the respiratory tree that is a complete ring.

Click **LAYER 3** *in the* **LAYER CONTROLS** *window, and you will see the following image:*

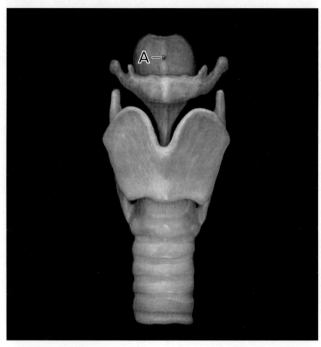

©McGraw-Hill Education

Mouse-over the pin on the screen to find the information necessary to identify the following structure:

A. _epiglottis_

EXERCISE 11.11:

Respiratory System—Larynx—Lateral View

SELECT TOPIC: Larynx **SELECT VIEW:** Lateral

Click **LAYER 1** *in the* **LAYER CONTROLS** *window, and you will see the following image:*

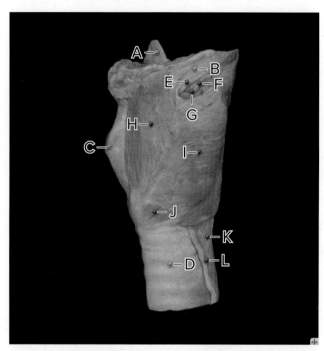

©McGraw-Hill Education

Mouse-over the pins on the screen to find the information necessary to identify the following structures:

A. _____
B. _____
C. _____
D. _____

Nonrespiratory System Structures (blue pins)

E. _____
F. _____
G. _____
H. _____
I. _____
J. _____
K. _____
L. _____

696 MODULE 11 The Respiratory System

CHECK POINT

Larynx, Lateral View

1. Name the cranial nerve that contributes to the internal laryngeal nerve.
2. Name the sensory innervation of the internal laryngeal nerve.
3. Name the companion artery and vein to the internal laryngeal nerve.

CHECK POINT

Larynx, Lateral View, *continued*

4. Name the superior projection of the posterior thyroid cartilage.
5. Name the inferior projection of the posterior thyroid cartilage.

Click **LAYER 2** *in the* **LAYER CONTROLS** *window, and you will see the following image:*

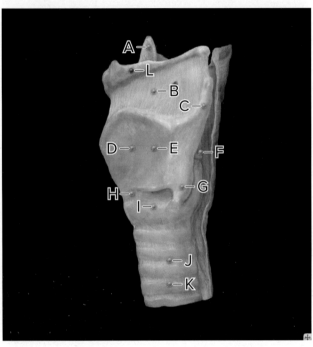

©McGraw-Hill Education

Mouse-over the pins on the screen to find the information necessary to identify the following structures:

A. _____
B. _____
C. _____
D. _____
E. _____
F. _____
G. _____
H. _____
I. _____
J. _____
K. _____

Nonrespiratory System Structure (blue pin)

L. _____

Click **LAYER 3** *in the* **LAYER CONTROLS** *window, and you will see the following image:*

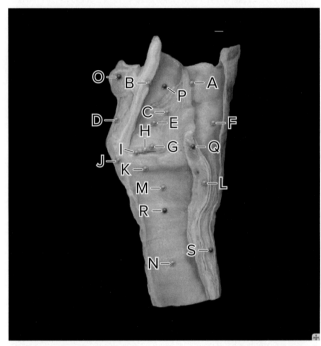

©McGraw-Hill Education

Mouse-over the pins on the screen to find the information necessary to identify the following structures:

A. _____
B. _____
C. _____
D. _____
E. _____
F. _____
G. _____
H. _____
I. _____
J. _____
K. _____
L. _____
M. _____
N. _____

Nonrespiratory System Structures (blue pins)

O. _____
P. _____
Q. _____
R. _____
S. _____

> **CHECK POINT**
>
> **Larynx, Lateral View,** *continued*
>
> 6. Name two structures with the major role in sound production.
> 7. Name the portion of the pharynx posterior to the larynx.
> 8. Name the sensory and motor innervations of the recurrent laryngeal nerve.

EXERCISE 11.12:
Respiratory System—Larynx—Posterior View

SELECT TOPIC	SELECT VIEW
Larynx	Posterior

Click **LAYER 1** *in the* **LAYER CONTROLS** *window, and you will see the following image:*

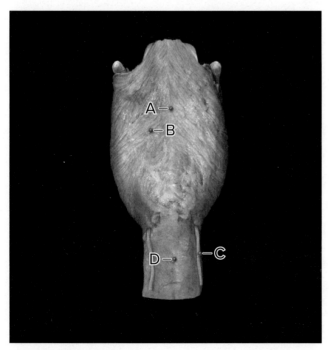

©McGraw-Hill Education

Mouse-over the pins on the screen to find the information necessary to identify the following nonrespiratory system structures:

A. _____
B. _____
C. _____
D. _____

> **CHECK POINT**
>
> **Larynx, Posterior View**
>
> 1. What is a raphe?
> 2. Name the innervation for the inferior pharyngeal constrictor muscle.
> 3. Name the nerve responsible for the innervation of all but one intrinsic laryngeal muscle.

Click **LAYER 2** *in the* **LAYER CONTROLS** *window, and you will see the following image:*

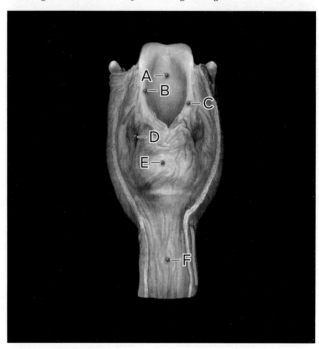

©McGraw-Hill Education

Mouse-over the pins on the screen to find the information necessary to identify the following structures:

A. _____
B. _____
C. _____
D. _____
E. _____

Nonrespiratory System Structure (blue pin)

F. _____

698 MODULE 11 The Respiratory System

CHECK POINT

Larynx, Posterior View, *continued*

4. Name the superior opening of the larynx.
5. When food is "caught in the throat," where is it usually lodged?
6. Name the mucous membrane fold that surrounds the structure in question 4.

Click **LAYER 3** *in the* **LAYER CONTROLS** *window, and you will see the following image:*

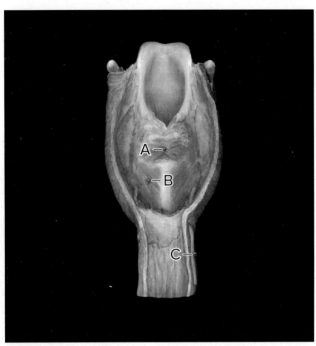

©McGraw-Hill Education

Mouse-over the pins on the screen to find the information necessary to identify the following nonrespiratory system structures:

A. _____
B. _____
C. _____

CHECK POINT

Larynx, Posterior View, *continued*

7. Name a muscle that abducts the vocal folds.
8. Name a nerve of the larynx that is a branch of the vagus nerve (CN X).

Click **LAYER 4** *in the* **LAYER CONTROLS** *window, and you will see the following image:*

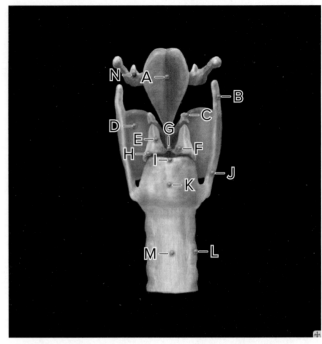

©McGraw-Hill Education

Mouse-over the pins on the screen to find the information necessary to identify the following structures:

A. _____
B. _____
C. _____
D. _____
E. _____
F. _____
G. _____
H. _____
I. _____
J. _____
K. _____
L. _____
M. _____

Nonrespiratory System Structure (blue pin)

N. _____

CHECK POINT

Larynx, Posterior View, *continued*

9. Name the process that is the posterior attachment site for the vocal ligament.
10. Name the cartilaginous structure that provides attachment for the lateral and posterior cricoarytenoid muscles.

IN REVIEW

What Have I Learned?

The following questions cover the material that you have just learned, the larynx. Apply what you have learned to answer these questions on a separate piece of paper.

1. Name three pharyngeal constrictors.

2. Name a muscle that tenses and elongates the vocal ligaments.

3. Describe the cricothyroid ligament.

4. Describe the thyrohyoid membrane.

5. Name the structure also known as the Adam's apple.

6. Name the U-shaped structures that provide a rigid skeleton for the trachea.

7. What are the corniculate cartilages? Where are they embedded?

SELECT ANIMATION
Respiration (Interactive)
Pulmonary Ventilation: Inspiration
PLAY

This Interactive Animation will walk you through the inspiration phase of breathing. Be sure to interact with the animation to glean all the information available to you.

Open the **Module 11—Respiratory System ANIMATION LIST**.

Under the second heading of **Respiration (Interactive)**, select **1. Pulmonary ventilation: Inspiration** and you will see the image to the right:

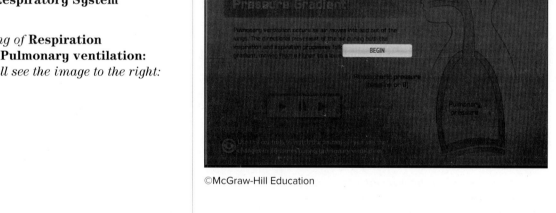
©McGraw-Hill Education

Select the **BEGIN** button in the middle, and interact with the seven images contained in this animation.

When finished with the interactions, click the **X** in the top-right corner to close the animation.

After reviewing the animation, answer the following questions:

1. _____ _____ occurs as _____ moves into and out of the _____.

2. The directional movement of the air during both the inspiration and expiration processes follows a _____ _____, moving from a _____ to a _____ _____.

3. _____ _____ is the difference between the _____ _____ found within the _____ and the _____ _____ found in the _____ surrounding the _____.

4. The _____ _____ helps the lungs to _____ in the _____ _____ during _____.

5. Although the pleural pressure changes _____ throughout _____ and _____, it continually remains more _____ than _____ _____, and is always _____ in comparison to the _____ pressure.

6. _____ law is described as the _____ relationship between _____ and _____ where as _____ increases, _____ decreases and vice versa.

7. Lung volume changes in response to changes in the overall _____ _____ volume.

8. This occurs in response to the _____ or _____ of the _____ _____ the _____ and _____ _____.

9. When these muscles _____, they _____ the overall _____ volume and in turn _____ the _____ volume.

10. For air to move _____ the lungs, the _____ _____ must be _____ _____ _____, or _____ in comparison.

SELECT ANIMATION
Respiration (Interactive)
Alveolar Gas Exchange:
Oxygen Diffuses into Blood
PLAY

This Interactive Animation will walk you through the gas exchange in the lungs. Be sure to interact with the animation to glean all the information available to you.

Open the **Module 11—Respiratory System ANIMATION LIST.**

Under the second heading of **Respiration (Interactive)**, select **2. Alveolar Gas Exchange: Oxygen Diffuses into Blood** and you will see the image to the right:

©McGraw-Hill Education

Select the **BEGIN** *button in the middle, and interact with the seven images contained in this animation.*

When finished with the interactions, click the **X** *in the top-right corner to close the animation.*

After reviewing the animation, answer the following questions:

1. During _____, air entering the body travels to the _____ through _____ _____ structures,, ultimately entering the _____ zone.

2. Here, _____ found within the _____ must cross the _____ membrane and enter into the _____ in order to be _____ throughout the _____.

3. Where does gas exchange occur? How many of these structures exist in the lungs?

4. The presence of alveoli indicates . . .

5. Describe Type I pneumocytes. Where are they found?

6. Describe Type II pneumocytes. Where are they found? What do they secrete?

7. What is the function of surfactant? What does this prevent?

8. The _____ membrane separates the _____ environment within the _____ from the _____ environment of the _____.

9. List the three layers of the respiratory membrane.

10. What does partial pressure refer to?

11. Oxygen crosses the air/liquid barrier at a rate . . .

12. This is referred to as _____ Law or _____ Law.

13. The difference in the _____ _____ of oxygen between the _____ air and the _____ _____ drives its movement out of the _____, across the _____ _____, and into the _____.

SELECT ANIMATION

Respiration (Interactive)
Gas Transport of Oxygen within the Blood

PLAY

This Interactive Animation will walk you through transport of oxygen in the blood. Be sure to interact with the animation to glean all the information available to you.

Open the **Module 11—Respiratory System ANIMATION LIST.**

Under the second heading of **Respiration (Interactive)**, *select* **3. Gas Transport of Oxygen within the Blood** *and you will see the image to the right:*

©McGraw-Hill Education

Select the **BEGIN** *button in the middle, and interact with the seven images contained in this animation.*

When finished with the interactions, click the **X** *in the top-right corner to close the animation.*

After reviewing the animation, answer the following questions:

1. When is oxygen in the air actually available to the cells of the body?

2. Once in the _____ _____, oxygen molecules enter the _____, also known as _____ _____ _____, by _____.

3. Here they encounter millions of _____.

4. Inside the red blood cells, oxygen binds to four _____ _____ _____ that each contain their own _____ region.

5. The binding results in a change to the _____ - _____ _____ of the _____, now called _____, and alters its _____ to reflect a _____ _____ coloration.

6. What do hemoglobin saturation levels reflect?

7. Why do hemoglobin saturation levels increase in increments of 25%?

8. What relationship is demonstrated by an Oxygen-Hemoglobin Dissociation Curve?

9. A reduction in the partial pressure of oxygen results in . . .

10. An increase in oxygen partial pressure results in . . .

11. The Y-axis of this graph represents . . .

12. As seen with an Oxygen-Hemoglobin Dissociation Curve, the nonlinear shape of the graph demonstrates . . .

13. The X-axis of this graph represents . . .

14. Where is the partial pressure of oxygen highest? What happens to the partial pressure of oxygen as it moves toward the tissues?

SELECT ANIMATION
Respiration (Interactive)
Systemic Gas Exchange: Oxygen into Systemic Cells
PLAY

This Interactive Animation will walk you through transport of oxygen from the capillaries to the tissues. Be sure to interact with the animation to glean all the information available to you.

Open the **Module 11—Respiratory System ANIMATION LIST.**

Under the second heading of **Respiration (Interactive)**, select **4. Systemic Gas Exchange: Oxygen into Systemic Cells**

and you will see the image to the right:

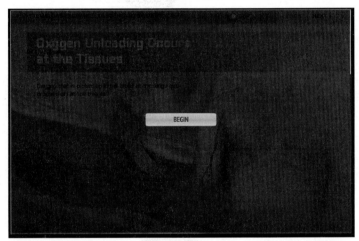

©McGraw-Hill Education

Select the **BEGIN** button in the middle, and interact with the five images contained in this animation.

When finished with the interactions, click the **X** in the top-right corner to close the animation.

After reviewing the animation, answer the following questions:

1. Where does oxygen that is picked up in the blood of the lungs get dropped off?

2. Where is the partial pressure of oxygen lowest?

3. This pressure gradient drives . . .

4. This ultimately increases the partial pressure of oxygen . . .

5. This ultimately decreases the partial pressure of oxygen . . .

6. The P_{O_2} = _____ in the capillaries.

7. The P_{O_2} = _____ in the tissues.

8. What causes increased oxygen demand in the tissues?

9. This increased oxygen demand stimulates . . .

10. What does this increased oxygen demand increase in the tissues?

11. Why does the steep slope region of the Oxygen-Hemoglobin Dissociation Curve occur?

12. What four phenomena also influence the saturation level of hemoglobin?

13. What is the Bohr Effect? What effect does the Bohr Effect have on the Oxygen-Hemoglobin Dissociation Curve?

14. What effect does this shift have on hemoglobin's affinity for oxygen? What does this allow?

SELECT ANIMATION
Respiration (Interactive)
Systemic Gas Exchange:
Carbon Dioxide into Blood
PLAY

This Interactive Animation will walk you through the movement of carbon dioxide into the blood. Be sure to interact with the animation to glean all the information available to you.

Open the **Module 11—Respiratory System ANIMATION LIST.**

Under the second heading of **Respiration (Interactive)**, select **5. Systemic Gas Exchange: Carbon Dioxide into Blood** and you will see the image to the right:

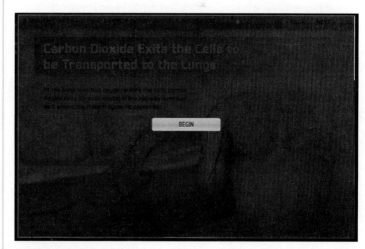
©McGraw-Hill Education

Select the **BEGIN** button in the middle, and interact with the three images contained in this animation.

When finished with the interactions, click the **X** in the top-right corner to close the animation.

After reviewing the animation, answer the following questions:

1. What gas exits the cells at the same time and in the opposite direction as oxygen?

2. What determines the directional movement of both carbon dioxide and oxygen?

3. What is the P_{CO_2} in the tissues? What is the P_{CO_2} of the blood in the capillaries? Which direction will the CO_2 move?

4. What increases the red blood cell's affinity for CO_2? This is referred to as the _____ Effect.

SELECT ANIMATION
Respiration (Interactive)
Gas Transport of Carbon Dioxide within the Blood
PLAY

This Interactive Animation will walk you through transport of carbon dioxide while it is in the blood. Be sure to interact with the animation to glean all the information available to you.

Open the **Module 11—Respiratory System ANIMATION LIST.**

Under the second heading of **Respiration (Interactive)**, select **6. Gas Transport of Carbon Dioxide within the Blood** and you will see the image to the right:

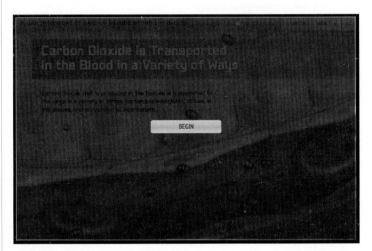

©McGraw-Hill Education

Select the **BEGIN** button in the middle, and interact with the six images contained in this animation.

When finished with the interactions, click the **X** in the top-right corner to close the animation.

After reviewing the animation, answer the following questions:

1. Carbon dioxide is transported to the lungs in a variety of forms. List three of them.

2. In what manner does CO_2 bind with hemoglobin?

3. What is carbaminohemoglobin? Approximately what percentage of the overall CO_2 in the blood is transported as carbaminohemoglobin?

4. Approximately what percentage of the overall CO_2 in the blood remains dissolved in the plasma?

5. Approximately what percentage of the overall CO_2 in the blood is transported is converted to bicarbonate?

6. Where is the enzyme carbonic anhydrase located? What is its function?

7. Carbonic acid will quickly dissociate into . . .

8. How do these ions affect the pH within the red blood cell?

9. Describe the chloride shift in the red blood cell. What does it maintain?

10. Where do the bicarbonate ions remain as they travel toward the lungs?

SELECT ANIMATION
Respiration (Interactive)
Alveolar Gas Exchange:
Carbon Dioxide into Alveoli
PLAY

This Interactive Animation will walk you through the movement of carbon dioxide from the blood to the lungs. Be sure to interact with the animation to glean all the information available to you.

Open the **Module 11—Respiratory System ANIMATION LIST.**

Under the second heading of **Respiration (Interactive)**, *select* **7. Alveolar Gas Exchange: Carbon Dioxide into Alveoli** *and you will see the image to the right:*

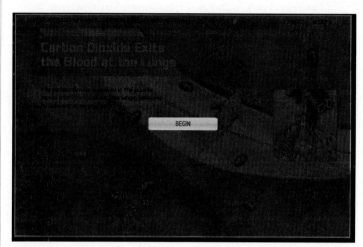
©McGraw-Hill Education

Select the **BEGIN** *button in the middle, and interact with the three images contained in this animation.*

When finished with the interactions, click the **X** *in the top-right corner to close the animation.*

After reviewing the animation, answer the following questions:

1. How does the carbon dioxide dissolved in the plasma move from the blood into the lungs?

2. Describe in detail the movement of bicarbonate, chloride, and hydrogen ions as they move in and out of the red blood cell.

3. How is carbonic acid formed in the red blood cell?

4. What is the function of carbonic anhydrase enzyme? Where do the end products diffuse?

5. As blood enters the capillaries of the lungs, what induces carbon dioxide to travel across the respiratory membrane?

SELECT ANIMATION
Respiration (Interactive)
Pulmonary Ventilation: Expiration — PLAY

This Interactive Animation will walk you through the expiration phase of breathing. Be sure to interact with the animation to glean all the information available to you.

Open the **Module 11—Respiratory System ANIMATION LIST**.

Under the second heading of **Respiration (Interactive)**, *select* **8. Pulmonary Ventilation: Expiration** *and you will see the image to the right:*

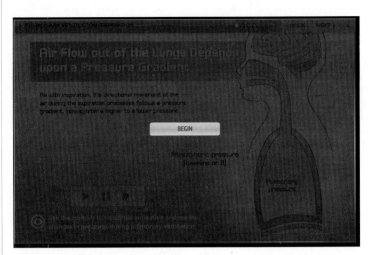
©McGraw-Hill Education

Select the **BEGIN** *button in the middle, and interact with the four images contained in this animation.*

When finished with the interactions, click the **X** *in the top-right corner to close the animation.*

After reviewing the animation, answer the following questions:

1. As with inspiration, the directional movement of the air during the expiration process follows a _____ _____, moving from a _____ to a _____ _____.

2. What events decrease the volume of the thoracic cavity and lungs?

3. What is the result of this volume decrease?

4. What is the result of the answer to question number 3?

5. The movement of the air into or out of the lungs depends upon a _____ _____, with the flow of air moving from a _____ to a _____ pressure.

6. How are changes in blood pH monitored? How do they maintain homeostasis?

7. What affect does a decrease in blood pH have on the respiratory system?

8. What is the overall result?

SELECT ANIMATION
Clinical Applications
Respiratory Overview/
Respiratory Failure (3D) — PLAY

After viewing the animation, answer the following questions:

1. The lungs are part of the _____ _____ _____, and contain a series of subdividing tubes, beginning with the _____.

2. Air passes through the _____ into smaller tubes, called _____, then into tiny elastic air sacs called _____, where _____ _____ occurs.

3. _____ _____ and _____ share a thin membrane through which _____ _____ occurs.

4. _____ diffuses from the _____ into the _____ for transport to _____ _____ and _____ _____, a _____ _____ of _____ _____, diffuses from the _____ into the _____ to be _____ from the body.

5. Shifts in _____ and _____ _____ _____ in the _____ can signal _____ _____, a condition in which gas exchange is _____, due to _____ involving the _____ or other organs.

6. In _____ _____ _____, too _____ _____ diffuses into the _____.

7. In _____ _____ _____, too much _____ _____ remains in the blood.

8. _____ _____ _____ _____, known as _____, and _____ _____ _____ _____, known as _____, commonly result in _____ _____.

9. _____ includes two main conditions— _____ _____ and _____.

10. _____ _____ involves _____ of the _____ _____—thick _____ accumulation along the _____ lining, and narrowing of the _____, called _____, which can result in blocked or narrowed _____.

11. In _____, some _____ walls are _____, leading to fewer, larger, formless sacks.

12. In addition, some of the _____ collapse, impeding _____ _____ out of the _____.

13. In contrast, _____ _____ _____ _____ involves _____ to one or both lungs, caused by factors such as _____, or blood infection called _____.

14. _____ leads to _____ _____ _____ and _____ _____, reducing the surface available for gas exchange.

15. List two common treatments for respiratory failure.

16. Other common treatments and their effects include . . .

SELECT ANIMATION
Clinical Applications
Asthma (3D)
PLAY

After viewing the animation, answer the following questions:

1. Asthma is a _____ _____ _____ that obstructs _____ _____ in and out of the _____ _____.

2. Normally as the _____ contracts and relaxes, _____ moves freely in and out of the _____ and _____ to the _____ and then to the _____, where _____ _____ takes place.

3. During this process, _____ _____ will _____ out of the _____ into the _____ while _____ will _____ from the _____ into the _____.

4. _____ _____ in the _____ walls is controlled by the _____ _____ _____.

5. _____ _____ relaxes _____ muscle and produces _____ when the air is _____, _____, and free of _____.

6. _____ _____ contracts _____ _____ and produces _____ when the air is _____, _____, or contains _____.

7. People with asthma have _____ _____ and _____ airways that are _____ to _____ that can trigger an _____ _____.

8. List the outdoor asthma triggers.

9. List the indoor asthma triggers.

10. List the food allergen asthma triggers.

11. List the physiological asthma triggers.

12. During an asthma attack, these triggers . . .

13. List the chemical mediators of inflammation.

14. What do these chemical mediators of inflammation precipitate?

15. What occurs during a bronchospasm?

16. What do the goblet cells contribute and what is the result?

17. What effect does this have on normal gas exchange?

18. List the symptoms of a bronchospasm.

19. How do medicines that treat asthma affect the structures involved in the bronchospasm?

20. What are the most important asthma medications? Give examples.

21. How do these drugs keep asthma under control?

22. What effect does this have on the airways?

23. What is the result of regular use of these asthma medications?

24. What are the two types of bronchodilator drugs? Give examples of each.

25. What effect do these have on bronchial smooth muscle?

26. Why is the regular use of long-acting maintenance medications critical?

27. What does this reduce?

28. When flare-ups do happen . . .

29. What is the purpose of short-acting rescue medications?

SELECT ANIMATION
Clinical Applications
COPD: Emphysema and Bronchitis (3D)
PLAY

After viewing the animation, answer the following questions:

1. What is COPD?

2. Describe the process of gas exchange.

3. Describe in detail how breathing begins.

4. Where does gas exchange take place?

5. Describe the normal airways and alveoli. How do they respond when air is inhaled and exhaled?

6. What are capillaries? What do they allow? Describe the interface where this occurs.

7. What structures control the size of the airway or bronchioles?

8. What protective layer covers the smooth muscle? How does it protect the tubes of the respiratory tree?

9. List the two main conditions included in COPD.

10. Describe emphysema. What does it destroy? What does this lead to? What does this reduce?

11. Describe chronic bronchitis. What accompanies it?

12. What is caused by the inflammation and mucus?

14. What symptoms do COPD patients experience?

15. What is the leading causative factor of COPD?

16. List other causative factors of COPD.

17. Describe the treatment capable of reversing the damage to the airways and lungs and for COPD.

18. List the treatments available and lifestyle changes that can manage and slow the disease while increasing the quality of life for the patient.

18. What is a bullectomy?

SELECT ANIMATION
Clinical Applications
Upper Respiratory Tract Infection (3D)
PLAY

After viewing the animation, answer the following questions:

1. List the structures that comprise the upper respiratory tract.

2. What barriers do these structures provide as we breathe?

3. In the nose, _____, _____, and other _____ stick to the mucus in the nasal passageway while the _____ _____ warms and humidifies the air.

4. The shape of the _____ forces air to take a _____ _____ _____ _____.

5. _____ _____ _____ can't make the turn so they get caught in the _____ of the posterior wall instead.

6. The _____ and _____ _____ contain _____ cells that _____ and _____ _____ that land on them.

7. Air then passes through the _____ on its way to the _____ _____ _____.

8. What is the most common viral upper respiratory tract infection? (list all three synonyms)

9. What is the most common cold virus?

10. How does this virus enter the body?

11. Where does the virus invade?

12. What is the result?

13. What are the possible results from inflammation of the nasal membranes?

14. What are the possible results from inflammation of the pharyngeal lining?

15. What is the possible results from inflammation of the larynx?

16. What is the cure for the common cold?

17. What do treatments focus on?

18. List common lifestyle modification treatments.

19. List the medical options for treating the common cold. How do they help?

20. Why are antibiotics not used to treat the common cold?

**SELECT ANIMATION
Clinical Applications
Acidosis (3D)
PLAY**

After viewing the animation, answer the following questions:

1. Define acid-base balance.

2. Define pH.

3. What is expressed as a pH value?

4. What is the range of normal blood pH?

5. Metabolic processes constantly release _____ which freely release _____ ions, resulting in . . .

6. What is the body's chemical and physiological response?

7. When does respiratory acidosis occur?

8. If respiration cannot keep pace with _____ _____ production, _____ _____ builds up in the _____.

9. Excess _____ _____ combines with _____ to produce _____ _____.

10. _____ _____ disassociates into _____ and _____ ions.

11. What lowers pH and causes acidosis?

12. Metabolic acidosis occurs when . . .

13. List the conditions causing metabolic acidosis.

14. This results in . . .

15. What is the first step in treatment for acidosis?

16. What may be used to treat metabolic acidosis?

17. What may be used to treat respiratory acidosis?

**SELECT ANIMATION
Clinical Applications
Alkalosis (3D)** — PLAY

After viewing the animation, answer the following questions:

1. Define acid-base balance.

2. Define pH.

3. What is expressed as a pH value?

4. What is the range of normal blood pH?

5. Metabolic processes constantly release _____ which freely release _____ ions, resulting in . . .

6. What is the body's chemical and physiological response?

7. When does respiratory alkalosis occur?

8. What causes a carbon dioxide deficit?

9. What is the result of a carbon dioxide deficit?

10. Less _____ _____ dissociates into fewer free _____ ions.

11. What causes alkalosis?

12. Metabolic alkalosis occurs when . . .

13. List the conditions causing metabolic alkalosis.

14. This results in . . .

15. Discuss the primary treatment for respiratory alkalosis.

16. Discuss the treatment for metabolic alkalosis.

17. When does the treatment of alkalosis include antiemetic drugs? How do they help?

MODULE 12

The Digestive System

Overview: The Digestive System

Think of everything that you have eaten today, or this week for that matter. After you swallowed it, where did it go? Out of sight, out of mind. We don't even think about food again until it's meal time or our stomach growls during class (or worse yet, during a test). What happens to the food that you eat, between the time you swallow it and the time that the waste is eliminated? What structures are involved with this transformation?

The nutrients in your food are mostly unavailable to your body when they are swallowed. The primary nutrients—proteins, carbohydrates, and fats must be broken down into their basic building blocks before they can be absorbed into the bloodstream.

Only then can your body build proteins, carbohydrates, and fats from these available building blocks. It's like moving a large desk into a small office. The desk won't fit through the door, so you need to disassemble it first into smaller parts that *will* fit through the door. Once the parts have passed through the door and inside the office, they can be reassembled into a desk. Likewise, proteins, for example, are too large to pass through the epithelium of the small intestine and into the capillaries located there.

Digestive enzymes must first break these proteins down into their smallest building blocks—amino acids. These amino acids can then pass through the epithelium of the small intestine and into the bloodstream. Then, when your body needs to make a protein molecule—say for example, your growing hair, it puts together the required amino acids and constructs hair. This is where the office desk analogy breaks down. When you reassembled the desk in the small office, it was more or less the same desk that you had taken apart. The only difference would be any missing pieces lost in the move. In your body, the proteins that you create to build structures are unique and completely unlike the original protein in your food. But the point is the same, they must be broken down into smaller pieces to make it through the "door" of your digestive system.

To learn the structures of the digestive system, we will begin with an animated overview of the entire system. Then, we will look at the anatomical structures of the alimentary canal sequentially, from proximal to distal. Next we will look at the accessory digestive organs, and we will conclude by learning how we get energy out of the food we eat.

From the **HOME** *screen, click the drop-down box on the* **MODULE** *menu.*

From the systems listed, click on **Digestive.**

SELECT ANIMATION
Digestive System Overview
PLAY

After viewing the animation, answer the following questions:

1. What are the four main functions of the digestive system?

2. Name the two types of digestion.

3. Where does digestion begin? How does this occur?

4. Another name for chewing is _____.

5. What is the function of the salivary glands? Of saliva?

6. What prevents food from entering the nasal cavity while swallowing?

7. What muscles push food particles into the pharynx?

8. Name the structure that prevents food from entering the respiratory system.

9. Name the structure that connects the pharynx to the stomach.

10. Once it has been swallowed, the food mass is called a _____.

11. What is the term for the involuntary wavelike contractions that propel the bolus to the stomach?

12. What are rugae? What are their functions?

13. The stomach cells secrete . . .

14. What effect do these secretions have on the bolus?

15. The bolus, mixed with stomach secretions, is now called _____.

16. _____ exits the stomach through the _____ _____ and enters the _____ _____.

17. Name the major site of nutrient absorption.

18. Name the three parts of the small intestine, from proximal to distal.

19. What digestive aids enter the duodenum? Where do they originate?

20. How are nutrients absorbed?

21. What is the destination of the chyme not absorbed in the small intestine?

22. List the sequence of structures the chyme passes through as it becomes feces. What has been absorbed from the chyme as it passes through the colon?

23. Where are the feces stored? What causes fecal elimination?

SELECT ANIMATION
Digestive System Anatomy & Physiology (3D) PLAY

After viewing the animation, answer the following questions:

1. The digestive system consists of specialized _____ and _____ that _____ _____ and _____ _____ to body cells.

2. The digestive organs form a continuous tube called the _____ _____.

3. In normal digestion, _____ _____ moves down the _____ and into the _____, where the food is _____ _____ into _____ _____.

4. From the stomach, these particles enter the _____ _____, where _____ from the _____, _____ and the _____ _____ break down the particles into _____ _____.

5. Along the lining of the _____ _____, cellular projections called _____ _____ the _____ as well as _____, _____ and _____.

6. _____ _____ _____ move into the _____ _____, which absorbs more _____ and _____.

7. The remaining material enters the _____, where it will be stored until it exits the body as solid waste called _____.

Self-Quiz
Take this opportunity to check your progress by taking the **QUIZ**. See the **Introduction Module** for a reminder on how to access the **QUIZ** for this Study Area.

The Oral Cavity

EXERCISE 12.1:
Digestive System—Oral Cavity and Pharynx—Lateral View

SELECT TOPIC	SELECT VIEW
Oral Cavity and Pharynx	Lateral

Click **LAYER 1** *in the* **LAYER CONTROLS** *window, and you will see the following image:*

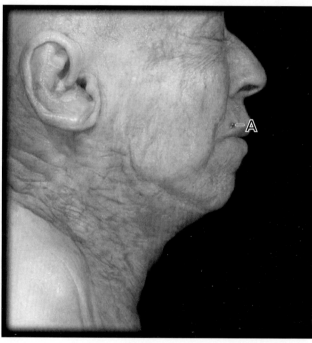

©McGraw-Hill Education

Mouse-over the pin on the screen to find the information necessary to identify the following structure:

A. _____lips_____

CHECK POINT
Oral Cavity and Pharynx, Lateral View

1. Name the fleshy folds surrounding the mouth.
2. Name the midline vertical groove of the upper lip.
3. Name the muscle contained in the lips.

Click **LAYER 2** *in the* **LAYER CONTROLS** *window, and you will see the following image:*

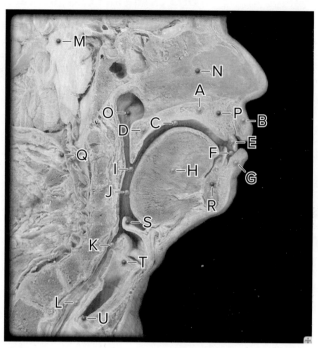

©McGraw-Hill Education

Mouse-over the pins on the screen to find the information necessary to identify the following structures:

A. _____hard palate_____
B. _____upper lip_____
C. _____oral cavity proper_____
D. _____soft palate_____
E. _____teeth_____
F. _____oral vestibule_____
G. _____lower lip_____
H. _____tongue_____
I. _____pharynx_____
J. _____oropharynx_____
K. _____laryngopharynx_____
L. _____esophagus_____

Nondigestive System Structures (blue pins)

M. _____brain_____
N. _____nasal septum with mucosa_____
O. _____nasopharynx_____
P. _____maxilla_____
Q. _____atlas_____
R. _____mandible_____
S. _____epiglottis_____
T. _____larynx_____
U. _____trachea_____

CHECK POINT

Oral Cavity and Pharynx, Lateral View, *continued*

4. Name the narrow space between the dental arches, lips, and cheeks.
5. What is its function?
6. Name the space bounded by the teeth, tongue, and hard palate.

IN REVIEW

What Have I Learned?

The following questions cover the material that you have just learned, the oral cavity. Apply what you have learned to answer the following questions on a separate piece of paper.

1. Name the three functions of the lips.

2. What is the function of the oral cavity?

3. What is the pharynx? What are the three subdivisions?

4. Name the divisions that are part of the respiratory system, the digestive system, or both.

5. Describe the structure of the tongue.

6. What is the tongue's function?

7. Name the structure involved in oral breathing.

8. The Latin term pharynx means _____.

9. Name the structure that separates the oropharynx from the nasopharynx.

10. Name the structure that separates the oral cavity from the nasal cavity. Which bones form this structure?

11. Name the tube-shaped structure located between the oropharynx and the esophagus.

The Salivary Glands and Teeth

EXERCISE 12.2:
Digestive System—Salivary Glands—Lateral View

SELECT TOPIC	SELECT VIEW
Salivary Glands	Lateral

Click **LAYER 1** *in the* **LAYER CONTROLS** *window, and you will see the following image:*

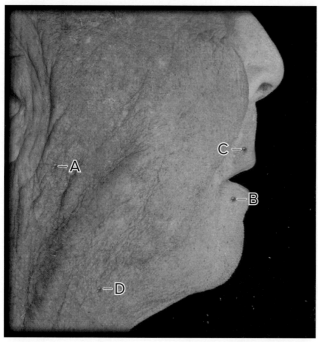

©McGraw-Hill Education

Mouse-over the pins on the screen to find the information necessary to identify the following structures:

A. _____

B. _____

C. _____

D. _____

CHECK POINT

Salivary Glands, Lateral View

1. The superficial part of which salivary gland is located anterior to the auricle?
2. Name the salivary gland located inferior and medial to the body of the mandible.
3. Name the three major paired salivary glands.

Click **LAYER 2** *in the* **LAYER CONTROLS** *window, and you will see the following image:*

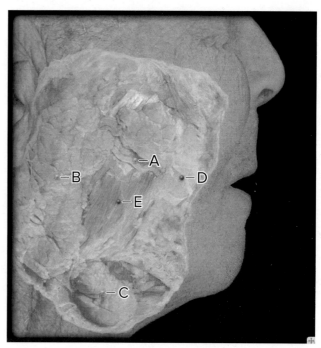

©McGraw-Hill Education

Mouse-over the pins on the screen to find the information necessary to identify the following structures:

A. _____

B. _____

C. _____

Nondigestive System Structures (blue pins)

D. _____

E. _____

CHECK POINT

Salivary Glands, Lateral View, *continued*

4. Which salivary gland produces 25–30 percent of your saliva? Where is it located?
5. Which salivary gland produces 60–70 percent of your saliva? Where is it located?
6. Name the salivary duct that ends in the vestibule of the oral cavity opposite the second maxillary molar.

Click **LAYER 3** *in the* **LAYER CONTROLS** *window, and you will see the following image:*

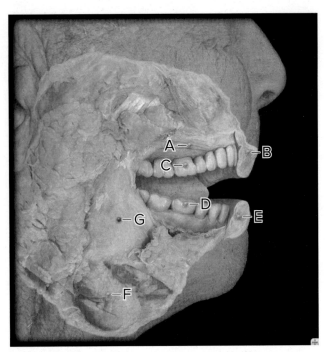

©McGraw-Hill Education

Mouse-over the pins on the screen to find the information necessary to identify the following structures:

A. _____

B. _____

C. _____

D. _____

E. _____

F. _____

Nondigestive System Structure (blue pin)

G. _____

CHECK POINT

Salivary Glands, Lateral View, *continued*

7. Describe the gingivae. What is another name for them? What is gingivitis?
8. Describe the permanent mandibular teeth. What is their function?
9. Describe the permanent maxillary teeth. Are they identical in number and function as the mandibular teeth? What do you suppose is the reason for this?

Click **LAYER 4** *in the* **LAYER CONTROLS** *window, and you will see the following image:*

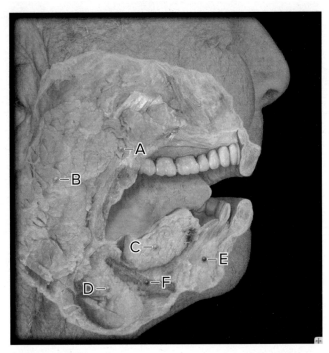

©McGraw-Hill Education

Mouse-over the pins on the screen to find the information necessary to identify the following structures:

A. _____

B. _____

C. _____

D. _____

Nondigestive System Structures (blue pins)

E. _____

F. _____

CHECK POINT

Salivary Glands, Lateral View, *continued*

10. Name the salivary gland inferior to the tongue. What percentage of your saliva does it secrete?
11. What is the sublingual fossa?
12. Name a muscle responsible for the elevation of the floor of the mouth.

MODULE 12   The Digestive System   719

Click **LAYER 5** *in the* **LAYER CONTROLS** *window, and you will see the following image:*

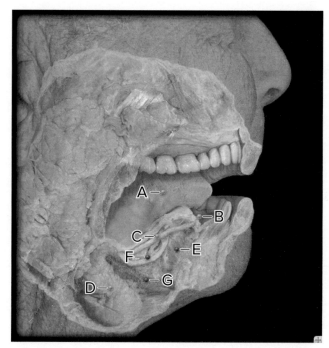

©McGraw-Hill Education

Mouse-over the pins on the screen to find the information necessary to identify the following structures:

A. _____
B. _____
C. _____
D. _____

Nondigestive System Structures (blue pins)

E. _____
F. _____
G. _____

CHECK POINT

Salivary Glands, Lateral View, *continued*

13. Name the three functions of the tongue.
14. Where are the taste buds located?
15. What gives the dorsal surface of the tongue a "feltlike" appearance?

EXERCISE 12.3:
Digestive System—Teeth—Superior and Inferior Views

SELECT TOPIC	SELECT VIEW
Teeth	Superior–Inferior

Click **LAYER 1** *in the* **LAYER CONTROLS** *window, and you will see the following image:*

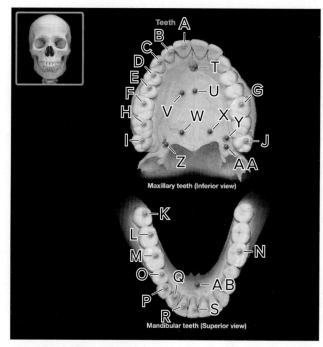

©McGraw-Hill Education

Mouse-over the pins on the screen to find the information necessary to identify the following structures:

A. _____
B. _____
C. _____
D. _____
E. _____
F. _____
G. _____
H. _____
I. _____
J. _____
K. _____
L. _____
M. _____
N. _____
O. _____

P. _____
Q. _____
R. _____
S. _____

Nondigestive System Structures (blue pins)

T. _____
U. _____
V. _____
W. _____
X. _____
Y. _____

Z. _____
AA. _____
AB. _____

CHECK POINT

Teeth, Superior and Inferior Views

1. Describe the Universal/National System used for numbering adult teeth.
2. Name the teeth, on both bones, important for biting and cutting.
3. Which teeth are the longest? What is their function?

IN REVIEW

What Have I Learned?

The following questions cover the material that you have just learned, the salivary glands and teeth. Apply what you have learned to answer the following questions on a separate piece of paper.

1. Name the globular, encapsulated fat body prominent in the cheeks of infants.

2. What is its function?

3. Name a muscle involved with elevation of the mandible.

4. Name the secretions of each extrinsic salivary gland.

5. What is mastication?

6. What is deglutition?

7. What is the name for the "sockets" of the teeth?

8. Describe the submandibular duct. What is its function?

9. What is the general and special sensory innervation of the lingual nerve?

10. The lingual frenulum has been mentioned several times in this topic. What is the lingual frenulum?

11. Describe the Universal/National System used for numbering adult teeth.

12. Name the teeth, on both bones, important for biting and cutting.

13. Which teeth are the longest? What is their function?

14. Name all teeth important for grinding and crushing.

15. What foramina transmit the lesser palatine nerves and blood vessels?

16. Name the largest molar. Hint: There is a pair on each bone.

17. Name the pairs of teeth known as the wisdom teeth.

18. What nerves and blood vessels are transmitted through the incisive fossa?

The Esophagus

EXERCISE 12.4:
Digestive System—Esophagus—Anterior View

SELECT TOPIC	SELECT VIEW
Esophagus	Anterior

Click **LAYER 1** *in the* **LAYER CONTROLS** *window, and you will see the following image:*

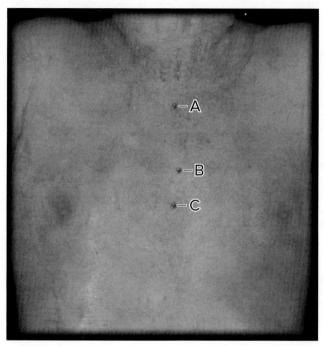

©McGraw-Hill Education

Mouse-over the pins on the screen to find the information necessary to identify the following nondigestive system structures:

A. _____

B. _____

C. _____

CHECK POINT

Esophagus, Anterior View

1. Where is the esophagus located?
2. Describe the esophagus.
3. What is its function?

Click **LAYER 2** *in the* **LAYER CONTROLS** *window, and you will see the following image:*

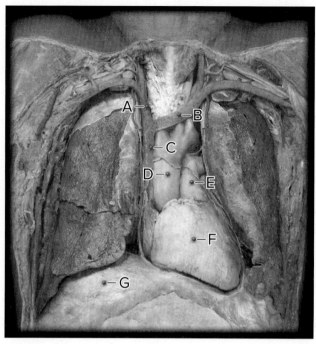

©McGraw-Hill Education

Mouse-over the pins on the screen to find the information necessary to identify the following nondigestive system structures:

A. _____

B. _____

C. _____

D. _____

E. _____

F. _____

G. _____

722 MODULE 12 The Digestive System

Click **LAYER 3** *in the* **LAYER CONTROLS** *window, and you will see the following image:*

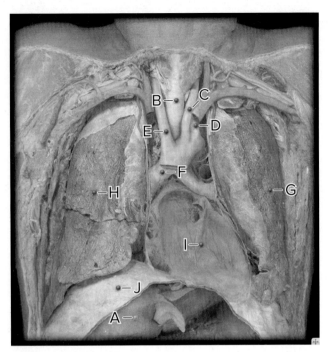

©McGraw-Hill Education

Mouse-over the pin on the screen to find the information necessary to identify the following structure:

A. _____

Nondigestive System Structures (blue pins)

B. _____
C. _____
D. _____
E. _____
F. _____
G. _____
H. _____
I. _____
J. _____

CHECK POINT

Esophagus, Anterior View, *continued*

4. List the functions of the liver.
5. What are hepatocytes?
6. What can lead to the destruction of hepatocytes?

Click **LAYER 4** *in the* **LAYER CONTROLS** *window, and you will see the following image:*

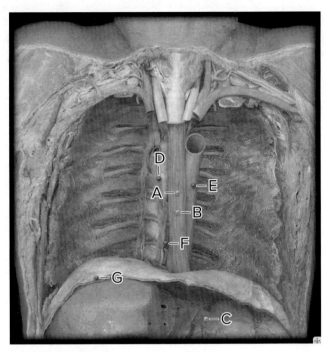

©McGraw-Hill Education

Mouse-over the pins on the screen to find the information necessary to identify the following structures:

A. _____
B. _____
C. _____

Nondigestive System Structures (blue pins)

D. _____
E. _____
F. _____
G. _____

CHECK POINT

Esophagus, Anterior View, *continued*

7. The esophagus is located in what three body regions?
8. Describe the structure of the esophagus.
9. What is peristalsis?

Click **LAYER 5** *in the* **LAYER CONTROLS** *window, and you will see the following image:*

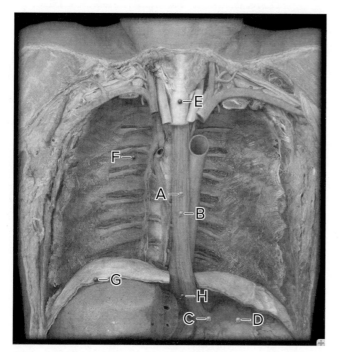

©McGraw-Hill Education

Mouse-over the pins on the screen to find the information necessary to identify the following structures:

A. _____
B. _____
C. _____
D. _____

Nondigestive System Structures (blue pins)

E. _____
F. _____
G. _____
H. _____

CHECK POINT

Esophagus, Anterior View, *continued*

10. The esophagus conveys food to which digestive organ?
11. What is reflux esophagitis?
12. Name the opening in the diaphragm for the passage of the esophagus.

EXERCISE 12.5a:
Digestive System—Esophagus—Histology

SELECT TOPIC	SELECT VIEW
Esophagus	LM: Low Magnification

Click the **TAGS ON/OFF** *button, and you will see the following image:*

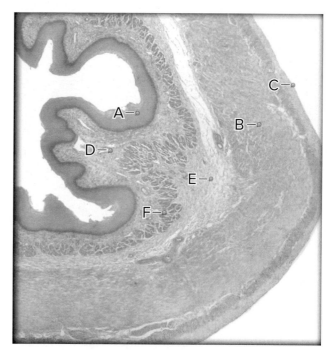

©McGraw-Hill Education/Greg Reeder

Mouse-over the pins on the screen to find the information necessary to identify the following structures:

A. _____
B. _____
C. _____
D. _____
E. _____
F. _____

724 MODULE 12 The Digestive System

EXERCISE 12.5b:
Digestive System—Esophagus

SELECT TOPIC	SELECT VIEW
Esophagus	LM: High Magnification

Click the **TAGS ON/OFF** *button, and you will see the following image:*

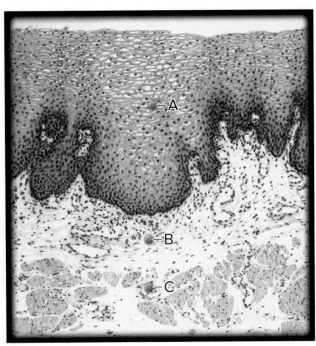

©McGraw-Hill Education/Al Telser

Mouse-over the pins on the screen to find the information necessary to identify the following structures:

A. _____
B. _____
C. _____

EXERCISE 12.5c:
Digestive System—Gastro-esophageal Junction

SELECT TOPIC	SELECT VIEW
Gastro-esophageal Junction	LM: High Magnification

Click the **TAGS ON/OFF** *button, and you will see the following image:*

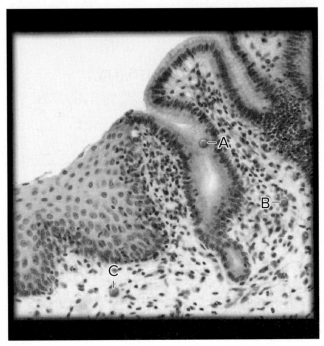

©McGraw-Hill Education/Al Telser

Mouse-over the pins on the screen to find the information necessary to identify the following structures:

A. _____
B. _____
C. _____

MODULE 12 The Digestive System 725

The Abdominal Cavity

EXERCISE 12.6:
Digestive System—Abdominal Cavity—Anterior View

SELECT TOPIC	SELECT VIEW
Abdominal Cavity	Anterior

Click **LAYER 1** in the **LAYER CONTROLS** window, and you will see the following image:

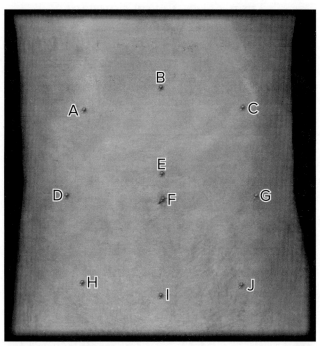

©McGraw-Hill Education

Mouse-over the pins on the screen to find the information necessary to identify the following nondigestive system structures:

A. right hypochondriac region
B. epigastric region
C. left hypochondriac region
D. right flank region
E. umbilical region
F. umbilicus
G. left flank region
H. right inguinal region
I. pubic region
J. left inguinal region

CHECK POINT
Abdominal Cavity, Anterior View

1. What is the umbilicus? Which abdominal region contains the umbilicus?
2. Name the abdominal region located superior to that region.
3. Name the abdominal region located inferior to the region in question 1.

Click **LAYER 2** in the **LAYER CONTROLS** window, and you will see the following image:

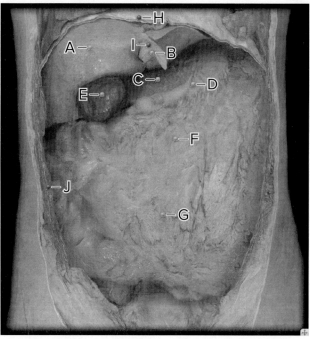

©McGraw-Hill Education

Mouse-over the pins on the screen to find the information necessary to identify the following structures:

A. liver
B. round ligament of liver
C. lesser omentum
D. stomach
E. gallbladder
F. transverse colon
G. greater omentum

Nondigestive System Structures (blue pins)

H. diaphragm
I. falciform ligament of liver
J. musculature of abdominal wall

CHECK POINT

Abdominal Cavity, Anterior View, *continued*

4. What is the greater omentum? What is its function?
5. What is the lesser omentum? Where is it located?
6. Describe the structure of the gallbladder. Where is it located?

Click **LAYER 3** *in the* **LAYER CONTROLS** *window, and you will see the following image:*

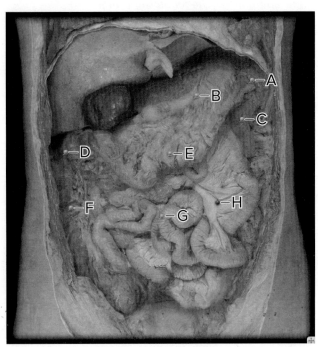

©McGraw-Hill Education

Mouse-over the pins on the screen to find the information necessary to identify the following structures:

A. left colic flexure
B. stomach
C. descending colon
D. right colic flexure
E. transverse colon
F. ascending colon
G. small intestine
H. mesentery of small intestine

CHECK POINT

Abdominal Cavity, Anterior View, *continued*

7. Name the four parts of the stomach from proximal to distal.
8. Name the location where the ascending colon joins the transverse colon. What is another name for this location? To what does this name refer?
9. Name the location where the transverse colon joins the descending colon. What is another name for this location? What does this name refer to?

Click **LAYER 4** *in the* **LAYER CONTROLS** *window, and you will see the following image:*

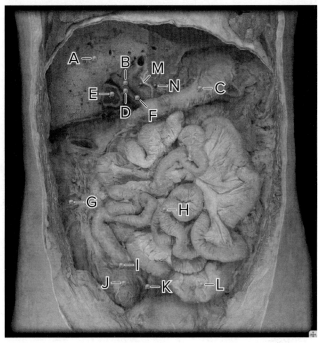

©McGraw-Hill Education

Mouse-over the pins on the screen to find the information necessary to identify the following structures:

A. liver
B. common hepatic duct
C. stomach
D. cystic duct
E. gallbladder
F. bile duct
G. ascending colon
H. small intestine
I. ileocecal junction
J. cecum
K. vermiform appendix
L. sigmoid colon

Nondigestive System Structures (blue pins)

M. _hepatic portal v._
N. _celiac a. and branches_

CHECK POINT

Abdominal Cavity, Anterior View, *continued*

10. Name the structure at the junction of the small intestine and large intestine. What part of each intestine is joined here?
11. Name the pouch of the large intestine into which the small intestine empties. What muscle regulates the flow from one to the other?
12. Name the slender hollow appendage attached to this pouch. What is its function? What structures does it contain? What is it called when the lumen of this structure becomes obstructed?

Click **LAYER 5** *in the* **LAYER CONTROLS** *window, and you will see the following image:*

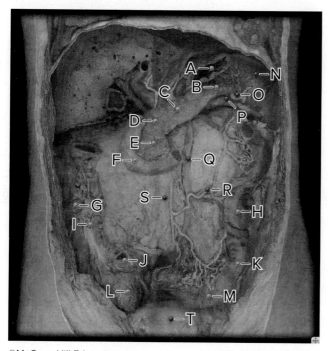

©McGraw-Hill Education

Mouse-over the pins on the screen to find the information necessary to identify the following structures:

A. _stomach_
B. _tail of pancreas_
C. _body of pancreas_
D. _neck of pancreas_
E. _head of pancreas_
F. _duodenum_
G. _ascending colon_
H. _descending colon_
I. _omental appendices_
J. _ileum_
K. _taeniae coli_
L. _cecum_
M. _sigmoid colon_

Nondigestive System Structures (blue pins)

N. _spleen_
O. _splenic a._
P. _splenic v._
Q. _inferior mesenteric v. & tributaries_
R. _inferior mesenteric a. & branches_
S. _abdominal aorta_
T. _urinary bladder with peritoneum_

CHECK POINT

Abdominal Cavity, Anterior View, *continued*

13. What are the taeniae coli? What is their function?
14. What are haustra?
15. Name the peritoneal appendages filled with fat. What is their function?

728 MODULE 12 The Digestive System

Click **LAYER 6** *in the* **LAYER CONTROLS** *window, and you will see the following image:*

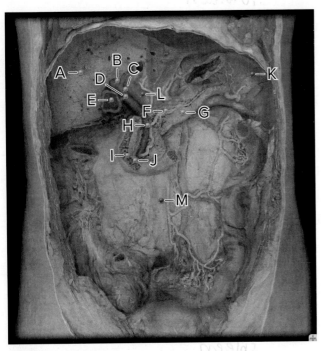

©McGraw-Hill Education

Mouse-over the pins on the screen to find the information necessary to identify the following structures:

A. liver
B. hepatic duct
C. common hepatic duct
D. cystic duct
E. gallbladder
F. pancreatic duct
G. pancreas
H. bile duct
I. duodenum
J. major duodenal papilla

Nondigestive System Structures (blue pins)

K. spleen
L. hepatic portal ✓
M. abdominal aorta

CHECK POINT

Abdominal Cavity, Anterior View, *continued*

16. Name and describe the largest lymphatic organ.
17. What is its function?

IN REVIEW

What Have I Learned?

The following questions cover the material that you have just learned, the abdominal cavity. Apply what you have learned to answer the following questions on a separate piece of paper.

1. Using the following figure, label the nine abdominal regions.

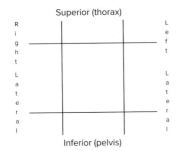

2. What is the function of the gallbladder?

3. Name the muscles of the abdominal wall from superficial to deep.

4. Name the four parts of the colon from proximal to distal.

5. Name the three parts of the small intestine, from proximal to distal.

6. Name the most mobile part of the large intestine.

7. Which parts of the colon absorb water? Which parts of the colon absorb electrolytes? Which parts of the colon store feces?

The Stomach

After viewing the animation, answer the following questions:

1. Where is the stomach located? Between which two organs?

2. What is the function of the stomach? What two processes contribute to this function?

3. What structures of the stomach form the superior and inferior borders?

4. Describe the four stomach regions.

5. How is the distal part of the stomach subdivided?

6. What is the function of the pyloric sphincter?

7. What are gastric rugae? What is their function?

8. Describe the four layers of the stomach.

9. Name the layers of the stomach muscularis. How does it compare to the rest of the digestive tract?

10. What substance is secreted by the mucous cells of the epithelium?

11. List three functions of gastric mucus.

12. Describe the gastric pits.

13. What are the four different cells found in gastric glands? What are their functions?

Self-Quiz

Take this opportunity to check your progress by taking the **QUIZ**. See the **Introduction Module** for a reminder on how to access the **QUIZ** for this Study Area.

EXERCISE 12.7a:
Digestive System—Stomach—Histology

 SELECT TOPIC: Stomach ▶ SELECT VIEW: LM: Low Magnification

Click the **TAGS ON/OFF** *button, and you will see the following image:*

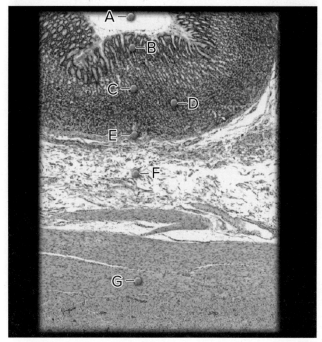

©McGraw-Hill Education/Al Telser

Mouse-over the pins on the screen to find the information necessary to identify the following structures:

A. _____
B. _____
C. _____
D. _____
E. _____
F. _____
G. _____

730 MODULE 12 The Digestive System

EXERCISE 12.7b:
Digestive System—Stomach—Histology

SELECT TOPIC	SELECT VIEW
Stomach	LM: Medium Magnification

Click the **TAGS ON/OFF** *button, and you will see the following image:*

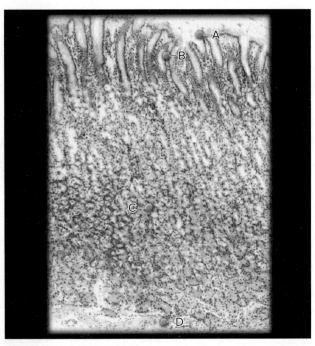

©McGraw-Hill Education/Al Telser

Mouse-over the pins on the screen to find the information necessary to identify the following structures:

A. _____

B. _____

C. _____

D. _____

EXERCISE 12.7c:
Digestive System—Stomach—Histology

SELECT TOPIC	SELECT VIEW
Stomach	LM: High Magnification

Click the **TAGS ON/OFF** *button, and you will see the following image:*

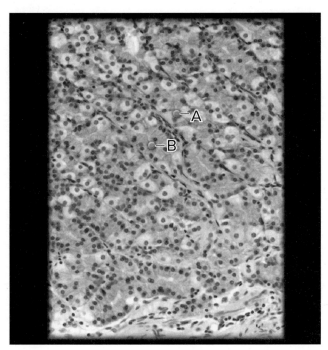

©McGraw-Hill Education/Al Telser

Mouse-over the pins on the screen to find the information necessary to identify the following structures:

A. _____

B. _____

EXERCISE 12.7d:
Digestive System—Stomach—Histology

SELECT TOPIC	SELECT VIEW
Stomach	SEM: Low Magnification

Click the **TAGS ON/OFF** *button, and you will see the following image:*

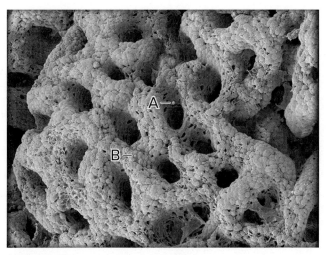

©Science Photo Library/Alamy Stock Photo

Mouse-over the pins on the screen to find the information necessary to identify the following structures:

A. _____

B. _____

CHECK POINT

Stomach, Histology

1. Describe the muscularis mucosae. What is its function?
2. Describe the gastric pits. What is their function?
3. Describe the gastric glands. What is their function?

EXERCISE 12.8:
Digestive System—Stomach and Duodenum— Anterior View

SELECT TOPIC	SELECT VIEW
Stomach and Duodenum	Anterior

Click **LAYER 1** *in the* **LAYER CONTROLS** *window, and you will see the following image:*

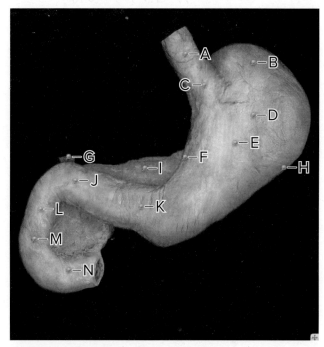

©McGraw-Hill Education

Mouse-over the pins on the screen to find the information necessary to identify the following structures:

A. _____
B. _____
C. _____
D. _____
E. _____
F. _____
G. _____
H. _____
I. _____
J. _____
K. _____
L. _____
M. _____
N. _____

MODULE 12 The Digestive System

CHECK POINT

Stomach and Duodenum, Anterior View

1. Name the part of the stomach located at the junction with the esophagus. What structure does it contain?
2. Name the dome-shaped superior part of the stomach. What is usually retained there?
3. Name the terminal part of the stomach. Name the ring of muscle located there. What is the function of this part of the stomach and the muscular ring?

Click **LAYER 2** *in the* **LAYER CONTROLS** *window, and you will see the following image:*

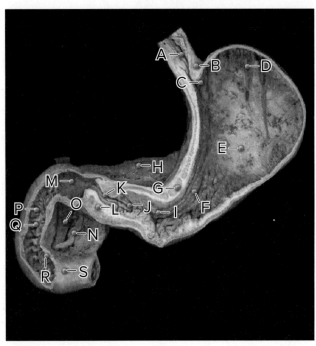

©McGraw-Hill Education

Mouse-over the pins on the screen to find the information necessary to identify the following structures:

A. _____
B. _____
C. _____
D. _____
E. _____
F. _____
G. _____
H. _____
I. _____
J. _____
K. _____
L. _____
M. _____
N. _____
O. _____
P. _____
Q. _____
R. _____
S. _____

CHECK POINT

Stomach and Duodenum, Anterior View, *continued*

4. Name the muscular structure that prevents reflux of stomach contents. Name two times that this muscle relaxes.
5. Name the structures that allow the stomach to expand as it fills.
6. What is the major duodenal papilla? What is its function?

SELECT ANIMATION

Gastric Secretion   PLAY

After viewing the animation, answer the following questions:

1. List the three phases of gastric secretion.

2. What initiates the cephalic phase?

3. Describe the parasympathetic response involved in gastric secretion.

4. What role does gastrin play?

5. What initiates the gastric phase?

6. Describe the parasympathetic reflex and the direct stimulation that results in this phase.

7. What is the net result of these nervous stimuli?

8. What initiates the intestinal phase?

9. What two events inhibit gastric secretion?

10. Describe the three steps involved in this inhibition.

Self-Quiz
Take this opportunity to check your progress by taking the **QUIZ**. See the **Introduction Module** for a reminder on how to access the **QUIZ** for this Study Area.

SELECT ANIMATION
HCl Production
PLAY

After viewing the animation, answer the following questions:

1. Which cells produce hydrochloric acid in the stomach? Where are these cells located?

2. What enzyme is involved in the formation of carbonic acid?

3. What two substrates are involved in this reaction?

4. Carbonic acid dissociates into _____.

5. Which ion is transported back to the bloodstream?

6. What exchange occurs at the ion exchange molecule in the plasma membrane?

7. How are the hydrogen ions transported into the duct of the gastric gland?

8. What movement occurs with the chloride ions?

9. What ions are transported into the parietal cell in exchange for the hydrogen ions?

10. What is the result of this movement?

Self-Quiz
Take this opportunity to check your progress by taking the **QUIZ**. See the **Introduction Module** for a reminder on how to access the **QUIZ** for this Study Area.

734 MODULE 12 The Digestive System

IN REVIEW

What Have I Learned?

The following questions cover the material that you have just learned, the stomach. Apply what you have learned to answer the following questions on a separate piece of paper.

1. Name each different type of gastric gland and the products that they produce.

2. Name the structure that conducts mucus and secretions from the gastric glands to the lumen of the stomach.

3. What is a lumen?

4. Name the elongated, nodular gland posterior to the stomach. Describe it.

5. What are the functions of this elongated, nodular gland?

6. Name the part of the small intestine to receive ingested material from the stomach. What is that ingested material called?

7. List the functions of the structure in question 6.

8. Name the four parts of this structure.

9. What is the function of the bile duct? Where is it located?

10. What is the function of the pancreatic duct? Where is it located?

The Small Intestine

EXERCISE 12.9:

Digestive System—Duodenum—Histology

SELECT TOPIC	SELECT VIEW
Duodenum	LM: Low Magnification

Click the **TAGS ON/OFF** button, and you will see the following image:

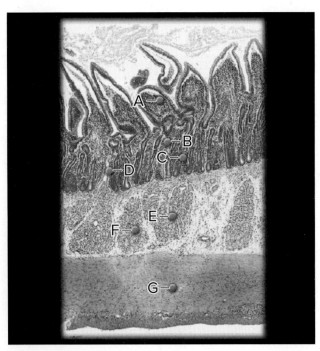

©McGraw-Hill Education/Dennis Strete

Mouse-over the pins on the screen to find the information necessary to identify the following structures:

A. _____
B. _____
C. _____
D. _____
E. _____
F. _____
G. _____

CHECK POINT

Small Intestine (Duodenum), Histology

1. Name and describe the structures that increase the surface area of the small intestine. What is their function?
2. Describe the structures also known as the crypts of Lieberkühn. What are their functions?
3. What are submucosa glands of the duodenum? Where are they located? What is their function?

MODULE 12 The Digestive System 735

EXERCISE 12.10a:
Digestive System—Jejunum—Histology

SELECT TOPIC: Jejunum ▸ **SELECT VIEW**: LM: Medium Magnification

Click the **TAGS ON/OFF** *button, and you will see the following image:*

©McGraw-Hill Education/Al Telser

Mouse-over the pins on the screen to find the information necessary to identify the following structures:

A. _____

B. _____

C. _____

D. _____

E. _____

F. _____

G. _____

EXERCISE 12.10b:
Digestive System—Jejunum—Histology

SELECT TOPIC: Jejunum ▸ **SELECT VIEW**: LM: High Magnification

Click the **TAGS ON/OFF** *button, and you will see the following image:*

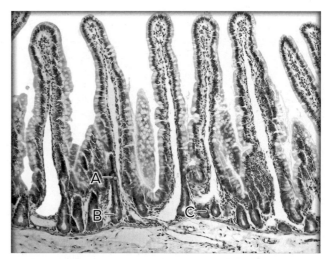

©McGraw-Hill Education/Al Telser

Mouse-over the pins on the screen to find the information necessary to identify the following structures:

A. _____

B. _____

C. _____

CHECK POINT

Jejunum, Histology

1. Name the cells that produce and release mucin into the intestinal lumen.
2. Where are these cells located?

736 MODULE 12   The Digestive System

EXERCISE 12.11a:
Digestive System—Ileum—Histology

SELECT TOPIC: Ileum → SELECT VIEW: LM: Low Magnification

Click the **TAGS ON/OFF** *button, and you will see the following image:*

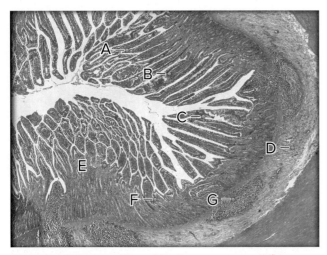

©McGraw-Hill Education/Steve Sullivan

Mouse-over the pins on the screen to find the information necessary to identify the following structures:

A. _____
B. _____
C. _____
D. _____
E. _____
F. _____
G. _____

EXERCISE 12.11b:
Digestive System—Ileum—Histology

SELECT TOPIC: Ileum → SELECT VIEW: LM: High Magnification

Click the **TAGS ON/OFF** *button, and you will see the following image:*

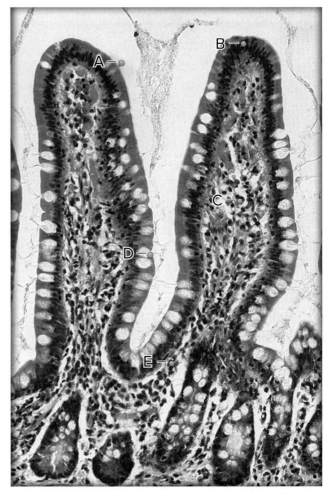

©Image Source/Getty Images

Mouse-over the pins on the screen to find the information necessary to identify the following structures:

A. _____
B. _____
C. _____
D. _____
E. _____

MODULE 12 The Digestive System 737

EXERCISE 12.11c:
Digestive System—Ileum—Histology

SELECT TOPIC	SELECT VIEW
Ileum	TEM: Medium Magnification

Click the **TAGS ON/OFF** *button, and you will see the following image:*

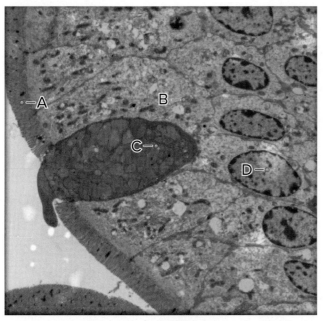

©EM Research Services, Newcastle University

Mouse-over the pins on the screen to find the information necessary to identify the following structures:

A. _____
B. _____
C. _____
D. _____

EXERCISE 12.12:
Digestive System—Small Intestine—Histology

SELECT TOPIC	SELECT VIEW
Small Intestine	SEM: Low Magnification

Click the **TAGS ON/OFF** *button, and you will see the following image:*

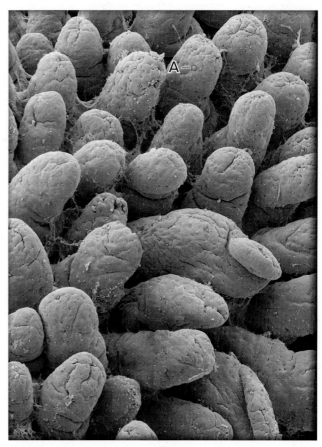

©Science Photo Library/Alamy Stock Photo

Mouse-over the pin on the screen to find the information necessary to identify the following structure:

A. _____

EXERCISE 12.13:
Digestive System—Peyer's Patch—Histology

SELECT TOPIC	SELECT VIEW
Peyer's Patch	LM: Medium Magnification

Click the **TAGS ON/OFF** button, and you will see the following image:

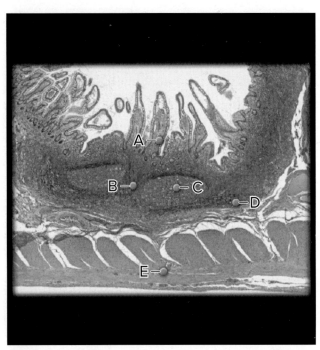

©McGraw-Hill Education/Al Telser

Mouse-over the pins on the screen to find the information necessary to identify the following structures:

A. _____
B. _____
C. _____
D. _____
E. _____

EXERCISE 12.14:
Digestive System—Intestinal Microvilli—Histology

SELECT TOPIC	SELECT VIEW
Intestinal Microvilli	TEM: High Magnification

Click the **TAGS ON/OFF** button, and you will see the following image:

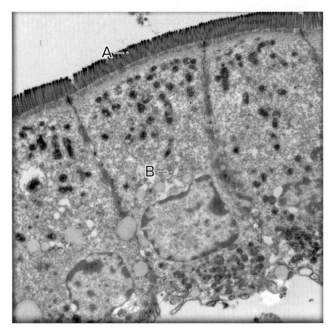

©EM Research Services, Newcastle University

Mouse-over the pins on the screen to find the information necessary to identify the following structures:

A. _____
B. _____

EXERCISE 12.15:
Imaging—Stomach and Small Intestine

SELECT TOPIC	SELECT VIEW
Stomach and Small Intestine	X Ray: Anterior–Posterior

Click the **TAGS ON/OFF** button, and you will see the following image:

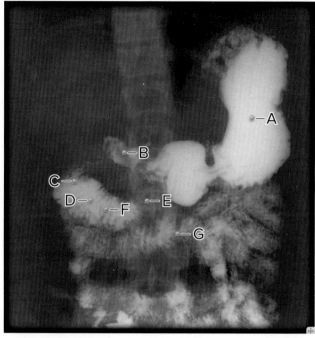

©McGraw-Hill Education

Mouse-over the pins on the screen to find the information necessary to identify the following structures:

A. _____
B. _____
C. _____
D. _____
E. _____
F. _____
G. _____

EXERCISE 12.16:
Imaging—Digestive—Small Intestine

SELECT TOPIC	SELECT VIEW
Small Intestine	CT: Axial

Click the **TAGS ON/OFF** button, and you will see the following image:

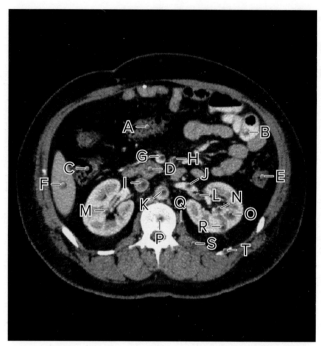

©McGraw-Hill Education

Mouse-over the pins on the screen to find the information necessary to identify the following structures:

A. _____
B. _____
C. _____
D. _____
E. _____
F. _____

Nondigestive System Structures (blue pins)

G. _____
H. _____
I. _____
J. _____
K. _____

740 MODULE 12 The Digestive System

L. _____
M. _____
N. _____
O. _____
P. _____

Q. _____
R. _____
S. _____
T. _____

IN REVIEW

What Have I Learned?

The following questions cover the material that you have just learned, the small intestine. Apply what you have learned to answer these questions on a separate piece of paper.

1. What are villi? What is their function?

2. Which layer regulates the length of the villi?

3. The contraction of which layer compresses the glands of the mucosa to expel their contents?

4. Name and describe the layer responsible for peristalsis.

5. Which layer contains blood and lymph vessels, lymphoid nodules, and nerve plexus?

6. What is the function of the submucosal glands?

7. What cell types are found on the epithelium of the small intestine?

8. What are the functions of these cells?

9. Name two phenomena that change the shape of the villi.

10. What layer of the small intestine contains Peyer's patches? In what part of the small intestine are they found? What is their function?

The Colon

EXERCISE 12.17a:
Digestive System—Colon—Histology

SELECT TOPIC	SELECT VIEW
Colon	LM: Low Magnification

*Click the **TAGS ON/OFF** button, and you will see the following image:*

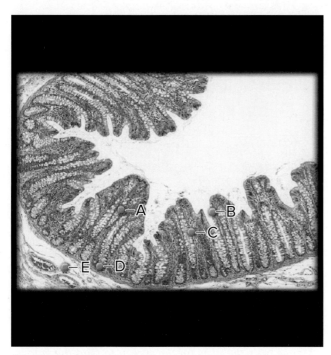

©Dr. Thomas Caceci, Virginia-Maryland Regional College of Veterinary Medicine

Mouse-over the pins on the screen to find the information necessary to identify the following structures:

A. _____
B. _____
C. _____
D. _____
E. _____

EXERCISE 12.17b:
Digestive System—Colon—Histology

SELECT TOPIC	SELECT VIEW
Colon	LM: Medium Magnification

*Click the **TAGS ON/OFF** button, and you will see the following image:*

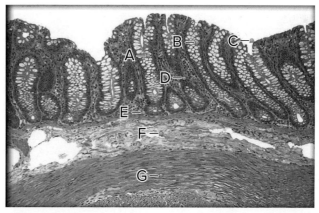

©Victor Eroschenko

Mouse-over the pins on the screen to find the information necessary to identify the following structures:

A. _____
B. _____
C. _____
D. _____
E. _____
F. _____
G. _____

CHECK POINT
Colon, Histology

1. Name and describe the tubular glands located in the colon.
2. What is their function?
3. What is the dominant cell-type in these glands?
4. Name the tissue-type that lines the colon. What is its function?
5. The contraction of which muscles compresses the glands of the mucosa and expels their contents?
6. Name the layer of the colon containing blood and lymph vessels.

EXERCISE 12.18a
Imaging—Digestive—Colon

SELECT TOPIC: Colon ▸ SELECT VIEW: CT: Axial

Click the **TAGS ON/OFF** button, and you will see the following image:

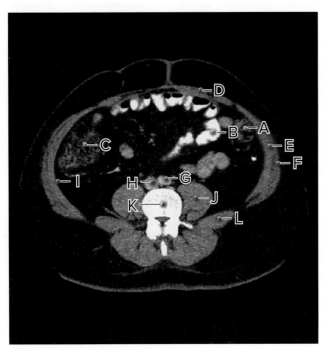

©Victor Eroschenko

Mouse-over the pins on the screen to find the information necessary to identify the following structures:

A. _____
B. _____
C. _____

Nondigestive System Structures (blue pins)

D. _____
E. _____
F. _____
G. _____
H. _____
I. _____
J. _____
K. _____
L. _____

EXERCISE 12.28b:
Imaging—Colon

SELECT TOPIC: Colon ▸ SELECT VIEW: X ray: Anterior–Posterior

Click the **TAGS ON/OFF** button, and you will see the following image:

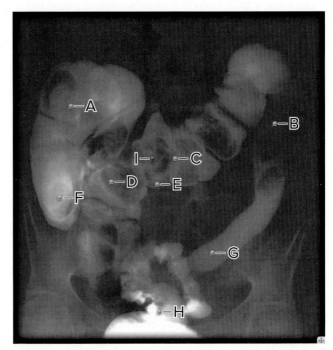

©McGraw-Hill Education

Mouse-over the pins on the screen to find the information necessary to identify the following structures:

A. _____
B. _____
C. _____
D. _____
E. _____
F. _____
G. _____
H. _____

Nondigestive System Structure (blue pin)

I. _____

IN REVIEW

What Have I Learned?

The following questions cover the material that you have just learned, the histology of the colon. Apply what you have learned to answer the following questions on a separate piece of paper.

1. Name and describe the tubular glands located in the colon.

2. What is their function?

3. What is the dominant cell-type in these glands?

4. Name the tissue-type that lines the colon. What is its function?

5. The contraction of which muscles compresses the glands of the mucosa and expels their contents?

6. Name the layer of the colon containing blood and lymph vessels.

The Peritoneum

EXERCISE 12.19:
Digestive System—Peritoneum—Midsagittal View

SELECT TOPIC: Peritoneum ▶ SELECT VIEW: Midsagittal

Click **LAYER 1** in the **LAYER CONTROLS** window, and you will see the following image:

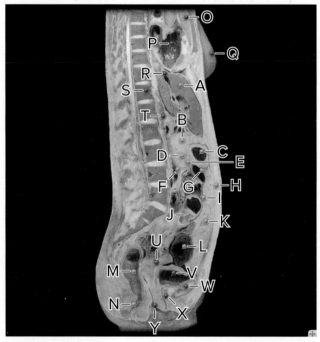

©McGraw-Hill Education

Mouse-over the pins on the screen to find the information necessary to identify the following structures:

A. _____
B. _____
C. _____
D. _____
E. _____
F. _____
G. _____
H. _____
I. _____
J. _____
K. _____
L. _____
M. _____
N. _____

Nondigestive System Structures (blue pins)

O. _____
P. _____
Q. _____
R. _____
S. _____
T. _____
U. _____
V. _____
W. _____
X. _____
Y. _____

MODULE 12 The Digestive System

CHECK POINT

Peritoneum, Midsagittal View

1. Name the single layer of serous membrane that coats the outer surface of many abdominal organs.
2. What is its function?
3. Name the single layer of serous membrane that lines the wall of the abdomen.

Self-Quiz

Take this opportunity to check your progress by taking the **QUIZ**. See the **Introduction Module** for a reminder on how to access the **QUIZ** for this Study Area.

IN REVIEW

What Have I Learned?

The following questions cover the material you have just learned, the peritoneum. Apply what you have learned to answer these questions on a separate piece of paper.

1. Name the single layer of serous membrane that coats the outer surface of many abdominal organs.

2. What is its function?

3. Name the single layer of serous membrane that lines the wall of the abdomen.

4. Describe the transverse mesocolon.

5. What is its function?

6. What is the peritoneal cavity? What does this term mean?

7. Name the double-layered fold of peritoneum capable of storing large amounts of fat.

8. How is it associated with the immune system?

The Liver

SELECT ANIMATION — Liver — **PLAY**

After viewing the animation, answer the following questions:

1. The liver is the _____ internal organ of the body.

2. Where is it located?

3. Name the four lobes of the liver. Which two are part of another lobe?

4. Name and describe the structure that separates the two anterior lobes.

5. What are the two fetal remnants found on the liver?

6. What structures are located in the porta hepatis?

7. Histologically, the liver is composed of functional units called _____ _____.

8. The hub of these functional units is a _____ _____.

9. Describe the portal triad.

10. Describe the sinusoids. How are they arranged?

11. How are the hepatocytes arranged?

12. What are the two basic functions of the liver?

13. Name the three categories of this task.

14. Describe these three categories of tasks.

Self-Quiz

Take this opportunity to check your progress by taking the **QUIZ**. See the **Introduction Module** for a reminder on how to access the **QUIZ** for this Study Area.

MODULE 12 The Digestive System 745

EXERCISE 12.20:
Digestive System—Liver—Anterior and Postero—inferior Views

SELECT TOPIC	SELECT VIEW
Liver	Anterior–Postero—inferior

Click **LAYER 1** in the **LAYER CONTROLS** window, and you will see the following image:

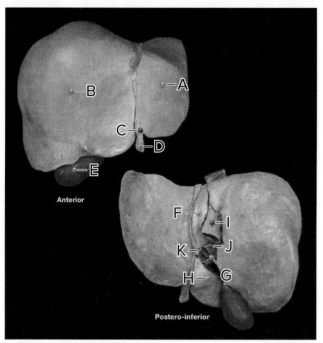

©McGraw-Hill Education

Mouse-over the pins on the screen to find the information necessary to identify the following structures:

A. left lobe of liver
B. right " " "
C. falciform ligament of liver
D. round ligament of liver
E. gallbladder
F. caudate lobe
G. common hepatic duct
H. quadrate lobe of liver

Nondigestive System Structures (blue pins)

I. inferior vena cava
J. hepatic portal v
K. hepatic a. proper

CHECK POINT
Liver, Anterior and Postero-inferior Views

1. Name the largest lobe of the liver. What two lobes make up this lobe?
2. Name the structures that separate the left and right lobes of the liver.
3. Name the functions of the liver.

EXERCISE 12.21a:
Digestive System—Liver—Histology

SELECT TOPIC	SELECT VIEW
Liver	LM: Low Magnification

Click the **TAGS ON/OFF** button, and you will see the following image:

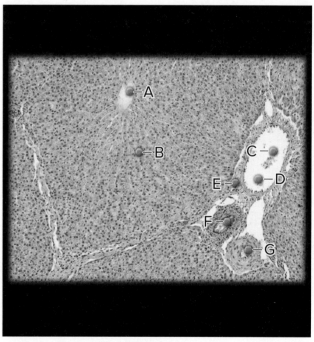

©McGraw-Hill Education/Dennis Strete

Mouse-over the pins on the screen to find the information necessary to identify the following structures:

A.
B.
C.
D.
E.
F.
G.

EXERCISE 12.21b:
Digestive System—Liver—Histology

SELECT TOPIC	SELECT VIEW
Liver	LM: Medium Magnification

Click the **TAGS ON/OFF** *button, and you will see the following image:*

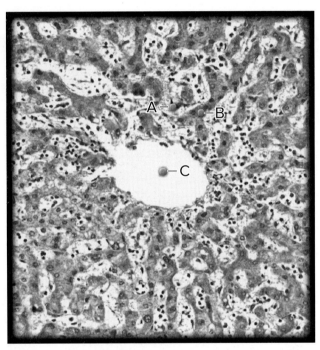

©McGraw-Hill Education/Al Telser

Mouse-over the pins on the screen to find the information necessary to identify the following structures:

A. _____

B. _____

C. _____

EXERCISE 12.21c:
Digestive System—Liver—Histology

SELECT TOPIC	SELECT VIEW
Liver	LM: High Magnification

Click the **TAGS ON/OFF** *button, and you will see the following image:*

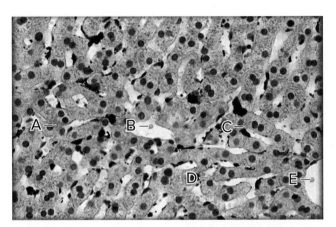

©Victor Eroschenko

Mouse-over the pins on the screen to find the information necessary to identify the following structures:

A. _____

B. _____

C. _____

D. _____

E. _____

EXERCISE 12.21d:
Digestive System—Liver—Histology

SELECT TOPIC	SELECT VIEW
Liver	TEM: Medium Magnification

Click the **TAGS ON/OFF** button, and you will see the following image:

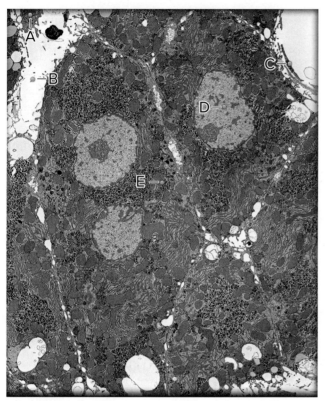

©SPL/Science Source

Mouse-over the pins on the screen to find the information necessary to identify the following structures:

A. _____
B. _____
C. _____
D. _____
E. _____

EXERCISE 12.22:
Digestive System—Hepatocyte—Histology

SELECT TOPIC	SELECT VIEW
Hepatocyte	TEM: High Magnification

Click the **TAGS ON/OFF** button, and you will see the following image:

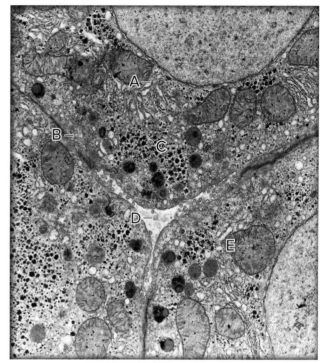

©Biophoto/Science Source

Mouse-over the pins on the screen to find the information necessary to identify the following structures:

A. _____
B. _____
C. _____
D. _____
E. _____

CHECK POINT

Liver, Histology

1. Name the liver cells arranged in cords within a liver lobule. What are their functions?
2. What are hepatic sinusoids? What is their function?
3. Name the structure that brings venous blood to the liver sinusoids.

EXERCISE 12.23:
Digestive System—Biliary Ducts—Anterior View

SELECT TOPIC: **Biliary Ducts** ▶ SELECT VIEW: **Anterior**

Click **LAYER 1** *in the* **LAYER CONTROLS** *window, and you will see the following image:*

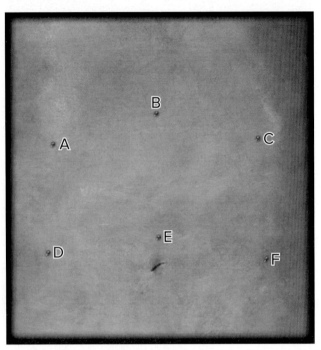

©McGraw-Hill Education

Mouse-over the pins on the screen to find the information necessary to identify the following nondigestive system structures:

A. right hypochondriac region
B. epigastric region
C. left hypochondriac region
D. right flank region
E. umbilical region
F. left flank region

Click **LAYER 2** *in the* **LAYER CONTROLS** *window, and you will see the following image:*

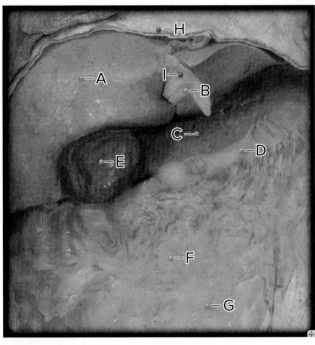

©McGraw-Hill Education

Mouse-over the pins on the screen to find the information necessary to identify the following structures:

A. liver
B. round ligament of liver
C. lesser omentum
D. stomach
E. gallbladder
F. transverse colon
G. greater omentum

Nondigestive System Structures (blue pins)

H. diaphragm
I. falciform ligament of liver

Click **LAYER 3** in the **LAYER CONTROLS** window, and you will see the following image:

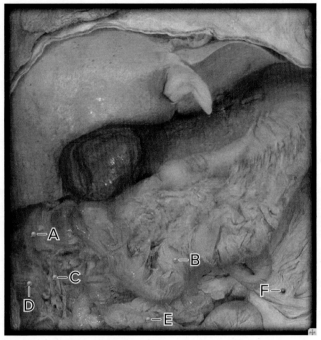

©McGraw-Hill Education

Mouse-over the pins on the screen to find the information necessary to identify the following structures:

A. right colic flexure
B. transverse colon
C. omental appendices
D. ascending colon
E. small intestine
F. mesentery of small intestine

Click **LAYER 4** in the **LAYER CONTROLS** window, and you will see the following image:

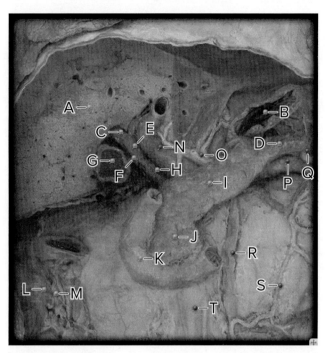

©McGraw-Hill Education

Mouse-over the pins on the screen to find the information necessary to identify the following structures:

A. liver
B. stomach
C. hepatic duct
D. tail of pancreas
E. common hepatic duct
F. cystic duct
G. gallbladder
H. bile duct
I. body of pancreas
J. head of pancreas
K. duodenum
L. omental appendices
M. ascending colon

Nondigestive System Structures (blue pins)

N. hepatic portal v.
O. celiac a. and branches
P. splenic v.
Q. splenic a.
R. inferior mesenteric v. & tributaries
S. inferior mesenteric a. & branches
T. abdominal aorta

MODULE 12 The Digestive System

CHECK POINT
Biliary Ducts, Anterior View

1. Name the two structures that receive bile from the liver. What structure do they form when they merge?
2. Bile passes from the _____ to the _____ for storage and from the _____ to the _____ for _____ of _____.
3. What two ducts unite to form the bile duct?

CHECK POINT
Biliary Ducts, Anterior View, *continued*

4. Name the structure that transmits pancreatic secretions.
5. Name the structure formed by the joining of this structure and the bile duct.
6. Into which section of the small intestine does the structure in question 5 empty?

Click **LAYER 5** *in the* **LAYER CONTROLS** *window, and you will see the following image:*

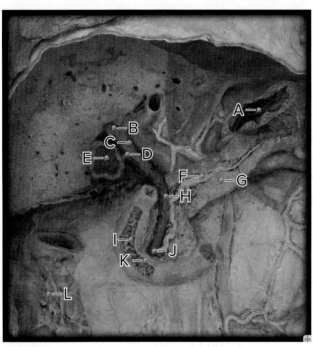

©McGraw-Hill Education

Mouse-over the pins on the screen to find the information necessary to identify the following structures:

A. stomach
B. hepatic duct
C. common hepatic duct
D. cystic duct
E. gallbladder
F. pancreatic duct
G. pancreas
H. bile duct
I. duodenum
J. hepatopancreatic ampulla
K. major duodenal papilla
L. ascending colon

Clinical Application

SELECT ANIMATION
Liver Function/Liver Failure (3D) PLAY

After viewing the animation, answer the following questions:

1. The liver is the _____ _____ organ in the body.

2. It consists of _____ lobes composed of hundreds of _____.

3. Each lobule, the _____ _____ element of the liver, metabolizes _____, _____, and _____.

4. Lobules process _____, the _____ that is the main _____ _____ for cells.

5. _____ from the digestive tract flows into the _____ where the _____, the primary cell type in the liver, store excess _____ as _____ and distribute _____ to the body.

6. _____ also secrete _____, which helps the body digest _____.

7. _____ cells reside in the liver and help to remove _____ and _____ from the blood.

8. Hepatocytes also produce _____ _____ _____ to prevent _____.

9. The other vital functions performed by the liver include . . .

10. Why do toxins accumulate in the liver? What can this accumulation cause?

11. What is the most common liver failure? How is it most often caused?

12. How does chronic liver failure damage the liver?

13. How are dying liver cells replaced? What do these form?

14. What does this scar tissue block? What is the result? What does this gradually diminish?

15. Is acute liver failure more or less common than chronic liver failure?

16. List common causes of acute liver failure.

17. What effect these have on the liver?

18. Is cirrhosis reversible?

19. Where does treatment for cirrhosis focus?

20. What is the only definitive treatment for chronic liver disease?

21. Which forms of hepatitis can be prevented with vaccines?

22. What medications block the replication of viral hepatitis?

23. How is acute liver failure caused by an overdose of acetaminophen treated?

24. How soon must Mucomyst be administered to help the liver safely excrete acetaminophen metabolites?

IN REVIEW

What Have I Learned?

The following questions cover the material that you have just learned, the liver. Apply what you have learned to answer the following questions on a separate piece of paper.

1. Name the structure that is the remnant of the umbilical vein of the fetus.

2. Name the blood vessel that carries absorbed products of digestion to the liver.

3. Name the organ that stores, concentrates, and releases bile.

4. Name the structure that brings arterial blood to the liver sinusoids.

5. Name the structure in the liver lobule that collects and transports bile toward the bile duct.

6. What is the hepatic portal triad?

7. Name the elevation on the wall of the duodenum that contains the hepatopancreatic ampulla.

8. What is the exocrine function of the pancreas?

9. What is the endocrine function of the pancreas?

10. What hormones are produced by the pancreas?

The Pancreas

EXERCISE 12.24:
Digestive System—Pancreas (Exocrine)—Histology

SELECT TOPIC	SELECT VIEW
Pancreas (Exocrine)	LM: High Magnification

Click the **TAGS ON/OFF** button, and you will see the following image:

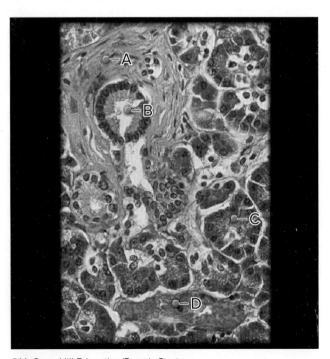

©McGraw-Hill Education/Dennis Strete

Mouse-over the pins on the screen to find the information necessary to identify the following structures:

A. _____

B. _____

C. _____

D. _____

CHECK POINT

Pancreas (Exocrine), Histology

1. Name the oval collection of secretory (acinar) cells in the exocrine pancreas.
2. What do these cells secrete? How are these secretions regulated?
3. Name the series of ducts that transport digestive enzymes through the exocrine pancreas. Where is the destination of these enzymes?
4. Name the structure that drains blood from, and contains hormones from, the endocrine pancreas.
5. Name the structure that separates the pancreatic lobules.

MODULE 12 The Digestive System 753

EXERCISE 12.25a:
Digestive System—Pancreas—Histology

SELECT TOPIC: Pancreas ▶ SELECT VIEW: LM: Low Magnification

Click the **TAGS ON/OFF** *button, and you will see the following image:*

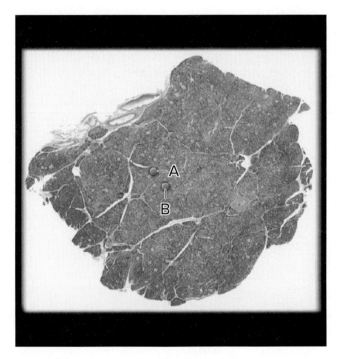

©Dr. Thomas Caceci, Virginia-Maryland Regional College of Veterinary Medicine

Mouse-over the pins on the screen to find the information necessary to identify the following structures:

A. _____

B. _____

EXERCISE 12.25b:
Digestive System—Pancreas—Histology

SELECT TOPIC: Pancreas ▶ SELECT VIEW: LM: Medium Magnification

Click the **TAGS ON/OFF** *button, and you will see the following image:*

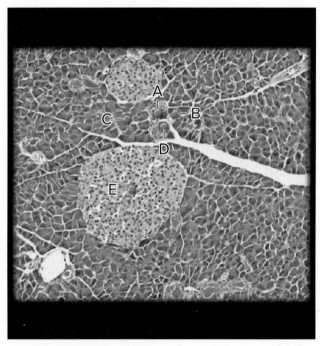

©McGraw-Hill Education/Al Telser

Mouse-over the pins on the screen to find the information necessary to identify the following structures:

A. _____

B. _____

C. _____

D. _____

E. _____

IN REVIEW

What Have I Learned?

The following questions cover the material that you have just learned, the histology of the pancreas. Apply what you have learned to answer the following questions on a separate piece of paper.

1. Name the oval collection of secretory (acinar) cells in the exocrine pancreas.

2. What do these cells secrete? How are these secretions regulated?

3. Name the series of ducts that transport digestive enzymes through the exocrine pancreas. Where is the destination of these enzymes?

4. Name the structure that drains blood from, and contains hormones from, the endocrine pancreas.

5. Name the structure that separates the pancreatic lobules.

ATP Synthesis

We have been discovering the structures involved in digestion—the process of breaking down our food into smaller, simple molecules that can be absorbed into the bloodstream. But that is not the end of the story. Amino acids will be recombined by your body to build proteins, fatty acids, and glycerol will be recombined into energy-storing fat molecules, and glucose will either be stored as glycogen molecules in your liver and muscle tissues or used to fuel the production of energy in the form of ATP. Any of these molecules can serve to supply the fuel to feed the ATP production process, but they must be converted into glucose first. The following animations will walk you through the dynamics of digesting the foods you eat. As you view these animations, answer the questions that follow each animation to check your comprehension of the information presented.

SELECT ANIMATION
Chemical Digestion and Absorption (3D)
PLAY

After viewing the animation, answer the following questions:

1. List the structures included in the digestive tract.

2. List the accessory organs of the digestive tract. What is their function?

3. List the functions of the digestive tract.

4. Where does digestion begin?

5. What is the function of saliva?

6. What is the function of amylase, an enzyme in saliva?

7. Food moves through the _____ and enters the _____, where _____ _____ and _____ _____ continue breaking it down.

8. The resultant breakdown product, called _____, contains _____, small _____, _____, _____, _____ and _____.

9. Chyme exits the stomach and enters the _____ _____.

10. The _____ _____ digests and absorbs each component of _____.

11. _____ _____ eventually break down _____ into several simple sugars called _____.

12. _____ _____ _____ are responsible for transporting _____ across the _____ _____ _____ into the _____ using _____ _____.

13. After transport into the cell, _____ _____ move the _____ out of the cell, and eventually into the _____ for use by the _____.

14. Enzymes called _____ break down proteins into _____ _____ like _____.

15. Amino acids such as _____ are _____ with _____ ions via the _____ _____ _____.

16. Various _____ _____ _____ then transfer _____ _____ into the _____ _____, where they are available to build _____ needed by the body.

17. Minerals such as _____ are _____ with _____ and _____ _____.

18. _____ _____ _____ emulsify _____.

19. Then, _____ and _____ _____ digest them into _____ _____ and _____.

20. Bile acid droplets called _____ absorb the _____ _____ and _____ as well as any _____ _____ _____ in the chyme and deliver them to the _____ _____ _____ for _____.

21. Within the cell, the _____ and _____ _____ _____ are packaged into _____, which are delivered to the _____ or _____, in the _____ _____ for transport to the _____ system and eventual return to the _____.

SELECT ANIMATION
Hydrolysis of Sucrose   PLAY

After viewing the animation, answer the following questions:

1. The enzyme sucrase breaks the disaccharide _____ into two monosaccharides: _____, or _____ sugar, and _____, or _____ sugar.

2. Where does this reaction occur?

3. For hydrolysis to occur, the sucrose must bind to what part of the sucrase enzyme?

4. What happens to the enzyme when this occurs? What happens to the sucrose?

5. What molecule breaks the bond? Do you see why the process is called hydrolysis? hydro = water and lysis = to break. The bond is broken by water.

6. After the bond is broken and the two monosaccharides are released, what happens to the enzyme?

7. How many times can this process be repeated?

8. What three events can occur to end this process?

Self-Quiz
Take this opportunity to check your progress by taking the **QUIZ**. See the **Introduction Module** for a reminder on how to access the **QUIZ** for this Study Area.

Clinical Application

SELECT ANIMATION
Lactose Intolerance (3D)   PLAY

After viewing the animation, answer the following questions:

1. Lactose intolerance is an example of _____ _____.

2. Lactose is the major _____ in _____ and _____ _____.

3. Normally, ingestion of lactose stimulates cells lining the _____ _____ to secrete a _____ _____ called _____.

4. _____ divides the _____ into _____ _____ _____ which are easily absorbed.

5. In lactose intolerance, _____ is either _____ or _____.

6. After ingestion of food containing lactose, no _____ is available to break it down.

7. As a result, _____ accumulates in the _____ _____ which disrupts normal water absorption causing _____.

8. The unabsorbed water and _____ _____ enter the _____ _____ where _____ metabolize the _____.

9. The bacteria break the lactose down into two simple sugars. What are they?

10. This process produces _____ that fills the _____ _____ causing _____ _____ and _____.

11. In addition, lactose in the _____ _____ causes _____ _____ producing watery feces known as _____.

12. Is there a cure for lactose intolerance?

13. List the treatments for lactose intolerance.

MODULE 12 The Digestive System

EXERCISE 12.26:
Coloring Exercise

Identify the organs of the digestive system. Then, with colored pens or pencils, color in these organs.

Note: Many organs are shown separated from one another. In life, many digestive organs overlap one another when viewed from this perspective.

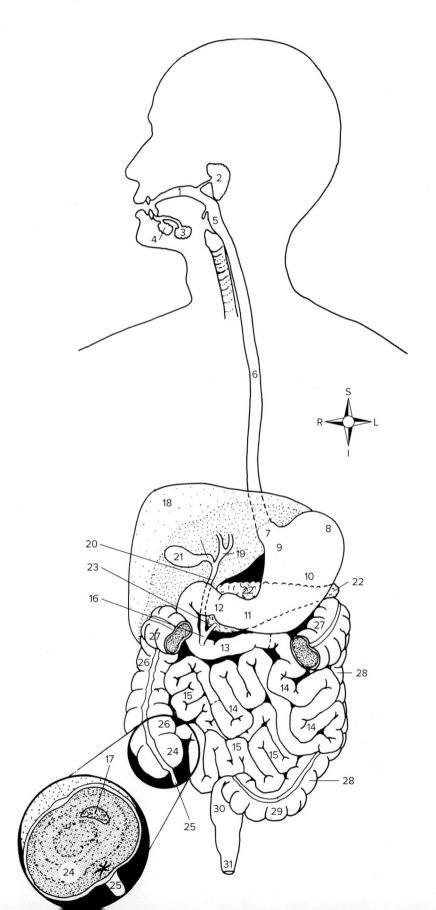

1.
2.
3.
4.
5.
6.
7.
8.
9.
10.
11.
12.
13.
14.
15.
16.
17.
18.
19.
20.
21.
22.
23.
24.
25.
26.
27.
28.
29.
30.
31.

MODULE 13
The Urinary System

Overview: The Urinary System

Metabolic wastes, toxins, drugs, hormones, salts, hydrogen ions, and water are all excreted by the urinary system of your body. To do this, it must filter your blood. All of your blood plasma is filtered 60 times per day. That's once every 24 minutes. The average plasma filtration rate is 125 mL per minute for adults with two kidneys. This produces 180 liters or 45 gallons of filtrate every day. Aren't you glad that you don't have to urinate all of that fluid? Fortunately, your body reabsorbs 99 percent of this fluid, so that you only excrete 1.8 liters or 0.45 gallons of urine daily.

The urinary system also regulates blood volume and pressure; regulates the amounts of dissolved solutes in body fluids—called osmolarity; helps to control blood pressure, red blood cell count and the blood's ability to carry oxygen, the pH of body fluids, calcium homeostasis, and, free radical detoxification; and it has the ability to convert amino acids to glucose during times of starvation. Talk about multitasking!

We will begin our study with an overview of the structures of the urinary system. We will then take a sequential look at the organs in great detail, from the kidneys through the urethra. Finally, we will look at how molecules move through the body and pass through the cell membrane.

From the **HOME** *screen, click the drop-down box on the* **MODULE** *menu.*

From the systems listed, click on **Urinary.**

   SELECT ANIMATION
Urinary System Overview   PLAY

After viewing the animation, answer the following questions:

1. What are the major functions of the urinary system?

2. What are the organs of the urinary system?

3. Where are the kidneys located?

4. What is the renal hilum?

5. Blood to be filtered is transported to the kidney by the _____ _____.

6. Filtered blood leaves the kidney via the _____ _____.

7. Name the structure that covers the outer surface of the kidney.

8. Describe the structure of the interior of a kidney.

9. Fluids from the _____ _____ ultimately are funneled into the _____ _____ of the _____.

10. What is the function of the ureters?

11. How is urine propelled through the ureters?

760 MODULE 13 The Urinary System

12. What is the urinary bladder? Where is it located?

    sigmoid colon

13. Where do the ureters drain into the urinary bladder?

14. Name the muscle of the urinary bladder wall.

15. Another name for urination is _____.

16. How is urine expelled from the urinary bladder?

17. Compare the functions of the male and female urethras.

18. What is the function of the internal urethral sphincter muscle? Is it under voluntary or involuntary control?

19. What is the function of the external urethral sphincter muscle? Is it under voluntary or involuntary control?

20. How are both of these sphincters involved with urination?

Self-Quiz
Take this opportunity to check your progress by taking the **QUIZ**. See the **Introduction Module** for a reminder on how to access the **QUIZ** for this Study Area.

 SELECT ANIMATION
Kidney Anatomy and Urine Production (3D) **PLAY**

After viewing the animation, answer the following questions:

1. Where are the kidneys located?

2. What structures sit on top of the kidneys?

3. What structures comprise the renal lobe?

4. What blood vessels supply each renal lobe?

5. Name the smallest branches of the renal artery. What structures do they supply?

6. Describe the structure of each nephron.

7. Name the first step of urine production. Where does it occur?

8. Describe the structures that comprise the renal corpuscle.

9. Compare the diameter of the afferent arteriole and the efferent arteriole.

10. This size difference produces . . .

11. What does this pressure force out of the blood? List these substances.

12. Where do they move? What do these substances pass through in the process of moving?

13. Which substances are too large to pass through the membrane? Where do they remain?

14. Name the second step of urine production. What occurs during this step?

15. The renal tubule consists of . . .

16. What occurs as the fluid flows through the renal tubule during the second step of urine formation?

17. What specific substances are reabsorbed by active transport?

18. What does the peritubular capillary reabsorb through osmosis?

19. Name the final step of urine formation. Where does this primarily occur?

20. What substances are extracted during the final step?

21. Where do these substances move from? Where are they secreted?

22. Where do the wastes go from here?

The Upper Urinary System

EXERCISE 13.1:
Urinary System—Upper Urinary—Anterior View

SELECT TOPIC	SELECT VIEW
Upper Urinary	Anterior

After clicking **LAYERS 1** through **3** to orient yourself to our location, click **LAYER 4** in the **LAYER CONTROLS** window, and you will see the following image:

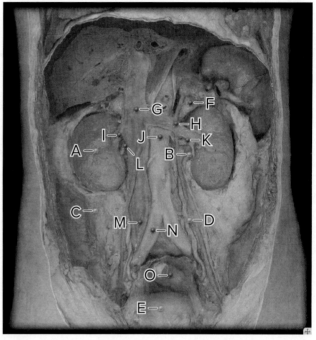

©McGraw-Hill Education

Mouse-over the pins on the screen to find the information necessary to identify the following structures:

A. Kidney
B. renal pelvis
C. pararenal fat
D. abdominal part of ureter
E. urinary bladder with peritoneum

Nonurinary System Structures (blue pins)

F. suprarenal gland
G. inferior vena cava
H. left renal v
I. right renal v
J. ~~~~~~~~~~ abdominal a
K. left renal a
L. right renal a

MODULE 13 The Urinary System

M. common iliac v
N. common iliac a
O. sigmoid colon

CHECK POINT

Upper Urinary, Anterior View

1. Describe the structure of the kidneys.
2. What is their function?
3. Name the structure that conveys urine from the renal pelvis.

Nonurinary System Structures (blue pins)

L. liver
M. spleen
N. inferior vena cava
O. celiac a
P. superior mesenteric a
Q. right renal a
R. left renal a
S. abdominal aorta
T. right renal v
U. inferior mesenteric a

CHECK POINT

Upper Urinary, Anterior View, *continued*

4. What is the function of the renal pyramids? How many are located in each kidney?
5. Name the tip of the renal pyramids. Where does it project?
6. What is a minor calyx? What is its function?

Click **LAYER 5** in the **LAYER CONTROLS** window, and you will see the following image:

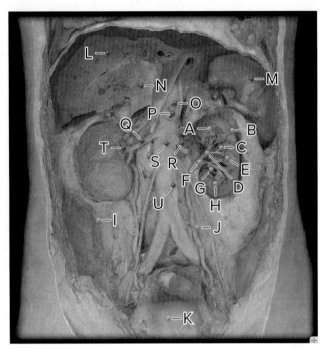

©McGraw-Hill Education

Click **LAYER 6** in the **LAYER CONTROLS** window, and you will see the following image:

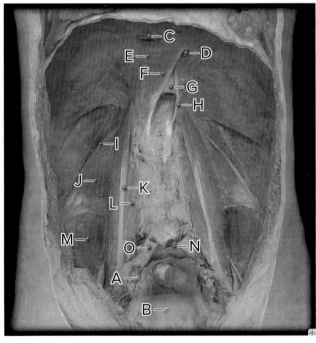

©McGraw-Hill Education

Mouse-over the pins on the screen to find the information necessary to identify the following structures:

A. renal medulla
B. renal cortex
C. minor calyx
D. major calyx
E. renal papilla
F. renal pelvis
G. renal sinus
H. renal pyramid
I. pararenal fat
J. abdominal part of ureter
K. urinary bladder with peritoneum

Mouse-over the pins on the screen to find the information necessary to identify the following structures:

A. ureter
B. urinary bladder w peritoneum

Nonurinary System Structures (blue pins)

C. inferior vena cava
D. esophagus
E. diaphragm
F. right crus of diaphragm
G. abdominal aorta
H. left crus of diaphragm
I. lumbar plexus & branches
J. quadratus lumborum m
K. psoas minor m
L. psoas major m
M. iliacus m
N. common iliac v
O. common iliac a

CHECK POINT

Upper Urinary, Anterior View, *continued*

7. What differences exist in the pathway of the ureter for males and females?
8. How long is the ureter?
9. List the pathway of urine from the kidney to the urethra.

Self-Quiz

Take this opportunity to check your progress by taking the **QUIZ**. See the **Introduction Module** for a reminder on how to access the **QUIZ** for this Study Area.

IN REVIEW

What Have I Learned?

The following questions cover the material that you have just learned, the upper urinary system. Apply what you have learned to answer these questions on a separate piece of paper.

1. Name the arteries that supply blood to the kidneys.
2. Name the veins that drain blood from the kidneys.
3. Which is anterior to the other?
4. Name the funnel-shaped structure that drains urine from the kidneys to the ureter.
5. Name the structure that conducts urine from the minor calyx to the renal pelvis.
6. How many of each calyx are located in each kidney?
7. Name the outer layer of the kidney. What are its extensions into the middle part of the kidney called? What structures are contained in this outer layer?
8. Name the inner layer of the kidney.
9. Name the part of the kidney where blood vessels and the renal pelvis enter and exit the kidney.
10. Name the part of the kidney between the structure in question 9 and the renal papillae.
11. Name the muscle in the walls of the urinary bladder.
12. What is the trigone? Name the openings that define the trigone.
13. The size and position of the urinary bladder varies with _____.
14. The volume of the urine also effects the position of _____.

The Kidney

 SELECT ANIMATION: Kidney—Gross Anatomy — PLAY

After viewing the animation, answer the following questions:

1. Where are the kidneys located?

2. What is the renal hilum? Name three structures that pass through the hilum.

3. The hilum is continuous with the . . .

4. How is the kidney tissue divided? Name the structures of the one tissue division that project into the other.

5. The base of each pyramid is located at the junction of these two tissues, the _____ _____ _____.

6. Name the structure at the apex of the renal pyramids. Where does it project?

7. Several _____ _____ merge to form the larger _____ _____, which merge to form a single, funnel-shaped _____ _____.

8. How many lobes are in each kidney? Each lobe consists of . . .

9. How is the blood carried to the kidneys for filtration?

10. List the series of branches that form from this artery up to and including the glomerulus. Give the locations where each artery is found.

11. Name the initial filtering component of the kidney. Name the functional filtration unit of the kidney. What are its components?

12. Name the arteriole that leaves each glomerulus. Upon leaving the glomerulus, this vessel enters . . .

13. For nephrons in the renal cortex, a _____ _____ _____ forms around the . . .

14. List the series of vessels that carry blood as it drains from these capillary networks.

15. What is the destination of the efferent arterioles associated with nephrons at the corticomedullary junction?

16. These capillaries are known as the _____ _____.

17. Name the sequence of vessels for the blood draining these capillaries.

Self-Quiz
Take this opportunity to check your progress by taking the **QUIZ**. See the **Introduction Module** for a reminder on how to access the **QUIZ** for this Study Area.

EXERCISE 13.2:
Urinary System—Kidney—Anterior View

SELECT TOPIC: Kidney ▸ **SELECT VIEW:** Anterior

Click **LAYER 1** *in the* **LAYER CONTROLS** *window, and you will see the following image:*

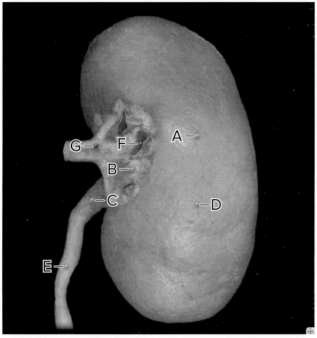

©McGraw-Hill Education

Mouse-over the pins on the screen to find the information necessary to identify the following structures:

A. _____
B. _____
C. _____
D. _____
E. _____

Nonurinary System Structures (blue pins)

F. _____
G. _____

CHECK POINT
Kidney, Anterior View

1. Name the connective tissue coat of the kidney. What is its function?
2. Describe the structure of the renal hilum. What is its function?
3. What is the function of the renal pelvis?

Click **LAYER 2** *in the* **LAYER CONTROLS** *window, and you will see the following image:*

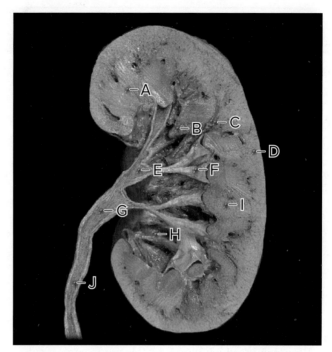

©McGraw-Hill Education

Mouse-over the pins on the screen to find the information necessary to identify the following structures:

A. _____
B. _____
C. _____
D. _____
E. _____
F. _____
G. _____
H. _____
I. _____
J. _____

CHECK POINT
Kidney, Anterior View, *continued*

4. Describe the structure of the renal pyramids.
5. Name the structures that separate the renal pyramids. These structures are an extension of what part of the kidney?
6. Which nephron parts are located in the renal cortex? Which are located in the renal medulla?

Clinical Application

SELECT ANIMATION
Kidney Anatomy/Renal Failure (3D) — PLAY

After viewing the animation, answer the following questions:

1. Where are the kidneys located?

2. List the functions of the kidneys.

3. What structures compose the renal lobe? It is referred to as the _____ _____ of the kidney.

4. Branches of which blood vessel supply each renal lobe?

5. What structure is the functional unit of the kidney?

6. What structures make up this functional unit?

7. Where does initial blood filtration occur?

8. Describe the glomerulus.

9. Describe the Bowman's capsule.

10. What is the purpose for the size difference between the afferent arteriole and efferent arteriole of the glomerulus?

11. What is the function of the renal tubule?

12. What percentage of the glomerular filtrate is reabsorbed as it passes through the renal tubule?

13. List the structures that compose the renal tubule.

14. Define acute renal failure.

15. List the causes of prerenal acute renal failure.

16. What processes result in less filtrate moving into the Bowman's capsule.

17. What causes intrarenal acute renal failure?

18. What situations can cause damage to the nephron?

19. List the causes of postrenal acute renal failure.

20. List the treatments for acute renal failure.

21. Define chronic renal failure.

22. List the causes for loss of nephron function.

23. List the treatments for chronic renal failure.

24. What treatment is required for end-stage renal failure?

I. _____
J. _____
K. _____
L. _____
M. _____
N. _____
O. _____
P. _____
Q. _____
R. _____
S. _____
T. _____

EXERCISE 13.3a:
Imaging—Kidney

SELECT TOPIC: Kidney ▸ SELECT VIEW: CT: Axial 1

Click the **TAGS ON/OFF** button, and you will see the following image:

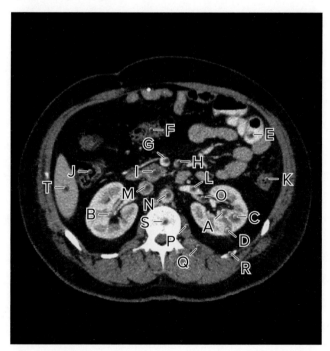

©McGraw-Hill Education

Mouse-over the pins on the screen to find the information necessary to identify the following structures:

A. _____
B. _____
C. _____
D. _____

Nonurinary System Structures (blue pins)

E. _____
F. _____
G. _____
H. _____

EXERCISE 13.3b:
Imaging—Kidney

SELECT TOPIC: Kidney ▸ SELECT VIEW: CT: Axial 2

Click the **TAGS ON/OFF** button, and you will see the following image:

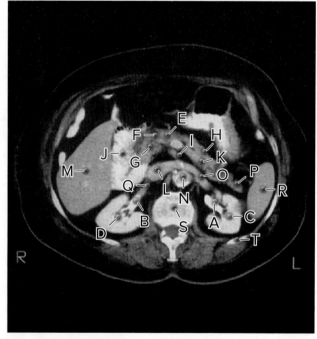

©McGraw-Hill Education

Mouse-over the pins on the screen to find the information necessary to identify the following structures:

A. _____
B. _____
C. _____
D. _____

Nonurinary System Structures (blue pins)

E. _____
F. _____
G. _____
H. _____
I. _____
J. _____
K. _____
L. _____
M. _____

N. _____
O. _____
P. _____
Q. _____
R. _____
S. _____
T. _____

Self-Quiz

Take this opportunity to check your progress by taking the **QUIZ**. See the **Introduction Module** for a reminder on how to access the **QUIZ** for this Study Area.

IN REVIEW

What Have I Learned?

The following questions cover the material that you have just learned, the kidney. Apply what you have learned to answer these questions on a separate piece of paper.

1. What is the distribution of the renal artery?

2. Describe its branching before it enters the kidney.

3. List the structures drained by the renal vein.

4. Describe the ureter. What is its function?

5. Trace the pathway of urine as it passes through and then out of the kidney through the urethra.

The Renal Corpuscle

EXERCISE 13.4:

Urinary System—Renal Corpuscle—Histology

SELECT TOPIC	SELECT VIEW
Renal Corpuscle	LM: High Magnification

Click the **TAGS ON/OFF** button, and you will see the following image:

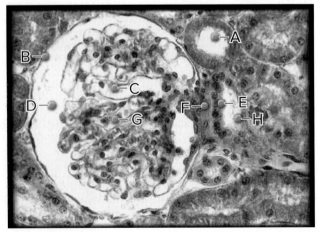

©Victor Eroschenko

Mouse-over the pins on the screen to find the information necessary to identify the following structures:

A. _____
B. _____
C. _____
D. _____
E. _____
F. _____
G. _____
H. _____

CHECK POINT

Renal Corpuscle, Histology

1. Name the two structures that make up the renal corpuscle.
2. What is the function of the renal corpuscle?

SELECT ANIMATION
Kidney—Microscopic Anatomy   PLAY

After viewing the animation, answer the following questions:

1. Name the functional filtration unit of the kidney. Approximately how many of these are located in each kidney?

2. Name the two parts of a nephron.

3. Describe the structure of the renal corpuscle. What are its two poles? What structures are located at each pole?

4. Describe the visceral layer of the glomerular capsule. What are podocytes and pedicels?

5. Describe the parietal layer of the glomerular capsule. What is the capsular space?

6. Name and describe the three components of the filtration membrane of the glomerulus.

7. Name and describe the three parts of the renal tubule.

8. What are the structural differences between the thick and thin segments of the nephron loop?

9. Several distal convoluted tubules drain into _____ _____ that pass through the _____ of the kidney. These merge to form _____ _____ that drain into a _____ _____.

10. What events occur as the fluid passes through the renal tubule?

11. Describe the two types of nephrons. What percent of the total number of nephrons consist of each type?

12. Describe the blood pathway to the vascular pole of the glomerulus. Name the apparatus located at this point.

13. What are the three parts of this apparatus? What is its function?

14. The efferent arteriole from each glomerulus enters a _____.

15. For cortical nephrons, where is the peritubular capillary network located? Where does it drain?

16. The efferent arterioles associated with juxtaglomerular nephrons enter capillaries known as the _____ _____ that surround . . . Where does this blood drain?

Self-Quiz
Take this opportunity to check your progress by taking the **QUIZ**. See the **Introduction Module** for a reminder on how to access the **QUIZ** for this Study Area.

SELECT ANIMATION
Urine Formation   PLAY

After viewing the animation, answer the following questions:

1. What are the primary functions of the kidneys?

2. In the process of these functions, they regulate . . . How?

3. Where is urine formed? What are the three processes in its formation?

4. Where does glomerular filtration occur? How does this occur?

5. What is tubular fluid? What occurs to it as it flows through the renal tubule?

6. What occurs to the fluid during tubular reabsorption?

7. How do the solutes move across the tubule wall into the interstitial fluid? Where do they go from there?

8. How does water move through the tubule wall? What percent of water is reabsorbed from each portion of the renal tubule?

9. What percent of the water in the glomerular filtrate returns to the bloodstream? What happens to the remaining water?

10. What event occurs during the process of tubular secretion?

11. Give some examples of waste products involved in this process.

12. What happens to the waste products as they are prepared for excretion?

13. Tubular fluid that enters the collecting ducts is called _____.

14. What determines whether the urine produced is diluted or concentrated?

15. Describe the urine produced when water intake is high and also when water intake is limited.

16. What are the two key factors that determine the kidneys' ability to concentrate urine?

Self-Quiz
Take this opportunity to check your progress by taking the **QUIZ**. See the **Introduction Module** for a reminder on how to access the **QUIZ** for this Study Area.

EXERCISE 13.5:
Urinary System—Renal Cortex—Histology

SELECT TOPIC	SELECT VIEW
Renal Cortex	LM: Medium Magnification

Click the **TAGS ON/OFF** *button, and you will see the following image:*

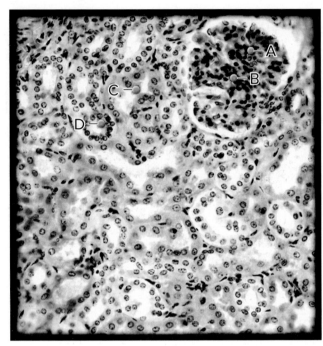

©McGraw-Hill Education/Al Telser

Mouse-over the pins on the screen to find the information necessary to identify the following structures:

A. _____

B. _____

C. _____

D. _____

MODULE 13 The Urinary System 771

EXERCISE 13.6:
Urinary System—Renal Medulla—Histology

SELECT TOPIC Renal Medulla ▸ **SELECT VIEW** LM: Medium Magnification

Click the **TAGS ON/OFF** *button, and you will see the following image:*

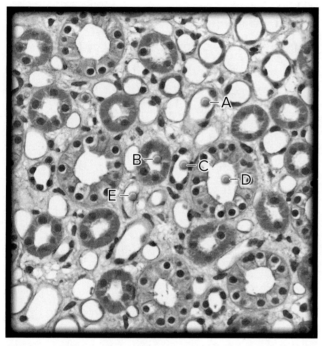

©McGraw-Hill Education/Al Telser

Mouse-over the pins on the screen to find the information necessary to identify the following structures:

A. _____

B. _____

C. _____

D. _____

E. _____

EXERCISE 13.7:
Urinary System—Renal Medulla (Longitudinal Section)

SELECT TOPIC Renal Medulla (longitudinal section) ▸ **SELECT VIEW** LM: Medium Magnification

Click the **TAGS ON/OFF** *button, and you will see the following image:*

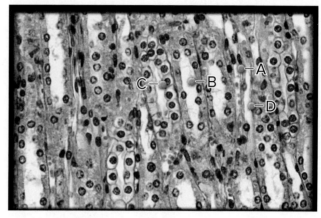

©Victor Eroschenko

Mouse-over the pins on the screen to find the information necessary to identify the following structures:

A. _____

B. _____

C. _____

D. _____

772 MODULE 13 The Urinary System

EXERCISE 13.8:
Urinary System—Podocyte—Histology

SELECT TOPIC	SELECT VIEW
Podocyte	SEM: High Magnification

Click the **TAGS ON/OFF** *button, and you will see the following image:*

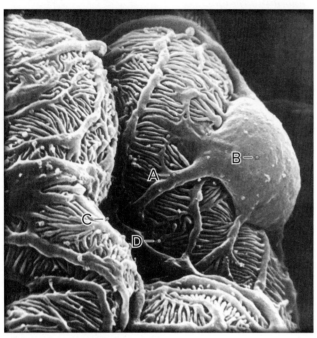

©Don Fawcett/Science Source

Mouse-over the pins on the screen to find the information necessary to identify the following structures:

A. _____

B. _____

C. _____

D. _____

EXERCISE 13.9:
Urinary System—Distal Convoluted Tubule—Histology

SELECT TOPIC	SELECT VIEW
Distal Convoluted Tubule	TEM: Low Magnification

Click the **TAGS ON/OFF** *button, and you will see the following image:*

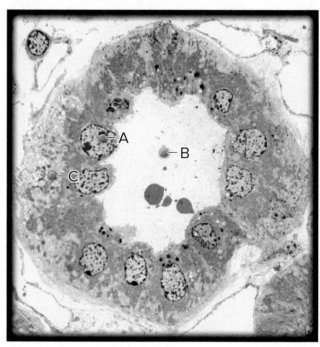

©EM Research Services, Newcastle University

Mouse-over the pins on the screen to find the information necessary to identify the following structures:

A. _____

B. _____

C. _____

MODULE 13   The Urinary System   773

EXERCISE 13.10:
Urinary System—Proximal Convoluted Tubule—Histology

SELECT TOPIC	SELECT VIEW
Proximal Convoluted Tubule	TEM: Low Magnification

Click the **TAGS ON/OFF** button, and you will see the following image:

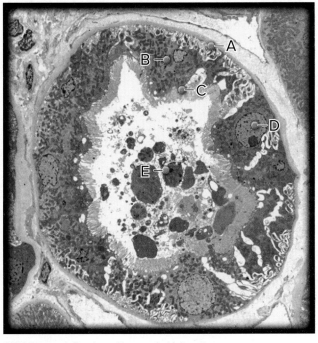

©EM Research Services, Newcastle University

Mouse-over the pins on the screen to find the information necessary to identify the following structures:

A. _____

B. _____

C. _____

D. _____

E. _____

IN REVIEW

What Have I Learned?

The following questions cover the material that you have just learned, the renal corpuscle. Apply what you have learned to answer these questions on a separate piece of paper.

1. Name the three parts of the renal tubule.

2. Name the specialized cells of the distal convoluted tubule that are part of the juxtaglomerular device.

3. What is the function of the juxtaglomerular device?

4. Name the structure that regulates the final volume and electrolyte content of the urine. What hormone influences this regulation?

5. What is the function of the capsular space?

SELECT ANIMATION

Glomerular Filtration and Regulation (Interactive)
Filtration Overview

PLAY

This Interactive Animation will walk you through the first step in urine formation. Be sure to interact with the animation to glean all the information available to you.

Open the **Module 13—Urinary System ANIMATION LIST**.

Under the second heading of **Glomerular Filtration and Regulation (Interactive)**, *select* **1. Filtration Overview** *and you will see the following image:*

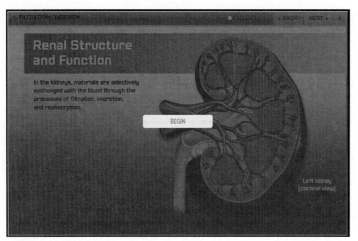

©McGraw-Hill Education

Select the **BEGIN** *button in the middle, and interact with the five images contained in this animation.*

When finished with the interactions, click the **X** *in the top-right corner to close the animation.*

After reviewing the animation, answer the following questions:

1. List the processes in the kidneys where materials are selectively exchanged.

2. Where in the kidneys are these tasks completed? How many of these structures exist in the kidneys?

3. What are the three ultimate outcomes of these tasks?

4. What drives the movement of blood from the afferent arteriole into the glomerulus?

5. What impacts this blood flow and how is it compensated?

6. Where does the filtrate collect?

7. What structure separates the blood in the glomerulus from the filtrate in the capsular space.

8. What structures combine to form the filtration membrane?

9. How are materials forced across the filtration membrane?

10. What five substances can cross the filtration membrane?

SELECT ANIMATION
Glomerular Filtration and Regulation (Interactive)
Glomerular Filtration
PLAY

This Interactive Animation will continue to walk you through the role of the glomerulus in the first step in urine formation. Be sure to interact with the animation to glean all the information available to you.

Open the **Module 13—Urinary System ANIMATION LIST.**

Under the second heading of **Glomerular Filtration and Regulation (Interactive),** *select* **2. Glomerular Filtration** *and you will see the following image:*

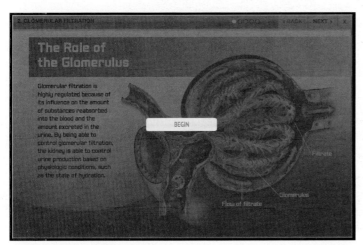

©McGraw-Hill Education

Select the **BEGIN** *button in the middle, and interact with the five images contained in this animation.*

When finished with the interactions, click the **X** *in the top-right corner to close the animation.*

After reviewing the animation, answer the following questions:

1. Why is glomerular filtration highly regulated?

2. By being able to control glomerular filtration, what basis allows the kidney to control urine production? Give an example.

3. What drives the outward movement of materials from the blood into the capsular space? How?

4. What is the actual glomerular hydrostatic pressure out?

5. What two forces oppose filtration?

6. When does capsular hydrostatic pressure occur? What is the result?

7. What processes are responsible for colloid hydrostatic pressure?

8. What is the actual capsular hydrostatic pressure in?

9. What is the actual blood colloid osmotic pressure in?

10. What is necessary for filtration to occur?

11. How is net filtration pressure determined? What is the actual net filtration pressure?

MODULE 13 The Urinary System

SELECT ANIMATION
Glomerular Filtration and Regulation (Interactive)
Control of Glomerular Filtration Rate

PLAY

This Interactive Animation will walk you through the process of regulating glomerular filtration. Be sure to interact with the animation to glean all the information available to you.

Open the **Module 13—Urinary System ANIMATION LIST.**

Under the second heading of **Glomerular Filtration and Regulation (Interactive)**, *select* **3. Control of Glomerular Filtration Rate** *and you will see the following image:*

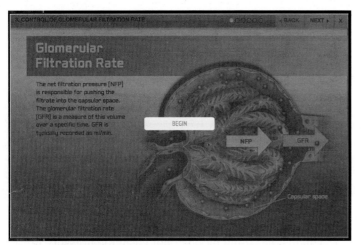

©McGraw-Hill Education

Select the **BEGIN** *button in the middle, and interact with the six images contained in this animation.*

When finished with the interactions, click the **X** *in the top-right corner to close the animation.*

After reviewing the animation, answer the following questions:

1. What is responsible for pushing the filtrate into the capsular space?

2. What is the glomerular filtration rate? How is it typically recorded?

3. What is the relationship between net filtration pressure and glomerular filtration rate? How are they positively correlated?

4. When is GFR regulated intrinsically within the kidneys?

5. What blood vessels adjust to regulate GFR? How do they accomplish this? What affect does it have on blood flow?

6. What are the exceptions to the relatively stable GFR maintained by autoregulation? What can result from these exceptions?

7. What is the GFR at a low mean arterial blood pressure?

8. What is the GFR at 180 mmHg mean arterial blood pressure?

9. What is the GFR above 180 mmHg mean arterial blood pressure?

10. List two extrinsic controls influencing GFR.

11. Describe the two ways that sympathetic nervous stimulation can reduce GFR.

12. How is Angiotensin II release stimulated? What does Angiotensin release cause? What is the result?

13. Using the interactive arrow to stimulate the kidneys, what affect does a stress or emergency stimulus have on glomerular surface area? What affect does this have on GFR?

14. What hormone does the heart release in response to increases in blood volume or blood pressure? What is the result?

15. List the three pathways.

16. Describe glomerular filtration.

17. What other systems play an important role in influencing glomerular filtration?

SELECT ANIMATION
Tubular Reabsorption and Secretion (Interactive)
Renal Tubule Structure and Function
PLAY

This Interactive Animation will walk you through an overview of the structure and function of the renal tubules of the nephron. Be sure to interact with the animation to glean all the information available to you.

Open the **Module 13—Urinary System ANIMATION LIST.**

Under the third heading of **Tubular Reabsorption and Secretion (Interactive),** *select* **1. Renal Tubule Structure and Function** *and you will see the following image:*

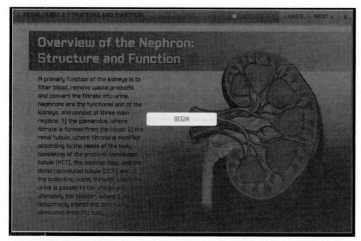

©McGraw-Hill Education

Select the **BEGIN** *button in the middle, and interact with the five images contained in this animation.*

When finished with the interactions, click the **X** *in the top-right corner to close the animation.*

After reviewing the animation, answer the following questions:

1. List three primary functions of the kidneys.

2. What are the functional units of the kidneys? Describe the three main regions of the functional units including their subdivisions and functions.

3. The _____ formed at the _____ _____ enters the _____ _____ at the _____, where it is now referred to as the _____ _____.

4. As the _____ _____ moves through each section of the _____ _____, adjustments in its _____ occur through the processes of _____ and _____.

5. List the selective barriers between the blood and the renal tubule that must be crossed by materials that are reabsorbed or secreted.

6. Describe reabsorption.

7. Describe the process of secretion.

8. Which materials ultimately remain in the blood supply? Which eventually become components of urine?

9. What must first happen for reabsorption to occur?

10. What method does potassium use to move through the tubule walls? Which materials use transcellular transport?

MODULE 13 The Urinary System

SELECT ANIMATION
Tubular Reabsorption and Secretion (Interactive) Proximal Convoluted Tubule — PLAY

This Interactive Animation will walk you through the function of the proximal end of the renal tubule. Be sure to interact with the animation to glean all the information available to you.

Open the **Module 13—Urinary System ANIMATION LIST**.

Under the third heading of **Tubular Reabsorption and Secretion (Interactive)**, *select* **2. Proximal Convoluted Tubule** *and you will see the following image:*

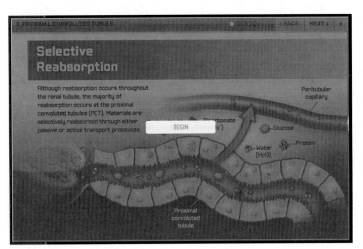

©McGraw-Hill Education

Select the **BEGIN** *button in the middle, and interact with the five images contained in this animation.*

When finished with the interactions, click the **X** *in the top-right corner to close the animation.*

After reviewing the animation, answer the following questions:

1. Although reabsorption occurs throughout the _____ _____, the majority of reabsorption occurs . . .

2. What two methods are employed to selectively reabsorb materials?

3. Where does the majority of sodium reabsorption occur?

4. How does sodium leave the tubular fluid?

5. What structure actively moves sodium out of the cell? Where are these structures located?

6. What does this allow and also maintain?

7. What percent of the cumulative water reabsorption occurs at the proximal convoluted tubule?

8. What are present in a set number in the luminal membrane? What do they enhance?

9. The movement of water follows that of _____ in this region, and is referred to as _____ _____ _____.

10. What materials enter the tubular cells through sodium-glucose symport channels? Where are these channels located? What does this allow?

11. How does glucose that has entered the tubular cells eventually exit? Where are these structures located? What process do they employ as they exit?

12. Within the tubular fluid, _____ [_____] and _____ ions [_____] combine to form _____ _____ [_____], which quickly dissociates into _____ _____ [_____] and _____.

13. _____ remains in the fluid while _____ diffuses into the _____ _____, where it combines with _____ to form _____. This then dissociates into _____ and _____. _____ is reabsorbed and _____ ions are secreted.

SELECT ANIMATION
Tubular Reabsorption and Secretion (Interactive)
The Nephron Loop
PLAY

This Interactive Animation will walk you through the function of the middle portion of the renal tubule. Be sure to interact with the animation to glean all the information available to you.

Open the **Module 13—Urinary System ANIMATION LIST**.

Under the third heading of **Tubular Reabsorption and Secretion (Interactive)**, *select* **3. The Nephron Loop** *and you will see the following image:*

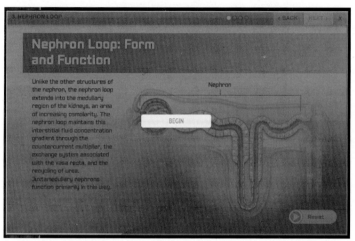

©McGraw-Hill Education

Select the **BEGIN** *button in the middle, and interact with the four images contained in this animation.*

When finished with the interactions, click the **X** *in the top-right corner to close the animation.*

After reviewing the animation, answer the following questions:

1. What makes the nephron loop unlike the other structures of the nephron?

2. What is different about this region of the kidneys?

3. Describe the countercurrent multiplier. What does the nephron loop maintain here?

4. Which nephrons function primarily with the countercurrent multiplier?

5. How do the descending and ascending limbs of the nephron loop vary in permeability? What substances are each of these permeable and impermeable to?

6. What is dictated by these differences in permeability?

7. What effect results deep in the medulla? What is this called?

8. What is the osmolarity of the descending limb of the nephron loop in the renal cortex?

9. What is the osmolarity of the nephron loop deep in the renal medulla?

10. What is the osmolarity of the ascending limb of the nephron loop in the renal cortex?

11. How does the blood flow in the vasa recta compare to the flow of tubular fluid? What is the result?

12. What does this opposing flow encourage and enhance?

13. The higher concentration of _____ in the medulla is largely influenced by the high concentration of _____ in this region.

14. Where does the urea within the medulla enter the tubular fluid?

15. Where does the urea continue to travel on its way to the collecting ducts?

16. Is the region of the collecting ducts permeable or impermeable to urea? What does this allow?

SELECT ANIMATION
Tubular Reabsorption and Secretion (Interactive) Distal Convoluted Tubule PLAY

This Interactive Animation will walk you through the function of the distal convoluted tubules. Be sure to interact with the animation to glean all the information available to you.

Open the **Module 13—Urinary System ANIMATION LIST.**

Under the third heading of **Tubular Reabsorption and Secretion (Interactive),** *select* **4. Distal Convoluted Tubule** *and you will see the following image:*

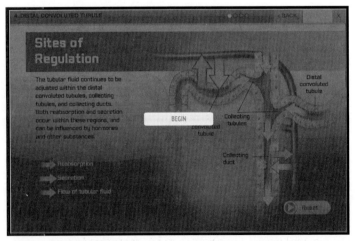
©McGraw-Hill Education

Select the **BEGIN** *button in the middle, and interact with the four images contained in this animation.*

When finished with the interactions, click the **X** *in the top-right corner to close the animation.*

After reviewing the animation, answer these questions:

1. Where does the tubular fluid continue to be adjusted?

2. List the two process that occur in these regions? What influences these processes?

3. The resorption of which two items is highly regulated? Why?

4. What is the function of aldosterone? How?

5. _____ follows the movement of _____ by _____, thus _____ indirectly enhances _____ _____.

6. What other hormone also increases water reabsorption? How?

7. _____ [_____] and _____ [_____] reabsorption are closely tied to _____ _____ levels.

8. At normal levels (_____ to _____ mg/dl], _____ percent of _____ and _____ percent of _____ in the filtrate is reabsorbed in the _____.

9. When plasma _____ levels decline, _____ _____ [_____] is released from the _____ _____ and changes the reabsorption rate of each ion.

10. _____ stimulates the opening of _____ channels in the _____, thus increasing the reabsorption of _____, while it inhibits _____ reabsorption in the _____.

11. Where does the majority of secretion occur?

12. Which ions within the blood secreted back into the tubular fluid?

13. List the other substances secreted at the distal convoluted tubule.

SELECT ANIMATION
Tubular Reabsorption and Secretion (Interactive)
Regulation and Elimination of Urine
PLAY

This Interactive Animation will walk you through the regulation of the composition and volume of urine. Be sure to interact with the animation to glean all the information available to you.

Open the **Module 13—Urinary System ANIMATION LIST.**

Under the third heading of **Tubular Reabsorption and Secretion (Interactive)**, select **5. Regulation and Elimination of Urine** and you will see the following image:

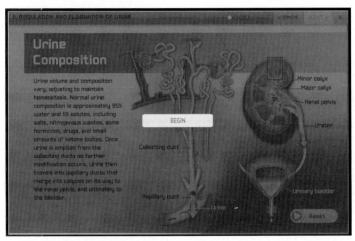

©McGraw-Hill Education

Select the **BEGIN** button in the middle, and interact with the four images contained in this animation.

When finished with the interactions, click the **X** in the top-right corner to close the animation.

After reviewing the animation, answer the following questions:

1. Why does the volume and composition of urine vary?

2. What is the normal urine composition?

3. List the solutes found in normal urine.

4. When is there no further modification of the urine?

5. Where does the urine then travel?

6. What is the normal average daily urine output?

7. What variables can cause variations in daily urine output?

8. List two other means of excreting fluids from the body.

9. What is the minimum amount of urine required per day to eliminate the wastes from the body?

10. What is the actual urine volume that results from normal atrial natriuretic peptide secretion?

11. What is the result of increased atrial natriuretic peptide secretion on urine volume?

12. What is the result of decreased atrial natriuretic peptide secretion on urine volume?

13. Compare the relationship between urine volume and normal, high, and low blood pressure.

14. Compare the relationship between urine volume and normal, high, and low fluid intake.

15. Compare the relationship between urine volume and normal, high, and low aldosterone levels.

16. Compare the relationship between urine volume and normal, high, and low antidiuretic hormone levels.

17. List two common measurements of urine output. What are the normal ranges for these two measurements?

18. If the specific gravity is too low, the urine contains _____ _____ and represents _____ _____.

19. A high specific gravity represents an _____ _____ of _____.

20. What pigment gives urine its yellowish color? When is this more apparent?

21. What is the specific gravity of the "hydrated" urine sample?

22. What is the specific gravity of the "dehydrated" urine sample?

23. What is the pH of the "hydrated" urine sample?

24. What is the pH of the "dehydrated" urine sample?

25. List the dynamic processes within the kidney. What are the results of these processes? What do they maintain?

26. What other role do the kidneys play? List two examples.

The Ureter

EXERCISE 13.11a:
Urinary System—Ureter—Histology

SELECT TOPIC: Ureter ▶ SELECT VIEW: LM: Low Magnification

Click the **TAGS ON/OFF** *button, and you will see the following image:*

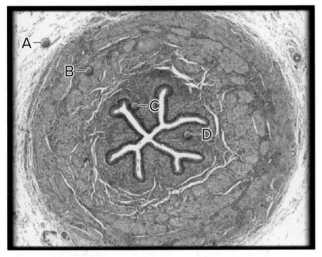

©McGraw-Hill Education/Al Telser

Mouse-over the pins on the screen to find the information necessary to identify the following structures:

A. _____
B. _____
C. _____
D. _____

EXERCISE 13.11b:
Urinary System—Ureter—Histology

SELECT TOPIC: Ureter ▶ SELECT VIEW: LM: High Magnification

Click the **TAGS ON/OFF** *button, and you will see the following image:*

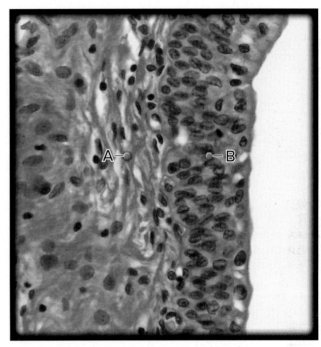

©McGraw-Hill Education/Al Telser

Mouse-over the pins on the screen to find the information necessary to identify the following structures:

A. _____
B. _____

CHECK POINT

Ureter, Histology

1. Name the tissue that loosely attaches the ureter to adjacent structures.
2. What structures are found in this tissue?
3. How does the muscularis change in composition along the length of the ureter? What is its function?
4. What is the mucosa? What is its function?
5. Describe the transitional epithelium. What is unique about its function?

784 MODULE 13 The Urinary System

EXERCISE 13.12a:
Imaging—Kidney, Ureter, and Urinary Bladder

SELECT TOPIC	SELECT VIEW
Kidney, Ureter, and Urinary Bladder	Pyelogram: Posterior

Click the **TAGS ON/OFF** *button, and you will see the following image:*

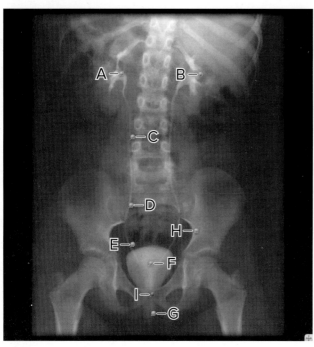

©McGraw-Hill Education

Mouse-over the pins on the screen to find the information necessary to identify the following structures:

A. _____
B. _____
C. _____
D. _____
E. _____
F. _____
G. _____

Nonurinary System Structures (blue pins)

H. _____
I. _____

EXERCISE 13.12b:
Imaging—Kidney, Ureter, and Urinary Bladder

SELECT TOPIC	SELECT VIEW
Kidney, Ureter, and Urinary Bladder	IVP: Posterior

Click the **TAGS ON/OFF** *button, and you will see the following image:*

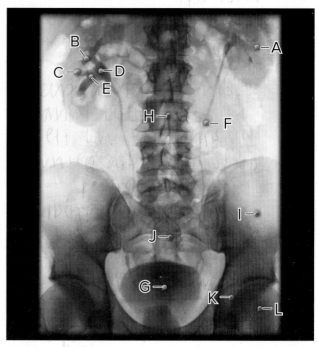

©McGraw-Hill Education

Mouse-over the pins on the screen to find the information necessary to identify the following structures:

A. _____
B. _____
C. _____
D. _____
E. _____
F. _____
G. _____

Nonurinary System Structures (blue pins)

H. _____
I. _____
J. _____
K. _____
L. _____

The Female Lower Urinary System

EXERCISE 13.13:
Urinary System—Lower Urinary—Female—Sagittal View

SELECT TOPIC Lower Urinary—Female ▸ **SELECT VIEW** Sagittal

Click **LAYER 1** in the **LAYER CONTROLS** window, and you will see the following image:

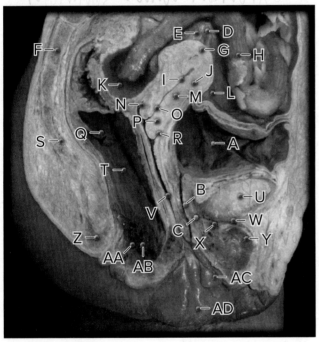

©McGraw-Hill Education

Mouse-over the pins on the screen to find the information necessary to identify the following structures:

A. urinary bladder
B. urethra
C. external urethral sphincter m

Nonurinary System Structures (blue pins)

D. uterine tube
E. ovary
F. sacrum

G. fundus of uterus
H. small intestine
I. uterine cavity
J. body of uterus
K. recto-uterine pouch
L. vesico-uterine pouch
M. uterus
N. posterior vaginal fornix
O. cervical canal
P. cervix of uterus
Q. transverse fold of rectum
R. anterior vaginal fornix
S. coccyx
T. rectum
U. pubic symphysis
V. vagina
W. deep dorsal v. of clitoris
X. crus of clitoris
Y. clitoris
Z. external anal sphincter
AA. anal canal
AB. internal rectal venous plexus
AC. labium minus
AD. labium majus

CHECK POINT

Lower Urinary—Female, Sagittal View

1. Describe the structure of the urinary bladder.
2. What are the two functions of this organ?
3. What is the function of the female urethra?
4. Give two reasons why urinary tract infections are more common in females.

Self-Quiz

Take this opportunity to check your progress by taking the **QUIZ**. See the **Introduction Module** for a reminder on how to access the **QUIZ** for this Study Area.

The Male Lower Urinary System

EXERCISE 13.14:
Urinary System—Lower Urinary—Male—Sagittal View

SELECT TOPIC: Lower Urinary—Male ▶ SELECT VIEW: Sagittal

Click **LAYER 1** in the **LAYER CONTROLS** window, and you will see the following image:

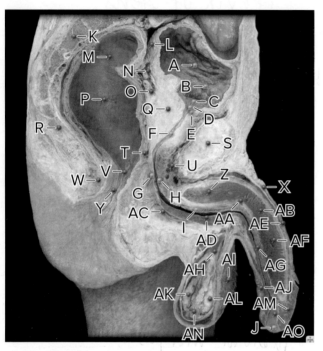

©McGraw-Hill Education

Mouse-over the pins on the screen to find the information necessary to identify the following structures:

A. urinary bladder
B. left ureteric orifice
C. internal urethral orifice
D. preprostatic urethra
E. internal urethral sphincter
F. prostatic urethra
G. external urethral sphincter m.
H. membranous urethra
I. spongy urethra
J. external urethral orifice

Nonurinary System Structures (blue pins)

K. sacrum
L. recto-vesical pouch
M. transverse fold of rectum

N. pelvic part of vas deferens
O. seminal vesicle
P. rectum
Q. prostate
R. coccyx
S. pubic symphysis
T. internal anal sphincter
U. prostatic venous plexus
V. anal column
W. external anal sphincter
X. deep fascia of penis
Y. anal canal
Z. left crus of penis
AA. tunica albuginea of penis
AB. penis
AC. bulbospongiosus m - male
AD. bulb of penis
AE. deep dorsal v. of penis
AF. corpus cavernosum penis
AG. corpus spongiosum penis
AH. spermatic fascia
AI. scrotum
AJ. remnant of prepuce of penis
AK. epididymis
AL. testis
AM. glans penis
AN. scrotal part of vas deferens
AO. navicular fossa

CHECK POINT
Lower Urinary—Male, Sagittal View

1. Name the structures that form the trigone of the urinary bladder.
2. Name the first part of the male urethra.
3. Where are sperm formed?
4. Name the four different parts of the male urethra. What is the function of the male urethra?
5. Name the external orifice of the urinary tract.

Self-Quiz
Take this opportunity to check your progress by taking the **QUIZ**. See the **Introduction Module** for a reminder on how to access the **QUIZ** for this Study Area.

IN REVIEW

What Have I Learned?

The following questions cover the material that you have just learned, the lower urinary tract. Apply what you have learned to answer these questions on a separate piece of paper.

1. Describe the structure of the urinary bladder.

2. What are the two functions of this organ?

3. What is the function of the female urethra?

4. Give two reasons why urinary tract infections are more common in females.

5. Name the structures that form the trigone of the urinary bladder.

6. Name the first part of the male urethra.

7. Where are sperm formed?

8. Name the four different parts of the male urethra. What is the function of the male urethra?

9. Name the external orifice of the urinary tract.

The Urinary Bladder

EXERCISE 13.15:
Urinary System—Urinary Bladder—Histology

SELECT TOPIC	SELECT VIEW
Urinary Bladder	LM: Low Magnification

Click the **TAGS ON/OFF** button, and you will see the following image:

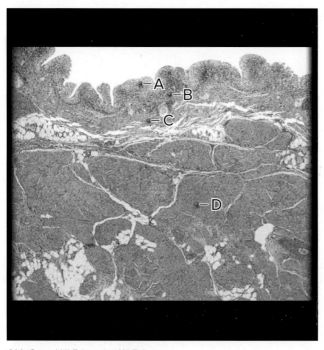

©McGraw-Hill Education/Al Telser

Mouse-over the pins on the screen to find the information necessary to identify the following structures:

A. _____

B. _____

C. _____

D. _____

CHECK POINT

Urinary Bladder, Histology (Low Magnification)

1. Name the connective tissue layer of the urinary bladder that contains blood, lymph vessels, and nerves.
2. Name the three-layered muscle of the urinary bladder.
3. What is this muscle's function?

EXERCISE 13.16:
Urinary System—Urinary Bladder—Histology

SELECT TOPIC: Urinary Bladder ▸ **SELECT VIEW:** LM: High Magnification

*Click the **TAGS ON/OFF** button, and you will see the following image:*

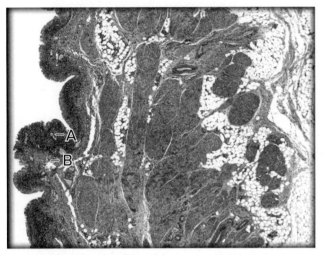

©McGraw-Hill Education/Al Telser

Mouse-over the pins on the screen to find the information necessary to identify the following structures:

A. _____

B. _____

CHECK POINT
Urinary Bladder, Histology (High Magnification)

1. Describe the transition epithelium of the urinary bladder.
2. What is the function of this epithelium?
3. Name the connective tissue layer that attaches the epithelium to the muscularis mucosa.

 SELECT ANIMATION Micturition Reflex PLAY

After viewing the animation, answer the following questions:

1. What type of reflex is the micturition reflex? What pathway do the impulses take?

2. How is the reflex coordinated? What signals can influence this reflex?

3. What causes the increase in the frequency of action potentials along this pathway?

4. What is the parasympathetic response to these action potentials? How does the urinary bladder respond to these stimuli?

5. What causes a conscious desire to urinate?

6. What happens if urination is not appropriate?

7. What causes the external urethral sphincter to remain contracted? What does this prevent?

8. When urination is desired, what stimulates the micturition reflex?

9. How does the brain then initiate urination?

Self-Quiz
Take this opportunity to check your progress by taking the **QUIZ**. See the **Introduction Module** for a reminder on how to access the **QUIZ** for this Study Area.

IN REVIEW

What Have I Learned?

The following questions cover the material that you have just learned, the urinary bladder. Apply what you have learned to answer these questions on a separate piece of paper.

1. Name the connective tissue layer of the urinary bladder that contains blood, lymph vessels, and nerves.

2. Name the three-layered muscle of the urinary bladder.

3. What is this muscle's function?

4. Describe the transition epithelium of the urinary bladder.

5. What is the function of this epithelium?

6. Name the connective tissue layer that attaches the epithelium to the muscularis mucosa.

Self-Quiz

Take this opportunity to check your progress by taking the **QUIZ**. See the **Introduction Module** for a reminder on how to access the **QUIZ** for this Study Area.

Clinical Application

SELECT ANIMATION
Fluid and Electrolyte Imbalances (3D)   **PLAY**

After viewing the animation, answer the following questions:

1. What substances mix together to comprise the body's fluids?

2. Describe the majority of solutes in the extracellular fluid.

3. What are electrolytes?

4. What is one of the most abundant electrolytes?

5. In order to sustain normal _____ functions and _____ volume, these elements are maintained within a narrow range in a process called _____.

6. What is the term for the sodium concentration in blood plasma? What does it indicate?

7. At what level is normal serum sodium maintained?

8. Define hyponatremia.

9. Which is the most common type of hyponatremia? What does it involve?

10. How does profuse sweating contribute to dilutional hyponatremia?

11. How does chronic kidney disease contribute to dilutional hyponatremia?

12. What affect does this abnormal process have?

13. How does water retention in the interstitial spaces lead to cell damage and inability to function normally?

14. Describe the treatments for hyponatremia.

15. What is hypertonic saline solution? How does it treat hyponatremia?

16. How do antidiuretic hormone inhibitor medications treat hyponatremia?

EXERCISE 13.17:
Coloring Exercise

Identify the features of the urinary system. Then color them in using colored pens or pencils.

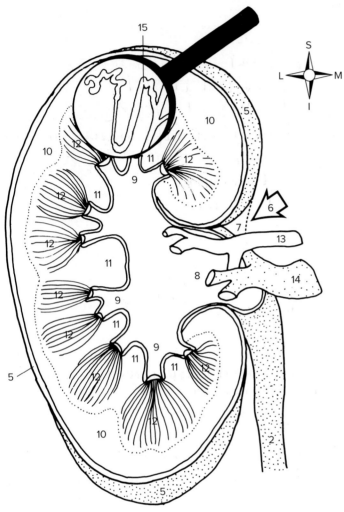

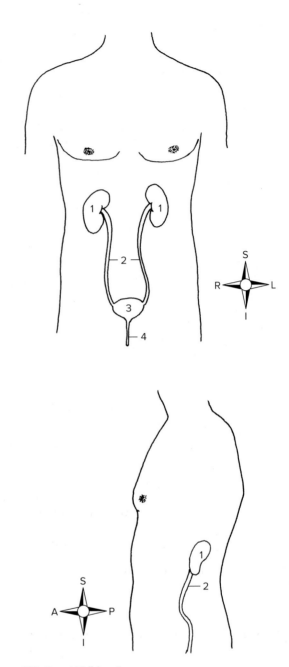

©McGraw-Hill Education

☐ 1. _____
☐ 2. _____
☐ 3. _____
☐ 4. _____
☐ 5. _____
☐ 6. _____
☐ 7. _____
☐ 8. _____

☐ 9. _____
☐ 10. _____
☐ 11. _____
☐ 12. _____
☐ 13. _____
☐ 14. _____
☐ 15. _____

MODULE 14

The Reproductive System

Overview: The Reproductive System

Without the reproductive system, you wouldn't be here! This is the one organ system that doesn't demand much attention until those wonderful years of puberty—and then it takes over like a raging fire. Hormone levels fluctuate and structures change as the body transitions from childhood into adulthood. This is a time when life takes on a whole different perspective.

Our perspective with *Anatomy & Physiology | Revealed*® will be to look at the reproductive system from a macroscopic as well as a microscopic level. We will begin with the female reproductive system, focusing on the structures unique to the human female. We will then explore how the body orchestrates her monthly reproductive cycles. We will conclude with the male reproductive system, which, although not as complex as the female's, is every bit the product of the same precise engineering.

From the **HOME** *screen, click the drop-down box on the* **MODULE** *menu.*

From the systems listed, click on **Reproductive.**

SELECT ANIMATION
Female Reproductive System Overview   PLAY

After viewing the animation, answer the following questions:

1. Name the internal organs of the female reproductive system.

2. Name the other organs of the female reproductive system.

3. Describe the structure of the ovaries.

4. Name the ligaments of the ovaries.

5. Name the three structures housed in each suspensory ligament.

6. What is the mesovarium? What is its function?

7. Name the capsule that surrounds each ovary. What tissue type is it?

8. Name the two regions of the ovary. Which region contains the ovarian follicles?

9. What is located within each follicle?

10. What event triggers development of some follicles?

11. What events occur during ovulation?

12. The remnant of the ruptured follicle is the _____ _____, which later degenerates into a _____ _____.

13. Where are the uterine tubes located? What is the mesosalpinx?

14. Into where do the uterine tubes directly open? Why?

15. Name the four segments of each uterine tube.

16. What are the fimbriae? What is their function?

17. Where does fertilization normally occur?

18. What is the shape and function of the uterus?

19. Name the three regions of the uterus.

20. The uterus' hollow lumen connects to the _____ _____ superiorly and opens into the vagina via the _____ _____.

21. Describe the three layers of the uterus.

22. Describe the two layers of the lamina propria. Which layer is shed during menstruation? How is it regenerated?

23. Name the three functions of the uterus.

24. Describe the three layers of the vaginal wall.

25. How do bacteria help inhibit the growth of pathogens in the vagina?

The Female Breast

EXERCISE 14.1:
Reproductive System—Breast—Female—Anterior View

SELECT TOPIC: Breast—Female ▸ SELECT VIEW: Anterior

Click **LAYER 1** *in the* **LAYER CONTROLS** *window, and you will see the following image:*

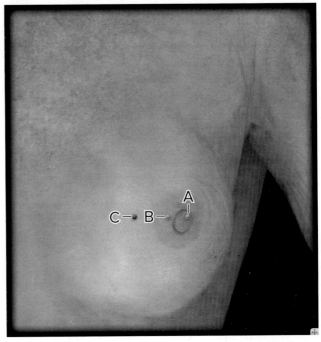

©McGraw-Hill Education

Mouse-over the pins on the screen to find the information necessary to identify the following structures:

A. nipple
B. areola
C. breast

CHECK POINT

Breast—Female, Anterior View

1. Name the elevation of skin containing lactiferous ducts for transferring milk during nursing.
2. Name the structure that maintains a permanent pigment increase during the first pregnancy.
3. What is the function of this structure?

Click **LAYER 2** *in the* **LAYER CONTROLS** *window, and you will see the following image:*

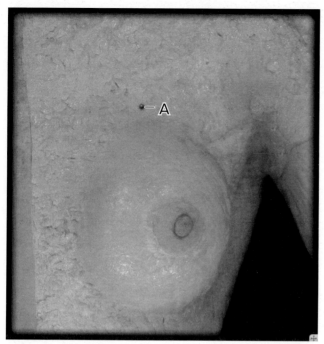

©McGraw-Hill Education

Mouse-over the pin on the screen to find the information necessary to identify the following structure:

A. pectoral superficial fascia

CHECK POINT

Breast—Female, Anterior View, *continued*

4. What tissue-types make up the pectoral superficial fascia?
5. What structures do they contain?
6. Name four functions of this structure.

Click **LAYER 3** *in the* **LAYER CONTROLS** *window, and you will see the following image:*

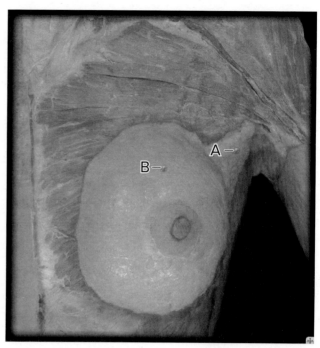

©McGraw-Hill Education

Mouse-over the pins on the screen to find the information necessary to identify the following structures:

A. axillary process
B. body

CHECK POINT

Breast—Female, Anterior View, *continued*

7. Where is the base of the breast located? What is located at its apex?
8. Where is the axillary process of the breast located?
9. What three variables affect the size and shape of the breast?

Click **LAYER 4** *in the* **LAYER CONTROLS** *window, and you will see the following image:*

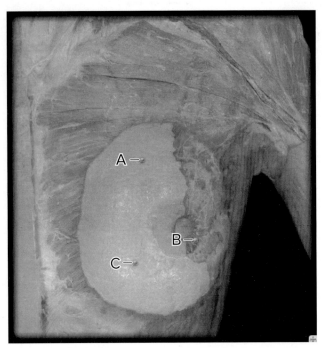

©McGraw-Hill Education

Mouse-over the pins on the screen to find the information necessary to identify the following structures:

A. suspensory ligament
B. lactiferous ducts
C. septum of breats

CHECK POINT

Breast—Female, Anterior View, *continued*

10. Name the structures that support the breast tissue.
11. What affect does age have on these structures?
12. Describe the lactiferous ducts.

Click **LAYER 5** *in the* **LAYER CONTROLS** *window, and you will see the following image:*

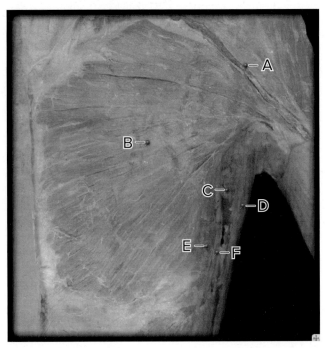

©McGraw-Hill Education

Mouse-over the pins on the screen to find the information necessary to identify the following nonreproductive system structures:

A. cephalic v
B. pectoralis major m
C. axillary lymph nodes
D. latrismus dorsi m
E. serratus anterior m
F. lateral thoracic v.

EXERCISE 14.2:

Imaging—Reproductive System—Mammogram

SELECT TOPIC: **Mammogram** ▶ SELECT VIEW: **X Ray: Lateral**

Click the **TAGS ON/OFF** button, and you will see the following image:

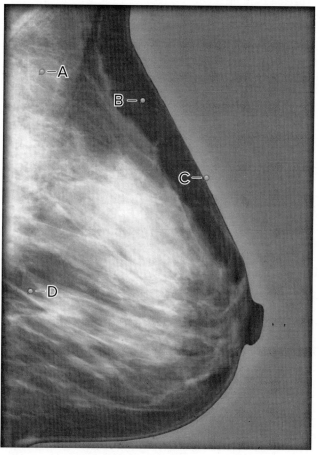

©Science Photo Library/Alamy Stock Photo

Mouse-over the pins on the screen to find the information necessary to identify the following structures:

A. _____

B. _____

C. _____

D. _____

Clinical Application

 SELECT ANIMATION: **Breast Anatomy/Breast Cancer (3D)** PLAY

After viewing the animation, answer the following questions:

1. What glands lie inside the breast? What muscle do they lie over? What two tissues surround them?

2. The mammary gland is composed of groups of _____ that form _____, which drain through _____ _____ to the _____.

3. What cells line the lactiferous ducts and lobules?

4. Describe the structure of the lobules.

5. How does the structure of the lobules change during pregnancy? What changes occur after birth?

6. What structures surround the breast radially? What is their function?

7. How does breast cancer begin?

8. Where do most breast cancers form?

9. What increases the chances for DNA damage to occur?

10. What type of growth does a cell with damaged DNA exhibit?

11. This mass of cells is a . . .

12. What is the stimulus for breast tissue growth? Where is this stimulus produced? During what time in a woman's life is it produced?

13. Therefore, the longer the time between _____ and _____, the greater the chance for developing _____ _____.

14. What are the sources of some breast cancers?

15. What is the normal function of the genes BRCA1 and BRCA2?

16. How do mutations to the genes BRCA1 and BRCA2 prevent their normal function? What is the result?

17. What may happen as tumors invade the surrounding tissue?

18. How do displaced tumor cells travel? What do they become elsewhere?

19. List other risks for breast cancer.

20. Can men get breast cancer?

21. List three factors that determine the treatment for breast cancer.

22. Describe a breast conserving lumpectomy. When is it performed?

23. Describe a modified radical mastectomy. When is it performed?

24. Describe a radical mastectomy. When is it performed?

25. When is radiation therapy performed?

26. Describe chemotherapy.

IN REVIEW

What Have I Learned?

The following questions cover the material that you have just learned, the breast. Apply what you have learned to answer these questions on a separate piece of paper.

1. What is the composition of the breast?

2. What muscle is deep to the base of the breast?

3. The "tail" of the breast extends into the _____.

4. Into what structures do the suspensory ligaments divide the breast?

5. What tissue partially replaces the breast fat tissue during pregnancy and lactation?

6. The _____ allows movement of the breast independent of the thoracic wall muscles.

7. What are the functions of the suspensory ligaments?

8. What is the function of lactiferous ducts?

9. Where does each duct open?

The Female Pelvis

EXERCISE 14.3:
Reproductive System—Pelvis—Female—Sagittal View

SELECT TOPIC	SELECT VIEW
Pelvis—Female	Sagittal

Click **LAYER 1** in the **LAYER CONTROLS** window, and you will see the following image:

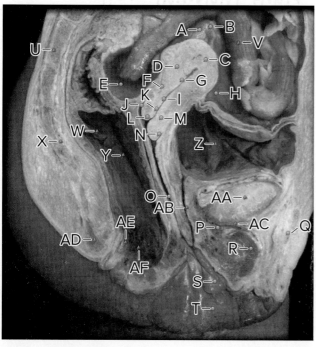

Mouse-over the pins on the screen to find the information necessary to identify the following structures:

A. ovary
B. uterine tube
C. fundus of uterus
D. body of uterus
E. recto-uterine pouch
F. uterus
G. uterine cavity
H. vesico-uterine pouch
I. internal os of uterus
J. posterior vaginal fornix
K. cervical canal
L. external os of uterus
M. cervix of uterus
N. anterior vaginal fornix
O. vagina
P. crus of clitoris
Q. mons pubis
R. clitoris
S. labium minus
T. labium majus

Nonreproductive System Structures (blue pins)

U. sacrum
V. small intestine
W. transverse fold of rectum
X. coccyx
Y. rectum
Z. urinary bladder
AA. pubic symphysis
AB. urethra
AC. deep dorsal v. of clitoris
AD. external anal sphincter
AE. anal canal
AF. internal rectal venous plexus

CHECK POINT

Pelvis—Female, Sagittal View

1. Name the primary organ of female sexual response. Describe its structures.
2. Describe the structure of the crus of the clitoris. What is its function?
3. Describe the structure of the labium minus. What is its function?

EXERCISE 14.4:
Reproductive System—Pelvis—Female—Superior View

SELECT TOPIC: Pelvis—Female
SELECT VIEW: Superior

Click on **LAYER 1** in the **LAYER CONTROLS** window, and you will see the following image:

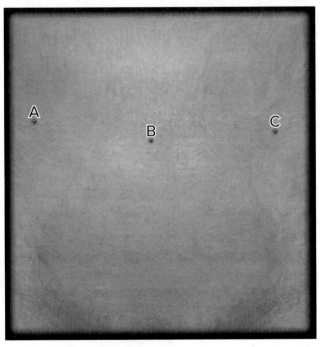

©McGraw-Hill Education

Mouse-over the pins on the screen to find the information necessary to identify the following nonreproductive system structures:

A. right inguinal region
B. pubic region
C. left inguinal region

CHECK POINT

Pelvis—Female, Superior View

1. Name the lower medial abdominal region.
2. What is another name for this region?
3. Name the two regions that flank this region on either side.

MODULE 14 The Reproductive System 799

Click **LAYER 2** *in the* **LAYER CONTROLS** *window, and you will see the following image:*

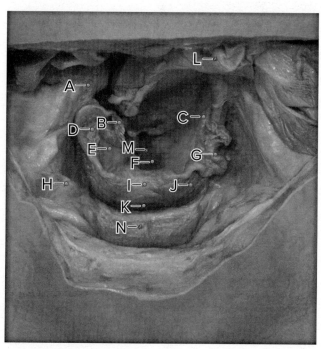

©McGraw-Hill Education

Mouse-over the pins on the screen to find the information necessary to identify the following structures:

A. Suspensory ligament of ovary
B. fimbriae of uterine tube
C. ovary
D. mesosalpinx
E. broad ligament of uterus
F. recto-uterine pouch
G. uterine tube
H. round ligament of uterus
I. uterus
J. mesometrium
K. vesico-uterine pouch

Nonreproductive System Structures (blue pins)

L. Sigmoid colon
M. rectum
N. urinary bladder with peritoneum

CHECK POINT

Pelvis—Female, Superior View, *continued*

4. What is a hysterectomy?
5. Name the blind recess that lies between the urinary bladder and the uterus.
6. Name the structure that is the embryonic remnant of the gubernaculum of the ovary. What was its embryonic function?

EXERCISE 14.5:

Imaging—Uterus and Vagina

SELECT TOPIC	SELECT VIEW
Uterus and Vagina	Sagittal

Click the **TAGS ON/OFF** *button, and you will see the following image:*

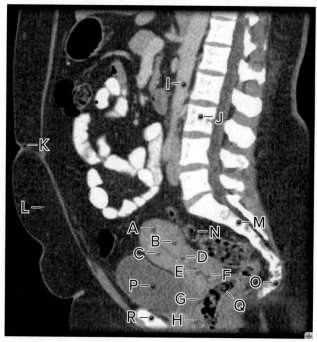

©McGraw-Hill Education

Mouse-over the pins on the screen to find the information necessary to identify the following structures:

A. _____
B. _____
C. _____
D. _____
E. _____
F. _____

G. _____
H. _____

Nonreproductive System Structures (blue pins)

I. _____
J. _____
K. _____
L. _____

M. _____
N. _____
O. _____
P. _____
Q. _____
R. _____

IN REVIEW

What Have I Learned?

The following questions cover the material that you have just learned, the female pelvis. Apply what you have learned to answer these questions on a separate piece of paper.

1. Name the region between the labia minora.

2. What is the urogenital triangle?

3. Describe the structure of the labium minus. What is its function?

4. Describe the structure of the labium majus. What is its function?

5. Describe the structure of the vagina. What is its function?

6. Define parturition. What is the standard measure for timing parturition?

7. Name the structure that varies its position with the fullness of the urinary bladder and rectum. What is the function of that structure?

8. The junction of the cervical canal and what structure forms the internal os? What is the function of that structure?

9. Name the female gonads. What is their function?

10. Name the location for a tubal ligation. What is this procedure?

11. Name the location where the cartilage softens in late pregnancy to allow a slight separation of the pubic bones.

12. Name the structure that conducts the ovarian vessels and nerves.

The Uterus

EXERCISE 14.6a:
Reproductive System—Uterus—Histology

SELECT TOPIC	SELECT VIEW
Uterus	LM: Low Magnification

Click the **TAGS ON/OFF** *button, and you will see the following image:*

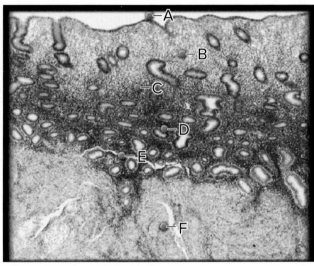

©McGraw-Hill Education/Al Telser

Mouse-over the pins on the screen to find the information necessary to identify the following structures:

A. _____
B. _____
C. _____
D. _____
E. _____
F. _____

EXERCISE 14.6b:
Reproductive System—Uterus—Histology

SELECT TOPIC	SELECT VIEW
Uterus	LM: High Magnification

Click the **TAGS ON/OFF** *button, and you will see the following image:*

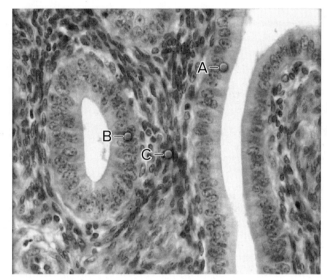

©Lutz Slomianka

Mouse-over the pins on the screen to find the information necessary to identify the following structures:

A. _____
B. _____
C. _____

CHECK POINT

Uterus, Histology

1. What is unique for the cervical endometrium compared to the endometrium of the body and fundus of the uterus?
2. Describe the endometrial glands.
3. The endometrium is formed from what two layers?
4. Name the thick, smooth muscle layer of the uterus.
5. Name the superficial part of the endometrium sloughed during menstruation.

EXERCISE 14.7:
Reproductive System—Ovary—Histology

SELECT TOPIC: Ovary ▸ SELECT VIEW: LM: Low Magnification

Click the **TAGS ON/OFF** button, and you will see the following image:

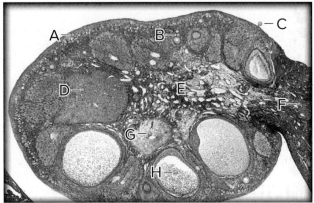

©Victor Eroschenko

Mouse-over the pins on the screen to find the information necessary to identify the following structures:

A. _____
B. _____
C. _____
D. _____
E. _____
F. _____
G. _____
H. _____

The Ovarian Follicle

SELECT ANIMATION: Comparison of Meiosis and Mitosis — PLAY

After viewing the animation, answer the following questions:

1. What is the normal human chromosome number? This number is referred to as _____.

2. There are _____ pairs of chromosomes, known as _____ pairs.

3. What process occurs before both meiosis and mitosis? What happens to the chromosomes afterward?

4. How many successive divisions occur in meiosis?

5. List the events of the first division.

6. List the events of the second division.

7. Define haploid.

8. How many cell divisions occur during mitosis? What is the result of mitosis?

SELECT ANIMATION: Meiosis (3D) — PLAY

After viewing the animation, answer the following questions:

1. What is meiosis? What is its purpose?

2. What are gametes?

3. How many divisions are there in meiosis? What is the result?

4. How does the number of chromosomes in the daughter cells compare to the number of chromosomes in the initial parent cell?

5. What is the term for having half the number of chromosomes as the parent cell? How many chromosomes do the daughter cells contain?

6. What must occur before the first cell division occurs?

7. What is the term for the first cell division? What is the term for the beginning phase?

8. What events occur in this phase?

9. What happens to the chromatids in each pair of chromosomes late in Prophase 1? What is this process called?

10. What does this process create?

11. What events occur during Metaphase 1?

12. What is independent assortment?

13. What events occur during Anaphase 1?

14. What events occur during Telophase 1? What is the result?

15. How does Meiosis 2 begin? What occurs during this phase?

16. What events occur during Metaphase 2?

17. What events occur during Anaphase 2? What is the result? What is each now considered?

18. What events occur during Telophase 2? What is the result?

19. What is the condition of the gametes once meiosis is complete? What is the ultimate result?

20. What occurs once these gametes meet? What is this cell called? How many chromosomes does this cell contain?

21. What process does this cell employ as it continues to develop?

SELECT ANIMATION
Unique Features of Meiosis
PLAY

After viewing the animation, answer the following questions:

1. Name the three unique features of meiosis.

2. Describe the events of synapsis.

3. Describe the events of crossing-over.

4. What is another name for crossing-over?

5. The name for the first nuclear division is a _____ _____.

6. What is the result of reduction division?

7. What effect does crossing-over have on the sister chromatids of each daughter cell?

EXERCISE 14.8:
Reproductive System—Primordial Follicle—Histology

SELECT TOPIC: Primordial Follicles ▸ SELECT VIEW: LM: High Magnification

*Click the **TAGS ON/OFF** button, and you will see the following image:*

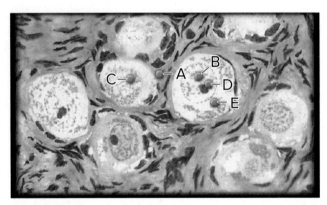

©McGraw-Hill Education/Al Telser

Mouse-over the pins on the screen to find the information necessary to identify the following structures:

A. _____

B. _____

C. _____

D. _____

E. _____

CHECK POINT
Primordial Follicle, Histology

1. A primordial follicle is one that has not responded to _____ to become a _____.
2. Where are these follicles located?
3. What structure surrounds the oocyte?

SELECT ANIMATION: Female Reproductive Cycles — PLAY

After viewing the animation, answer the following questions:

1. Female reproductive cycles are initiated usually between the ages of _____ and _____, a period known as _____.

2. How long is the average female reproductive cycle? Where do the changes occur during the cycle?

3. During each cycle, the hypothalamus releases _____ _____, which stimulates the _____ _____ to release two _____ hormones, _____ _____ _____ and _____ _____.

4. Name the target organ for these hormones.

5. Name the two components of the female reproductive cycle.

6. Ovulation occurs on day _____ of the _____-day *ovarian* cycle.

7. What are the 14 days prior to ovulation called in the ovarian cycle? The 14 days following ovulation?

8. What regulates the ovarian cycle?

9. Ovulation occurs on day _____ of the _____-day *uterine* cycle.

10. Describe the two subdivisions of the uterine cycle that occur before ovulation. And the 14 days following ovulation.

11. What controls the uterine cycle?

12. Describe the first five days of the follicular stage.

13. What occurs during this time in the uterine cycle? This is the _____ phase, commonly referred to as . . .

14. Describe the ovarian events during days 6–13 of the follicular phase.

15. Describe the ovarian events two days before ovulation. What hormones influence these events?

16. What causes the final maturation of the follicle? How soon before ovulation does this occur? What is another name for the mature follicle?

17. What event occurs in the primary oocyte just prior to ovulation? What does this form?

18. What event occurs in the uterus during the proliferative stage? Name the hormone responsible for this event and where that hormone is produced.

19. Describe the events of ovulation. What hormone peaks at this time?

20. During the luteal stage, what structure is formed from the remaining luteal cells? What is the function of this structure and the hormones that it produces?

21. After the ovum is fertilized and implants in the uterine wall, what hormone is produced by the cells of the implantation site? What is the function of this hormone?

22. How long does the corpus luteum continue hormone production? What transition takes place after this time? What structures produce hormones in the place of the corpus luteum?

23. If fertilization does not occur, when does the corpus luteum become the corpus albicans? What effect does this have on hormone levels? What events do these changes initiate?

EXERCISE 14.9:

Reproductive System—Primary Follicle—Histology

SELECT TOPIC	SELECT VIEW
Primary Follicle	LM: High Magnification

Click the **TAGS ON/OFF** button, and you will see the following image:

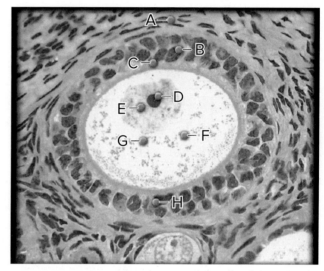

©McGraw-Hill Education/Al Telser

Mouse-over the pins on the screen to find the information necessary to identify the following structures:

A. _____

B. _____

C. _____

D. _____

E. _____

F. _____

G. _____

H. _____

CHECK POINT

Primary Follicle, Histology

1. Name the cells responsible for estrogen secretion. Where are they located?
2. Name the glycoprotein produced by the oocyte.
3. Is the primary oocyte diploid or haploid? What do these terms mean?
4. Name and describe the structure that surrounds the primary oocyte.

EXERCISE 14.10:
Reproductive System—Secondary Follicle—Histology

SELECT TOPIC	SELECT VIEW
Secondary Follicle	LM: High Magnification

Click the **TAGS ON/OFF** button, and you will see the following image:

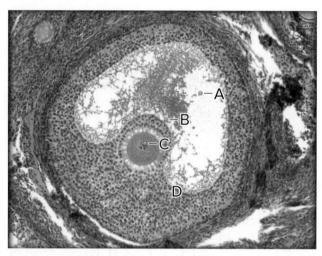

©McGraw-Hill Education/Al Telser

Mouse-over the pins on the screen to find the information necessary to identify the following structures:

A. _____
B. _____
C. _____
D. _____

CHECK POINT

Secondary Follicle, Histology

1. Name the fluid-filled cavity rich in estrogen. Where is it located?
2. Name the innermost layer of granulosa cells. These cells are in contact with which layer?

SELECT ANIMATION
Ovulation Through Implantation — PLAY

After viewing the animation, answer the following questions:

1. What are oocytes? Where are they produced? What is the process of their production?

2. How many primary oocytes are produced before birth? Are they diploid or haploid?

3. How many primary oocytes survive until puberty? Of these, how many will become secondary oocytes?

4. What meiotic stage do the primary oocytes remain in until just before ovulation?

5. Describe the structure of the follicle.

6. What events occur in the primary oocyte just before ovulation? What do these events form?

7. How do the ovaries alternate between ovulations?

8. What events occur during ovulation? Name the structures that receive the ovulated secondary oocyte.

9. What processes move the secondary oocyte toward the uterus?

10. Where does fertilization occur?

11. How many sperm are deposited in the vagina during ejaculation? How many reach the ampulla? How soon do they arrive?

12. How many sperm penetrate the secondary oocyte? What does this event trigger?

13. What events occur on a cellular level during fertilization? What cell do they form? How many chromosomes does this cell contain? Is it diploid or haploid?

14. What is the term for a fertilized oocyte (ovum)? What term describes the mitotic divisions of this fertilized secondary oocyte? Where is the fertilized ovum located while these mitotic divisions occur?

15. What is a morula? What is the term for the hollow sphere of cells into which the morula transforms? Where is the morula located when this transition occurs?

16. What are trophoblasts and embryoblasts? What structures are formed from each?

17. What role do the hormones estrogen and progesterone play in the implantation of the blastocyst?

18. What is implantation? When does it occur?

19. What changes occur in the trophoblast as implantation occurs? What are the functions of these new tissues?

20. When does the blastocyst become embedded superficially in the endometrium?

EXERCISE 14.11:
Reproductive System—Corpus Luteum—Histology

SELECT TOPIC	SELECT VIEW
Corpus Luteum	LM: High Magnification

Click the **TAGS ON/OFF** *button, and you will see the following image:*

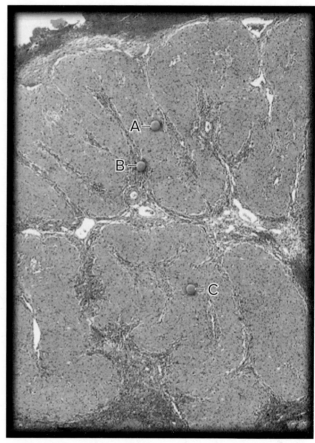

©McGraw-Hill Education/Al Telser

Mouse-over the pins on the screen to find the information necessary to identify the following structures:

A. _____

B. _____

C. _____

EXERCISE 14.12:
Reproductive System—Corpus Albicans—Histology

SELECT TOPIC: Corpus Albicans ▸ **SELECT VIEW**: LM: High Magnification

Click the **TAGS ON/OFF** *button, and you will see the following image:*

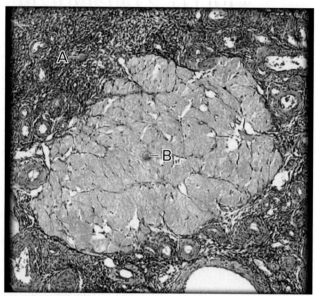

©McGraw-Hill Education/Al Telser

Mouse-over the pins on the screen to find the information necessary to identify the following structures:

A. _____

B. _____

CHECK POINT
Corpus Albicans, Histology

1. The corpus albicans is a connective tissue _____ at the surface of the _____.
2. The corpus albicans is a remnant of _____.
3. What does it identify?

Did you know that the name corpus albicans means "white body" and the name corpus luteum means "yellow body"? These are descriptive terms for their visual appearance.

The Uterine Tube

EXERCISE 14.13a:
Reproductive System—Uterine Tube—Histology

SELECT TOPIC: Uterine Tube ▸ **SELECT VIEW**: LM: Low Magnification

Click the **TAGS ON/OFF** *button, and you will see the following image:*

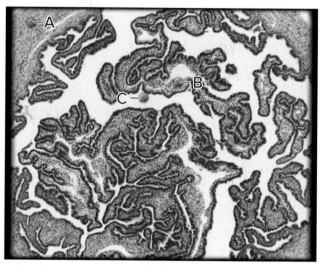

©McGraw-Hill Education/Al Telser

Mouse-over the pins on the screen to find the information necessary to identify the following structures:

A. _____

B. _____

C. _____

CHECK POINT
Uterine Tube, Histology (Low Magnification)

1. What is the central tubular cavity of the uterine tube? What is its function?
2. What is the lining of the uterine tube? It consists of which tissue type?
3. Name the smooth muscle layers in the wall of the uterine tube. What is their function?

EXERCISE 14.13b:
Reproductive System—Uterine Tube—Histology

SELECT TOPIC	SELECT VIEW
Uterine Tube	LM: High Magnification

Click the **TAGS ON/OFF** *button, and you will see the following image:*

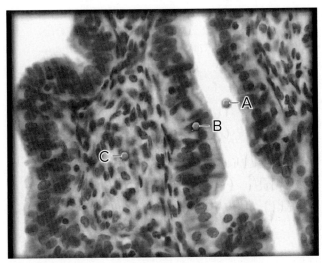

©McGraw-Hill Education/Al Telser

Mouse-over the pins on the screen to find the information necessary to identify the following structures:

A. _____

B. _____

C. _____

CHECK POINT

Uterine Tube, Histology (High Magnification)

1. Name the tissue type lining the uterine tubes.
2. What is the function of this tissue?
3. Name the layer that attaches the epithelium to the muscularis of the uterine tube. Name two structural layers not present in the uterine tube.

EXERCISE 14.13c:
Reproductive System—Uterine Tube—Histology

SELECT TOPIC	SELECT VIEW
Uterine Tube	SEM: Low Magnification

Click the **TAGS ON/OFF** *button, and you will see the following image:*

©Steve Gschmeissner/Science Source

Mouse-over the pins on the screen to find the information necessary to identify the following structures:

A. _____

B. _____

810 MODULE 14 The Reproductive System

The Female Perineum

EXERCISE 14.14:
Reproductive System—Perineum—Female—Inferior View

SELECT TOPIC	SELECT VIEW
Perineum—Female	Inferior

Click on **LAYER 1** *in the* **LAYER CONTROLS** *window, and you will see the following image:*

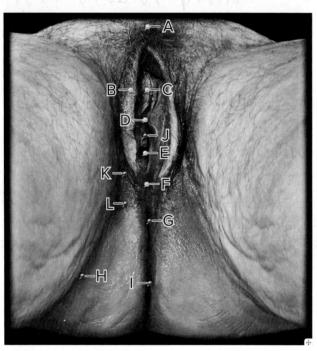

©McGraw-Hill Education

Mouse-over the pins on the screen to find the information necessary to identify the following structures:

A. _____
B. _____
C. _____
D. _____
E. _____
F. _____
G. _____
H. _____
I. _____

Nonreproductive System Structures (blue pins)

J. _____
K. _____
L. _____

CHECK POINT
Perineum—Female, Inferior View

1. Name the subcutaneous fat pad anterior to the pubic symphysis and pubic bones. When does the fat layer increase and decrease?
2. Name the limiting structures of the urogenital triangle. What is contained in this area?
3. What is the hymen? What is its function?

Click on **LAYER 2** *in the* **LAYER CONTROLS** *window, and you will see the following image:*

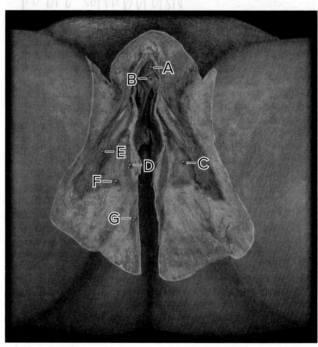

©McGraw-Hill Education

Mouse-over the pins on the screen to find the information necessary to identify the following structures:

A. _____
B. _____
C. _____

Nonreproductive System Structures (blue pins)

D. _____
E. _____
F. _____
G. _____

CHECK POINT
Perineum—Female, Inferior View, *continued*

4. Name the hood of thin skin over the glans clitoris. What are its functions?
5. Describe the structure of the glans clitoris. What is its function?

MODULE 14   The Reproductive System   811

Click **LAYER 3** *in the* **LAYER CONTROLS** *window, and you will see the following image:*

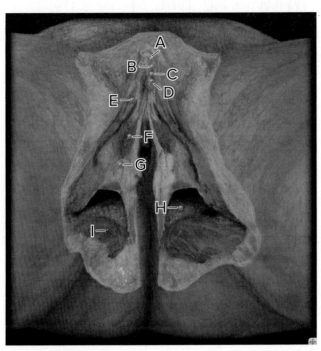

©McGraw-Hill Education

Mouse-over the pins on the screen to find the information necessary to identify the following structures:

A. _____
B. _____
C. _____
D. _____
E. _____
F. _____
G. _____
H. _____

Nonreproductive System Structure (blue pin)

I. _____

CHECK POINT

Perineum—Female, Inferior View, *continued*

6. What is the function of the clitoris? What occurs within the clitoris during sexual arousal?
7. Name the three sections of the clitoris.
8. What is the function of the bulb of the vestibule?

EXERCISE 14.15:
Reproductive System—Vagina—Histology

SELECT TOPIC	SELECT VIEW
Vagina	LM: Low Magnification

Click the **TAGS ON/OFF** *button, and you will see the following image:*

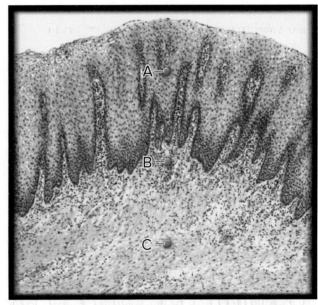

©McGraw-Hill Education/Al Telser

Mouse-over the pins on the screen to find the information necessary to identify the following structures:

A. _____
B. _____
C. _____

IN REVIEW

What Have I Learned?

The following questions cover the material that you have just learned, the female perineum. Apply what you have learned to answer these questions on a separate piece of paper.

1. What is an episiotomy? Why is this procedure done?

2. What is the anal triangle?

3. What is the intergluteal cleft?

4. What is the gluteal fold?

5. Name the potential space between the labia minora. Name the contents of this space from anterior to posterior.

6. Name the inferior opening of the vagina.

7. Name the structure that protects the external genitalia.

8. What are the functions of the labia minus?

9. What are the actions of the ischiocavernosus muscles?

10. What are the actions of the bulbospongiosus muscles?

11. Name the paired mucus-producing glands lateral to the vaginal opening. What is their function?

12. Describe the structures of the clitoris. What is the function of each?

The Male Pelvis

SELECT ANIMATION — Male Reproductive System Overview — **PLAY**

After viewing the animation, answer the following questions:

1. The primary male sex organs, or __gonads__ are the __testes__

2. List the pathway of sperm from the primary sex organs through the penis.
   __testes, epididymis, vas deferens, ejaculatory duct, urethra, penis__

3. Name the accessory sex glands. What is their function?
   __seminal vesicles, protate, burowoethal gland,__

4. Describe the location of the testes. How does the temperature there compare to core body temperature? Why? __inbetween thighs, 3°C below human body temp — optimal sperm production__

5. How is the temperature of the testes regulated? What muscles are involved? What are the actions of these muscles when the temperature is too warm *and* too cold? __changing position of testes in scrotum & decreasing temp of blood reaching testes.__

6. Describe the counter-current heat exchanger mechanism. What effect does this mechanism have on the arterial blood temperature as it reaches the testes?

   __~pic~__

7. The testes are the site of what events?
   __spermatogenisis__
   __production of male sex hormone__

8. Name the two connective tissue coats of the testes.
   __inner tunica albuginea__
   __outer tunica vaginallis__

9. Internally, the testes are divided into __250-300__ wedge-shaped __lobules__. Each __lobule__ contains up to __3__ ducts called __semineferous tubules__. What event occurs in these ducts?

   __where spermatogenisis occurs__

10. What is equivalent to the combined length of these ducts in both testes?

    15 football fields

11. How many sperm are in the average ejaculate?

    one half billion

12. Describe the epithelium that lines the seminiferous tubule. germ cells, sustentacular cells sperm cells

13. Name the cells responsible for testosterone production. Where are they located?

    intersticial (leydig)

14. Describe the pathway of the sperm as they leave the seminiferous tubule, mature, and are stored.

    leave testes through rete testis, efferent ductules, epididymis

15. How many sperm form each day?

    60 million

16. What structure is continuous with the tail of the epididymis? Where is this structure located?

    vas deferens, leaves scrotum in spermatic cord & enters pelvic cavity to urinary

17. The __vas deferens__ joins the duct of the __seminal vesicle__ to form the __ejaculatory duct__, which passes through the __prostate__ to empty into the __prostatic__ part of the __urethra__.

18. Name the accessory gland that produces 60 percent of the semen volume. Describe the secretion produced.

    seminal vesicles

19. Name the accessory gland that produces 30 percent of the semen volume. Describe the secretion produced.

    prostate
    thin, milkly, slightly acidic

20. Name the accessory gland that produces mucus that neutralizes the acidic urethra. What percent of the total semen volume consists of this fluid?

    bulburoethral glands   5%

21. What are the dual functions of the penis?

    lower urinary tract
    male organ of copulation

22. What are the divisions of the penis?

    root / shaft / glans

23. Describe the erectile bodies of the penis.

    root cons   pic

24. Which structure encloses the spongy urethra? What is the terminal end of this structure?

    bulb, corpus spongiosum

25. How is the size and shape of the penis determined?

    erectile tissue

26. What events occur in the process of an erection?

    ET fill with blood

EXERCISE 14.16:
Reproductive System—Pelvis—Male—Sagittal View

SELECT TOPIC: Pelvis—Male   SELECT VIEW: Sagittal

Click **LAYER 1** in the **LAYER CONTROLS** window, and you will see the following image:

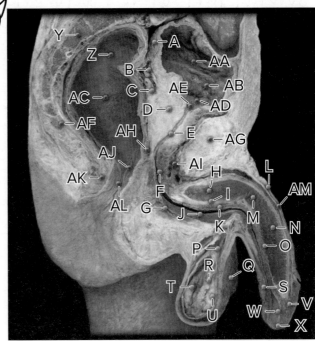

©McGraw-Hill Education

Mouse-over the pins on the screen to find the information necessary to identify the following structures:

A. recto-vesical pouch
B. pelvic part of vas deferens
C. seminal vesicle
D. prostate
E. prostatic urethra
F. membranous urethra
G. bulbospongiosus m. – male
H. left crus of penis
I. tunica albuginea of penis
J. bulb of penis
K. spongy urethra
L. deep fascia of penis
M. penis
N. corpus cavernosum penis
O. corpus spongiosum penis
P. spermatic fascia
Q. scrotum
R. remnant of prepuce of penis
S. scrotal part of vas deferens
T. epididymis
U. testis
V. glans penis
W. navicular fossa
X. external urethral orifice

Nonreproductive System Structures (blue pins)

Y. sacrum
Z. transverse fold of rectum
AA. urinary bladder
AB. left ureteric orifice
AC. rectum
AD. internal urethral orifice
AE. preprostatic urethra
AF. coccyx
AG. pubic symphysis
AH. internal anal sphincter
AI. prostatic venous plexus
AJ. anal columns
AK. external anal sphincter
AL. anal canal
AM. deep dorsal v. of penis

CHECK POINT

Pelvis—Male, Sagittal View

1. Name the paired and unpaired erectile tissues of the penis. What is their function?
2. What structure is responsible for most of the increased size and rigidity of the penis during an erection?
3. What is circumcision? How does the structure of the prepuce compare between circumcised and uncircumcised males?

IN REVIEW

What Have I Learned?

The following questions cover the material that you have just learned, the male pelvis. Apply what you have learned to answer these questions on a separate piece of paper.

1. Describe the structure of the crus of the penis. What is its function?

2. Describe the structure of the tunica albuginea of the penis. What is its function?

3. Describe the structure of the bulb of the penis. What is its function?

4. What are the actions of the bulbospongiosus muscle in the male? What are the results of these actions?

5. Describe the parts of the male urethra from proximal to distal. What is the function of the male urethra?

6. Name the accessory reproductive organ that contributes 30 percent of the semen volume. What is BPH?

7. Name the accessory reproductive organ that contributes 60 percent of the semen volume. What is the pH of this contribution?

8. Name the network of veins that drain the erectile tissues of the penis. What other structures are drained by these veins?

The Prostate

EXERCISE 14.17:

Reproductive System—Prostate—Histology

| SELECT TOPIC | SELECT VIEW |
| Prostate Gland | LM: Medium Magnification |

Click the **TAGS ON/OFF** button, and you will see the following image:

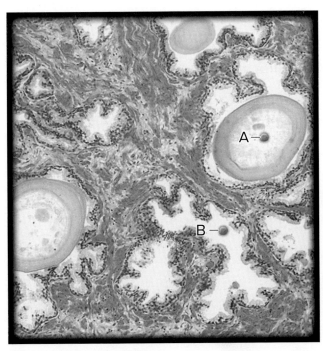

©McGraw-Hill Education/Al Telser

Mouse-over the pins on the screen to find the information necessary to identify the following structures:

A. _____

B. _____

CHECK POINT

Prostate, Histology

1. Describe the structure of the tubuloalveolar glands of the prostate.
2. What is the function of prostatic secretions?

Clinical Application

| SELECT ANIMATION | |
| Prostate/BPH/Cancer (3D) | PLAY |

After viewing the animation, answer the following questions:

1. The prostate is part of the _____ _____ _____.

2. The prostate produces _____ _____, a component of _____, that carries _____ during _____.

3. Describe the urethra. What structure does it pass through?

4. Prostate tissue is divided into several _____ _____, _____ zone, the _____ zone, and _____ zone.

5. The prostate consists of _____ _____ and _____ surrounded by _____ tissue called the _____.

6. What hormone influences prostate growth and development throughout life?

7. What enzyme do prostatic stromal cells express? What is the function of this enzyme?

8. _____ binds to _____ cells, releasing _____ _____ that stimulate _____ _____ _____ and _____.

9. What happens with this hormone as men age?

10. What happens to circulating testosterone levels?

11. What other hormone levels rise?

12. What can these hormonal changes lead to?

13. These changes are characteristic of two situations. What are they?

14. Describe benign prostatic hyperplasia.

15. What does benign prostatic hyperplasia cause over time?

16. In which zone does prostate cancer occur?

17. Describe the process of prostate cancer growth. What two structures develop?

18. How do prostate tumors disrupt the urine stream?

19. Describe metastasis.

20. List the treatments for benign prostatic hyperplasia. How do they work? What is the result?

21. What is a radical prostatectomy?

22. List other treatments for prostate cancer.

The Male Perineum

EXERCISE 14.18:
Reproductive System—Perineum—Male—Inferior View

SELECT TOPIC	SELECT VIEW
Perineum—Male	Inferior

Click **LAYER 1** *in the* **LAYER CONTROLS** *window, and you will see the following image:*

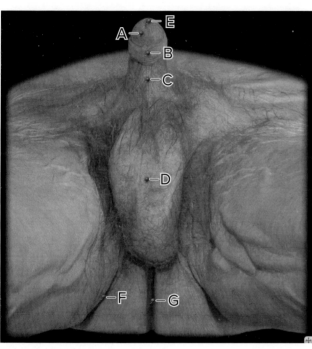

©McGraw-Hill Education

Mouse-over the pins on the screen to find the information necessary to identify the following structures:

A. glans penis
B. remnant of prepuce of penis
C. body of penis
D. scrotum

Nonreproductive System Structures (blue pins)

E. external urethral orifice
F. gluteal fold
G. intergluteal cleft

CHECK POINT

Perineum—Male—Inferior View

1. Name the pouch of skin that contains the testes. What muscle is in the wall of the pouch?
2. What changes occur when this muscle contracts? Why?
3. The size of this pouch varies with _____.

Click **LAYER 2** *in the* **LAYER CONTROLS** *window, and you will see the following image:*

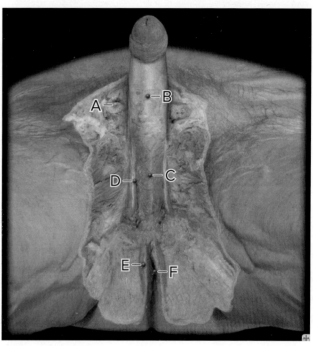

©McGraw-Hill Education

Mouse-over the pins on the screen to find the information necessary to identify the following structures:

A. spermatic cord
B. deep fascia of penis

Nonreproductive System Structures (blue pins)

C. bulbospongiosus m. - male
D. ischiocavernosus m.
E. external anal sphincter
F. anus

CHECK POINT

Perineum—Male, Inferior View, *continued*

4. Describe the actions of the ischiocavernosus muscle in the male.
5. Describe the actions of the bulbospongiosus muscle in the male. What are the results of these actions?
6. What is the function of the deep fascia of the penis?

Click **LAYER 3** *in the* **LAYER CONTROLS** *window, and you will see the following image:*

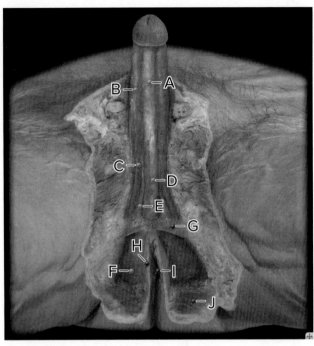

©McGraw-Hill Education

Mouse-over the pins on the screen to find the information necessary to identify the following structures:

A. corpus spongiosum penis
B. corpus cavernosum penis
C. crus of penis
D. bulb of penis
E. perineal membrane
F. ischioanal fossa

Nonreproductive System Structures (blue pins)

G. superficial transverse perineal m.
H. external anal sphincter
I. anus
J. gluteus maximus m.

CHECK POINT

Perineum—Male, Inferior View, *continued*

7. Name the single erectile body on the ventral aspect of the penis. This is a continuation of what structure?
8. Name the paired erectile body on the dorsal aspect of the penis. This is a continuation of what structure?
9. What is the function of these erectile tissues?

IN REVIEW

What Have I Learned?

The following questions cover the material that you have just learned, the male perineum. Apply what you have learned to answer these questions on a separate piece of paper.

1. How does the scrotal temperature compare with body temperature? Why?

2. How is the scrotum divided?

The Penis and Scrotum

EXERCISE 14.19:
Reproductive System—Penis and Scrotum—Anterior View

SELECT TOPIC: Penis and Scrotum **SELECT VIEW:** Anterior

Click **LAYER 1** *in the* **LAYER CONTROLS** *window, and you will see the following image:*

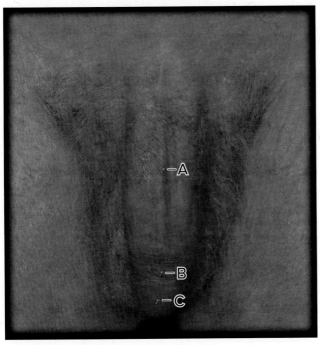

©McGraw-Hill Education

Mouse-over the pins on the screen to find the information necessary to identify the following structures:

A. body of penis
B. remnant of prepuce of penis
C. glans penis

CHECK POINT

Penis and Scrotum, Anterior View

1. Name the unattached portion of the penis.
2. Name the distal expansion of the penis.
3. Define flaccid. How does the dorsal surface compare when the penis is flaccid and when it is erect?

Click **LAYER 2** *in the* **LAYER CONTROLS** *window, and you will see the following image:*

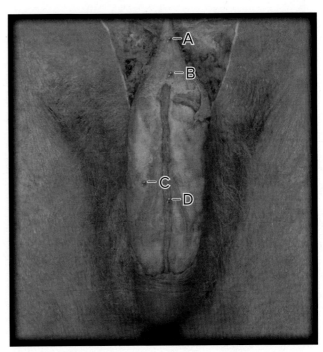

©McGraw-Hill Education

Mouse-over the pins on the screen to find the information necessary to identify the following structures:

A. suspensory ligament of penis
B. superficial fascia of penis
C. deep fascia of penis

Nonreproductive System Structure (blue pin)

D. superficial dorsal v. of penis

CHECK POINT

Penis and Scrotum, Anterior View, *continued*

4. Name the large vein on the dorsal surface of the penis that is *not* involved in an erection.
5. Describe the deep fascia of the penis. What is its function?
6. Name the sling of deep fascia that suspends the body of the penis.

Click **LAYER 3** *in the* **LAYER CONTROLS** *window, and you will see the following image:*

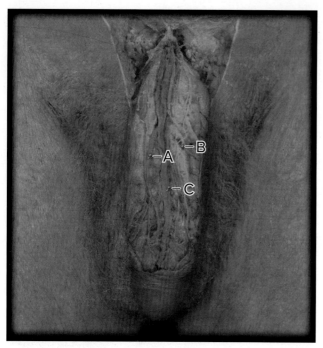

©McGraw-Hill Education

dorsal a. of penis ↙

Mouse-over the pins on the screen to find the information necessary to identify the following structures:

A. ~~deep fascia of penis~~
B. dorsal n. of penis
C. deep dorsal v. of penis

CHECK POINT

Penis and Scrotum, Anterior View, *continued*

7. What prevents blood from escaping the erectile tissue?
8. Name the paired arteries on the dorsal side of the penis that distribute to the skin and glans of the penis.
9. Name the nerve that supplies innervation to the body of the penis.

Click **LAYER 4** *in the* **LAYER CONTROLS** *window, and you will see the following image:*

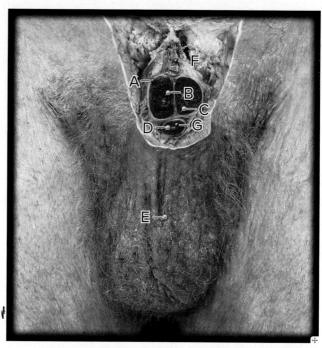

©McGraw-Hill Education

Mouse-over the pins on the screen to find the information necessary to identify the following structures:

A. tunica albuginea of penis
B. corpus cavernosum penis
C. body of penis
D. corpus spongiosum penis
E. scrotum

Nonreproductive System Structures (blue pins)

F. deep dorsal v. of penis
G. spongy urethra

CHECK POINT

Penis and Scrotum, Anterior View, *continued*

10. Name the distal end cap of the corpora cavernosa.
11. Name the distal end of the corpus spongiosum.
12. Name the vein that drains the erectile tissues of the body of the penis.

Click **LAYER 5** *in the* **LAYER CONTROLS** *window, and you will see the following image:*

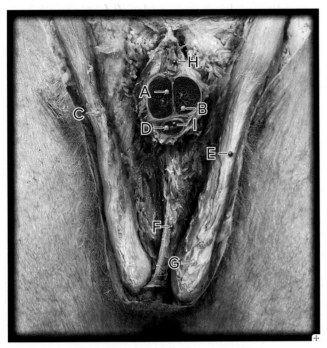

©McGraw-Hill Education

Mouse-over the pins on the screen to find the information necessary to identify the following structures:

A. corpus cavernosum penis
B. body of penis
C. spermatic cord
D. corpus spongiosum penis
E. spermatic fascia
F. septum of scrotum
G. testis with spermatic fascia

Nonreproductive System Structures (blue pins)

H. deep dorsal v. of penis
I. spongy urethra

CHECK POINT

Penis and Scrotum, Anterior View, *continued*

13. Name the structures located in the spermatic cord. What muscle is located in the middle fascial layer? What is the function of this muscle?
14. Name the structure that separates the right and left testes.
15. Name the structures that enclose the spermatic cord. What are the three layers that make up these structures?

Click **LAYER 6** *in the* **LAYER CONTROLS** *window, and you will see the following image:*

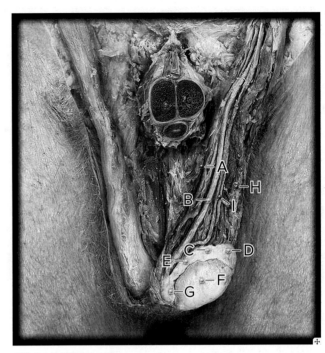

©McGraw-Hill Education

Mouse-over the pins on the screen to find the information necessary to identify the following structures:

A. spermatic fascia
B. vas deferens
C. epididymis
D. head of epididymis
E. body of epididymis
F. testis
G. tail of epididymis

Nonreproductive System Structures (blue pins)

H. pampiniform venous plexus
I. testicular a. and branches

CHECK POINT

Penis and Scrotum, Anterior View, *continued*

16. Name the paired male gonads. What are their functions?
17. Name the structure for maturation and storage of spermatozoa. How long can spermatozoa be stored?
18. What are the three parts of the structure in question 17?

IN REVIEW

What Have I Learned?

The following questions cover the material that you have just learned, the penis and scrotum. Apply what you have learned to answer these questions on a separate piece of paper.

1. What is emission? Name the structure that conducts sperm and testicular fluid during emission.

2. Name the arteries located within the spermatic cord. What is their distribution?

3. Name the network of veins that ascend within the spermatic cord. Where do these drain? (Hint: Right is different from left.)

The Seminiferous Tubule

EXERCISE 14.20a:
Reproductive System—Seminiferous Tubule—Histology

SELECT TOPIC	SELECT VIEW
Seminiferous Tubule	LM: High Magnification

Click the **TAGS ON/OFF** button, and you will see the following image:

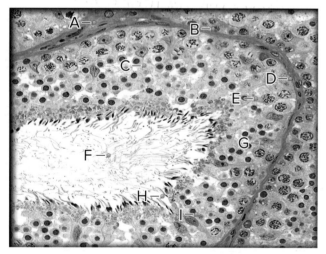

©Dr. Thomas Caceci, Virginia-Maryland Regional College of Veterinary Medicine

Mouse-over the pins on the screen to find the information necessary to identify the following structures:

A. _____
B. _____
C. _____
D. _____
E. _____
F. _____
G. _____
H. _____
I. _____

CHECK POINT

Seminiferous Tubule, Histology

1. Name the site of spermatogenesis. Where is this site located? How many are in each location?
2. Name the central tubular cavity of the seminiferous tubules. What is usually contained in this cavity?
3. What are the functions of the sustentacular cells? What other names are they known by?

MODULE 14   The Reproductive System   823

EXERCISE 14.20b:
Reproductive System—Seminiferous Tubule—Histology

SELECT TOPIC: Seminiferous Tubule → **SELECT VIEW**: SEM: Low Magnification

Click the **TAGS ON/OFF** button, and you will see the following image:

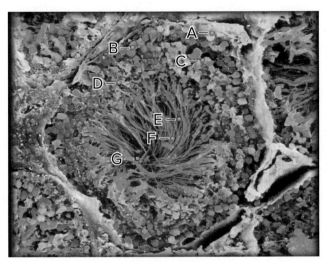

©Prof. P.M. Motta/Univ. "La Sapienza", Rome/Science Source

Mouse-over the pins on the screen to find the information necessary to identify the following structures:

A. _____
B. _____
C. _____
D. _____
E. _____
F. _____
G. _____

SELECT ANIMATION: Spermatogenesis   PLAY

After viewing the animation, answer the following questions:

1. Where are sperm formed?

2. Describe the structure and lining of the seminiferous tubules. What specialized cell types are located there?

3. Name the germ cells from which sperm cells arise.

4. What differences occur in the daughter cells of the spermatogonia? Are they haploid or diploid? How many chromosomes do they each contain?

5. The primary spermatocytes divide by _____ to form _____ _____ spermatocytes, each containing _____ chromosomes. The _____ _____ divide again to form _____.

6. What is spermiogenesis? Where does it occur? What changes occur during this process?

7. Sperm cells are _____ when they leave the seminiferous tubules and testis and mature in the _____.

Self-Quiz
Take this opportunity to check your progress by taking the **QUIZ**. See the **Introduction Module** for a reminder on how to access the **QUIZ** for this Study Area.

IN REVIEW

What Have I Learned?

The following questions cover the material that you have just learned, the seminiferous tubule. Apply what you have learned to answer these questions on a separate piece of paper.

1. Name the most primitive cell in the male germ line. Diploid or haploid?

2. Name the diploid cells derived from the cells in question 1. These cells give rise to _____, which are (haploid/diploid). These cells in turn give rise to _____, which are (haploid/diploid).

3. Name and describe the structure of the male germ cells. What is their function?

824 MODULE 14 The Reproductive System

EXERCISE 14.21:
Coloring Exercise

Identify the features of the female reproductive system. Then color in the features with colored pens or pencils.

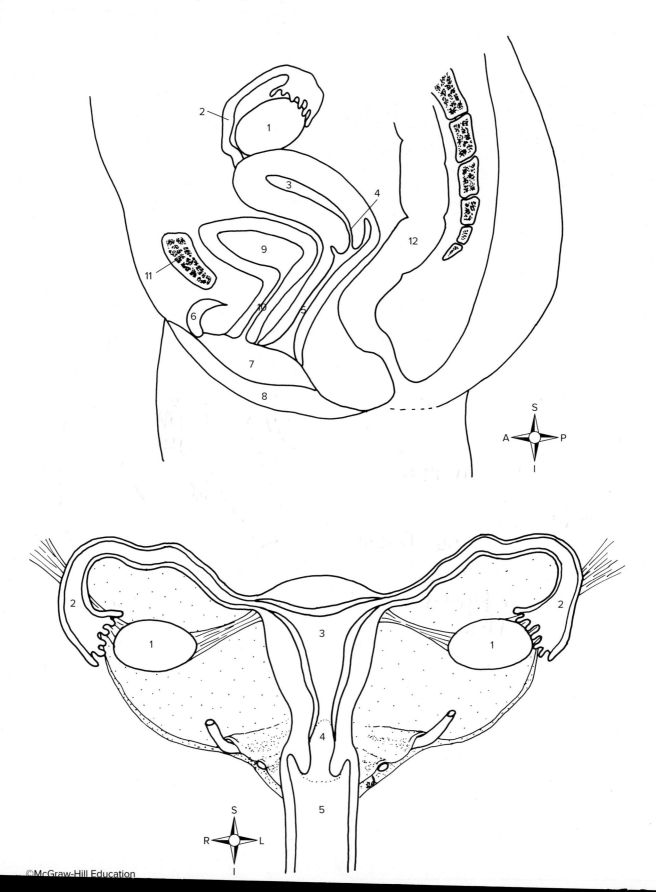

MODULE 14 The Reproductive System **825**

☐ 1. _____
☐ 2. _____
☐ 3. _____
☐ 4. _____
☐ 5. _____
☐ 6. _____
☐ 7. _____
☐ 8. _____
☐ 9. _____
☐ 10. _____
☐ 11. _____
☐ 12. _____

EXERCISE 14.22:

Coloring Exercise

Identify the features of the male reproductive system. Then color in the features with colored pens or pencils.

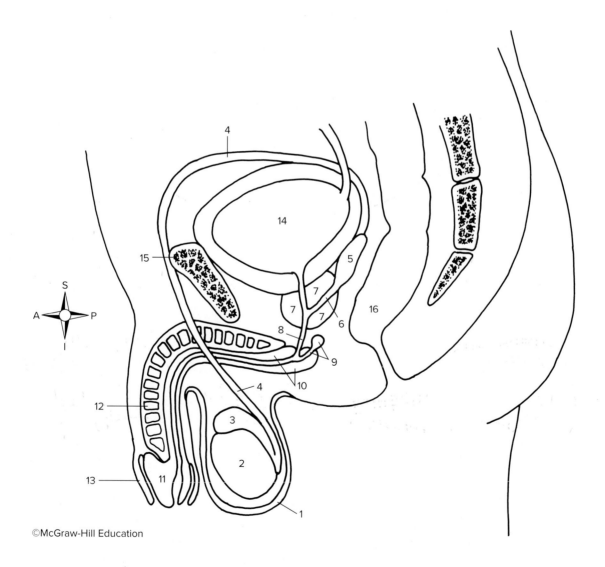

©McGraw-Hill Education

MODULE 14 The Reproductive System

☐ 1. _____
☐ 2. _____
☐ 3. _____
☐ 4. _____
☐ 5. _____
☐ 6. _____
☐ 7. _____
☐ 8. _____

☐ 9. _____
☐ 10. _____
☐ 11. _____
☐ 12. _____
☐ 13. _____
☐ 14. _____
☐ 15. _____
☐ 16. _____